Horner

NATO ASI Series

Advanced Science Institutes Series

A series presenting the results of activities sponsored by the NATO Science Committee, which aims at the dissemination of advanced scientific and technological knowledge, with a view to strengthening links between scientific communities.

The Series is published by an international board of publishers in conjunction with the NATO Scientific Affairs Division

A Life Sciences B Physics	Plenum Publishing Corporation London and New York
C Mathematical and Physical Sciences D Behavioural and Social Sciences E Applied Sciences	Kluwer Academic Publishers Dordrecht, Boston and London
F Computer and Systems Sciences G Ecological Sciences H Cell Biology I Global Environmental Change	Springer-Verlag Berlin Heidelberg New York London Paris Tokyo Hong Kong Barcelona Budapest

PARTNERSHIP SUB-SERIES

1. Disarmament Technologies	Kluwer Academic Publishers
2. Environment	Springer-Verlag/Kluwer Acad. Publishers
3. High Technology	Kluwer Academic Publishers
4. Science and Technology Policy	Kluwer Academic Publishers
5. Computer Networking	Kluwer Academic Publishers

The Partnership Sub-Series incorporates activities undertaken in collaboration with NATO's Cooperation Partners, the countries of the CIS and Central and Eastern Europe, in Priority Areas of concern to those countries.

NATO-PCO DATABASE

The electronic index to the NATO ASI Series provides full bibliographical references (with keywords and/or abstracts) to about 50000 contributions from international scientists published in all sections of the NATO ASI Series. Access to the NATO-PCO DATABASE is possible via a CD-ROM "NATO Science & Technology Disk" with user-friendly retrieval software in English, French and German (© WTV GmbH and DATAWARE Technologies Inc. 1992).

The CD-ROM can be ordered through any member of the Board of Publishers or through NATO-PCO, Overijse, Belgium.

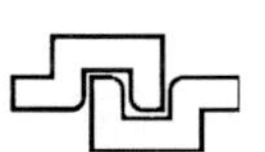

Series G: Ecological Sciences, Vol. 41

Springer
Berlin
Heidelberg
New York
Barcelona
Budapest
Hong Kong
London
Milan
Paris
Santa Clara
Singapore
Tokyo

Physiological Ecology of Harmful Algal Blooms

Edited by

Donald M. Anderson

Biology Department
Woods Hole Oceanographic Institution
Woods Hole, MA 02543, USA

Allan D. Cembella

Institute for Marine Biosciences
Halifax, Nova Scotia, B3H 3Z1, Canada

Gustaaf M. Hallegraeff

Department of Plant Science
University of Tasmania
Hobart, Tasmania 7001, Australia

With 201 Figures (1 Colour Plate) and 45 Tables

Proceedings of the NATO Advanced Study Institute on "The Physiological Ecology of Harmful Algal Blooms", held at the Bermuda Biological Station for Research, Bermuda, May 27–June 6, 1996

Library of Congress Cataloging-in-Publication Data

Physiological ecology of harmful algal blooms / edited by Donald M. Anderson, Allan D. Cembella, Gustaaf M. Hallegraeff.
p. cm. – (NATO ASI series. Series G, Ecological sciences; vol. 41)
"Proceedings of the NATO Advanced Study Institute on The Physiological Ecology of Harmful Algal Blooms, held at the Bermuda Biological Station for Research, Bermuda, USA, May 27–June 6, 1996" – T.p. verso.
Includes bibliographical references and index.
ISBN 3-540-64117-3 (hard cover)
1. Toxic marine algae–Ecophysiology–Congresses. 2. Algal blooms–Congresses. 3. Dinoflagellate blooms–Congresses.
I. Anderson, Donald M. (Donald Mark) II. Cembella, Allan D., 1952– . III. Hallegraeff, Gustaaf M. IV. NATO Advanced Study Institute on Physiological Ecology of Harmful Algal Blooms (1996: Bermuda Islands) V. Series; NATO ASI series. Series G, Ecological sciences; no. 41.
QK568.T67P48 1998 579.8'165–dc21 98-5063 CIP

ISSN 0258-1256
ISBN 3-540-64117-3 Springer-Verlag Berlin Heidelberg New York

Printed in Germany

Typesetting: Camera ready by authors/editors
Printed on acid-free paper
SPIN 10525523 31/3136 - 5 4 3 2 1 0

Preface

Throughout the world, coastal countries are heavily impacted by the phenomena we now term "harmful algal blooms" or HABs. Many of these phenomena were called red tides in the past because of the intense discoloration of the water by the pigments in the algae involved. That term is too general, however, as it includes many blooms which discolor the water but cause no harm, and ignores blooms of highly toxic cells which cause problems at very low (and essentially invisible), cell densities. HABs take many forms and have equally diverse effects. The common thread linking these phenomena is that they cause harm. Some algal species produce potent neurotoxins which accumulate in shellfish that feed on those algae, resulting in poisoning syndromes in human consumers called paralytic, diarrhetic, amnesic, and neurotoxic shellfish poisoning (PSP, DSP, ASP, and NSP respectively). Ciguatera fish poisoning (CFP) occurs when algal toxins accumulate in tropical fish flesh. Planktonic "blooms" are not involved in CFP since the causative algae are benthic or epiphytic, but the origin of the toxin is in microscopic algae and the manner in which the toxins move through and affect the food web is similar to other HAB phenomena. The diverse array of HAB toxins can alter marine ecosystem structure and function, affecting fecundity and survival at all levels. Some toxic blooms kill wild and farmed fish populations, while non-toxic algal species can cause problems through biomass effects - shading of submerged vegetation, disruption of food web dynamics and structure, and oxygen depletion as the blooms decay.

HAB phenomena are not new. Harmful "red tides" are recorded in the Bible and other early documents, and the fossil record includes forms identical to species known to cause toxic outbreaks today. What is new is the apparent proliferation of these phenomena over the last several decades. Scientists argue about the causes of this expansion. Some view it as a global "epidemic" resulting from pollution and other changes in the nature of coastal ecosystems due to human activities. Others argue that much of the expansion can be explained by increases in the number and skill of scientists studying the problem, by advances in toxin detection methodologies, and by the proliferation of aquaculture and other fisheries activities that require careful monitoring of product quality. In one sense, this latter viewpoint argues that the problem may not be any worse than before - we are simply more aware of its nature and extent.

Whatever the explanation, there is a HAB problem and it is significant and the impacts are growing. There has been a steady increase in the number of known toxic species, algal toxins, toxic outbreaks, and fisheries and ecosystem impacts, all at a time when human reliance on the coastal zone for food, recreation, and commerce is rapidly expanding. Recognizing this growing problem, a Working Group was convened by the Scientific Committee for Oceanic Research (SCOR) and the Intergovernmental Oceanographic Commission (IOC) of UNESCO to investigate the Physiological Ecology of Harmful Algal Blooms. The terms of reference of Working Group #97 were:

> 1) To review and analyze data on the physiological ecology and biochemical aspects of harmful algal blooms, especially those resulting in toxic episodes, paying particular attention to nutritional, environmental and physiological factors;
>
> 2) To assemble within two years the Working Group's findings and submit for publication a report, summarizing the state of knowledge and identifying priority areas for future research.

At a meeting immediately after the Sixth International Conference on Toxic Marine Phytoplankton in Nantes, Working Group members decided that the most efficient means of accomplishing these goals was to convene an Advanced Study Institute (ASI). A proposal was written to NATO and funding was obtained. ASIs are technical workshops in which participants and lecturers meet together for 10 days of focused oral presentations, demonstrations, and discussions on particular topics. NATO provided the bulk of the funding for this ASI, but significant contributions were also provided by SCOR and the IOC.

The working group then met in Tokyo before the 7th International Conference on Toxic Phytoplankton in Sendai and chose a series of lecture topics and individuals to deliver those lectures. The ASI was advertised in major journals as well as through direct mailings, electronic mailing lists and society newsletters.

The difficult task of selecting participants from the hundreds of applications that were received was the job of the International Organizing Committee. Working within constraints specified by the co-sponsors, the Organizing Committee accepted and solicited applicants from a variety of countries and disciplines, attempting to broaden the scientific and geographic coverage of the meeting. Special efforts were made to provide opportunities for postdoctoral investigators and graduate students to attend. We deeply regret having had to decline the applications of many fully-qualified individuals, and hope that those colleagues understand the scientific and political constraints of the decision process.

The ASI was convened at the Bermuda Biological Station for Research (BBSR) from May 27 - June 6, 1996. This proved to be an excellent location, as the facilities of the BBSR included a modern and spacious lecture hall, several teaching laboratories and seminar rooms, and on-site housing and dining facilities for all participants. The ASI format was very different from most other scientific meetings, in that it involved ten days of lectures, technical demonstrations, and discussions. The days were long but varied, with lectures in the morning and sometimes in the afternoon, and demonstrations and discussions in the afternoon and evening, often lasting well into the night. The lectures were divided into two theme areas: 1) The ecology of critical groups of harmful phytoplankton (= "Autecology"); and (2) The ecophysiological processes and mechanisms that affect harmful bloom formation and the production of phycotoxins (="Ecophysiological Processes and Mechanisms"). A typical day involved several invited presentations on a common topic, followed by an hour or more of open discussion about the major issues raised by those talks. Posters relating to either the Autecology or the Ecophysiology themes were also on display throughout the meeting.

A major feature of the ASI was a series of demonstrations of advanced techniques and equipment useful in ecophysiological studies. The topics included:1) molecular probes for HAB species; 2) mixotrophy; 3) neuroreceptor assays; 4) remote sensing and ocean optics; 5) new cyst techniques; 6) flow cytometry and image analysis; 7) cytotoxicity; and 8) rheology and small scale processes. The ASI also included discussion sessions on a number of topics, sometimes to explore new issues, and other times to expand upon ideas raised during the lectures. Discussion groups were convened on:1) grazing estimates; 2) culturing bottlenecks; 3) advanced imaging techniques; 4) eutrophication and HABs; 5) rheology and small scale processes; 6) cell cycles; and 7) autecology and the species concept. Once again, participation in these discussions groups was excellent, with spirited discussions lasting well into the evening hours, despite the many distractions of Bermuda.

Overall, the ASI was a great success. The format of the meeting, the location at the BBSR, and the dedication and energy of the organizing committee, the lecturers, and the discussion and demonstration leaders all combined to produce a stimulating and memorable time. It is indeed unfortunate that many colleagues applied but could not be invited to the meeting, but we hope that the publication of this book for the ASI will allow those who could not attend to benefit from the scientific presentations. I am exceedingly grateful to NATO, IOC, and SCOR for their financial support, and to all ASI participants for helping to make the meeting so productive and enjoyable. Special thanks also go to E. Le Fave, J. Ridge, and L. Simons for superb administrative support and to Allan Cembella and Gustaaf Hallegraeff for their tireless and meticulous efforts as session chairs and co-editors of this volume.

Donald M. Anderson
ASI Director

Taxonomic Notes

Harmful algal blooms are often almost monospecific events. Correctly assessing the precise taxonomic identity of the causative organism thus becomes crucial in deciding whether knowledge on toxicology, physiology and ecology gained from similar blooms can be reliably applied to the species at hand. Resolution of the species concept in harmful algae therefore has become a profound issue of discussion at all major conferences dealing with toxic phytoplankton (see Chapters by Scholin and Gallagher). The resulting name changes always cause concern and confusion to non-specialists, but reflect the ever developing scientific understanding of natural relationships among organisms. The scientific history and taxonomic justification behind the now widely accepted name change of the toxic dinoflagellate *Gonyaulax tamarensis* Lebour, via competing allocations within the genera *Alexandrium* Halim, *Gessnerium* Halim or *Protogonyaulax* Taylor, but now to be called *Alexandrium tamarense* (Lebour) Balech, is described in the Chapter by Taylor. Similarly, the history of the name change of the toxic diatom *Nitzschia pungens* f.*multiseries* Hasle to *Pseudo-nitzschia multiseries* (Hasle) Hasle is described in the Chapter by Bates *et al.*. For a few other toxigenic algal bloom species discussed during the Bermuda workshop, such taxonomic disputes have not been formally settled and some of authors have chosen to use one name while others have used a different name. As editors, we chose not to arbitrarilly alter the selected species designation in these cases where the disputes are not fully resolved.

One example, is the correct species designation of the fish-killing raphidophyte flagellate *Heterosigma akashiwo.* This species has often been confused with the benthic flagellate *Olisthodicus luteus* N. Carter, which is a non-toxic organism with a different cell shape, colour and swimming pattern. Hulburt (1965) was the first to describe the toxigenic plankton flagellate under the Botanical Code of Nomenclature using the name *Olisthodiscus carterae*, but he failed to indicate a holotype of the species. This latter technical detail also invalidates the new combination *Heterosigma carterae* (Hulburt) F.J.R. Taylor proposed by Taylor (1992). Two years after Hulburt's description, Hada (1967) independently described a similar protozoan under the Zoological Code of Nomenclature as *Entomosigma akashiwo* Hada which he later transferred to the new genus *Heterosigma* as *H. akashiwo* (Hada) Hada 1968. Sournia (1973) formally typified the genus *Heterosigma* (type species: *H. inlandica* Hada) while Hara and Chihara (1987) formally designated a holotype for *H. akashiwo* (see Throndsen 1996, and references therein). A taxonomic workshop, led by Ø. Moestrup, during the 8th International Conference on Harmful Algae, Vigo, Spain, in June 1997 has recommended to adopt the name *Heterosigma akashiwo* (Hada) Hada ex Sournia.

Another problematic species complex is what in the Bermuda workshop we have referred to as the fish-killing dinoflagellate *Gymnodinium mikimotoi* (see chapters by Gentien and Steidinger *et al.*). This species designation includes the European fish killer *Gyrodinium aureolum* Hulburt, sensu Braarud and Heimdal (1970). However, the true *Gyrodinium aureolum* Hulburt (1957) clearly is a different species with a different position of the nucleus, dorsoventral flattening of the cell, number of chloroplasts and pyrenoids, and posssibly lacking an apical groove. In Japan this fish-killing dinoflagellate was described in 1984 as *Gymnodinium nagasakiense* Takayama et Adachi , but these Japanese authors eventually withdrew their proposed species designation once an earlier Japanese description as *Gymnodinium mikimotoi* Miyake et Kominami ex Oda 1935 was recognised. The Pacific *G. mikimotoi* and the European *Gyrodinium aureolum* are morphologically similar and are regarded as conspecific although genetic differences between European and Pacific populations do exist (see Partensky *et al* . 1988, and references therein).

Another controversial name change refers to an attempt to separate non-toxic planktonic *Prorocentrum* species (type species *P. micans* Ehrenberg) from toxigenic benthic species (see Chapter by Tindall and Morton), for which a reinstatement of the poorly defined genus *Exuviaella* (type species *E. marina* Cienkowski) has been proposed (McLachlan *et al.* 1997). Even though the proposed division of the genus *Prorocentrum* has merit, a taxonomic workshop during the 8th International Conference on Harmful Algae, Vigo, Spain, held in June 1997, concluded that the species *E. marina* is so poorly defined (it has been used for at least 6 different taxa) that the name *Exuviaella* best be completely discarded.

Other taxonomic questions which remain to be solved concern the dinoflagellate *Noctiluca,* of which there is likely to exist more than one species (see Chapter by Elbrachter and Qi), the number of valid species of the haptophyte *Phaeocystis* (see Chapter by Lancelot *et al.*) and the PSP dinoflagellate *Gymnodinium catenatum* Graham, for which increasing evidence suggests that the nontoxic, North-European form may represent a different taxon (see Chapter by Hallegraeff and Fraga).

To alleviate the problems of ever-changing taxonomy of harmful phytoplankton, it is recommended: (1) to study type-material or, if this is not available, collect and re-examine material from the type locality; (2) establish and curate type specimen collections using permanent mounts, photomicrographs (rather than drawings), video tapes and preferably living cultures; and (3) incorporate life cycle features e.g. cysts in species descriptions. Original names as much as possible should be retained until complete information is available on the existing available and valid genera.

References

McLachlan, J.L., Boalch, G.T. and Jahn, R. (1997). Reinstatement of the genus *Exuviaella* (Dinophyceae, Prorocentrophycidae) and an assessment of *Prorocentrum lima. Phycologia* 36: 38-46.

Partensky, F., Vaulot, D., Couté, A., Sournia, A. (1988). Morphological and nuclear analysis of the bloom-forming dinoflagellates *Gyrodinium* cf.*aureolum* and *Gymnodinium nagasakiense. J. Phycol.* 24: 408-415.

Sournia, A. (1973). Catalogue des espèces et taxons infraspecifiques de Dinoflagellés marins actuels. 1. Dinoflagellés libres. *Beih. Nova Hedwigia* 48: 1-92.

Throndsen, J. (1996). Note on the taxonomy of *Heterosigma akashiwo* (Raphidophyceae). *Phycologia* 35, 367

Taylor, F.J.R. (1992) The taxonomy of harmful marine phytoplankton. *Gior. Bot. Ital.* 126: 209-219.

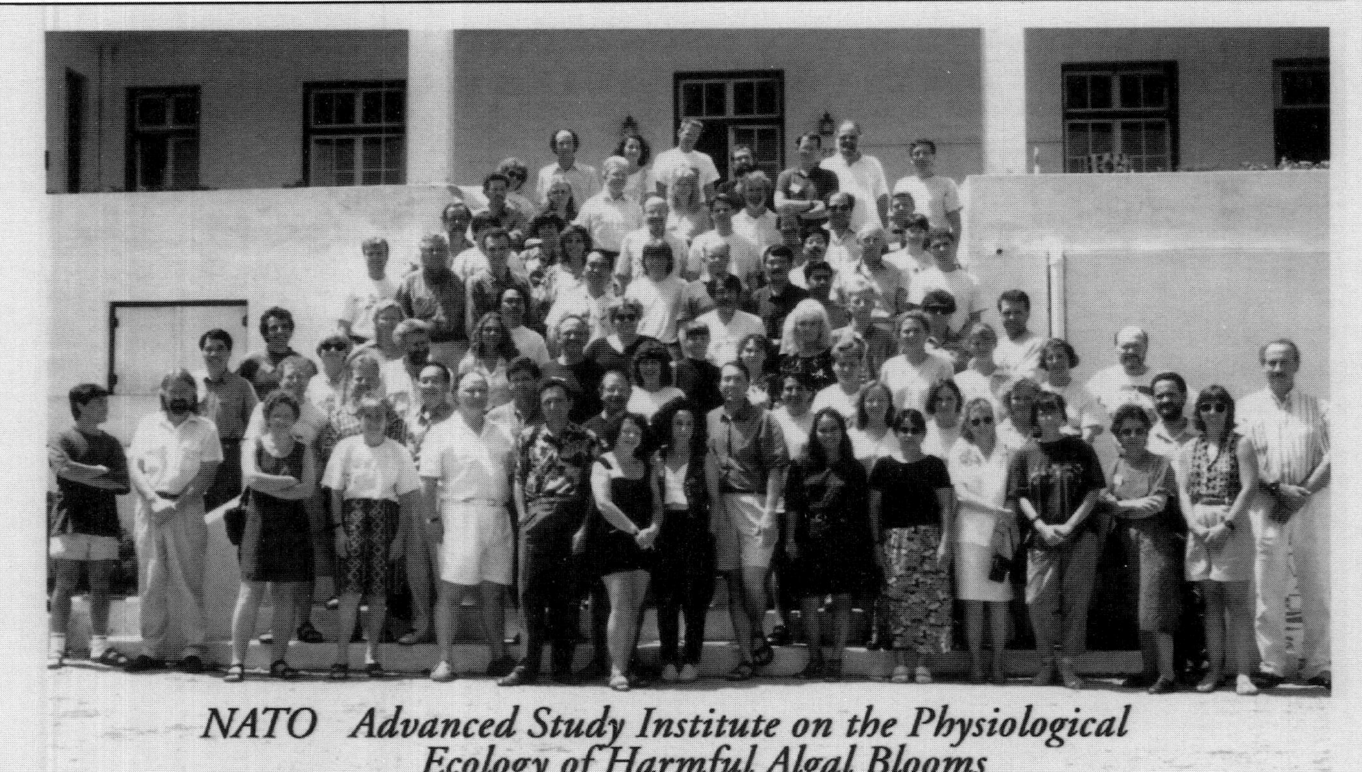

NATO Advanced Study Institute on the Physiological Ecology of Harmful Algal Blooms
May 27 - June 6, 1996
Bermuda Biological Station for Research

1 Alejandro CLEMENT; **2** Gustaff HALLEGRAEFF; **3** Hanne KAAS; **4** Marta ESTRADA; **5** Ed BLACK; **6** Serge MAESTRINI; **7** Edna GRANÉLI; **8** Isabel BRAVO; **9** Don ANDERSON; **10** Rhodora CORALLES; **11** Ann ANTON; **12** Susan GALLACHER; **13** Isabel REYERO FERNANDEZ; **14** Maria Antonia SAMPAYO; **15** Fran VAN DOLAH; **16** Don TINDALL; **17** Youlian PAN; **18** Steve BATES; **19** Greg DOUCETTE; **20** Yuzaburo ISHIDA; **21** William COCHLAN; **22** F.J.R. "Max" TAYLOR; **23** JoAnn BURKHOLDER; **24** Janice LAWRENCE; **25** Elisa BERDALET; **26** Manuel ZAPATA; **27** Els MAAS; **28** Tracey McDONNELL; **29** Eurgain Haf JOHN; **30** Linda MEDLIN; **31** Santiago FRAGA; **32** Raffael JOVINE; **33** Chris SCHOLIN; **34** John CULLEN; **35** Virginia GARCIA; **36** Charles TRICK; **37** Catherine LEGRAND; **38** Masaaki KODAMA; **39** Maija BALODE; **40** Gareth WOOD; **41** Sonya DYHRMAN; **42** Pat TESTER; **43** Bente EDVARDSEN; **44** Grant PITCHER; **45** Susana FRANCA; **46** Wayne CARMICHAEL; **47** Per CARLSSON; **48** Rosella PISTOCCHI; **49** Harry PELETIER; **50** Gaspar TARONCHER OLDENBURG; **51** Gires USUP; **52** Silvia MENDEZ; **53** Ichiro IMAI; **54** Giorgio HONSELL; **55** Chris PARRISH; **56** F. Hoe CHANG; **57** Ahmet KIDEYS; **58** Yasuo FUKUYO; **59** Bernd LUCKAS; **60** Jane LEWIS; **61** Alexander VERSHININ; **62** Patrick GENTIEN; **63** Allan CEMBELLA; **64** Sue BLACKBURN; **65** Jürgen LENZ; **66** Ted SMAYDA; **67** Jane GALLAGHER; **68** Per Juel HANSEN; **69** Kevin SELLNER; **70** D. Wayne COATS; **71** José CORDOVA; **72** Dan KAMYKOWSKI; **73** Maurice LEVASSEUR; **74** Maureen KELLER; **75** Quay DORTCH; **76** Larry BRAND; **77** Esther GARCÉS I PIERES; **78** Roel RIEGMAN; **79** Gerry PLUMLEY; **80** Greg BOYER; **81** Malte ELBRÄCHTER; **82** Kristinn GUDMUNDSSON **not pictured:** Jeff TURNER, Ian JENKINSON, Jean LINCOLN, Mike SIERACKI.

Table of Contents

The *Alexandrium* Complex and Related Species

F. J. R. Taylor and Y. Fukuyo
The Neurotoxigenic Dinoflagellate Genus *Alexandrium* Halim: General Introduction 3

C. A. Scholin
Morphological, Genetic and Biogeographic Relationships of Toxic Dinoflagellates *Alexandrium tamarense*, *A. catenella* and *A. fundyense* 13

D. M. Anderson
Physiology and Bloom Dynamics of Toxic *Alexandrium* Species, with Emphasis on Life Cycle Transitions 29

Y. Ishida, A. Uchida, and Y. Sako
Genetic and Biochemical Approaches to PSP Toxin Production of Toxic Dinoflagellates 49

G. M. Hallegraeff and S. Fraga
Bloom Dynamics of the Toxic Dinoflagellate *Gymnodinium catenatum*, with Emphasis on Tasmanian and Spanish Coastal Waters 59

G. Usup and R. V. Azanza
Physiology and Bloom Dynamics of the Tropical Dinoflagellate *Pyrodinium bahamense* 81

Fish-killing Species

I. Imai, M. Yamaguchi, and M. Wantanabe
Ecophysiology, Life Cycle, and Bloom Dynamics of *Chattonella* in the Seto Inland Sea, Japan 95

T. J. Smayda
Ecophysiology and Bloom Dynamics of *Heterosigma akashiwo* (Raphidophyceae) 113

K. A. Steidinger, G. A. Vargo, P. A. Tester, and C. R. Tomas
Bloom Dynamics and Physiology of *Gymnodinium breve* with Emphasis on the Gulf of Mexico 133

P. Gentien
Bloom Dynamics and Ecophysiology of the *Gymnodinium mikimotoi* Species Complex 155

J. M. Burkholder, H. B. Glasgow, Jr., and A. J. Lewitus
Physiological Ecology of *Pfiesteria piscicida*, with General Comments on "Ambush-Predator" Dinoflagellates 175

B. Edvardsen and E. Paasche
Bloom Dynamics and Physiology of *Prymnesium* and *Chrysochromulina* 193

C. Lancelot, M. D. Keller, V. Rousseau, W. O. Smith, Jr., and S. Mathot
Autecology of the Marine Haptophyte *Phaeocystis* sp. 209

J. C. Gallagher
Genetic Variation in Harmful Algal Bloom Species: An Evolutionary Ecology Approach 225

Other HAB Species

S. Y. Maestrini
Bloom Dynamics and Ecophysiology of *Dinophysis* spp. 243

S. S. Bates, D. L. Garrison, and R. A. Horner
Bloom Dynamics and Physiology of Domoic-Acid-Producing *Pseudo-nitzschia* Species 267

D. R. Tindall and S. L., Morton
Community Dynamics and Physiology of Epiphytic/Benthic Dinoflagellates Associated with Ciguatera 293

M. Elbrächter and Y. Qi
Aspects of *Noctiluca* (Dinophyceae) Population Dynamics 315

C. A. Scholin
Development of Nucleic Acid Probe-Based Diagnostics for Identifying and Enumerating Harmful Algal Bloom Species 337

M. Elbrächter and E. Schnepf
Parasites of Harmful Algae 351

Autecology: Synthesis

G. M. Hallegraeff
Concluding Remarks on the Autecology of Harmful Algal Blooms 371

Theme 2 - Ecophysiological Processes and Mechanisms

Ecophysiological Role of Toxin Production

A. D. Cembella
Ecophysiology and Metabolism of Paralytic Shellfish Toxins in Marine Microalgae 381

S. S. Bates
Ecophysiology and Metabolism of ASP Toxin Production 405

J. L. C. Wright and A. D. Cembella
Ecophysiology and Biosynthesis of Polyether Marine Biotoxins 427

Zooplankton Grazing

J. T. Turner, P. A. Tester, and P. J. Hansen
Interactions Between Toxic Marine Phytoplankton and Metazoan and Protistan Grazers 453

Eutrophication and Nutrient Supply

R. Riegman
Species Composition of Harmful Algal Blooms in Relation to Macronutrient Dynamics 475

G. L. Boyer and L. E. Brand
Trace Elements and Harmful Algal Blooms 489

P. Carlsson and E. Granéli
Utilization of Dissolved Organic Matter (DOM) by Phytoplankton, Including Harmful Species 509

The Role of Mixotrophy in Harmful Algal Bloom Nutrition

P. J. Hansen
Phagotrophic Mechanisms and Prey Selection in Mixotrophic Phytoflagellates 525

E. Granéli and P. Carlsson
The Ecological Significance of Phagotrophy in Photosynthetic Flagellates 539

Swimming Behavior Buoyancy and Small-Scale Physical Processes

J. J. Cullen and J. G. MacIntyre
Behavior, Physiology and the Niche of Depth-Regulating Phytoplankton 559

D. Kamykowski, H. Yamazaki, A. K. Yamazaki, and G. J. Kirkpatrick
A Comparison of How Different Orientation Behaviors Influence Dinoflagellate Trajectories and Photoresponses in Turbulent Water Columns 581

M. Estrada and E. Berdalet
Effects of Turbulence on Phytoplankton 601

Bacterial Interactions With Harmful Algal Bloom Species

G. J. Doucette, M. Kodama, S. Franca, and S. Gallacher
Bacterial Interactions with Harmful Algal Bloom Species: Bloom Ecology, Toxigenesis, and Cytology 619

A. D. Cembella
Ecopysiological Processes and Mechanisms: Towards Common Paradigms for Harmful Algal Blooms Microalgae 649

ASI Participants List 653

Index 655

Autecology

The Neurotoxigenic Dinoflagellate Genus *Alexandrium* Halim: General Introduction

F.J.R. "Max" Taylor[1] and Yasuwo Fukuyo[2]

[1] Oceanography, Department of Earth and Ocean Sciences, University of British Columbia, 6270 University Boulevard, Vancouver, B.C., Canada V6T 1Z4

[2] Centre for Asian Natural Environmental Sciences, University of Tokyo, 1-1-1 Yayoi, Bunkyyo-Ku, Tokyo 113, Japan.

1. Introduction

Members of the genus *Alexandrium* Halim are gonyaulacoid dinoflagellates of the family Goniodomaceae, subfamily Helgolandinioideae (Fensome *et al.* 1993). All are marine, thecate and photosynthetic, being strongly pigmented in the motile state. Some are bioluminescent, and the genus contains more neurotoxic members (seven to nine, depending on taxonomic acceptance) than any other harmful algal genus although some are non- or weakly toxic. In North America, the common name "red tide" is associated with outbreaks of PSP on both east and west coasts, even though it is rare for the water to be visibly discolored during outbreaks, the species preferring subsurface levels. Observed from the air an intense bloom of *A. catenella* in Sequim Bay, Washington, was a dark reddish-brown color (Taylor, 1968). It was this bloom that was found to be heavily infected with the parasite *Amoebophrya ceratii,* leading to the suggestion that this might be used as a control organism. This seemingly naive proposal is now being taken seriously by several groups.

Alexandrium catenella was the first dinoflagellate to be linked to PSP, initially by the positive correlation of dinoflagellate concentrations in seawater with mussel toxicity (Sommer *et al.* 1937; Sommer and Meyer, 1937) and later by extraction and purification of the toxins from cultured cells (Schantz et al. 1966). Woloszynska and Conrad (1939) attributed an outbreak of mussel poisoning in Belgium in 1938 to a new species, *Pyrodinium phoneus* (thought by Steidinger, 1973 and Taylor, 1975 to be *A. tamarense* but Balech, 1995 thought it was more like *A. ostenfeldii*). Needler (1949) implicated *A. tamarense* (confirmed by Lebour) as the culprit on the east coast of North America using coincident bloom and mussel poisoning data. Prakash and Taylor (1966) added *A. acatenella* as a west coast toxin source in the first event in which human fatality and a bloom coincided.

Subsequently other species have been added to the list (name changes complicate the history for the uninitiated: see below). The toxins, which are blockers of neural sodium channels, are of the saxitoxin family, and are the commonest cause of paralytic shellfish poisoning (PSP) in temperate and cold waters. Additional species of bivalves, lobsters, crabs and other invertebrates continue to be added to the PSP vector list (see Shumway, 1995 for a recent review). The closely related *Pyrodinium*

NATO ASI Series, Vol. G 41
Physiological Ecology of Harmful Algal Blooms
Edited by D. M. Anderson, A. D. Cembella and G. M. Hallegraeff

bahamense var. *compressum* can be densely packed into the gut of tropical planktivorous fish which, when eaten whole, can kill humans in the phenomenon known as clupeotoxicity (Halstead, 1978, toxin source unknown at the time). Shumway (1995) has suggested that the death of humans from eating squid in such regions may be an extension of the clupeotoxicity food chain. At least one species, *A. monilatum*, is a fish killer (Connell and Cross, 1950) and harm has been suggested to arise from the consumption of zooplankton made toxic by their diet of larval fish (White, 1981).

An important development in the history of the genus was the discovery of benthic resting cysts as the mechanism for seeding blooms (Wall, 1975), bodies similar to those in sediments having been seen previously in cultures of *A. tamarense* by Braarud (1945) and by numerous later authors, e.g. Dale (1977), Fukuyo (1979). Other dinoflagellates had been found to use this strategy for surviving unfavorable planktonic conditions, largely through the pioneering work of Wall and Dale (1968). Subsequently, Turpin *et al.*. (1978) showed that the resting cyst was formed as part of the sexual process, again similar to other dinoflagellates (Pfiester and Anderson, 1987).

Alexandrium catenella was the first member of this genus to also be found in the southern hemisphere, first on the west coast of South Africa and then in the southern part of Chile. Since then it and other species of the genus have been found in several other southern hemisphere locations as might be expected from the biogeography of other dinoflagellate species (reviewed by Taylor, 1987a).

2. Taxonomy

It is essential that the species under investigation in physiological and ecological studies be clearly identified and conceptually understood. Name changes have had potentially confusing effects, particularly if the underlying reasons are not understood. Here the history and reasons are outlined.

The *Alexandrium* genus, with 29 named species, has been taxonomically monographed by Balech (1995) and briefer descriptions of the known toxic members have been provided by Taylor et al. (1996). Because of its involvement in toxic events some of its members, notably *A. tamarense* (under various names) and *A. catenella*, have been subject to intensive physiological and biochemical study. This has raised interesting and fundamental issues involving the concept of species, population genetics and biogeography, addressed by others in this volume.

Most of the earliest described species were placed in the genus *Gonyaulax*. This was because they were similar in general form, although their right-handed girdle displacement (usually close to one girdle width) is less than the typical members of that genus. Their tabulation, particularly the antapical pattern, seemed to be similar, although significant differences are now recognized. Their surfaces were much more delicately developed than typical *Gonyaulax* but this was not considered sufficient to remove them from that genus.

The early descriptive history is full of errors although they were not unusual for the time. Technically, *A. ostenfeldii* was the first to be described (as a species of very closely related genus *Goniodoma*) but the tabulational figures of Paulsen (1904, 1949) were poor and contradictory. Considering that the importance of tabulation was not fully appreciated at the time, this is not surprising. The species is now recognized as redefined much later by Balech and Tangen (1985), a common practice. The next to be described was *A. tamarense*, assigned to *Gonyaulax*. However, Lebour (1925), in her figures, optically reversed the epithecal pattern. Taylor (1975) showed that it was a mirror image of the real pattern. This probably came about by viewing a detached epitheca from the inner face, an easy error to make (Halim did the same in his description of *Gessnerium*.). However, she carried this error over to a ventral view, through wrong reconstruction. It is interesting that this error was not noticed for nearly fifty years, even though the species was recorded on various occasions in the interim!

The descriptions of *A. catenella* and *A. acatenella* by Whedon and Kofoid (1936), considered to differ from *A. tamarense* partly because of the error in the tabulation of the latter by Lebour, rounded out most of the major temperate species of the group, their descriptions erring only in minor plate detail. The distinction of *A.acatenella* is based primarily on body shape distinction and this relates to the issue of species distinction discussed below. Other early species descriptions and names, such as *Pyrodinium phoneus* and *Gonyaulax excavata*, are no longer viewed as distinct.

A new element arose with the description of *Alexandrium minutum* by Halim (1960). His figure is quite clear, even including some sulcal details. Its apical tabulation seemed sufficiently distinct to warrant separation from the others assigned (then) to *Gonyaulax*. Loeblich and Loeblich (1979) later advocated rejection of the genus due to the inadequacy of the description but few would agree, including Balech, who re-examined and redescribed the species from the type locality.

Howell (1953) described a fish-killing species and named it *Gonyaulax monilata*. Unfortunately Halim (1967) unwittingly renamed it *Gessnerium mochimaensis*, its apparent generic difference being due to the same error Lebour had made: drawing the inside of the epitheca instead of the outside. Loeblich and Loeblich (1979) corrected this error and validated the genus but its corrected pattern was fundamentally the same as *Alexandrium*. In a description of another species with a tabulation like *Alexandrium* although attributed to *Gonyaulax*, Steidinger (1971) noted that she could recognize a "*catenella* group" that was distinct from most members of *Gonyaulax*. Subsequently Taylor (1979) recognized these species as members of two distinct genera: a new genus *Protogonyaulax* for those in which the first apical plate made visible contact with the apical pore complex (APC) and *Alexandrium* in which its homologue did not. This attempt to recognize two closely related genera was abandoned when the first apical plate of the type species of *Alexandrium, A.minutum,* was shown to sometimes contact the APC. All thus became members of *Alexandrium.*

2.1 Species and Infraspecific Units

Balech's exhaustive monograph (1995) will undoubtedly be the definitive reference for the taxonomy of this genus for some time to come. However, it must be noted that there are other interpretations of the taxa that need to be recognized and discussed, particularly as other types of data raise doubts as to the separation of some of the species. His approach was primarily that of a classical zoologist with a fine eye for morphological detail. Many of the species distinctions rest on very small details, such as the presence or absence of the ventral pore, as long as they are non-variable. No infraspecific taxa, such as variety, form, or subspecies are recognized. The first two categories are not available to zoologists and the latter, subspecies, is based on a degree of geographic isolation which is hard to apply to marine dinoflagellates at this time. Latin diagnoses are not provided because they are not required by the International Code of Zoological Nomenclature (ICZN).

One of us has argued (Taylor, 1992) that morphologically-based species, "morphospecies", are the most pragmatically useful basis for named units at present, with full recognition of the potential for considerable biochemical/genetic variation within them (see contributions by Scholin and Gallagher, this volume). This genetically-based variability makes it essential to designate which strain is being studied experimentally. This morphologically-based taxonomy supports Balech's system. However, there are several constraints to this approach.

In the first place, there must be an agreement as to the degree of difference required for recognition at the levels of genus, species or infraspecific taxon. Is it reasonable to use differences for species that are much smaller in one genus than in another? This is certainly the case with *Alexandrium.* The traditional criteria for the recognition of botanical varieties and forms have been described by Taylor, 1976) although they need to be revised in the light of contemporary genetic knowledge. Several "species" in the Balechian treatment could well be considered to be varieties, such as those differing by the presence or absence of a ventral pore only, especially if they are interfertile with the type variety. An example of this would be the distinction of *A. fundyense* from *A. tamarense.* Schmidt and Loeblich (1979) distinguished *A. excavatum,* which they characterized as lacking a pore, from *A. tamarense* (similar but with a pore) at the varietal level. Initially they had hoped that such appearance differences might coincide with differences in toxicity and luminescence, but found that the latter were not tightly linked to morphology.

Second, all stages of the life cycle should be considered in recognizing taxa. It has been argued that cyst morphology, or differences in phase dominance, should be significant factors in the recognition, not only of species, but of genera also (Taylor, 1987b; Fensome *et al.* 1993). In this connection it is notable that *A. pseudogoniaulax* (and probably other species in the same "group" *sensu* Balech, 1995) is significantly different from other *Alexandrium* species in that its normal mode of vegetative division is by ecdysis followed by division in a temporary division cyst (eleutheroschisis) whereas most divide by desmoschisis (see Pfiester and Anderson, 1987 for more details concerning these types of reproduction). At present the mode of division in the closely related *Goniodoma* (see below) is not known. If it is by eleutheroschisis a case could be made for moving *A. pseudogoniaulax* back into *Goniodoma* from whence it came.

Third, the system should be "robust", that is, other characters should covary consistently with morphology unless differences are otherwise explicable. Many other features, mostly biochemical, such as allozyme or toxin content, or ribosomal DNA sequences, have now been examined and are amenable to being expressed in the form of "trees" representing the relationships of the various forms *of that character.* One would like to believe that these patterns arise as part of the evolutionary process and that they reflect the true, relatively recent phylogeny of the group. However, such patterns do not agree with one another in detail nor are they fully concordant with morphology. *A. fundyense* does not form a coherent clade in the isozyme or rRNA trees, being interspersed with *A. tamarense.* However, in the LSU tree of Scholin *et al.* (1995) Japanese *A. catenella* and *A. tamarense* group closely together and far from the North American members of these morphospecies, presumably because of long divergence time. These problems are detailed and discussed fully in the papers by Scholin and Gallagher in this volume.

3. Relationships

Members of the genus *Alexandrium* all have an identical plate pattern, allowing for variations in the size and contacts of the first apical plate. All also have similar smooth-walled cysts although a seemingly paratabulate surface has been reported for a cyst population in the Adriatic (Montresor *et al.* 1991). As has been noted previously (Taylor, 1979; Balech, 1985; Fensome *et al.* 1993; Taylor *et al..*, 1995; Balech, 1995), *Pyrodinium* also has an identical tabulation pattern although it has a cyst which has long spines and a distinctive way of opening (Wall and Dale, 1969). This alone would be enough to distinguish it at the generic level but it also differs in that all three of the small platelets in sector I (Ii,Im,Iu) of the Taylor-Evitt homology model (see Fensome et al. 1993 for extensive application of the model) are enclosed by large sulcal lists in the sulcal depression whereas Iu is usually not so in *Alexandrium* (and so was often referred to as the first postcingular plate and not a sulcal: Balech, 1995, weakens this distinction by terming the plate the S.s.a. in *Alexandrium*). *Pyrodinium* also has a heavily sculptured theca with apical and antapical spines, Balech using this to bolster separation of the genera. Given the apparent close relationship, the production of saxitoxin-like compounds in *Pyrodinium* is unremarkable.

Another genus with very similar tabulation is *Goniodoma* which Balech asserts is indistinguishable on the basis of tabulation although in Taylor's (1976) and Fensome *et al.*'s (1993, text Fig.105) interpretation the plate homologous to posterior sulcal plate of *Alexandrium* is (Z in the T-E system) completely out of the sulcus in *Goniodoma* and would be considered an antapical in the traditional Kofoid system. Several *Alexandrium* species, including *A. pseudogoniaulax,* seem to be intermediate in this regard. Only the Im platelet is in the sulcal groove in the type species. The cyst of *Goniodoma,* if it has one, is unknown at present. Balech points to the heaviness of the theca (like *Pyrodinium*) as another distinction and asserts that the apical pore plate is inclined transversely, an interpretation with which we cannot agree (scanning electron micrographs clearly show that it is inclined roughly 45° to the mid-dorsoventral axis). Toxicity is not yet known for this genus. The resemblances in tabulation led Fensome *et al.*(1993)to include these genera in the same family.

It is interesting to note that, although the toxins they produce are very different, mostly polyethers, the benthic genera *Gambierdiscus, Ostreopsis and Coolia* also have a fundamentally similar tabulation although the resemblance is not east to see at first due to plate size and shape distortions. Fensome *et al.* (1993) classified them in a separate subfamily (Gambierdiscoideae) of the same family (Goniodomaceae).

Other genera placed in the Goniodomaceae by Fensome *et al.* (1993) include *Helgolandinium, Fragilidium* (considered synonymous by some authors) and *Pyrophacus.* The cyst of the latter is very unusual. Another tabulationally similar genus is *Pyrocystis*, although this was placed in a separate family because of a major life-cycle difference (the thecate stage is ephemeral, with most of the time spent as a floating vegetative cyst). Although none of these latter genera are known to produce toxins their apparently close affinity suggests that this deserves scrutiny.

4. Conclusions

It is likely that this important genus will continue to be on the cutting edge of harmful algal studies. The taxonomy should pay more regard to life-cycle and genetic information so that the patterns observed are compatible. The use of varieties and forms should reduce the number of taxa distinguished by very small differences, indicating affinities at an infraspecific level. In this publication, no formal changes were made, but it is our view that this should be the next taxonomic revision. In view of the widespread distribution of the morphospecies, population genetics become ever more important in understanding the underlying genetic polymorphisms and, particularly, to attempts to distinguish anthropogenic from natural dispersals.

5. References

Balech, E. (1985). The genus *Alexandrium* or *Gonyaulax* of the tamarense group, pp. 33-38. In *Toxic Dinoflagellates.*, Anderson, D.M., White, A.W. Baden, D.G. (eds.), Elsevier/North Holland, New York.

Balech, E. (1995). *The genus Alexandrium Halim (Dinoflagellata).* Sherkin Island Mar. Stat., Special Publs.

Balech, E., Tangen, K. (1985). Morphology and taxonomy of toxic species in the tamarensis group (Dinophyceae) *Alexandrium excavatum* (Braarud) *comb.nov.* and *Alexandrium ostenfeldii* (Paulsen) *comb. nov.*. *Sarsia* 70: 333-343.

Braarud, T. (1945). Morphological observations on marine dinoflagellate cultures (*Porella perforata, Gonyaulax tamarensis, Protoceratium reticulatum*). *Arch. norske Vidensk.Akad. Oslo, Mat.-Nat.Kl.* 1944, 11:1-18.

Connell, C.H., Cross, J.B. (1950). Mass mortality of fish associated with the protozoan *Gonyaulax* in the Gulf of Mexico. *Science* 112: 359-363.

Dale, B. (1977). Cysts of the toxic red-tide dinoflagellate *Gonyaulax excavata* (Braarud) Balech from Oslofjorden, Norway. *Sarsia* 63:29-34

Fensome, R.A., Taylor, F.J.R., Norris, G., Sarjeant, W.A.S.,Wharton, D.I., Williams, G.J. (1993). *A classification of living and fossil dinoflagellates.* Micropaleontology, Amer. Mus. Nat. Hist., Special Publ. 7, 351 pp.

Fukuyo (1979). Theca and cyst of *Gonyaulax excavata* (Braarud) Balech found at Ofunato Bay, Pacific coast of northern Japan. pp.61-64, In *Toxic Dinoflagellate Blooms*, Taylor, D.L., Seliger, H.H. (eds.), Elsevier North Holland, N.Y.

Halim, Y. (1960). *Alexandrium minutum , n.gen n.sp.*, dinoflagelle provocant des eaux rouges. *Vie Milieu* 11: 102-105.

Halim, Y. (1967). Dinoflagellates of the South-East Caribbean Sea (East Venezuela). *Int. Rev. Ges. Hydrobiol.* 52: 701-755.

Halstead (1978). *Poisonous and Venemous Marine Animals of the World*, Revised Edit.. Darwin Press, Princeton, N.J.. 1042 pp.

Howell, J.F. (1953). *Gonyaulax monilata sp. nov.,* the causative dinoflagellate of a red tide in the east coast of Florida in August-September 1951. *Trans. Am. Micr. Soc.* 72: 153-156.

Loeblich, A.R. and Loeblich, L.A. (1979). The systematics of *Gonyaulax* with special reference to the toxic species, pp. 41-46. In *Toxic Dinoflagellate Blooms*, Taylor, D.L., Seliger, H.H. (eds.), Elsevier, New York.

Montresor, A., Zingone, A., Marino, D.(1993). The paratabulate resting cyst of *Alexandrium pseudogoniualax* (Dinophyceae). pp. 159-164. In *Toxic Phytoplankton Blooms in the Sea*, Smayda, T.J., Shimizu, Y. (eds.), Elsevier, New York.

Needler, A.B.(1949). Paralytic shellfish poisoning and *Gonyaulax tamarensis. J.Fish.Res.Bd. Canada* 7: 490-504.

Paulsen, O. (1904). Plankton investigation in the water around Iceland in 1903. *Medd. Kom. Havunders. Kobenhaven, Ser. Plankton* 1: 1-40.

Paulsen, O. (1949). Observations on dinoflagellates. *K. Dan. Vidensk. Selsk., Biol. Skr.* 6: 1-67.

Pfeister, L., Anderson, D. (1987). Dinoflagellate reproduction. Chpt. 14, pp. 611-648 In *The Biology of Dinoflagell;ates*; Taylor, F.J.R. (ed.), Blackwell, Oxford.

Prakash, A., Taylor, F.J.R. (1966). A 'red water' bloom of *Gonyaulax acatenella* in the Strait of Georgia and its relation to paralytic shellfish toxicity. *J. Fish. res. Bd. Canada* 23: 1265-1270.

Schantz, E.J., Lynch, J.M., Vayvada, G., Matsumoto, K., Rapoport, H.(1966). The purification and charcterization of the poison produced by *Gonyaulax catenella* in axenic culture. *Biochem.* 5: 1191-1195.

Scholin, C.A., Hallegraeff, G.M., Anderson, D.M. (1995) Molecular evolution of the *Alexandrium tamarense* 'species complex' (Dinophyceae): dispersal in the North American and West Pacific regions. *Phycologia* 34: 472-485.

Shumway, S. E. (1995). Phycotox-related shellfish poisoning: bivalve molluscs are not the only vectors. *Reviews in Fisheries Science* 3: 1-31.

Sommer, H., Meyer, K.F. (1937). Paralytic shellfish poisoning. *Arch. Pathol.* 24: 560-598.

Sommer, H., Whedon, W.H., Kofoid, C.A., Stohler, R. (1937). Relation of paralytic shellfish poison to plankton organisms of the genus *Gonyaulax*. *Arch. Pathol.* 24: 537-559.

Steidinger, K.A. (1973). Phytoplankton research: a conceptual review based on eastern Gulf of Mexico research. *CRC Rev. Microb.* 3: 49-68.

Taylor, F.J.R. (1968). Parasitism of the toxin-producing dinoflagellate by the endoparasitic dinoflagellate *Amoebophrya ceratii*. *J. Fish. Res. Bd. Canada* 25: 2241-2245.

Taylor, F.J.R. (1975). Taxonomic difficulties in red tide and paralytic shellfish poison studies: the "*Tamarensis* complex" of *Gonyaulax*. *Env. Letters* 9: 103-109.

Taylor, F.J.R. (1976). Dinoflagellates from the International Indian Ocean Expedition. *Bibliotheca Bot.* 132: 1-234+46pls.

Taylor, F.J.R. (1979). The toxigenic gonyaulacoid dinoflagellates.pp. 45-56. In *Toxic Dinoflagellate Blooms,* Taylor, D.L., Seliger, H.H.(eds.), . Elsevier, North Holland, N.Y.

Taylor, F.J.R.(1987a). Ecology of marine dinoflagellates. General and marine ecosystems. pp. 398- 502, In *The Biology of Dinoflagellates,* Taylor, F.J.R. (ed.), Blackwell, Oxford.

Taylor, F.J.R. (1987b). Taxonomy and classification. Appendix, pp.723-731, In *The Biology of Dinoflagellates,* Taylor, F.J.R. (ed.), Blackewll, Oxford

Taylort, F.J.R. (1992). The taxonomy of harmful phytoplankton. *Giorn. Bot. Ital.* 126: :

Taylor, F.J.R., Fukuyo, Y.,Larsen, J. (1995). Taxonomy of harmful dinoflagellates. pp. 283-317, In *Manual on Harmful Microalgae*, Hallegraeff, G., Anderson, D.M., Cembella, A.D. Enevoldsen, H. (eds.), IOC UNESCO, Paris.

Turpin, D.H., Dobell, P.E.R., Taylor, F.J.R. (1978). Sexuality and cyst formation in Pacific strains of the toxic dinoflagellate *Gonyaulax tamarensis*. *J. Phycol.* 14: 235-238.

Wall D. (1975). Taxonomy and cysts of red tide dinoflagellates, pp. 242-255. In *Toxic Dinoflagellate Blooms*., LoCicero, V.R. (ed.), Mass. Sci. and Technol. Found., Wakefield, Mass.

Wall, D., Dale, B. (1969). The "hystrichosphaerid" resting spore of the dinoflagellate *Pyrodinium bahamense* Plate, 1906. *J. Phycol.* 5: 140-149.

Whedon, W.F., Kofoid, C.A. (1936). Dinoflagellates of the San Francisco region. 1. On the skeletal morphology of two new species, *Gonyaulax catenella* and *Gonyaulax acatenella* . *Univ. Cal. Publ. Zool.* 41: 25-34.

White, A. W. (1981). Sensitivity of marine fishes to toxins from the red-tide dinoflagellate *Gonyaulax excavata* and implications for fish kills. *Mar. Biol.* 65: 255-260.

Woloszynska, J., Conrad,W. (1939). *Pyrodinium phoneus*, n.sp., agent de la toxicité des moules du canal maritime de Bruges a Zeebrugge. *Bull. Mus. Roy. Belge* 15: 1-5.

Morphological, Genetic, and Biogeographic Relationships of the Toxic Dinoflagellates *Alexandrium tamarense*, *A. catenella*, and *A. fundyense*

Christopher A. Scholin

Monterey Bay Aquarium Research Institute
P.O. Box 628
Moss Landing CA 95039 USA

1. Introduction

Marine dinoflagellates of the genus *Alexandrium* are arguably the best characterized harmful algal species known. Some, but not all, representatives of this diverse group produce potent toxins responsible for paralytic shellfish poisoning (PSP), a neurological affliction that has caused human illness for centuries and claimed hundreds of lives (Quayle 1969; Prakash *et al.* 1971). Not surprisingly, *Alexandrium* spp. have captured the imagination of researchers, with investigations spanning morphology, biochemistry, toxicity, genetics, bloom dynamics, population biogeography, dispersal and evolution. All of these themes of research depend on a unified systematic scheme to define inter and intra-specific boundaries. At present, the only internationally accepted standard for distinguishing between such groups rests largely on cells' morphological features and life histories (e.g., Taylor 1985). In some cases, however, these features may belie an underlying genetic diversity, an understanding of which is crucial to a wide range of investigations.

In an effort to better define species boundaries and to assess intraspecific genetic divergence among representatives of this genus, researchers have increasingly turned to comparisons using both morphological and subcellular criteria (Table 1). Results of these studies have made it clear that extensive genetic diversity may exist among a group of organisms that share an overall conserved set of morphological and life history characteristics. A problem thus arises when groups defined by traditional morphological-based descriptions ("morphospecies") do not coincide with groups identifiable by subcellular criteria ("genospecies"). How does one reconcile the differences? What do these differences tell us about the biology of the organisms in question? Confusing relationships between morphotypic and genotypic associations can confound numerous research initiatives. Nowhere is this situation better illustrated than with representatives of the so-called *Alexandrium* "*tamarensis* complex". The definition of species boundaries, the genetic basis of toxin production and means to distinguish between endemic and introduced flora of this group have all been debated with respect to the morphospecies concept (e.g., Taylor 1975, 1984, 1985; Balech and Tangen 1985; Steidinger 1990).

For at least some members of the *tamarensis* complex, the confusion over morphotypes and their relationship to cells' subcellular characteristics appears to be rooted in the organisms evolutionary history, global population biogeography and dispersal. This paper

NATO ASI Series, Vol. G 41
Physiological Ecology of Harmful Algal Blooms
Edited by D. M. Anderson, A. D. Cembella and
G. M. Hallegraeff

Table 1. Comparison of the different types of criteria applied to delineate representatives of *A. tamarense*, *A. catenella* and *A. fundyense*. The list is not exhaustive, but rather is a representation of the type of work done over the past decade.

Criterion	Example reference
Morphology	Taylor 1984, 1985 Fukuo 1985 Steidinger 1990 Steidinger and Moestrup 1990 Balech 1985, 1995
Isozyme Electrophoresis	Cembella et al. 1988 Sako et al. 1990 Hayhome et al. 1989
toxin profile	Oshima et al. 1982 Maranda et al. 1985 Cembella et al. 1987 Kim et al. 1993 Anderson et al. 1994
immunogenicity	Sako et al. 1993 Adachi et al. 1993
sexual crosses	Destombe and Cembella 1990 Sako et al. 1990 Sako et al. 1992
chloroplast RFLP patterns RRNA/rDNA sequences	Boczar et al. 1991 Destrombe et al. 1992 Adachi et al. 1994, 1996 Scholin et al. 1994 Scholin & Anderson 1996

examines this situation with an emphasis on globally distributed representatives of *A. tamarense*, *A. catenella* and *A. fundyense*. An evolutionary model is advanced to explain how the patterns observed today may have developed over the course of millions of years (Scholin *et al.* 1995). Given this global perspective it is possible to reconcile the apparent incongruence between morphospecies and genospecies designations without invalidating the morphospecies paradigm. Lastly, dispersal hypotheses are considered in light of the proposed evolutionary scheme with particular emphasis on North American, Japanese and Australian regional populations.

2 Relationships Between Morphological Features and Subcelluar Criteria

Alexandrium taxonomy is founded upon detailed descriptions of thecal plate morphology (Balech 1995). Continual re-evaluation of these characters has altered the group's generic and species concepts, leaving a legacy of confusing taxonomic designations. Detailed histories of these changes and the rationale behind them are found elsewhere and are not considered here (e.g., Taylor 1984, 1985; Balech 1985; Steidinger 1990). A consensus to use the *Alexandrium* genus designation was reached in 1989 (Steidinger and Moestrup 1990), but some debate over appropriate species and "strain" assignments continues (e.g., Anderson *et al.* 1994). *Alexandrium tamarense*, *A. catenella* and *A. fundyense* perhaps best illustrate this situation.

For years taxonomic authorities have agreed that *A. tamarense*, *A. catenella* and *A. fundyense* are closely related. The first two species are distributed widely throughout much of the world's coastal regions whereas *A. fundyense* is found primarily in eastern North America (Hallegraeff 1993, Balech 1995). The *A. tamarense* designation harbors both toxic and non-toxic strains. In contrast, all *A. catenella* and *A. fundyense* known to date are toxic. Cells' inherent toxicity or "potency" can vary significantly within and between regional populations (e.g., Oshima *et al.* 1992; Maranda *et al.* 1985; Cembella *et al.* 1987; Kim *et al.* 1993; Anderson *et al.* 1994). The distinction of these organisms as unique species rests primarily on the presence or absence of a ventral pore in the 1' apical plate, the overall shape of the cells and the propensity to form chains (Taylor 1984; Fukuyo 1985). However, morphological "intermediates" between the extremes are known to occur and may arise by changing environmental conditions and/or natural genetic variation (Sako et al. 1990). The question is, are *A. tamarense, A. catenella*, and *A. fundyense* distinct species or instead a continuum of strains or varieties of a single species? Secondly, what are the morphotypic and genetic relationships between different regional populations of these groups?

Different views as to the taxonomic positions of *A. tamarense*, *A. catenella* and *A. fundyense* and an interest in the organisms' population biogeography sparked years of research, examples of which are compiled in Table 1. Numerous investigations have focused on "testing" the morphospecies concept by defining groups of organisms using morphological criteria and then comparing those same organisms using some type of subcellular criteria. Isozyme electrophoretic patterns, toxin profiles, reaction towards poly and monoclonal antibodies, mating compatibility, chloroplast DNA restriction restriction enzyme patterns, and ribosomal RNA (rRNA) gene (rDNA) sequences have been documented for numerous isolates of *Alexandrium, A. tamarense* and *A. catenella* in particular. Today, the same themes are still under investigation in laboratories throughout the world, with new attention focused on developing rapid diagnostic tests to identify species collected from cultured or natural samples.

The intensive effort spent trying to reconcile the debate over morphospecies and genospecies concepts begs the question raised long ago: Are *A. tamarense*, *A. catenella* and *A. fundyense* distinct species? Ironically, in our quest to answer this question we have uncovered a much more fundamental and far older question: What is a species? Unfortunately, there is no simple answer to this deceptively complex question because the definition of "species" is inherently subjective (see Gallagher, this volume). It

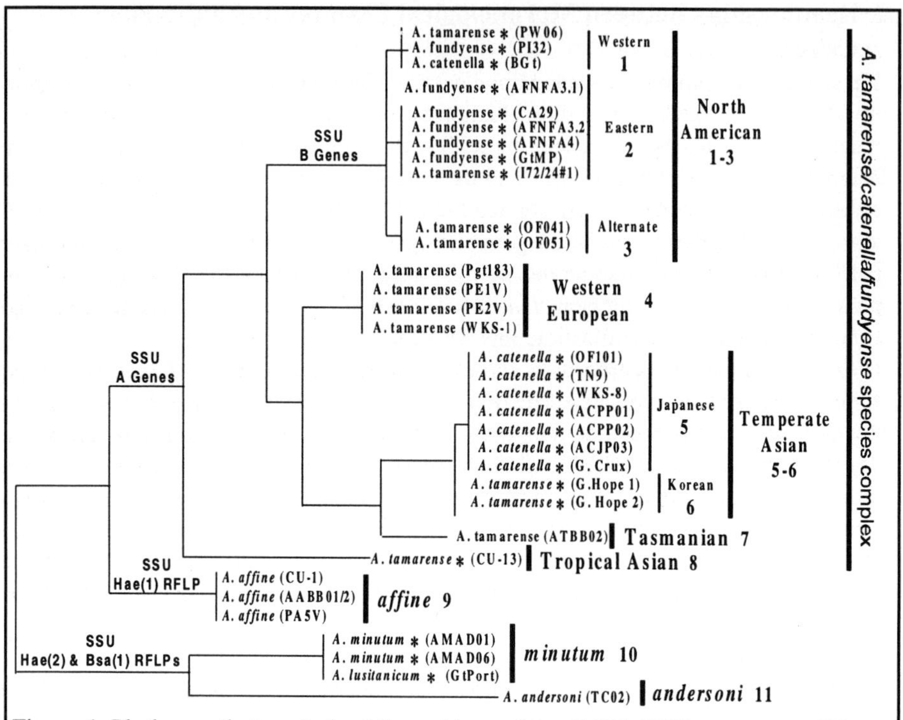

Figure 1. Phylogenetic tree derived from *Alexandrium* LSU rDNA sequences. SSU rDNA RFLP characteristics are also shown on appropriate branches (see Table 2; reprinted from Scholin et al. 1994)

is worth noting that in the vast majority of cases, species designations based on morphological criteria provide a remarkably accurate picture of distinct taxonomic units. For example, data from numerous laboratories shows clearly that *Alexandrium* are distinct from other dinoflagellate genera and that the *A. tamarense/catenella/fundyense* group as a whole is distinct from *A. affine*, *A. minutum*, *A. andersoni*, *A. insuetum* and *A. pseudo-gonyaulax*. In these cases, groups defined by morphological criteria agree with those defined by subcellular characteristics. The morphospecies paradigm "works" (e.g., see Adachi *et al.* 1994, 1996; Scholin *et al.* 1994). The utility of that approach only comes in to question when it fails to resolve satisfactorily the organisms in question. *Alexandrium tamarense*, *A. catenella* and *A. fundyense* exemplify the latter situation because taxonomists are faced with choosing species names in an historically correct context, while at the same time preserving important information about the divergence of these organisms and their genetically distinct populations.

Returning to the question at hand: Are *A. tamarense*, *A. catenella* and *A. fundyense* distinct species? Apparently not, at least when one considers both morphology and some measure of the organisms genetic divergence. In nature the three morphotypes encompass multiple evolutionary lineages, all of which appear to spring from a common ancestor (Cembella *et al.* 1988; Scholin *et al.* 1995). Among these groups there are no strict relationships between "morphospecies" and "genospecies." That is, morphological features used to define the three species may in fact appear to be positively *or* negatively correlated with subcellular criteria (see refs. in Table 1). Based on rDNA sequences, *A. tamarense*, *A. catenella* and *A. fundyense* comprise at least five major groups, two of which harbor distinct subgroups (Fig. 1; see also Adachi *et al.* 1994, 1996). Scholin *et al.* (1994) have noted that these divisions agree with and further resolve clades defined by isozyme electrophoresis, immunogenecity, and cells' overall capacity to produce toxin (see also Scholin and Anderson 1996).

The extensive genetic diversity among isolates of *A. tamarense* and *A. catenella* was first recognized in the mid 1980's. Further characterization of these lineages has compelled researchers to ascribe particular names to the groups in order to simplify discussions of their existence and their relatedness. Already a number of new, non-traditional nomenclatures have been advanced, some of which use different designations for the same group of organisms (Table 2). This author has labeled phylogenetic clades of *Alexandrium* as "ribotypes" and "subribotypes" using geographic locales to denote lineages of *tamarensis* complex representatives, or species designations for other congeners (Fig. 1). At present there are no standards for integrating results of molecular characterizations with accepted morphological descriptions. It is reasonable to assume that use of strain designations will continue, even expand, and therefore some consideration should be given soon as to how this information can be best integrated into the existing taxonomic scheme.

3. Population Biogeography and the Monophyletic Dispersal Hypothesis

Cembella *et al.* (1988) were the first to suggest that confusion over *A. tamarense* and *A. catenella* morphotypes and their relationship to cells' subcellular characteristics was rooted in the organisms evolutionary history. Cladistic analysis of electrophoretic variants led those authors to conclude that "there is little evidence that the morphotypic extremes (*catenella* versus *tamarensis*) are derived from distinctly different ancestral lines". Subsequent analyses have unanimously supported this notion, with further recognition that observed relationships between morphotypes and cells' genetic similarities can depend on the geographic origin of isolates in question (Sako *et al.* 1990, Cembella *et al.* 1988, Hayhome *et al.* 1989).

For example, some *A. tamarense, A. catenella* and *A. fundyense* isolated from North America share a high degree of genetic identity despite their morphological differences. In contrast, some *A. tamarense* isolated from western Europe and North America are genetically distinct even though they are morphologically indiscriminable. Analogous situations are found among other *A. tamarense* and *A. catenella* isolated from Australia, Asia, North America and Korea (Scholin and Anderson 1994; Scholin *et al.* 1994). Recognition of phenotypic plasticity within a genetically similar population (e.g., North American *A. tamarense* and *A. catenella*) and phenotypic overlap between genetically distinct populations (e.g., western European and North American *A. tamarense*) provides a framework from which to view the evolution and dispersal of these organisms.

The possibility that *A. tamarense* and *A. catenella* are distinct species versus varieties of the same species may be considered in the context of two alternative evolutionary histories, or phylogenetic trees, labeled here as "Polyphyletic Radiation" (Fig. 2a)

Table 2. A comparison of different nomenclatures used to define varieties of *A. tamarense, A. catenella* and *A. fundyense* collected from globally distributed regional populations.

SSU RFLP Group[a]	ITS-Type[b]	LSU ribotype[c]	Example Strain[d]	Toxic?[e]	Species Designation	Isolation locale[f]
I	*tamarense*	North American				
		Eastern*				
			AFNFA3	Yes	*A. fundyense*	Newfoundland
			AFNFA4	Yes	*A. fundyense*	Newfoundland
			GtCA29	Yes	*A. fundyense*	Cape Ann, MA
			GtMP	Yes	*A. fundyense*	Orleans, MA
			I72/24#1	Yes	*A. tamarense*	Ballast water (Muroran, Japan)**
II		Western*				
			PW06	Yes	*A. tamarense*	Port Benny, AK
			PI32	Yes	*A. fundyense*	Porpoise Isl., AK
			BGt1	Yes	*A. catenella*	Russian River, CA
		Alternate*				
			OF041	Yes	*A. tamarense*	Ofunato Bay, Japan
			OF051	Yes	*A. tamarense*	Ofunato Bay, Japan
III	WKS1	Western European				
			Pgt183	No	*A. tamarense*	Plymouth, U.K.
			PE1V	No†	*A. tamarense*	Galicia, Spain
			PE2V	No	*A. tamarense*	Galicia, Spain
			WKS-1	No	*A. tamarense*	Tanabe Bay, Japan
	catenella	Temperate Asian				
		Japanese*				
			OF101	Yes	*A. catenella*	Ofunato Bay, Japan
			TN9	Yes	*A. catenella*	Tanabe Bay, Japan
			WKS-8	Yes	*A. catenella*	Tanabe Bay, Japan
			ACPP01	Yes	*A. catenella*	Port Phillip, Australia
			ACPP02	Yes	*A. catenella*	Port Phillip, Australia
			ACJP03	Yes	*A. catenella*	Ballast water (Kashima, Japan)
			G. Crux	Yes	*A. catenella*	Ballast water (Singapore)***
		Korean*				
			G. Hope 1	Yes	*A. tamarense*	Ballast water (Samchonpo, S. Korea)**
			G. Hope 2	Yes	*A. tamarense*	Ballast water (Samchonpo, S. Korea)**
		Tasmanian				
			ATBB01	No†	*A. tamarense*	Bell Bay, Tasmania
	Thai	Tropical Asian				
			CU13	Yes	*A. tamarense*	Gulf of Thailand

[a]Based on results of the A/B gene restriction tests (Scholin and Anderson 1994).
[b]Subdivisions based on analysis of rDNA ITS regions (Adachi et al. 1994, 1996).
[c]Subdivisions based on LSU rDNA phylogeny (Fig. 1); *"subribotype" designations based on fine-scale LSU rDNA sequence variation (Scholin et al. 1994).
[d]See Scholin et al. 1994
[e]Determined by mouse bioassy and/or HPLC analysis; †= may contain trace amounts of toxin (D. Kulis, pers.comm.).
[f]**Origin of ballast water (Hallegraeff and Bolch 1992);
***Hailing port of vessel; origin of ballast water uncertain (Hallegraeff and Bolch 1992).

and "Monophyletic Radiation" (Fig. 2b; Scholin *et al.* 1995). For simplicity, the models consider the tamarensoid and catenelloid morphotypes only. Primordial populations of *A. tamarense* and *A. catenella* ("protocat" and "prototam", respectively) are presumed to have dispersed to various regions of the world over millions of years and to have subsequently diverged giving rise to morphologically similar, genetically distinct, regional populations (numbered 1-3). Present-day phylogenetic relationships of the two morphotypes and their correspondence to geographic populations are hypothesized to depend on whether or not the organisms arose from distinct ancestral lines (Polyphyletic Radiation, Fig. 2a), or instead radiated from a single ancestor that included or gave rise to the morphotypes we observed today (Monophyletic Radiation, Fig 2b). Scholin *et al.* (1995) examined these two scenarios from the perspective of hypothetical phylogenetic trees and compared those predictions to actual phylogenies derived from SSU and LSU rDNA data. A summary of that work is described here.

For both models it is assumed that the immediate ancestor(s) of *A. tamarense* and *A. catenella* arose sometime in the ancient past, perhaps tens of millions of years ago. As time went by these organisms radiated naturally to many regions of the world by way of water mass movements and perhaps episodically by migratory water birds or violent storms, etc. (R. Scheltema and A. Gibor, pers. com.) At the same time, coastlines of spreading continents changed, sea ways opened and closed, and the earth saw dramatic alterations in climate. The combination of these events, over millions of years, are predicted to have given rise to sexually isolated, regional populations scattered throughout the world's oceans.

Scholin *et al.* (1995) proposed that prolonged isolation and independent evolution of regional groups could be "recorded" in the genomes of their descendants given the processes of genetic drift and selection, and that rDNA sequences were among those portions of the genome where this record might be found. If this were the case, then different regional populations of either morphotype could harbor divergent rDNA sequences (e.g., populations 1-3 in Fig. 2). However, if *A. tamarense* and *A. catenella* arose polyphyletically, then their overall phylogeny should reveal at least two different lineages as exemplified in Fig. 2a. For simplicity, the polyphyletic scenario originally put forward assumed that the two lineages also corresponded to two distinct morphotypes. In reality, however, two different lineages could harbor representatives with overlapping morphologies, this being a case of "convergent evolution". Regardless of whether the tamarensoid and catenlloid morphotypes arose through processes of divergence or convergence, the polyphyletic radiation model yields the same prediction: that a reconstruction of their evolution history by means of a molecular phylogeny will reveal at least two different, unrelated (or at least distantly related) ancestral stocks (see also Avise 1994).

In contrast, the monophyletic radiation model (Fig. 2b) predicts a molecular phylogeny whose termini ascribe genetically distinct regional populations. Each of these populations represent divergent strains or varieties that arose from the *same* primordial stock, a common ancestor that either included or gave rise to the spectrum of morphotypes presently observed. If this were the case, then representatives of *A. tamarense* and *A. catenella* that descended from the same population will appear most closely related genetically, whereas those from different populations will appear divergent; regardless of their morphotypic affinities (Fig. 2b; Avise 1994).

Scholin *et al.* (1995) concluded that the phylogeny shown in Fig. 1 and the extensive body evidence preceding it (Table 1) are most consistent with the monophyletic dispersal hypothesis. This conclusion was based on several facts, including that the *A. tamarense/catenella/fundyense* group is monophyletic with respect to other dinoflagellate genera as well as the closely related congeners *A. affine*, *A. minutum* and *A. andersoni*, that distinct yet co-occurring morphospecies can have similar (or identical) rDNA sequences, and that the overall phylogeny is one of geographic populations not distantly related, independent lineages that share a fortuitous convergent morphology.

With respect to *A. tamarense*, *A. catenella* and *A. fundyense*, how does the monophyletic dispersal hypothesis answer the question as to whether or not the three are distinct species *without* invalidating the morphospecies paradigm? From an operational perspective, *tamarense*, *catenella* and *fundyense* are "good" morphospecies because they ascribe three groups of organisms that are recognizable on the basis of a defined set of morphological characteristics (e.g., Balech 1995). Eliminating those designations would clearly result in a loss of valuable information (Anderson *et al.* 1994). Given the monophyletic dispersal model, however, one can also view those distinct groups in a broader context of their evolution and population biogeography, and thereby provide an explanation as to why "tests" of morphological-based taxonomic schemes have yielded conflicting results (Scholin *et al.* 1994).

4. Evidence for Natural and Human-Assisted Dispersal

Like other harmful phytoplankton, *Alexandrium* spp. are thought by some to be dispersing to regions of the world thought to be previously free of their presence. Several theories have been put forward to explain this apparent trend including, increased growth of endemic species, natural dispersal via ocean currents, human-assisted dispersal resulting from ships' ballast water or exchange of shellfish stocks, or a combination of all the above (Anderson 1989; Smayda 1990; Hallegraeff and Bolch 1992). The difficulty in distinguishing between dispersal hypotheses lies in differentiating endemic and introduced flora.

A specific prediction of the monophyletic radiation hypothesis is that *A. tamarense* and *A. catenella* descended from the same regional population are genetically similar, whereas those of different populations are genetically distinct, irrespective of morphotype (Fig. 2b). In the absence of dispersal, genetically distinct populations of *A. tamarense* and *A. catenella* should be limited in their global distribution. A further prediction is that each endemic, regional population of these species can be defined by a unique set of characters that include both morphological *and* subcellular criteria. If dispersal occurs after populations have diverged, however, then the established pattern of genetically distinct, geographically separate populations will change: dispersed populations will harbor the morphotypic and genetic signatures indicative of the region from which they descended but inhabit a different location than that of their parent population. Patterns of indigenous and introduced flora may thus be inferred by defining phylogenetic and morphotypic relationships of extant populations, and by viewing those continuities or discontinuities in the context of geography, the historic record, and potential natural or human-assisted dispersal mechanisms (see Brooks and McLennan 1991; Avise 1994).

a. Polyphyletic Radiation

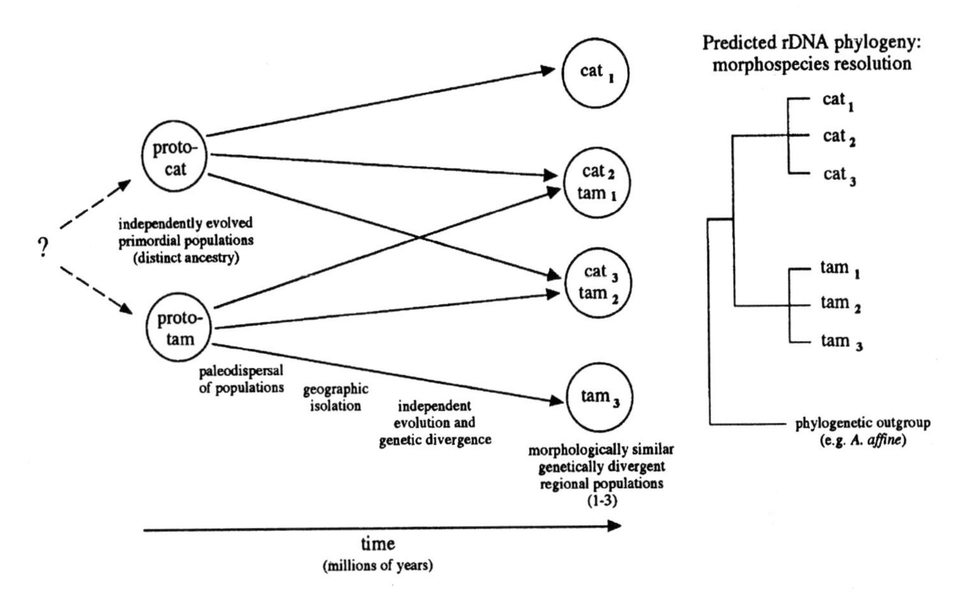

b. Monophyletic Radiation

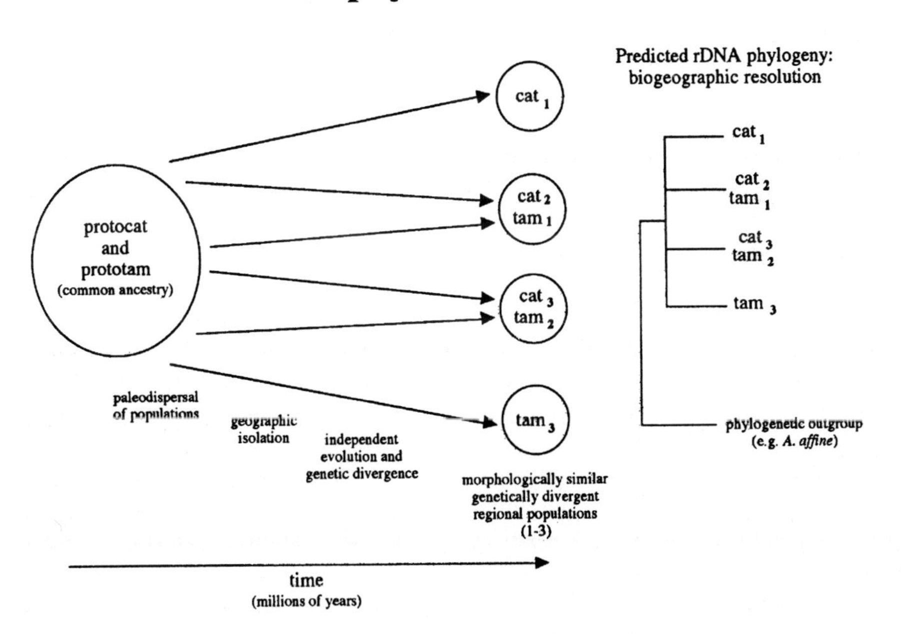

Figure 2. Hypothetical models that describe alternative evolutionary histories of *A. tamarense/catenella* and their corresponding phylogenies. See text for details (reprinted from Scholin et al. 1995).

Scholin *et al.* (1995) were the first to apply this concept in an attempt to explain the biogeography of genetically distinct representatives of *A. tamarense*, *A. catenella* and *A. fundyense* found in North America, Japan and Australia (summarized in Fig. 3). The authors concluded that eastern and western North American populations examined thus far on the basis of rDNA sequences were descended from a common population which at one time may have extended from the Bering Strait to the Labrador Sea. The fact that the two populations are distinguishable genetically today was taken as further evidence of a natural dispersal that occurred in the recent *geological* past, perhaps by way of the Arctic Ocean. Indeed, *A. tamarense* observed in plankton samples from the Beaufort, Labrador and Greenland Seas indicates that this species occurs at high latitudes (Taylor pers. comm.). The natural dispersal scenario predicts that those organisms should share a high degree of genetic similarity with their counterparts in eastern and western North America. The extensive genetic diversity found among *A. tamarense* and *A. catenella* from western North America (e.g., Cembella et al. 1988) is also indicative of dispersal, but in this case is predicted to have resulted from secondary contact of regional populations previously isolated for millions of years. An inference of whether or not the latter dispersal took place in ancient or modern times awaits further study (see Scholin 1993).

The remarkable genetic diversity among *A. tamarense* and *A. catenella* found in Japan suggests dispersal has occurred in this region of the world too. As above, secondary contact of regional populations is a good explanation for this pattern. At least four different lineages of the *tamarensis* complex occur in Japan: three groups of *A. tamarense* and one group of *A. catenella.* These groups are defined on the basis of isozymes, reactivity towards monoclonal antibodies, toxicity, and rDNA sequences (Table 1). Using this author's nomenclature, these organisms represent toxic *A. tamarense* (eastern and alternate North American ribotypes), non-toxic *A. tamarense* (western European ribotype) and toxic *A. catenella* (temperate Asian ribotype; Fig. 1, Table 2).

A combination of natural and human-mediated dispersals may be responsible for the association of morphotypes and corresponding genotypes of *A. tamarense/catenella* found in Japan today. As with a possible exchange between eastern and western North America, a natural dispersal of *A. tamarense* between North America and Japan is possible and may have occurred over a period of time similar to that described above. If this were the case, then extant populations of *A. tamarense* between these two regions should share a high degree of genetic and morphological identity. Thus far, limited tests support this notion as *A. tamarense* isolated from the Kamchatka Peninsula share genetic and morphotypic signatures indicative of those found in both Japan and North America (Scholin *et al.*, unpubl. obsv.). There is also good reason to believe that *A. tamarense* was introduced to Japan by human activity in modern times. The exchange of shellfish stocks and increased intercontinental shipping between Japan and other countries of the world are potential mechanisms whereby such transfers could occur (Fig. 3; e.g., Hallegraeff and Bolch 1992; Taylor, pers. comm.).

Similar to Japanese populations of *A. tamarense* and *A. catenella*, those found in Australia may have arrived through a combination of natural and human-assisted means. For example, until recently a non-toxic variety of *A. tamarense* was not known to occur in Australia. Increased awareness of the phytoplankton inhabiting the waters of that country is the likely reason that species was discovered. Indeed, the non-toxic *A. tamarense* from Tasmania is unique, being the sole representative of the Tasmanian ribotype (Fig.

1). This lineage is characteristic of what one might call a “hidden flora” given that it could have gone unnoticed for centuries and that it harbors a unique genetic signature not known anywhere else in the world.

In contrast, *A. catenella* isolated from Australia and Japan are genetically similar. One would predict, therefore, that Japanese and Australian *A. catenella* have descended from the same parental stock. As in the examples above, natural dispersal of this species between Asia and Australia cannot be ruled out. For example, reductions of sea level and equatorial sea surface temperatures during the last Pleistocene glacial maximum (ca. 18,000 years ago) may have provided a means by which this could have occurred (e.g., Fleminger 1985; van Oppen *et al.* 1994). Once again, a natural dispersal scenario predicts a genetic continuity of extant *A. catenella* populations along the proposed route (Fig. 3). *A. catenella* recently isolated from the coast of China agree with this prediction (Scholin *et al.*, unpubl. obsv.).

Human-mediated dispersals by way of ships’ ballast water are also a good explanation for the genetic continuity between Asian and Australian populations of *A. catenella*. In at least one instance a cargo vessel ballasted in Japan during an *A. catenella* bloom arrived in Australia containing viable *A. catenella* cysts that are genetically indiscriminable (by rDNA) to known populations of this species in both countries (Scholin *et al.* 1994). In two other separate cases, ships ballasted during *A. tamarense* blooms in Japan and South Korea also

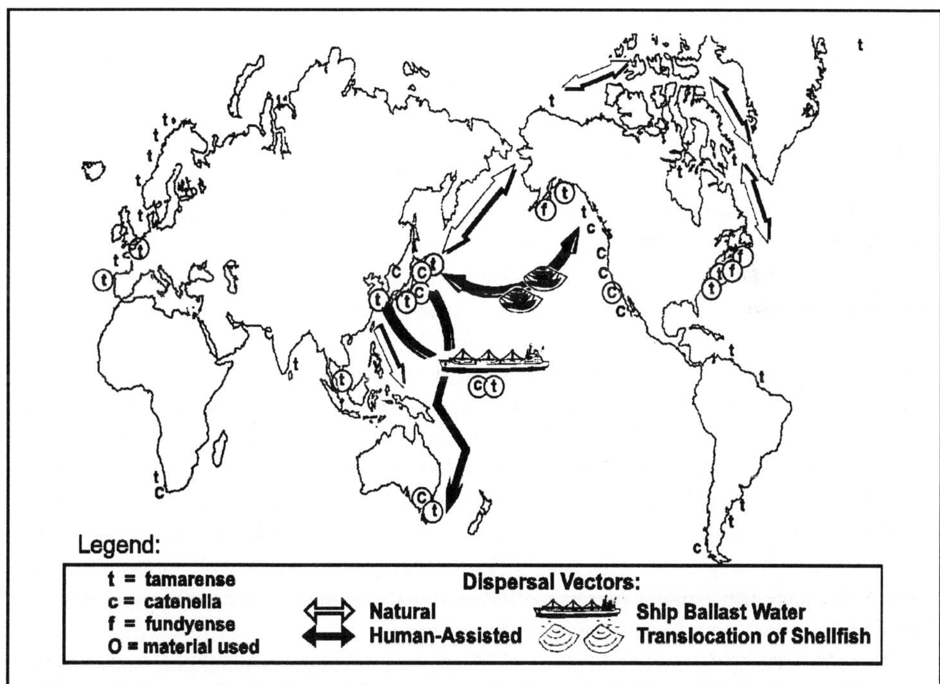

Figure 3. Known distributions of *A. tamarense* (t), *A. catenella* (c), and *A. fundyense* (f), and representatives brought into culture and used to construct the phylogenetic tree shown in Figure 1 (circled). Hypothesized routes of natural (hollow arrows) and human-assisted (solid arrows) disperal are shown (from Scholin et al. 1995).

inadvertently transported viable resting cysts to Australia. However, these cysts proved to be genetically distinct from the *A. tamarense/catenella* previously found in Australia, and genetically distinct from each other as well (Scholin *et al.* 1994). The "Australian inocula" of these species thus includes those of non-toxic *tamarense* (Tasmanian ribotype), toxic *catenella* (Japanese temperate Asian ribotype), and two varieties of toxic *tamarense* (Korean temperate Asian and eastern North American ribotype; Fig. 1). In Australian waters the latter two are known as cysts from ships' ballast water only. If those strains were introduced successfully, then one would predict that those areas inoculated with cysts could experience toxic blooms of cells that will appear of the tamarensoid morphotype with one of two known genotypes (eastern North American or Korean temperate Asian). However, to date, this has not been observed.

While human-mediated transport of viable *Alexandrium* resting cysts have been documented, and regional dispersal by natural means seems possible, proof that such events have in fact resulted in successful colonizations remains elusive. The uncertainty of dispersal timing stems from the relatively slow rate at which rDNA evolves and a lack of fossil evidence (Scholin *et al.* 1995). Nevertheless, further inference of dispersal events and their relative timing may be found by analysis of portions of genomes that may undergo more rapid rates of evolution (Avise 1994) and by mapping those groups identified onto the existing and growing rDNA-based phylogeny.

5. Conclusions

Alexandrium tamarense, *A. catenella* and *A. fundyense* can be defined as a suite of divergent lineages (strains or varieties) that share overlapping morphological characteristics. It is probable that the organisms arose from a common ancestor in the geological past and have since colonized many regions of the earth, thereby establishing isolated regional populations. Subsequent natural and human-mediated dispersals have disrupted this primordial pattern giving rise to the organisms' morphotypic, genetic and biogeographic characteristics observed today. The monophyletic dispersal hypothesis provides a theoretical perspective from which to view such events, and one means for distinguishing between endemic and introduced species.

There are a variety of antibody and nucleic acid probe-based tests now available that allow researchers to define particular groups of the *tamarensis* complex rapidly (Sako *et al.* 1993; Adachi *et al.* 1994; Scholin and Anderson 1996). This information can be cross referenced with a wide variety of other data such as morphology, toxicity, isozyme electrophoretic patterns and breeding group affinities. Application of rapid techniques for identifying particular strains of the *tamarensis* group will improve our understanding of the biogeography and genetic diversity of populations of this important group of organisms. Screening isolates representative of global populations not yet examined using molecular criteria (e.g., South America, South Africa) are now in progress with the goals of reconstructing dispersal pathways and identifying novel groups that may exist but are not yet known.

6. Acknowledgements

F.J.R. Taylor is thanked for reviewing early versions of this chapter and for providing many helpful comments to improve its contents. Cyndi Stubbs and Kelly Burgess are thanked for their expert technical assistance. This work was supported by the Monterey Bay Aquarium Research Institute through funds from the David and Lucile Packard Foundation.

7. References

Adachi, M., Sako, Y., Ishida, Y., Anderson, D.M., Reguera, B. (1993). Cross-reactivity of five monoclonal antibodies to various isolates of *Alexandrium* as determined by an indirect immunofluorescence method. *Nippon Suisan Gakkaishi* 59:1807.

Adachi, M., Sako, Y., Ishida, Y. (1994). Restriction fragment length polymorphism of ribosomal DNA internal transcribed spacer and 5.8S regions in Japanese *Alexandrium* species (Dinophyceae). *J. Phycol.* 30:857-863.

Adachi, M., Sako, Y., Isida, Y. (1996). Analysis of *Alexandrium* (Donophyceae) species using sequences of the 5.8S ribosomal DNA and internal transcribed spacer regions. J.Phycol. 32:424-432

Anderson, D.M. (1989). Toxic algal blooms and red tides: a global perspective. *In:* Okaichi,T., Anderson, D.M., Nemoto, T. [Eds.] *Red Tides: Biology, Environmental Science and Toxicology*. Elsevier, New York, pp. 11-20.

Anderson, D.M., Kulis, D.M., Doucette, G.J., Gallagher, J.C., Balech, E. (1994). Biogeography of toxic dinoflagellates in the genus *Alexandrium* from the northeastern United States and Canada. *Mar. Biol. (Berl.)* 120: 467-478.

Avise, J.C. (1994). *Molecular Markers, Natural History and Evolution*. Chapman and Hall, New York. 511 pp.

Balech, E. (1985). The genus *Alexandrium* or *Gonyaulax* of the tamarensis group. *In*: Anderson, D.M., White, A.W., Baden, D.G. [Eds.] *Toxic Dinoflagellates.* Elsevier, New York, pp. 33-38.

Balech, E. (1995). *The Genus Alexandrium Halim (Dinoflagellata).* Sherkin Island Marine Station, Cork, Ireland. 151 pp.

Balech, E., Tangen, K. (1985). Morphology and taxonomy of toxic species in the *tamarensis* group (Dinophyceae): *Alexandrium excavatum* (Braarud) comb. nov. and *Alexandrium ostenfeldii* (Paulsen) comb. nov. *Sarsia* 70: 333-43.

Boczar, B.A., Liston, J., Cattolico, R.A. (1991). Characterization of Satellite DAN from three marine dinoflagellates (Dinophyceae): *Glenodinium* sp. and two members of the toxic genus, *Protogonyaulax*. *Plant Physiol.* 97:613-618.

Brooks, D.R., McLennan, D.A. (1991). *Phylogeny, Ecology and Behavior. A Research Program in Comparative Biology*. University of Chicago Press, Chicago, London. 434 pp.

Cembella, A.D., Sullivan, J.J., Boyer, G.L., Taylor, F.J.R., Andersen, R.J. (1987). Variation in paralytic shellfish toxin composition within the *Protogonyaulax tamarensis/ catenella* species complex: red tide dinoflagellates. *Biochem. Syst. Ecol.* 15: 171-86.

Cembella, A.D., Taylor, F.J.R., Therriault, J-C. (1988). Cladistic analysis of electrophoretic variants within the toxic dinoflagellate genus *Protogonyaulax*. *Bot. Mar.* 31: 39-51.

Destombe, C., Cembella, A.D. (1990). Mating-type determination, gametic recognition and reproductive success in *Alexandrium excavatum* (Gonyaulacales, Dinophyta), a toxic red-tide dinoflagellate. *Phycologia* 29: 316-325.

Destombe, C., Cembella, A.D., Murphy, C.A., Ragan, M.A. (1992). Nucleotide sequence of the 18S ribosomal RNA genes form the marine dinoflagellate *Alexandrium tamarense* (Gonyaulacales, Dinophyta). *Phycologia* 31: 121-4.

Fleminger, A. (1985). The Pleistocene equatorial barrier between the Indian and Pacific Oceans and a likely cause for Wallace's line. In: Pierot-Bults, A.C., van der Spoel, S., Zahuranec B.J., Johnson, R.K. [Eds.] *Pelagic Biogeography. Proceedings of an International Conference. The Netherlands. Unesco Technical Papers in Marine Science* 49:84-97.

Fukuyo, Y. (1985). Morphology of *Protogonyaulax tamarensis* (Lebour) Taylor and *Protogonyaulax catenella* (Whedon and Kofoid) Taylor from Japanese coastal waters. *Bull. Mar. Sci.* 37: 529-537.

Hallegraeff, G., Bolch, C.J. (1992). Transport of toxic dinoflagellate cysts via ship's ballast water: implications for plankton biogeography and aquaculture. *J. Plankton Res.* 14: 1067-1084.

Hallegraeff, G. (1993). A review of harmful algal blooms and their apparent global increase. *Phycologia* 32: 79-99.

Hayhome, B.A., Anderson, D.M., Kulis, D.M., Whitten, D.J. (1989). Variation among congeneric dinoflagellates from the northeastern United States and Canada I. Enzyme electrophoresis. *Mar. Biol.* (*Berl.*) 101: 427-435.

Kim, C-H., Sako, Y., Ishida Y. (1993). Comparison of toxin composition between populations of *Alexandrium* spp. from geographically distant areas. *Nippon Suisan Gakkaishi* 59: 641-646.

Maranda, L., Anderson, D.M., Shimizu, Y. (1985). Comparison of toxicity between populations of *Gonyaulax tamarensis* of eastern North American waters. *Est. Coastal Shelf Sci.* 21: 401-410.

Oshima, Y., Hayakawa, M., Hashimoto, M., Kotaki, Y., Yasumoto, T. (1982). Classification of *Protogonyaulax tamarensis* from North Japan into three strains by toxin composition. *Nippon Suisan Gakkaishi* 48: 851-854.

Prakash, A., Medcof, J.C., and Tennant, A.D. (1971). Paralytic shellfish poisoning in eastern Canada. *Bull. Fish. Res. Board Can.* 177: 1-87.

Quayle, D.B. (1969). Paralytic shellfish poisoning in British Columbia. *Bull. Fish. Res. Board Can.* 168: 1-68.

Sako, Y., Adachi, M., Ishida, Y. (1993). Preparation and characterization of monoclonal antibodies to *Alexandrium* species. *In:* Smayda, T.J., Shimizu, Y. [Eds.] *Toxic Phytoplanton Blooms in the Sea.* Elsevier, New York, pp. 87-93.

Sako, Y., Uchida, A., Ishida, Y. (1985). Electrophoretic analysis of isozymes in red tide dinoflagellates (*Gymodinium nagasakiense, Protogonyaulax tamarensis,* and *Peridinium bipes*). *In*: Okaichi, T., Anderson, D. M., Nemotò, T. [Eds.] *Red Tides: Biology, Environmental Science and Toxicology*. Elsevier, New York, pp. 325-328

Sako, Y., Kim, C.H., Ninomiya, H., Adachi, M., Ishida, Y. (1990). Isozyme and cross analysis of mating populations in the *Alexandrium catenella/tamarense* species complex. *In:* Graneli, E. Sundstrom, B., Edler, L., Anderson, D.M. [Eds.] *Toxic Marine Phytoplankton.* Elsevier, New York, pp. 320-323.

Scholin, C.A. (1993). Analysis of toxic and non-toxic *Alexandrium* (Dinophyceae) species using ribosomal RNA gene sequences. Ph.D. thesis, Massachusetts Institute of Technology/Woods Hole Oceanographic Institution, WHOI-93-08, 251 pp.

Scholin, C.A., Anderson, D.M. (1994). Identification of species and strain-specific genetic markers for globally distributed *Alexandrium* (Dinophyceae). I. RFLP analysis of SSU rRNA genes. *J.Phycol.* 30:744-754.

Scholin, C.A., Anderson, D.M. (1996). LSU rDNA-based assays for discriminating species and strains of *Alexandrium* (Dinophyceae). *J. Phycol. 32:1022-1035.*

Scholin, C.A., Herzog, M., Sogin, M.L., Anderson, D.M., (1994). Identification of group- and strain-specific genetic markers for globally distributed *Alexandrium* (Dinophyceae). II. Sequence analysis of a fragment of the LSU rRNA gene. *J. Phycol.* 30:999-1011.

Scholin, C.A., Hallegraeff, G.M., Anderson, D.M. (1995). Molecular evolution of the *Alexandrium tamarense* "species complex" (Dinophyceae): dispersal in the North American and West Pacific regions. *Phycologia* 34: 472-485.

Smayda, T.J. (1990). Novel and nuisance phytoplankton blooms in the sea:evidence for a global epidemic. *In:* Graneli, E. Sundstrom, B., Edler, L., Anderson, D.M. [Eds.] *Toxic Marine Phytoplankton.* Elsevier, New York, pp. 29-40.

Steidinger, K.A. (1990). Species of the *tamarensis/catenella* group of *Gonyaulax* and the fucoxanthin derivative-containing Gymnodiniods. *In:* Graneli, E. Sundstrom, B., Edler, L., Anderson, D.M. [Eds.] *Toxic Marine Phytoplankton.* Elsevier, New York, pp. 11-16.

Steidinger, K.A., Moestrup, Ø. (1990). The taxonomy of *Gonyaulax*, *Pyrodinium*, *Alexandrium*, *Gessnerium*, *Protogonyaulax* and *Goniodoma*. *In:* Graneli, E. Sundstrom, B., Edler, L., Anderson, D.M. [Eds.] *Toxic Marine Phytoplankton.* Elsevier, New York, pp. 522-523.

Taylor, F.J.R. (1975). Taxonomic difficulties in red tide and paralytic shellfish poison studies: the "*tamarensis* complex" of *Gonyaulax. Env. Letters* 9: 103-119.

Taylor, F.J.R. (1984). Toxic dinoflagellates: taxonomic and biogeographic aspects with emphasis on *Protogonyaulax. In:* Ragelis, E.P. [Ed.] *Seafood Toxins.* American Chemical Society Symposium Series No. 262, Washington, D.C., pp. 77-97.

Taylor, F.J.R. (1985). The taxonomy and relationships of red tide dinoflagellates. *In:* Anderson, D.M., White, A.W., Baden D.G. [Eds.] *Toxic Dinoflagellates.* Elsevier, New York, pp. 11-26.

van Oppen, M.J.H., Diekmann, O.E., Wiencke, C., Stam, W.T., Olsen, J.L. (1994). Tracking dispersal routes: phylogeography of the Arctic-Antarctic disjunct seaweed *Acrosiphonia arcta* (Chlorophyta). *J. Phycol.* 30: 67-80.

Physiology and Bloom Dynamics of Toxic *Alexandrium* Species, with Emphasis on Life Cycle Transitions

Donald M. Anderson

Biology Department, Woods Hole Oceanographic Institution, Woods Hole MA 02543 USA

1. Introduction

Of the many genera of phytoplankton associated with harmful algal blooms (HABs), the dinoflagellate genus *Alexandrium* includes the largest number of toxic species. At least 8 species in this genus (*A. acatenella, A. catenella, A. cohorticula, A. fundyense, A. ostenfeldii, A. minutum, A. tamarense* (= *A. excavatum*) and *A. tamiyavanichi)* produce saxitoxin - the suite of compounds associated with paralytic shellfish poisoning (PSP) in humans (Cembella, this volume). An eighth species, *A. monilatum*, produces a poorly characterized toxin capable of killing fish but which appears to be unrelated to the PSP toxins (Aldrich *et al.* 1967).

This diversity of toxic species is matched by a diversity among strains of those species with respect to temperature requirements, toxicity, bioluminescence, and genetics. Some, such as *Alexandrium tamarense*, can be found in sub-Arctic and temperate zones (e.g., Taylor 1984; Cembella *et al.* 1988) as well as in tropical regions (Reyes-Vasquez *et al.* 1995). Although some of these globally distributed strains possess the same external morphology, they are genetically different. A comprehensive overview of the biogeography and population biology of the *Alexandrium* genus is provided elsewhere (Scholin (this volume), Scholin *et al.*, (1995), and Gallagher (this volume)). Here the focus will be on the autecological features that underlie many *Alexandrium* blooms, based predominantly on the small number of species that have been well-studied in the laboratory and the field. This effort will necessarily emphasize life cycle transformations and their quantitative effect on bloom dynamics, for it is in this specific area that *Alexandrium* blooms have been especially well-characterized and where differences from other HAB species become apparent. A concept that will recur throughout this discussion is that *Alexandrium* blooms have a "life-span" - a relatively short period of time in which these species are in the water column as motile vegetative cells. *Alexandrium* cells do not persist throughout the year, as do those of species such as *Gymnodinium breve* (Steidinger *et al.* this volume). Most *Alexandrium* species can be considered "background" bloom species, in that they are often outnumbered by co-occurring phytoplankton. High-biomass, monospecific *Alexandrium* blooms that discolor the water do occur, such as those of *A. minutum* in south Australia (Hallegraeff *et al.* 1988), but these are the exceptions rather than the rule.

2. The *Alexandrium* life cycle

The life histories of most *Alexandrium* species that have been studied involve an

NATO ASI Series, Vol. G 41
Physiological Ecology of Harmful Algal Blooms
Edited by D. M. Anderson, A. D. Cembella and G. M. Hallegraeff

alternation between asexual and sexual reproduction (Fig. 1). Repeated divisions (binary fission) lead to the proliferation of motile, vegetative cells as blooms develop. This is an asexual process that terminates when sexuality is induced. Sexuality begins with the formation of gametes which fuse to form swimming zygotes (planozygotes) which in turn become dormant, resting cysts (hypnozygotes). A useful compilation of the morphology of known *Alexandrium* cysts can be found in Bolch *et al.* (1991). Hereafter, the term "cyst" will refer to hypnozygotes formed through sexuality. Most species also produce another resting stage called a "temporary cyst" when motile, vegetative cells are exposed to unfavorable conditions such as mechanical shock or a sudden change of temperature or salinity. When conditions become favorable again, temporary cysts quickly re-establish a vegetative, motile existence. The temporary resting state thus allows the cells to withstand short-term environmental fluctuations.

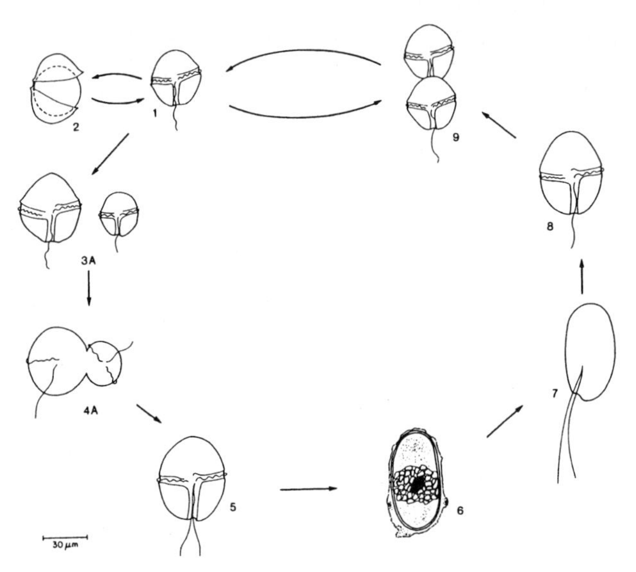

Fig. 1. Life cycle diagram of *Alexandrium tamarense*. Stages are identified as follows: (1) vegetative, motile cell; (2)temporary or pellicle cyst; (3) anisogamous "female" and "male" gametes; (4) fusing gametes; (5) swimming zygote or planozygote; (6) resting cyst or hypnozygote; (7&8) motile, germinated cell or planomeiocyte; and (9) pair of vegetative cells following division. Adapted from Anderson *et al.* 1996.

2.1 Mating mechanisms

In *A. tamarense*, both unequal-sized and equal-sized gametes have been reported (anisogamy and isogamy; Turpin *et al.* 1978; Anderson 1980), and generally, two different mating types are needed for a successful mating. As demonstrated by Destombe and Cembella (1990), however, self-recognition and fusion of gametes of *A. tamarense* clones does occur, although the zygote produced is not viable. These authors also demonstrated that there were not just two mating types in an assemblage of *A. tamarense* clones (i.e. that mating types appear not to be strictly bi-polar but rather are part of a continuum of mating affinity). Fusion between different "species" within *Alexandrium* is also possible, depending of course on one's definition of a species. Anderson *et al.* (1994) showed that *A. fundyense* and *A. tamarense* could produce viable cysts, but they also argue that these two species should be considered variants of a single species. Attempts to mate *A. tamarense* with *A. catenella* have not been successful to date (Y. Ishida, pers. comm.).

Little is known of the manner in which gametes locate each other and fuse. Sawayama *et al.* (1993) showed that concanavalin A (a lectin) and tunamycin (an inhibitor of glycoprotein synthesis) prevented sexual attachment in *A. catenella.* A cell-to-cell recognition system involving agglutinins similar to those found in *Chlamydomonas* is thus a possibility for *Alexandrium.* The location of these recognition sites on the cell surface and the details of the biochemical interaction between receptor and ligand remain unknown.

2.2 Induction of sexuality

In *Alexandrium* cultures, sexuality has been induced by the imposition of nutrient limitation - typically a decrease in nitrogen or phosphorus (Turpin *et al.* 1978; Anderson *et al.* 1984; Anderson and Lindquist 1985). Non-optimal temperature or light, and cessation of growth in nutrient-replete batch cultures due to over-crowding or carbon limitation do not result in cyst formation to any significant extent, yet there are occasional reports of spontaneous cyst formation in high nutrient cultures of other dinoflagellates (Morey-Gaines and Ruse 1980; S. Blackburn, pers. comm.). Yoshimatsu (1981) performed crossing experiments with *A. catenella* in nutrient-replete medium containing 1,400 μM nitrate and 11 μM phosphate, and observed zygote formation. The latter value is well above levels where Anderson *et al.* (1984) found virtually no cyst formation in *A. tamarense* cultures.

Numerous issues concerning the induction of sexuality remain to be resolved. The nutrient levels at which sexuality is induced are not known, nor is it known whether ambient concentrations or the size of internal nutrient pools triggers the transition. Regulatory mechanisms also remain a mystery. The pathways involved are especially perplexing since nitrogen-, phosphorus-, and iron-limitation can all induce sexuality (Anderson *et al.* 1984; Doucette *et al.* 1989). Finally, if gametes are formed when nutrients are depleted, how then does the zygote obtain sufficient nutrients to support prolonged dormancy, quiescence, germination and growth? One possibility is that gametes form before internal pools are completely exhausted. Another is that nutrient depletion is not required, but that sexuality is instead controlled by an endogenous clock similar to that shown to regulate excystment in *A. fundyense* (Anderson and Keafer 1987). Alternatively, perhaps nutrient uptake occurs during the planozygote stage. The latter suggestion derives from nutrient-limited laboratory batch cultures in which nutrients are exhausted very quickly. A large proportion of the cells fuse and become planozygotes but are arrested at that stage and never become cysts (Anderson *et al.* 1985; Anderson and Lindquist 1985). It is not known why this occurs, but one possibility is that only the first planozygotes to form are able to complete the transition to cysts, perhaps because they are able to take up nutrients before concentrations become too low in the batch cultures to permit significant uptake.

2.3. Dormancy and quiescence

The planozygotes that develop after the fusion of gametes swim for up to a week before falling to the sediment as resting cysts to begin dormancy. "Dormancy" is defined as the suspension of growth by active endogenous inhibition, and "quiescence" as the suspension of growth by unfavorable environmental (i.e. exogenous) conditions. Thus dormant cysts cannot germinate, even under optimal environmental conditions, whereas quiescent cysts are competent to germinate, but are inhibited from doing so by some environmental factor. *Alexandrium* cysts typically proceed through a mandatory dormancy period before they are capable of germination. The duration of this interval, which is generally considered a time for physiological "maturation", varies considerably among dinoflagellate species (12 hrs to 6 months; Anderson *et al.* 1996). For a single species, this dormancy can vary with storage temperature (Anderson 1980). Despite its critical importance to the ecology of all cyst-forming *Alexandrium* species, the duration of this interval is known only for *A. tamarense.* Cysts of *A. tamarense* stored at 4 °C mature in 4-6 months, whereas storage at warmer temperatures shortens the mandatory interval to 1-3 months (Anderson 1980). The duration of this process can have a significant effect on the timing of recurrent blooms, as cysts with a long maturation requirement may only seed one or two blooms per year, whereas those that can germinate in less time may cycle repeatedly between the plankton and the benthos and contribute to multiple blooms in a single season. Hallegraeff (this volume) argues that the short maturation time of *Gymnodinium catenatum* reduces the effect of cysts on motile cell dynamics, as the cysts germinate gradually throughout the year rather than as a single, synchronized pulse.

Once a cyst is mature and the dormancy interval is complete, the resting state will continue if external conditions are unfavorable for growth. Thus a quiescent cyst cannot germinate until an applied external constraint (such as cold temperature) is removed. The factors that break the quiescence of mature cysts are not known for most *Alexandrium* species. A primary stimulus for excystment of *A. tamarense* from temperate waters (Anderson and Morel 1979) and *A. minutum* from Australia (Bolch *et al.* 1991) is a shift in temperature to favorable levels, as occurs in seasonal warming or cooling. Cysts stored at cold temperatures remain quiescent until the temperature is increased. Likewise, cysts held at high temperatures maintain quiescence and germinate only when temperatures decrease to a favorable level (Anderson and Morel 1979; Anderson 1980). The absence of germination at these cold and warm extremes defines a permissive temperature "window" within which quiescent cysts will germinate, but outside of which they will continue their resting state. For *A. tamarense* from Cape Cod, the window ranges from 5 to 21 °C (Fig. 2). This does much to explain the occurrence of two discrete *A. tamarense* blooms each year in shallow embayments on Cape Cod, one in the spring and one in the fall (Anderson and Morel 1979). Mature, over-wintering cysts germinate when temperatures warm in the spring and a bloom results, depositing new cysts in the sediments. Those cysts need a month or more to mature, at which time the salt pond water is above 21 °C and thus is too warm to permit germination. As water temperatures decrease in the fall, germination is again possible, leading to another bloom. Temperature thus can maintain quiescence for extended periods, determine the duration of dormancy after cyst formation, synchronize or entrain cyst populations for more uniform germination, and initiate the excystment process. It is thus a major factor in the dynamics of dormancy, quiescence and germination of *Alexandrium* species, or at least for temperate species which have been the only ones studied in this regard. Species or strains from tropical waters where temperature fluctuations are less dramatic might not be as reliant on temperature cues, but this speculation awaits further research. In support of this concept, *A. catenella* strains from subtropical Sydney Harbor (winter temperature >15 °C) have a dormancy period of only 1-2 weeks (G. Hallegraeff, pers. comm.).

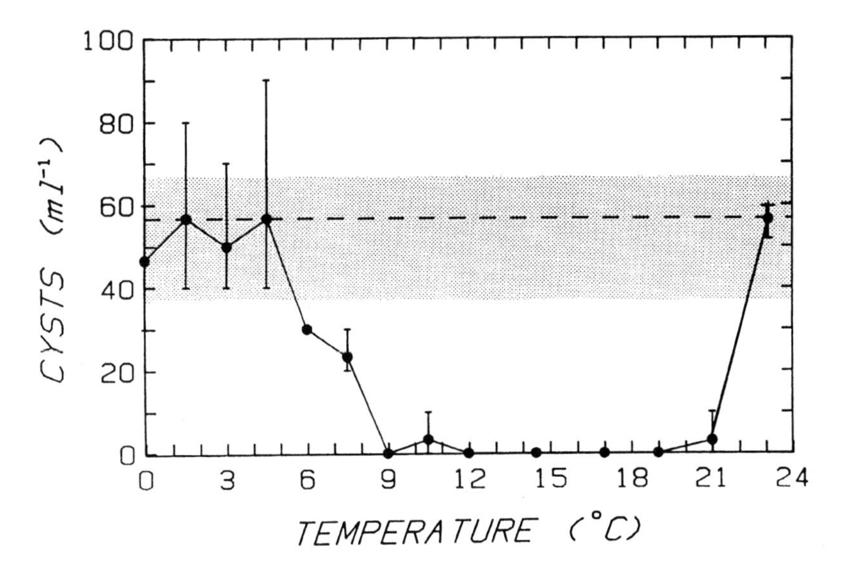

Fig. 2. Temperature "window" for *A. tamarense* cyst germination. Sediment from Perch Pond, (Cape Cod, MA USA) was incubated for three weeks at the indicated temperatures and germination determined by counting the remaining intact cysts. The means of three replicate counts of ungerminated cysts are plotted, and error bars indicate the range of those replicates. Dashed line and shaded area represent mean of initial counts ± SE. Differences between the initial mean and the final counts are assumed to represent germinated cysts.

The effects of other environmental factors on dormancy and excystment are less-studied, not just for *Alexandrium*, but for all dinoflagellate species. Nutrient concentrations do not seem to affect the success or rate of germination (Cannon 1993), but light, salinity, and oxygen are important to varying degrees. *Alexandrium tamarense* cysts did not germinate after 7 weeks of incubation in total darkness, although the presence of chlorophyll fluorescence in those cysts suggests that germination would eventually have occurred (Anderson *et al.* 1987). Light is thus not required for germination, but does accelerate the process. Likewise, very few Australian *A. minutum* cysts germinated in complete darkness, but high rates were observed at light levels as low as 20 $\mu E\ m^{-2} sec^{-1}$ (Cannon 1993) Germination of *A. minutum* occurred between 14-26 PSU, the salinity range of the waters in which the organism occurs in south Australia.

Oxygen (or the lack thereof) can have a dramatic effect on cyst germination. Among *Alexandrium* species, only *A. tamarense* has been tested thus far (Anderson *et al.* 1987), but it and most other dinoflagellate species examined have an absolute requirement for oxygen during germination. Cysts that are buried deep in the sediment can thus remain quiescent for years, their fate being either eventual death if anoxia persists, or germination should they be transported to the sediment surface or overlying water. The longevity of buried cysts is difficult to determine, but quantitative cyst profiles, radioisotope measurements, and a simple model suggest that the half-life of *A. tamarense* in anoxic sediments is approximately 5 years (Keafer *et al.* 1992).

2.3.1 Endogenous clock regulation

The dormancy and germination story can be more complicated than indicated above. Some dinoflagellate species can alternate between dormancy and quiescence through time, with the interval when germination is possible being determined by an endogenous, annual clock. Mature, quiescent cysts of *A. tamarense* from sediments in the Gulf of Maine did not germinate to any significant extent during fall and winter months, but did excyst in high numbers during the spring and summer (Anderson and Keafer 1987). An endogenous clock of this type would allow cysts to germinate even when deposited in deep waters where seasonal environmental cues such as temperature or daylength are small or nonexistent. In contrast, germination of *A. tamarense* cysts from shallow Cape Cod salt ponds showed no sign of clock control, suggesting that strains inhabiting those waters were regulated by the external environment (Anderson and Keafer 1987). This strategy makes sense in hindsight, given the variability in shallow, estuarine waters where a long winter or early spring could be detrimental to an organism restricted by an internal clock to a fixed time for germination.

The existence of endogenous annual clocks has not yet been confirmed in other strains or species of *Alexandrium*, but several studies hint that such control might exist. Kim (1994) suggest that endogenous clock-controlled germination can explain the variability in germination success seen in cysts from Jinhae Bay, Korea. In the St. Lawrence estuary in Canada, a study of cysts isolated from a sediment sample kept in the laboratory under constant conditions again suggests that there is an endogenous germination rhythm in those cysts, that the period is about one year, and that the germination "window" lasts about two months (Perez *et al.* in press).

It is noteworthy that in the Perez *et al.* study, as well as in those of Kim (1994) and Anderson and Keafer (1987), a small percentage (10 - 20%) of the cysts germinated during the interval when the clock control was inhibiting germination of the bulk of the cyst population. Two possibilities for this duality in response are suggested. First, the cyst population in the stored sediment samples may not represent a single genotype, but instead may include at least two strains with different germination physiology. Alternatively, clock control may not be operative in all cells within a population, perhaps as an adaptive strategy to ensure that germination is variable within that population. Work is clearly needed to better characterize the nature of endogenous control of *Alexandrium* germination, given the importance of excysted cells to overall bloom dynamics.

3. Bloom dynamics

Studies of the bloom dynamics of *Alexandrium* species are remarkably few, despite the importance of toxic species within this genus. The "tamarense" group of *Alexandrium* has been best-studied in this regard, but even there, most studies are descriptive and lack autecological detail. The following discussion will therefore focus on two hydrographic systems which are commonly associated with *Alexandrium* blooms and which demonstrate the extent to which life cycle transformations influence bloom dynamics in different habitats. The two regimes to be addressed are shallow, restricted embayments with localized blooms, and open coastal waters with widespread blooms.

3.1. Localized blooms in salt ponds, bays, and lagoons

Alexandrium blooms frequently occur in shallow salt ponds and coastal bays (e.g., Anderson *et al.* 1983; Su *et al.* 1993; Ho, and Hodgkiss 1993; Takeuchi *et al.* 1995;

Giacobbe *et al.* 1996). In some instances, blooms in these waters are simply a nearshore manifestation of large-scale coastal blooms, but in many cases, such as those discussed below, the blooms are localized, "point-source" events, where the coupling between cysts in the sediments and blooms in overlying waters is direct.

3.1.2. Seedbeds and excystment dynamics

A common assumption is that cyst "seedbeds" provide the inoculum for many *Alexandrium* blooms. The concept of a discrete seedbed is not appropriate in many locations, however, due to widespread distribution of cysts and the likelihood that germination will occur over a large area. The strongest evidence for localized cyst accumulations being linked to subsequent blooms comes from salt ponds on Cape Cod, MA., U.S.A. (Anderson *et al.* 1983), where blooms and PSP are confined to the immediate vicinity of the embayments where cyst are abundant (Anderson *et al.* 1982). Cysts are also important in larger bays, but the linkage is more difficult to quantify.

Encystment and excystment dynamics are determined by the interplay between physiological processes (such as maturation) and environmental controls on dormancy and quiescence. In the temperate Cape Cod salt ponds, *A. tamarense* cysts remain quiescent during the winter months since bottom temperatures are often near 0 °C and are thus below the lower limit of the permissive temperature "window" for germination (Fig. 2). Germination is possible once waters warm to 5-6 °C (Fig. 3), but the only cysts which excyst are those at the sediment surface where oxygen is available (Anderson *et al.* 1987). This temperature threshold is first marked by the appearance of red fluorescence due to chlorophyll that is synthesized in germinating cysts (Anderson and Keafer 1985), and subsequently by the appearance of distinctive germling cells (planomeiocytes) in the water column (Anderson *et al.* 1983). No fluorescence is observed in cysts from anoxic sediments (Fig. 3).

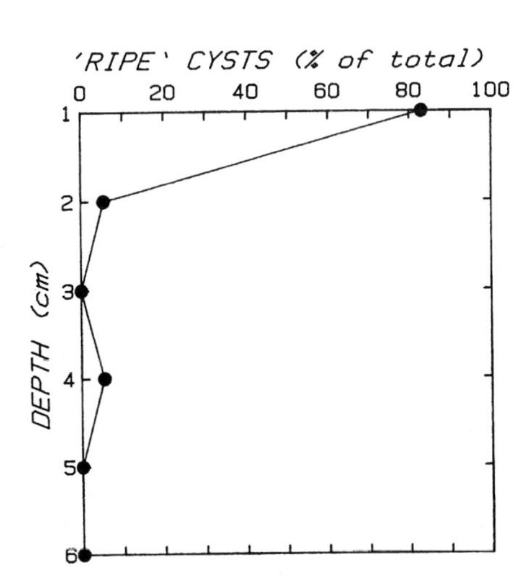

Fig. 3. Vertical profile of *Alexandrium tamarense* cyst fluorescence in Perch Pond. Cysts showing visible red chlorophyll fluoresence are plotted as a percentage of the total *A. tamarense* cysts present at each depth in the core.

The number of cysts that contribute to the bloom initiation process is small relative to the total number in the sediments. One reason is shown in Fig. 4 which depicts a set of vertical profiles of *A. tamarense* cysts from different locations in Perch Pond, a Cape Cod salt pond. Although the vertical distributions differ throughout the embayment, one pattern is clear - on average, only 20% of the cysts are in the top cm where oxygen is available. As seen in Fig. 3 from the same embayment, "ripe" or fluorescing cysts were only observed in the surface layer. These values, in conjunction with areal estimates of *A. tamarense* cyst abundance in Perch Pond, allow a calculation of the input or inoculum from cyst germination. Given an average density of 4.5 x 10^7 cysts m^{-2} (in the top 6 cm of sediment; Anderson *et al.* 1982) over an area of approximately 30,000 m^2, there would be 1.35 x 10^{12} total cysts in the sediments. Only 20% are in the oxygenated surface layer and only 80% of those will germinate based on their fluorescence (Fig. 3). The total input of planomeiocytes from germination would thus 2 x 10^{11} cells. If all cysts germinated simultaneously, this would yield a concentration of planomeiocytes in the water column of 3,500 cells l^{-1}, which would be a significant inoculum indeed. Germination is not simultaneous, however, but occurs over about a one-month interval in Perch Pond (Fig. 5). Thus the input of planomeiocytes would be about 100 cells l^{-1} per day, on average.

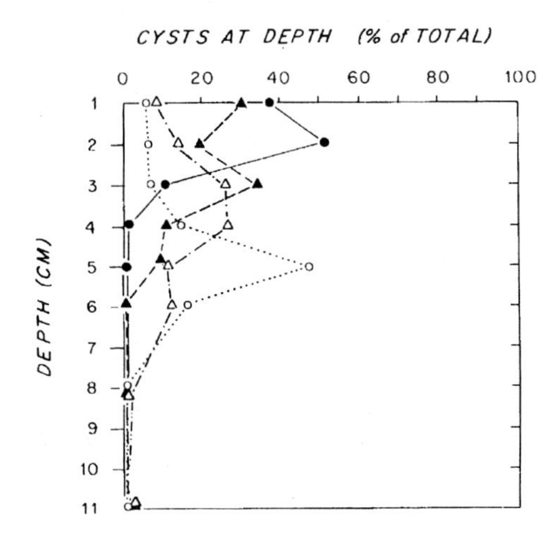

Fig 4. Vertical distribution of *Alexandrium tamarense* cysts from four different locations in Perch Pond.

Direct counts of the number of planomeiocytes inoculated into the water column at the onset of blooms are difficult to obtain since planomeiocytes divide to produce vegetative cells and thus the sustained inoculum of new cells from the sediment is not be easy to detect after a few generations. Anderson *et al.* (1983) observed up to 90 planomeiocytes l^{-1} (30% of the motile cell population) early in a Perch Pond bloom of *A. tamarense*. Takeuchi et al (1995) suggest that *A. catenella* cysts in Tanabe Bay (Japan) sediments germinate to yield an inoculum of 10-100 planomeiocytes l^{-1}. The consistency between these initial planomeiocyte concentrations in Tanabe Bay and the observed and calculated concentrations for Perch Pond is noteworthy, but extrapolation to other systems should be approached with caution.

Anderson *et al.* (1983) concluded that factors which lead to "bloom" versus "non-bloom" years within Cape Cod salt ponds depend more on the growth of the planktonic population, than on the size of the cyst inoculum. This may be true for that study, but the conclusion probably does not apply to all situations. A low inoculum of 10 cells l^{-1} from excystment would require three extra divisions and several more weeks of growth to achieve a bloom with biomass equivalent to one started from 100 cells l^{-1} at the onset of a growth season. The quantitative dynamics of cyst populations clearly requires further study.

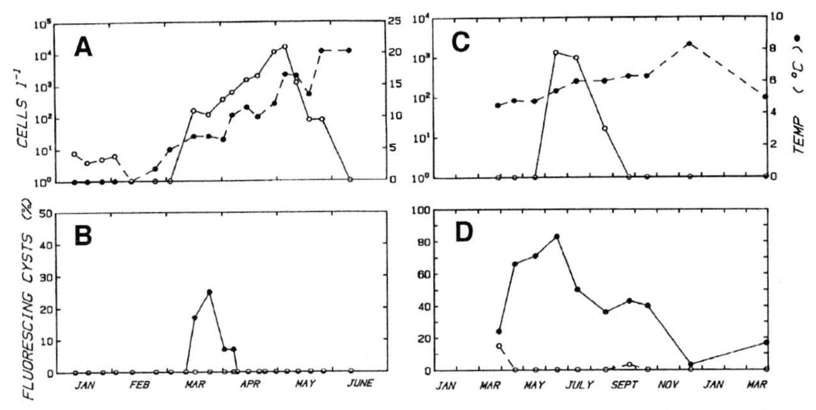

Fig 5 Motile cell and cyst dynamics two different hydrographic and environmental regimes - Perch Pond, a shallow Cape Cod embayment (A,B), and Sta. 29 near Cape Ann, a 160 m deep station in the Gulf of Maine (C,D). (A) Perch Pond motile cells (o) and temperature (●); (B) Perch Pond cyst fluorescence in the top cm of sediment (●) and at 5-6 cm depths in anoxic sediment (o); (C) Gulf of Maine motile cells (o) and temperature (●); and (D) Gulf of Maine cyst fluorescence in the top cm of sediment (●) and at 5-6 cm depths in anoxic sediment (o). From Anderson and Keafer (1985).

3.1.3. Bloom development and encystment

No effort will be made to generalize on the bloom dynamics of motile *Alexandrium* populations in shallow embayments other than to emphasize that such blooms are heavily dependent upon local hydrography and the manner in which it interacts with cell behavior, especially vertical migration. Shimada *et al.* (1996) demonstrated how *A. tamarense* blooms in Funka Bay, Japan are initiated from *in situ* cyst populations, with the distribution and abundance of the motile cells being markedly affected by the timing and strength of the Oyashio current as it flows into the bay. Studies of *A. minutum* in a Mediterranean lagoon by Giacobbe *et al.* (1996) demonstrated that the spring appearance of the species coincided with enhanced rainfall and freshwater runoff, and with stabilization of the water column. Watras *et al.* 1982 conducted laboratory growth studies and used the results to paramaterize a simple model which indicated that for the blooms in Cape Cod salt ponds, the development of *Alexandrium* populations depends solely on salinity-dependent temperature regulation of cell division rates. The same model produced a poor prediction of Bay of Fundy *Alexandrium* bloom dynamics, presumably because physical forcings are more influential in population accumulation in those open, tidally stirred waters.

Another example of physical/biological coupling and the importance of stratification in embayments was seen in Perch Pond, a salt pond which has a shallow entrance sill that restrict outflowing water to the low density surface layer (Anderson and Stolzenbach 1985). The diel vertical migration pattern of *A. tamarense* kept cells below that depth during the night, and even when the cells migrated close to the surface during the day, they remained deep enough to avoid transport out of the embayment with the outflowing surface layer (optimal light levels of 100-150 μE m^{-2}

sec^{-1} occur at 1.5 m or deeper). A density-driven exchange mechanism rapidly flushes water from these salt ponds, but the residence time of the *Alexandrium* cells is much longer due to the limited vertical extent of the migration. This coupling between organism behavior and the hydrography of the system restricts the extent to which cells and cysts can colonize adjacent waters and allows *Alexandrium* populations to accumulate to levels where toxicity becomes dangerous.

The duration of the blooms that have been followed in bays and salt ponds is generally two to three months or less (Anderson *et al.* 1983; Han *et al.* 1992; Shimada *et al.* 1996; Takeuchi *et al.* 1995). Giacobbe *et al.* (1996) describe an *A. minutum* bloom in a Mediterranean lagoon over a six month period, but the concentrations were at bloom levels for only two months - April and May. In Cape Cod, most of the bloom development occurs at water temperatures which are non-optimal for rapid growth of vegetative cells. Perch Pond isolates of *Alexandrium* grow fastest at 15-20 °C in the laboratory, but once the water reaches those temperatures in the field, blooms are typically on the decline and new cysts are already falling to the sediments (Anderson *et al.* 1983). Similarly, Han *et al.* (1992) found that *A. tamarense* disappears from the water column of Chinhae Bay, Korea at temperatures well below those that support optimal growth in the laboratory. The implication is that the induction of sexuality precludes the long-term persistence of vegetative *Alexandrium* cells in the plankton. This is in contrast to the long duration of *Gymnodinium breve* and *G. nagasakiense* populations (Steidinger *et al.* this volume; Gentien, this volume), for example, which remain in the water column for extended periods. The encystment/excystment cycle thus restricts the longevity of the *Alexandrium* vegetative cell population and appears not to be optimized for rapid or sustained vegetative growth. Whether this generalization from Cape Cod salt ponds applies to other areas of the world remains to be seen, but if so, it highlights an important aspect of *Alexandrium* autecology.

Laboratory studies suggest that the induction of sexuality in *Alexandrium* occurs as a result of nutrient limitation, yet this is not well-supported by field measurements. One problem in this regard is that gametes are not easily distinguished from vegetative cells in natural populations, and fusing gametes, though distinctive, are rarely observed. Gametes have thus never been enumerated in field studies. However, it is possible to recognize large, darkly-pigmented planozygotes (Anderson 1980) and to tabulate their abundance through time. Salt ponds are once again ideal for this type of time-series measurement. Studies in three Cape Cod salt ponds over two bloom seasons demonstrated that planozygote formation did not coincide with an obvious decrease in ambient nutrients (Anderson *et al.* 1983). In fact, planozygotes in the plankton and new cysts at the sediment surface were first observed when external nutrients were at or above concentrations equivalent to those measured during the earlier stages of bloom development when vegetative growth was rapid. It may be that as the ambient temperature increased during the blooms, the rates of uptake and metabolism of nutrients increased as well. Thus nutrient concentrations that were sufficient for balanced (but slow) growth at colder, early-bloom temperatures may not have been sufficient to maintain balanced growth when waters warmed and the *A. tamarense* growth rate increased. A gradual decrease in internal nutrient pools would thus occur, leading to nutrient limitation. Another possibility is that micronutrient limitation occurred, but was not measured (e.g. iron stress; Doucette *et al.* 1989). Alternatively, given the discovery of endogenous control of cyst germination for *A.*

tamarense (Anderson and Keafer 1987), endogenous or "clock"-regulated sexuality must also be considered. A final explanation is suggested by the remarkable consistency in the number of net cell divisions that occurred between the time the first *A. tamarense* cells were observed and the time that planozygotes appeared in the salt pond studies of Anderson *et al.* (1983). Despite a one-month differential in the onset of the blooms in the three salt ponds, zygotes appeared approximately six divisions after the first vegetative cells were seen (neglecting advection and grazing losses). This regularity is consistent with the gradual depletion of a stored product that eventually triggers sexuality, with replenishment of that stored reserve during the non-dividing planozygote stage. If verified, this again points to a temporal limit to *Alexandrium* blooms, perhaps under endogenous control.

Only two studies have attempted to enumerate *Alexandrium* planozygotes during blooms in order to quantify the importance of encystment in bloom decline (Anderson *et al.* 1983; Takeuchi *et al.* 1995). Both show that sexuality is induced well before the size of the bloom population peaks, and that during this late stage of bloom development, planozygotes can comprise 20-40% of the motile population. This number underestimates the total percentage of cells that become cysts, however, since it cannot account for the dynamic nature of the zygote sub-population. Each day, some planozygotes fall to the sediments as cysts, but new planozygotes appear following gamete fusion. The estimates do suggest that a large fraction of the bloom population encysts, and thus that bloom decline may be linked more to life cycle transitions than to grazing or other loss factors.

3.2 Coastal blooms

Another prominent habitat for *Alexandrium* blooms is in open coastal waters or large estuaries. The two regions that have been best studied in this regard are the St. Lawrence estuary in Canada and the Gulf of Maine in the U.S.

3.2.1. Seedbeds and excystment dynamics

Discrete seedbeds are difficult to identify in open coastal waters, since mapping surveys typically document widely distributed cyst populations. Quantitative cyst maps are available for *A. tamarense* (e.g., White and Lewis 1982; Anderson and Keafer 1985; Cembella *et al.* 1988), *A. catenella* (e.g., Yamaguchi *et al.* 1995), *A. minutum* (Erard *et al.* 1993) and *A. ostenfeldii* (Mackenzie *et al.* 1996), but few investigators have been able to obtain the data needed to demonstrate that a single location provides the bulk of the motile cell inoculum for a regional bloom. In fact, cyst distribution and abundance did not correlate with shellfish toxicity patterns along the coast of Maine (Thayer *et al.* 1983). Examples of discrete cyst seedbeds that lead to large-scale blooms do exist, however. For example, Cembella *et al.* (1988b) argue that *A. tamarense* cysts along the northern shore of the St. Lawrence estuary near the Manicouagan and Aux-Outardes rivers are responsible for toxic blooms which cause PSP on the south shore and further downstream in the estuary. On the northeast coast of Britain, *A. tamarense* cyst accumulations in the Firth of Forth have been linked to toxic blooms in the adjacent coastal waters to the north (Lewis *et al.* 1995). Evidence for the existence of a regional seedbed is also found in studies in the southwestern Gulf of Maine where *A. tamarense* blooms are confined to a buoyant coastal current formed from the outflows of two rivers in southern Maine (see below). Cyst surveys document a widespread distribution both nearshore and offshore (e.g., Lewis *et al.*

1979; Thayer *et al.* 1983; Anderson and Keafer 1985), but a shallow water source region or "initiation zone" has been identified in the eastern Casco Bay region just "downstream" from the mouths of the rivers that produce the coastal current waters. In two successive years (1993 and 1994), *A. tamarense* cells were generally absent in early spring at all stations in a large study area except those near the Casco Bay region (D. Anderson, unpub. data). Initial concentrations were low (approximately 50-100 cells l^{-1}), but quickly increased within the low salinity coastal current.

The size of the inoculum from cysts in coastal or estuarine systems is not known, as the large scales of the blooms have prevented detailed study, and planomeiocytes are rarely seen. Given the widespread cyst distribution typical of coastal areas, two scenarios are suggested with respect to bloom initiation. One involves the synchronized germination of cysts throughout the region, with only those cells which emerge into favorable growth conditions being responsible for blooms. Alternatively, rapid and synchronized germination in localized areas (e.g., shallow bays) might seed the blooms, leaving cysts that germinate more gradually in other areas with little quantitative impact. One indication of the low magnitude and protracted nature of the inoculum from deep water cyst germination is seen in the chlorophyll fluorescence of cysts in sediments 160 m deep in the Gulf of Maine (Anderson and Keafer 1985). As in the shallow salt ponds and embayments, the majority of the cysts were buried below the sediment surface, and thus were prevented from germinating by anoxia (Keafer *et al.* 1992). Germination of these deep water cysts appears to be controlled by an endogenous annual clock (Anderson and Keafer 1987). The percentage of fluorescing cysts in surface sediments increases sharply in the spring and then decreases during the summer and fall, remaining positive and significant for nearly 8 months (Fig. 3). Deep water cysts thus germinate over a much longer interval than the duration of the coastal bloom in those waters (~ two months), and much longer than the germination interval in the shallow salt ponds described above. Clearly, the cold temperatures and darkness of deep waters extend the excystment process, with only a fraction of the viable cysts in surface sediments actually participating in the bloom. For coastal blooms in the Gulf of Maine, we now suspect that cysts in shallow coastal waters or bays provide a larger and more synchronized inoculum than do the offshore cyst deposits, which may only be sinks where most cysts accumulate and eventually die, with little effect on overall bloom dynamics.

3.2.2. Bloom development and encystment

The complexities of *Alexandrium* blooms in dynamic coastal or estuarine systems are far from understood. One common characteristic of such phenomena is that physical forcings play a significant role in both bloom dynamics and the patterns of toxicity. The coupling between physics and biological "behavior" such as swimming, vertical migration, or physiological adaptation holds the key for understanding these phenomena, yet this is perhaps where our knowledge of this genus is weakest.

Once vegetative cells enter the water following cyst germination, their net growth and transport are heavily affected by circulation, nutrients, stratification, and other chemical or physical factors. Although many of these interactions remain uncharacterized, blooms of several *Alexandrium* species have been linked to particular water masses. In the St. Lawrence estuary, for example, patterns of PSP toxicity and *A. tamarense* cell distributions have been linked to the plume produced by the Manicouagan and Aux-Outardes rivers (Therriault *et al.* 1985). This flow generates a frontal zone extending into the estuary which is associated with high phytoplankton production, particularly of dinoflagellates. Examples of the importance of fronts in HAB bloom dynamics are many (e.g., Pingree *et al.* 1975; Simpson *et al.* 1979). The key issue here is that the freshwater plume generates a highly stratified water column which favors proliferation and retention of vertically migrating phytoplankton such as *Alexandrium.* A fraction of the vegetative cells retained in that zone are subject to transport across to the south shore of the estuary, where they are entrained into the Gaspé current which carries them towards the Gulf of St.. Lawrence. The frontal system at the Manicouagan and Aux-Outardes plume thus serves as an initiation zone and the Gaspé current as a transport pathway. Similar initiation zones and coastal current transport have also been indentified for *Alexandrium* blooms in the western Gulf of Maine (Franks and Anderson 1992a; see below).

The physical system is not the entire story, however. Although the Manicouagan and Aux-Outardes plume is essential for *A. tamarense*, this species is most abundant during mid- to late-summer, even though the characteristics of the plume and the front are well-established for a much longer interval. Clearly, other factors are regulating *A. tamarense* dynamics. Therriault *et al.* (1985) suggest that *A. tamarense* blooms in the St. Lawrence develop only when the proper combination of meteorological and hydrodynamic factors coincide to produce high temperatures, maximum water column stability, low nutrients, and low winds.

The frontal zone of the Manicouagan and Aux-Outardes rivers is also a site of enhanced cyst deposition (Cembella *et al.* 1988b). Unfortunately, no information is available on the timing or magnitude of cyst formation or the mechanisms underlying sexual induction in these waters.

Another example of the importance of freshwater in *Alexandrium* bloom dynamics is found in the southwestern Gulf of Maine, where the temporal and spatial pattern of persistent PSP outbreaks have been linked to a buoyant plume or coastal current originating in several rivers that empty into the Gulf near Casco Bay (Franks and Anderson 1992a; Anderson in press). Concentrations of the toxic dinoflagellate *Alexandrium tamarense* are much higher within the lower salinity waters of the plume than without, and toxicity in coastal shellfish rises and falls with the movement of the plume. This is in turn driven by the local wind stress, by rainfall and snow melt patterns, and by the general circulation of the Gulf. Downwelling conditions are conducive to toxicity development, since such conditions trap the plume and its associated cells tightly against the coast and accelerate them to the south. In contrast, upwelling favorable winds push the plume offshore, spreading it out laterally and dramatically decreasing the concentration of *A. tamarense* in nearshore waters due to the upwelling of deeper, saltier waters that contain no toxic cells.

The early season bloom dynamics of that area are not the only issue, however. Given a persistent southward flow of the coastal current, the red tide problem "downstream" should gradually diminish year after year as the sediments near the origin of the coastal current are depleted of cysts. In other words, once the *A. tamarense* cells leave the area and travel south in the coastal current, there is no hydrographic pathway that will bring them (or their cysts) back to the bloom initiation zone. Since the toxic blooms have been an annually recurrent event for over 25 years (Franks and Anderson 1992b), there must be a mechanism by which the cyst seedbeds are replenished. One possible explanation is suggested by the observation of late-season populations of *Alexandrium* in Casco Bay in 1993 and 1994, long after the initial pulses of cells had initiated a bloom in the early spring (unpub. data). With much-reduced rainfall and no snowmelt to drive the coastal current at that time, transport out of the region was limited and localized blooms (and presumably cyst deposition) occurred. Much remains to be clarified concerning the existence and dynamics of this putative initiation zone or seedbed.

As in other open coastal systems, no estimates are available on the extent to which encystment contributes to the decline of the blooms in the coastal current, nor is it known how sexuality is induced. Nutrient measurements during blooms are few, and planozygotes and newly formed cysts are rarely observed. As in other stratified systems, the high nutrient concentrations below the pycnocline (e.g. Franks and Anderson 1992a) would be accessible to vertically migrating *A. tamarense* cells, so it is not clear whether nutrient limitation actually occurs while the cells are associated with the buoyant coastal current.

Another important unknown in the coastal blooms concerns the possible stimulation of *A. tamarense* growth by the unique chemistry of the freshwater plumes. More cells are typically found within the low salinity plumes (e.g., Therriault *et al.* 1985; Franks and Anderson 1992a), but this could simply be a result of small-scale physics interacting with the cells migration behavior, or it could be a reflection of higher growth rates within the plume. Freshwater runoff from the heavily forested watershed contains significant levels of dissolved and particulate organic matter as well as metals and other micronutrients. It appears likely that some component of this mixture could be critical to the rapid growth of *A. tamarense* cells. Iron is a likely candidate for a stimulatory micronutrient, as Wells *et al.* (1991) showed that bioavailable iron was elevated in nearshore waters characteristic of the coastal current, and depleted offshore in the Gulf of Maine. The measured iron levels were within the range of those that stimulated or limited *A. tamarense* growth in laboratory cultures.

3.2.3. General environmental forcings

The large number of *Alexandrium* species involved in harmful events throughout the world makes it difficult to generalize about general environmental controls of bloom dynamics. The nutrition of these organisms is not unusual, although mixotrophy has been reported for some *Alexandrium* species (Jacobson and Anderson 1996) and more are probably capable of this strategy. Like most phytoplankton species, *Alexandrium* will respond to anthropogenic nutrient inputs, but there is no evidence that these species are preferentially stimulated compared to other phytoplankters, nor is there compelling evidence of any increase in *Alexandrium* bloom magnitude or frequency as a direct result of pollution. Indeed, *Alexandrium* blooms, including many toxic ones, occur in remote and relatively pristine waters, such as those in Alaska (Hall 1982) or southern Argentina (Benavides *et al.* 1995). A strong association with freshwater inputs is often seen (e.g. Franks and Anderson 1992a; Therriault et al. 1985), but this presumably reflects the importance of stratification and the supply of natural humic substances, trace elements, and other materials that might serve as growth stimulants.

On a larger scale, some workers have attempted to discern the influence of mesoscale weather patterns and lunar forcings on *Alexandrium* bloom timing. Balch (1981) observed a synchrony between seasonal *A. tamarense* blooms off the coast of Maine and maximum tidal ranges of major spring tides. Spring tides increase the height of the bottom mixed layer, erode the seasonal thermocline, and inject nutrients into surface waters, each of which could influence cyst germination and motile cell growth and accumulation. White (1987) examined patterns of PSP toxicity caused by *A. fundyense* for over 40 years and looked for correlations with environmental factors during pre-bloom months and summer toxicity episodes. A correlation with an 18.6 year cycle of lunar tidal modulation was observed, as was a relationship with salinity, windspeed and tidal energy dissipation. A possible relationship between El Niño/Southern Oscillation (ENSO) events and *Alexandrium* blooms on the west coast of the U.S. was suggested by Erickson and Nishitani (1985). These events affect sea surface temperature, winds, and sunlight over a large scale, and hence might be expected to influence the vertical stability of the water column, and thus dinoflagellate blooms. Clearly, local and regional *Alexandrium* blooms can be influenced by mesoscale circulation patterns, and thus basin-scale phenomena such as ENSO events can be important in the patterns of toxicity. A careful evaluation of these factors in the context of PSP dynamics would likely be a fruitful exercise in both Pacific and Atlantic waters.

3.2.4 Population biology

Much of the foregoing discussion treats large-scale, regional populations of *Alexandrium* as if they were composed of individual strains or genotypes, yet this is probably not the case. *Alexandrium* cells in the western Gulf of Maine coastal current or the plume of the Manicouagan and Aux Outardes rivers would appear to represent discrete, genetically uniform populations, yet biochemical and genetic studies are now revealing that considerable heterogeneity exists. (Details are provided elsewhere in this volume by Scholin and Gallagher and by Anderson *et al.* 1994; Cembella and Destombe 1996). It is now clear, for example, that two and perhaps three genetically distinct strains of *A. tamarense* occur within the Gulf of Maine and areas to the south. If these strains have different growth characteristics and physiology, fine-scale autecological understanding of the details of bloom dynamics will not be possible until the different genotypes within a region are characterized. For now, we must recognize that considerable genetic variability can exist within regional populations of *Alexandrium*, and that this variability is not easily detected in routine monitoring surveys. A major priority for future research should be to characterize the nature of the genetic variability within *Alexandrium* cells in a given region, and to determine under what conditions the different genotypes become dominant.

4.0 Summary

The ability of *Alexandrium* species to colonize multiple habitats and to persist over large regions through time is testimony to the adaptability and resilience of this important organism *Alexandrium* species are not known for rapid or "explosive" growth rates. Maximal rates in laboratory cultures are typically 0.5 to 0.7 divisions day^{-1} (e.g., Anderson *et al.* 1984; Su et al. 1993; Yamamoto and Tarutani 1996; Chang and McClean 1997), although rates near one division day^{-1} are reported for Australian *A. minutum* (G. Hallegraeff, pers. comm.). Population growth is typically not reflected in monospecific blooms but rather in moderate biomass levels and co-occurrence with other species. Blooms are not particularly long-lasting, and seem restricted in time by life cycle transitions. The cyst stage is clearly important in *Alexandrium* population dynamics, both with respect to bloom initiation and termination, but the nature of this linkage varies among habitats. In shallow embayments, cysts and motile cell blooms are tightly coupled, whereas in large temperate estuaries and open coastal waters, the linkage is more difficult to define and quantify. In both of these habitats, most of the cysts in the sediments do not germinate due to bioturbation, burial, and inhibition of germination by anoxia. Even when only the cysts in surface sediments are considered, the bulk of the widely distributed cysts in deeper waters may germinate too slowly or too far from suitable growth conditions to be a factor in coastal blooms. In one sense, *Alexandrium* species appear to use a type of r-selection strategy, producing many "offspring" in the form of cysts, only a few of which ever germinate to inoculate blooms. On the other hand, a complex life history and a low growth rate are often considered K-strategies. This group of dinoflagellates does not easily fit into such fixed categories.

Estimates of the inoculum size from excystment are small - on the order of tens to hundreds of cells l^{-1}, suggesting that major blooms require multiple, sustained vegetative divisions that in turn depend greatly on environmental conditions affecting motile cells. Nevertheless, the size of an excystment inoculum can have a bearing on the magnitude of a bloom, especially if that bloom is limited temporally due to seasonal temperatures or to some form of endogenous regulation of excystment and encystment. In small-scale blooms in embayments and in widespread coastal blooms, physical/biological coupling is a critical feature of population accumulation, growth, and dispersal. Behavioral adaptations such as vertical migration are important features in this regard. Bloom termination is clearly linked to life cycle transitions, although the relative importance of encystment relative to grazing or other loss factors has not been explicitly investigated. Overall, the *Alexandrium* species that have been studied in detail have proven to be remarkably resilient and capable of colonizing a spectrum of habitats and hydrographic regimes. It is thus of no surprise that the biogeographic range of these species has expanded considerably in recent times (Anderson *et al.* 1994; Scholin, this volume) and that PSP outbreaks remain a significant global problem.

5. Acknowledgments

Special thanks to T. Takeuchi, Y. Fukuyo, and T. Tekiguchi for providing unpublished data. This research was supported by grants from the National Science Foundation (OCE-9415536), the National Oceanic and Atmospheric Administration (NA36RM0190) through the Gulf of Maine Regional Marine Research Program, the National Sea Grant College Program Office, Department of Commerce (NA90-AA-D-SG480; WHOI Sea Grant Project R/B-121), and the Massachusetts Water Resources Authority. Contribution No. 9550 from the Woods Hole Oceanographic Institution.

6. References

Aldrich, D.V., Ray, S.M., Wilson,W.B. (1967). *Gonyaulax monilata:* Population growth and development of toxicity in cultures. *J. Protozool.* 14:636-639.

Anderson, D. M. (1980). Effects of temperature conditioning on development and germination of *Gonyaulax tamarensis* (Dinophyceae) hypozygotes. *J. Phycol.* 16:166-172.

Anderson, D.M. (in press). Bloom dynamics of toxic *Alexandrium* species in the northeastern United States. *Limnol.and Oceanogr.*

Anderson, D. M., Aubrey, D.G., Tyler, M. A., Coats, D. W. (1982). Vertical and horizontal distributions of dinoflagellate cysts in sediments. *Limnol. and Oceanogr.* 27: 757-765.

Anderson, D. M., Chisholm, S. W.,Watras, C. J. (1983). The importance of life cycle events in the population dynamics of *Gonyaulax tamarensis. Mar. Biol.* 76:179-190.

Anderson, D.M., Fukuyo, Y., Matsuoka, K. (1996). Cyst Methodologies. pp. 229-249, In: *Manual on Harmful Marine Microalgae.* Hallegraeff, G.M., Anderson, D.M., Cembella, A.E. (eds.). UNESCO, Paris.

Anderson, D. M., Keafer, B. A. (1985). Dinoflagellate cyst dynamics in coastal and estuarine waters. pp. 219-224, In: D. M. Anderson, White, A. W., Baden, D. G. (eds.) *Toxic dinoflagellates.* Proc. 3rd Int'l. Conf., Elsevier, New York.

Anderson, D. M., Keafer, B. A. (1987). An endogenous annual clock in the toxic marine dinoflagelate *Gonyaulax tamarensis. Nature* 325:616-617.

Anderson, D. M., Kulis, D. M., Binder, B. J. (1984). Sexuality and cyst formation in the dinoflagellate *Gonyaulax tamarensis*: Cyst yield in batch cultures. *J. Phycol.* 20:418-425.

Anderson, D. M., Kulis, D. M., Doucette, G. J., Gallagher, J.C., Balech, E. (1994). Biogeography of toxic dinoflagellates in the genus *Alexandrium* from the northeastern United States and Canada. *Mar. Biol.* 120:467-478.

Anderson, D. M., Lindquist, N. L. (1985). Time-course measurements of phosphorus depletion and cyst formation in the dinoflagellate *Gonyaulax tamarensis* (Lebour). *J. Exp. Mar. Biol. Ecol.* 86:1-13.

Anderson, D. M., Morel, F. M. M. (1979). The seeding of two red tide blooms by the germination of benthic *Gonyaulax tamarensis* hypnocysts. *Est. Coast. Mar. Sci.* 8: 279-293.

Anderson, D. M., Stolzenbach, K. D. (1985). Selective retention of two dinoflagellates in a well-mixed estuarine embayment: The importance of diel vertical migration and surface avoidance. *Mar. Ecol. Prog. Ser.* 25: 39-50.

Balch, W. M. (1981). An apparent lunar tidal cycle of phytoplankton blooming and community succession in the Gulf of Maine. *J. Exp. Mar. Biol. Ecol.* 55: 65-77.

Benavides, H., Prado, L., Diaz, S., Carreto, J.J. (1995). An exceptional bloom of *Alexandrium catenella* in the Beagle Channel, Argentina. pp. 113-119, In: *Harmful Marine Algal Blooms,* Lassus, P., Arzul, G., Erard, E., Gentien, P., Marcaillou, C. (eds). Lavoiser, Paris.

Bolch, C. J.,Blackburn, S. I.,Cannon, J. A., Hallegraeff, G. M. (1991). The resting cyst of the red tide dinoflagellate *Alexandrium minutum* (Dinophyceae). *Phycologia* 30:215-219.

Cannon, J. (1993). Germination of the toxic dinoflagellate, *Alexandrium minutum,* from sediments in the Port River, South Australia. pp. 103-107, In: *Toxic Phytoplankton Blooms in the Sea,* T. Smayda, Y. Shimizu (eds.), Elsevier.

Cembella, A. D., Destombe, C. (1996). Genetic differentiation among *Alexandrium* populations from Eastern Canada. pp. 447-450, In: Yasumoto, T., Oshima, Y., Fukuyo, Y. (eds.) *Harmful and Toxic Algal Blooms.* Intergovernmental Oceanographic Commission of UNESCO, Paris.

Cembella, A. D., Turgeon, J., Therriault, J.C., Beland, P. (1988). Spatial distribution of *Protogonyaulax tamarensis* resting cysts in nearshore sediments along the north coast of the lower St. Lawrence estuary. *J. Shellfish Res.* 7:597-610.

Chang, F. H., McClean, M. (1997). Growth responses of *Alexandrium minutum* (Dinophyceae) as a function of three different nitrogen sources and irradiance. *New Zeal. J. Mar. Fresh. Res.* 31:1-7.

Destombe, C., Cembella, A. (1990). Mating-type determination, gametic recognition and reproductive success in *Alexandrium excavatum* (Gonyaulacales, Dinophyta), a toxic, red tide dinoflagellate. *Phycologia* 29:316-325.

Doucette, G. J. , Cembella, A. D., Boyer, G. L. (1989). Cyst formation in the red tide dinoflagellate *Alexandrium tamarense* (Dinophyceae): effects of iron stress. *J. Phycol.* 25: 721-731.

Erard-Le-Denn, E., Desbruyeres, E., Olu, K. (1993). *Alexandrium minutum* resting cyst distribution in the sediments collected along the Brittany Coast, France. pp. 763-768, In: *Toxic Phytoplankton Blooms in the Sea,* T. Smayda, Shimizu, Y. (eds.), Elsevier, Amsterdam.

Erickson, G. and Nishitani, L. (1985). The possible relationship of El Nino/Southern Oscillation events to interannual variations in *Gonyaulax* populations as shown by records of shellfish toxicity. pp. 283-290, In : Wooster, W.S. and Fluharty, D.L. (eds) , *Proc. Meeting on El Nino effects in the Eastern Subarctic Pacific*, Washington Sea Grant Program, University of Washington.

Franks, P. J. S., Anderson, D. M. (1992a). Alongshore transport of a toxic phytoplankton bloom in a buoyancy current: *Alexandrium tamarense* in the Gulf of Maine. *Mar. Biol.* 112:153-164.

Franks, P. J. S., Anderson, D. M. (1992b). Toxic phytoplankton blooms in the southwestern Gulf of Maine: testing hypotheses of physical control using historical data. *Mar. Biol.* 112:165-174.

Giacobbe, M. G., Oliva, F. D., Maimone, G. (1996). Environmental factors and seasonal occurrence of the dinoflagellate *Alexandrium minutum*, a PSP potential producer, in a Mediterranean Lagoon. *Est. Coast. Shelf Sci.* 42:539-549.

Hall, S. (l982). T*oxins and toxicity of Protogonyaulax from the northeast Pacific*. Ph.D. Thesis, Univ. of Alaska.

Hallegraeff, G. M., Steffensen, D. A., Wetherbee, R. (1988). Three estuarine Australian dinoflagellates that can produce paralytic shellfish toxins. *J. Plank. Res.* 10:533-541.

Han, M. S., Jeon, J. K., Kim, Y. O. (1992). Occurrence of dinoflagellate *Alexandrium tamarense*, a causative organism of paralytic shellfish poisoning in Chinhae Bay, Korea. *J. Plank. Res.* 14:1581-1592.

Ho, K.C., Hodgkiss, I.J. (1993). Characteristics of red tides caused by *Alexandrium catenella* (Whedon & Kofoid) Balech in Hong Kong. pp. 263-268, In: *Toxic Phytoplankton Blooms in the Sea*, T. Smayda, Shimizu, Y. (eds.), Elsevier, Amsterdam.

Jacobson, D. M., Anderson, D.M. (1996) Widespread phagocytosis of ciliates and other protists by marine mixotrophic and heterotrophic dinoflagellates. *J. Phycol.* 32:279-285.

Keafer, B.A., Buesseler, K.O., Anderson, D.M. (1992). Burial of living dinoflagellate cysts in estuarine and nearshore sediments. *Mar. Micropaleont.* 20: 147-161.

Kim, C. H. (1994). Germinability of resting cysts associated with occurrence of toxic dinoflagellate *Alexandrium* species. *J. Aquacult.* 7:251-264.

Lewis, C. M., Yentsch, C. M., Dale. , B. (1979). Distribution of Gonyaulax excavata resting cysts in the sediments of Gulf of Maine. pp. 235-238, In: D. L. Taylor, Seliger, H.H. (eds.). *Toxic Dinoflagellates Blooms.* Proc.Int. Conf. (2nd). Elsevier, North Holland.

Mackenzie L., White, D., Oshima, Y., Kapa, J. (1996). The resting cyst and toxicity of *Alexandrium ostenfeldii* (Dinophyceae) in New Zealand. *Phycologia* 35 (2): 148-155

Morey-Gaines, G., Ruse, R. H. (1980). Encystment and reproduction of the predatory dinoflagellate *Polykrikos kofoidii* Chatton (Gymnodiniales). *Phycologia* 19:230-236.

Pingree, R., Pugh, P., Holligan, P., Forster, G. (1975). Summer phytoplankton blooms and red tides along tidal fronts in the approaches to the English Channel. *Nature* 258:672-277.

Perez, C. C., Roy, S., Levasseur, M., Anderson, D.M. (In press). Control of germination of *Alexandrium tamarense* cysts from the lower St. Lawrence estuary (Canada). *J. Phycol.*

Reyes-Vasquez, G., Ferraz-Reyes, E., Vasquez, E. (1979). Toxic dinoflagellate blooms in northeastern Venezuela during 1977. pp. 191-194, In: D. L. Taylor, Seliger, H.H. (eds.). *Toxic Dinoflagellates Blooms.* Proc.Int. Conf. (2nd). Elsevier, North Holland.

Sawayama, S., Sako, Y., Ishida, Y. (1993). Inhibitory effects of concanavalin A and tunicamycin on sexual attachment of *Alexandrium catenella* (Dinophyceae). *J. Phycol.* 29:189-190.

Scholin, C., G. M., Hallegraeff, D. M. Anderson. (1995). Molecular evolution of the *Alexandrium tamarense* "species complex" (Dinophyceae): dispersal in the North American and West Pacific regions. *Phycologia* 34:472-485.

Shimada, H., Hayashi, T., Mizushima, T. (1996). Spatial distribution of *Alexandrium tamarense* in Funka Bay, southwestern Hokkaido. pp. 219-221, In: Yasumoto, T., Oshima, Y., Fukuyo, Y. (eds.) *Harmful and Toxic Algal Blooms.* Intergovernmental Oceanographic Commission of UNESCO, Paris.

Simpson, J. H., Edelstein, D. J., Edwards, A.,Morris, N. C. G., Tett, P. B. (1979). The Islay Front: Physical structure and phytoplankton distribution. *Estuar. Coast. Mar. Sci.* 9:713-726.

Su, H. M., Liao, I. C., Chiang, Y. M. (1993). Mass mortality of prawn caused by *Alexandrium tamarense* in a culture pond in southern Taiwan. pp. 329-333, In: *Toxic Phytoplankton Blooms in the Sea*, T. Smayda, Shimizu, Y. (eds.), Elsevier, Amsterdam.

Takeuchi, T., Kokubo, T., Fukuyo, Y., Matsuoka, K. (1995). Quantitative relationship among vegetative cells, planozygotes, and hypnozygotes of *Alexandrium catenella* (Dinophyceae) in its blooming season at Tanabe Bay, Central Japan. Abstract, 7th Int'l. Conf. on Toxic Phytoplankton. Sendai, Japan.

Taylor, F.J.R. (1984). Toxic dinflagellates: taxonomic and biogeographic aspects with emphasis on Protogonyaulax. pp. 77-97, In: *Seafood Toxins*, E. Ragelis (ed.) Amer. Chem. Soc. Symposium Series. Washington, D.C.

Thayer, P. E., J. W. Hurst, C. M. Lewis, R. Selvin, C. M. Yentsch. (1983). Distribution of resting cysts of *Gonyaulax tamarensis* var. *excavata* and shellfish toxicity. *Can. J. Fish. Aquat. Sci.* 40:1308-1314.

Therriault, J. C., Painchaud, J., Levasseur, M. (1985). Controlling the occurrence of *Protogonyaulax tamarensis* and shellfish toxicity in the St. Lawrence Estuary: Freshwater runoff and the stability of the water column. pp. 141-146, In: D. M. Anderson, White, A.W., Baden, D.G. (eds.). *Toxic Dinoflagellates*, Elsevier, New York.

Turpin, D. H., Dobel, P. E. R., Taylor, F. J. R. (1978). Sexuality and cyst formation in Pacific strains of the toxic dinoflagellate *Gonyaulax tamarensis*. *J. Phycol.* 14:235-238.

Watras, C. J., Chisholm, S. W., Anderson, D.M. (1982). Regulation of growth in an estuarine clone of *Gonyaulax tamarensis*: Salinity-dependent temperature responses. *J. Exp. Mar. Biol. Ecol.* 62:25-37.

Wells, M.L., Mayer, L.M., Guillard, R.R. L. (1991). Evaluation of iron as a triggering factor for red tide blooms. *Mar. Ecol. Prog. Ser.* 69:93-102.

White, A. W. (1987). Relationships of environmental factors to toxic dinoflagellate blooms in the Bay of Fundy. *Rapp. P.-v. Réun. Cons. int. Explor. Mer* 187:38-46.

White, A. W., Lewis, C. M. (1982). Resting cysts of the toxic, red tide dinoflagellate *Gonyaulax excavata* in Bay of Fundy sediments. *Can. J. Fish. Aquat. Sci.* 39: 1185-1194.

Yamaguchi, M., Itakura, S., Imai., I. (1985). Vertical and horizontal distribution and abundance of resting cysts of the toxic dinoflagellates *Alexandrium tamarense* and *Alexandrium. catenella* in sediments of Hiroshima Bay, the Seto Inland Sea, Japan. *Nippon suisan Gakkaishi* 61:700-706.

Yamamoto, T., Tarutani, K. (1996). Growth and phosphate uptake kinetics of *Alexandrium tamarense* from Mikawa Bay, Japan. pp. 293-296, In: Yasumoto, T., Oshima, Y., Fukuyo, Y. (eds.) *Harmful and Toxic Algal Blooms*. Intergovernmental Oceanographic Commission of UNESCO, Paris.

Yoshimatsu, S. (1981). Sexual reproduction of *Protogonyaulax catenella* in culture. Heterothallism. *Bull. Plankton Soc. Japan* 28:131-139.

Genetic and Biochemical Approaches to PSP Toxin Production of Toxic Dinoflagellates

Yuzaburo Ishida[1], Aritsune Uchida[2] and Yoshihiko Sako[2]

[1] Department of Marine Biotechnology, Faculty of Engineering, Fukuyama University, Fukuyama City, Hiroshima 729-02, Japan
[2] Division of Applied Biosciences, Graduate School of Agriculture, Kyoto University, Kyoto 606, Japan

1. Introduction

Silva (1982) and Franca *et al.* (1995) suggested that dinoflagellate toxin production may be due to bacteria living in association with *A. lusitanicum.* Kodama and Ogata (1988) also pointed out the possible association of intracellular bacteria with toxin production by *A. tamarense,* and suggested that toxin production is not a hereditary characteristic. We maintain that the report on *A. tamarense* has not as yet been supported by sufficient evidence (Sako *et al.* 1992; Ishida *et al.* 1993). It is not clear whether the enzymes for PSP toxins are encoded by chromosomal, chloroplast, or mitochondrial DNA, or other DNA sources such as bacteria, viruses or plasmids. One approach to the resolution of this controversy is through studies of the genetics of toxin production. Detailed genetics of dinoflagellates is such a new field that the information available is far from extensive except for three dinoflagellates species+ *Gloeodinium montanum, Symbiodinium microadriaticum, and Crypthecodonium cohnii,* which can grow on agar. Another dinoflagellate *Gonyaulax polyedr*a, contains a luciferin-binding protein gene which represents the first dinoflagellate gene that has been cloned and sequenced at both cDNA and genomic levels (Lee *et al*. 1993).

In this chapter, we will deal with genetic analysis of PSP toxin production by the mating reaction, and biochemical and molecular analysis of PSP toxin synthesizing enzymes in *Alexandrium spp.* and related dinoflagellates.

2. Genetic analysis of paralytic shellfish poisoning (PSP) toxin production using mating reactions.

Hypotheses on the method of PSP toxin production in *A. catenella* and other dinoflagellates are divided and controversial (Kodama and Ogata,1988; Silva 1982; Franca *et al.* 1995; Sako *et al.* 1992; Ishida *et al.* 1993; Ishida 1993; Kim *et al.* 1993a; Sako *et al.* 1995). Therefore, we tried to determine whether the source of PSP toxins in toxic dinoflagellates is encoded by chromosomal DNA, chloroplast DNA, mitochondrial DNA, or contaminating bacteria,viruses or plasmids. The inheritance of toxin composition by F_1 cells from parents with different toxin compositions was studied. The proportion of PSP toxin composition was used as a phenotypic marker, because toxin composition is a stable property in *Alexandrium* spp. (Cembella *et*

NATO ASI Series, Vol. G 41
Physiological Ecology of Harmful Algal Blooms
Edited by D. M. Anderson, A. D. Cembella and G. M. Hallegraeff

al. 1987; Ogata *et al.* 1987; Kim *et al.* 1993b), as shown in Fig. 1(Sako *et al.* 1992, Sako *et al.* 1995).

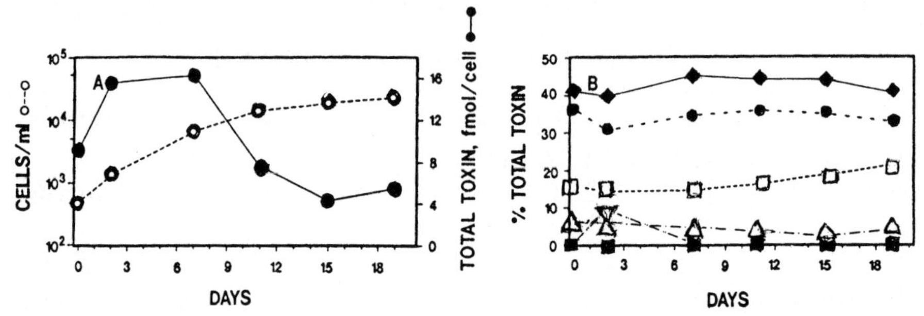

Fig. 1 Growth curve, PSP toxin content and toxin composition of *A. catenella* TNX22 as a function of incubation time.
Symbols: (A)○- - -○cell density (cells/ml); ●---●total toxins(f mol/cell).
(B)□GTX1+4; ●GTX5; ◆C1+2;△C4; × neoSTX; ▽STX

Crosses of parental mating types (mt^+ and mt^-) of *A. tamarense* and *A. catenella* with different PSP toxin composition were made and the toxin composition patterns of the F_1 cells were compared with those of the parents (Table 1; Sako *et al.* 1995). When a character is coded on chromosomal DNA in crosses, it is inherited in a 2:2 Mendelian pattern (Gillham 1978), whereas if it is located on the chloroplast DNA or mitochondrial DNA, it is inherited uniparentally from the mt^+ or mt^-. Other symbiotic or contaminant factors such as bacteria, viruses and plasmids will be also inherited uniparentally or at random.

Table 1 shows all four F_1 progeny of two $mt^+$and two mt^- produced by crossing *A. tamarense* OFX181 (mt^+) and OFY184 (mt^-) with similar toxin profiles had the same toxin composition as their parents. However, when for *A. tamarense* from Ofunato Bay, the parent strain OFY152 (mt^-,which has gonyautoxin GTX1+4 as its major toxin component), and the other parent OFX191 (mt^+, which has C1+C2, neosaxitoxin (neoSTX) and GTX1+4), were crossed, two F_1 progeny of mt^+ and mt^- showed the same toxin composition as the parental OFY152, and the other two F_1 progeny of mt^+ and mt^- showed the same toxin composition as parental OFX191(Ishida *et al.* 1993).

In pairing *A. catenella* TNX22 (mt^+) from Tanabe Bay with OFY101 (mt^-) from Ofunato Bay, having different toxin composition, F_1 progeny of mt^- and mt^+ showed the same toxin composition patterns as parents TNX22 and OFY101, respectively. Furthermore, the toxin composition of both F_1 and F_2 progeny was the same. The toxins of all F_1 progeny were inherited biparentally. From these inheritance patterns, we conclude that the suite of PSP toxins in *Alexandrium* is inherited in a 2:2 Mendelian pattern, and not in a uniparental pattern, nor randomly (Sako *et al.* 1992).

Differing from the normal Mendelian pattern with biparent type mentioned above, when parents TNY7 (mt^+) from Tanabe Bay and OFX072 (mt^-) from Ofunato Bay were crossed, F_1 and F_2 cells had a toxin composition different from both parents at a high frequency. In the case of the parent strains OFX072 (mt^+) and TNY7 (mt^-), the former produced neoSTX and STX, and the latter GTX4 and C4. Among 38 F_1 progeny, 9 clones (mt^+) and 8 clones (mt^-) showed the same toxin composition (A) as the parental TNY7 (mt^-), and 8 clones (mt^+) and 9 clones (mt^-) showed the same toxin composition (C) as the parental OFX072 (mt^+). The four remaining F_1 progenies from the parentsTNY7 and OFX072 showed toxin profiles (A' and C') different from those of the parents (Table 1).

Table 1 Relationship between toxin type and mating type of parent and F_1 strains in *A. catenella* and *A. tamarense* (Sako et al. 1995)

Parentage	Morph-type g)	ITS ribotype h)	Mating type	Toxin type	No. of cysts	No. of F_1 progeny	Mating type	Toxin type
A. catenella								
TNX 22	*catenella*	*catenella*	mt+	A^a	2	1	mt-	A
						1	mt+	A
						1	mt-	B
OFY 101	*catenella*	*catenella*	mt-	B^b		1	mt+	B
						9	mt+	A
TNY 7	*catenella*	*catenella*	mt-	A	17	8	mt-	A
						8	mt+	C
						9	mt-	C
OFX 072	*catenella*	*catenella*	mt+	C^c		2	mt+	A'
						2	mt-	C'
A. tamarense								
OFX 181	*tamarense*	*tamarense*	mt+	D^d	4	4	mt+	D
						4	mt-	D
						4	mt+	D
OFY 184	*tamarense*	*tamarense*	mt-	D		4	mt-	D
OFY 152	*tamarense*	*tamarense*	mt-	E^e	4	2	mt+	E
						2	mt-	E
						2	mt+	F
OFX 191	*tamarense*	*tamarense*	mt+	F^f		1	mt-	F

Main components are : a) GTX 4, 5 and C2 ; b) C2 and neo STX ; c) C2, neoSTX and STX ; d) GTX4 and C2 ; e) GTX1, 4 ; f) GTX4,C2 andneoSTX 。

g) By the criteria of Fukyo et al. (1985).

h) Grouping based on 5.8SrDNA and internal transcribed spacer(ITS)1,2 phylogeny (Adachi et al. 1996).

Next, backcrosses of parent TNY7 (mt^-) and F_1 progeny 707-4 (mt^+, which was generated from parents TNY7 and OFX072 and has the same composition as the parent OFX072) produced two B_1 cells (B_1; first backcross generation) having the same composition as F_1 progeny 707-4, and two B_1 cells having the same composition as the parent TNY7. But when the parent OFX072 (mt^+) and F_1 progeny 707-6 (mt^-, which was generated from parents TNY7 and OFX072 and has the same composition as the parent TNY7) were backcrossed, each of two cells among the four B_1 cells produced had the same composition as the parent OFX072 and F_1 707-6, respectively. Two other B_1 cells did not inherit either the parent or F_1 toxin composition. From this result, it is assumed that a general recombination in toxin genomes possibly occurs, and that the parent OFX072 is especially apt to produce a recombinant F_1 (Ishida *et al.* 1993). The mating types (mt^+ and mt^-) were not necessarily associated with toxin inheritance, as shown in Table 1. Genes synthesizing PSP toxins are not linked with the mating gene (Sako *et al.*1992; Ishida *et al.* 1993).

These results support the hypothesis that the genes for the enzymes that produce PSP toxin derivatives must be coded in the chromosomal DNA of *Alexandrium* and other toxic dinoflagellates.

3. Biochemical and molecular analysis of PSP toxins production in *A. catenella* and *G. catenatum* .

In order to obtain direct evidence for the hypothesis mentioned above, we worked to detect and purify enzymes associated with N-1 hydroxylation, C-13 carbamoylation, C-11 sulfation and N-21 sulfation of these derivatives, to determine the N-terminal amino acid sequences of these enzymes and to isolate the genes encoding them from *G. catenatum* and *A. catenella.*

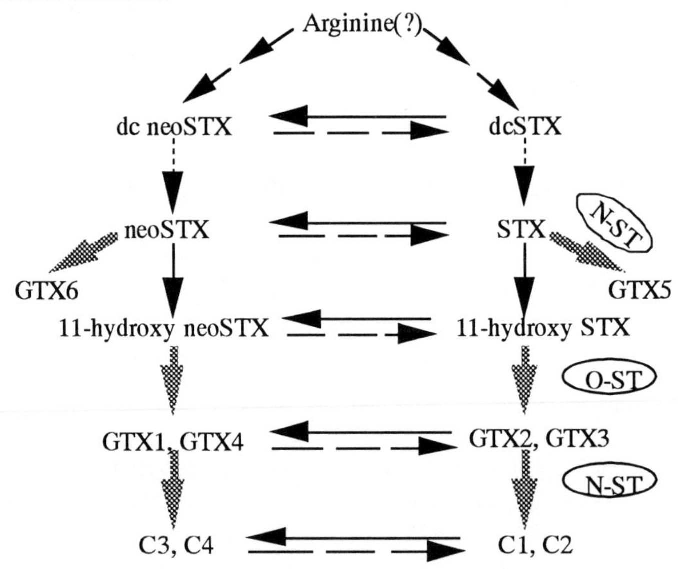

Fig. 2 Putative pathway of PSP biosynthesis in *G. catenatum*
sulfation, reduction,
oxidation, carbamoylation

First, we focused on sulfation referring to the putative pathway of PSP biosynthesis in *G. catenatum* (Fig. 2) because sulfated toxins predominate in the toxin composition of some dinoflagellates. Moreover, sulfation is known to be an important step in the metabolism of bioactive compounds, i.e. hormones and drugs, in mammals. Sulfation was catalyzed by sulfotransferase (ST) which transfers a sulfate group of 3'-phosphoadenosine 5'-phosphosulfate (PAPS) to these compounds (Mulder and Jakoby 1990; Saidha and Schiff 1994; Homma *et al.* 1992; Barnes *et al.* 1989; Hondoh *et a*l. 1993).

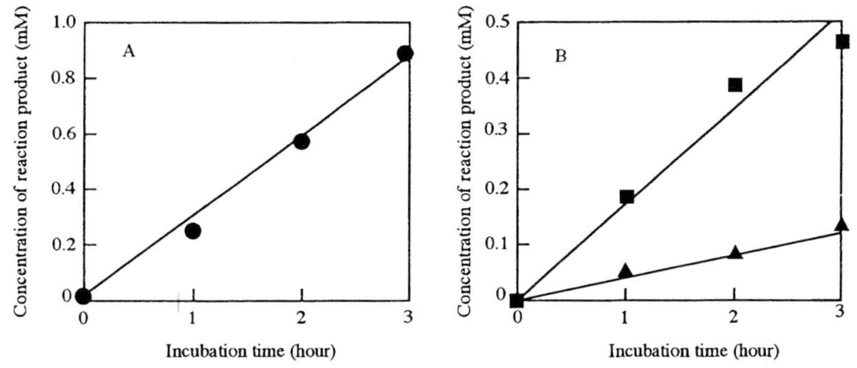

Fig. 3 Time-course of reaction products by N-sulfotransferase(N-ST). N-ST was incubated with PAPS as a sulfate donor and with STX (A) and GTX2 + 3 (B). A; ●GTX5 was produced. B; ■C1 and ▲C2 were produced.

We first detected a PSP toxin sulfotransferase from *G. catenatum* GC21V from Ria de Vigo, Spain. This enzyme transfers the sulfate group of PAPS to N-21 of STX and GTX2+3 and produces GTX5 and C1+2, respectively (Fig. 3). We purified this enzyme (N-sulfotransferase) as shown in Table 2. N-sulfotransferase enzyme was purified by $(NH_4)_2SO_4$ precipitation, and DEAE cellulose Blue Toyopearl, Mono Q and Superose 6 column chromatography. The purification procedure for N-sulfotransferase from *G. catenatum* is summarized in Table 2. This purified enzyme showed a single

Table 2 Summary of purification of N-sulfotransferase from*G. catenatum*

Preparation	Total protein (mg)	Total activity (units)	Yield (%)	Specific activity (units/mg)	Purification factor (fold)
Crude extract	239	-	-	-	-
$(NH_4)_2SO_4$	213	-	-	-	-
DEAE-cellulose	55.7	69.6	100	1.2	1
Blue Toyopearl	2.2	101	145	46	38
Mono Q	0.33	2.9	4.2	8.8	7
Superose 6	0.015	1.0	1.4	67	54

- , not analyzed.

band with Mr of 59 kDa by SDS-PAGE electrophoresis as well as by Superose 6 gel chromatography. Thus this enzyme seemed to be a monomer with a larger molecular mass than those of other mammalian sulfotransferases which have Mr of 30 - 35 kDa (Homma *et al.* 1992; Barnes *et al.* 1989; Hondoh *et al.* 1993), and phenol sulfotransferase of *Euglena* with Mr of 26 kDa (Saidha *et al.* 1994). This enzyme was active toward STX and GTX2+3 but not N-1 hydroxy toxins, neoSTX and

GTX1+4. The activity for STX or GTX2+3 was not inhibited by addition of neoSTX, GTX1+4 and other analogues of PSP toxins. This enzyme required PAPS as a sulfate donor, but not adenosine 5'-phosphosulfate (APS) and $MgSO_4$. The activity under standard conditions with PAPS was inhibited by addition of the PAPS analogue, 3''-phosphoadenosine 5'-phosphate (PAP), which is known to be a competitive inhibitor of sulfotransferase. Thectivity was stimulated by Mg^{2+} or Co^{2+}, but was inhibited by Ca^{2+}, Cu^{2+}, Fe^{2+}, Mn^{2+}, Ni^{2+} or Zn^{2+}. In the pH range of 4 to 10, the highest activity was observed at pH 6. This enzyme has an optimum temperature of 25°C (Table3).

Table 3 Comparison between properties of N-sulfotransferase (N-ST) and O-sulfotransferase(O-ST) in *G. catenatum*

	N-ST		O-ST
Substrate	STX	GTX2+3	11- α,β -hydroxy STX
	↓	↓	↓
Products	GTX5	C1+C2	GTX2+3
SO_4 donor	PAPS*		PAPS*
Opt. pH	6		6
Opt. Temp.	25 °C		35 °C
Metal co-factors	Mg^{2+}, Co^{2+}		no

* 3'-phosphoadenosine 5'-phosphosulfate

We also partially purified theN-sulfotransferases of *G. catenatum* MZ5 and MZ12 isolated from Miyazu Bay, Kyoto, Japan, respectively. Characteristics of both enzymes were similar to the enzyme from the GC21V strain, and *A. catenella* Acko5 from Uranouchi Bay, Japan which produced mainly C1+2 also had a similar N-sulfotransferase from *G. catenatum*. The enzyme from *A. catenella* produced C1+2 and GTX5 from GTX2+3 and STX, respectively. It was optimally active at 15°C. The activity was not stimulated by any metals, and was inhibited not only by Ca^{2+}, Cu^{2+}, Fe^{2+}, Mn^{2+}, Ni^{2+} or Zn^{2+}, but also by Co^{2+} and Mg^{2+}. These results suggest that the properties of N-sulfotransferase from *G. catenatum* are a little different from those of *A. catenella*.

The other sulfotransferase (O-sulfotransferase; O-ST) associated with conversion of 11- α,β -hydroxySTX into GTX2+3 was detected in a crude extract of *G. catenatum* GC21V, through a fraction bound to the DEAE-cellulose column. When this fraction was applied to a Blue-Toypearl column, an active fraction of O-sulfotransferase was eluted without binding and separated from N-sulfotransferase. This enzyme had an optimum pH of 6 and an optimum temperature of 35°C, and did not require divalentcation for activity (Table 3). These results suggest that the O-sulfotransferase specific to C-11 of 11- α,β -hydroxy STX was different from N-sulfotransferase(Yoshida et al., 1996). Oshima (1995) showed that crude extracts of *A. tamarense* transformed GTX2+3 to GTX1+4 while those of *G. catenatum* JP02 transformed GTX2+3 to C1+2, and that the extracts of both toxic (DE05) and non-toxic (DE09) isolates from Tasmania, Australia, produced the same reaction.

At present, we are trying to detect the genes encoding these enzymes in a *G. catenatum* cDNA library, using a DNA nucleotide sequence determined from N-terminal amino acid sequencing. In the near future, we are confident that the series of genes encoding the enzymes in *G. catenatum* and *A. catenella* will be found to be distributed on a portion of the chromosome genome which is not linked to the mating type locus. These findings will contribute to forecasting the occurrence of PSP toxin-producing dinoflagellates by *in situ* hybridization with DNA probes for the genes encoding these enzymes and preventing paralytic shellfish poisoning.

4. DNA analysis of some genes in *Alexandrium* spp. and related dinoflagellates

We have shown that the PSP toxin composition of F_1 progenies in *A. tamarense* and *A. catenella* is inherited in a Mendelian pattern. This fact suggests that molecular analysis based on genotypic characters of sulfotransferase enzymes specific to PSP production could be useful for identifying toxic dinoflagellates. Furthermore, to verify the presence of the toxin genes in the dinoflagellates, we will try to exploit promoters specific to dinoflagellates obtained from ferredoxin and elongation factor genes (FE-α)(Uchida *et al.,* unpublished data). They will be used for transformation of toxic genes, because dinoflagellate promoters will be needed for gene transfer and expression in dinoflagellates.

In addition, molecular approaches to identification of toxic *Alexandrium* spp. and related dinoflagellates have been attempted. Protein-based comparisons at the species level relied on enzyme electrophoresis and immunological methods (monoclonal antibodies). The former were used to investigate taxonomic problems in *A. catenella, A. tamarense* and *A. fundyense* (Taylor 1984; Cembella and Taylor 1986; Hayhome *et al.* 1989; Sako *et al.* 1990) . Based on isozyme analysis, all isolates of *A. catenella* from two geographically different areas were grouped into two populations, but the mating types were not discriminated (Sako *et al.* 1990). Antibody probes provided an estimate of similarity that can be used to delineate taxonomic boundaries (Sako *et al.* 1993; Adachi *et al.* 1993a, b and c). However, the cell surface antigens recognized by these antibodies may be subject to changes with differing environmental conditions.

To define genetic markers useful for classifying these organisms, Lenaers *et al.* (1991) and Scholin and Anderson (1993) recently used the small subunit (SS) and large subunit (LS) ribosomal DNA sequences as phylogenetic and taxonomic indicators for several species of the genus *Alexandrium*, and clarified their phylogenetic relationships. We targeted the regions containing the 5.8S rDNA and internal transcribed spacers (ITSs) for discrimination of *Alexandrium* at the intra-specific level (Adachi *et al.* 1994, 1995 and 1996a and b; Ishida et al. 1995). Unfortunately, the distance between isolates of *A. catenella* of two mating types (mt+ and mt-) was extremely small. By using RFLP (restriction fragment length polymorphism) analysis of the 5.8S rDNA and ITS regions, we could not discriminate between mt^+ and mt^- in *A. catenella* and *A. tamarense.*

The nucleotide sequences of 16S rDNA of chloroplast DNA (cpDNA), DNA of chloroplast-type ferredoxin, major DNA-binding protein(HCc) and elongation factor (FE- α), were determined for *A. tamarense* and compared with those of *Peridinium bipes* and other microalgae. We found that these genes were useful only for discrimination at the genus level or higher. Clearly, investigations of the genetics and molecular biology of *Alexandrium* species are in their infancy, and considerable work is needed before we can use these approaches to uncover the mysteries of toxin production and species dispersal.

5. Acknowledgments

We are indebted to Dr I. Yoshinaga, Dr C-H. Kim, Dr M. Adachi, Dr T. Yoshikawa and Mr T. Yoshida for valuable discussions and useful comments. This work was partly supported by a Grant-in-Aid for Scientific Research (04404014, 07456096) and by a grant from the Ministry of Agriculture, Forestry and Fisheries, Japan.

6. References

Adachi, M., Sako, Y. and Ishida, Y. (1993a) Application of monoclonal antibodies to field samples of *Alexandrium* species. *Nippon Suisan Gakkaishi* **59:** 1171-1175.

Adachi, M., Sako, Y. and Ishida, Y. (1993b) The identification of conspecific dinoflagellate *Alexandrium tamarense* from Japan and Thailand by monoclonal antibodies. *Nippon Suisan Gakkaishi* **59:** 327-332.

Adachi, M., Sako, Y., Ishida, Y., Anderson, D. M. and Reguera, B. (1993c) Cross-reactivity of five monoclonal antibodies to various isolates of *Alexandrium. Nippon Suisan Gakkaishi* **59:** 1807.

Adachi, M., Sako, Y. and Ishida, Y. (1994) Restriction fragment length polymorphism of ribosomal DNA internal transcribed spacer and 5.8S regions in Japanese. *Alexandrium* species (Dinophyceae). *J. Phycol.* **30:** 857-863.

Adachi, M., Sako, Y., Uchida, A. and Ishida, Y. (1995) Ribosomal DNA internal transcribed spacer regions (ITS) define species of the genus *Alexandrium.* In *Harmful Marine Algal Blooms*, Lassus, P., Arzul, F., Denn, E. E-L., Gentien, P. and Marcaillou-LeBaut, C. (eds.) Lavoisier Publ. Inc., Paris, pp. 15-20.

Adachi, M., Sako, Y. and Ishida, Y. (1996a) Identification of the toxic dinoflagellates *Alexandrium catenella* and *A. tamarense* (Dinophyceae) using DNA probes and whole-cell hybridization. J. Phycol. **32:** 1049-1052.

Adachi, M., Sako, Y. and Ishida, Y. (1996b) Analysis of *Alexandrium* (Dinophyceae) species using sequences of the 5.8S ribosomal DNA and internal transcribed spacer regions. J. Phycol. **32:** 424-432.

Barnes, S., Buchina, E. S., King, R. J., McBurnett, T.and Taylar, K. B. (1989) Bile acid sulfotransferase 1 from rat river sulfates bile acids and 3-hydroxy steroids: purification, N-terminal amino acid sequence, and kinetic properties. *J. Lipid. Res.* **30:** 529 - 540.

Cembella, A. E., Sullivan, J. J., Boyer, G. L., Taylor, F. J. R. and Anderson, R. J. (1987) Variation in paralytic shellfish toxin composition within the *Protogonyaulax tamarensis/catenella* species complex: Red tide dinoflagellates. *Biochem. Syst. Ecol.* **15:** 171 - 186.

Cembella, A. D. and Taylor, F. J. R. (1986) Electrophoretic variability within the *Protogonyaulax tamarensis/catenella* species complex: pyridine linked dehydrogenases. *Biochem. Syst .Ecol.* **14:** 311-323.

Franca, S., Viega,s S., Mascarenhas, V., Pinto, L. and Doucette, G. J. (1995) Prokaryotes in association with a toxic *Alexandrium lusitanicum* in culture. In *Harmful Marine Algal Blooms,* Lassus, P., Arzul, C., Erard-Le Denn, E., Gentien, P. and Marcaillou-LeBaut, C. (eds.) Lavoisier Publ. Inc., Paris, pp. 45-51.

Fukuyo, Y.(1985) Morphology of *Protogonyaulax tamarensis* (Lebour) TAYLOR and *Protogonyaulax catenella* (Whedon and Kofoid) TAYLOR) from Japanese coastal waters. Bull. Mar. Sci. **37:**529-537.

Gillham, N. W. (1978) In *Organella Heredity*, Raven Press, N Y., 602 pp.

Hallegraeff, G. M. (1993) A review of harmful algal blooms and their apparent global increase. *Phycologia* **32:** 79-99.

Hayhome, B., Anderson, D. M., Kulis, D. M. and Whitten, D. J. (1989) Variation among congeneric dinoflagellates from the north eastern United States and Canada *Mar. Biol.* **101:** 427-435.

Homma, H., Nakagome, I., Kamakura, M. and Matsui, M. (1992) Immnochemical characterization of developmental changes in rat hepatic hydroxysteroid sulfotransferase. *Biochem. Biophys. Act.* **1121:** 1169- 1174.

Hondoh, T., Suzuki, T., Hirato, K., Saitoh, H., Kadofuku, T., Sato, T. and Yanaihara, T. (1993) Purification and properties of estrogen sulfotransferase of human fetal liver. *Biomedical Res.* **14:** 129-136.

Ishida, Y. (1993) Who produces PSP? In *Harmful Algal News,* No.**7** :1-2.

Ishida, Y., Kim, C-H., Sako, Y., Hirooka, N. and Uchida, A. (1993) PSP toxin production is chromosome dependent in *Alexandrium* spp. In *Toxic Phytoplankton Blooms in the Sea*, Smayda, T. J. and Shimizu, Y. (eds.) Elsevier Sci. Publ. Amsterdam, pp. 881-887.

Ishida, Y., Sako, Y. and Adachi, M. (1995) Molecular approaches to identification of toxic marine dinoflagellates. In *MolecularApproaches to Food Safty Issuses Involving Toxic Microorganisms*, Eklund, M, Richard, J. L. and Mise, K. (eds.) Alaken Inc., Colorado, pp. 39-48.

Kim, C-H., Sako, Y. and Ishida, Y. (1993a) Variation of toxin production and composition in axenic cultures of *Alexandrium catenella* and *A. tamarense*. *Nippon Suisan Gakkaishi* **59:** 633-639.

Kim, C-H., Sako, Y. and Ishida, Y. (1993b) Comparison of toxin composition between populations of *Alexandrium* spp. from geographically distant areas. *Nippon Suisan Gakkaishi* **59:** 641-646.

Kodama, M. and Ogata, T. (1988) New insights into shellfish toxins. *Mar. Pollut. Bull.* **19:** 559-564.

Lee, D-H., Mittag, M., Sczekan, S., Morse, D. and Hastings, J. W. (1993) Molecular cloning and genomic organization of a gene for luciferin-binding protein from the dinoflagellate *Gonyaulax polyedra. J. Biol. Chem.* **268:** 8842-8850.

Lenaers, G., Scholin, C. A., Bhaud, Y., Saint-Hilaire, D. and Herzog, M. (1991) A molecular phylogeny of dinoflagellate protists (Pyrrhophyta) inferred from the sequence of the 24S rRNA divergent domains D1 and D8. *J. Mol. Evol*. **32:** 53-63.

Mulder, G. J. and Jakoby, W. B. (1990) Sulfation. In *Conjugation Reactions in Drug Metabolism*, Mulder, G. J. (ed.) Talor & Francis, London, pp. 107-161.

Ogata, T., Ishimaru, T. and Kodama, M. (1987) Effect of water temperature and light intensity on growth rate and toxicity change in *Protogonyaulax tamarensis. Mar. Biol.* **95:** 217-22.

Oshima, Y. (1995) Chemical and enzymatic transformation of paralytic shellfish toxins in marine organisms. In *Harmful Marine Algal Blooms,* Lassus, P., Arzul, F., Denn, E. E-L., Gentien, P. and Marcaillou-LeBaut, C. (eds.) Lavoisier Publ. Inc., Paris, pp. 475-480.

Saidha, T. and Schiff, J. A. (1994) Purification and properties of a sulfotransferase from *Euglena* using L-tyrosine as substrate. *Biochem. J.* **298:** 45-50.

Sako, Y., Kim, C-H., Ninomiya, H., Adachi, M. and Ishida, Y. (1990) Isozyme and cross analysis of mating populations on the *Alexandrium catenella/tamarense* species complex. In *Toxic Marine Phytoplankton*, Graneli, E., Sundstrom, B., Edler, L. and Anderson, D. M. (eds.) Elsevier Sci. Publ., New York, pp. 320-323.

Sako, Y., Kim, C-H. and Ishida, Y. (1992) Mendelian inheritance of paralytic shellfish poisoning toxin in the marine dinoflagellate *Alexandrium catenella. Biosci. Biotech. Biochem.* **56:** 692-694.

Sako, Y., Naya, N., Yoshida, T., Kim, C-H., Uchida, A. and Ishida, Y. (1995) Studies onstability and heredity of PSP toxin composition in the toxic dinoflagellate *Alexandrium*. In *Harmful Marine Algal Blooms*, Lassus, P., Arzul, F., Denn, E. E-L., Gentien, P. and Marcaillou-LeBaut, C. (eds.) Lavoisier Publ. Inc., Paris, pp. 345-350.

Scholin, C. A. and Anderson, D. M. (1993) Population analysis of toxic and nontoxic *Alexandrium* species using ribosomal RNA signature sequences. In *Toxic Phytoplankton Blooms in the Sea*, Smayda, T. J. and Shimizu, Y. (eds.) Elsevier Sci. Publ., Amsterdam, pp. 95-102.

Silva, E. S. (1982) Relationship between dinoflagellates and intracellular bacteria. In *Marine Algae in Pharmaceutical Science 2*, Hoppe, L. (ed.) Walter de Gruyter & Co., Berlin, pp. 269-288.

Taylor, F. J. R. (1984) Toxic dinoflagellates: taxonomic and biogeographic aspects with emphasis on *Protogonyaulax.* In *Seafood Toxins*, Ragelis, E. P. (ed.) ACS Symposium series **262**, Amer. Chem. Soc., Washington D. C., pp. 77-97.

Yoshida, T., Sako, Y., Uchida, A., Ishida, Y., Arakawa, O. and Noguchi, T. (1996) Purification and properties of paralytic shellfish poisoning sulfotransferase from toxic dinoflagellate Gymnodinium catenatum, In *Harmful and Toxic Algal Blooms*,Yasumoto, T., Oshima, Y. and Fukuyo, Y. (eds.) Intergovern. Oceanogr. Comm., UNESCO, pp. 499-502.

Bloom Dynamics of the Toxic Dinoflagellate *Gymnodinium catenatum*, with Emphasis on Tasmanian and Spanish Coastal Waters

G.M.Hallegraeff[1] and S. Fraga[2]

[1] Department of Plant Science, University of Tasmania, GPO Box 252-55, Hobart, Tasmania 7001, Australia

[2] Instituto Espanol de Oceanografia, Apdo. 1552, 36280 Vigo, Spain

1.Introduction

1.1. Taxonomic affinities

The chain-forming dinoflagellate *Gymnodinium catenatum* Graham 1943 is the only gymnodinioid dinoflagellate known which produces paralytic shellfish poisoning (PSP). Early speculation that *G. catenatum* may be a mutant of the armoured, chain-forming dinoflagellate *Alexandrium catenella* (Taylor 1980, Morey-Gaines 1982) has not been substantiated by comparative ultrastructural studies of the morphology of the nucleus, pyrenoid, pusule and flagellar root system (Rees and Hallegraeff 1991) nor molecular genetic studies (Zardoya *et al.* 1995). Similarly, the unique microreticulate brown hypnozygote of *G. catenatum* (Anderson *et al.* 1988) is unlike either the mucoid cyst type of *Alexandrium* or the spinose cyst of *Pyrodinium.* On present evidence, the taxonomic affinities of *G. catenatum* are within the Gymnodiniales, albeit it in a somewhat isolated position.

1.2. Global distribution

From a relatively obscure species which was reported only twice in the period 1940 to 1970, this dinoflagellate has been reported with increasing frequency starting in the late 1970s and early 1980s (summarised in Table 1). Its association with toxic PSP episodes was first reported from NW Spain in 1976, the Pacific coast of Mexico in 1979 and Australia (Tasmania) , Southern Japan and Portugal all in 1986. More recent records are also available from Venezuela, Uruguay , Morocco, and surprisingly from the tropical Indo-West Pacific in the Philippines, the Gulf of Thailand, Palau and Malaysia. While *G. catenatum* can easily be overlooked or misidentified when present as single cells (Blackburn *et al.* 1989; Yuki and Yoshimatsu 1989), when growing under optimal conditions its chain-forming habit (usually 8-16 cells, sometimes up to 64 cells long) makes it a very conspicuous organism which can readily be collected by plankton nets and they preserve even in harsh fixatives such as formalin. While increased scientific awareness may be partly responsible for its apparent increase in global distribution, one cannot escape the impression that this toxic dinoflagellate could be increasing in response to coastal eutrophication, ballast water translocations or even global warming (Fraga and Bakun 1993; Hallegraeff 1993; Nehring 1995).

NATO ASI Series, Vol. G 41
Physiological Ecology of Harmful Algal Blooms
Edited by D. M. Anderson, A. D. Cembella and G. M. Hallegraeff

Table 1. Chronology of first plankton records of *Gymnodinium catenatum* in different parts of the world

Year	Location	Reference
1939	Gulf of California	Graham 1943
1962	Mar del Plata , Argentina	Balech 1964
1976	NW Spain	Estrada *et al.* 1984
1979	Pacific coast of Mexico	Mee *et al.* 1986
1985	Tasmania, Australia	Hallegraeff and Sumner 1986
1986	Southern Japan	Ikeda *et al.* 1989; Nishioka *et al.* 1993
1986	Portugal	Franca and Almeida 1989
1988	Venezuela	La Barbera-Sanchez *et al.* 1993
1989	Gulf of Thailand	Matsuoka and Fukuyo 1994
1990	Phillipines	Fukuyo *et al.* 1993; Corrales *et al.* 1996
1990	Palau	Hallegraeff and Oshima, unpublished
1992	Uruguay	Mendez and Brazeiro 1993
1994	Morocco	Tagmouti *et al.* 1995
1994	Malaysia	Anton and Mohamad-Noor, this volume

Considering that *Gymnodinium catenatum* produces such a distinctive fossilisable cyst (Anderson *et al.* 1988), it is surprising that the cyst has been detected so rarely by micropaleontologists. The only unambiguous subfossil record refers to the Kattegat-Skagerrak in Northern Europe (Dale *et al.* 1993; Dale and Nordberg 1993), where a bloom period *ca.* 2000 to 500 years ago (4000 cysts/g sediment) coincided with a minor warming in climate. Viable cysts detected in recent years in Denmark (Ellegaard *et al.* 1993), the German Bight (North Sea) and Kiel Bight (Baltic Sea; maximum 17 cysts/ cm^3) (Nehring 1993) could possibly represent a continuation of a sparse population which managed to persist in the area under suboptimal environmental conditions. Even though this North European ecotype is non-chain-forming, it is surprising that the motile plankton cells have never been detected there in nature. Paulmier (1992) reported an organism similar to *G. catenatum* off Normandy (France) , which judging from its size and short chain length (2-5 cells) looks similar to the Danish strains. Personal observations by the first author on Danish cultures established from cysts by M. Ellegaard revealed that this "cold-water" form exhibits fast whirling swimming behaviour, distinct from the slow snake-like movement of its chain-forming counterparts. The pale yellow-green pigmentation of Danish cultures also is distinct from the golden-brown pigmentation of Tasmanian cultures. Attention is also drawn to discrepancies which exist between the cyst sizes reported from Spain, Tasmania and Japan (45-63 μm) compared to the smaller (30-40 μm) North European cysts (Table 2; Fig.3). The nature of a still smaller microreticulate cyst type (17-22 μm) known from Tasmanian and mainland Australian waters remains to be elucidated (Bolch and Hallegraeff 1990). Bimodal cyst size distributions have also been observed off Spain (Bravo, pers.comm.). No fossil cyst records are known from Japan, where modern cyst distributions (<10 cysts/mL sediment) are widespread in warm temperate waters in Southern Japan (Yatsushiro Sea to Seto Inland Sea, and Wakasa Bay;

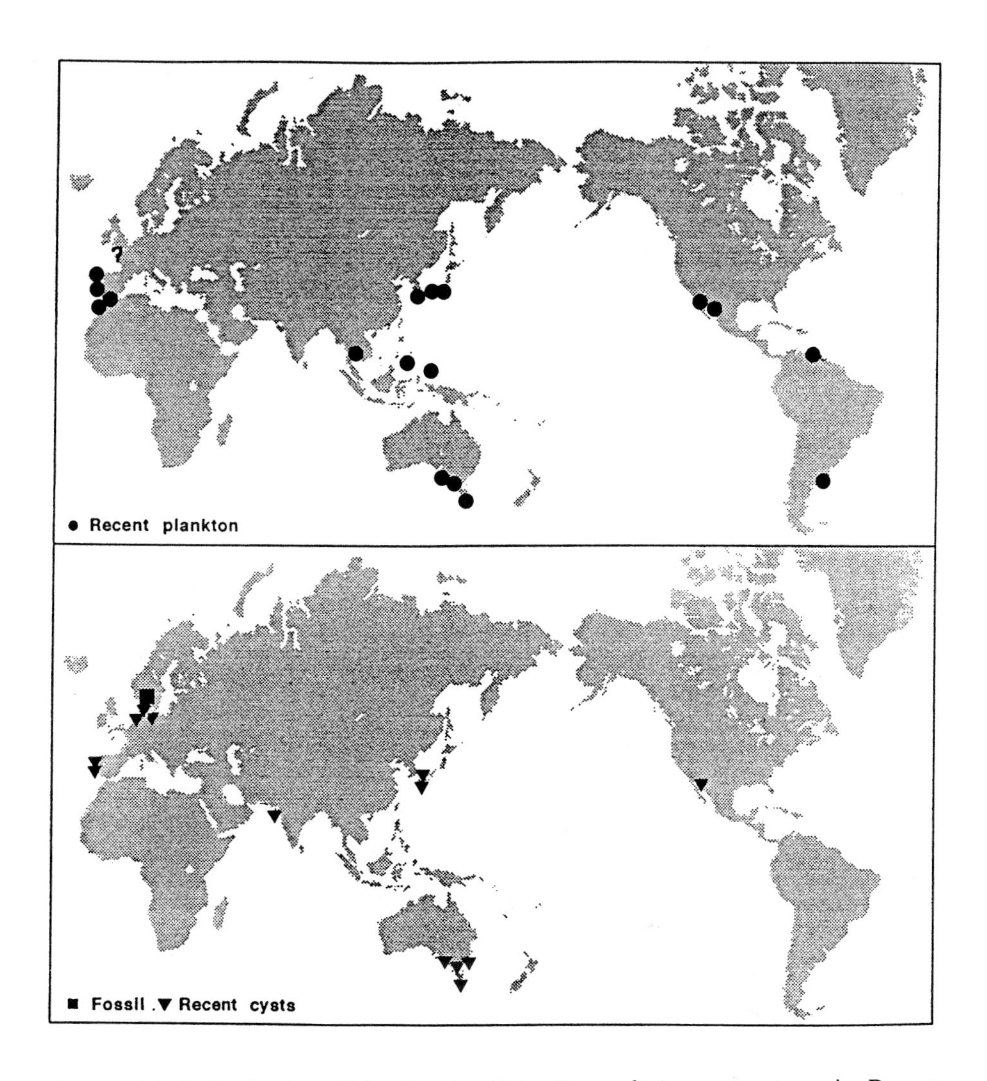

Fig.1. Global distribution of the dinoflagellate *Gymnodinium catenatum* in Recent plankton (top) and of the benthic cyst in Recent and fossil sediments (modified after Hallegraeff 1993 and Matsuoka and Fukuyo 1994).

Matsuoka and Fukuyo 1994) and neighbouring Korean waters (Kim *et al.* 1995). Similarly, no fossil records are known from the Australian region, where cyst surveys in dated sediment depths cores have unambiguously demonstrated the appearence of this species in Southern Tasmanian waters in *ca.* 1970 (McMinn *et al.* 1997). This evidence has fuelled speculation about the possible introduction of this organism via ship ballast water discharge or translocation of shellfish (Hallegraeff and Bolch 1992). In recent years, sparse populations of *G. catenatum* have also been detected on the mainland of Australia, off Port Lincoln in South Australia (Hallegraeff and Bolch, unpublished) and off Melbourne (Sonneman and Hill, in press). Modern cyst records are also known from NW Spain, especially in the Ria de Arosa (Blanco 1995b) and the outer region of Ria de Vigo (310 cysts/cm^3; Bravo and Anderson 1994), in the

South of Spain (Malaga) and off Portugal (Bravo, pers.comm.), the Gulf of California (Wrenn, pers. comm.), Uruguay (Mendez, pers.comm.), Dapeng Bay in China (Qi Yu-Zao, unpublished) and the Arabian Sea off India (K. Zonneveld, pers.comm.) (see map in Fig.1).

Table 2. Cyst diameter (μm) in different parts of the world

Origin	Mean size	Reference
subfossil Kattegat	30.5 ±2.5	Dale and Nordberg 1993; Thorsen *et al.* 1995
Denmark (Oresund)	30-40	Ellegaard *et al.* 1993
Germany	30-38	Nehring 1993
Japan	45-63	Matsuoka and Fukuyo 1994
Spain	38-60	Anderson *et al.* 1988; Blanco 1995b
Tasmania	37-55 17-22 (small form)	Blackburn *et al.* 1989; Bolch and Hallegraeff 1990

1.3. Confusion with morphologically similar species

In reviewing the literature on *G. catenatum* bloom dynamics, it is imperative to ensure that reported bloom events indeed refer to this taxon and have not been confused with similar chain-forming taxa. Fraga *et al* . (1995) elucidated the existence of a morphologically similar , nontoxic species *Gyrodinium impudicum*, which is generally smaller (17 μm diameter), produces shorter chains (mostly 4 cells long), has a larger cingular displacement and an apical groove (acrobase) which is an extension of the sulcus (sulcus and acrobase are clearly separated in *G. catenatum* ; Fig. 2). *G. impudicum* is known from Spanish (Ria de Vigo) and Portuguese Atlantic coastal waters, the Mediterranean coast of Spain (Fraga *et al* . 1995) , Fusaro lagoon (Italy; Carrada *et al* . 1991) , the Aegean Sea (Fraga *et al* . 1995), and Japanese (Iwasaki 1971), Australian (Melbourne, Sydney) and New Zealand waters (D.Hill, G.Hallegraeff and L.Mackenzie, unpublished observations). This species generally blooms in warm (22-28 °C), saline (36 to 38 ‰) and polluted waters and in culture grows much faster than *G.catenatum.* It is also possible that the chainforming *Cochlodinium catenatum* (descending girdle with 1.5 to 4 turns around the cell) occasionally may have been mistaken for *G. catenatum.* To date, only Tasmanian (Rees and Hallegraeff 1991) , Danish (Ellegaard *et al.* 1993), Portuguese (Franca 1987) and Mexican strains of *G. catenatum* (Morey-Gaines 1982) have been subjected to detailed ultrastructural characterisation using transmission electron microscopy. A closer examination of tropical strains is especially recommended.

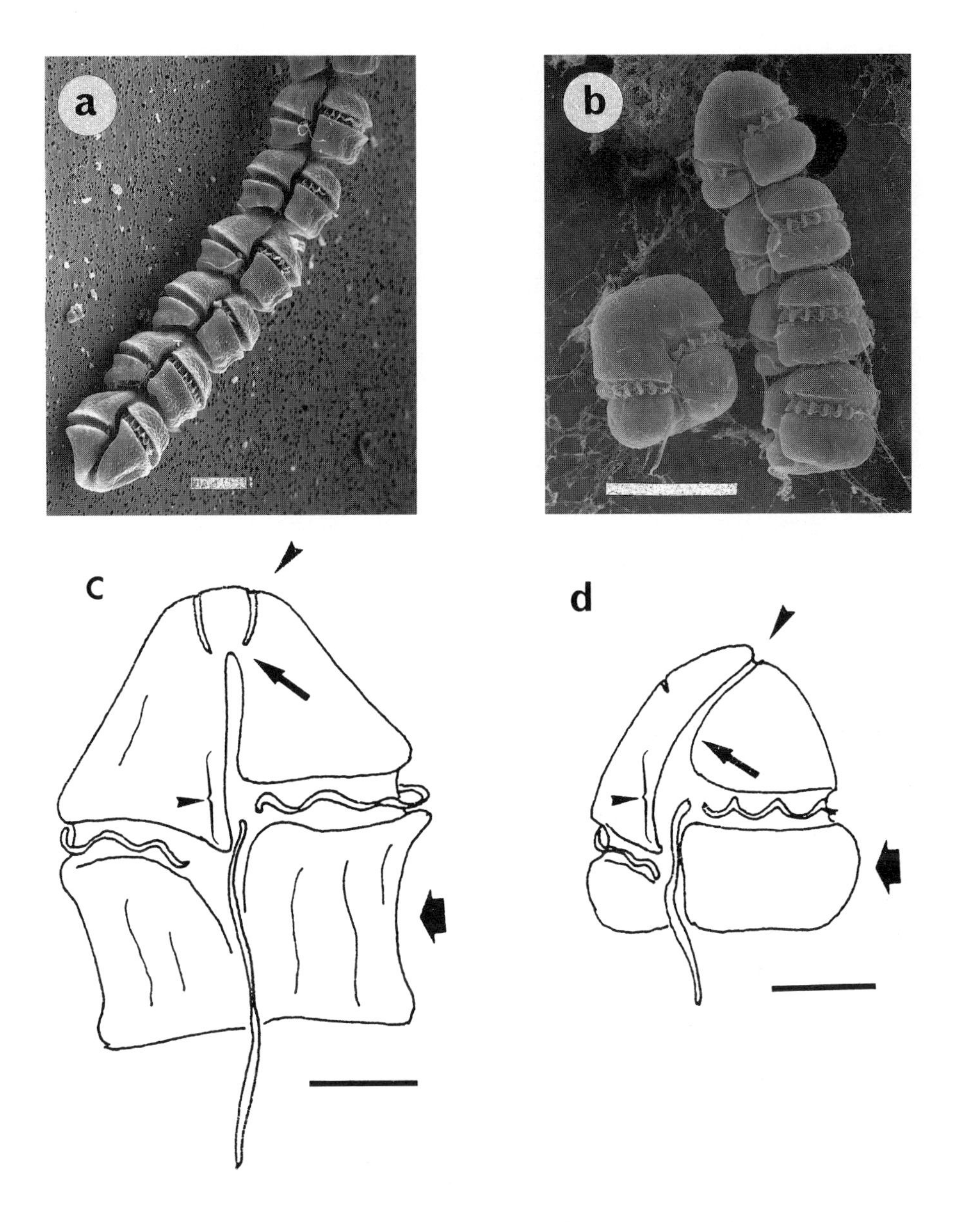

Fig. 2. Comparison between *Gymnodinium catenatum* (Figs 2 a, c) and the morphologically similar *Gyrodinium impudicum* (Figs 2 b, d). The sulcus and acrobase are clearly separated in *G. catenatum,* whereas the acrobase is an extension of the sulcus in *G. impudicum.* The latter species is also considerably smaller, has a larger cingular displacement and a rounder hypocone silhouette . Scale bars : 20 μm in Figs 2 a, b; 10 μm in Figs 2 c, d. (Micrographs: C.J. Bolch (Fig.2a); M. Delgado (Fig.2b); Line drawings in Figs 2 c, d after Fraga *et al.* 1995).

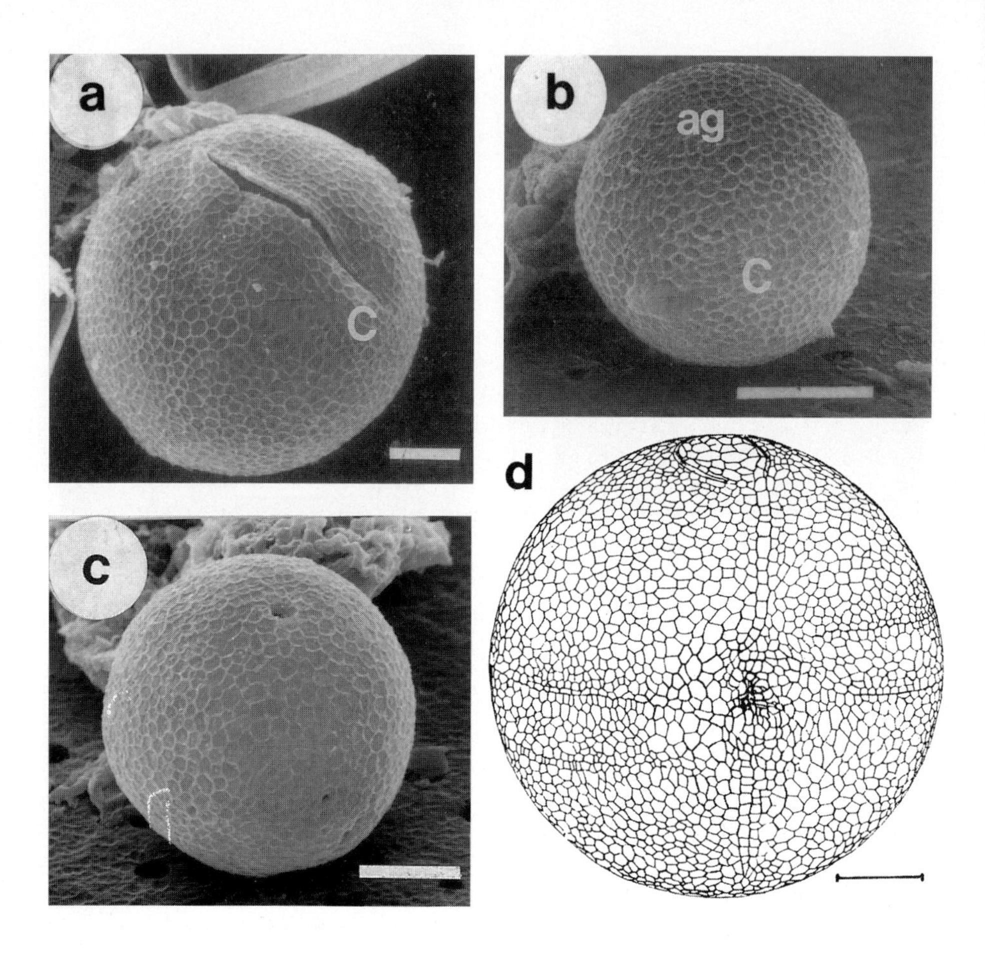

Fig.3. Scanning electron micrographs and sketch of the microreticulate resting cyst of *Gymnodinium catenatum* and related species: (a) typical cyst from Tasmanian waters, 43-55 μm diameter; (b) smaller rare cyst type from Tasmanian waters, 17-22 μm diameter ; (c) fossil cyst type from Kattegat, 30 μm diameter (after Bolch and Hallegraeff 1990; Hallegraeff 1993); (d) sketch of ventral surface of Spanish cyst (after Anderson *et al.* 1988). All scale bars 10 μm.

1.4. Recognition of *Gymnodinium catenatum* as a "species complex"

The map in Fig. 1 summarises the known global distribution of unambiguous taxonomic records of *G. catenatum* motile plankton dinoflagellates as well as of its microreticulate benthic resting cyst in Recent and fossil sediments. What at first impression may appear to be a widespread species in warm-temperate and tropical regions, we interpret to represent a "species complex". The Northern European (Denmark, Germany) and tropical forms (Thailand, Phillipines, Palau) appear to be different ecotypes compared to the well-studied Tasmanian, Spanish and Japanese populations. In culture, Tasmanian, Spanish and Japanese strains have broadly similar temperature optima of 12-18°C, exhibiting no growth at temperatures below 11-12.5 °C nor at temperatures as high as 30°C (Blackburn *et al.* 1989, Bravo and Anderson 1994, Ogata *et al.* 1989; Table 3). Tropical populations from Palau, the Phillippines and Thailand thrive at temperatures of 29-30 °C and therefore must represent at least different temperature strains (compare Guillard and Kilham 1977 regarding temperature strains of diatoms). In culture, the Danish strains exhibited a preference for slightly higher temperatures of 20-25°C (Ellegaard *et al.* 1993), which contrasts with the unfavourable ambient temperatures of North European waters (max. 12-18°C in summer), explaining the apparent absence of motile dinoflagellates in the water column. The Tasmanian strains are tolerant towards salinities in the range of 15 to 35°/oo (Blackburn *et al.* 1989).

Preliminary allozyme electrophoretic analyses on Tasmanian, Spanish and Japanese strains failed to show any differences (Blackburn, unpublished). Similarly, LSU rDNA sequences of Tasmanian and Spanish strains were identical, but those of Danish cultures differed at 3 positions (Ellegaard, pers.comm.) Further molecular genetic studies on *G. catenatum* are in progress trying to discriminate populations and identify global dispersal routes on the basis of RAPD and microsatellite molecular markers (C. Bolch and G. Hallegraeff, in progress). Sexual crossing experiments indicated mating compatibility between Tasmanian, Spanish and Japanese strains (Blackburn *et al.* 1989), but long-term viability of the progeny of some of these international crosses was limited (Blackburn *et al*, unpublished). Crosses between Danish and Spanish, and between Danish and Australian strains have been unsuccessful (Ellegaard, pers.comm.). At the same time, attempts to use PSP toxin fingerprints appear to distinguish between Tasmanian strains (producing 13-deoxydecarbamoyl toxins), Spanish (larger proportions of GTX5 and GTX6) and Japanese strains (complete absence of C3 and C4 toxins) (Oshima *et al.* 1993). The toxin profile of a Philippines culture was somewhat similar to that of Spanish strains (Fukuyo *et al.* 1993) while Danish strains were consistently non-toxic (Ellegaard and Oshima 1995). The comparative value of allozymes, rDNA sequences, RAPD and microsatellite molecular markers, mating compatibility and toxin finger prints as signatures of *G. catenatum* populations is the subject of continued investigation.

2. Life cycle

Bravo (1986) and Anderson *et al.* (1988) produced *G. catenatum* cysts by laboratory incubation of natural Spanish plankton net samples for 20 days in f/2 medium deficient in nitrate and phosphate. Blackburn *et al.* (1989) succeeded in inducing cyst production in unialgal Tasmanian cultures provided compatible opposite mating types were combined (heterothallism). After 3 days of incubation in a medium deficient in both nitrate and phosphate, small dancing groups of gametes and fusing gamete pairs

Table 3. Temperature optima of different populations

Location	Temperature (°C)		Authority
	culture	**nature**	
Tasmania	14.5-20 range; max 17.5-20; no growth at 30 nor <12.5	12-18	Blackburn *et al.* 1989; Hallegraeff *et al.* 1995
Denmark	15-29 range; max 20-25; no growth at 9	absent	Ellegaard *et al.* 1993
Spain	14-29 range; max. 22-28; no growth at 30; negligible at 11	12-18	Bravo and Anderson 1994
Japan	15-25 in culture	6-15; >20	Ogata *et al.* 1989; Ikeda *et al.* 1993; Nishioka *et al.* 1993
Palau	not available	29.6-29.9	Hallegraeff and Oshima, unpublished
Thailand	not available	>25	Matsuoka, unpublished
Argentina	not available	>17	Balech 1964; Carretto *et al.* 1995
Mexico	not available	14-17; >20	Graham 1943; Mee *et al.* 1986
Venezuela	no data	23-26	La Barbera-Sanchez, unpublished
Philippines	25	26-29	Fukuyo *et al.* 1993; Corrales *et al.* 1996

(with the girdles joined equatorially or perpendicular to each other) could be observed for up to 28 days after initiation of intercrossing experiments. The resulting planozygote (with double longitudinal flagellum) over a period of 14 days changed shape to become subspherical, then lost motility, contracted its cell contents and developed a thick cyst wall (Fig. 4). Hypnozygotes exhibited a very short dormancy (maturation) period of approximately 2 weeks at 17.5-20 °C. They showed no evidence of quiescence (inhibition of growth by unfavorable conditions), with germination possible (albeit retarded) when kept in the dark for 3 months and when stored at 4 - 5°C (Blackburn *et al.* 1989; Bravo and Anderson 1994). Growth media seemed to have no effect on excystment rates , and the optimal range of environmental conditions for cyst germination was similar to that of vegetative growth (Bravo and Anderson 1994). The most striking feature of the *G. catenatum* life cycle is that newly formed cysts can germinate rapidly (within 2 weeks of encystment) under a wide range of environmental conditions , providing inocula for a nearly continuous motile population of dinoflagellate cells. In Tasmanian waters *G. catenatum* is a resident of the water column throughout the year and planomeiocytes (the first cells resulting from cyst germination) have only rarely been detected in field samples during ten years of plankton monitoring. Cyst germination therefore is thought to play a minimal role in seasonal bloom dynamics in Tasmania, and constraints on the growth of the dinoflagellates in the water column are considered to determine seasonal and interannual bloom variations (Hallegraeff *et al* . 1995). In the Ria de Vigo, the water column is devoid of *G. catenatum* for many months in the year, and Bravo and Anderson (1994) argue that increased bottom temperatures after the upwelling period, and perhaps resuspension of cysts to higher light levels, could account for high cyst germination and increased vegetative growth rates in autumn. However, when considering the Iberian coastline as a whole, *G. catenatum* is present throughout the whole year between Cape Finisterre to the Alboran Sea (Fraga 1996; Fig.11; see section 4.2), which is considered to be the central distribution zone for this species , while Ria de Vigo represents only the northern limit of distribution. Benthic cyst stages of this dinoflagellate are important in sustaining this species through long periods in marginal environments when water column conditions are unfavourable for bloom formation. The development of immunological or molecular probes that can readily detect planomeiocytes in field samples would be of considerable value to more precisely estimate the contribution of cyst germination to dinoflagellate bloom dynamics . The application of a "germinating cell trap sampler" to this cyst species as developed by Ishikawa *et al.* (1995) would also be of considerable interest.

3. Nutrients and Vertical Migration

In culture experiments, growth of *G.catenatum* could be optimised (mean doubling time 3-4 days) by including a soil extract in the culture medium (Blackburn *et al.* 1989). It is not yet clear whether this is due to the chelating effect of soil extract or its stimulation of cellular metabolism; this question is presently being addressed by bioassays with individual humic fractions isolated from Tasmanian waters (M.Doblin, University of Tasmania). Preliminary results suggest that humic fractions do not markedly affect the growth rate but prolong the stationary phase, allowing the species to achieve higher biomass levels. In Tasmania, no relationship between *G.catenatum* blooms and macronutrient concentrations in the water column (5 to 15 $\mu g\ L^{-1}$ NO_3^- nitrogen, 30 to 50 $\mu g\ L^{-1}$ NH_4^+ nitrogen, 5 to 15 $\mu g\ L^{-1}$ $PO4^{3-}$ and 0.2 to 3 mg L^{-1} $Si(OH)_4$ could be detected. In Ria de Vigo in Spain, *G.catenatum* blooms occur at NO_3^- concentrations of 0.2-0.8 μM and NH_4^+ concentrations of 0.25-0.50 μM (Figueiras and Fraga 1990). The ability of this dinoflagellate to perform diurnal vertical migrations enables this species to capitalise on available deep water nutrients at night, even when nutrients in surface water become depleted . Figueiras and Fraga (1990) demonstrate this principle during a *G.catenatum* bloom in 1986 in Ria de Vigo, when a small upwelling event delivered deep (10m) nutrients only accessible to this strongly vertically migrating species. Fraga *et al.* (1992) demonstrated the uptake of deep nitrate nutrients during vertical migration of *G. catenatum* using a chemical balance approach assessing different forms of N, P and C in the water column. During the day, the cells photosynthesized carbohydrates in surface waters, but at night they used carbohydrates to take up N and P in deeper waters in order to synthesize proteins.

In Tasmanian waters *G.catenatum* also performs diurnal vertical migrations, rising close to surface at 1 m depth in the early morning and descending down to 3-6 m depth in the evening (Fig.5). Fraga *et al.* (1989) recorded swimming speeds of 300-400 $\mu m/sec$ (1.5 m/h) under laboratory conditions, which is slightly higher than field estimates from Tasmanian waters (4 m/9 hrs = 125 $\mu m/sec$) . In the Ria de Vigo

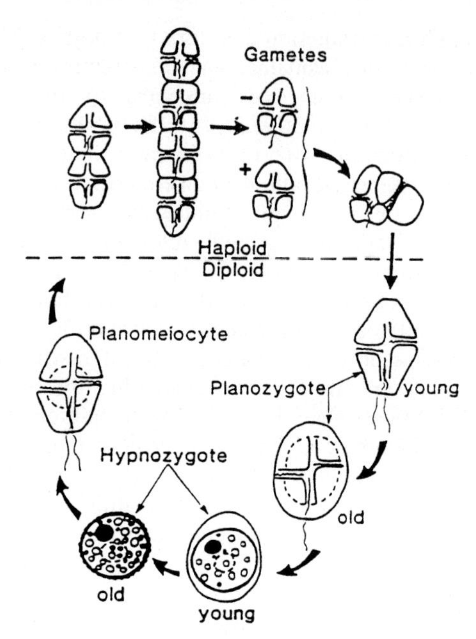

Fig.4. Diagrammatic summary of the sexual life cycle of *Gymnodinium catenatum*. Motile vegetative cells divide (by mitosis) to form chains, producing two types of gametes (heterothallism), pairs of which fuse to give a planozygote. This cell loses motility to form a benthic resting cyst (hypnozygote). Excystment produces a planomeiocyte, similar to a planozygote, which divides (by meiosis) to re-establish the planktonic vegetative stage (from Blackburn *et al.* 1989).

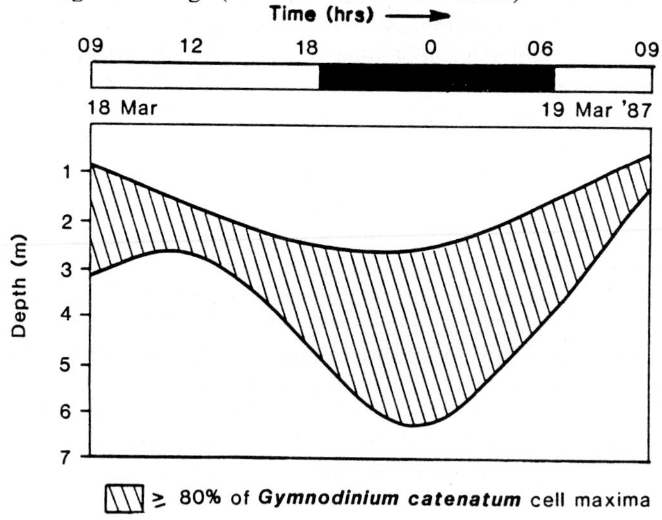

Fig.5. Diurnal vertical migration behaviour of *Gymnodinium catenatum* in southern Tasmanian waters, Derwent River, 18- 19 March 1987. The dinoflagellates descend and spread out at night down to 3-6 m depth and rise and accumulate at 1 m depth in the early morning (Hallegraeff and Bolch, unpublished data).

(Spain), *G. catenatum* can overcome downwelling velocities of 10m/day. It has been widely observed in culture experiments with *G. catenatum* which require the regular taking of subsamples (John *et al.* this conference; Doblin and Hallegraeff, unpublished), that this species is extremely sensitive to turbulence. This was confirmed by field observations on Tasmanian blooms which only occur following continuous periods of low windspeed (<5 m s^{-1}) for 5 days or more (Hallegraeff *et al* .1995). While in Spain, *G. catenatum* blooms paradoxically follow strong winds, it is possible that the deeper part of the water column remains stable; this remains to be investigated (Estrada, this volume).

4. Bloom dynamics

At present, the only long-term (10 year) ecological data sets available are for Tasmanian and Spanish/Portuguese blooms, while insufficient information yet exists to interpret Argentinian, Japanese , Mexican and Philippines bloom occurrences.

4.1. Tasmanian case history

Based on an analysis of historic plankton samples and cysts in sediment depth cores, McMinn *et al.* (1997) claim that this species was introduced into the Tasmanian region after 1970. The first bloom event was recorded in the Derwent estuary in 1980, but monitoring of bivalve toxicity was not instituted until 1986. The presence of extensive cyst beds in southern Tasmania (Bolch and Hallegraeff 1990) suggests that autochthonous bloom events originate from within the Huon and Derwent rivers (average depth 10-20 m), whereas dinoflagellate cells that are flushed out of the estuaries into oceanic waters always appear moribund. Blooms of *G. catenatum* exhibit significant interannual variability , with high shellfish toxicity in 1986, 1991 and 1993 (more than 8,000 µg PSP / 100 g shellfish meat) also coinciding with the greatest spatial extent of shellfish toxicity. In other years, such as 1989, 1990, 1992, 1995 and 1996 this dinoflagellate has been present in Tasmanian waters only in very low concentrations resulting in insignificant shellfish toxicity (Fig.7). Seasonal variations in dinoflagellate abundance are primarily controlled by water temperature (Hallegraeff *et al.* 1989), while interannual variations are thought to be related to differences in water temperature, rainfall patterns and wind stress (Hallegraeff *et al.* 1995). Surface water temperatures in southern Tasmanian waters exhibit maxima in February to March (15-18°C) and minima in July-August (10-11°C). *G. catenatum* blooms in Tasmania tend to occur mainly in the period December to June , when water temperatures range from 12 to 18 °C and salinities range from 28 to 34 °/oo. Within Tasmania, *G. catenatum* blooms are mainly confined to the humus-laden Huon and Derwent rivers and appear to require a rainfall event as a trigger. In most years the January-March period is characterised by drought conditions and the first sign of dinoflagellate blooms occurs 2-3 wks following the first major rainfall event. In 1990 the rainfall trigger occurred in mid-May during decreased water temperatures (14°C) resulting in an insignificant dinoflagellate population, whereas bloom events in 1986, 1991 and 1993 triggered in February-March at peak water temperatures (16°C) led to significant shellfish toxicities. Once a bloom has started, there appears to be no continued requirement for high river flow. The significant shellfish toxicity in April-July 1991, and especially the 1993 bloom were all associated with calm weather conditions, *i.e.* continuous periods of low windspeed (<5 m s^{-1}) for 5 days or

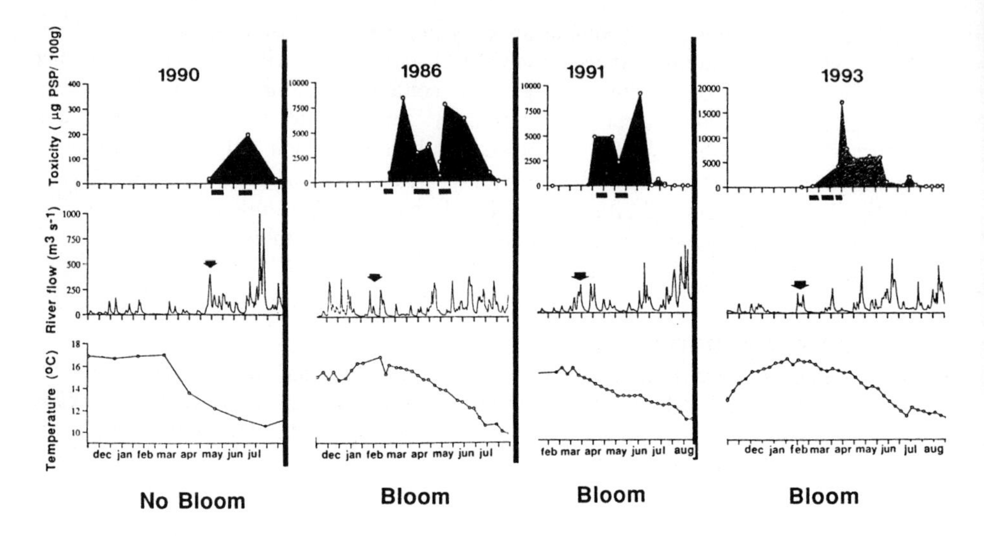

Fig.6. Environmental conditions in Tasmanian waters leading up to the 1990, 1986, 1991 and 1993 *Gymnodinium catenatum* population increases : (a) dinoflagellate abundance as reflected in PSP toxin levels in Huon estuary longline mussels; (b) Huon River flow. The arrow marks the increase in river flow associated with the initiation of the bloom; (c) Water temperature at 10m depth . Black bars underneath the shellfish toxicity graphs indicate periods of low windstress (<5 m s^{-1}); this explains the incidence in some years of multiple toxicity peaks (From Hallegraeff *et al.* 1995).

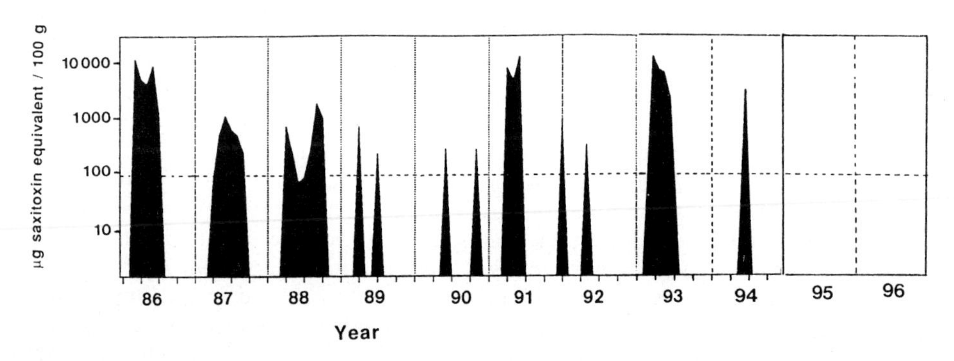

Fig.7. Interannual variability in the concentration of *Gymnodinium catenatum* toxins (µg saxitoxin equivalent per 100 g shellfish meat) in Tasmanian mussels in the period 1986 to 1996. A logarithmic scale of shellfish toxicity has been used to emphasize seasonal bloom patterns. The dashed line indicates the USFDA quarantine level of 80 µg per 100 g, above which shellfish farm closures need to be instituted. Note the absence of bloom events in 1989, 1990, 1992, 1995 and 1996 (From Hallegraeff *et al.* 1995).

more. The incidence in some years of multiple toxicity peaks also relates to cycles of bloom development and wind disturbance (Fig.6).

4.2. Spanish case history

Gymnodinium catenatum was first observed in North Atlantic waters of the rias of Galicia in the autumn of 1976 (Estrada *et al* . 1984), coincident with a PSP outbreak that affected several European countries to which cultured Spanish mussels had been exported (Luthy 1979). Although intensive phytoplankton studies had been carried out in Ria de Vigo during the 1950s by Margalef and coworkers (1955, 1956), this conspicuous chain-forming species had never been detected. Similarly, there exist no records of PSP in Galicia prior to 1976. As a consequence of the 1976 *G. catenatum* PSP outbreak, a phytoplankton and shellfish toxicity monitoring program was instigated in Galicia, but *G. catenatum* was not observed again until 1981 (perhaps due to an inadequate sampling strategy). In that year, it occurred off the Southwest coast of Portugal at the end of summer (Estrada 1995) and appeared more than a month later in the Ria de Vigo, which is located more than 400 km further north along the west coast of the Iberian Peninsula. Since then, this dinoflagellate has been commonly observed in the area resulting in shellfish toxicity and shellfish quarantine closures in many years . An in-depth analysis of long-term PSP toxicity data in the Spanish region has not yet been carried out.

Because of the significance of PSP for public health and the economic interests of the aquaculture industry in the Galician rias (annual production of cultured mussels is approximately 200,000 metric tons), the main research objective has been to elucidate the triggering mechanisms and develop a predictive capacity of blooms to minimize the damage to the aquaculture industry. Initially, a good correlation between *G. catenatum* blooms in the rias and upwelling relaxation was observed (Fraga *et al.* 1988). The hydrography of the west coast of the Iberian Peninsula is characterized by seasonal wind driven coastal upwelling which can be considered to be a northern extension of the NW Africa upwelling system (Wooster *et al* . 1976). A strong high pressure system over the North Atlantic in summer causes alongshore northerly winds off Portugal and Galicia which force upwelling events from April through September. *G. catenatum* was commonly oberved inside the rias at the end of summer , when the upwelling favorable winds relax, or when they reverse in their direction and become southerlies. The precise origin of these populations is not clear. The observed rapid increases in dinoflagellate cell numbers can not be simply explained by vegetative growth only; moreover, due to the action of wind the water of the rias is exchanged in a few days. Two alternative explanations for these blooms have been put forward : (a) advection of populations from outside the ria or (b) a massive excystment of benthic resting cysts from local sediments (Blanco 1995a). Fraga *et al* (1988), in the first complete description of the 1985 bloom event of *G. catenatum* in Ria de Vigo, correlated upwelling relaxation (Fig.8) with the introduction of warm offshore surface waters into the ria as reflected in the occurrence of a mixed phytoplankton assemblage of warm-water oceanic and coastal dinoflagellates. These authors considered that a massive cyst germination event seemed unlikely since the distinctive germling cells (planomeiocytes) which emerge from cysts were never detected in water samples (compare with similar observations in Tasmania). Figueiras and Pazos (1991), however, argued that resuspension of bottom cysts was a more realistic hypothesis to explain a major *G. catenatum* bloom in Ria deVigo in 1986, because they could not observe a dinoflagellate bloom at the oceanic stations before it appeared in the ria. The

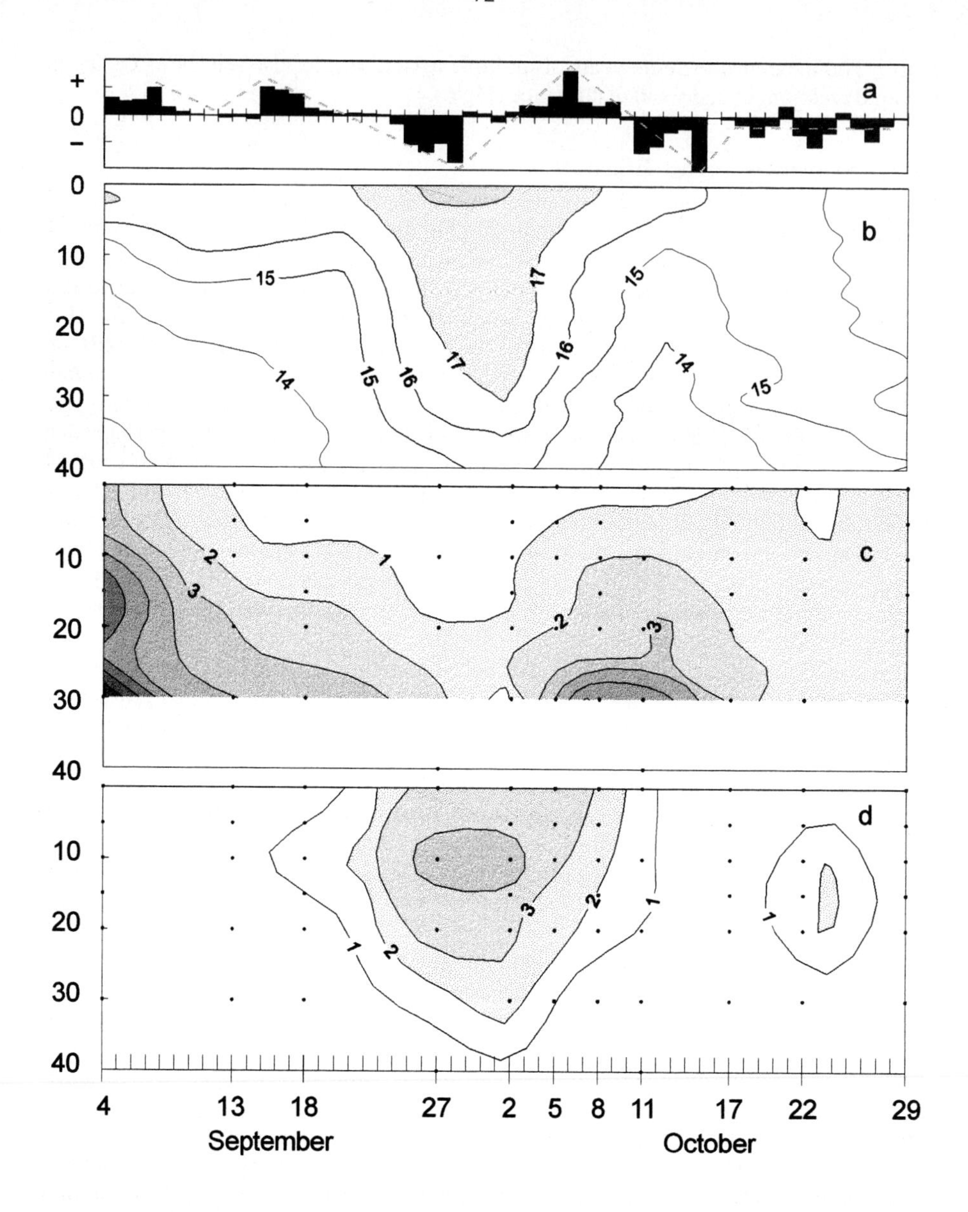

Fig.8. Relationship between *Gymnodinium catenatum* blooms in the Ria de Vigo and upwelling relaxation during 1990; (a) upwelling index for a station at 43°N 11°W; (b) CTD measurements of water temperature (°C) at a station near the mouth of the Ria de Vigo; c) nitrate concentrations (μmol kg^{-1}) at the same station; d) *Gymnodinium catenatum* cell concentrations in log (x+1) cells L^{-1} at the same station (after Fraga et *al* 1993).

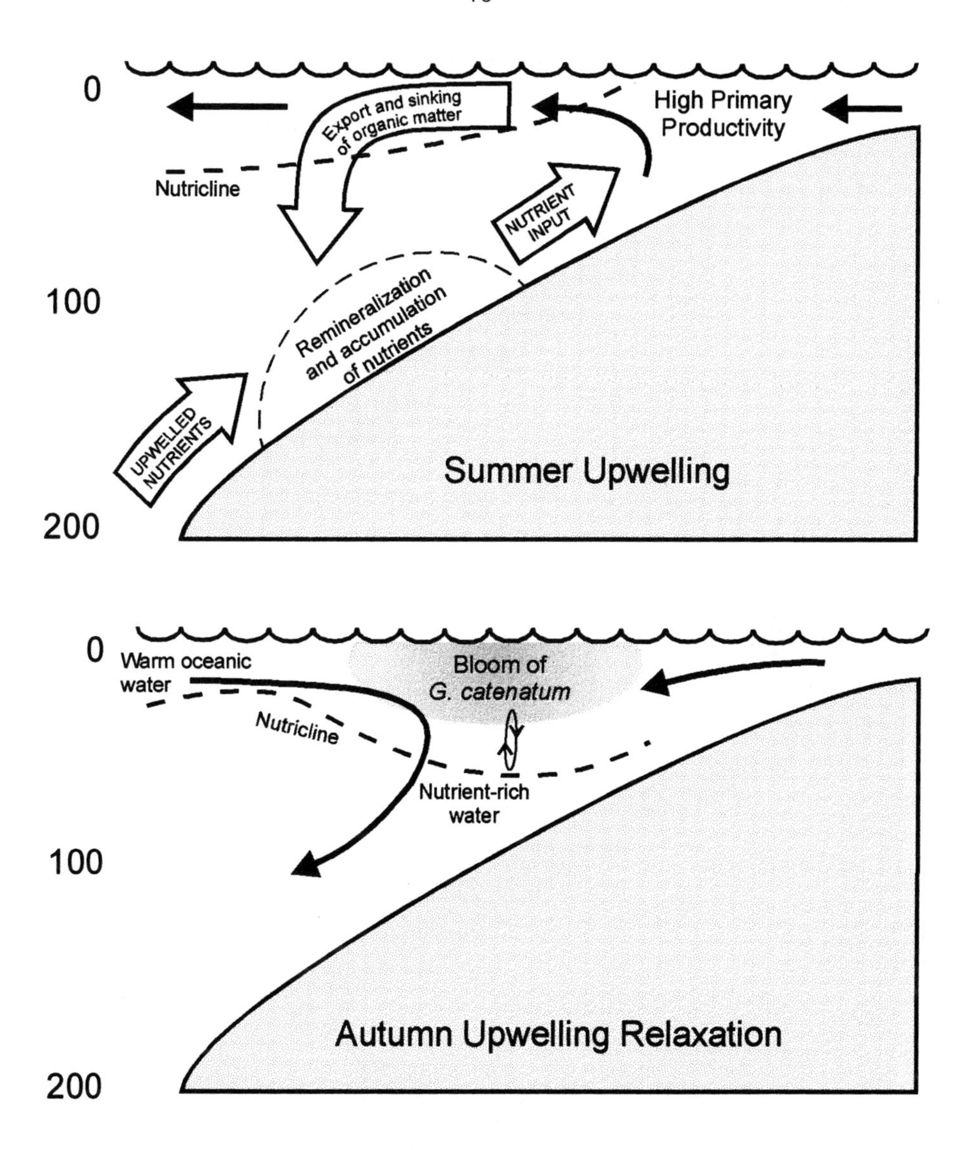

Fig.9. Hydrography of the Ria de Vigo during (a) summer upwelling and (b) autumn upwelling relaxation. Note the shallow nutricline in (a) allowing for diatom blooms, while the advection of warm oceanic water in (b) leads to a deepening of the nutricline, providing a competitive advantage to migrating dinoflagellates (from Fraga and Bakun 1993).

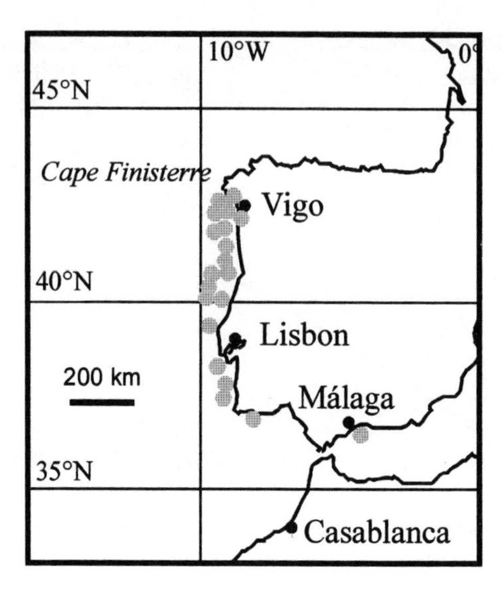

Fig.10. Distribution of *Gymnodinium catenatum* cysts in coastal waters off SW Europe, based on published and unpublished data by I. Bravo, J. Blanco and T. Moita.

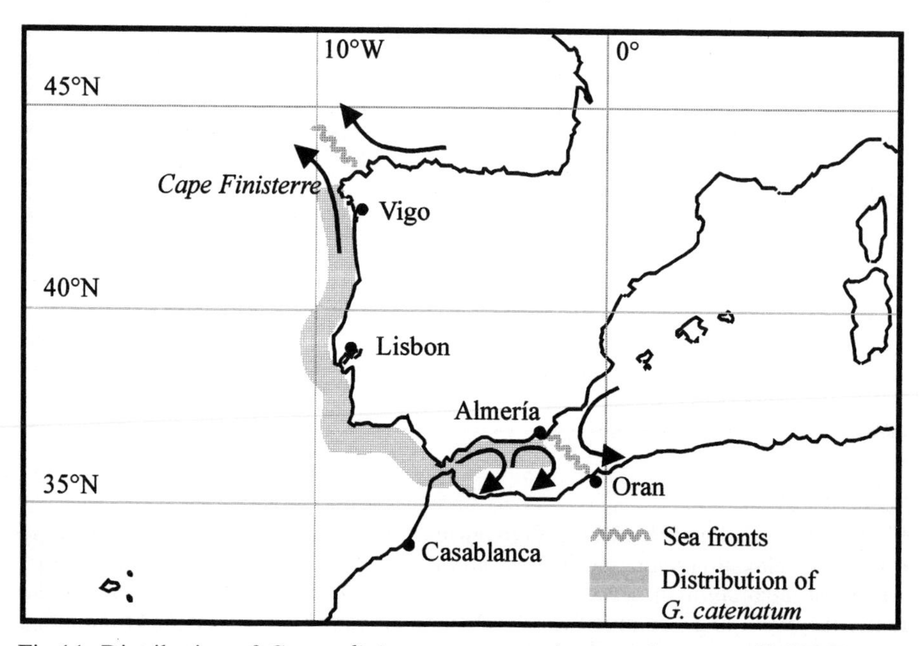

Fig.11. Distribution of *Gymnodinium catenatum* in coastal waters off SW Europe During oceanic cruise Morena 0593 in May 1993 the dinoflagellate was observed in oceanic waters off Galicia and North Portugal, with blooms in Ria de Vigo occurring 10 days after it was found off Portugal (from Fraga 1996).

presence of low concentrations of *G. catenatum* resting cysts inside the rias has now been confirmed (Bravo and Anderson, 1994; Blanco 1995b) but their presence still cannot readily explain the triggering of a massive bloom event , even when taking into account that bottom water temperatures are highest and most suitable for cyst germination at the time of year when blooms usually take place (Bravo and Anderson, 1994). Gomez *et al.* (1996) studying the short-time scale development of *G. catenatum* blooms in the Ria de Vigo also support the concept of shoreward advection of offshore populations, with intrusions of coastal waters being induced by westerly winds. Upwelling relaxation therefore not only results in a deepening of the nutricline, but it also allows for the advection of warm (17°C) water, the introduction of an offshore inoculum and the further concentration of growing cells in convergence zones (Fig.9).

Until recently, the lack of suitable samples from oceanic waters of the Iberian Peninsula has been a stumbling block to further the hypothesis of an offshore origin of *G.catenatum* in the rias. Fraga (1996) documented the presence of *G. catenatum* in oceanic waters off the west coast of the Iberian peninsula throughout all seasons, including winter (Fig.11). The dinoflagellate was observed in May 1993 during an oceanic cruise, when it was the dominant species in net samples taken over the southernmost part of the shelfbreak off Portugal at a time when it was not observed in the Rias Baixas of Galicia. The most northern station at which *G. catenatum* was observed was 100 km south of Ria de Vigo. After the completion of the cruise, strong southerly winds persisted and *G. catenatum* was observed in Ria de Vigo 10 days after it was found off Portugal. This appearance in the mouth of Ria de Vigo during May coincided with an increase in bottom temperature of about 3°C . During a second cruise nine months later, *G. catenatum* was observed in winter west of Lisbon . During a third cruise in March 1995 the dinoflagellate was also observed in offshore deep sea oceanic waters in front of Portugal. The salinity and temperature characteristics of the water in which *G. catenatum* was observed during the winter cruise match that of the central portion of Eastern North Atlantic Central Water (ENCW) (Alvarez-Salgado *et al.* 1993), which is the upwelling water of the west coast of Galicia that has a subtropical origin. To improve prediction of Spanish bloom events, it is recommended to focus future research efforts on bloom dynamics of the shelfbreak off Portugal (Moita 1993; Sousa *et al.* 1995; Vilarinho and Moita 1996), where cyst beds have been found all along the coast (Fig.10) . Bloom dynamics off Portugal also relate to upwelling patterns, where an offshore origin of *G. catenatum* in the Azores frontal system is currently being explored .

5. Conclusions

• Temperature determines the biogeography of different ecotypes of *G. catenatum* , but it can also play a role in creating an environmental window for seasonal plankton blooms in warm-temperate waters (in Tasmania and Spain). Subfossil blooms in the Kattegat-Skagerrak of an organism similar to *G. catenatum* also coincided with a minor warming in climate. The molecular genetic, ultrastructural, biochemical and physiological characteristics of different global populations of this species remains to be elucidated.

• Benthic cyst stages of this dinoflagellate appear not to play a role in seasonal bloom dynamics, but their major function is to sustain this species through long periods when water column conditions are unfavorable for bloom formation (e.g. in Northern

Europe). The quantitative contribution of cyst germination to bloom dynamics (e.g. in Spain) remains to be investigated.

• The inoculum for *G. catenatum* blooms can derive from overwintering vegetative cells (Tasmania, Spain) but germination of benthic cysts is also possible. Both local estuarine origin (Tasmania) and alongshore transport of offshore populations (Spain) have been claimed as inoculation sources.

•The strong vertical migration ability of *G. catenatum* allows this organism to establish predominance over diatoms under calm stable water column conditions (Tasmania) or under conditions where the nutricline is deeper than 10m (Spain). *G.catenatum* commonly co-occurs with other migrating chainforming (*Alexandrium affine* in Spain; *Pyrodinium bahamense* in Palau and Mexico) and non-chain-forming dinoflagellate species (*Ceratium tripos, C. furca, C. fusus* in Tasmania). This dinoflagellate species is extremely sensitive to turbulence in laboratory experiments, the precise physiological basis for which remains to be elucidated.

6. Acknowledgements

We gratefully acknowledge the following colleagues for helpful discussions and / or permission to quote their unpublished data, S. Blackburn, J. Blanco, I. Bravo, C. Bolch, J. Carreto, M. Ellegaard, A. La Barbera, K. Matsuoka, S. Mendez , T. Moita , J. Wrenn and K. Zonneveld.

7. References

Alvarez-Salgado, X.A., Roson, G , Perez, F.F., Pazos,Y. (1993). Hydrographic variability off the Rias Baixas (NW Spain) during the upwelling season. *J. Geophys. Res.* 98 (C8) :14447-14455.

Anderson, D.M., Jacobson, D.M., Bravo, I., Wrenn, J.H. (1988). The unique, microreticulate cyst of the naked dinoflagellate *Gymnodinium catenatum. J. Phycol.* 24: 255-262.

Anton, A., Mohamad-Noor, N. (1996). Harmful Algal Bloom (HAB) species in the straits of Melaka, Malaysia. NATO-ASI conference, Bermuda, poster abstract E1.

Balech, E. (1964). El plancton de Mar del Plata durante el periodo 1961-1962. *Bol.Inst.Biol. Mar.Univ.Nac.Buenos Aires, Mar del Plata* 4: 1-49.

Blackburn, S.I., Hallegraeff, G.M., Bolch, C.J. (1989).Vegetative reproduction and sexual life cycle of the toxic dinoflagellate *Gymnodinium catenatum* from Tasmania, Australia. *J. Phycol.* 25: 577-590.

Blanco, J. (1995a) A model of the effect of cyst germination on the development of the *Gymnodinium catenatum* populations on the west coast of the Iberian peninsula. In: Lassus, P., Arzul, G., Erard-Le Denn, E., Gentien, P., Marcaillou-Le Baut, C. (eds.) *Harmful Marine Algal Blooms*, Lavoisier, Paris, pp. 563-566.

Blanco, J. (1995b). The distribution of dinoflagellate cysts along the Galician (NW Spain) coast. *J. Plankton Res.* 17: 283-302.

Bolch, C.J., Hallegraeff, G.M. (1990).Dinoflagellate cysts from Recent marine sediments of Tasmania, Australia. *Bot. Mar.* 33: 173-192.

Bravo, I. (1986). Germinacion de quistes, cultivo y enquistamiento de *Gymnodinium catenatum* Graham. *Inv.Pesq.* 50: 313-321.

Bravo, I., Anderson, D.M. (1994). The effects of temperature, growth medium and darkness on excystment and growth of the toxic dinoflagellate *Gymnodinium catenatum* from northwest Spain. *J. Plankton Res.* 16: 513-525.

Carrada, G.C., Casotti, R., Modigh, M., Saggiomo (1991). Presence of *Gymnodinium catenatum* (Dinophyceae) in a coastal Mediterranean lagoon. *J. Plankton Res.* 13: 229-238

Carreto, J.I., Lutz, V.A., Carignan, M.O.,Cucchi Colleoni, A.D, De Marcos, S.G.(1995). Hydrography and chlorophyll *a* in a transect from the coast to the shelf-break in the Argentinian Sea.*Continental Shelf. Res.* 15(2/3): 315-336.

Corrales, R.A., Gonzales, C. , Roman, R. (1996). March 1996 *Gymnodinium catenatum* bloom: First record for Manila Bay, Philippines. Phycotoxins@biome.bio.dfo.ca, 15 March 1996.

Dale, B., Madsen, A , Nordberg, K., Thorsen, T.A. (1993). Evidence for prehistoric and historic "blooms" of the toxic dinoflagellate *Gymnodinium catenatum* in the Kattegat -Skagerrak region of Scandinavia. In: Smayda, T.J., Shimizu, Y. (eds.), *Toxic Phytoplankton Blooms in the Sea,* Elsevier Science Publishers, pp. 47-52.

Dale, B., Nordberg, K. (1993). Possible environmental factors regulating prehistoric and historic "blooms" of the toxic dinoflagellate *Gymnodinium catenatum* in the Kattegat -Skagerrak region of Scandinavia. In: Smayda, T.J., Shimizu, Y.(eds.), *Toxic Phytoplankton Blooms in the Sea,* . Elsevier Science Publishers,.pp. 53-57

Ellegaard, M., Christensen, N.F., Moestrup, Ø (1993). Temperature and salinity effects on growth of a non-chain-forming strain of *Gymnodinium catenatum* (Dinophyceae) established from a cyst from Recent sediments in the Sound (Øresund), Denmark. *J. Phycol.* 29: 418-426.

Ellegaard, M. , Oshima, Y. (1995). The dinoflagellate *Gymnodinium catenatum* Graham 1943 (Dinophyceae) from sediments in Northern Europe. Abstracts, 7th Int. Conf. Toxic Phytoplankton, Sendai, Japan, p.63

Estrada, M., Sanchez, F.J., Fraga, S. (1984). *Gymnodinium catenatum* (Graham) en las rias gallegas (NO de Espana). *Inv.Pesq.* 48: 31-40.

Estrada M. (1995) Dinoflagellate assemblages in the Iberian upwelling area. In : Lassus, P., Arzul, G., Erard, E.,Gentien, P., Marcaillou, C. (eds) *Harmful Marine Algal Blooms* .Lavoisier, Paris pp. 157-162

Figueiras, F.G., Fraga, F. (1990). Vertical nutrient transport during proliferation of *Gymnodinium catenatum* Graham in Ria de Vigo, Northwest Spain. In: Graneli, E., Sundstrom, B., Edler, L., Anderson, D.M. (eds). *Toxic Marine Phytoplankton,* Elsevier Science Publishing Co., pp. 144-148.

Figueiras, F.G, Pazos, Y (1991) Hydrography and phytoplankton of Ria de Vigo before and during a red tide of *Gymnodinium catenatum* Graham. *J. Plankton Res.* 13:589-608

Fraga , F, Perez, F.F, Figueiras,F.G., Rios, A.F.(1992) Stoichiometric variations of N, P, C and O2 during a *Gymnodinium catenatum* red tide and their interpretation. *Mar. Ecol. Prog. Ser.* 87: 123-134

Fraga, S.(1996). Wintering of *Gymnodinium catenatum* Graham (Dinophyceae) in Iberian waters. In: Yasumoto, T., Oshima, Y. and Fukuyo, Y. (eds), *Harmful and toxic algal blooms*, IOC, Paris., pp. 211-214.

Fraga, S., Anderson, D.M., Bravo, I., Reguera, B., Steidinger, K.A., Yentsch, C.M. (1988). Influence of upwelling relaxation on dinoflagellates and shellfish toxicity in Ria de Vigo, Spain. *Est. Coastal Mar. Sc.* 27: 349-361.

Fraga, S., Bakun, A. (1993). Global climate change and harmful algal blooms: the example of *Gymnodinium catenatum* on the Galician coast. In: Smayda, T.J., Shimizu, Y. (eds), *Toxic Phytoplankton Blooms in the Sea.*, Elsevier Science Publishers. B.V., pp. 59-65.

Fraga, S., Bravo, I., Delgado, M., Franco, J.M. , Zapata, M. (1995). *Gyrodinium impudicum* sp. nov. (Dinophyceae), a non toxic, chain-forming, red tide dinoflagellate. *Phycologia* 34: 514-521.

Fraga, S., Gallager, S.M., Anderson, D.M. (1989). Chain-forming dinoflagellates: an adaptation to red tides. In: Okaichi, T., Anderson, D.M., Nemoto, T. (eds), *Red Tides: Biology, Environmental Science, and Toxicology*, Elsevier Science Publishing Co., pp. 281-284.

Franca, S. (1987). Ultrastructural study of *Gymnodinium catenatum*, toxic dinoflagellate -preliminary results. *Abstr.XXII. Reuniao Anual da Sociedade Portuguesa de Microscopia Electronica, Evora*, abstract 14.

Franca, S., Almeida, J.F. (1989). Paralytic shellfish poisons in bivalve molluscs on the Portuguese coast caused by a bloom of the dinoflagellate *Gymnodinium catenatum* . In: Okaichi, T., Anderson, D.M., Nemoto, T. (eds), *Red Tides: Biology, Environmental Science , and Toxicology*, Elsevier Science Publishing Co., pp. 93-96.

Fukuyo, Y., Kodama, M., Ogata, T., Ishimaru, T., Matsuoka, K., Okaichi, T., Maala, A.M., Ordones, J.A. (1993). Occurrence of *Gymnodinium catenatum* in Manila Bay, the Philippines. In: Smayda, T.J., Shimizu , Y. (eds), *Toxic Phytoplankton Blooms in the Sea.*, Elsevier Science Publishers.B.V., pp. 875-880.

Gomez Fermin, E., Figueiras, F.G., Arbones, B. , Villarino, M.L. (1996). Short-time scale development of a *Gymnodinium catenatum* population in the Ria de Vigo (NW Spain). *J. Phycol.* 32: 212-221.

Graham, H.W. (1943). *Gymnodinium catenatum*, a new dinoflagellate from the Gulf of California. *Trans.Am. Microsc.Soc.* 62: 259-261.

Guillard, R.R.L., Kilham, P. (1977). The ecology of marine plankton diatoms. In: Werner , D. (ed), The biology of diatoms. *Botanical Monographs* 13: 372-469.

Hallegraeff, G.M. (1993). A review of harmful algal blooms and their apparent global increase. *Phycologia* 32: 79-99.

Hallegraeff, G.M. and Bolch, C.J. (1992). Transport of diatom and dinoflagellate resting spores in ships' ballast water : Implications for plankton biogeography and aquaculture. *J. Plankton Res.* 14 : 1067 - 1084.

Hallegraeff, G.M., McCausland, M.A., Brown, R.K. (1995). Early warning of toxic dinoflagellate blooms of *Gymnodinium catenatum* in southern Tasmanian waters. *J. Plankton Res.* 17: 1163-1176.

Hallegraeff, G.M., Sumner, C.E. (1986). Toxic plankton blooms affect shellfish farms. *Australian Fisheries* 45: 15-18.

Hallegraeff, G.M., Stanley, S.O., Bolch, C.J., Blackburn, S.I. (1989). *Gymnodinium catenatum* blooms and shellfish toxicity in Southern Tasmania, Australia. In : Okaichi, T., Anderson, D.M., Nemoto, T. (eds), *Red tides: Biology, Environmental Science and Toxicology*, Elsevier, pp. 75-78. .

Ikeda, T., Matsuno, S., Sato, S., Ogata, T., Kodama, M., Fukuyo, Y., Takayama, H. (1989). First report on paralytic shellfish poisoning caused by *Gymnodinium catenatum* Graham (Dinophyceae) in Japan. In: Okaichi, T., Anderson , D.M., Nemoto, T. (eds), *Red Tides: Biology, Environmental Science , and Toxicology,* Elsevier, pp. 411-414.

Ishikawa, A., Fujita, N., Taniguchi, A. (1995). A sampling device to measure *in situ* germination rates of dinoflagellate cysts in surface sediments. *J. Plankton Res.* 17: 647-651

Iwasaki, H. (1971). Studies on the red tide dinoflagellates-V. On *Polykrikos schwartzi*, Butschli. *Bull. Japan. Soc.Sci.Fish.* 37: 606-609. (in Japanese)

John, E.H., Flynn, K.J., Flynn, K., Reguera, B., Reyero, M.I. , Franco, J. (1996). Response of the toxic dinoflagellate *Gymnodinium catenatum* to changes in inorganic nutrients and salinity. NATO-ASI workshop, Bermuda.Poster abstract P5

Kim, H.G., Matsuoka, K., Lee, S.G. (1995). New findings of a toxic dinoflagellate *Gymnodinium catenatum* from Chinhae Bay, Korea. Abstracts, 7th Int. Conf. Toxic Phytoplankton, Sendai, Japan, p.38

La Barbera-Sanchez, A., Hall, S., Ferraz-Reyes, E. (1993). *Alexandrium* sp., *Gymnodinium catenatum* and PSP in Venezuela. In: Smayda, T.J., Shimizu, Y.(eds), *Toxic Phytoplankton Blooms in the Sea*, Elsevier Science Publishers.B.V., pp. 281-285.

Luthy, J. (1979). Epidemic paralytic shellfish poisoning in Western Europe, 1976. In: Taylor, D.L., Seliger, H.H. (eds.), *Toxic Dinoflagellate Blooms*, Elsevier, New York, pp. 15-22.

Margalef, R. (1956). Estructura y dinamica de la "purga de mar" en la R¡a de Vigo. *Inv. Pesq.* 5: 113-134.

Margalef, R., Duran, M., Saiz, F. (1955). El fitoplancton de la ria deVigo de enero de 1953 a marzo de 1954. *Inv. Pesq.* 2: 85-129.

Matsuoka, K., Fukuyo, Y. (1994). Geographical distribution of the toxic dinoflagellate *Gymnodinium catenatum* Graham in Japanese coastal waters. *Bot. Mar.* 37: 495-503

McMinn, A., Hallegraeff, G.M., Thomson, P., Short, S., Heijnis, H. (1997). Microfossil evidence for the Recent introduction of the toxic dinoflagellate *Gymnodinium catenatum* into Tasmanian waters. *Mar. Ecol. Progr. Ser.* (submitted).

Mee, L.D., Espinosa, M., Diaz, G. (1986). Paralytic shellfish poisoning with *Gymnodinium catenatum* red tide on the Pacific coast of Mexico. *Mar. Environm, Res.* 19: 77-92

Mendez, S. , Brazeiro, A. (1993). *Gymnodinium catenatum* and *Alexandrium fraterculus* associated with a toxic period in Urugay. Proc.6th Int.Conf.Toxic Phytoplankton, Nantes , France, abstract p.139

Moita, M.T. (1993). Development of toxic dinoflagellates in relation to upwelling patterns off Portugal. In: Smayda, T.J., Shimizu , Y. (eds), *Toxic Phytoplankton Blooms in the Sea.*, Elsevier Science Publishers.B.V., pp. 299-304.

Morey-Gaines, G. (1982). *Gymnodinium catenatum* Graham (Dinophyceae): morphology and affinities with armoured forms. *Phycologia* 21:154-163

Nehring, S. (1995). *Gymnodinium catenatum* Graham (Dinophyceae) in Europe: a growing problem? *J. Plankton Res.* 17: 85-102.

Nishioka, J., Wada, Y., Imanishi, Y.. (1993). On the occurrences of *Gymnodinium catenatum* (Dinophyceae) in Kumihama Bay. *Bull. Kyoto Inst. Ocean. Fish. Sci.* 16:43-49.

Ogata, T., Kodama, M. , Ishimaru, T. (1989). Effect of water temperature and light intensity on growth rate and toxin production of dinoflagellates. In: Okaichi, T., Anderson, D.M., Nemoto, T. (eds), *Red Tides: Biology, Environmental Science, and Toxicology,* pp. 423-426.

Oshima, Y., Blackburn, S.I. , Hallegraeff, G.M. (1993). Comparative study on paralytic shellfish toxin profiles of the dinoflagellate *Gymnodinium catenatum* from three different countries. *Mar. Biol.* 116: 471-476.

Paulmier, G. (1992). Catalogue illustre des microphytes planctoniques et benthiques des Cotes Normandes. Rapp. Intern. Dir. Resour.Viv. IFREMER, DRV-92.007-RH/LE Robert, pp. 1-107.

Rees, A.J.J. , Hallegraeff, G.M. (1991). Ultrastructure of the toxic, chain-forming dinoflagellate *Gymnodinium catenatum* (Dinophyceae). *Phycologia* 30: 90-105.

Sonneman, J.A., Hill, D.R.A. (1997). A taxonomic survey of cyst-producing dinoflagellates from the coastal waters of Victoria, Australia. *Bot.Mar.* (in press).

Sousa, I., Alvito, P., Franca, S., de Sampayo, M.A., Martinez, A.G., Rodriguez-Vazquez (1995). Data on paralytic shellfish toxins related to recent *Gymnodinium catenatum* blooms in Portugal coastal waters. In: Lassus, P., Arzul, G., Erard, E., Gentien, P., Marcaillou, C. (eds), *Harmful Marine Algal Blooms,* Lavoisier, Intercept Ltd., pp. 825-829.

Tagmouti, F., Chafak, H., Fellat-Zarrouk, R., Talbi, M., Blaghen, M., Mikou , Guittet, E. (1995). Detection of toxins in bivalves of Moroccan coasts. Abstracts, 7th Int. Conf. Toxic Phytoplankton, Sendai, Japan, p.40

Taylor, F.J.R. (1980). On dinoflagellate evolution. *BioSystems* 13: 65-108.

Thorsen, T.A., Dale, B., Nordberg, K. (1995). "Blooms" of the toxic dinoflagellate *Gymnodinium catenatum* as evidence of climatic fluctuations in the late Holocene of southwestern Scandinavia. *The Holocene* 5: 435-446.

Yuki, K., Yoshimatsu, S. (1989). Morphology of the athecate dinoflagellate *Gymnodinium catenatum* in culture. *Bull. Plankton Soc. Japan* 34: 109-117.

Vilarinho, M.G., Moita , M.T. (1996). Resultados preliminares do estudo da variabilidade interanual do fitoplankton toxico na costa portuguesa. In: Matamoros, E., Delgado, M. (eds). *IV. Reunion Iberica sobre Fitoplancton toxico y biotoxinas. Actas de la Reunion.* Generalitat de Catalunya, pp. 69-76

Wooster, W.S., Bakun, A., McLain, D.R. (1976). The seasonal upwelling cycle along the eastern boundary of the North Atlantic. *J. Mar.Res.* 34: 131-141

Zardoya, R., Costas, E., Lopez-Rodas, V., Garrido-Pertierra, A., Bautista, J.M. (1995). Revised dinoflagellate phylogeny inferred from molecular analysis of lsu rRNA gene sequences. *J. Mol. Evolution* 41: 637-645.

Physiology and Bloom Dynamics of the Tropical Dinoflagellate *Pyrodinium bahamense*

Gires Usup
Jabatan Sains Laut, Universiti Kebangsaan Malaysia, 43600 Bangi, Selangor, Malaysia

Rhodora V. Azanza
Marine Science Institute, University of the Philippines, Diliman, Quezon City 1101, Philippines

1. Introduction

The thecate dinoflagellate *Pyrodinium bahamense* has probably existed since the Eocene (McMinn 1989). Fossil cysts of the species, referred to as *Polysphaeridium zoharyii* (Wall and Dale 1969) have been recorded from many geographical locations, as far north as Japan and the British Isles and as far south as Australia. These fossil cysts also seemed to have been widely distributed in the Arabian and Mediterranean seas. In contrast, modern, extant cysts of *P. bahamense* seem to have a more limited distribution, mainly in Southeast Asia, the Caribbean and in western Europe (Ferreira and Dale 1997). It is uncertain whether the modern cysts are truly absent from other regions since data is lacking. The present-day distribution of vegetative cells likewise seems to be confined to the tropical Indo-west Pacific and the tropical Atlantic (Hallegraeff 1993).

1.1 Taxonomy, morphology and life cycle of *P. bahamense*

Pyrodinium bahamense was first described from the Bahamas in the Atlantic Ocean by Plate (1906). In 1931 Böhm described a compressed form of the species based on specimens from the Persian Gulf. Since then, the genus *Pyrodinium* has undergone several revisions (Taylor and Fukuyo 1989). Steidinger *et al.* (1980) reexamined the morphology of vegetative cells collected from the tropical Pacific and Atlantic and established two varieties, var. *bahamense* and var. *compressum*. Balech (1985) did not find enough morphological differences between the two varieties to warrant separation. Morphologically both varieties are similar, with a Kofoid plate pattern of (P_o, P_i), 4', 6'', 6c, 8s, 6''', 2''''. *Pyrodinium bahamense* var. *compressum*, however, forms chains and in this configuration the cells appear flattened anterior-posteriorly. The variety *bahamense* has never been reported to exist in chains although duplets are formed following cell division (Buchanan 1968). Interestingly, the variety *bahamense* at present seems to be confined to the tropical Atlantic, while the variety *compressum* is found only in the tropical Pacific. A more significant difference between the two forms is that all toxic red tides of the species to date are due to the variety *compressum* while the variety *bahamense* has never been reported as toxic.

A first description of the asexual life cycle of *P. bahamense* was provided by Buchanan (1968) based on observations of vegetative cells in Oyster Bay, Jamaica. One of the stages in the postulated cycle involves an ecdysed, spherical pellicle cyst stage, from

NATO ASI Series, Vol. G 41
Physiological Ecology of Harmful Algal Blooms
Edited by D. M. Anderson, A. D. Cembella and G. M. Hallegraeff

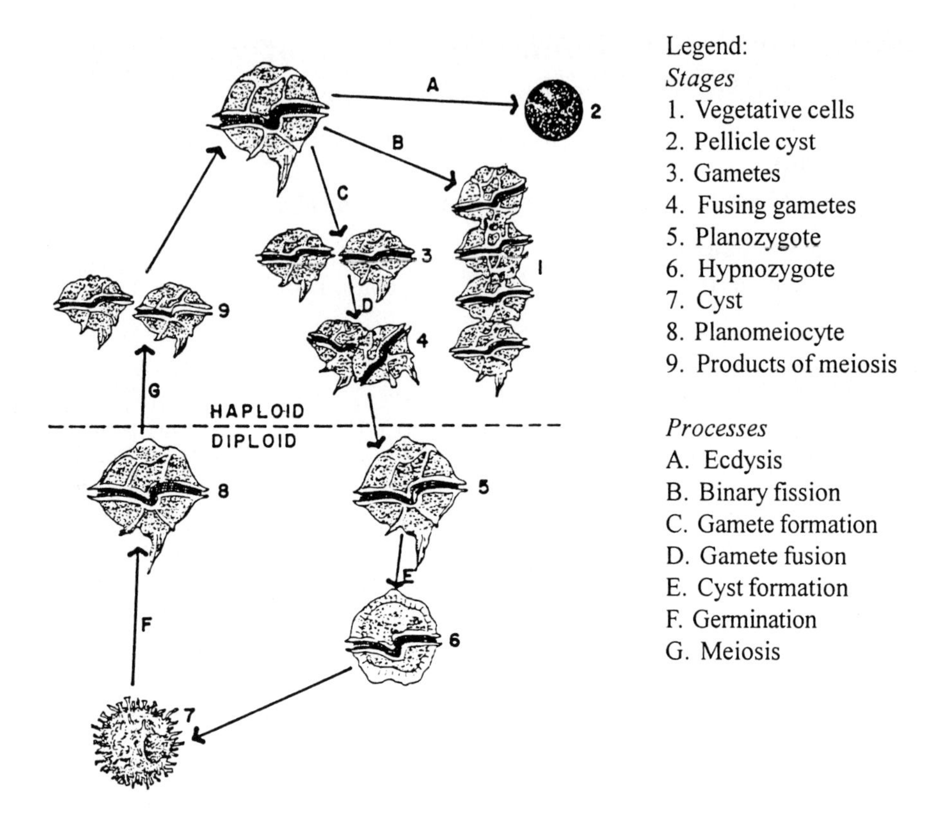

Fig. 1. The life cycle of *Pyrodinium bahamense*. (Original illustration by Rhodora V. Azanza).

which thecate cells subsequently redevelop. The availability of laboratory cultures have enabled a more complete description of the life cycle of *P. bahamense* var. *compressum* to be made (Fig. 1). Vegetative cells undergo asexual reproduction by oblique division without ecdysis. In cultures it was observed that cells divided with the onset of the light cycle, taking ca. 2 h for division to be completed (Usup 1995). This is similar to what have been observed in field samples (Buchanan 1968; Maclean 1977).

The presence of a sexual cycle in the life-history of the species was suggested by the discovery of resting cysts in sediment samples (Wall and Dale 1969; Matsuoka *et al.* 1989; Usup *et al.* 1989). Subsequently, resting cysts were produced in culture (Corrales *et al.* 1995) and also in concentrated live samples collected from a bloom patch and allowed to stand overnight. There is strong evidence to suggest that *P. bahamense* is heterothallic (Corrales *et al.* 1995; Usup 1995). The gametes are small and seem to be produced in and released from the larger vegetative cells (Azanza and Larsen 1997). Fusing gametes are of the same size and their thecal plates disintegrate at the point of fusion (Corrales *et al.* 1995). The resulting planozygote is larger than vegetative cells and the hypnozygotes contain 1-2 distinct red-colored chromatophores. Germination experiments on laboratory-produced cysts suggest a dormancy period of 2.5 - 3 months (Corrales *et al.* 1995).

1.2 *Pyrodinium bahamense* as a harmful algal bloom species

Early interest in *P. bahamense* were mainly due to its bioluminescence (Seliger *et al.* 1971). Bahia Posforescente in Puerto Rico, for example, is named for the spectacular bioluminescence of *Pyrodinium* blooms (McLaughlin and Zahl 1961). Prior to 1972, *P. bahamense* was not regarded as a species of economic or public health importance. In 1972 the first paralytic shellfish poisoning (PSP) cases linked to *P. bahamense* var. *compressum* blooms were reported in Papua New Guinea (Maclean 1973; Worth *et al.* 1975). In March of 1976 PSP cases due to *P. bahamense* were reported from Brunei Bay in Borneo (Beales 1976; Roy 1977). This was followed in 1983 by outbreaks in Manila Bay, the Philippines (Estudillo and Gonzales 1984), in 1987 on the Pacific coast of Guatemala in Central America (Rosales-Loessener *et al.* 1989) and more recently in Mexico (Orellana-Cepeda *et al.* 1997). *Pyrodinium bahamense* continues to be the major HAB problem in these countries. Even though *P. bahamense* var. *compressum* blooms have limited geographical coverage, the species has caused more fatal PSPs than any other species. In most cases the vector for the toxins to humans have been bivalves, but planktivorous fish (e.g. *Sardinella* sp. and *Rastrelliger* sp.) have also been implicated (Jaafar and Subramaniam 1984).

2. *Pyrodinium bahamense* bloom dynamics

2.1 Bloom initiation

Resting cysts have been proven as the sources of seed populations in *Alexandrium* and *Gymnodinium* blooms, particularly in temperate regions (Anderson 1984; Hallegraeff 1993). In the case of *P. bahamense* the relative importance of resting cysts as opposed to vegetative cells (background population) in initiating blooms is still uncertain. Corrales and Crisostomo (1996) conducted detailed mapping of resting cyst distribution in Manila Bay, the Philippines and found densities to be highest in Bataan and Cavite (Fig. 2). Interestingly these were also the two areas where blooms normally appear first and were most persistent in the bay. Villanoy *et al.* (1996) found that in Manila Bay the highest density of *P. bahamense* cysts in the water column occurred during the northeast monsoon, when vertical mixing was most intense. They proposed that in a relatively shallow, semi-enclosed area like Manila Bay, the appearance of *P. bahamense* blooms is related to the resuspension of resting cysts by turbulence. In the open coastal waters of Sabah, Malaysia, no such cyst beds have been discovered, suggesting that blooms may be initiated by vegetative cells present in low numbers in the plankton.

Blooms that develop in a particular area could also result from horizontal advection of seed populations from elsewhere. Sotto and Young (1995) hypothesized, for example, that the *Pyrodinium* bloom observed in Dec 1992 - Jan 1993 around Cebu Island in the Philippines originated from the northeast (Leyte and Samar) where blooms occurred in Nov 1992. The probability that the blooms in Cebu originated locally could not, however,

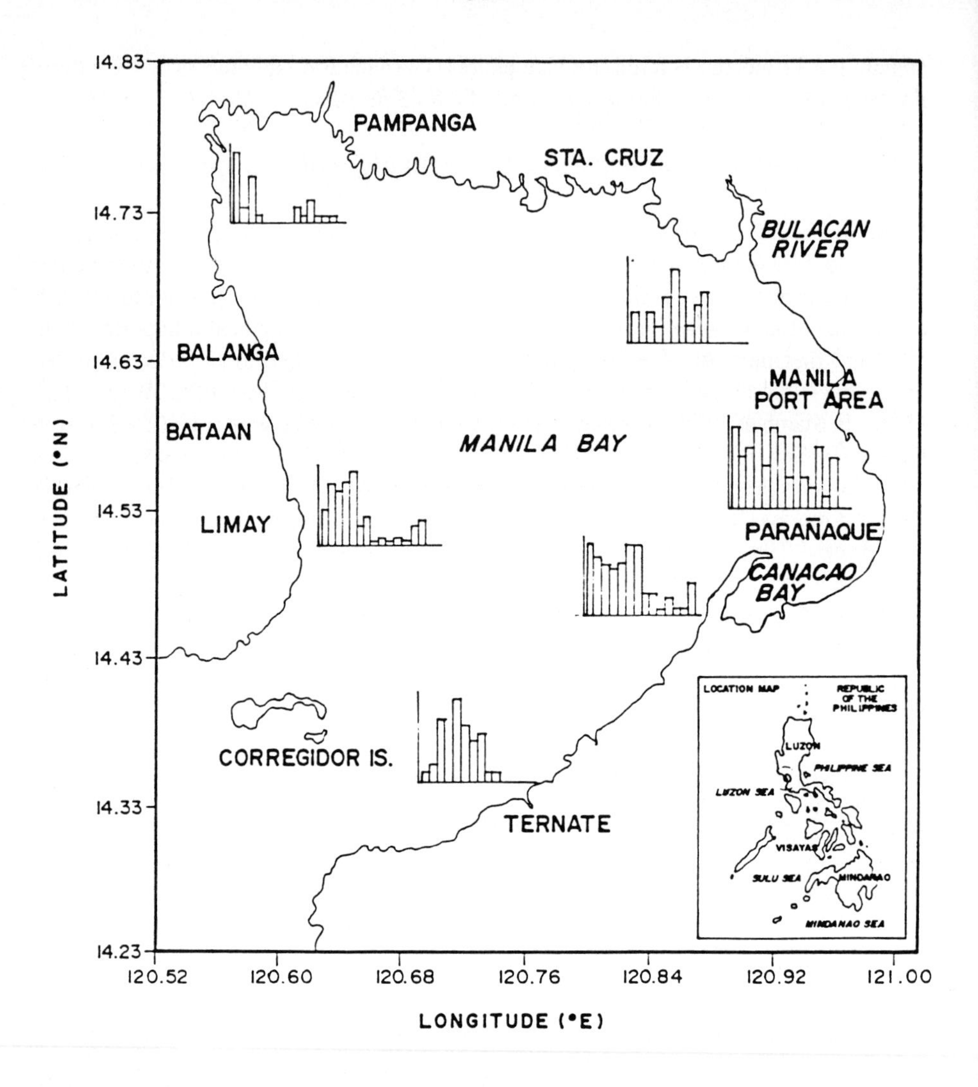

Fig. 2. The distribution of *Pyrodinium bahamense* resting cysts in the surface sediment of Manila Bay from August 1993 to May 1994. Cyst densities (cysts cm^{-3} sediment) for each region were as follows: Bataan 0-450; Pampanga 0-70; Bulacan 0-40; Paranaque 0-60; Canacao 0-300; Ternate 0-80. (From Corrales and Crisostomo 1996).

be discounted since cyst distribution studies were not carried out. Seliger (1989) also hypothesized that horizontal advection was responsible for the spread of *P. bahamense* red tides from the northeast to the southwest along the coastline of Sabah, Malaysia. The seeding of *P. bahamense* var. *compressum* red tide events is clearly a subject that requires more understanding as this knowledge may impact the siting of aquaculture projects, particularly shellfish farms.

2.2 Bloom development

Regardless of the source of seed population, the factors responsible for the development of *P. bahamense* blooms remain poorly understood. Typical peak cell density in a *P. bahamense* red tide patch is on the order of 10^6 cells L^{-1}. At present it is uncertain if such high cell densities are due to enhanced *in situ* growth, physical accumulation, or both. Observations that *P. bahamense* red tides tend to be patchy in both the horizontal and vertical dimensions could not discount either mechanism. Maclean (1977) estimated that the growth rate of *P. bahamense* populations during a red tide in Papua New Guinea, after the bloom reached its peak, was ca. 0.3 divisions d^{-1}, comparable to the maximum of ca. 0.4 divisions d^{-1} obtained in laboratory batch cultures (Usup 1995).

Marked horizontal and vertical patchiness of *P. bahamense* cells in a bloom suggest that seed populations were delivered into a water mass containing sufficient concentrations of a limiting nutrient, which could be of terrestrial origin. This may help explain the observations that *P. bahamense* red tides tend to occur after periods of heavy rain (Usup and Lung 1991), close to shore and in embayments. A potential source of growth-enhancing nutrient that has been suggested is mangrove areas (Maclean 1989), which are common in locations where blooms occur. More research is clearly needed to elucidate the relative importance of biological and physical factors in the establishment and maintenance of these blooms.

2.3 Bloom decline

The disappearance of P. bahamense blooms is just as sudden as its appearance. In Sabah, Malaysia, patches of *P. bahamense* red tides remained intact for a period of 10-14 days (Usup *et al.* 1989). Factors that lead to the decline of these blooms are still unknown. It has been observed from field samples that at least one species of tintinnid, *Favella* sp. is able to feed on *P. bahamense* (Usup *et al.* 1989) but the importance of grazing in the disappearance of blooms needs more investigation. Dissipation of cells by turbulence could be more important. This is another area where more research is needed.

Studies by Villanoy *et al.* (1996) and Corrales and Crisostomo (1996) in Manila Bay indicate that sexual reproduction and production of resting cysts are important events at the end of *P. bahamense* blooms. Factors that promote the sexual cycle are still unknown, although work by Corrales *et al.* (1995) suggest that nutrient limitation is an important factor.

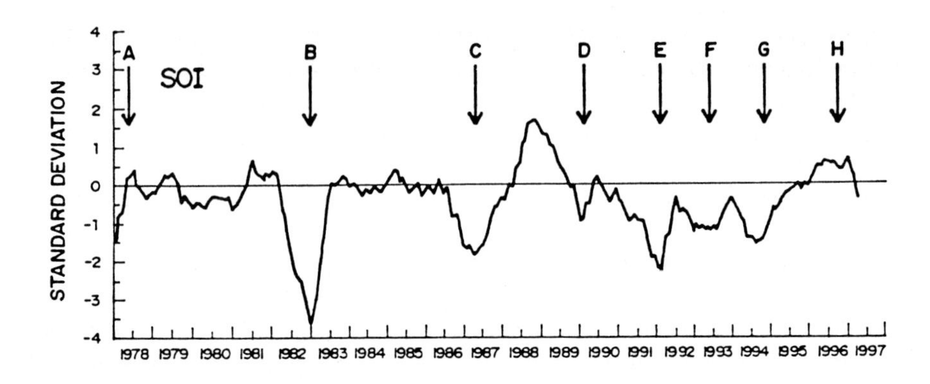

Fig. 3. Relationship between ENSO events and major toxic *Pyrodinium* red tides in the western Pacific region for the period 1978 - 1997. The Southern Oscillation Index (SOI) values were based on readings taken in Tahiti, Hawaii and Darwin, Australia. Arrows indicate years when *P. bahamense* red tides occurred in the Philippines and Malaysia. Red tides also occurred in 1972 and 1976, which were also ENSO event years. (Data provided by the ASEAN-Canada Red Tide Network and NOAA through the Philippines Weather Bureau).

2.4 Periodicity of *P. bahamense* red tides

Outbreaks of *P. bahamense* red tides have been largely aperiodic and unpredictable. Attempts have been made to correlate these events with periodic cycles of weather and meteorology. On a large scale, Maclean (1989) suggested that outbreaks of *P. bahamense* red tides in the western Pacific region coincide with the El-Nino Southern Oscillation (ENSO) events (Fig. 3). It was suggested that ENSO events probably affect different parts of the western Pacific to varying degrees on each occasion, creating a short- or long-term environment suitable for toxic blooms in one locality or another. The mechanisms underlying this association between *Pyrodinium* blooms and ENSO events are unclear.

In Sabah, Malaysia from 1976 to 1986 *P. bahamense* red tides seem to have coincided with the inter-monsoon period, with most outbreaks occurring in July and Dec-Jan (Usup and Lung 1991). In the Philippines, for the period of 1991-1995, blooms have occurred in June-July (early southwest monsoon) while highest cyst densities in the sediment were highest during the northeast monsoon. These observations suggest a relationship between *P. bahamense* red tides and local weather events. In Papua New Guinea, Maclean (1976) noted that *P. bahamense* red tides seemed to coincide with upwelling events. As more research is carried out, it may become evident that these meteorological events may modify the physical and chemical properties of the water column leading to favorable conditions for *P. bahamense* growth. An important aspect of this modification may be the supply of limiting nutrient(s) into coastal waters through runoff from land sources.

3. The physiology of *P. bahamense*

Most of the available data on the physiology of *P. bahamense* have been obtained from laboratory studies on clonal cultures of isolates from Malaysia (Usup 1995; unpublished data). The potential problems associated with extrapolation of laboratory data to *in situ* situations are recognized. Nevertheless, some of the laboratory data do confirm earlier observations made on natural populations.

3.1 Salinity, temperature and light requirements

Data obtained from the field during *P. bahamense* red tides suggested that the species prefers water of high salinity. In Malaysia, blooms have typically been in waters of salinities 30 ‰ or higher (Usup *et al.* 1989). In Papua New Guinea blooms occurred in water of salinities 28 ‰ or higher (Maclean 1976) while in the Philippines the blooms occurred in water of 31 ‰ or higher salinity (Corrales and Hall 1993). Wall and Dale (1969) reported that the optimum salinity for *P. bahamense* var. *bahamense* was 35 ‰. The high salinity requirement of *P. bahamense* was also evident from earlier efforts to culture the species. Oshima *et al.* (1985) and McLaughlin and Zahl (1961) found that diluting seawater by 10 % resulted in poor growth of their cultures. Usup (1995) found that an isolate from Malaysia was able to grow at salinities of 20 - 36 ‰ and did not survive at 16 ‰ (Fig. 4a).

Seawater temperature in the natural habitat of *P. bahamense* ranges from 25 to 31° C. Laboratory studies on an isolate from Malaysia (Usup 1995) showed that the temperature limits for growth are 22 to 34° C, with optimum growth at 28° C (Fig. 4b). There is thus very good correspondence between temperature tolerance of *P. bahamense* in cultures and the temperature range in its natural habitat.

Not much data is currently available on effects of irradiance on the physiology of *P. bahamense*. During previous red tides in Sabah, living cells of *P. bahamense* were found in high density (~ 10^5 cells L^{-1}) at depths of 20 m and more, deeper than the euphotic zone (Usup *et al.* 1989). Downwelling irradiance in a bloom patch would also be significantly reduced as a result of self-shading. At present the length of time in which *P. bahamense* could remain viable under low light conditions is still not known. In laboratory cultures, *P. bahamense* was able to grow well at an irradiance value of 50 $\mu E\ m^{-2}\ s^{-1}$. Compensation for low light conditions as evidenced by an increase in cellular chlorophyll *a* content, however, occurred at 90 $\mu E\ m^{-2}\ s^{-1}$ (Fig. 4c). There is also evidence that *P. bahamense* can grow well under continuous illumination, so apparently a dark phase is not required in its cell cycle (Usup 1995). This is similar to some of the other dinoflagellates that have been studied, for example *Amphidinium carterae* and *Scrippsiella trochoidea* (Dixon and Syrett 1988).

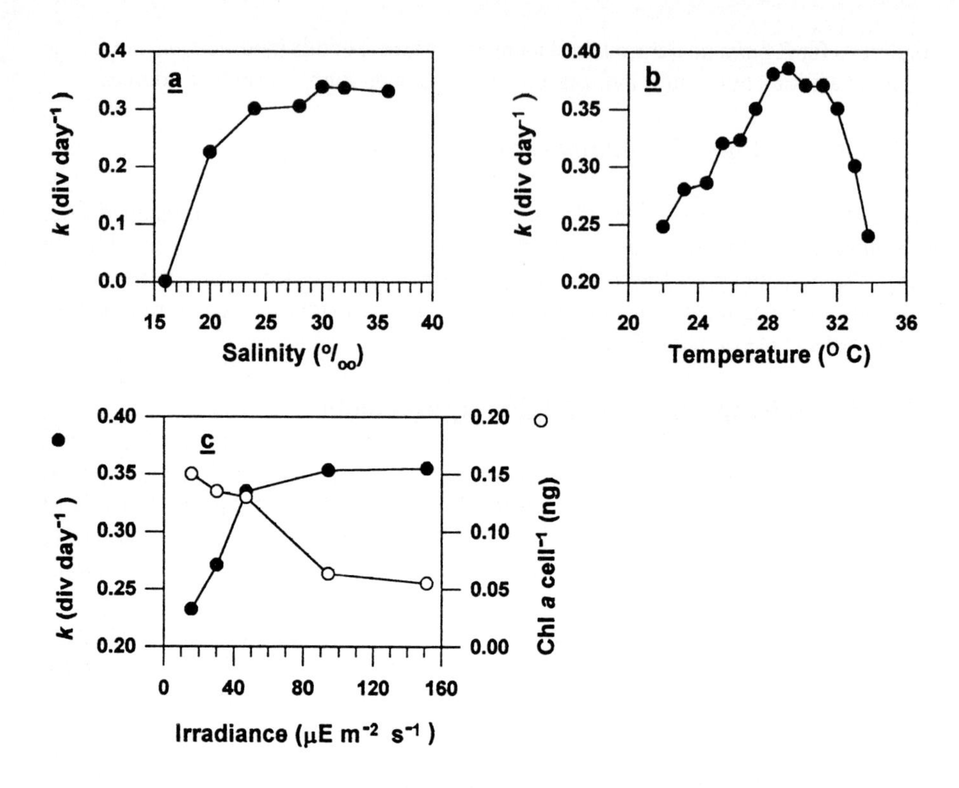

Fig. 4. Effects of (a) salinity, (b) temperature, and (c) irradiance on the growth of *P. bahamense* var. *compressum* in modified ES medium batch cultures. (From Usup 1995).

3.2 Nutrient requirements

It has been suggested that *P. bahamense* may have fastidious nutrient requirements, judging by the difficulty in establishing laboratory cultures of the species (Blackburn and Oshima 1989). The few instances whereby the species has been successfully cultured seemed to support this contention. Difficulties in establishing laboratory cultures, until quite recently, have been the main reason for lack of physiological data on the species. The major pattern that emerged from instances where culturing were successful was the requirement for soil extract supplement, regardless of the medium employed. This was true for cultures of isolates from Palau (Oshima *et al.* 1985), Malaysia (Usup 1995) and the Caribbean (McLaughlin and Zahl 1961). In the study of Usup (1995) it was found that soil extract supplement did not significantly affect division rates, but resulted in prolongation of the exponential phase of growth. As a result densities achieved in batch cultures increased from 1600 cells mL^{-1} without soil extract supplement to 4700 cells mL^{-1} when 10 mL L^{-1} soil extract was added to the medium (Fig. 5). These results suggest the presence

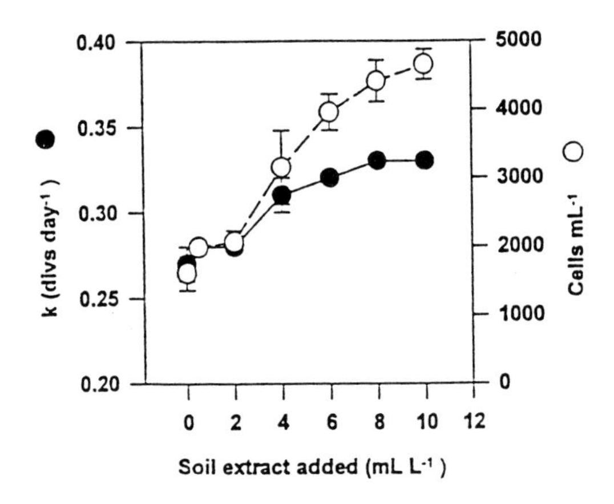

Fig. 5. Effects of soil extract supplement on the growth of *P. bahamense* var. *compressum* in modified ES medium batch cultures. (From Usup 1995).

of a growth factor in the soil extract which is required in low concentration but continuous availability by *P. bahamense*. These findings are of significance because it has been suggested that material of mangrove origin may be a factor that triggers blooms of *P. bahamense* in tropical coastal waters (Maclean 1989).

Soil extract has previously been shown to stimulate the growth of some dinoflagellates and diatoms (Prakash and Rashid 1968; Prakash *et al.* 1973; Gedziorowska and Plinski 1990). This stimulation is thought to result from several mechanisms including chelation of trace metals by organic ligands (Prakash and Rashid 1968; Mackey 1984; Sunda 1988; Bruland *et al.* 1991), and provision of nutrients, particularly nitrogen, which otherwise may be growth-limiting (Graneli *et al.* 1989; Carlsson *et al.* 1993). Usup (1995) investigated the potential benefits of soil extract to *P. bahamense* growth in cultures and found strong evidence that the soil extract serves as a source of selenium (Fig. 6). Similar to soil extract, selenium supplement did not significantly affect division rates but resulted in enhanced culture yields. The highest yield obtained in Se-supplemented cultures (ca. 6000 cells mL^{-1}) was comparable to or even better than yields obtained in soil extract supplemented cultures. It is possible, however, that growth was still limited by other factors since densities obtained were lower compared to cultures of other similar-sized dinoflagellates. Furthermore, the *P. bahamense* clones were unable to grow in an artificial seawater-based medium even when supplemented with Se. Results from the same study also indicated that *P. bahamense* could utilize selenite (Se-IV) and organic selenide but not selenate. This is similar to results that have been obtained for other phytoplankton species (Price *et al.* 1987; Vandermeulen and Foda 1988). The relevance of these findings to *P. bahamense* growth in its natural habitat remains to be tested but nevertheless the data indicate the potential importance of land-derived nutrients in promoting blooms of the species in coastal waters.

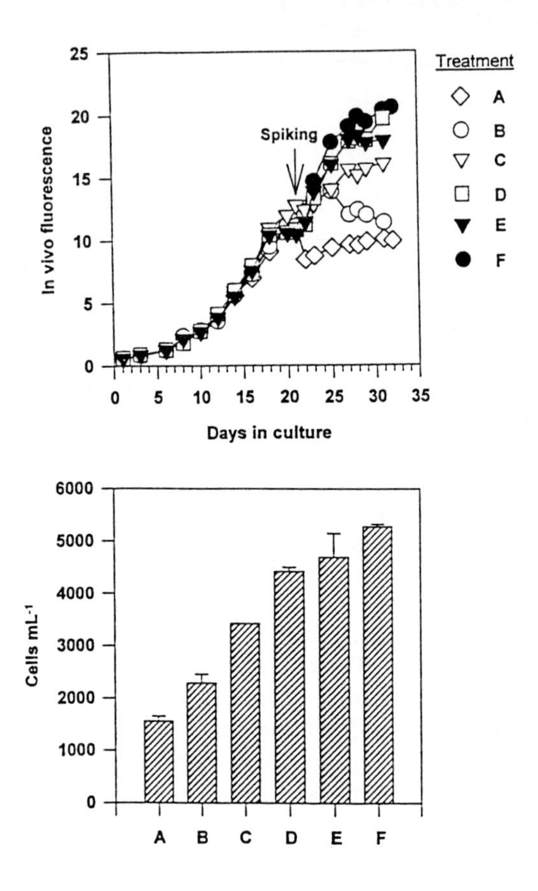

Fig. 6. Effects of selenite supplement on *P. bahamense* var. *compressum* growth. Upper panel - *in vivo* fluorescence profiles of batch cultures spiked at early stationary phase with various nutrient combinations. Lower panel - highest yields obtained from the cultures. Spiking combinations were: A-no spiking; B-10 mL L^{-1} modified ES medium nutrients mix; C-4 mL L^{-1} soil extract; D-10^{-7} M selenite; E-10^{-7} M selenite, 200 mg L^{-1} bicarbonate, 10 mL L^{-1} medium nutrients mix; F-10^{-7} M selenite, 10 mL L^{-1} medium nutrients mix. (From Usup 1995).

In modified ES medium (Usup 1995) *P. bahamense* was able to utilize nitrate as the nitrogen source. There is also evidence that the species is able to utilize urea, while its tolerance to ammonia seems to be low. Its ability to utilize organic may be very limited since clones could not grow on alanine, arginine or histidine in culture. This is an important aspect since coastal waters in which the species is found receive considerable inputs of organic wastes. It is also evident that *P. bahamense* is able to utilize both inorganic and organic phosphorus (Usup 1995). In modified ES medium sodium glycerophosphate is supplied as a source for phosphorus. There is also evidence that the species has alkaline phosphatase activity (Gonzalez-Gil *et al.* in press).

4. Summary

It has been more than two decades since *Pyrodinium bahamense* was first recognized as an important HAB species. There is evidence showing that the species has a wider geographical occurrence than previously thought, and it may pose significant PSP problems in more countries. More data on the species have emerged from studies that have been carried out. Its life cycle is now well understood, as are its basic physiological requirements. Several aspects pertaining to its toxicity have also been elucidated (discussed elsewhere in this volume). Nevertheless there are still several studies that need to be carried out, particularly pertaining to the following aspects: 1) bloom dynamics of the species, ideally identifying the factors that promote formation of the blooms, 2) direct/indirect effects of *P. bahamense* on the physiology and well-being of other marine species, particularly on the larval stages of commercially important fish and shellfish, 3) comparative physiology of *P. bahamense* isolates from the different regions where the species occur, 4) phylogenetic relationship between *P. bahamense* var. *bahamense* and var. *compressum* by molecular biology methods, and 5) the biogeography of *P. bahamense.*

5. Acknowledgments

We thank the Intergovernmental Oceanographic Commission and Universiti Kebangsaan Malaysia for travel grants. Research supported in part by grants from the Malaysia government, the US National Science Foundation (INT-9224474) the ASEAN-Canada Cooperative Programme in Marine Science, the Philippine Department of Science and Technology - PCAMRD and the Canadian International Development Agency (CIDA) through the ACPMS II.

6. References

Anderson, D.M. (1984). Shellfish toxicity and dormant cysts in toxic dinoflagellate blooms. pp. 125-138, In: *Seafood Toxins*, E.P. Ragelis (ed.), American Chemical Society. Washington, D.C.

Azanza, R., Larsen, J. (1997). Variation in nutrient concentration: effects on *Pyrodinium* cells in culture. Abstract. *VIIIth International Conference on Harmful Algae*. Vigo, Spain,

Balech, E. (1985). A redescription of *Pyrodinium bahamense* Plate (Dinoflagellate). *Rev. Paleobot. Palyn.* 45:17-34.

Beales, R.W. (1976). A red tide in Brunei's coastal waters. *Brunei Museum J.* 3:167-182.

Blackburn, S.I., Oshima, Y. (1989). Review of culture methods for *Pyrodinium bahamense.* pp. 257-266, In: *Biology, Epidemiology and Management of Pyrodinium Red Tides*, G.M. Hallegraeff, Maclean, J.L. (eds.), ICLARM, Manila.

Bruland, K.W., Donat, J.R., Hutchins, D.A. (1991). Interactive influences of bioactive trace metals on biological production in oceanic waters. *Limnol. Oceanogr.* 36:1555-1577.

Buchanan, R.J. (1968). Studies at Oyster Bay in Jamaica, West Indies, IV. Observations an the morphology and asexual cycle of *Pyrodinium bahamense* Plate. *J. Phycol.* 4:271-277.

Carlsson, P., Segatto, A.Z., Graneli, E. (1993). Nitrogen bound to humic matter of terrestrial origin - a nitrogen pool for coastal phytoplankton? *Mar. Ecol. Prog. Ser.* 97:105-116

Corrales, R.A., Hall, S. (1993). Isolation and culture of *Pyrodinium bahamense* var. *compressum* from the Philippines. pp. 725-730, In: *Toxic Phytoplankton Blooms in the Sea*, T.J. Smayda Shimizu, Y. (eds.) Elsevier, Amsterdam.

Corrales, R.A., Crisostomo, R.C. (1996). Variation of *Pyrodinium* cyst density. pp. 181-184, In: *Harmful and Toxic Algal Blooms*, T. Yasumoto, Oshima, Y., Fukuyo, Y. (eds.), IOC-UNESCO, Paris.

Corrales, R.A., Reyes, M., Martin, M. (1995). Notes on the encystment and excystment of *Pyrodinium bahamense* var. *compressum in vitro*. pp. 573-578, In: *Harmful Marine Algal Blooms*, P. Lassus, Arzul, G., Erard-Le Denn, E., Gentien, P., Marcaillou-Le Baut, C. (eds.), Lavoisier Science Publ., Paris.

Dixon, G.K., Syrett, P.J. (1988). The growth of dinoflagellates in laboratory cultures. *New Phytol.* 109:297-302.

Estudillo, R.A., Gonzales, C.L. (1984). Red tides and paralytic shellfish poisoning in the Philippines. pp. 52-79, In: *Toxic Red Tides and Shellfish Toxicity in Southeast Asia*, A.W. White, Anraku, M., Hooi, K.K. (eds.), SEAFDEC/IDRC, Ottawa.

Ferreira, A.A., Dale, B. (1997). Distribution of cysts from toxic or potentially toxic dinoflagellates along the Portuguese coast. Abstract. *VIIIth International Conference on Harmful Algae.* Vigo, Spain.

Gedziorowska, D., Plinski, M. (1990). Humic compounds and growth response of phytoplankton dominated by dinoflagellates (late spring bloom): observation in coastal waters of the southern Baltic. pp. 155-160, In: *Toxic Marine Phytoplankton*, E. Graneli, Sundstrom, B., Edler, L., Anderson, D.M. (eds.), Elsevier, Amsterdam.

Gonzalez-Gil, S., Keafer, B.A., Jovine, R.V.M., Anderson, D.M. In press. Detection and quantification of alkaline phosphatase in single cells of phosphorus-starved marine phytoplankton. *Mar. Ecol. Prog. Ser.*

Graneli, E., Ollson, P., Sundstrom, B., Edler, L. (1989). *In situ* studies on the effects of humic acids on dinoflagellates and diatoms. pp. 209-212, In: *Red Tides: Biology, Environmental Science and Toxicology*, T. Okaichi, Anderson, D.M., Nemoto, T. (eds.), Elsevier, New York.

Hallegraeff, G.M. (1993). A review of harmful algal blooms and their apparent global increase. *Phycologia* 32:79-99.

Jaafar, M., Subramaniam, S. (1984). Occurrences of red tide in Brunei Darussalam and methods of monitoring and surveillance. pp. 17-24, In: *Toxic Red Tides and Shellfish Toxicity in Southeast Asia*, A.W. White, Anraku, M., Hooi, K.K. (eds.), SEAFDEC/ IDRC, Ottawa.

Mackey, D. (1984). Trace metals and the productivity of shelf waters of North West Australia. *Australian J. Mar. Freshwat. Res.* 35:505-516.

Maclean, J.L. (1973). Red tide and paralytic shellfish poisoning in Papua New Guinea. *Papua New Guinea Agric. J.* 24:131-138.

Maclean, J.L. (1977). Observations on *Pyrodinium bahamense* Plate, a toxic dinoflagellate in Papua New Guinea. *Limnol. Oceanogr.* 22:234-254.

Maclean, J.L. (1989). Indo-Pacific red tides, 1985-1988. *Mar. Poll. Bull.* 20:304-310.

Matsuoka, K., Fukuyo, Y., Gonzales, C.L. (1989). A new discovery of cysts of *Pyrodinium bahamense* var. *compressum*. pp. 301-304, In: *Red Tides:Biology, Environmental Science and Toxicology*, T. Okaichi, Anderson, D.M., Nemoto, T. (eds.), Elsevier, New York.

McLaughlin, J.J.A., Zahl, P.A. (1961). *In vitro* culture of *Pyrodinium*. *Science* 229:144-146.

McMinn, A. (1989). Late Pleistocene dinoflagellate cysts from Botany Bay, New South Wales, Australia. *Micropaleontology* 35:1-9.

Orellana-Cepeda, E., Martinez-Romero, E., Munoz-Cabrera, L., Lopez-Ramirez, P., Cabrera-Mancilla, E., Ramirez-Camarena, C. (1997). The toxicity of *Pyrodinium bahamense* var. *compressum* blooms on the shellfish on South West coast of Mexico. Abstract. *VIIIth International Conference on Harmful Algae*. Vigo, Spain.

Oshima, Y., Harada, T., Yasumoto, T. (1985). Paralytic shellfish poisoning. pp. 20-31, In: *Studies on Tropical Fish and Shellfish Infested by Toxic Dinoflagellates*, T. Yasumoto (ed.), Tohuku Univ., Japan.

Plate, L. (1906). *Pyrodinium bahamense* n.g., n. sp. die Leucht-Peridinee des "Feuersees" von Nassau, Bahamas. *Arch. Protistenk.* 7:411-429.

Prakash, A., Rashid, M.A. (1968). Influence of humic substances on the growth of marine phytoplankton: dinoflagellates. *Limnol. Oceanogr.* 13:598-606.

Prakash, A., Rashid, M.A., Jensen, A., Subba Rao, D.V. (1973). Influence of humic substances on the growth of marine phytoplankton: diatoms. *Limnol. Oceanogr.* 18:516-524.

Price, N.M., Thompson, P.A., Harrison, P.J. (1987). Selenium: an essential element for growth of the coastal marine diatom *Thalassiosira pseudonana* (Bacillariophyceae). *J. Phycol.* 23:1-9.

Rosales-Loessener, F., Porras, E.D., Dix, M.W. (1989). Toxic shellfish poisoning in Guatemala. pp. 113-116, In: *Red Tides: Biology, Environmental Science and Toxicology*, T. Okaichi, Anderson, D.M., Nemoto, T. (eds.), Elsevier, New York.

Roy, R.N. (1977). Red tide and outbreak of paralytic shellfish poisoning in Sabah. *Med. J. Malays.* 31:247-251.

Seliger, H.H. (1989). Mechanisms for red tides of *Pyrodinium bahamense* var. *compressum* in Papua New Guinea, Sabah and Brunei Darussalam. pp. 53-71, In: *Biology, Epidemiology and Management of Pyrodinium Red Tides*, G.M. Hallegraeff, Maclean, J.L. (eds.), ICLARM, Manila.

Seliger, H.H., Carpenter, J.H., Loftus, M., Biggley, N.H., McElroy, W.D. (1971). Bioluminescence and phytoplankton successions in Bahia Fosforescente, Puerto Rico. *Limnol. Oceanogr.* 16:608-622.

Sotto, F.B., Young, J.G. (1995). *Pyrodinium bahamense* var. *compressum* cells in the Visayan Sea, Philippines: transport by water currents. pp. 243-247, In: *Harmful Marine Algal Blooms*, P. Lassus, Arzul, G., Erard-Le Denn, E., Gentien, P., Marcaillou-Le Baut, C. (eds.), Lavoisier Science Publ., Paris.

Steidinger, K., Tester, L.S., Taylor, F.J.R. (1980). A redescription of *Pyrodinium bahamense* var. *compressa* (Böhm) stat. nov. from Pacific red tides. *Phycologia* 19:329-337.

Sunda, W.G. (1988). Trace metal interactions with marine phytoplankton. *Biol. Oceanogr.* 6:411-442.

Taylor, F.J.R., Fukuyo, Y. (1989). Morphological features of the motile cell of *Pyrodinium bahamense*. pp. 207-217, In: *Biology, Epidemiology and Management of Pyrodinium*

Red Tides, G.M. Hallegraeff, Maclean, J.L. (eds.), ICLARM, Manila.

Usup, G. (1995). *The Physiology and Toxicity of the Red Tide Dinoflagellate Pyrodinium bahamense var. compressum*. Ph. D. Thesis. Boston University, U.S.A.

Usup, G., Ahmad, A., Ismail, N. (1989). *Pyrodinium bahamense* var. *compressum* red tide studies in Sabah, Malaysia. pp. 97-110, In: *Biology, Epidemiology and Management of Pyrodinium Red Tides*, G.M. Hallegraeff, Maclean, J.L. (eds.), ICLARM, Manila.

Usup, G., Lung, Y.K. (1991). Effects of meteorological factors on toxic red tide events in Sabah, Malaysia. *Mar. Ecol.* 12:331-339.

Vandermeulen, J.H., Foda, A. (1988). Cycling of selenite and selenate in marine phytoplankton. *Mar. Biol.* 98:115-123.

Villanoy, C.L., Corrales, R.A., Jacinto, G.S., Cuaresma Jr, N.T., Crisostomo, R.P. (1996). Towards the development of a cyst-based model for *Pyrodinium* red tides in Manila Bay, Philippines. pp. 189-192, In: *Harmful and Toxic Algal Blooms*, T. Yasumoto, Oshima, Y., Fukuyo, Y. (eds.), IOC-UNESCO.

Wall, D., Dale, B. (1969). The hystrichospaerid resting spore of the dinoflagellate *Pyrodinium bahamense* Plate 1906. *J. Phycol.* 5:140-149.

Worth, G.K., Maclean, J.L., Price, M.J. (1975). Paralytic shellfish poisoning in Papua New Guinea. *Pac. Sci.* 29:1-5.

Ecophysiology, Life Cycle, and Bloom Dynamics of *Chattonella* in the Seto Inland Sea, Japan

Ichiro Imai[1], Mineo Yamaguchi[2], and Masataka Watanabe[3]

[1]Laboratory of Marine Environmental Microbiology, Division of Applied Biosciences, Graduate School of Agriculture, Kyoto University, Kyoto 606-01, Japan
[2]Red Tide Biology Section, Red Tide Research Division, Nansei National Fisheries Research Institute, Ohno, Hiroshima 739-04, Japan
[3]Water and Soil Environment Division, National Institute for Environmental Studies, Tsukuba, Ibaraki 305, Japan

1. Introduction

In Japanese coastal waters, the first red tide of *Chattonella* occurred in Hiroshima Bay in 1969. In the summer of 1972, *C. antiqua* formed a red tide, and caused mass mortality of about 14 million cultured yellowtail (*Seriola quinqueradiata*) worth over ¥ 7 billion in Harima-Nada, eastern Seto Inland Sea, Japan (Okaichi 1989). This was the most severe fishery damage caused by red tides in Japan. *Chattonella* red tides have subsequently occurred until the present in the western coastal waters.

Subrahmanyan (1954) reported a red tide of *C. marina* (*Hornellia marina*) accompanying mass mortality of fish in Malabar Coast in India, and Tseng *et al.* (1993) recorded the first red tide of *C. marina* in Dapeng Bay, southern China, in 1991. Species of *Chattonella* were also found in southeast Asia, southern Brazil (Odebrecht and Abreu 1995), south Australia, and in Dutch coastal waters (Vrieling *et al.* 1995). Thus, *Chattonella* spp. are thought to be widely distributed in temperate and subtropical/tropical areas of the world.

According to Hara *et al.* (1994), the genus *Chattonella* comprises seven species, i.e. *C. subsalsa*, *C. antiqua*, *C. marina*, *C. globosa*, *C. minima*, *C. ovata*, and *C. verruculosa*. Amongst these species, *C. antiqua* and *C. marina* (Fig.1) are known to be the most harmful fish-killing species. *C. verruculosa* also kills cultured fish(Yamamoto and Tanaka 1990, Baba *et al.* 1995). In this review, the ecophysiology, life cycle, and population dynamics of *Chattonella* (*C. antiqua* and *C. marina*) are summarized and discussed with respect to their occurrences in the Seto Inland Sea.

Chattonella antiqua ***Chattonella marina***

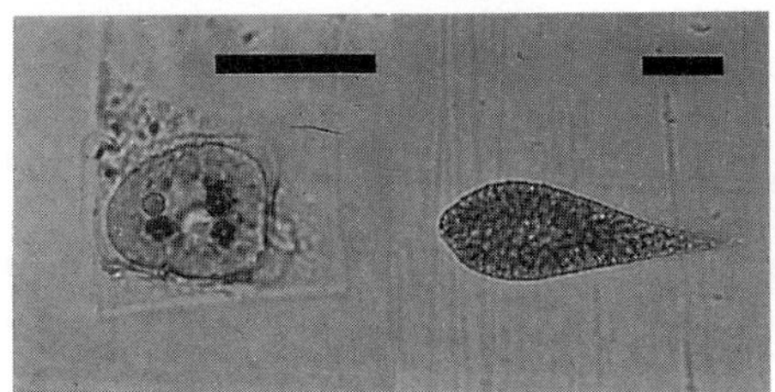

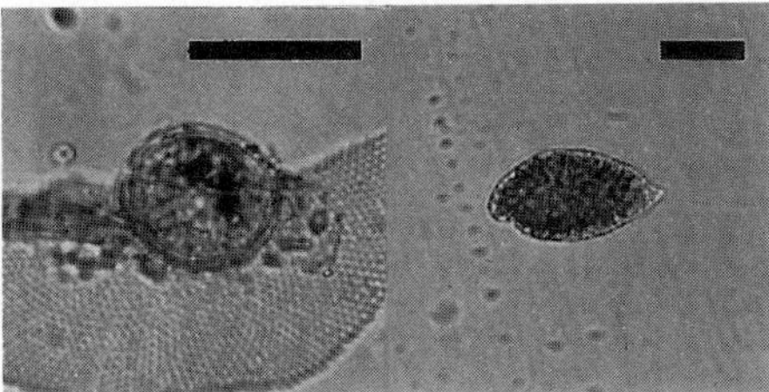

Fig.1. Cyst (left) and vegetative cell (right) of *Chattonella antiqua* and *C. marina*. Scale bar, 30 μm. (After Imai and Itoh 1988)

NATO ASI Series, Vol. G 41
Physiological Ecology of Harmful Algal Blooms
Edited by D. M. Anderson, A. D. Cembella and G. M. Hallegraeff

2. Growth responses of *Chattonella antiqua* and *C. marina* to environmental factors

A massive increase in cell number is essential for occurrences of red tides. The vegetative cells of *C. antiqua* and *C. marina* multiply by binary fission. Therefore, to understand the mechanisms of red tide occurrences it is important to know the effects of environmental factors on the growth of *C. antiqua* and *C. marina*. Irradiance, water temperature, salinity, and nutrients are the most significant factors affecting the growth of *Chattonella*.

Growth of *C. antiqua* and *C. marina* is observed at an irradiance of 30 μmol m^{-2} sec^{-1} or more, with saturation at 110 μmol m^{-2} sec^{-1} (Nakamura and Watanabe 1983a, Yamaguchi *et al.* 1991). The maximal growth rates of both species are about 1 div. day^{-1} (Yamaguchi *et al.* 1991). Cell division in *C. antiqua* and *C. marina* can be synchronously induced under light-dark regimes, and division occurs in the dark period (Nemoto and Furuya 1985, Ono 1988).

The growth responses of *Chattonella* were examined at 30 combinations of different temperatures (10° - 30°C) and salinities (10 - 35 psu) under a saturating light intensity of 120 μmol m^{-2} sec^{-1} (Yamaguchi *et al.* 1991). *Chattonella* grew at temperatures from 15° to 30°C and salinity from 10 to 35 psu (Fig.2). Maximal growth rates of *C. antiqua* and *C. marina* were obtained with the combination of 25°C and 25 psu, and 25°C and 20 psu, respectively. The most suitable temperature and salinity for *C. antiqua* are identical to those obtained by Nakamura and Watanabe (1983a). Yamochi (1984) reported that survival ranges were 13° to 31°C for both species of *Chattonella*. In Harima-Nada in the Seto Inland Sea, vegetative cells of *C. antiqua* and *C. marina* were observed in the temperature ranges of 19.2° - 28.8°C and 18.8° - 28.0°C, respectively (Yoshimatsu and Ono 1986). Blooms of *Chattonella* generally occurred during summer at temperatures higher than 23°C. The water temperature drops to around 10°C or below in the Seto Inland Sea during winter, so vegetative cells of *Chattonella* cannot overwinter in the water column.

Growth and uptake kinetics of nitrate and phosphate were examined in detail for *C. antiqua*. According to Nakamura and Watanabe (1983b), the uptake kinetics of *C. antiqua* followed the Michaelis-Menten equation, and the half-saturation constants (Ks) were 3.0 μM for nitrate and 1.9 μM for phosphate, respectively. *C. antiqua* could take up these nutrients in the dark at a rate almost equal (83 - 93 %) to that in the light. The half-saturation constants for the growth of *C. antiqua* for nitrate and phosphate were calculated to be 1.0 μM and 0.11 μM, respectively (Nakamura *et al.* 1988). Under stable and stratified conditions without mixing events in the summer, nutrient levels in the Seto Inland Sea such as at Harima-Nada are considered to be too low (compared with the above values) to support the growth of *Chattonella* in the surface water above the thermocline (nutricline).

Nutrient requirements have been investigated for *C. antiqua*. This species uses nitrate, ammonium, and urea, but not amino acids (glycine, alanine, and glutamate) as nitrogen sources (Nakamura and Watanabe 1983c). Inorganic phosphate, and occasionally glycerophosphate, can serve as phosphorus sources (Iwasaki 1973, Nakamura and Watanabe 1983c). For micro nutrients, *C. antiqua* needs Fe^{3+} and vitamin B_{12} (Iwasaki 1973, Nakamura and Watanabe 1983c, Nishijima and Hata 1986).

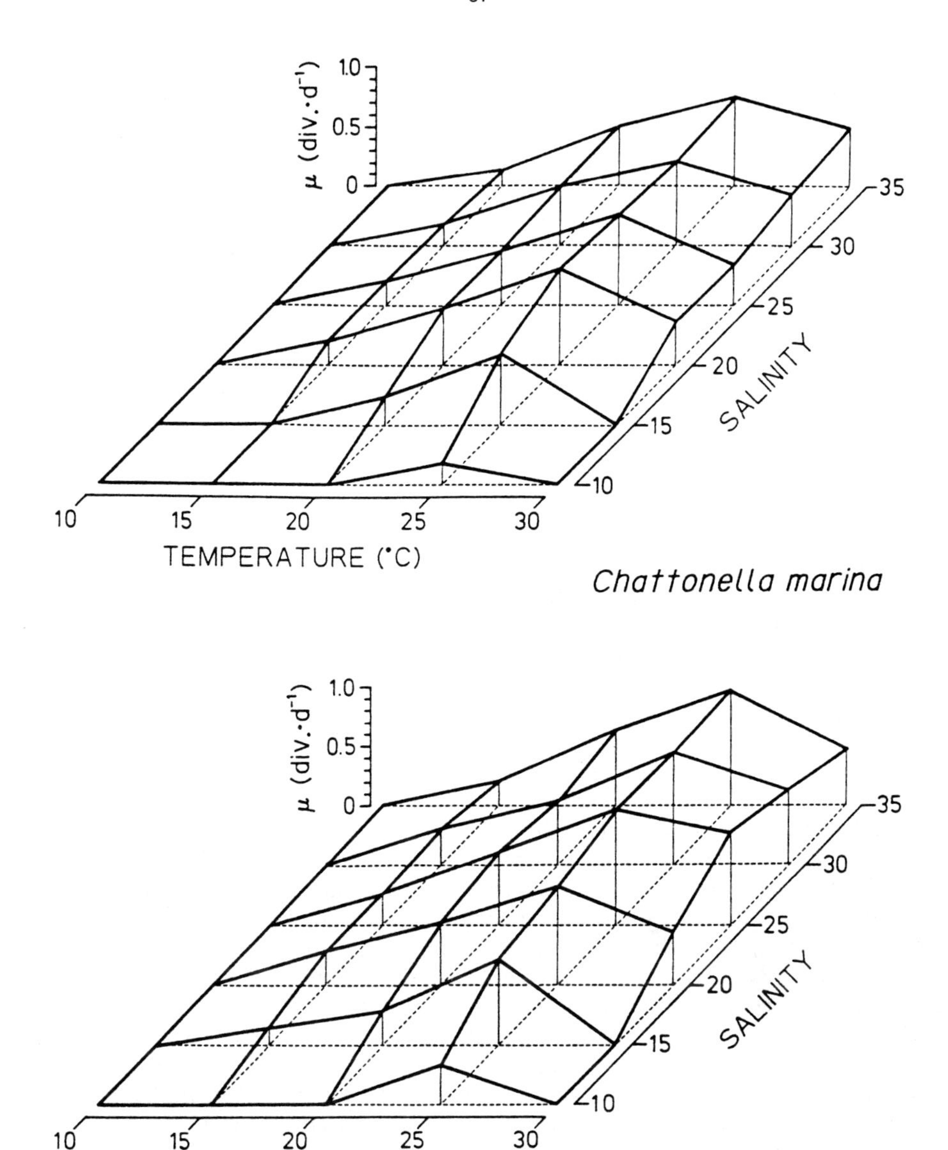

Fig.2. Multivariate response surfaces of growth rates in *Chattonella marina* and *C. antiqua* as a function of temperature and salinity. (After Yamaguchi *et al.* 1991)

3. Life history of *Chattonella*

Information on the life history of *Chattonella* was limited before 1986 when their cysts were first identified from bottom sediments of the Seto Inland Sea of Japan (Imai and Itoh 1986). The cysts of *C. antiqua* and *C. marina* (Fig.1) are typically hemispherical, 25 - 35 μm in diameter and 15 - 25 μm in height, and usually adhere to solid surfaces such as diatom frustules and sand grains. They are yellow-greenish to brownish in color and have several spots of dark brown or black material. Discrimination between the cysts of *C. antiqua* and *C. marina* is not possible on the basis of their morphological characteristics (Fig.1, Imai and Itoh 1988). Live cysts have spherical chloroplasts smaller than those of vegetative cells. There is a structure for future germination on the cyst wall, and an opening with a diameter of about 7 μm is observed on the wall of empty cysts after germination (Imai and Itoh 1988).

In both *C. antiqua* and *C. marina*, cyst formation has been observed in culture under laboratory conditions (Imai 1989, 1990, Nakamura *et al.* 1990). The cysts of both species formed in culture display morphological characteristics quite similar to natural cysts observed in the bottom sediments collected from the Seto Inland Sea. The combination of factors such as nutrient depletion (especially nitrogen limitation), adherence to solid surfaces (glass beads or glass slide), and low irradiance of about 15 μmol m^{-2} sec^{-1} or below (or darkness) was effective for cyst formation (Imai 1989, 1990, Nakamura and Umemori 1991).

It was unknown until recently whether or not cysts of *Chattonella* were formed through sexual reproduction. Yamaguchi and Imai (1994) determined the nuclear DNA content at various stages in the life history of *C. antiqua* and *C. marina* employing an epifluorescence microscopy-based fluorometry system (Yamaguchi 1992), and found that vegetative cells of both *C. antiqua* and *C. marina* were diploid and that their cysts were haploid, *i.e.,* they have a diplontic life history.

Fig.3 shows a schematic representation of the diplontic life history of *Chattonella antiqua* and *C. marina* based on DNA microfluorometry (Yamaguchi and Imai 1994). The life history is divided into two distinct phases: a vegetative propagation phase and a non-motile dormant phase. The vegetative cells grow by binary fission under normal growth conditions. The cells are diplont (2C to 4C). In nutrient-depleted conditions, small pre-encystment cells (1C) are produced after meiosis. The small pre-encystment cells of *C. antiqua* and *C. marina* are quite similar in size and morphology; they were also observed in natural seawater of Hiroshima Bay, the Seto Inland Sea, at the final stage of a red tide (Imai *et al.* 1993). The pre-encystment cells change into cysts under low light conditions. The ploidy level of the cyst is 1C. After a dormant period, a small vegetative cell (1C) germinates from each cyst and undergoes diploidization. The diplont small cell enlarges into a normal vegetative cell (2C) without cell fusion. At present we have very little information about how meiosis and diploidization occur.

4. Annual life cycle of *Chattonella* in the Seto Inland Sea

Red tides of *C. antiqua* and *C. marina* are caused by the motile, planktonic stage in their life history. However, both species of *Chattonella* are coastal neritic species and their life cycles incorporate benthic cyst stages. This cyst stage is not directly responsible for the red tide, but the cysts may play an important role in the total ecology of the neritic species *Chattonella* by serving several important functions as is the case in toxic dinoflagellates (Wall 1971, 1975, Anderson and Wall 1978, Fukuyo *et al.* 1982, Anderson *et al.* 1983, Dale 1983).

In the Seto Inland Sea, water temperature fluctuates seasonally. Temperatures of around 10°C or below are common in winter, and of 25°C or higher in summer. From laboratory experiments using sediment samples collected from the Seto Inland Sea, temperature was confirmed to be a principal factor affecting the physiology of the cysts of *Chattonella* (Imai *et al.* 1984a, 1989, 1991, Imai and Itoh 1987).

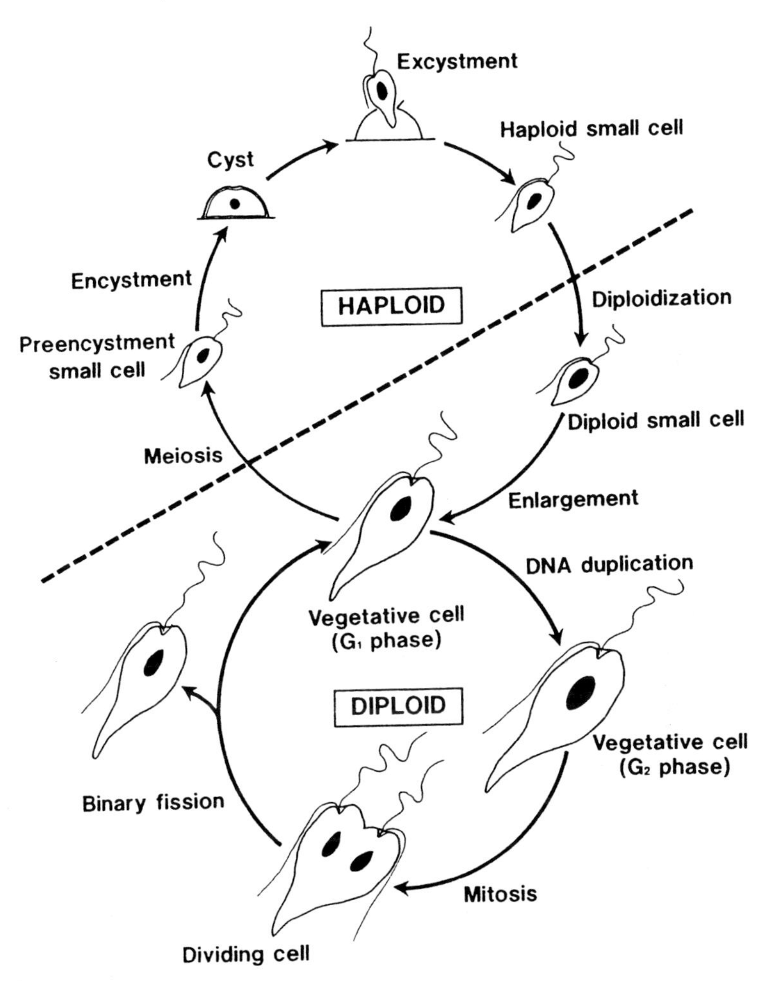

Fig.3. Schematic representation of the life history of *Chattonella antiqua* and *C. marina*, based on DNA microfluorometry. (After Yamaguchi and Imai 1994)

Fig.4 shows the effects of incubation temperature on the germination of matured cysts, and of storage temperature on the maturation (acquisition of germination ability) of cysts in sediments. Germination of cysts was not observed at 10°C, and was very low at 15° and 18°C, whereas the rate increased at 20°C with maxima at 22° and 25°C, and then decreased markedly at 30°C. The maturation and dormancy of *Chattonella* cysts are also significantly affected by temperature. For maturation, storage temperatures of 11°C or below for more than 4 months are needed, whereas no maturation is observed at 20°C or higher. Temperatures of 15° and 18°C are critical for maturation. Mature cysts in sediments maintained germination ability at the storage temperature of 11°C or below, they lost it progressively at 15° and 18°C, and rapidly at 20°C or higher (Imai *et al.* 1989).

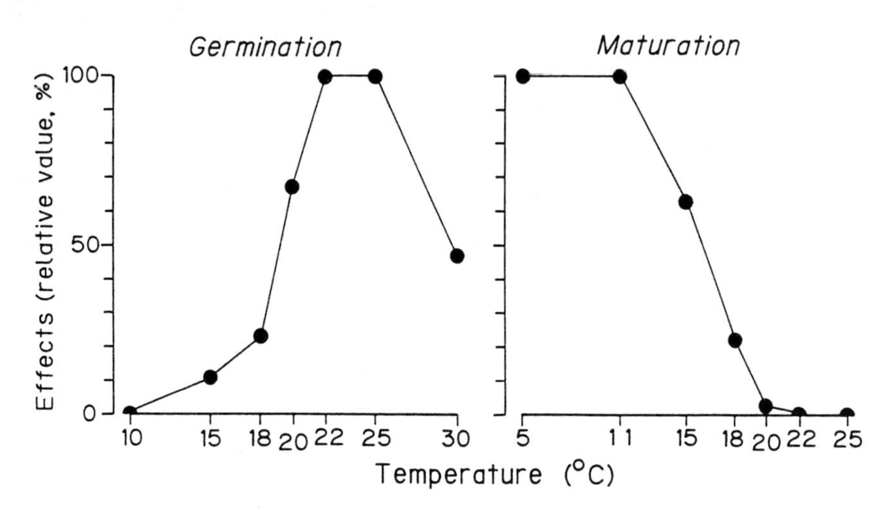

Fig.4. Effects of incubation temperature on the germination of cysts of *Chattonella*, and of storage temperature on maturation of the cysts in sediments. The ordinates show the percentage of maximum value obtained in experiments. (After Imai *et al.* 1991)

Using freshly collected bottom sediments of the Seto Inland Sea, the seasonality of germination ability in cysts was investigated for *Chattonella* (Imai *et al.* 1984b, Imai and Itoh 1987). Germination ability was low from autumn to early winter, then increased gradually to a high level, which was maintained between spring and early summer, and again decreased rapidly during summer.

The annual life cycle of *Chattonella* is summarized in Fig.5 (Imai and Itoh 1987, Imai *et al.* 1991). In the Seto Inland Sea, vegetative cells of *C. antiqua* and *C. marina* are generally observed from June to September, and occasionally form red tides in July and August. The vegetative cells originate from the germination of the cysts in bottom sediments in early summer, when the bottom temperature rises to a threshold level of around 20°C (Imai *et al.* 1984a). The vegetative cells multiply asexually in summer, and produce small pre-encystment cells under unfavorable conditions such as nutrient depletion. These cells then sink to the sea bottom (Imai *et al.* 1993), and encystment is completed there after the adhesion to solid surfaces such as diatom frustules and sand grains. The cysts spend a period of spontaneous (genetically regulated) dormancy there until next spring. They never germinate in autumn even when the bottom water temperature descends to the optimal range for germination, 20° to 22°C. The maturation of cysts progresses in sediments during winter. In spring, many cysts have completed the period of spontaneous dormancy, and have acquired the capacity for germination. From spring to early summer, they must spend a period of post dormancy (quiescence) due to a temperature too low for germination. Vegetative cells thereafter appear in the water column via cyst germination. Most of the cyst populations, however, remain in sediments without germinating, and they are carried over to the next year or year to year via the secondary dormancy which is induced by high temperature during summer (Imai *et al.* 1989). The secondary dormancy was also reported in the resting spores of the diatom *Eunotia soleirolii* (von Stosch and Fecher 1979).

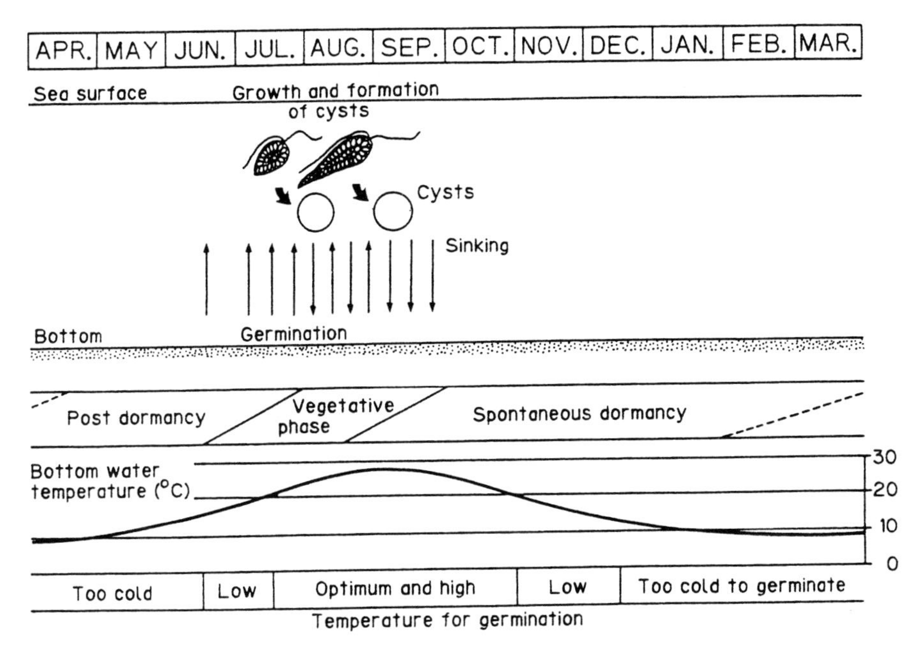

Fig.5. Schematic representation of the annual life cycle of *Chattonella* in the Seto Inland Sea, including vegetative and cyst phases. The seasonal fluctuation of bottom water temperature is also shown. (After Imai and Itoh 1987, Imai *et al.* 1991)

Since the vegetative cells of *Chattonella* are usually observed only in the summer, *C. antiqua* and *C. marina* presumably spend most of their life cycle as cysts in sediments. In summary, the life cycle of *Chattonella* is considered to be well adapted to the temperature fluctuations in temperate seas such as the Seto Inland Sea. Moreover, alternation between benthic and planktonic stages is probably unconstrained, because most part of the Seto Inland Sea is shallower than 50 m (Imai and Itoh 1987, Imai *et al.* 1991).

5. Bloom dynamics of *Chattonella* in Suo-Nada, western Seto Inland Sea

The dynamics of cysts and vegetative cells of *Chattonella* were investigated together with the environmental factors in Suo-Nada, the Seto Inland Sea. Fig.6 shows the distribution of vegetative cells of *Chattonella* in surface water (mean of 0- and 5-m samples), and surface and bottom water (1 m above the bottom) temperatures, in Suo-Nada during a period from July 12 to 14 in 1984 (Imai *et al.* 1986). The bottom water temperature, which affects the germination of *Chattonella* cysts (Fig.4), was optimum for germination (≥ 20°C) in the western coastal area, but lower in

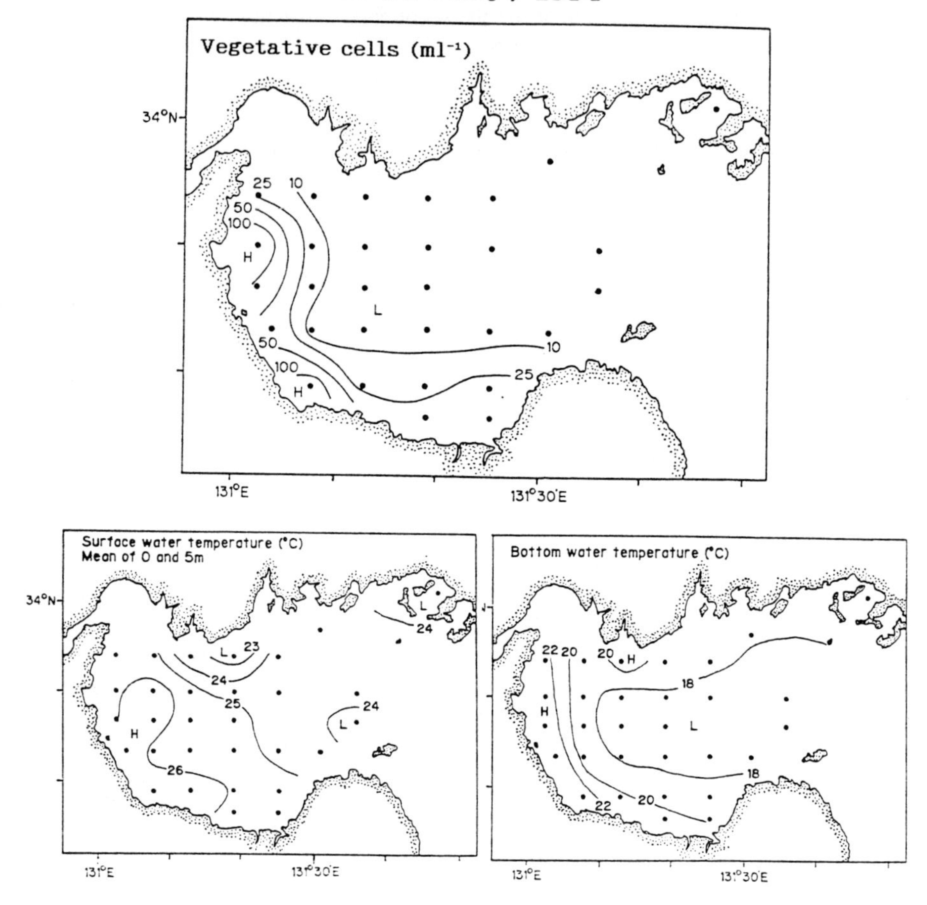

Fig.6. Distribution of vegetative cells of *Chattonella* in surface water (mean of 0 and 5m samples), and temperatures of surface and bottom water during a period of July 12 to 14, 1984. (After Imai *et al.* 1986)

the central deep water area during this period. The surface water temperature (22.3° - 26.8°C) was in an optimum range for the growth of vegetative cells (20° - 30°C) in the whole area. The density of *Chattonella* vegetative cells (*C. antiqua* and *C. marina*) was higher in the coastal area (maximum 248 cells ml^{-1}) than in the offshore water. The *Chattonella* populations subsequently developed into a "red tide" in late July and caused fish mortality. The weather was fair and sunny, and the winds blowing to the western coastal area from offshore constantly prevailed in daytime during July of 1984, which presumably concentrated *Chattonella* there because *Chattonella* are accumulated in the surface layer by the vertical migration in daytime (Watanabe *et al.* 1995).

From this result and the subsequent field observations in Suo-Nada, a conceptual model of a *Chattonella* red tide outbreak is presented in Fig.7. We inferred that the cysts started to germinate from the coastal shallow area where the bottom water temperature was relatively higher than the deeper area and reached the optimum

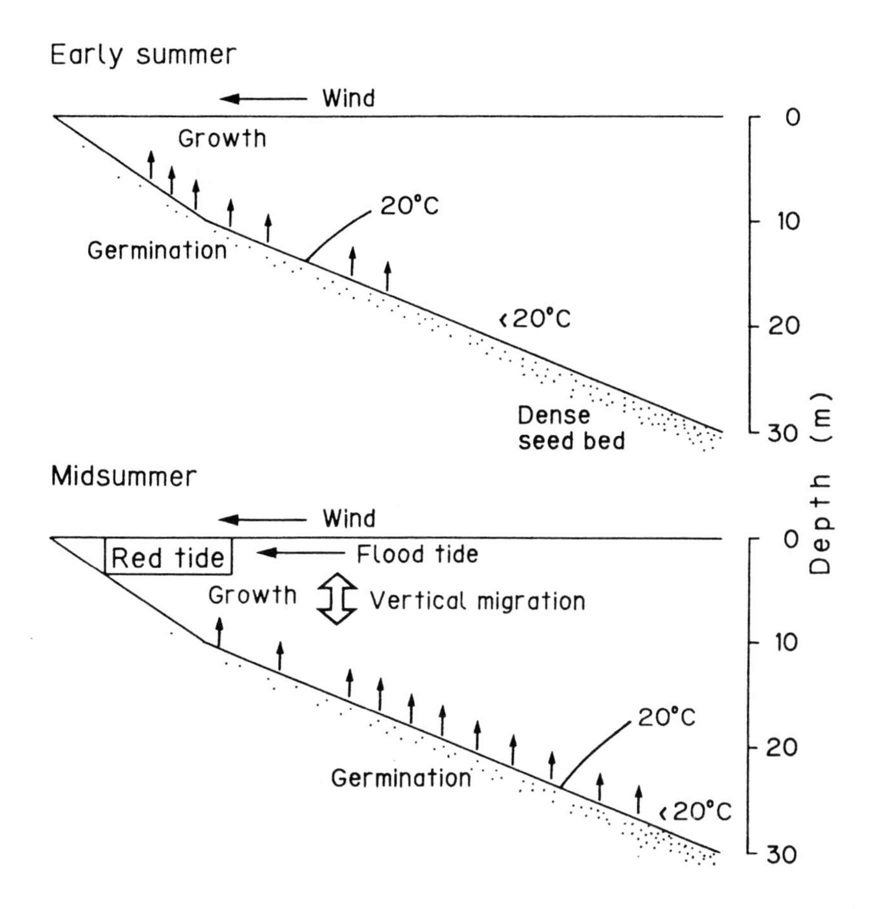

Fig.7. A schematic representation of the outbreak of *Chattonella* red tide in the western coastal area of Suo-Nada, western Seto Inland Sea. (After Imai *et al.* 1986, Imai 1990)

level (*ca.* 20°C) for germination in early June. The primary vegetative populations first appear as a result of the germination of cysts in south-western coastal shallow area in Suo-Nada despite the lower densities of cysts, and start to multiply in surface waters. Since the bottom water temperature rises gradually from the coastal shallow area to deeper areas, the germination of *Chattonella* cysts presumably continues for a rather long period (more than 2 months). In mid summer (Fig.7), the cysts in deeper waters germinate, and new vegetative cells start to multiply in surface waters. Accumulation of vegetative cells by the winds mentioned earlier might play an important role in supporting the population density in a high level, because *Chattonella* cells vertically migrate by swimming to the upper layer during daytime (Watanabe *et al.* 1995).

The temporal changes in the total live *Chattonella* cysts and those with germination ability in freshly collected sediment samples (top 1 cm) are shown at 3 stations in Suo-Nada during the summer of 1987 (Fig.8). The population density of *Chattonella* was low (≤ 5 cells $ml^{-1)}$ in late June, reached a high level (maximum 212 cells ml^{-1}) in mid July, and stayed there (maximum 81 cells ml^{-1}) until early August (Imai 1990).

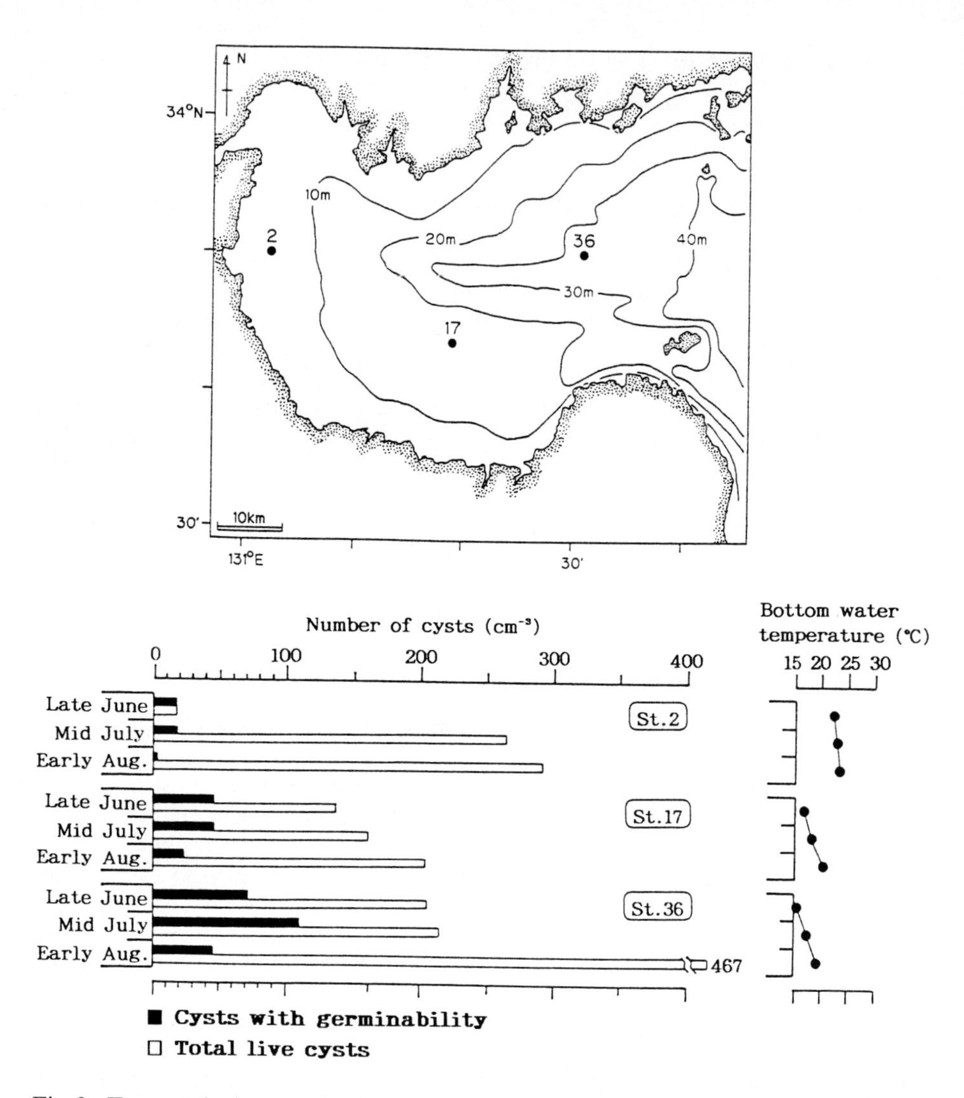

Fig.8. Temporal changes in the total live cysts of *Chattonella* and those with germinability in freshly collected sediment samples at 3 stations in Suo-Nada, during the summer of 1987. The changes of bottom water temperature are also shown. (After Imai 1990, modified)

The number of cysts with germination ability did not change markedly between late June and mid July, although the number of vegetative cells in the water column drastically increased about 2 orders of magnitude. This implies that the number of cysts that actually germinate is much less than previously expected. Thus a small portion of cyst populations may actually germinate, and the growth of the resultant vegetative populations probably hold the key to the subsequent development to the red tide. Hence *Chattonella* appear to adopt the inoculum strategy of "long period germination" which allows *Chattonella* more chances for the population development during summer. The occurrence of blooms then supplies new cysts to the bottom sediments of Suo-Nada (Fig.8). An increase in cyst numbers after *Chattonella* blooms was also observed in northern Hiroshima Bay (Imai *et al.* 1993) and Harima-Nada (Imai *et al.* unpublished).

6. Marine ecosystem succession and *C. antiqua* bloom in a mesocosm

Diel vertical migration and nocturnal uptake of PO_4^{3-} and N by *C. antiqua* in the nutrient-rich lower layer was demonstrated in a 1.5-m-tall axenic culture tank, in which nutrients were vertically stratified analogous to nutrients observed in a natural *Chattonella* red tide (Watanabe *et al.* 1991). However, diel vertical migration of *C. antiqua* has not been observed in the field, and its migratory depth is unknown.

In 5 years of observation at stations in northern Harima-Nada, eastern Seto Inland Sea, the presence of a shallow P-cline (5 - 7 m) was most characteristic in 1987, a year with a *C. antiqua* outbreak, compared to other years (1984, 1985, 1986, 1988) in which P-clines were usually found at 12 - 15 m (Nakamura *et al.* 1989, Watanabe *et al.* 1991). Watanabe *et al.* (1995) used an *in situ* mesocosm to show that *C. antiqua* undergoes vertical migration at night to reach deep water that has ample nutrients, and that a bloom of *C. antiqua* occurred when there was a stable nutricline at a shallow depth within the range of vertical migration of *C. antiqua.*

Mesocosm experiments were conducted at a cove off the Ieshima Islands in northern Harima-Nada, from 21 July to 14 August 1989. The mesocosm was 18 m deep, 5-m diameter and had a volume of ~ 350 m^3. Details are described in Watanabe *et al.* 1995. At the beginning of the experiment on 21 July 1989 temperature was 21.5°C at the bottom which was almost optimum for excystment of *C. antiqua* (22°C, Imai and Itoh 1987, 1988). During the experiment, temperature at the surface was in the range 24.5° - 27.5°C, optimum for growth of *C. antiqua.* Immediately after enclosure on 21 July, N and P were enriched throughout the water column.

After the first nutrient enrichment, centric diatoms, dinoflagellates, and others showed rapid growth in the 0 - 10-m zone. N, P, and Si in the 0 - 5 m zone were taken up rapidly within a few days. N, P, and Si were thus stratified at 5 - 6 m. Centric diatoms could not grow in the surface layer after 27 July due to limitation of not only N and P but also Si.

After 2 August, dinoflagellates and other flagellates increased in number until vertical circulation ended on 8 August. When vertical circulation stopped, there was a large accumulation of dinoflagellates at the water surface as a result of vertical migration. On 10 August, most dinoflagellates, in particular *Scrippsiella trochoidea*, suddenly disappeared from the surface layer.

From 4 to 7 August, *C. antiqua* occurred at ~ 0.5 cells ml^{-1}, however, when vertical circulation ended, *C. antiqua* was present at 2 cells ml^{-1} (at 5 m at 0900 hours on 8 August). From 9 to 13 August, the population of *C. antiqua* at the surface (at 0900 hours) increased from 8 to 116 cells ml^{-1}, and there was a *C. antiqua* red tide.

Between 1500 and 1800 hours on 12 August, *C. antiqua* cells began to migrate downward (Fig.9). At 0300 hours on 13 August, most cells (229 cells ml^{-1}) migrated to a depth of 7.5 m, which was within the nutrient-rich zone. Between 0300 and 0600 hours, the cells began to migrate upward, and at 1200 hours, most cells reached the surface (421 cells ml^{-1}). The migration speed (as group velocity) was calculated to be ~ 0.8 m h^{-1}, with upward and downward migration speeds the same.

This *in situ* mesocosm study showed that oceanographic conditions leading the marine ecosystem in the Seto Inland Sea toward a *C. antiqua* red tide involved a

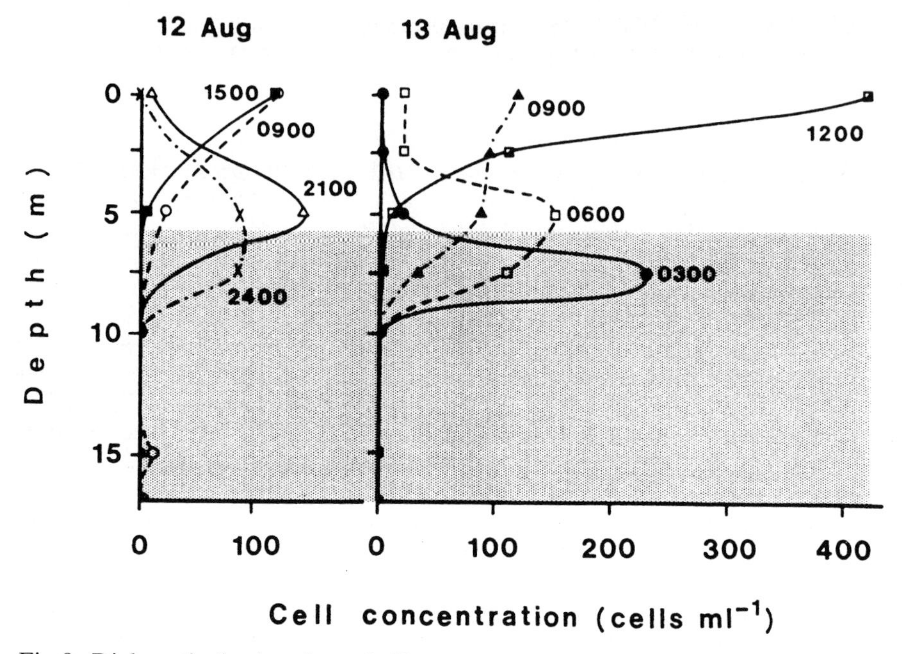

Fig.9. Diel vertical migration of *Chattonella antiqua* observed in the mesocosm between 12 and 13 August 1989. Artificial nutrient enrichment was conducted in the shaded zone (6-18 m). (After Watanabe *et al.* 1995)

complex combination of factors: excess of N and P compared with Si, initial bottom temperature of 20° - 22°C (optimum for excystment to inoculate *C. antiqua*), surface temperature of 25° - 27°C optimum for growth of *C. antiqua*), stable shallow nutrient stratification, absence of copper toxicity, cessation of vertical circulation, selective grazing toward small-sized phytoplankton by zooplankton, and diel vertical migration by *C. antiqua* (Watanabe *et al.* 1995).

7. A hypothesis on the mechanisms of *Chattonella* red tide

When vegetative cells of *Chattonella* appear in the water column after the germination of cysts, environmental factors such as nutrients and competitors (mainly diatoms) may crucially affect the development of *Chattonella* populations thereafter. A hypothesis (Fig.10) on the occurrence mechanisms of *Chattonella* red tide is presented based on the results of a research project conducted by the Nansei National Fisheries Research Institute of Fisheries Agency (Imai 1995).

It is empirically known that *Chattonella* red tides have occurred when diatoms are scarce (the order of 10^2 cells ml^{-1} or lower) in surface water (*e.g.* Yoshimatsu and Ono 1986, Montani *et al.* 1989, Nakamura *et al.* 1989, Imai 1990). Moreover, *Chattonella* often forms red tides even in the presence of rather high concentrations of $Si(OH)_4$, in which the growth of diatoms may not be limited (Montani *et al.* 1989, Nakamura *et al.* 1989).

As competitors, the diatoms (mainly Centrales) seem to dominate over *Chattonella* because diatoms generally have higher growth rates (Eppley 1977, Yamaguchi 1994). However, the diatoms form resting stage cells under conditions of nutrient limitation

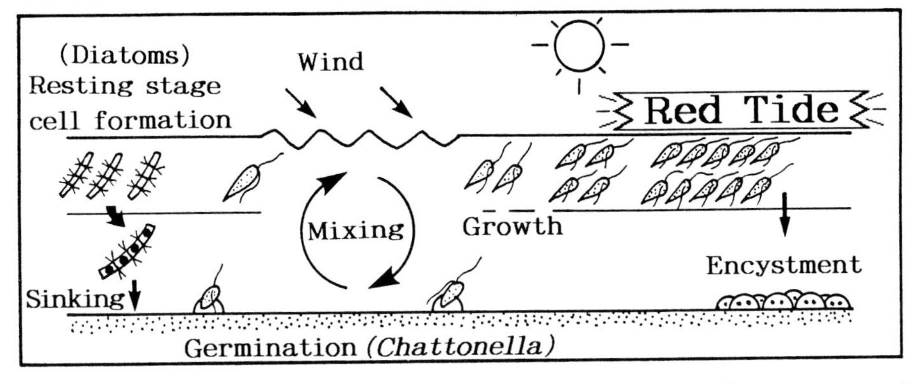

Fig.10. A schematic representation of the hypothetic scenario on the occurrence of *Chattonella* red tides in the Seto Inland Sea in summer. (After Imai 1995, modified)

especially of nitrogen (Davis *et al.* 1980, French and Hargraves 1980, Hargraves and French 1983, Garrison 1984, Smetacek 1985, Itakura *et al.* 1993). Itakura and Imai (1994) found that resting spores of *Chaetoceros* spp. were observed more frequently in areas at low ambient DIN concentrations (<1μM) in Harima-Nada, eastern Seto Inland Sea. Resting stage cells rapidly sink to the sea bottom, and exist at densities of 10^3 - 10^6 g^{-1} wet sediment (Imai *et al.* 1990). Accordingly, the cell concentrations of diatoms can be reduced in water columns via resting stage cell formation and sinking, after a strong stratification and the resultant exhaustion of nutrients in the surface layer. The diatom resting stage cells cannot generally germinate and/or rejuvenate at low light intensities or in the dark (Hollibaugh *et al.* 1981, Hargraves and French 1983, Garrison 1984, Imai *et al.* 1996), whereas *Chattonella* cysts can germinate in the dark (Imai *et al.* 1984a). This allows the continuous and selective seeding of *Chattonella* populations to the surface water. The diatoms which survive in surface water might be in a physiologically less active state induced by nutrient depletion (Kuwata and Takahashi 1990). Moreover, the capacity for vertical migration could aid in the growth and accumulation of *Chattonella* populations in a stratified water column with a shallow nutricline (thermocline) (Watanabe *et al.* 1995).

Because of the decrease of diatoms in the surface water, *Chattonella* spp. obtain a competitive advantage, and thus can become dominant despite lower growth rates of at most 1 div. day^{-1}, after mixing events supply nutrients to the surface layer. Yanagi (1989) previously pointed out by the analysis of weather and marine conditions that mixing events by strong winds frequently contribute to the occurrences of *Chattonella* red tides in Harima-Nada. If the nitrogen concentration increased 2 μM by a mixing event and *C. antiqua* utilized it exclusively, the cell increase was estimated to be 260 cells ml^{-1} based on the minimum cell quota (q_0) of *C. antiqua* (q_0^N = 7.7 mol $cell^{-1}$, q_0^P = 0.6 pmol $cell^{-1}$, Nakamura 1985). *C. antiqua* can theoretically reach this level from 2 cells ml^{-1} within 1 week with a growth rate of 1 div. day^{-1}. The timely mixing event (= nutrient supply to the surface layer), which must occur after an appropriate stratification accompanied the nutrient exhaustion and sinking and/or inactivation of diatoms, is thought to be essential for the occurrence of *Chattonella* red tides. The existence of primary vegetative populations of *Chattonella* seeded from the germination of cysts is of course the prerequisite in this scenario.

8. Acknowledgements

We are very grateful to Drs. Y. Ishida, M. Anraku, K. Itoh, T. Onbé, H. Iwasaki, Y. Yoshida, and T. Honjo for their helpful advice and encouragements during the course of study. We thank Drs. S. Itakura and Y. Matsuo for their fruitful collaborations. We also wish to thank the officers and crew of R/V Shirafuji-Maru, Nansei National Fisheries Research Institute, for their kind cooperation at sea. The research was supported by grants from the Fisheries Agency and Environment Agency of Japan.

9. References

Anderson, D.M., Chisholm, S.W., Watras, C.J. (1983). Importance of life cycle events in the population dynamics of *Gonyaulax tamarensis*. *Mar. Biol.* 76: 179-189.

Anderson, D.M., Wall, D. (1978). Potential importance of benthic cysts of *Gonyaulax tamarensis* and *G. excavata* in initiating toxic dinoflagellate blooms. *J. Phycol.* 14: 224-234.

Baba, T., Momoyama, K., Hiraoka, M. (1995). A harmful flagellated plankton increased in Tokuyama Bay. *Bull. Yamaguchi Pref. Naikai Fish. exp. Stn.* No.24: 1121-122 (in Japanese).

Dale, B. (1983). Dinoflagellate resting cysts : "benthic plankton". In: Fryxell, G.A. (ed.) *Survival strategies of algae*. Cambridge Univ. Press, Cambridge, p.69-136

Davis, C.O., Hollibaugh, J.T., Seibert, D.L.R., Thomas, W.H., Harrison, P.J. (1980). Formation of resting spores by *Leptocylindrus danicus* (Bacillariophyceae) in a controlled experimental ecosystems. *J. Phycol.* 16: 296-302.

Eppley, R.W. (1977). The growth and culture of diatoms. In: Werner, D. (ed.) *The biology of diatoms*. Blackwell, Oxford, p.24-64.

French, F.W., Hargraves, P.E. (1980). Physiological characteristics of plankton diatom resting spores. *Mar. Biol. Lett.* 1: 185-195.

Fukuyo, Y., Watanabe, M.M., Watanabe, M. (1982). Encystment and excystment of red-tide flagellates II. Seasonality on excystment of *Protogonyaulax tamarensis* and *P. catenella*. *Res. Rep. natl. Inst. environ. Stud.* No.30: 43-52 (in Japanese with English abstract).

Garrison, D.L. (1984). Plankton diatoms. In: Steidinger, K.A., Walker, L.M. (eds) *Marine plankton life cycle strategies*. CRC Press, Boca Raton, Florida, p.1-17.

Hara, Y., Doi, K., Chihara, M. (1994). Four new species of *Chattonella* (Raphidophyceae, Chromophyta) from Japan. *Jpn. J. Phycol.* 42: 407-420.

Hargraves, P.E., French, F.W. (1983). Diatom resting spores: significance and strategies. In: Fryxell, G.A. (ed.) *Survival strategies of the algae*. Cambridge Univ. Press, Cambridge, p.49-68.

Hollibaugh, J.T., Seibert, D.L.R., Thomas, W.H. (1981). Observations on the survival and germination of resting spores of three *Chaetoceros* (Bacillariophyceae) species. *J. Phycol.* 17: 1-9.

Imai, I. (1989). Cyst formation of the noxious red tide flagellate *Chattonella marina* (Raphidophyceae) in culture. *Mar. Biol.* 103: 235-239.

Imai, I. (1990). Physiology, morphology, and ecology of cysts of *Chattonella* (Raphidophyceae), causative flagellates of noxious red tides in the Inland Sea of Japan. *Bull. Nansei natl. Fish. Res. Inst.* No.23: 63-166 (in Japanese with English abstract).

Imai, I. (1995). Ecological control of *Chattonella* red tides by diatoms. *Kaiyo Monthly*, 27: 603-612 (in Japanese).

Imai, I., Itakura, S., Itoh, K. (1990). Distribution of diatom resting cells in sediments of Harima-Nada and northern Hiroshima Bay, the Seto Inland Sea, Japan. *Bull.. coast. Oceanogr.* 28: 75-84 (in Japanese with English abstract).

Imai, I., Itakura, S., Itoh, K. (1991). Life cycle strategies of the red tide causing flagellates *Chattonella* (Raphidophyceae) in the Seto Inland Sea. *Mar. Poll. Bull.* 23: 165-170.

Imai, I., Itakura, S., Ouchi, A. (1993). Occurrence of *Chattonella* red tide and cyst dynamics in sediments in northern Hiroshima Bay, the Seto Inland Sea, Japan. *Nippon Suisan Gakkaishi* 59: 1-6 (in Japanese with English abstract).

Imai, I., Itakura, S., Yamaguchi, M., Honjo, T. (1996). Selective germination of *Heterosigma akashiwo* (Raphidophyceae) cysts in bottom sediments under low light conditions: A possible mechanism of the red tide initiation. In: Yasumoto, T., Oshima, Y., Fukuyo, Y. (eds.) *Harmful and toxic algal blooms.* IOC-UNESCO, p.197-200.

Imai, I., Itoh, K. (1986). A preliminary note on the cysts of *Chattonella* (Raphidophyceae), red tide flagellates, found in bottom sediment in Suo-Nada, western Seto Inland Sea, Japan. *Bull. Plankton Soc. Japan* 33: 61-63 (in Japanese with English abstract).

Imai, I., Itoh, K. (1987). Annual life cycle of *Chattonella* spp., causative flagellates of noxious red tides in the Inland Sea of Japan. *Mar. Biol.* 94: 287-292.

Imai, I., Itoh, K. (1988). Cysts of *Chattonella antiqua* and *C. marina* (Raphidophyceae) in sediments of the Inland Sea of Japan. *Bull. Plankton Soc. Japan* 35: 35-44.

Imai, I., Itoh, K., Anraku, M. (1984a). Distribution of dormant cells of *Chattonella* in Harima-Nada, eastern Seto Inland Sea, and temperature characteristics of germination. *Bull. Plankton Soc. Japan* 31: 35-42 (in Japanese with English abstract).

Imai, I., Itoh, K., Anraku, M. (1984b). Extinction dilution method for enumeration of dormant cells of red tide organisms in marine sediments. *Bull. Plankton Soc. Japan* 31:123-124.

Imai, I., Itoh, K., Anraku, M. (1989). Dormancy and maturation in the cysts of *Chattonella* spp. (Raphidophyceae), red tide flagellates in the Inland Sea of Japan. In: Okaichi, T., Anderson, D.M., Nemoto, T. (eds.) *Red tides: Biology, environmental science, and toxicology.* Elsevier, New York, p.289-292.

Imai, I., Itoh, K., Terada, K., Kamizono, M. (1986). Distribution of dormant cells of *Chattonella* (Raphidophyceae) and occurrence of summer red tide in Suo-Nada, western Seto Inland Sea. *Bull. Jpn. Soc. sci. Fish.* 52: 1665-1671 (in Japanese with English abstract).

Itakura, S. Imai, I. (1994). Distribution of *Chaetoceros* (Bacillariophyceae) resting spores observed in the surface water of Harima-Nada, in the summer of 1991, with reference to the oceanographic conditions. *Bull. Jpn. Soc. Fish. Oceanogr.* 58: 29-42 (in Japanese with English abstract).

Itakura, S., Yamaguchi, M., Imai, I. (1993). Resting spore formation and germination of *Chaetoceros didymus* var. *protuberans* (Bacillariophyceae) in clonal culture. *Nippon Suisan Gakkaishi* 59: 807-813 (in Japanese with English abstract).

Iwasaki, H. (1973). The physiological characteristics of neritic red tide flagellates. *Bull. Plankton Soc. Japan* 19: 46-56. (in Japanese with English abstract).

Kuwata, A., Takahashi, M. (1990). Life-form population responses of a marine planktonic diatom, *Chaetoceros pseudocurvisetus*, to oligotrophication in regionally upwelled water. *Mar. Biol.* 107: 503-512.

Montani, S., Tokuyasu, M., Okaichi, T. (1989). Occurrence and biomass estimation of *Chattonella marina* red tides in Harima-Nada, the Seto Inland Sea, Japan. In: Okaichi, T., Anderson, D.M., Nemoto, T. (eds.) *Red tides: Biology, environmental science, and toxicology.* Elsevier, New York, p.197-200.

Nakamura, Y. (1985). Kinetics of nitrogen- or phosphorus-limited growth and effects of growth conditions on nutrient uptake in *Chattonella antiqua. J. oceanogr. Soc. Japan* 41: 381-387.

Nakamura, Y., Takashima, J., Watanabe, M. (1988). Chemical environment for red tides due to *Chattonella antiqua* in the Seto Inland Sea, Japan Part 1. Growth bioassay of the seawater and dependence of growth rate on nutrient concentrations. *J. oceanogr. Soc. Japan*, 44: 113-124.

Nakamura, Y., Umemori, T. (1991). Encystment of the red tide flagellate *Chattonella antiqua* (Raphidophyceae): cyst yield in batch cultures and cyst flux in the field. *Mar. Ecol. Prog. Ser.* 78: 273-284.

Nakamura, Y., Umemori, T., Watanabe, M. (1989). Chemical environment for red tides due to *Chattonella antiqua* Part 2. Daily monitoring of the marine environment throughout the outbreak period. *J. oceanogr. Soc. Japan* 45: 116-128.

Nakamura, Y., Umemori, T., Watanabe, M., Kulis, D.M., Anderson, D.M. (1990). Encystment of *Chattonella antiqua* in laboratory cultures. *J. oceanogr. Soc. Japan* 46: 35-43.

Nakamura, Y., Watanabe, M.M. (1983a). Growth characteristics of *Chattonella antiqua* (Raphidophyceae) Part 1. Effects of temperature, salinity, light intensity and pH on growth. *J. oceanogr. Soc. Japan* 39: 110-114.

Nakamura, Y., Watanabe, M.M. (1983b). Nitrate and phosphate uptake kinetics of *Chattonella antiqua* in light/dark cycles. *J. oceanogr. Soc. Japan* 39: 167-170.

Nakamura, Y., Watanabe, M.M. (1983c). Growth characteristics of *Chattonella antiqua* Part 2. Effects of nutrients on growth. *J. oceanogr. Soc. Japan* 39: 151-155.

Nemoto, Y., Furuya, M. (1985). Inductive and inhibitory effects of light on cell division in *Chattonella antiqua. Plant Cell Physiol.* 26: 669-674.

Nishijima, T., Hata, Y. (1986). Physiological ecology of *Chattonella antiqua* (Hada) Ono on B group vitamin requirements. *Bull. Jpn. Soc. sci. Fish.* 52: 181-186.

Odebrecht, C., Abreu, P.C. (1995). Raphidophycean in southern Brazil. *Harmful Algae News* No.12/13: 4.

Okaichi, T. (1989). Red tide problems in the Seto Inland Sea, Japan. In: Okaichi, T., Anderson, D.M., Nemoto, T. (eds.) *Red tides: Biology, environmental science, and toxicology*. Elsevier, New York, p.137-142.

Ono, C. (1988). Cell cycles and growth rates of red tide organisms in Harima Nada area, eastern part of Seto Inland Sea, Japan. *Bull. Akashiwo Res. Inst. Kagawa Pref.* No.3: 1-67 (in Japanese with English abstract).

Smetacek, V.S. (1985). Role of sinking in diatom life-history cycle: ecological, evolutionary and geological significance. *Mar. Biol.* 84: 239-251.

Subrahmanyan, R. (1954). On the life-history and ecology of *Hornellia marina* gen. et sp. nov., (Chloromonadineae), causing green discoloration of the sea and mortality among marine organisms off the Malabar Coast. *Indian J. Fish.* 1: 182-203.

Tseng, C.K., Zhou, M.J., Zou, J.Z. (1993). Toxic phytoplankton studies in China. In: Smayda, T.J., Shimizu, Y. (eds.) *Toxic phytoplankton blooms in the sea.* Elsevier, New York, p.347-352.

von Stosch, H.A., Fecher, K. (1979). "Internal thecae" of *Eunotia soleirolii* (Bacillariophyceae): Development, structure and function as resting spores. *J. Phycol.* 15: 233-243.

Vrieling, E.G., Koeman, R.P.T., Nagasaki, K., Ishida, Y., Peperzak, L., Gieskes, W.W.C., Veenhuis, M. (1995). *Chattonella* and *Fibrocapsa* (Raphidophyceae): First observation of, potentially harmful, red tide organisms in Dutch coastal waters. *Neth. J. Sea Res.* 33: 183-191.

Wall, D. (1971). Biological problems concerning fossilizable dinoflagellates. *Geoscience and Man* 3: 1-15.

Wall, D. (1975). Taxonomy and cysts of red-tide dinoflagellates. In: LoCicero, V.R. (eds.) *Toxic dinoflagellate blooms*. Mass. Sci. Tech. Found., Wakefield, p.249-255.

Watanabe, M., Kohata, K., Kimura, T. (1991). Diel vertical migration and noctur-nal up-take of nutrients by *Chattonella antiqua* under stable stratification. *Limnol. Oceanogr.* 36: 593-602.

Watanabe, M., Kohata, K., Kimura, T., Takamatsu, T., Yamaguchi, S., Ioriya, T. (1995). Generation of a *Chattonella antiqua* bloom by imposing a shallow nutricline in a mesocosm. *Limnol. Oceanogr.* 40: 1447-1460.

Yamaguchi, M. (1992). DNA synthesis and cell cycle in the noxious red-tide dinoflagellate *Gymnodinium nagasakiense*. *Mar. Biol.* 112: 191-198.

Yamaguchi, M. (1994). Physiological ecology of the red tide flagellate *Gymnodinium nagasakiense* (Dinophyceae) - Mechanism of the red tide occurrence and its prediction. *Bull. Nansei natl. Fish. Res. Inst.* No.27: 251-394 (in Japanese with English abstract).

Yamaguchi, M., Imai, I. (1994). A microfluorometric analysis of nuclear DNA at different stages in the life history of *Chattonella antiqua* and *Chattonella marina* (Raphidophyceae). *Phycologia,* 33: 163-170.

Yamaguchi, M., Imai, I., Honjo, T. (1991). Effects of temperature, salinity and irradiance on the growth rates of the noxious red tide flagellates *Chattonella antiqua* and *C. marina* (Raphidophyceae). *Nippon Suisan Gakkaishi* 57: 1277-1284 (in Japanese with English abstract).

Yamamoto, C., Tanaka, Y. (1990). Two species of harmful red tide plankton increased in Fukuoka Bay. *Bull. Fukuoka Fish. exp. Stn.* No.16: 43-44 (in Japanese).

Yamochi, S. (1984). Effects of temperature on the growth of six species of red-tide flagellates occurring in Osaka Bay. *Bull. Plankton Soc. Japan* 31: 15-22 (in Japanese with English abstract).

Yanagi, T. (1989). Physical parameters of forecasting red tide in Harima-Nada, Japan. In: Okaichi, T., Anderson, D.M., Nemoto, T. (eds.) *Red tides: Biology, environmental science, and toxicology.* Elsevier, New York, p.149-152.

Yoshimatsu, S., Ono, C. (1986). The seasonal appearance of red tide organisms and flagellates in the southern Harima-Nada, Inland Sea of Seto. *Bull. Akashiwo Res. Inst. Kagawa Pref.* No.2: 1-42 (in Japanese with English abstract).

Ecophysiology and Bloom Dynamics of *Heterosigma akashiwo* (Raphidophyceae)

Theodore J. Smayda

Graduate School of Oceanography, University of Rhode Island,
Kingston RI 02881 USA

1. Introduction

In situ dynamics of *Heterosigma akashiwo*, renown for its ichthyotoxic blooms (Honjo 1994), are shaped by a combination of genetically fixed and environmentally flexible autecological parameters modified by variable synecological processes. Ecophysiology of *Heterosigma akashiwo* is evaluated based on some key aspects of its cellular, population and community behavior selected from the >300 publications available. Not all salient ecophysiological characteristics are considered because of space limitations. Honjo's (1992, 1993, 1994) important reviews should also be consulted.

2. Taxonomy

The taxonomic status of *Heterosigma akashiwo* has been confused by its close resemblance to *Olisthodiscus luteus*, first described from a brackish habitat (Carter 1937), and now considered a member of the benthic psammon. The strongly flattened *O. luteus* cell is probably an adaptation to interstitial microhabitats (Larsen and Moestrup 1989). Most pelagic blooms attributed to *O. luteus* are almost certainly those of *H. akashiwo.* Their synonymy, taxonomic confusion and controversial phylogenetic position have received considerable attention (Hulburt 1965; Loeblich and Fine 1977; Hara and Chihara 1987; Taylor 1992; Throndsen 1996). Throndsen's conclusion that the epithet *akashiwo* has priority over *carterae* is applied here. I have also treated pelagic blooms attributed to *Olisthodiscus* as *Heterosigma* blooms. Key sources of morphological and ultrastructural information on *H. akashiwo* include Hara *et al.* (1985), Hara and Chihara (1987), Leadbeater (1969), Vesk and Moestrup (1987) and Parra *et al.* (1991).

3. Cellular Characteristics

3.1 Cell size, shape, pigmentation and chloroplasts

Hetersigma akashiwo is unicellular, 10-25 μm in length, 8-15 μm in breadth, and 8-10 μm in thickness. Its cellular shape varies from spheroidal to ovoid to oblong dependent upon the conditions and stages of growth (Hara and Chihara 1987). Cell volume ranges about 3-5-fold, from *ca.* 500 μm^3 up to 1775 μm^3 (Table 1). Cellular carbon, nitrogen, protein and chlorophyll contents are within the range of the allometric regression lines found for 33 microalgal species (Montagnes *et al.* 1994). The size and shape of nanoplanktonic *H. akashiwo* cells favor a high specific absorption coefficient (m^2 mg chl a^{-1}), with a photon absorption rate of 1.95 to 3.8 x

NATO ASI Series, Vol. G 41
Physiological Ecology of Harmful Algal Blooms
Edited by D. M. Anderson, A. D. Cembella and G. M. Hallegraeff

10^{11} quanta s^{-1} $cell^{-1}$ (Watanabe and Miyazaki 1982). The main light-harvesting pigments are chlorophyll *a* + *c*, with fucoxanthin the dominant (74%) carotenoid component, and violaxanthin also important (Latasa *et al.* 1992). The ratios of chl *a*:*c* (6.6:1 w/w), chl *a*: fucoxanthin (3.3:1), and chl *a*: violoxanthin (~ 12:1) are similar to those for *Chattonella antiqua* (Latasa et al. 1992), but these ratios are unlikely to be constant. However, unlike *C. antiqua* which ceases pigment synthesis in the dark (Kohata and Watanabe 1989), there is continuous light (L) and dark (D) phase synthesis of pigments by *H. akashiwo.*

Three remarkable features characterize the strategy of photon-capture by *H. akashiwo*: a very high number of chloroplasts, high cellular chloroplast volume, and uncoupling of chloroplast replication from cell division. Photon-capture ability varies greatly among individual cells and within populations: from 4 to 95 chloroplasts per cell have been recorded (Satoh *et al.* 1977; Cattolico 1978). *Heterosigma* populations can maintain a specific mean chloroplast number per cell, but the chloroplast number shifts with changes in growth conditions. During exponential growth, from 20 to 25 chloroplasts per cell are common; lag phase populations have a lower average number (Cattolico *et al.* 1976; Cattolico 1978; Hara and Chihara 1987), and cysts have only 4 to 11 chloroplasts per cell (Imai *et al.* 1993). Chloroplast number varies with irradiance, duration of the photoperiod light phase, diurnally, and nutrient availability (Cattolico *et al.* 1976; Cattolico 1978; Satoh *et al.* 1977). Of special interest to bloom inception, at low irradiance an increase in photon flux density from *ca.* 40 to 80 $\mu E\ m^{-2}s^{-1}$ increased chloroplast numbers by 120%, which doubled when the photoperiod increased from 6 to 12h L (Cattolico 1978).

Cells of *H. akashiwo* allocate a significant portion of the total cell volume (44 to 69%) to accommodate the high chloroplast numbers (Cattolico 1978). However, the ratio of the absolute amounts of chloroplast and nuclear DNA remains relatively constant: chloroplast DNA is *ca.* 5% of total levels (Cattolico 1978). Since the total chloroplast DNA level is relatively constant, individual chloroplast DNA levels are determined by the number of chloroplasts present. Cellular chlorophyll levels range from 3.3 to 5.9 pg $cell^{-1}$ between 5 - 30°C, with maximal levels at 15°C and low irradiance (Table 1; Tomas 1980a; Montagnes *et al.* 1994). The ratios of C:chl*a* (Table 1) are similar (19:1) to those in *Chattonella antiqua* (Kohata and Watanabe 1989).

Chloroplast replication and cell division in *H. akashiwo* are uncoupled, which is unique among microalgae (Cattolico *et al.* 1976; Satoh *et al.* 1977), and apparently in response to different irradiance signals. Chloroplast replication, dependent upon onset of the dark period, begins significantly in advance of cell division, and is regulated by onset of the photoperiod L phase. The two processes overlap by only 3 h (Wada *et al.* 1987). The autecological benefits of this asynchronous replication are obscure, but would appear to favor "surge uptake of photons" via an increased chloroplast number stimulated by improved growth conditions.

3.2 Glycocalyx and mucus secretion

Copious amounts of mucus secreted by *H. akashiwo* cells surround the cysts (Imai *et al.* 1993), encapsulate its masses of non-motile cells (Tomas 1978b), and form pelagic filaments during blooms (Pratt 1966). This mucus has been implicated in fish kills by sticking to gill lamellae leading to respiratory and osmoregulatory failure (see Chang *et al.* 1990). Numerous muciferous bodies (= mucocysts) surrounding the chloroplasts are the discharge sites (Leadbeater 1969; Hara and Chihara 1987). Motile cells (Honjo 1993; Yokote *et al.* 1985), as in *Chattonella*, are also enveloped by an external mucus cover (1-2 μm) termed the glycocalyx. The glycocalyx in *Heterosigma* is composed of acidic, complex carbohydrates (probably hyaluronic acid) and a neutral protein-carbohydrate complex of glycoproteins. Its functional role in the autecology of *H. akashiwo* is unknown; in higher plants, glycoproteins are involved in cell recognition, reception of chemical information, ion-exchange and selective absorption (Honjo 1993). An affinity for colloidal iron is also claimed for the glycocalyx (Honjo 1994). Given the complex life history and marked allelochemical and allelopathic traits of *H. akashiwo*, and apparent importance of iron during its blooms, the glycolcalyx may play a role in these cellular processes.

3.3. Flagella, motility, diel migration, nutrient retrieval

Heterosigma akashiwo has two heterodynamic flagella: rapid beating of the anterior flagellum provides the motile force; a shorter, rigid flagellum trails posteriorly (Hara *et al.* 1985; Vesk and Moestrup 1987). The cell swims spirally (Hara and Chihara 1987; Imai *et al.* 1993), gyrating 1-2 times per second (Throndsen 1973), whereas *O. luteus* cells glide smoothly, without rotation (Hara *et al.* 1985). The direction of swimming is variable (Throndsen 1973), and may reflect shifts in the cellular center of gravity and altered buoyancy (Wada *et al.* 1985). Diel variations in specific gravity (r') occur, from 1.100 to >1.146 gm cm^{-3}, being maximal during the photoperiod dark phase and minimal during cell division (Watanabe 1982). The r' values (1.055) of cells accumulating in the surface layer of experimental flasks were lower than in cells remaining at the bottom (1.105) (Wada *et al.* 1985). These diel variations in r' may be linked to variations in refractive fatty particles (triacylglycerols), which exhibit a daily cycle of production and resorption (Wada *et al.* 1987).

Cells of *H. akashiwo* exhibit diel vertical migrations characterized by daytime-ascent and dark-descent (Hatano *et al.* 1983; Wada *et al.* 1985, 1987; Watanabe *et al.* 1988; MacKenzie 1991). A diel migration in the Seto Inland Sea covered a distance of 10-15 m (Watanabe *et al.* 1988). Migration was triggered by a photon flux density of *ca.* 0.3 μE m^{-2} s^{-1} at 20°C and 14L:10D photoperiod (Takahashi and Hara 1989). Upward and downward migrational speeds vary. At 20°C and 14L:10D, descent initiated 30 min prior to the D-phase was rated at 139 μm s^{-1}, and ascent initiated 2 h before L-phase equalled 10 μm s^{-1}, yielding an descent:ascent ratio of 14:1 (Hatano *et al.* 1983). These descent and ascent swimming rates correspond to the laboratory rates reported by Throndsen (1973) of 130 to 160 μm s^{-1}, and Bauerfeind *et al.* (1986) of 20 μm s^{-1}, and may explain the discrepancy in their experimental results. The mean rate of 140 μm s^{-1} at a body length of 15 μm reported by Throndsen corresponds to a swimming speed of *ca.* 10 body lengths s^{-1}. Thus, *H. akashiwo* cells can search and respond to the physical and chemical micro-structure of the habitat over an extensive spatial domain relatively rapidly in seeking favorable growth conditions, and independent of their vertical migration capability.

This "swim" strategy is exhibited by flagellates generally in contrast to the "sink" strategy exhibited by non-motile species (Smayda 1997). *Heterosigma akashiwo* is aided in this by an ability to swim through significant vertical temperature and salinity gradients; cells can cross gradients of 6.5°C and 5.7‰ (Yamochi *et al.* 1982). Thus, motility of the cells is not a series of random motions, but an important, auto-regulated behavioral aspect significant to the autecology and bloom dynamics of *H.*

akashiwo. Bioconvective accumulations leading to density inversion and "falling finger" descent of *H. akashiwo* (see Fig. 3 in Tomas 1978b) are also significant *in situ* (Watanabe and Harashima 1982; Harashima *et al.* 1985).

Nutrient status influences vertical migration in *H. akashiwo.* Nitrogen-depleted cells cease vertical migration, but regain this ability when provided with nitrogen (Hatano et al. 1983; Takahashi and Fukazawa 1982; Takahashi and Ikawa 1987; Yamochi and Abe 1984). This resumption following N re-supply can be rapid: populations N-limited for 10 days resumed migration within 45 min after N enrichment, and in about 10 hr when N-deprived for 28 days (Hatano e*t al.* 1983). This rapid recovery contrasts markedly with the sluggish adaptive ability which *H. akashiwo* exhibits in response to light and nutrient stresses. Cells exposed to a disrupted photoperiod sequence of 6L:6D took 3-5 days to re-acclimate and entrain their migrational pattern to this new light phasing (Hatano *et al.* 1983). When exposed to continuous light, this diel migrational pattern persisted for 3 days, and for 4 days in continuous darkness. During N-limitation, migration persisted for 6 days, and during $P0_4$ limitation migration persisted for 9 days. This influence of cellular nutrient status on migration of nutrient-limited *H. akashiwo* reflects an evolved behavior which may generally characterize flagellates. Their migrations into nutrient-rich subsurface layers access nutrients which are limiting within the euphotic zone (see Smayda 1997). Following dark assimilation of these nutrients, resumed upward migration into the photic layer leads to photosynthetic incorporation of the stored nutrients into biochemical pathways. Such nutrient-retrieval migrations of P-deficient *H. akashiwo*, with dark uptake capability, have been demonstrated experimentally (Watanabe *et al.* 1983, 1988), and has been proposed for N- and Fe-depleted cells (Watanabe *et al.* 1983; Yamochi *et al.* 1982).

3.4 Photosynthesis, growth, and nutrition

Heterosigma akashiwo fixes $C0_2$ via the C_3 carbon reduction pathway (Takahashi and Ikawa, 1987). d-Mannitol has been considered the principal soluble storage and excreted product of photosynthesis (Bidwell 1957; Hellebust 1963), but this is challenged by Takahashi and Ikawa (1987). The latter authors suggest that mannitol is more important in osmo-regulation than as a storage product, claiming that ß-1-3 glucan is the main storage product, with mannitol excretion only 1.5% of photosynthate, whereas Hellebust reported a 10% rate. Photosynthesis-irradiance (PI) curves show P_{max} increases with temperature between 5 to 30°C (Tomas 1980a). Langdon (1987, 1988) has quantified the photosynthesis-growth-irradiance relationships (P-G-I) of *H. akashiwo* and *Skeletonema costatum*, competitive in their overlapping seasonal and geographical distributional ranges, and compared these with *Alexandrium tamarense*, against which *H. akashiwo* also competes. At 15°C and 14L:10D photoperiod, the photosynthetic quotients ($0_2/C0_2$ = 1.43 ± 0.09) and photosynthetic efficiency (2.3 ± 0.10 x 10^{-3} µmol 0_2 (µg chl *a*)$^{-1}$h^{-1}(µE m^{-2} s^{-1})$^{-1}$ are similar. They differ considerably in other P-G-I parameters, however, with those for *H. akashiwo* being intermediate. The compensation intensity (I_c = 9 µE m^{-2} s^{-1}) of *H. akashiwo* is 4-fold lower than for *Alexandrium,* and 8-fold greater than for *Skeletonema*. The maintenance respiration rate based on chlorophyll is 9-fold higher, and 4-fold lower than for *Skeletonema* and *Alexandrium*, respectively. Based on carbon (R_{OC} = 0.47 µmol 0_2 µg^{-1}C h^{-1}), the differences are 4- and 2.5-fold, respectively. The ratio (0.13) of dark respiration to light-saturated gross photosynthesis (R_{max}:P_{max}) of *H. akashiwo* cells is close to the classical ratio of 0.10, but about 2.5-fold lower than for dinoflagellates generally, and slightly below the average (0.16) for diatoms and haptophytes (Langdon 1993). Net growth efficiency (NG_e) of *H. akashiwo* is also intermediate. The maximal growth rate of 0.88 d^{-1} for *H. akashiwo* in Langdon's experiments is considerably below the 2.0 d^{-1} reported by Tomas (1978a) and the exceptional rates ranging from 2.0 to 5.0 d^{-1} in outdoor mesocosms and 3.3 d^{-1} in laboratory incubations (Honjo and Tabata 1985). Irradiance also influences cell volume, carbon quota and chlorophyll concentration of *H. akashiwo*, effects further modified by nutrient availability (Langdon 1987; Thompson *et al.* 1991).

Heterosigma akashiwo has relatively high nutrient requirements and K_s constants for uptake (Table 1), with marked ability for luxury uptake of PO_4 and storage of polyphosphate (up to 30 mM depots) for subsequent use, such as an energy source (Watanabe *et al.* 1987, 1988, 1989) for photophosphorylation of PO_4 assimilated during nutrient-retrieval migrations. Polyphosphate storage particularly benefits the Narragansett Bay (Tomas 1979) and Osaka Bay (Watanabe *et al.* 1982) strains which are unable to synthesize alkaline phosphatase (Nakamura 1985) needed to assimilate organic P. Ability to synthesize alkaline phosphatase appears to be a variable feature, however, given evidence that the Fukuyama and Gokasho strains of *H. akashiwo* utilize glycerophosphate (Hosaka 1992; Watanabe *et al.* 1982). Nygaard and Tobiesen (1993) report that bacterivorous feeding is induced in P-limited *H. akashiwo,* whose ingestion of bacterial P can then exceed maintenance requirements and support cell division. Confirmation of this alternative mechanism of phosphorus acquisition is needed.

Large variations in cell quotas of metals (Table 1) accompany variations in luxury PO_4 uptake and polyphosphate storage, and are characterized by complex excretion/uptake kinetics which possibly maintain cellular charge balance (Watanabe *et al.* 1989). Of the essential metals, Fe and Mn are particularly implicated in *H. akashiwo* dynamics. Numerous experiments suggest natural populations frequently depend upon enrichment of Fe and, secondarily, Mn for bloom continuance (Takahashi and Fuzukawa 1982; Yamochi 1983). This response is generally consistent with its high Fe requirement relative to other flagellates (Brand 1991; Watanabe and Nakamura 1984). *Heterosigma akashiwo*, *Chattonella antiqua* and *Prorocentrum minimum* have similar K_s constants (24 - 28 $\mu g\ L^{-1}$) for Fe uptake, but when Fe-enriched the raphidophytes achieve maximal uptake rates much faster than *P. minimum.* Okaichi *et al.* (1989) attributed this to the glycocalyx thought to be involved in Fe uptake.

The inorganic nitrogen nutrition of *H. akashiwo* and its interactions with irradiance which provides reductant are disputed. Wood and Flynn (1995) challenge experimental evidence that NO_3 is the preferred inorganic N substrate, particularly at higher irradiance (Iwasaki and Sasada 1969; Chang and Page 1995). These authors believe that NH_4 is more important than NO_3, and interpret biochemical data to indicate that NO_3-reared populations are more N-stressed than when grown on other N sources. Chang and Page, in contrast, report that at low irradiance high NH_4 and urea enrichments are toxic and repress growth. *Heterosigma akashiwo* becomes N-limited at relatively high N concentrations below 20°C, but exhibits surge uptake of NH_4 which progressively increases with N-limitation (French and Smayda 1995).

Table 1. Some morphological and physiological properties of *Heterosigma akashiwo*. DNA, chloroplast number and chlorophyll are per cell; V_{max} as pmol cell^{-1} min^{-1}, except for Fe given as fg cell^{-1} h^{-1}; I_c and I_{sat} are compensation and saturation intensities, respectively; PQ = photosynthetic quotient.

CELL CHARACTERISTICS		***SOURCE***
Cell Size	491 - 1763 μm^3	(12)
DNA	2.2 - 2.6 ng	(3)
RNA:DNA	4.7 - 6.6	(2)
Chloroplasts	4 - 95	(4)
Chlorophyll	3.3 - 5.9 pg	(16)

CELL QUOTAS (q_0)		
C	119 - 429 pg cell^{-1}	(12)
N	24 "	(6)
P	1.82 "	(6)
Fe	5.37 "	(19)
Zn	37.9 fg cell^{-1}	(19)
Mn	12.64 "	(19)
Co	0.94 "	(19)
B_{12}	0.09 "	(10)
B_{12}	24 molecules μm^{-3}	(10)
C:N	5.1 - 7.5:1 (log)	(15)
C:N	7.1 - 9:1 (stationary)	(15)
N:P	11 - 25:1	(15)
C:Chl	19 - 41:1	(7,15)

NUTRIENT UPTAKE		***SOURCE***
NH_4 -	K_S = 1.99 - 2.45 μM	(15)
	V_{max} = 0.34 - 0.41	(15)
PO_4 -	K_S = 1.0 - 1.98 μM	(15)
	V_{max} = 0.38 - 0.72	(15)
Fe -	K_S = 23.5 μg L^{-1}	(11)
	V_{max} = 1.4 fg	(11)
B_{12} -	K_S = 0.37 - 0.60 ng L^{-1}	(10)

PHOTOSYNTHESIS - GROWTH		
I_c	9 μE m^{-2} s^{-1}	(8)
I_{sat}	0.28 ly min^{-1}	(16)
PQ	1.43	(8)
μ_{max}	2.0 - ~ 5.0 d^{-1}	(5)
Resp:Ps	0.13	(9)

DIVERSE PROPERTIES		
°C optimum	20 °C	(14,18)
range:	< 2 - > 30 °C	(14,20)
S‰ optimum	15-25‰	(14,18)
range:	~ 1 - >50‰	(14,18)
Motility : min	20 μm s^{-1}	(1)
max	1250 "	(17)
other:	130 - 150 "	(13)

Sources: 1. Bauerfeind *et al.* 1986; 2. Berdalet *et al.* 1992; 3. Cattolico 1978; 4. Cattolico *et al.* 1976; 5. Honjo and Tabata 1985; 6. Hosaka 1992; 7. Kohata and Watanabe,1989; 8. Langdon 1987; 9. Langdon 1993; 10. Nishijima and Hata,1984; 11. Okaichi *et al.* 1989; 12. Thompson *et al.* 1991; 13. Throndsen 1973; 14. Tomas 1978a; 15. Tomas 1979; 16. Tomas 1980a; 17. Watanabe and Harashima 1986; 18. Watanabe *et al.* 1982; 19. Watanabe *et al.* 1989; 20. Yamochi 1989.

Chang and Page suggest that *H. akashiwo* achieves maximum growth *in situ* when well-illuminated and with NO_3 as the major N-source. This may partly account for its bloom-events during post-upwelling relaxations, but NH_4 would appear to be the more abundant inorganic N species in nutrient-enriched habitats and at fish-farming sites where *Heterosigma* commonly blooms. Based on the frequent association of *H. akashiwo* blooms with nutrient-enriched watermasses and laboratory experiments, Iwasaki (1973, 1979) suggested that organic growth factors such as purines and pyrimidines stimulate blooms. Such capability remains unresolved (see Mahoney and McLaughlin 1977); in fact, there is considerable variation in the ability of geographically isolated populations of *H. akashiwo* to utilize even the simple compound of urea (see Watanabe *et al.* 1982). This variable nutritional capacity among strains and their different responses to temperature and salinity suggest that *H.*

akashiwo comprises at least three different physiological races (see Watanabe *et al.* 1982).

3.5 Temperature and salinity requirements; growth rate

Heterosigma akashiwo is euryhaline and eurythermal. Cells tolerate a salinity range of 2 to >50‰ (Tomas 1978a), whereas clonal differences exist in response to lower salinities (Hosaka 1992; Honjo 1992). Salinity can select for blooms of *H. akashiwo* over more stenohaline flagellates (Mahoney and McLaughlin 1979). The optimal salinity for growth of a Narragansett Bay clone varied with temperature (Tomas 1978a): 5 - 50‰ at 5°C, and 10 - 20‰ at 20°C. Salinity also influences motility and survival of the resting stage (Tomas 1978a), and vegetative cells are very sensitive to Ca and K concentrations and Ca:Mg ratio (McLachlan 1964). Temperature tolerance, in culture, ranges from <5° to >30°C (Tomas 1978; Yamochi 1989), with mortality occurring at 33°C (Yamochi 1984a). Thermal strains also occur (Watanabe *et al.* 1982); an optimal growth-temperature of 15 - 25°C generally characterizes *H. akashiwo*. For a Narragansett Bay clone, growth rate significantly increased above 10°C - *ca.* 3-fold between 10° and 20°C (Fig.4 in Tomas 1978a). Natural populations of *H. akashiwo* occur over a wide temperature range. At the extremes, it has been found at *ca.* 3° to 4°C in Narragansett Bay and Norwegian coastal waters, (Tomas 1980b; Throndsen 1969), at 1° to 3°C in Kamchatka embayments (Konovalova 1995), with blooms occurring at 30°C in Hakata Bay (Honjo 1974). There is a remarkably persistent relationship between temperature and blooms of *H. akashiwo* throughout its distributional range (Fig. 1). Temperature must be at least 15°C, with blooms clustering within the narrow range between 15° - 20°C. This temperature-bloom relationship has been reported for upwelling habitats and coastal lagoons along the Iberian Peninsula (Figueras and Rios 1993; Pazos *et al.* 1995; Sampayo and Moita 1984), and in Oslofjord (Throndsen 1976), Masan Bay, Korea (Park *et al.* 1989), and the coastal waters of British Columbia (Taylor and Haigh 1993; Taylor *et al.* 1994), New Zealand (Chang *et al.* 1990; Rhodes *et al.* 1993), Chile (Clement and Lembeye 1993) and Peru (Rojas de Mendiola 1979). In Narragansett Bay, a 17-year trend showed a bimodal bloom cycle linked to temperature. The annual bloom begins soon after 15°C is reached, with a first maximum occurring at *ca.* 20°C, followed by a lull at 22°C, and then a second maximum when temperature drops to between 20° and 15°C (Tomas 1980b). In certain Japanese coastal waters, *Heterosigma* blooms diverge from this bloom-temperature relationship. They usually begin at >18°C and often occur between *ca.* 22° - <31°C (Honjo 1974; Yamochi 1983, 1989). A bloom occurred at >25°C in Kastela Bay, Adriatic Sea (Marasovic and Pucher-Petkovic 1985). Temperature is clearly a major factor in the growth and seasonal bloom cycle of *H. akashiwo* (Fig. 2).

4. Life cycle

Meroplanktonic *H. akashiwo* has two distinct benthic overwintering stages: non-motile, vegetative cells agglutinated into plasmodial masses of large numbers encapsulated by mucilage (Tomas 1978b) and bonafide resting cysts (Imai *et al.* 1993). The Osaka Bay population survived winter conditions as both vegetative and benthic stages (Yamochi and Joh 1986). The cysts are smaller and have fewer chloroplasts than plasmodial and flagellated cells. Similar to *Chattonella, Heterosigma* cysts tolerate low sedimentary oxygen and high sulfide levels (Montani *et al.* 1995). Seed banks of up to *ca.* 3 x 10^4 cysts g^{-1} of wet sediment have been found (Imai and Itakura 1991; Yamochi 1989), with the cysts capable of dark survival for at least eight months at 11°C (Imai *et al.* 1996). Cyst densities of *Heterosigma* are often *ca.* 100-fold lower than those of *S. costatum* within the same sediments, and against whose excystment of potential bloom inocula *Heterosigma* must often compete. Unlike *Skeletonema* (Imai *et al.* 1996), cysts of *H. akashiwo* germinate in the dark at 2° to 25°C (Yamochi 1984b), with successful germination rates suddenly increasing to 85 - 100% above 14°C. This may favor *H. akashiwo* initially, but subsequent irradiance characteristics at the sediment-water interface determine whether this seeding by excysted cells will favor bloom initiation, or that of *S. costatum* which has the advantage of greater seed bed populations and faster growth rates. The plasmodial benthic phase releases motile cells between 10 - 30°C (Tomas 1978b). Their sudden releases (= swarming) may be an important alternative mechanism by which *H. akashiwo* emerges from a resting phase to initiate blooms.

5. Distribution

Figure 2 updates Honjo's (1992) distributional map of *H. akashiwo;* it includes unpublished observations of occurrence in Australian, Namibian and Southwest African waters, in Florida Bay and Tampa Bay, U.S. Isolated occurrences reported for the tropical/subtropical waters of Taiwan (Shen and Chiang 1971), Singapore (Taylor 1990) and Bermuda (Tomas 1980b) need confirmation. The distributional pattern in both hemispheres suggests a temperate zone biogeography, with three main centers of occurrence in the northern hemisphere: in U.S. coastal waters north of 40°N; European coastal waters between *ca.* 40° (Portugal) and 55°N (Oslofjord), and in the Pacific region from ca. 55° to 30°N extending from the Kamchatka Peninsula (Konovalova 1995) throughout Japanese (Honjo 1993), South Korean (Park 1991) and Chinese coastal waters (Qi *et al.* 1993). In the southern hemisphere, *H. akashiwo* has bloomed along the Peruvian coast (Rojas de Mendiola 1979); between ca. 35° to 45°S in Chile (Clement and Lembeye 1993); in the coastal waters of Namibia and Southwest Africa, and in Australia and New Zealand (Chang *et al.* 1990; Rhodes *et al.* 1993). Subtropical and tropical habitats would not appear to be hostile, given frequent blooms of *H. akashiwo* at 20° to 25°C and growth, in culture, even at 30°C (Yamochi 1984a, 1989). Temperature may play a restrictive role in limiting distribution within subtropical and tropical habitats, however, if the seasonal temperature minimum lies above that needed in "over-wintering" quiescence. For temperate populations, the required minimum appears to be <10°C (Yamochi and Joh 1986; Itakura *et al.* 1996). The occurrences of *H. akashiwo* in a lake in Tennessee, U.S. (see Throndsen 1969) and saline lake in Kamchatka (Konovalova 1995) are unique (Fig. 1).

Hetersigma akashiwo has *three* distinct regional populations which bloom in habitats characterized by one of three different primary features: physically, chemically and biologically dominated habitats. These ecotypes contrast with those in "normal" habitats defined as those relatively pristine in nutrient enrichment, and in which upwelling and aquacultural activities are absent, e.g., Narragansett Bay (Pratt 1966) and Sechelt Inlet, British Columbia (Taylor *et al.* 1994). Neither locality is nutrient-enriched; Narragansett Bay is well-mixed year-round and exhibits tight benthic-pelagic coupling; Sechelt Inlet is influenced by Fraser River discharge. Contrary to claims by Taylor *et al.* (1994), stratification is neither essential for *H. akashiwo* to bloom, nor is the Sechelt Inlet bloom-site generally representative of its bloom habitats.

Physically dominated systems in which *Heterosigma* blooms occur are regions of intermittent upwelling: the Peruvian system (Rojas de Mendiola 1979), Spanish Galician rias (Pazos *et al.* 1995; Tilstone *et al.* 1994), Chloe Archipelago, Chile (Clement and Lembeye 1993) and, presumably, in Namibian coastal waters (Fig. 1). While these episodic blooms follow post-upwelling relaxations, the indigenous populations must be adapted to habitat turbulence. Field and experimental evidence reveals that *H. akashiwo* has a greater tolerance for turbulence than dinoflagellates (Berdalet and Estrada 1993; Pazos *et al.* 1995). Climatological disturbances during an El Niño event induced pronounced upwelling - relaxation sequences and associated wind, rain and runoff events, leading to dramatic blooms of H. akashiwo in New Zealand coastal waters (Rhodes *et al.* 1993; Chang *et al.* 1995). This, too, is a physically stimulated bloom response.

Chemically dominated habitats also support significant *H. akashiwo* blooms. Precipitous blooms in nutrient-enriched habitats are often the first recordings of the presence of this species in those regions: inner Oslofjord (Braarud 1969), Kastela Bay [Adriatic Sea] (Marasovic and Pucher-Petkovic 1985), inner New York Bight (Mahoney and McLaughlin 1977). This species also blooms in nutrient-enriched Tokyo Bay (Han *et al.* 1989), sub-regions of the Seto Inland Sea (Hada 1974), Hakata Bay (Honjo 1974) and Cascais Bay, Portugal (Sampayo and Moita 1984). As in Masan Bay, Korea (Park *et al.* 1989), a buildup of domestic/industrial waste-nutrients preceded its bloom occurrences. Experiments show the need to distinguish between macro- (N, P) and micro-nutrient (Fe, other trace metals) effects in stimulating *H. akashiwo* blooms in nutrient-enriched habitats. Bloom continuance, or triggering requires Fe and/or Mn enrichment (Honjo 1974; Takahashi and Fukazawa 1982; Yamochi 1983, 1989). An apparent dependency of *H. akashiwo* bloom events on river runoff in various regions (Honjo, 1974; MacKenzie 1991; Taylor and Haigh 1993) has been linked to Fe delivery. This was considered to be "crucial" to blooms in Osaka Bay (Yamochi 1989). A similar dependence on Fe has been reported for the related ichthyotoxic raphidophyte, *Chattonella* (Okaichi *et al.* 1989). Vigorous growth of *H. akashiwo* following exposure to sediment extracts has also been cited as evidence that micro-nutrients and complexation ligands are significant bloom-factors (Honjo 1974). Macro-nutrients might therefore primarily determine bloom magnitude and frequency, rather than limit *H. akashiwo* nutritionally. For example, uncoupling of bloom dependency on macro-nutrients would be facilitated by nutrient-retrieval migrations. However, based on cell quota (Table 1) and residual *in situ* nutrient levels, Watanabe *et al.* (1988) concluded P and N could limit *H. akashiwo* blooms in some regions of the Seto Inland Sea. P-limitation may also occur in inner Tokyo Bay (Hosaka 1992).

Biologically dominated habitats stimulatory to *H. akashiwo* blooms refer specifically to aquacultural sites, particularly salmonid fish-farming sites. The presence of this species was often revealed only after initiation of fish-farming which seemingly triggers fish-killing blooms, e.g., salmonid farming in British Columbia (Taylor and Haigh 1993), Chile (Clement and Lembeye 1993), New Zealand (Chang *et al.* 1990; MacKenzie 1991), Scotland (Ayres *et al.* 1982), Spain (see Chang *et al.* 1990), and United States (Taylor and Haigh 1993); yellowtail and sea bream culture in Japan (Honjo 1993); and milkfish culture in Taiwan (Shen and Chiang 1971). *Hetersigma akashiwo* blooms began in 1981 in South Korean waters after initiation of large-scale shellfish aquaculture (Park *et al.* 1989). Since the farmed-fish are pen-fed, altered grazing on the local food-web favoring *H. akashiwo* blooms seems

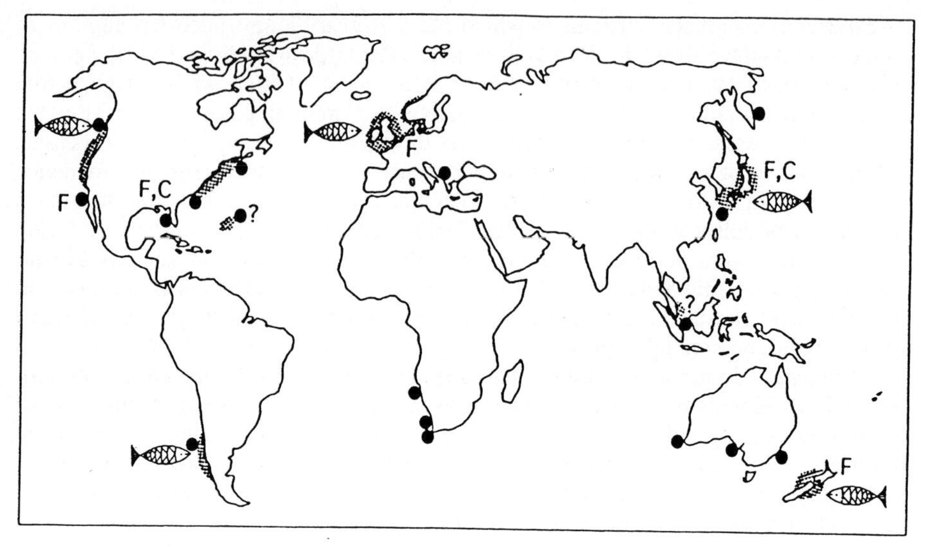

Figure 1. Distribution of *Heterosigma akashiwo*, including unpublished observations of Hallegraeff, Pitcher and Tomas. Fish icons indicate sites where farmed-fish kills have occurred during *Heterosigma akashiwo* blooms (modified from Honjo 1992). Symbols C and F indicate *Chattonella* spp. and *Fibrocapsa japonica*, respectively, also occur.

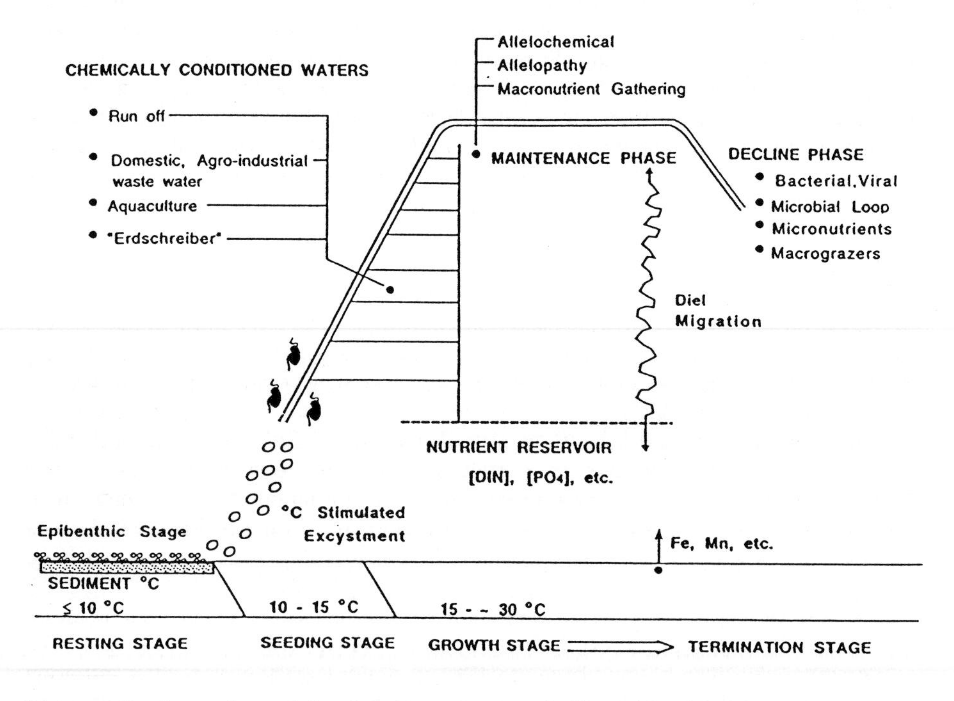

Figure 2. General model of key processes influencing bloom outbreaks of *Heterosigma akashiwo*; see text for details.

unlikely, whereas this may be a factor at shellfish culture sites. Chemical modification of the local habitat resulting from nutrient excretion and decomposition of undigested prey fed to the penned fish seems a more likely stimulus for these *Heterosigma* blooms. Nutrients may not significantly exceed pre-aquacultural levels, but is there some component of fish excreta favorable to *H. akashiwo* blooms? Growth of *H. akashiwo* is stimulated by several organic growth factors (Iwasaki 1979) possibly present in fish excreta. Decomposing oyster faeces also has been reported to stimulate growth of *H. akashiwo* (Iwasaki 1973).

The distributional features and the different ecotypic populations of *H. akashiwo* reveal that it has a very wide niche: it blooms 1) both in nutrient-enriched and at moderate nutrient levels; 2) in highly stratified and moderately mixed watermasses; 3) in habitats heavily influenced by runoff; 4) in regions of relatively constant high salinity; 5).in habitats of naturally established trophodynamics, and 6) in habitats in which the natural grazing and associated nutrient turnover have become unbalanced by fish-farming and shellfish aquacultural activities. This apparent occurrence of regional ecotypes of *H. akashiwo* isolated into discrete, yet distinctive habitats may reflect an evolving speciation consistent with the three physiological and ecologically different strains of *H. akashiwo* distinguished by Watanabe *et al.* (1982).

A noteworthy feature of the increased prominence of *H. akashiwo* blooms is the frequent co-occurrence (Fig. 1) with other raphidophytes: *Fibrocapsa japonica* and one or more *Chattonella* spp. All are strongly ichthyotoxic and share numerous ecophysiological features. A raphidophyte niche may be increasingly opening up, particularly in chemically modified habitats and at fish-farming sites, and possibly in "competition" with the dinoflagellate life-form niche. Thus, all three taxa co-occur in Japanese, Korean and Chinese coastal waters, and in Tampa Bay and Florida Bay, U.S. (Tomas, pers. comm.). *Fibrocapsa* and *Chattonella*, along with a problematic occurrence of *H. akashiwo*, have recently been reported from the Dutch Wadden Sea (Vrieling *et al.* 1995), with *Fibrocapsa* also reported in the contiguous waters extending from Germany to France. *Heterosigma* and *Fibrocapsa* co-occur in San Diego Bay (Lackey and Clendenning 1965), as found recently in New Zealand coastal waters (Rhodes *et al.* 1993) and Narragansett Bay (pers. obs.).

6. Bloom model

Figure 2 presents a general model of bloom formation by *H. akashiwo* based on the ecophysiological data considered here. The relative importance of these factors undoubtedly varies among specific bloom sites. This aspect and the model itself will be discussed in greater detail elsewhere. Space constraints allow for only a summary here. *H. akashiwo* must overcome three basic impediments for bloom formation: a temperature restraint, a chemical restraint, and interspecific competition. Since *H. akashiwo* is meroplanktonic, its dormant stages must provide the seed population for bloom initiation. Temperature is the key factor at this stage of the bloom; it must exceed 10°C for excystment and viable cell release from its epibenthic plasmodial and resting cyst stages. However, successful release does not assure bloom initiation. This partly depends upon the magnitude of release relative to inocula of other species also excysting. For example, at this stage of the bloom cycle, competition between *Heterosigma* and *Skeletonema costatum*, its primary competitor throughout much of its distributional range, is light-regulated. Darkened sediments during the seeding stage favor *Heterosigma*; illumination favors germination of *Skeletonema*. Upon exposure of the liberated *Heterosigma* cells to water column irradiance, significant increases in cellular chloroplast numbers occur favoring their photosynthesis and population growth. This growth is likewise temperature-dependent; the habitat must be minimally 15°C for growth to be vigorous. If the environment meets these temperature criteria, a *chemical* obstacle seemingly must then be overcome for progression of *H. akashiwo* to the next stage of its bloom cycle. At the cellular level, such a restraint is suggested by its low nutrient uptake affinity (high K_S), apparent high micronutrient (Fe) requirement, and possible dependence on growth promoters. At the population level, the frequent association of *Heterosigma* blooms in watermasses chemically conditioned by river runoff, domestic and agro-industrial waste water, fish-farming sites (excreta), and growth stimulation during *in situ* admixture of sediments with overlying waters (= *Erdschreiber* solutions) also suggest that a chemical barrier must be overcome. Given suitable chemical conditioning of the habitat, interspecific competition then becomes the key factor in bloom formation. Of the three basic requirements that *H. akashiwo* must meet to bloom, this capability would appear to be intrinsically more attainable, and likely, than overcoming the temperature and chemical barriers to cellular and population growth. The marked allelochemical (Pratt 1966; Honjo *et al.* 1978; Craigie and McLachlan 1964) and allelopathic (see Tomas and Deason 1981; Uye and Takamatsu 1990) effectiveness of *H. akashiwo* against competing species and grazers is probably the best known and broadest based among phytoplankton. During the period of waning macronutrient levels, nutrient retrieval migrations probably facilitate bloom formation, but it is unresolved whether nutrient limitation is the leading cause of bloom termination. There is much better evidence that bacterial and viral infections (Imai *et al.* 1995; Nagasaki *et al.* 1994a,b) commonly accompany declining *H. akashiwo* populations. The broad spectrum allelopathy of *H. akashiwo* against microbial loop and macrograzer components suggests that their predation is not a significant factor in bloom-termination.

Bloom regulation of *H. akashiwo* is, thus, under multifactorial control, occurring as a series of habitat impediments that must be sequentially overcome in order to bloom. This probably applies generally to phytoplankton blooms. However, although *H. akashiwo* shares many ecophysiological traits in common with harmful algal species generally (Smayda, 1997), the remarkably high degree of broad spectrum allelochemical and allelopathic antagonisms exhibited against competing species and potential grazers distinguishes this species from others. The causes of unpredictable blooms and interannual variations within given areas are obscure, but may be related to temperature and a chemically ("water quality", nutrient) regulated selection for the bloom species and population growth. If these favor *H. akashiwo*, then its allelochemical and allelopathic capacities significantly enhance the likelihood that it will bloom. This distinguishes *H. akashiwo* from other harmful species whose blooms appear to be correlated more with increased nutrient levels, seasonal reduction in turbulence, frontal system development, or periodic seedings from coastal currents (Smayda and Reynolds, *in review*).

7. References

Ayres, P.A., Seaton, D.D., Tett, P.B. (1982). Plankton blooms of economic importance to fisheries in UK waters 1968-1982. ICES CM 1982/L:38, 25 p.

Bauerfeind, M., Elbrächter, M., Steiner, R., Throndsen, J. (1986). Application of laser doppler spectroscopy (LDS) in determining swimming velocities of motile phytoplankton. *Mar. Biol.* 93: 323-327.

Berdalet, M., Estrada, M. (1993). Effects of turbulence on several dinoflagellate species. *In*: Smayda, T.J., Shimizu, Y. (eds.) *Toxic Phytoplankton Blooms In The Sea.* Elsevier, Amsterdam, pp. 737-740.

Berdalet, M., Latasa, M., Estrada, M. (1992). Variations in biochemical parameters of *Heterocapsa* sp. and *Olisthodiscus luteus* grown in 12:12 light:dark cycles. I. Cell cycle and nucleic acid composition. *Hydrobiologia* 238: 139-147.

Bidwell, R.G.S. (1957). Photosynthesis and metabolism of marine algae. I. Photosynthesis of two marine flagellates compared with *Chlorella. Can. J. Bot.* 35: 945-950.

Brand, L. (1991). Minimum iron requirements of marine phytoplankton and the implications for biogeochemical control of new production. *Limnol. Oceanogr.* 36: 1756-1771.

Braarud, T. (1969). Pollution effect upon the phytoplankton of Oslofjord. ICES CM 1969/L:15, 23 p.

Carter, N. (1937). New or interesting algae from brackish water. *Arch. Protistenk.* 90: 1-68.

Cattolico, R.A. (1978). Variation in plastid number. Effect on chloroplast and nuclear deoxyribonucleic acid complement in the unicellular alga *Olisthodiscus luteus. Plant Physiol.* 62: 558-562.

Cattolico, R.A., Boothroyd, J.C., Gibbs, S.P. (1976). Synchronous growth and plastid replication in the naturally wall-less alga *Olisthodiscus luteus. Plant Physiol.* 57: 497-503.

Chang, F.H., Page, M. (1995). Influence of light and three nitrogen sources on growth of *Heterosigma carterae* (Raphidophyceae). *N.Z. J. Mar. Freshw. Res.* 29: 299-304.

Chang, F.H., Anderson, C., Boustead, N.C. (1990). First record of *Heterosigma* (Raphidophycean) bloom with associated mortality of cage-reared salmon in Big Glory Bay, New Zealand. *N.Z. J. Mar. Freshw. Res.* 24: 461-469.

Chang, F.H., MacKenzie, L., Till, D. Hannah, D., Rhodes, L. (1995). The first toxic shellfish outbreaks and the associated phytoplankton blooms in early 1993 in New Zealand. *In*: Lassus, P., Arzul, G., Erard-LeDenn, E., Gentien, P., Marcaillou-LeBaut, C. (eds.) *Harmful Marine Algal Blooms.* Lavoisier, Paris, pp. 145-150.

Clement, A., Lembeye, G. (1993). Phytoplankton monitoring program in the fish farming region of South Chile. *In*: Smayda, T.J., Shimizu, Y. (eds.) *Toxic Phytoplankton Blooms In The Sea.* Elsevier, Amsterdam, pp. 223-228.

Craigie, J.S., McLachlan, J. (1964). Excretion of colored ultraviolet-absorbing substances by marine algae. *Can. J. Bot.* 42: 23-33.

Figueiras, F.G., Rios, A. F. (1993). Phytoplankton succession, red tides and the hydrographic regime in the Rias Bajas of Galicia. *In*: Smayda, T.J., Shimizu, Y. (eds.) *Toxic Phytoplankton Blooms In The Sea.* Elsevier, Amsterdam, pp. 239-244.

French, D.P., Smayda, T.J. (1995). Temperature regulated responses of nitrogen limited *Heterosigma akashiwo,* with relevance to its blooms. *In*: Lassus, P., Arzul, G., Erard-LeDenn, E., Gentien, P., Marcaillou-LeBaut, C. (eds.) *Harmful Marine Algal Blooms.* Lavoisier, Paris, pp. 585-590.

Han, M.-S., Furuya, K., Nemoto, T. (1989). Species-specific photosynthesis of red tide phytoplankton in Tokyo Bay. *In*: Okaichi, T., Anderson, D.M., Nemoto, T. (eds.) *Red tides: Biology, Environmental Science and Toxicology.* Elsevier, Amsterdam, pp. 213-216.

Hada, Y. (1974). The flagellata examined from polluted water of the Inland Sea, Setonaikai. *Bull. Plankton Soc. Jpn.* 20: 112-125.

Hara, Y., Chihara, M. 1987. Morphology, ultrastructure, and taxonomy of the Raphidophycean alga *Heterosigma akashiwo. Bot. Mag. Tokyo* 100: 151-163.

Hara, Y., Inouye, I., Chihara, M. (1985). Morphology and ultrastructure of *Olisthodiscus luteus* (Raphidophyceae) with special reference to the taxonomy. *Bot. Mag. Tokyo* 98: 251-262.

Harashima, A., Watanabe, M., Fujishiro, I. (1985). A numerical experiment on the bioconvection in the culture of a flagellate. *Res. Rep. Natl. Inst. Environ. Study* 80: 103-118.

Hatano, S., Hara, Y., Takahashi, M. (1983). Preliminary study on the effects of photoperiod and nutrients on the vertical migratory behavior of a red tide flagellate, *Heterosigma akashiwo. Jap. J. Phycol.* 31: 263-269.

Hellebust, J. (1965). Excretion of some organic compounds by marine phytoplankton. *Limnol. Oceanogr.* 10: 192-206.

Honjo, T. (1974). Studies on the mechanisms of red tide occurrence in Hakata Bay. IV. Environmental conditions during the blooming season and essential factors of red tide occurrence. *Bull. Tokai Reg. Fish. Res. Lab.* 79: 77-121.

Honjo, T. (1992). Harmful red tides of *Heterosigma akashiwo. NOAA Tech. Rept. NMFS* 111: 27-32.

Honjo, T. (1993). Overview on bloom dynamics and physiological ecology of *Heterosigma akashiwo. In*: Smayda, T.J., Shimizu, Y. (eds.) *Toxic Phytoplankton Blooms In The Sea.* Elsevier, Amsterdam, pp. 33-41.

Honjo, T. (1994). The biology and prediction of representative red tides associated with fish kills in Japan. *Rev. Fish. Sci.* 2: 225-253.

Honjo, T., Shinmouse, T., Ueda, N., Hanaoka, T. (1978). Changes of phytoplankton composition and its characteristics during red tide season. *Bull. Plankton Soc. Jpn.* 25: 13-19.

Honjo, T., Tabata, K. (1985). Growth dynamics of *Olisthodiscus luteus i*n outdoor tanks with flowing coastal water and in small vessels. *Limnol. Oceanogr.* 30: 653-664.

Hosaka, M. (1992). Growth characteristics of a strain of *Heterosigma akashiwo* (Hada) Hada isolated from Tokyo Bay, Japan. *Bull. Plankton Soc. Jpn.* 39: 49-58.

Hulburt, E.M. (1965). Flagellates from brackish waters in the vicinity of Woods Hole, Massachusetts. *J. Phycol.* 1: 87-94.

Imai, I., Itakura, S. (1991). Densities of dormant cells of the red tide flagellate *Heterosigma akashiwo* (Raphidophyceae) in bottom sediments of northern Hiroshima Bay, Japan. *Bull. Jap. Soc. Microb. Ecol.* 6: 1-7.

Imai, I., Itakura, S., Itoh, K. (1993). Cysts of the red tide flagellate *Heterosigma akashiwo*, Raphidophyceae, found in bottom sediments of northern Hiroshima Bay, Japan. *Nippon Suisan Gakkaishi* 59: 1669-1673.

Imai, I., Itakura, S., Yamaguchi, M., Honjo, T. (1996). Selective germination of *Heterosigma akashiwo* (Raphidophyceae) cysts in bottom sediments under low light conditions: A possible mechanism of the red tide initiation. *In*: Yasumoto, T., Oshima,Y., Fukuyo, Y. (eds.) *Harmful and Toxic Algal Blooms*. Intergovernmental Oceanographic Commission of UNESCO 1996, Paris, pp. 197-200.

Imai, I., Ishida, Y., Sakaguchi, K., Hata, Y. (1995). Algicidal marine bacteria isolated from northern Hiroshima Bay, Japan. *Fish. Sci.* 61: 628-636.

Itakura, S., Nagasaki, K., Yamaguchi, M., Imai, I. (1996). Cyst formation in the red tide flagellate *Heterosigma akashiwo* (Raphidophyceae). *J. Plankton Res.* 18: 1975-1979.

Iwasaki, H. (1973). The physiological characteristics of neritic red tide flagellates. Bay. Bull. *Plankton Soc. Jpn.* 19: 46-56.

Iwasaki, H. (1979). Physiological ecology of red tide flagellates. *In*: Levandowsky, M., Hutner, S.H. (eds.) *Biochemistry and Physiology of Protozoa.* Vol. 1, 2nd Edition, Academic Press, NY, pp. 357-393.

Iwasaki, H., Sasada, K. (1969). Studies on the red tide dinoflagellates. II. On *Heterosigma inlandica* appeared in Gokasho Bay, Shima Peninsula. *Bull. Jpn. Soc. Scient. Fish.* 35: 943-947.

Kohata, K., Watanabe, M. (1989). Diel changes in composition of photosynthetic pigments in *Chattonella antiqua* and *Heterosigma akashiwo* (Raphidophyceae). *In*: Okaichi, T., Anderson, D.M., Nemoto, T. (eds.) *Red tides: Biology, Environmental Science and Toxicology.* Elsevier, Amsterdam, pp. 329-332.

Konovalova, G. (1995). The dominant and potentially dangerous species of phytoflagellates in the coastal waters of East Kamchatka. *In*: Lassus, P., Arzul, G., Erard-LeDenn, E., Gentien, P., Marcaillou-LeBaut, C. (eds.) *Harmful Marine Algal Blooms.* Lavoisier, Paris, pp. 169-174.

Lackey, J .B., Clendenning, K.A. (1965). Ecology of the microbiota of San Diego Bay, California. *Trans. San Diego Soc. Nat. Hist.* 14: 9-40.

Langdon, C. (1987). On the causes of interspecific differences in the growth-irradiance relationship for phytoplankton. Part 1. A comparative study of the growth-irradiance relationship of three marine phytoplankton species: *Skeletonema costatum, Olisthodiscus luteus* and *Gonyaulax tamarensis. J. Plankton Res.* 9: 459-482.

Langdon, C. (1988). On the causes of interspecific differences in the growth-irradiance relationship for phytoplankton. Part 2. A general review. *J. Plankton Res.* 10: 1291-1312.

Langdon, C. (1993). The significance of respiration in production measurements based on oxygen. *ICES mar. Sci. Symp.* 197: 69-78.

Larsen, J., Moestrup, Ø. (1989). *Guide to Toxic and Potentially Toxic Marine Algae.* Fish Inspection Svc., Minist. Fisheries, Copenhagen.

Latasa, M., Berdalet, E., Estrada, M. (1992). Variations in biochemical parameters of *Heterocapsa* sp. and *Olisthodiscus luteus* grown in 12:12 light:dark cycles. II. Changes in pigment composition. *Hydrobiologia* 238: 149-157.

Leadbeater, B.S.C. (1969). A fine structural study of *Olisthodiscus luteus* Carter. *Br. Phycol. J.* 4: 3-17.

Loeblich III, A.R., Fine, K.E. (1977). Marine chloromonads more widely distributed in neritic environments than previously thought. *Proc. Biol. Soc. Wash.* 90: 388-399.

MacKenzie, L. (1991). Toxic and noxious phytoplankton in Big Glory Bay, Stewart Island, New Zealand. *J. Appl. Phycol.* 3: 19-34.

Mahoney, J.B., McLaughlin, J.J.A. (1977). The association of phytoflagellate blooms in lower New York Bay with hypereutrophication. *J. expl. mar. Biol. Ecol.* 28: 53-65.

Mahoney, J.B., McLaughlin, J.J.A. (1979). Salinity influence on the ecology of phytoflagellate blooms in lower New York Bay and adjacent waters. *J. expl. mar. Biol. Ecol.* 37: 213-223.

Marasovic, I., Pucher-Petkovic, T. (1985). Effects of eutrophication on the structure of the coastal phytoplankton community. *Rapp. Comm. int. Mer Médit.* 29: 137-139.

McLachlan, J. (1964). Some considerations of the growth of marine algae in artificial media. *Can. J. Microb.* 10: 769-782.

Montagnes, D.J., Berges, J.A., Harrison, P.J., Taylor, F.J.R. (1994). Estimating carbon, nitrogen, protein and chlorophyll *a* from volume in marine phytoplankton. *Limnol. Oceanogr.* 39: 1044-1060.

Montani, S., Ichimi, K., Meksumpun, S., Okaichi, T. (1995). The effects of dissolved oxygen and sulfide on germination of the cysts of different phytoflagellates. *In*: Lassus, P., Arzul, G., Erard-LeDenn, E., Gentien, P., Marcaillou-LeBaut, C. (eds.) *Harmful Marine Algal Blooms.* Lavoisier, Paris, pp. 627-632.

Nagasaki, K., Ando, M., Imai, I., Itakura, S. Ishida, Y. (1994a). Virus-like particles in *Heterosigma akashiwo* (Raphidophyceae): a possible red tide disintegration mechanism. *Mar. Biol.* 119: 307-312.

Nagasaki,K, Ando, M., Itakura, S., Imai, I., Ishida, Y. (1994b). Viral mortality in the final stage of *Heterosigma akashiwo* (Raphidophyceae) red tide. *J. Plankton Res.* 16: 1595-1599.

Nakamura, Y. (1985). Alkaline phosphatase activity of *Chattonella antiqua* and *Heterosigma akashiwo*. *Res. Rep. Natl. Inst. Environ. Study, Jpn.* 80: 67-72.

Nishijima,T., Hata, Y. (1984). Physiological ecology of *Heterosigma akashiwo* Hada on B group vitamin requirements. *Bull. Jpn. Soc. Scient. Fish.* 50: 1505-1510.

Nygaard, K., Tobiesen, A. (1993). Bacterivory in algae: A survival strategy during nutrient limitation. *Limnol. Oceanogr.* 38: 273-279.

Okaichi, T., Montani, S. Hiragi, J., Hasui, A. (1989). The role of iron in the outbreaks of *Chattonella* red tide. *In*: Okaichi, T., Anderson, D.M., Nemoto, T. (eds.) *Red tides: Biology, Environmental Science and Toxicology.* Elsevier, Amsterdam, pp. 353-356.

Park, J.S. (1991). Red tide occurrence and countermeasure in Korea. *In*: Park, J.S., Kim, J.G. (eds.) *Proceedings 1990 Korean-French Seminar on Red Tides.* Natl. Fish. Dev. Agency, ROK, pp. 1-24.

Park, J.S., Kim, H.G., Lee, S.G. (1989). Studies on red tide phenomena in Korean coastal waters. *In*: Okaichi, T., Anderson, D.M., Nemoto, T. (eds.) *Red tides: Biology, Environmental Science and Toxicology.* Elsevier, Amsterdam, pp. 37-40.

Parra, O., Rivera, P.R., Foyd, G.L.,.Wilcox, L.W. (1991). Cultivo, morfologia, ultraestructura y taxonomia de un fitoflagelado asociado a mareas rojas en Chile: *Heterosigma akashiwo* (Hada) Hada. *Gayana Bot.* 48: 101-110.

Pazos, Y., Figueiras, F.G., Avarez-Salgado, X.A., Roson, G. (1995). The control of succession in red tide species in the Ría de Arousa (NW Spain) by upwelling and stability. *In*: Lassus, P., Arzul, G., Erard-LeDenn, E., Gentien, P., Marcaillou-LeBaut, C. (eds.) *Harmful Marine Algal Blooms.* Lavoisier, Paris, pp. 645-650.

Pratt, D.M. (1966). Competition between *Skeletonema costatum* and *Olisthodiscus luteus* in Narragansett Bay and in culture. *Limnol. Oceanogr.* 11: 447-455.

Qi, Y., Zhang, Z., Hong, Y., Lu, S. Zhu, C., Li, Y. (1993). Occurrence of red tides on the coasts of China. *In*: Smayda, T.J., Shimizu, Y. (eds.) *Toxic Phytoplankton Blooms In The Sea.* Elsevier, Amsterdam, pp. 43-46.

Rhodes, L.L., Haywood, A.J., Ballantine, W.J., MacKenzie, A.L. (1993). Algal blooms and climate anomalies in north-east New Zealand, August-December 1992. *N.Z. J. Mar. Freshw. Res.* 27: 419-430.

Rojas de Mendiola, B. (1979). Red tide along the Peruvian coast. *In*: Taylor, D.L., Seliger, H.H. (eds.) *Toxic Dinoflagellate Blooms*. Elsevier, Amsterdam, pp. 183-190.

Sampayo, M., Moita, M.T. (1984). An *Olisthodiscus luteus* red water: Its dynamics during 24 hours. ICES CM 1984/B:12, 11 p.

Satoh, E, Watanabe, M.M., Fujii, T. (1987). Photoperiodic regulation of cell division and chloroplast replication in *Heterosigma akashiwo*. Plant Cell Physiol. 28: 1093-1099.

Shen, S-I, Chiang, Y-M. (1971). Notes on organism causing brown water in milkfish ponds. *Fishery Ser. China-Amer. jt. Comm. rural Reconstr.* 11: 84-86.

Smayda, T.J. (1997). Harmful phytoplankton blooms: their ecophysiology and general relevance to phytoplankton blooms in the sea. Limnol. Oceanogr. *In press.*

Smayda, T.J., Reynolds, C.S. (In review). Community assembly in marine phytoplankton: application of recent models to harmful dinoflagellate blooms. *J. Plankton Res.*

Takahashi, K., Fuzukawa, N. (1982). A mechanism of "red tide" formation. II. Effect of selective nutrient stimulation on the growth of different phytoplankton species in natural water. *Mar. Biol.* 70: 267-273.

Takahashi, K., Ikawa, T. (1987). Effect of nitrogen starvation on photosynthetic carbon metabolism in the marine Raphidophycean flagellate *Heterosigma akashiwo*. *Plant Cell Physiol.* 28: 840-852.

Takahashi, M., Hara, Y. (1989). Control of diel migration and cell division rhythms of *Heterosigma akashiwo* by day and night cycles. *In*: Okaichi, T., Anderson, D.M., Nemoto, T. (eds.) *Red tides: Biology, Environmental Science and Toxicology*. Elsevier, Amsterdam, pp. 265-268.

Taylor, F.J.R. (1990). Red tides, brown tides and other harmful algal blooms: the view into the 1990's. *In*: Granéli, E., Sundström, B., Edler, L., Anderson, D.M. (eds.) *Toxic Marine Phytoplankton.* Elsevier, Amsterdam, pp. 527-533.

Taylor, F.J.R. (1992). The taxonomy of harmful marine phytoplankton. *Giorn. Bot. Ital.* 126: 209-219.

Taylor, F.J.R., Haigh, R. (1993). The ecology of fish-killing blooms of the chloromonad flagellate *Heterosigma* in the Strait of Georgia and adjacent waters. *In*: Smayda, T.J., Shimizu, Y. (eds.) *Toxic Phytoplankton Blooms In The Sea.* Elsevier, Amsterdam, pp. 705-710.

Taylor, F.J.R., Haigh, R., Sutherland, T.F. (1994). Phytoplankton ecology of Sechelt Inlet, a fjord system on the British Columbia coast. II. Potentially harmful species. *Mar. Ecol. Prog. Ser.* 103: 151-164.

Thompson, P.A., Guo, M., Harrison, P.J., Parslow, J.S. (1991). Influence of irradiance on cell volume and carbon quota for ten species of marine phytoplankton. *J. Phycol.* 27: 351-360.

Throndsen, J. (1969). Flagellates of Norwegian coastal waters. *Nytt Mag. Bot.* 16: 161-216.

Throndsen, J. (1973). Motility in some marine nanoplankton flagellates. *Norw. J. Zool.* 21: 193-200.

Throndsen, J. (1976). Occurrence and productivity of small marine flagellates. *Norw. J. Bot.* 23: 269-293.

Throndsen, J. (1996). Note on the taxonomy of *Heterosigma akashiwo* (Raphidophyceae). *Phycologia* 35: 367.

Tilstone, G.H., Figueiras, F.G., Fraga, S. (1994). Upwelling-downwelling sequences in the generation of red tides in a coastal upwelling system. *Mar. Ecol. Prog. Ser.* 112: 241-253.

Tomas, C.R. (1978a). *Olisthodiscus luteus* (Chrysophyceae) I. Effects of salinity and temperature on growth, motility and survival. *J. Phycol.* 14: 309-313.

Tomas, C.R. (1978b). *Olisthodiscus luteus* (Chrysophyceae) II. Formation and survival of a benthic stage. *J. Phycol.* 14: 314-319.

Tomas, C.R. (1979). *Olisthodiscus luteus* (Chrysophyceae) III. Uptake and utilization of nitrogen and phosphorus. *J. Phycol.* 15: 5-12.

Tomas, C.R. (1980a). *Olisthodiscus luteus* (Chrysophyceae) IV. Effects of light intensity and temperature on photosynthesis and cellular composition. *J. Phycol.* 16: 149-156.

Tomas, C.R. (1980b). *Olisthodiscus luteus* (Chrysophyceae) V. Its occurrence, abundance and dynamics in Narragansett Bay. *J. Phycol.* 16: 157-166.

Tomas, C.R., Deason, E.E. (1981). The influence of grazing by two *Acartia* species on *Olisthodiscus luteus* Carter. *P.S.Z.N.I. Mar. Ecol.* 2: 215-223.

Uye, S., Takamatsu, K. 1990. Feeding interactions between planktonic copepods and red-tide flagellates from Japanese coastal waters. *Mar. Ecol. Prog. Ser.* 59: 97-197.

Vesk, M., Moestrup, Ø. (1987). The flagellar root system in *Heterosigma akashiwo* (Raphidophyceae). *Protoplasma* 137: 15-28.

Vrieling, E.G., Koeman, R.P.T., Nagasaki, K., Ishida, Y., Peperzak, L., Gieskes, W.W. C., Veenhuis, M. (1995). *Chattonella* and *Fibrocapsa* (Raphidophyceae): Novel, potentially harmful red tide organisms in Dutch coastal waters. *Neth. J. Sea Res.* 33: 183-191.

Wada, M., Hara, Y., Kato, M., Yamada, M., Fujii, T. (1987). Diurnal appearance, fine structure and chemical composition of fatty particles in *Heterosigma akashiwo* (Raphidophyceae). *Protoplasma* 137: 134-139.

Wada, M., Miyazaki, A., Fujii, T. (1985). On the mechanisms of diurnal vertical migration behavior of *Heterosigma akashiwo* (Raphidophyceae). *Plant Cell Physiol.* 26: 431-436.

Watanabe, M. (1982). The diurnal variations in the cell densities of *Olisthodiscus luteus* and *Skeletonema costatum*. *Res. Rep. Natl. Inst. Environ. Study* 30: 143-154.

Watanabe, M., Harashima, A. (1982). Bioconvection in culture of *Olisthodiscus luteus* and Rayleigh-Taylor instability. *Res. Rep. Natl. Inst. Environ. Study* 30: 155-173.

Watanabe, M., Miyazaki, T. (1982). Accumulation of *Heterosigma akashiwo* at the water surface due to vertical migration and absorption coefficient in microcosm. *Res. Rep. Natl. Inst. Environ. Study Jpn* 80: 13-22.

Watanabe, M., Nakamura, Y. (1984). Growth characteristics of a red tide flagellate *Heterosigma akashiwo* Hada. 2. The utilization of nutrients. *Res. Rep. Natl. Inst. Environ. Study Jpn.* 63: 59-68.

Watanabe, M., Kohata, K., Kunugi, M. (1987). ^{31}P nuclear magnetic resonance study of intracellular phosphate pools and polyphosphate metabolism in *Heterosigma akashiwo* (Hada) Hada (Raphidophyceae). *J. Phycol.* 23: 54-62.

Watanabe,M., Kohata, K., Kunugi, M. (1988). Phosphate accumulation and metabolism by *Heterosigma akashiwo* (Raphidophyceae) during diel vertical migration in a stratified microcosm. *J. Phycol.* 24: 22-28.

Watanabe,M.M., Nakamura, Y., Kohata, K. (1983). Diurnal vertical migration and dark uptake of nitrate and phosphate of the red tide flagellates, *Heterosigma akashiwo* Hada and *Chattonella antiqua* (Hada) Ono (Raphidophyceae). *Jap. J. Phycol.* 31: 161-166.

Watanabe, M., Nakamura, Y., Mori, S., Yamochi, S. (1982). Effects of physico-chemical factors and nutrients on the growth of *Heterosigma akashiwo* Hada from Osaka Bay, Japan. *Jap. J. Phycol.* 30: 279-288.

Watanabe, M., Takamatsu, T., Kohata, K., Kunugi, M., Kawashima, M., Koyama, M. (1989). Luxury phosphate uptake and variation of intracellular metal concentrations in *Heterosigma carterae* (Raphidophyceae). *J. Phycol.* 25: 428-436.

Wood, G.J., Flynn, K.J. (1995). Growth of *Heterosigma carterae* (Raphidophyceae) on nitrate and ammonium at three photon flux densities: evidence for N stress in nitrate-growing cells. *J. Phycol.* 31: 859-867.

Yamochi, S. (1983). Mechanisms for outbreak of *Heterosigma akashiwo* red tide in Osaka Bay, Japan. Part 1. Nutrient factors involved in controlling the growth of *Heterosigma akashiwo* Hada. *J. Ocean. Soc. Jpn.* 39: 311-316.

Yamochi, S. (1984a). Effects of temperature on the growth of six species of red-tide flagellates occurring in Osaka Bay. *Bull. Plankton Soc. Jpn.* 31: 15-22.

Yamochi, S. (1984b). Mechanisms for outbreak of *Heterosigma akashiwo* red tide in Osaka Bay, Japan. Part 3. Release of vegetative cells from bottom mud. *J. Ocean. Soc. Jpn.* 40: 433-348.

Yamochi, S. (1989). Mechanisms for outbreak of *Heterosigma akashiwo* red tide in Osaka Bay, Japan. *In*: Okaichi, T., Anderson, D.M., Nemoto, T. (eds.) *Red tides: Biology, Environmental Science and Toxicology.* Elsevier, Amsterdam, pp. 253-256.

Yamochi,S., Abe, T. (1984). Mechanisms to initiate a *Heterosigma akashiwo* red tide in Osaka Bay. II. Diel migration. *Mar. Biol.* 83: 255-261.

Yamochi,S., Abe, T., Joh, H. (1982). Study on the mechanisms of red water blooms by *Olisthodiscus luteus* at Tanigawa Fishing Port, Osaka Bay - Characteristics in occurrence of *O. luteus* and its diurnal vertical migration. *Res. Per. Natl. Inst. Environ. Study Jpn.* 30: 191-214.

Yamochi, S., Joh, H. (1986). Effects of temperature on the vegetative cell liberation of seven species of red-tide algae from the bottom mud in Osaka Bay. *J. Ocean. Soc. Jpn.* 42: 266-275.

Yokote, M., Honjo, T., Asakawa, M. (1985). Histochemical demonstration of a glycocalyx on the cell surface of *Heterosigma akashiwo. Mar. Biol.* 88: 295-299.

Bloom Dynamics and Physiology of *Gymnodinium breve* with Emphasis on the Gulf of Mexico

Karen A. Steidinger[1], Gabriel A. Vargo[2], Patricia A. Tester[3], and Carmelo R. Tomas[1]

[1] Florida Marine Research Institute, Florida Department of Environmental Protection, 100 Eighth Avenue S.E., St. Petersburg, FL 33701
[2] Department of Marine Science, University of South Florida, 140 Seventh Avenue S., St. Petersburg, FL 33701
[3] National Marine Fisheries Service, NOAA, Southeast Fisheries Center Beaufort Laboratory, 101 Pivers Island Road, Beaufort, NC 28516

1. Introduction

The Gulf of Mexico has over 30 toxic microalgal species, but only one consistently produces dramatic fish kills and presents human health risks. The causative microalga is *Gymnodinium breve,* an unarmored dinoflagellate. It is one of possibly two or three gymnodinioid species that produce brevetoxins which cause fish kills and other marine mortalities, cause filter-feeding animals such as oysters and clams to become toxic to humans when consumed (Neurotoxic Shellfish Poisoning or NSP), and cause an airborne toxin (particulate irritant) in seaspray.

In the Gulf of Mexico, the first reports of fish kills and discolored water date back to 1530. The Spanish explorer Alvar Nuñez Cabeza de Vaca landed at Tampa Bay, Florida, in 1528 and spent several years in this area as part of an expedition to establish a Spanish settlement. In his book published in 1542, he reported that the local Indians chronicled their activities by the occurrence of fish kills and red water. It was not until the 1800s that other incidents of fish kills, toxic shellfish, and discolored water were reported. In the 1880s, reports of fish kills on sailing routes were common, and vessel captains would frequently notice that fish in their live wells were dying even before the vessel encountered discolored water. They referred to this phenomenon as "poisoned water" (Rounsefell and Nelson 1966).

1.1. Taxonomy and world-wide distribution

The taxonomy of morphospecies in the fucoxanthin derivative-containing gymnodinioids, e.g., *Gymnodinium breve* Davis, *G. breve*-like forms, *G. mikimotoi* Miyake et Kominami ex Oda, *G.* cf. *mikimotoi* NZ, *Gyrodinium aureolum* Hulburt, *G.* cf. *aureolum* EUR, *G.* cf. *aureolum* NZ, is confused (Steidinger 1990). New gymnodinioid isolates with new toxic compounds (Chang 1996; Haywood *et al.* 1996; Mackenzie *et al.* 1996) and harmful algal bloom event records of gymnodinioids, e.g., Japan (Steidinger *et al.* 1989) and Israel (Kimor 1996), are confounding the problem. In the Gulf of Mexico, *Gymnodinium breve* and *G. mikimotoi* can co-occur during bloom events. Because both species are extremely pleomorphic, it is sometimes difficult to separate

NATO ASI Series, Vol. G 41
Physiological Ecology of Harmful Algal Blooms
Edited by D. M. Anderson, A. D. Cembella and G. M. Hallegraeff

the two using bright-field microscopy. However, field specimens of *G. mikimotoi* can be identified by the dorso-ventral compression and the features of the apical groove and sulcal intrusion on the epitheca, which are clearly visible using differential interference microscopy (Steidinger *et al.* 1989). In culture (CCCP 429), stressed *G. mikimotoi* can have a cone-shaped epitheca and even an apical carina (personal observation). It can resemble *G. breve* except that specimens still have a large, elongated nucleus in the left portion of the cell extending from the hypotheca into the epitheca, and it has the typical ventral ridge.

The shape and width of the sulcal intrusion onto the epitheca in *G. breve* can vary with the size and shape of the cell; smaller and less ventrally concave cells can have a wider anterior sulcal intrusion. The most conservative character in Gulf of Mexico *G. breve* and *G. mikimotoi* is the apical groove length in relation to the epitheca and sulcal intrusion.

Gymnodinium breve morphs have been described from Spain, Japan, Greece, France, Israel, and New Zealand. They range from the morphotype, like Florida and Texas isolates of *G. breve* (typically 20 to 40 µm), to those resembling *G. mikimotoi* or *G.* cf. *aureolum* and include various forms referred to as "butterfly" types (Fraga and Sanchez 1985; Haywood *et al.* 1996) up to 100 µm wide, or just large *G. breve* cells about 50 µm wide. Some are heavily pigmented, some are barely pigmented, and all have a spherical nucleus, usually in the lower left quadrant. In one large Florida butterfly type of 70 to 90 µm that occurs offshore, the nucleus is difficult to detect without staining. This butterfly type probably represents an undescribed species. Although Dragovich (1967) reported that this butterfly type reverted to the standard *G. breve* form in culture, the cultures were not clonal. Haywood *et al.* (1996) also reported that their large butterfly isolate gave way to the 50 µm cells resembling Wilson's strain of *G. breve* but did not specify whether their cultures were clonal.

In all probability, we are dealing with several new species that need to be characterized morphologically, cytologically, biochemically, and genetically. Haywood (personal communication) is taking this approach, and her work may help clarify species with affinities to *G. breve, G. mikimotoi,* and *Gyrodinium aureolum.*

The type locality for *Gymnodinium breve* is Florida (Davis 1948). Increased population size (blooms) above background levels that cause discoloration (2×10^6), fish kills ($1–2.5 \times 10^5$), NSP ($>5 \times 10^3$), or respiratory irritation ($>1 \times 10^3$) have been recorded off Mexico, Texas, Louisiana, Mississippi, Alabama, Florida, South Carolina, and North Carolina. The species has also been observed in the Gulf Stream off the Chesapeake Bay and may even have been carried to Spain by the Gulf Stream Current (Fraga and Sanchez 1985). Whether this species causes harmful algal blooms in other parts of the world awaits clarification of the taxonomy of *G. breve*-like species. If one of the *G. breve*-like forms isolated from New Zealand waters is genetically and biochemically identical to Florida *G. breve* isolates, then the question becomes, How did Florida *G. breve* get to New Zealand?

Gymnodinium breve is not the only flagellate that produces brevetoxins. Recently, *Chattonella marina, C. antiqua,* and *Fibrocapsa japonica* have been reported to produce this class of polyether toxins (Ahmed *et al.* 1995; Khan *et al.* 1996a, 1996b). Whether or not these raphidophytes cause NSP or produce airborne toxins and therefore present human health risks is not documented. The fact that these raphidophytes bloom and produce brevetoxins could confuse the interpretation of NSP events, particularly in the presence of *G. breve*-like forms.

2. First bloom stage: initiation/onset

Prior to 1973, the initiation of Florida red tides caused by *G. breve* was thought to be inshore around passes (Rounsefell and Nelson 1966) because that is where visible signs, e.g. fish kills, discolored seawater, and human respiratory irritation, were most obvious. However, a review of historical federal and state cruise track data from 1957–1959, 1964–1965, and 1967 (Steidinger 1975a) indicated that *G. breve* blooms originated offshore between 18 and 74 km on the mid West Florida Shelf. This information, when combined with data from other harmful algal bloom events around the world, indicates that there are common denominators to microalgal bloom development that can be divided into several stages: initiation, growth, maintenance, and termination/dissipation (Steidinger and Vargo 1988). Offshore initiation was verified with subsequent offshore cruise track data showing increased *G. breve* populations above background appearing offshore and then several weeks later inshore (Tester and Steidinger 1997). Offshore population increases were first noted by Dragovich and Kelly (1966). Over 70% of documented blooms since 1878 originated in late summer–fall months and about 17% originated in winter–spring (Rounsefell and Nelson 1966, Steidinger and Roberts, unpublished).

The exact timing of and location of bloom inception is difficult to predict at this time without full knowledge of the West Florida Shelf physical dynamics and complementary remote sensing data, but the sequence of events that results in blooms can be examined to assess the potential of prediction. Physical-chemical processes such as hydrodynamic conditions and oceanic fronts, circulation patterns, meteorological forcing functions, and upwelling/downwelling events act to produce gradients and ecotones of environmental conditions that affect physiological rate processes of microalgal blooms. The end product of growth and photosynthesis may be separated spatially from the region where the process originated or achieves maximum rates. It is critical to identify and understand the sequence of events responsible for blooms and the conditions that foreshadow them. It is not enough to examine the manifestations of blooms such as discolored water and toxic shellfish.

2.1. Occurrence and abundance of *G. breve* on the shelf

Gymnodinium breve is a common unarmored dinoflagellate in the Gulf of Mexico at background concentrations of 1 to 1000 cells L^{-1} (Dragovich and Kelly 1966; Steidinger 1975b; Geesey and Tester 1993). Although cell concentrations are higher at the surface, the species can be found down to >50 m. This principally neritic species does not tolerate salinities in Florida bays and lagoons of less than 24‰. However, it has been found at above background levels in lower salinity water in Mississippi (C. Montcreiff, personal communication). The highest inshore impact area along Florida's coast is Tarpon Springs/Clearwater to Sanibel, yet red tides can impact all coastal areas of Florida because of transport by currents and wind-driven circulation to more northern and southern locations. The high-impact area has had a red tide in 21 of the past 22 years, whereas areas north and south have had 4 to 10 red tides in the same period.

Gymnodinium breve blooms can even be transported from the Gulf of Mexico to the western North Atlantic. Although this is a subtropical–warm temperate species, there are no convincing reports of blooms in the Caribbean Sea, the source waters for the Gulf of Mexico (Tester and Steidinger 1997). Even regions within the Gulf of Mexico, other than the west Florida continental shelf, do not experience *G. breve* blooms frequently nor are the conditions in the Atlantic normally favorable for blooms. But there has been a long history of *G. breve* blooms along the central and southwest Florida coast. Does this region serve as the source for other blooms in the Gulf of Mexico and Atlantic? Do combinations of basin topography, hydrography, circulation, and chemistry on the West Florida Shelf provide an optimal habitat for *G. breve*?

2.2. Life cycle and source of seed stock

The asexual and sexual life cycle phases of dinoflagellate bloom species often represent successful adaptation strategies for survival and dispersal. A sexual life cycle with a benthic resting stage or a motile diploid nondividing stage can represent a reservoir or seed population for bloom development as suggested by Steidinger (1975a, 1975b). *Gymnodinium breve* reproduces asexually by oblique binary division with an observed rate in culture of 0.2 to 1 division day^{-1} (Wilson 1966, Shanley 1985, Shanley and Vargo 1993). Its sexual cycle has been only partially elucidated through the planozygote stage by Walker (1982). It is a haplont with heterothallic and homothallic strains (Steidinger, unpublished). Gamete production in the homothallic strain (clonal) appears to involve the nucleus of a vegetative cell moving from the left lower quadrant to the middle of the cell as the cell rounds out. The nucleus then divides into two nuclei, and the cell divides producing two easily identifiable gametes of different morphology and cytology than the vegetative cell (Steidinger, unpublished). The size of the gametes is $^1/_3$ to $^1/_2$ the size of a vegetative cell. Fusion of the homothallic strain gametes is as described for the heterothallic strain (Walker 1982), and they both produce a planozygote with two longitudinal flagella. Neither strain has been induced to produce hypnozygotes, although layered, thick-walled round cells that form within the plasmalemma of a *G. breve* cell have been observed in natural seawater samples (see Fig. 1 for life cycle).

In the 1994–1996 Florida red tide, palmelloid stages of *G. breve* were documented in seawater samples that were kept at ambient temperature for 3 months. Rounded cells in mucous sheaths were aggregated in various numbers. The round cells had numerous oil globules around the periphery and divided within the sheaths producing two cells, each with a sheath. Some clumps or aggregates had over 100 cells. The sheaths eventually dissolved, and the released round cells formed clumped, typical motile *G. breve* cells but with pointed epithecae. Clumped cells of similar appearance were found in fresh seawater samples (Steidinger, unpublished). This requires further investigation to determine whether the palmelloid stage is a temporary resting cyst or a stressed stage responding to adverse environmental conditions.

2.3. Influence of fronts and winds in initiation/onset of blooms

Blooms manifest in the mid-shelf region between 18 and 74 km off the west Florida coast (Steidinger 1975b) in 12–37-m water depth, frequently enough that offshore initiation is validated (Steidinger and Haddad 1981, Tester and Steidinger 1997). Observations and data sets from the Florida Department of Environmental Protection's

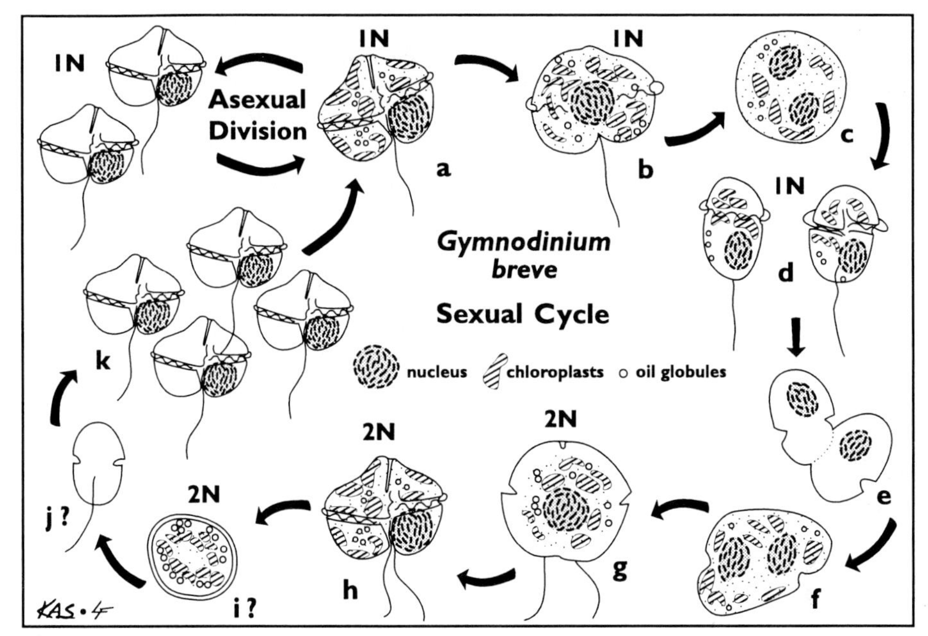

Fig. 1 Life cycle of *Gymnodinium breve*. (**a**) Vegetative 1N cell that is capable of mitotic division or producing gametes. (**b**) First stage in gamete formation with movement of nucleus to center. (**c**) Second stage of gamete formation with division of nucleus. (**d**) Third stage with production of isogametes. In heterothallic strains, there are + and – mating types that fuse. (**e**) In homothallic strains, gametes are the same mating type. (**f**) Cytoplasmic fusion precedes nuclear fusion. (**g**) Planozygote with 2N nucleus and 2 longitudinal flagella. (**h**) Motile planozygote. (**i**) Unconfirmed presumed hypnozygote. (**j**) Unconfirmed meiocyte. (**k**) Resultant 4 1N cells for meiosis II.

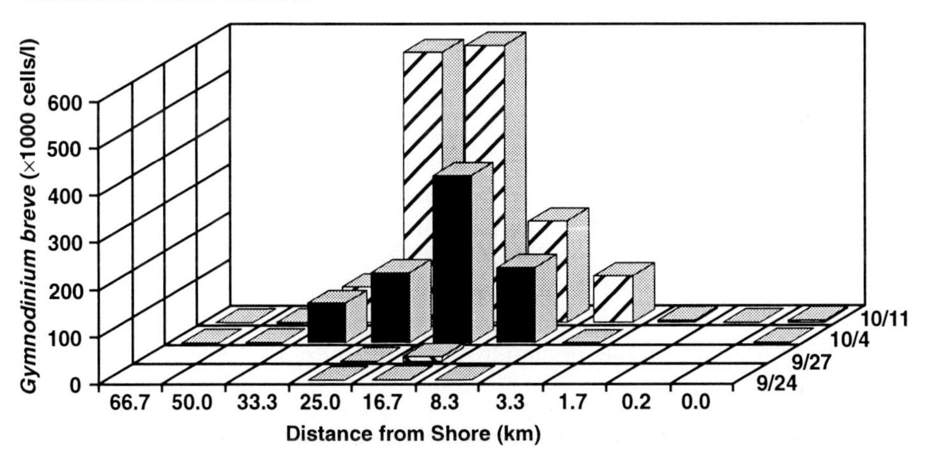

Fig. 2 Late summer 1976 red tide that initiated offshore Midnight Pass, Florida, as illustrated by *G. breve* counts along offshore–onshore transects over a time sequence.

Florida Marine Research Institute red tide logs (>30 years) help substantiate the theory of offshore initiation. Early distribution and abundance studies, based on cross-shelf transects, found the highest occurrence (measured as % of samples containing *G. breve* at the most offshore stations (25 km) and the lowest at the inshore stations (8 km) (Dragovich and Kelly 1966). Tester and Steidinger (1997) document three

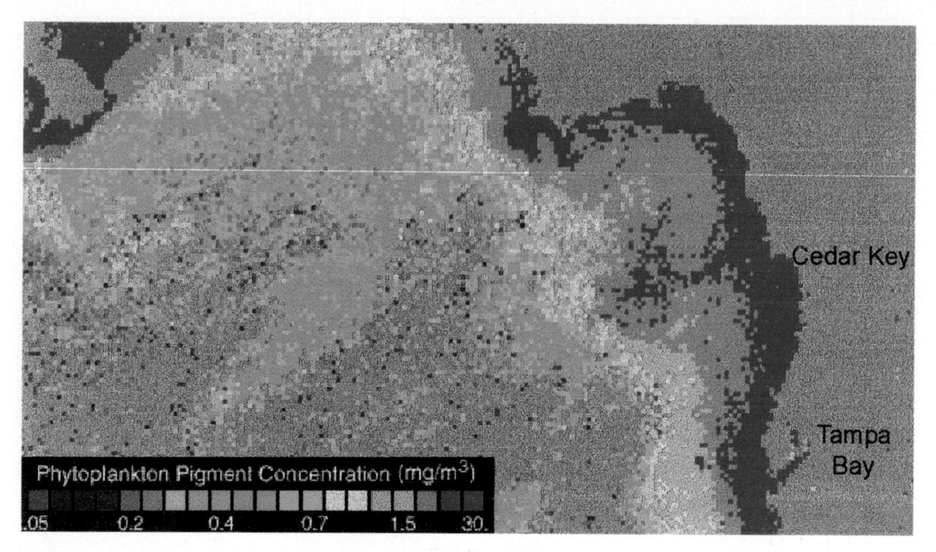

Fig. 3 Coastal Zone Color Scanner image from 15 November 1979 depicting higher chlorophyll in band off Cedar Key, FL; represents *G. breve* red tide.

blooms: late summer 1976 (Fig. 2), November–December 1979 (Fig. 3, CZCS image of 15 November 1979), and September–November 1985, all of which had higher cell numbers in the offshore to mid-shelf stations prior to cells being detected in nearshore waters.

Haddad and Carder (1979) and Haddad (1982) note that although wind speeds are not generally sufficient to drive upwelling at the shelf edge, except in winter, the penetration of the Gulf Loop Current water (>36.5‰, <24°C) onto the shelf has been documented during or prior to *G. breve* blooms in 6 out of 7 years between 1967 and 1977 (Table 1). Spin-off eddies (>20 to 250 km) from the Gulf Loop Current have been documented by Maul (1977), Sturges (1994), and Vukovich (1995). These eddies move shoreward across the West Florida Shelf at ca. 0.5 km day^{-1}, exchanging heat and momentum and suspending bottom particles (Haddad and Carder 1979). They are considered part of the annual cycle of front variability (growth, decay, and northward penetration) of the Gulf Loop Current (mean 11 ± 3 mo) (Vukovich 1995). See Fig. 4 for general current patterns in the Gulf of Mexico. Near-bottom intrusions of dense, relatively cool, high-salinity water are identifiable at the 40-m isobath as a mid-shelf front; Eastern Gulf water of oceanic origin (36.4‰ salinity) can be detected even further shoreward over the shelf.

Table 1 Association of oceanic intrusions with red tides. Example: EGW had penetrated to within 37 km by September 1977 and in early October surfaced nearshore where a sudden bloom of *G. breve* was observed (data from Haddad and Carder 1979).

Bloom year	Loop current water	Eastern Gulf water (oceanic)
1967	Within 93 km of shore	Within 56 km of shore
May 1971	Within 93 km of shore	Within 28–56 km of shore
August 1971	Within 56 km of shore	Within 28–56 km of shore
September 1972	Within 112 km of shore	Within 75 km of shore
September 1977	On shelf	Within 37–74 km of shore
October 1977		<20 km of shore

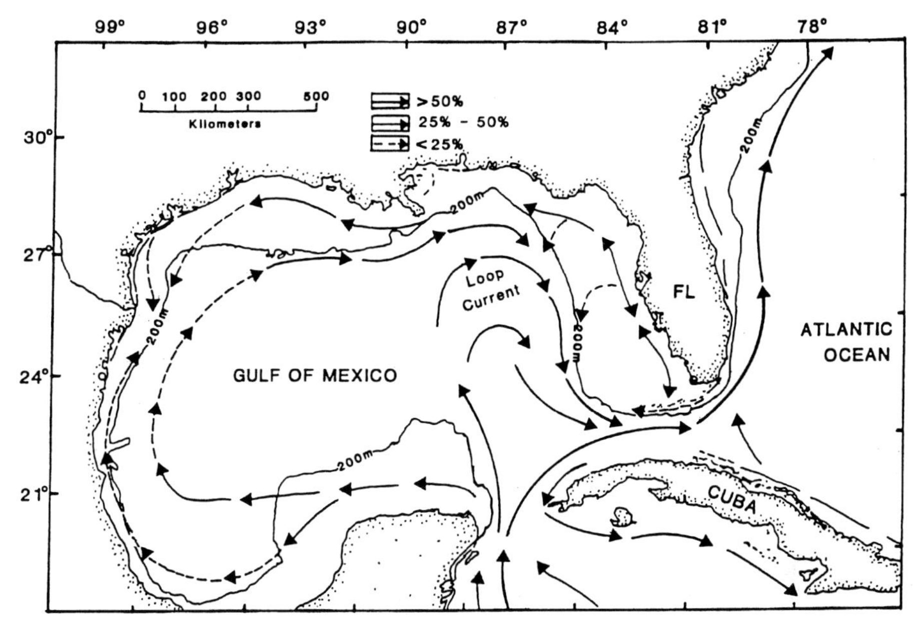

Fig. 4 Generalized current patterns in the Gulf of Mexico. Redrawn from UNEP/CEPAL Caribbean Environment Project #1037.

3. Second bloom stage: population growth on the shelf

Population growth rates must exceed loss rates, such as those for predation and advection, for blooms to develop. Steidinger and Vargo (1988) stated, "The study of cell adaptations for the utilization of light and nutrients, for the production of alleopathic growth factors, and for migratory behavior provides us with clues as to how dinoflagellates may take advantage of environmental conditions to out-compete other species to form near-monospecific blooms with high population densities." Initial growth occurs offshore where populations above 5000 cells L^{-1} reach 10^5 and 10^6 cells L^{-1} within 2 to 4 weeks at division rates of <1 div. day^{-1}.

3.1. Physical-chemical tolerances

Field- and laboratory-derived limits for the temperature range of *G. breve* range from 9°C to 33°C with optimum growth occurring between 22°C and 28°C (Rounsefell and Nelson 1966; Steidinger and Ingle 1972; Eng-Wilmot *et al.* 1977). Aldrich (1959) determined it would not survive below 7°C, whereas Eng-Wilmot *et al.* (1977) found that laboratory cultures died at 4°C and growth decreased below 17°C. However, in earlier laboratory studies, Wilson (1955) found that *G. breve* would survive and remain active at 8°C to 10°C if the temperature was lowered slowly at a rate of 1°C per hour. The upper limit of 33°C is based upon the presence of a bloom (10^5 to 10^6 cells L^{-1}) in Tampa Bay during 1971.

Laboratory studies on the salinity tolerance of a Florida isolate of *G. breve* show that optimum growth occurs between salinities of 27 to 37‰ with reduced growth below 24‰ (Aldrich and Wilson 1960; Wilson 1966). Field records examined by Rounsefell and Nelson (1966) support the laboratory observations with maximum abundance found between 31 and 37‰ and lowest abundance at salinities below 30.5‰. *G. breve*, therefore, is relatively stenohaline with a preference for salinities characteristic of coastal and offshore waters. Interestingly, red tides have initiated during drought conditions and terminated following heavy rainfall and delivery of freshwater to inshore systems.

The causal factors for the occurrence of red tide (*G. breve*) blooms have been the central question in the study of this phenomenon since the first reports of fish kills and discolored seawater. Feinstein (1956) was one of the earliest investigators who attempted to relate the occurrence of blooms with external factors. Based on regression analysis of red tide occurrence between 1844 and 1955, no significant relationships were found between the frequency and intensity of blooms and hurricane and tropical storm passages during that time frame or with rainfall using monthly averages of precipitation data from 1922 through 1955, whether rainfall was calculated over the calendar year or just the rainy season. Similarly, she found no significant relationships with river flow from the Peace River drainage system and the Apalachicola river system, either considered separately or combined. The highest correlation coefficient obtained for any relationship was an r^2 of 0.46 when she combined river flow from the Peace and Apalachicola rivers with water year precipitation data for the entire state of Florida; however, the relationship was still not statistically significant. It is of interest that she refers to a manuscript by Chew in which he states that the runoff from the Apalachicola River "seems to retain its identity *at least* [emphasis by Feinstein 1956] as far south as Tampa Bay at some seasons." This fact was used as a basis for formulating the hypothesis that river flow from the Apalachicola was related to red tides along the central sections of the West Florida Shelf. As indicated earlier, no significant relationships were found. It would be of interest to re-evaluate some of the information over a longer time span using different statistical evaluation techniques.

3.2. Nutrients and growth factors in the second stage of bloom development

While environmental parameters such as temperature and salinity may determine the distribution or presence/absence of a species within a region, the nutrient regime will determine its ability to grow and the levels of biomass or numerical abundance it can attain. *Gymnodinium breve* is very efficient at using macronutrients such as phosphorus and nitrogen. Under laboratory and field conditions it can utilize both inorganic and organic forms, but it has chelated trace metal and vitamin requirements.

With very few exceptions, all of the earlier studies on nitrogen (N) and phosphorus (P) requirements for *G. breve* found that it grew well, but at relatively low growth rates, at low levels of both macronutrients. Increasing the amount of available P did not enhance growth rates, although it was highly efficient in producing and maintaining high populations at low N and P concentrations, and it had the capacity to utilize a variety of organic N and P compounds to support growth. All of these characteristics could be used to describe it as a K-selected species—one that is adapted to low nutrient, oligotrophic environments. The nutrient regime on the West Florida Shelf is oligotrophic. Inorganic N and P levels rarely exceed 0.5 µg-at L^{-1} and are more commonly 0.1 to 0.2 µg-at L^{-1} within 2–4 km of the shore (Dragovich *et al.* 1961, 1963; Vargo and Shanley 1985; Vargo 1988; Vargo and Howard-Shamblott 1990).

Table 2 Standing stock of Soluble Reactive Phosphorus (SRP) and Total Hydrolizable Dissolved Organic Phosphorus (THDOP), their total, the fraction of that required for daily photosynthesis; growth estimates and the number of days supply available for the Photosynthetic Requirements for Phosphorus (PNS) from the standing stock.

	mg P M^{-2}			P required mg M^{-2} D^{-1}		% MET		Days supply
Station	SRP	THDOP	Total	Growth	PNS	Growth	PNS	
RT1	35.8	22.8	58.6	10.9	24.4	100	100	2.4
RT2	62.0	58.9	120.9	13.4	24.9	100	100	4.8
RT3	32.4	0	32.4	9.6	17.9	100	100	1.8
RT4	27.1	7.8	34.9	28.6	50.9	100	68.5	0.7
RT5	14.7	13.2	27.9	30.2	52.6	92.4	53	0.5
RT6	24.9	15.7	40.6	22.4	38.8	100	100	1.0
RT7	48.1	19.5	67.6	19.3	22.3	100	100	3.0
RT8	21.7	29.5	51.2	No *G. breve*	–	–	–	–

All studies conducted to date support the concept that *G. breve* does not require high nutrient levels to support normal growth rates and relatively high population abundance. Wilson and Ray (1958) determined that phosphate concentrations between 0.1 and 1 μM supported maximum growth; increased phosphate concentrations above the minimum required did not yield increased biomass (i.e., cell number). Field data also support this conclusion. Wilson (1966) reported that Dragovich *et al.* (1963) concluded that high P levels were not required for *G. breve* blooms based on the analysis of 5 years of total phosphate (TP) data. Relatively few cells were found in waters with concentrations of TP <6 μg L^{-1} (0.2 μg-at L^{-1}) or greater than 120 μg L^{-1} (~4 μg-at L^{-1}). These values were confirmed in laboratory culture studies by Wilson (1966) and complement earlier findings by Odum *et al.* (1955) and Bien (1957) that P is not a limiting nutrient for *G. breve* blooms. Odum further stated, "in general P levels of coastal water seem similar in a red tide year to those of a non-red tide year," and Bien observed that "waters of the west coast of Florida maintain a sufficient P concentration to support a red tide at all times of the year...."

More recent studies by Vargo (1988) confirm these earlier conclusions. P requirements for growth, photosynthesis, and cellular P-turnover rates were calculated for a 1986 bloom off the mouth of Tampa Bay. The measured standing stock of inorganic and hydrolizable organic phosphate ranged from 28 to 121 mg/m^2. Based on estimated growth rates of 0.2 div. day^{-1} and primary production rates calculated from relationships developed by Shanley (1985), sufficient P was available in the water column to meet the daily requirements of this bloom (Table 2). The two stations with the highest population density would, however, deplete the water column supply within 1 day. Therefore, except at high population levels (biomass), sufficient P is available to maintain blooms (population levels in this bloom ranged from 6×10^4 to 6.5×10^5 cells L^{-1}) in coastal to mid-shelf waters.

G. breve is highly efficient in the acquisition and utilization of available inorganic phosphate. Vargo and Howard-Shamblott (1990) found that P-deficient cells took up phosphate 2–4-fold faster than P-sufficient cells and increased cellular stores 3-fold within 2 hours after the addition of an organic phosphorus source. Growth rates were saturated at approximately 1 μg-at L^{-1} P (similar to the value reported by Wilson and Ray 1958) with a K_s for growth of 0.18 μg-at L^{-1}. This low K_s value is similar to that reported for the oceanic dinoflagellate, *Pyrocystis* (Rivkin and Swift 1985) and suggests that *G. breve* is adapted for growth at the low P concentrations commonly found in coastal waters. It is also of interest to note that the K_s value is equivalent to 5.6 μg P L^{-1}—a value close to the value of 6 μg L^{-1} determined by Wilson (1966) as the minimum concentration required to maintain populations.

Vargo and Howard-Shamblott (1990) also determined the yield per unit P produced by *G. breve*. A range of 2 to 9 × 10^6 cells, depending upon size, are produced per μg-at of P. This range is similar to that originally calculated by Wilson (1966) and confirms that this species is highly efficient in the utilization of available P. When this high efficiency for P use is combined with the low K_s for growth and the enhanced uptake rates, we have the characteristics of a species that is capable of growth and maintenance at low P levels. Phosphorus enrichment may support some additional biomass, but it is not required to maintain *G. breve* blooms at fish-killing levels.

Nitrogen, rather than phosphorus, is most often considered to be limiting in marine waters (Hecky and Kilham 1988). It is curious, therefore, that there is far less information available about N metabolism for *G. breve* than P metabolism. Wilson (1966) determined that it would use a variety of vitamins, amino acids, and other organic N compounds as N sources. The use of amino acids was later confirmed by Baden and Mende (1979). Initially, Wilson (1966) suggested that *G. breve* did not use nitrate and grew solely on ammonia and organic N. However, his experiments were all carried out in NH-15 medium that contained added ammonia, EDTA, TRIS, vitamins, and other organic sources. Subsequent work by Doig (1973) confirmed use of ammonia as an N-source for growth. The notion that nitrate was not utilized probably came from measurements by Dragovich *et al.* (1961) that indicated there was never sufficient nitrate + nitrite in the water column to support a bloom. Less than 1% of their samples in neritic waters (>4000) had nitrate levels of 14 μg NO_3 + NO_2 (1 μg-at L^{-1}). Their ammonia levels were always higher than nitrate, especially after rains. This leads to the suggestion that ammonia would be the primary N-source and that direct input of ammonia from rain could be an important source.

If, as suggested by Wilson (1966), *G. breve* requires approximately 15 μg P to support 10^6 cells, then based on a ratio of N:P of 8:1, 10^6 cells would require approximately 120 μg N (8.6 μg-at)— a concentration rarely found in coastal or shelf waters (see above). Odum *et al.* (1955) did a similar calculation using a lower N:P ratio and estimated a value of 45 μg N L^{-1} (3 μg-at L^{-1}), a more realistic value.

Shimizu and Wrensford (1993) and Shimizu *et al.* (1995) showed substantial increased cell yield of *G. breve* with organic N sources, urea, glycine, leucine, and aspartic acid. These organic substrates also influenced the production of brevetoxins. However, there are no values for uptake and growth rates for *G. breve* as a function of nitrate, nitrite, ammonia, or urea. Based on one series of preliminary experiments by Vargo (unpublished), growth rates ranged from 0.16 to 0.195 div. day^{-1} and were independent of ammonia or urea concentration over a range of 0.5 to 7 μg-at L^{-1}. A hyperbolic curve was fitted to the data, and K_s values of 0.47 and 1.07 μg-at L^{-1} were estimated for growth on ammonia and urea, respectively. These are low relative to other estuarine dinoflagellates and more in line with open-water species (Hecky and Kilham 1988). A single nitrate uptake experiment over a range of 1 to 10 μg-at L^{-1} was also done using replicate cultures. The disappearance of nitrate over a 2-hour incubation was used as a measure of uptake. The calculated K_s value of 0.42 (Fig. 5) is equiva-

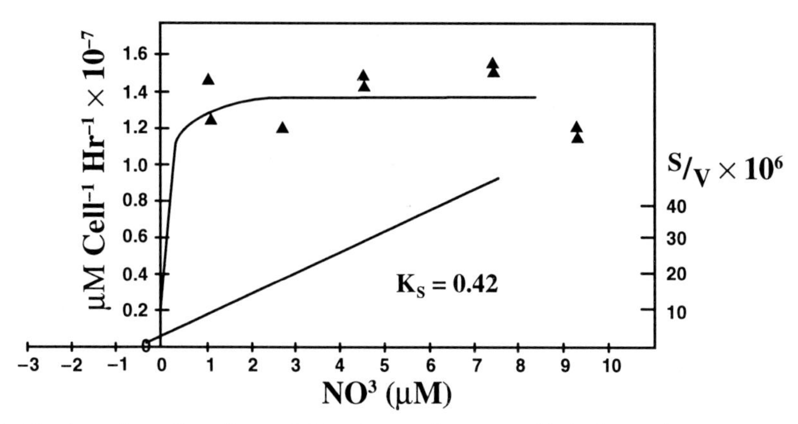

Fig. 5 Two-hour nitrate (1 to 10 μg-at L^{-1}) uptake experiment run with duplicate cultures.

lent to that for growth on ammonia. Both values are indicative of a species with a high affinity for inorganic nitrogen and, when combined with the low K_S for phosphate, reinforces the concept that *G. breve* is a species adapted for growth in low-nutrient environments.

Between November and March, offshore *G. breve* blooms occur in a West Florida Shelf environment that is well mixed down to about 100 m, during which time nutrients past the 10-m isobath are very low. Ammonia is in the order of 1–3 μM while nitrate is lower, between 0.5 and 1 μM, and phosphorus is either undetectable or <0.5 μM. Below 100 m, there is a pronounced nitrocline with nitrate nearing 20–30 μM. Ammonia and phosphorus are in the range of <0.5 μM and 1–5 μM, respectively. This appears to be associated with nutrient-rich Caribbean water that comes in with the Loop Current. Except for intrusions of the Loop Current, these nutrients do not appear to get mixed into the euphotic zone (Tomas, unpublished). To supply nutrients for offshore *G. breve* blooms, the source must come from upwelled offshore water with some minor additions from rainfall. Outflow from the Mississippi River and 10 major Florida rivers in the Big Bend area does not produce a nutrient signal in the critical areas of *G. breve* bloom initiation and growth (Tomas, unpublished).

There is some evidence to suggest that *G. breve*, and its associated bacterial flora, conditions its own medium, whether it be artificial culture media or natural seawater. Like other phytoplankton in culture, *G. breve* has a lag phase, an exponential growth phase, and a stationary phase. *G. breve* has maintained itself in culture for more than a year without the addition of fresh medium, but this occurs in non-axenic cultures. *G. breve* cannot be successfully maintained in axenic culture for more than several months (Wilson 1966).

3.3. Growth and light

Early studies by Wilson (1955), Aldrich (1960), and Wilson *et al.* (1975) suggested that *G. breve* was capable of maintaining "normal" growth at low light levels. Aldrich (1960) determined that growth was not limited by light at 200 ft. candles (approx. 2.2% of full sunlight), whereas Wilson *et al.* (1975) measured growth rates of 0.18 to 0.45 div. day^{-1} over a 12-week period in a culture grown at 750 ft. candles of continuous light.

More recent experiments by Shanley (1985) and Shanley and Vargo (1993) demonstrated that growth is a hyperbolic function of irradiance over a range of 24 to 160 μE m^{-2} sec^{-1} (approx. 1.5 to 10% of full sunlight). Growth was saturated at 45 μE with a calculated I_{comp} (compensation intensity) of 6 μE, which is indicative of a low-light adapted species. However, when adapted to 24 μE and subsequently transferred to 160 μE, growth rates increased from 0.08 to 0.31 div. day^{-1} within 4 days. This was 2-fold faster than the time required to reduce cellular chlorophyll levels. The fact that *G. breve* responds as a low-light species is somewhat enigmatic given the wealth of *in situ* observations of extremely high population levels at the surface. But with self-shading, bioconvection, and chromatic adaptation (see Millie *et al.* 1995), perhaps individual cells, although exposed to varying light intensities, are protected from high light. This is certainly an area that requires further investigation.

Gymnodinium breve is photosynthetically efficient. Net photosynthesis rates measured by carbon-14 uptake were equivalent to rates determined by oxygen evolution. This indicates that there was minimal carbon loss over the range of light levels tested. Measured respiration rates ranged from 15% to 29% of net photosynthesis, but the variation was in the photosynthesis rates since respiration was invariant at approximately 4 pgC $cell^{-1}$ hr^{-1} over the light gradient used (Shanley and Vargo 1993). Because net photosynthesis increased over the light gradient from 10 to 25 pgC $cell^{-1}$ hr^{-1}, the P:R ratio approximately doubled. This net increase in photosynthesis was reflected in realized growth rates that were linearly related to chlorophyll-specific and cell-specific photosynthetic capacity. Thus *G. breve* has the capability of maintaining a relatively constant respiration rate while translating most of its photosynthetic capacity into growth—a highly desirable physiological adaptation for a species.

3.4. Migratory behavior

Vertical migratory behavior in *G. breve* may be helpful in understanding some of the patterns displayed by *in situ* populations. Daytime distributions of natural populations appear to be concentrated at the surface, but there is no discrete population maximum at depth during darkness, as would be expected if this species had a classic vertical migration pattern. During the 1971 bloom, wide areas of the patch reef community off Tampa Bay and Sarasota were killed when high *G. breve* populations were trapped beneath a thermocline. Oxygen depletion via cell respiration and bacterial degradation of dead fish were responsible for the massive die-off of benthic organisms (Smith 1975, 1979). Why did the *G. breve* populations remain at depth rather than exhibiting their vertical migration to the surface during daylight hours?

Laboratory experiments in vertical columns suggests that the vertical behavior of *G. breve* is due to the interactions of phototaxis and endogenous rhythm, and possibly negative geotaxis (Heil 1986). The vertical movement of this species is independent of light intensity over a range of 50 to 250 μE m^{-2} sec^{-1}. Populations actually start movement toward the surface prior to the onset of the light period and begin dispersal by randomly swimming from the surface prior to the dark period, indicating that *G. breve* may possess a circadian rhythm (Heil 1986). Maintenance of the rhythm in continuous light and in continuous darkness for several days reinforces this assumption. Heil (1986) also exposed populations to a 5°C thermocline in the surface layer of the vertical column. Cells migrated upward into the surface layer while the thermocline was forming, but when the temperature differential reached its maximum, downward movement was restricted and cells were trapped above the thermocline. This result suggests that the absolute rate of change of temperature over the thermocline may be the determining factor. If this is the case, then cells would be trapped below a thermocline as they are trapped above it, which gives a possible explanation for the impact on the benthic fauna during the 1971 bloom. The response to a thermocline may also explain the extremely high populations observed in the surface microlayer during very calm periods. If sufficient daytime heating occurred during calm weather, vertically migrating cells would accumulate until the maximum thermal differential was reached, after which cells would be trapped until nighttime cooling or wind-mixing broke the discontinuity.

4. Third bloom stage: maintenance of blooms

Blooms that occur in the nutrient-depleted, oligotrophic, offshore shelf waters must be maintained by regenerated nutrients. Other than wet and dry deposition and N-fixation, there are few sources of "new" nutrients (sensu Dugdale and Goering 1967) for maintenance. Conversely, maintenance of blooms in nearshore, enriched environments is a result of a multitude of potential sources of "new" nutrients (i.e., point and non-point sources and atmospheric deposition). Offshore maintenance occurs in a nutrient-sufficient but unenriched environment of shelf water. For this to occur, turnover rates must be high. Inshore maintenance often occurs in an enriched environment. When blooms are established in bays and lagoons with poor flushing rates and exchange, population densities can increase and blooms can be prolonged. Yet blooms in enriched environments also terminate in the same environment, irrespective of nutrient inputs. Factors such as benthic predation, changing nutrient ratios, limiting nutrient availability, and dilution of water masses and growth conditioners have been suggested as precursors to bloom termination.

4.1. Influence of water masses, currents, and winds in maintenance of blooms

Cross-shelf movement of blooms from the mid-shelf region is generally influenced by local circulation patterns and winds. There is evidence for a mean southward flow over the shelf when the Gulf Loop Current is at the West Florida Shelf edge (Sturges and Evans 1983). Drift bottles released during the summer Hourglass cruises (Williams *et al.* 1977) resulted in high returns from the lower west coast (Tampa to Ft. Myers), whereas winter releases resulted in the greatest number of returns from the Florida Keys and Atlantic. Wind circulation patterns over the eastern Gulf of Mexico are southerly during spring, summer, and early fall with wind speeds of <12 km hr^{-1} more than 50% of the time between June and August (Haddad and Carder 1978). Lee and Willams (1988) identified a slightly different annual cycle of wind stress, northward in the summer and southward in the fall, resulting in coastal upwelling or downwelling that may serve to concentrate or disperse blooms. Certainly the onshore and southerly movement of late summer–fall blooms has been observed repeatedly (Murphy *et al.* 1975, Haddad and Carder 1979, Roberts 1979, Tester and Steidinger 1997), but the direct coupling of transport of blooms from the mid-shelf front shoreward and the rate of transport are poorly understood. The influence of wind stress in coastal waters on stratification, seasonal circulation patterns, sea surface height anomalies, and the timing and persistence of upwelling/downwelling cycles are prime topics for research. Weisberg *et al.* (1996) suggest there is a seasonal pattern to monthly mean current velocities throughout the water column at one of their mooring sites on the mid-shelf. Based on Acoustic Doppler Current Profiler (ADCP) information, average currents tend to be southerly during winter and spring with northerly flows during late summer (Sept–Oct) and fall. Nearshore upwelling and associated shoreward flow also occurs when southerly and westerly winds are maintained (Yang *et al.,* submitted). The upwelling process occurs in relatively shallow nearshore waters and therefore is not associated with nutrient enrichment. However, shoreward flows would provide a mechanism for seeding nearshore blooms from offshore initiation zones.

Inherent in these research questions are those related to reinoculation. If a red tide remains offshore when inshore concentrations of *G. breve* decrease to below background levels, the offshore population can reinoculate inshore areas and red tide conditions may be reestablished. Persistence of reintroduction of cells from offshore and longevity of a red tide event are complementary (Steidinger 1975, Steidinger *et al.* 1995).

4.2. Nutrients and maintenance of blooms

Transport mechanisms can concentrate populations in shallow, nearshore waters where they may be present for days to months. How can such high populations be supported for these time frames in what are normally considered to be oligotrophic waters? Since *G. breve* growth rates are low, and if we assume that uptake is reflected by the cellular N:P ratio, then N and P fluxes need not be high to maintain elevated population levels for long periods. However, *G. breve* has also adopted other strategies to cope with low inorganic N and P availability, e.g., the ability to utilize organic forms of both N and P. Wilson (1966) lists 6 organic P compounds that will support growth equivalent to orthophosphate. Vargo and Shanley (1985) confirmed the ability to produce a cell-surface alkaline phosphatase that, unlike some other species, was not repressed by low (0.5 to 1 μg-at L^{-1}) concentrations of orthophosphate. Thus *G. breve* can utilize inorganic and organic P simultaneously at low *in situ* inorganic P levels.

Baden and Mende (1979) confirmed and added to Wilson's (1966) finding that *G. breve* could use amino acids as a nitrogen source. Their study demonstrated that *G. breve*, unlike other species, utilized the entire compound rather than just stripping off the amine for uptake, thus acquiring a carbon chain as well as the nitrogen. It is unclear whether the carbon chains from the amino acids could be utilized directly for carbohydrate synthesis, but *G. breve* does have the ability to use glucose as an organic carbon source in the light (Baden and Mende 1978). Glucose uptake displays saturation kinetics and is done by facilitative transport processes. Although the uptake rate is low, they speculate that it could provide sufficient carbon for cell maintenance.

Finally, some indication of the degree of support for *G. breve* maintenance in nearshore waters from estuarine input can be estimated from the P requirements calculated by Vargo (1988) and outlined in an earlier section. To support growth and photosynthesis at population levels of about 10^6 cells L^{-1} and a growth rate of 0.2 div. day^{-1}, the P pool would have to turn over twice a day (Table 2). Because the water column standing stock is essentially constant in coastal waters at 0.1 to 0.2 μg-at L^{-1} within 2 km of shore, input must essentially be the same to maintain these concentrations. Based on inorganic and hydrolizable organic P measurements in Tampa Bay, shelf–Bay exchange rates must be on the order of 1% to 3% per day. This flux rate would easily support a population of >10^5 cells per liter at P sufficiency and thus represents a significant source of P for bloom maintenance.

Tampa Bay and Charlotte Harbor are the two largest estuaries on the west coast of Florida. Both are phosphate-enriched as a result of input from their respective drainage basins, which include the Hawthorne phosphatic deposits that occur to the east of Tampa Bay. Although inputs to these estuaries are high, they display a steep gradient with relatively low concentrations near their mouth (Vargo and Shanley 1985, Stoker 1986, Vargo 1988, KEA 1992). As noted above, inorganic P levels are generally less than 0.5 μg-at L^{-1} within 2 km of the shore, suggesting a relatively low flux into coastal waters from these two sources. One could argue that the P is immediately taken up at the mouth of the estuary; however, the concentration is relatively constant year round (at least for Tampa Bay, KEA 1992), whether red tides or diatom blooms are present. Therefore, while P flux from these estuaries may be a source for bloom maintenance and possibly enhanced biomass once populations are transported into nearshore waters, the estuaries are not a source of N and P for bloom initiation. *G. breve* blooms develop in mid-shelf, oligotrophic waters for which the species is ecologically well adapted.

5. Fourth bloom stage: dissipation/termination

The theory of an appropriate water mass or stratified water column as a necessary condition for dinoflagellate blooms is found throughout the harmful algal bloom literature. What causes the demise of blooms? Frequently, the low nutrient levels are cited as an important factor. This is less clear for *G. breve* because it is adapted for low- nutrient environments.

Many of the blooms on the West Florida Shelf subside before they reach nearshore waters (Steidinger and Haddad 1981). Are growth conditions insufficient, or does the cross-shelf transport mechanism fail? Do shifts in wind patterns cause short-term, alternating upwelling-downwelling events that provide needed nutrients but allow no net onshore-offshore movement, so the bloom persists at a mid-shelf front? The account of the 1987–1988 *G. breve* bloom off North Carolina attributes its collapse to low water temperature (<10°C) and a series of strong wind events that both mixed the water column and disrupted the water mass, moving it out of Onslow Bay, NC (Tester *et al.* 1991). Steidinger and Haddad (1981) theorized that the integrity of the water mass in which *G. breve* grows is critical to the continuance of a red tide and that once that integrity is compromised, e.g., diluted by mixing, the bloom cannot be maintained. This theory needs to be investigated from an integrated physical-biological-chemical approach.

5.1. Offshore entrainment and transport

Florida red tides can remain offshore or be transported inshore by winds or currents. If they remain offshore, high population centers can be entrained in offshore fronts and transported downstream.

Between 1878 and 1996, at least eight red tide-related fish kills have been reported from southwest Florida between Cape Romano and the Dry Tortugas off the Florida Keys (Rounsefell and Nelson 1966, Steidinger *et al.* 1995). Such events can be accompanied by an atypical configuration of the Gulf Loop Current and southward transport of a large volume of inner-shelf water (Murphy *et al.* 1975). Reports of sustained *G. breve* blooms, e.g., 1 to 2 months, became easier to understand with the recent description of an eddy system operative between the Dry Tortugas and upper Keys (Lee *et al.* 1994, personal communication). The Tortugas Gyre is a cyclonic, recirculating feature (100–180 km) with a 40–108-day duration. It is dependent on a well-developed Gulf Loop Current and the consequent offshore position of the Florida Current and has a strong influence on the transport and retention of plankton in the Florida Keys (Lee *et al.* 1994). Between episodes of gyre recirculation, there are 20–30 days of intense eastward flow, an ideal transport mechanism to move *G. breve* from the Gulf to the Atlantic.

Prior to 1972, there were no reports of blooms of *G. breve* in the Atlantic, and except for Lackey's (1969) observation of a single cell, this species was not known outside the Gulf of Mexico. Marshall (1982) has since recorded *G. breve* in the Gulf Stream off Virginia. However, in studies initiated in the Gulf of Mexico and along the southeast US coast after an unusual bloom during fall–winter 1987 off Cape Lookout, NC, *G. breve* was found to be ubiquitous in low numbers throughout its range (Geesey and Tester 1993, Tester *et al.* 1993). Cell concentrations of >0.02 to 1000 × 10^6 cells L^{-1} were documented from Miami to Jacksonville, FL, in 1972, 1977, 1980, 1983, and again in 1995. In three of these blooms (1983, 1987, and 1995), the Gulf Stream was the documented transport agent for higher-than-background cell concentrations into appropriate receiving waters where cell division could be supported (Tester *et al.* 1991, Steidinger *et al.* 1995). Transport of an entrained bloom out of an area represents one form of dissipation.

6. Conclusions

The unarmored, toxic dinoflagellate *Gymnodinium breve* is ubiquitous in the Gulf of Mexico at background concentrations (<1000 cells L^{-1}), but it frequently blooms in fall months on the West Florida Shelf. Higher-than-background concentrations are first detected offshore on the mid shelf, and the initiation zone has been characterized as 18 to 74 km off Clearwater to Sanibel, Florida. It takes several weeks for the bloom to increase in cell density and be transported inshore by currents and winds. *Gymnodinium breve* blooms develop and are even maintained in oligotrophic coastal waters, albeit initial growth and transport are associated with frontal systems. We do not know whether growth and physiological rate processes are the same along the spatial gradient from initiation or onset to termination. What we do know is that the adaptive strategies of *G. breve* are successful. It can out-compete or exclude other plankters to form nearly monospecific blooms that can cover thousands of km^2 and can last for months. What are its adaptive strategies? It is adapted to high-salinity water with wide temperature ranges. It is very efficient at using inorganic nitrogen and phosphorus but can easily use organic N and P; and typically at cell concentrations of 10^4 to 10^6 L^{-1}, daily population requirements for macronutrients are met. The species has a high affinity for inorganic nitrogen and a low K_s for phosphate. Its growth rate as measured *in situ* is 0.2 to 0.3 div. day^{-1}, and it is photosynthetically efficient over varying light levels as it maintains a relatively constant respiration rate. Its photobiology and behavior protect it as it concentrates and disperses in surface waters in daylight. Its morphology and toxins protect it from certain zooplankton predators. Its life cycle ensures that it occupies the same niche year after year; however, we have not determined the complexity or simplicity of this cycle, nor have we determined the complexity or simplicity of its other adaptations to the point of being able to predict bloom dynamics.

7. References

Ahmed, M.S., Arakawa, O., Onoue, Y. (1995). Toxicity of cultured *Chattonella marina.* In: Lassus, P., Arzul, G., Erard-Le Denn, E., Gentien, P., Marcaillou-Le Baut, C. (eds.) *Harmful Marine Algal Blooms.* Lavoisier, Paris, pp. 499–504.

Aldrich, D.V. (1959). Physiological studies of red-tide. U.S. Fish Wildl. Serv. Circ. 62, pp. 69–71.

Aldrich, D.V. (1960). Physiology of the Florida red tide organism. U.S. Fish Wildl. Serv. Circ. 92, pp. 40–42.

Aldrich, D.V., Wilson, W.B. (1960). The effects of salinity on growth of *Gymnodinium breve* Davis. *Biol. Bull.* 119: 57–64.

Baden, D.G., Mende, T.J. (1978). Glucose transport and metabolism in *Gymnodinium breve. Phytochemistry* 17: 1553–1558.

Baden, D.G., Mende, T.J. (1979). Amino acid utilization by *Gymnodinium breve. Phytochemistry* 18: 247–251.

Bein, S.J. (1957). The relationship of total phosphorus concentration in sea water to red tide blooms. *Bull. Mar. Sci. Gulf Caribb.* 7: 316–329.

Chang, F.H. (1996). A review of knowledge of a group of closely related, economically important toxic *Gymnodinium/Gyrodinium* (Dinophyceae) species in New Zealand. *J. Royal Soc. N.Z.* 26: 381–394.

Davis, C.C. (1948). *Gymnodinium brevis* sp. nov., a cause of discolored water and animal mortality in the Gulf of Mexico. *Bot. Gaz.* 109: 358–360.

Doig, M.T. (1973). The growth and toxicity of the Florida red tide organism, *Gymnodinium breve.* PhD. Dissertation, Univ. South Florida, Tampa. 80 p.

Dragovich, A. (1967). Morphological variations of *Gymnodinium breve* in situ. *Q. J. Fla. Acad. Sci.* 30: 245–249.

Dragovich, A., Kelly, J.A Jr. (1966). Distribution and occurrence of *Gymnodinium breve* on the west coast of Florida, 1964–65. U.S. Fish Wildl. Serv. Spec. Sci. Rep. Fish. 541. 15 p.

Dragovich, A., Finucane, J.H., Mays, B.Z. (1961). Counts of red tide organisms, *Gymnodinium breve* and associated oceanographic data from Florida west coast, 1957–1959. U.S. Fish Wildl. Serv. Spec. Sci. Rep. Fish. 369. 175 p.

Dragovich, A., Finucane, J.H., Kelly, J.H. Jr., Mays, B.Z. (1963). Counts of red tide organisms, *Gymnodinium breve* and associated oceanographic data from Florida west coast, 1960–1961. U.S. Fish Wildl. Serv. Spec. Sci. Rep. Fish. 455. 40 p.

Dugdale, R.C., Goering, J.J. (1967). Uptake of new and regenerated forms of nitrogen in primary productivity. *Limnol. Oceanogr.* 12: 196–206.

Eng-Wilmot, D.L., Hitchcock, W.S., Martin, D.F. (1977). Effects of temperature on the proliferation of *Gymnodinium breve* and *Gomphosphaeria aponina. Mar. Biol.* 41: 71–77.

Feinstein, A. (1956). Correlations of various ambient phenomena with red tide outbreaks on the Florida West shelf. *Bull. Mar. Sci. Gulf Caribb.* 6: 209–232.

Fraga, S., Sanchez, F.J. (1985). Toxic and potentially toxic dinoflagellates found in Galician rias (NW Spain). In: Anderson, D.M., White, A.W., Baden, D.G. (eds.) *Toxic Dinoflagellates.* Elsevier, NY, pp. 51–54.

Geesey, M., Tester, P.A. (1993). *Gymnodinium breve*: ubiquitous in Gulf of Mexico waters? In: Smayda, T.J., Shimizu, Y. (eds.) *Toxic Phytoplankton Blooms in the Sea.* Elsevier, NY, pp. 251–255.

Haddad, K.D. (1982). Hydrographic factor associated with west Florida toxic red tide blooms: an assessment for satellite predictions and monitoring. M.S. Thesis, Univ. South Florida. 155 p.

Haddad, K.D., Carder, K.L. (1979). Oceanic intrusion: one possible initiation mechanism of red tide blooms on the west coast of Florida. In: Taylor, D.L., Seliger, H.H. (eds.) *Toxic Dinoflagellate Blooms.* Elsevier, NY, pp. 269–274.

Haywood, A., Mackenzie, L., Garthwaite, I., Towers, N. (1996). *Gymnodinium breve* 'look-alikes': three *Gymnodinium* isolates from New Zealand. In: Yasumoto, T., Oshima, Y. and Fukuyo, Y. (eds.) *Harmful and Toxic Algal Blooms.* Intergovernmental Oceanographic Commission of UNESCO. Sendai Kyodo Printing Co. Ltd, pp. 227–230.

Heckey, R.E., Kilham, P. (1988). Nutrient limitation of phytoplankton in freshwater and marine environments: A review of recent evidence on the effects of enrichment. *Limnol. Oceanogr.* 33: 796–822.

Heil, C. (1986). Vertical migration of *Ptychodiscus brevis* (Davis) Steidinger. M.S. Thesis, Univ. South Florida. 118 p.

Khan, S., Arakawa, O., Onoue, Y. (1996a). A toxicological study of the marine phytoflagellate, *Chattonella antiqua* (Raphidophyceae). *Phycologia* 35: 239–244.

Khan, S., Arakawa, O., Onoue, Y. (1996b). Neurotoxin production by a chloromonad *Fibrocapsa japonica* (Raphidophyceae). *J. World Aquacult. Soc.* 27: 254–263.

KEA (1992). Review and synthesis of historical Tampa Bay water quality data. Tampa Bay National Estuary Program Tech. Rep. No. 7-92.

Kimor, B. (1996). A short-lived toxic phytoplankton bloom in Haifa Bay (Israel). *EOS, Trans. Am. Geophys. Un.* 76 (suppl.): OS34.

Lackey, J.B. (1969). Microbial studies in the FWPCA project area with comparisons to other subtropical and tropical areas. In: McAllister, R.F. (ed.) *Demonstration of the Limitations and Effects of Waste Disposal on an Ocean Shelf.* Fla. Ocean Sci. Inst. Rep. AR-69-2, pp. 52–58.

Lee, T.N., Williams, E. (1988). Wind-forced transport fluctuations of the Florida Current. *J. Phys. Oceanogr.* 18: 937–946.

Lee, T.N., Clarke, M.E., Williams, E., Szmant, A.F., Berger, T. (1994). Evolution of the Tortugas gyre and its influence on recruitment in the Florida Keys. *Bull. Mar. Sci.* 54: 621–646.

Mackenzie, L., Haywood, A., Adamson, J., Truman, P., Till, D., Satake, M., Yasumoto, T. (1996). Gymnodimine contamination of shellfish in New Zealand. In: Yasumoto, T., Oshima, Y. and Fukuyo, Y. (eds.) *Harmful and Toxic Algal Blooms.* Intergovernmental Oceanographic Commission of UNESCO. Sendai Kyodo Printing Co. Ltd., pp. 97–100.

Marshall, H. G. (1982). The composition of phytoplankton within the Chesapeake Bay plume and adjacent waters off the Virginia coast, U.S.A. *Estuarine Coastal Shelf Sci.* 15: 29–43.

Maul, G.A. (1977). The annual cycle of the Gulf Loop. Part I: Observations during a one-year time series. *J. Mar. Res.* 35: 219–247.

Millie, D.F., Kirkpatrick, G.J., Vinyard, B.T. (1995). Relating photosynthetic pigments and *in vivo* optical density spectra to irradiance for the Florida red-tide dinoflagellate *Gymnodinium breve. Mar Ecol. Prog. Ser.* 120: 65–75.

Murphy, E.B., Steidinger, K.A., Roberts, B.S., Williams, J., Jolley, J.W. (1975). An explanation for the Florida east coast *Gymnodinium breve* red tide of November, 1972. *Limnol. Oceanogr.* 20: 481–486.

Odum, H.T., Lackey, J.B., Hynes, J., Marshall, N. (1955). Some red tide characteristics during 1952–1954. *Bull. Mar. Sci. Gulf Caribb.* 5: 247–258.

Rivkin, R.B., Swift, E. (1985). Phosphorus metabolism of oceanic dinoflagellates: phosphorus uptake, chemical composition, and growth of *Pyrocystis noctiluca. Mar. Biol.* 88: 189–198.

Roberts, B.S. (1979). Occurrence of *Gymnodinium breve* red tides along the west and east coasts of Florida during 1976 and 1977. In: Taylor, D.L., Seliger, H.H. (eds.) *Toxic Dinoflagellate Blooms.* Elsevier, NY, pp. 199–202.

Rounsefell, G.A., Nelson, W.R. (1966). Red-tide research summarized to 1964 including an annotated bibliography. U.S. Fish Wildl. Serv. Spec. Sci. Rep. Fish. 535. 85 p.

Shanley, E. (1985). Photoadaptation in the red-tide dinoflagellate *Ptychodiscus brevis.* M.S. Thesis, Univ. South Florida. 122 p.

Shanley, E., Vargo, G.A. (1993). Cellular composition, growth, photosynthesis, and respiration rates of *Gymnodinium breve* under varying light levels. In: Smayda, T.J., Shimizu, Y. (eds.) *Toxic Phytoplankton Blooms in the Sea.* Elsevier, NY, pp. 831–836.

Shimizu, Y., Wrensford, G. (1993). Peculiarities in the biosynthesis of brevetoxins and metabolism of *Gymnodinium breve.* In: Smayda, T.J., Shimizu, Y. (eds.) *Toxic Phytoplankton Blooms in the Sea.* Elsevier, NY, pp. 919–923.

Shimizu, Y., Watanabe, N., Wrensford, G. (1995). Biosynthesis of brevetoxis and heterotrophic metabolism in *Gymnodinium breve.* In: Lassus, P., Arzul, G., Erard-Le Denn, E., Gentien, P., Marcaillou-Le Baut, C. (eds.) *Harmful Marine Algal Blooms.* Lavoisier, Paris, pp. 351–357.

Smith, G.B. (1975). The 1971 red tide and its impact on certain reef-fish communities in the eastern Gulf of Mexico. *Environ. Lett.* 9: 141–152.

Smith, G.B. (1979). Relationship of eastern Gulf of Mexico reef-fish communities to the equilibrium theory of insular biogeography. *J. Biog.* 6: 49–61.

Stoker, Y.E. (1986). Water quality of the Charlotte Harbor estuarine system, Florida, November 1982–1984. U.S.G.S. Open-File Rep. 85-563.

Steidinger, K.A. (1975a). Basic factors influencing red tides. In: LoCicero, V.R. (ed.) Proceedings of the First International Conference on Toxic Dinoflagellate Blooms. Mass. Sci. Tech. Found., Wakefield, MA, pp. 153–162.

Steidinger, K.A. (1975b). Implications of dinoflagellate life cycles on initiation of *Gymnodinium breve* red tides. *Environ. Lett.* 9: 129–139.

Steidinger, K.A. (1990). Species of the tamarensis/catenella group of *Gonyaulax* and the fucoxanthin derivative-containing gymnodinioids. In: Graneli, E., Sundstrom, B., Edler, L., Anderson, D.M. (eds.) *Toxic Marine Phytoplankton.* Elsevier, NY, pp. 11–16.

Steidinger, K.A., Ingle, R.M. (1972). Observations on the 1971 red tide in Tampa Bay, Florida. *Environ. Lett.* 3: 271–278.

Steidinger, K.A., Haddad, K.D. (1981). Biologic and hydrologic aspects of red tides. *Bioscience* 31: 814–819.

Steidinger, K.A., Vargo, G.A. (1988). Marine dinoflagellate blooms: dynamics and impacts. In: Lembi, C., Waaland, J.R. (eds.) *Algae and Human Affairs.* Cambridge Univ. Press, NY, pp. 373–401.

Steidinger, K.A., Babcock, C., Mahmoudi, B., Tomas, C., Truby, E. (1989). Conservative taxonomic characters in toxic dinoflagellate species identification. In: Okaichi, T., Anderson, D.M., Nemoto, T. (eds.) *Red Tides: Biology, Environmental Science, and Toxicology.* Elsevier, NY, pp. 285–288.

Steidinger, K.A., Roberts, B.S., Tester, P.A. (1995). Florida red tides. Harmful Algal News 12/13: 1–3.

Sturges, W.A. (1994). The frequency of ring separations from the Loop Current. *J. Phys. Oceanogr.* 24: 1647–1651.

Sturges, W.A., Evans, J.C. (1983). On the variability of the Loop Current in the Gulf of Mexico. *J. Mar. Res.* 41: 639–653.

Tester, P.A., Steidinger, K.A. (1997). *Gymnodinium breve* red tide blooms: initiation, transport and consequences of surface circulation. *Limnol. Oceanogr.* (in press)

Tester, P.A., Stumpf, R.P., Vukovich, F.M., Fowler, P.K., Turner, J.T. (1991). An expatriate red tide bloom: transport, distribution, and persistence. *Limnol. Oceanogr.* 36: 1053–1061.

Tester, P.A., Geesey, M.E., Vukovich, F.M. (1993). *Gymnodinium breve* and global warming: What are the possibilities? In: Smayda, T.J., Shimizu, Y. (eds.) *Toxic Phytoplankton Blooms in the Sea.* Elsevier, NY, pp. 67–72.

Vargo, G.A. (1988). Phosphorus dynamics in red-tide blooms on the West Florida Shelf: P requirements for growth and photosynthesis. Abstract, 1988 fall AGU/ASLO meeting, San Francisco, CA.

Vargo, G.A., Howard-Shamblott, D. (1990). Phosphorus dynamics in *Ptychodiscus brevis:* cell phosphorus, uptake and growth requirements. In: Graneli, E., (eds.) *Toxic Marine Phytoplankton.* Elsevier, NY, pp. 324–329.

Vargo, G.A., Shanley, E. (1985). Alkaline phosphatase activity in the red-tide dinoflagellate, *Ptychodiscus brevis. PSZNI Mar. Ecol.* 6: 251–264.

Vukovich, F.M. (1995). An updated evaluation of Loop Current eddy-shedding frequency. *J. Geophys. Res.* 100C5: 8655–8659.

Walker, L.M. (1982). Evidence for a sexual cycle in the Florida red tide dinoflagellate *Ptychodiscus brevis* (=*Gymnodinium breve*). *Trans. Am. Microsc. Soc.* 101: 287–293.

Weisberg, R.H., Black, B.D., Yang, H. (1996). Seasonal modulation of the west Florida continental shelf circulation. *Geophys. Res. Ltr.* (in press)

Williams, J., Grey, W.F., Murphy, E.B., Crane, J.J. (1977). Drift analyses of eastern Gulf of Mexico surface circulation. Mem. Hourglass Cruises 4(3): 1–134.

Wilson, W.B. (1955). Laboratory studies of *Gymnodinium brevis*. 14 p. Mimeographed address to Am. Assoc. Adv. Sci.

Wilson, W.B. (1966). The suitability of sea-water for survival and growth of *Gymnodinium breve* Davis; and some effects of phosphorus and nitrogen on its growth. Fla. Board Conserv. Mar. Lab. Prof. Pap. Ser. No. 7. 42 p.

Wilson, W.B. (1967). Forms of the dinoflagellate *Gymnodinium breve* Davis in cultures. *Contrib. Mar. Sci.* 12: 120–134.

Wilson, W.B., Ray, S. (1958). Nutrition of red tide organisms. U.S. Fish Wildl. Serv. Gulf Fish. Invest. Annu. Rep., pp. 62–65.

Wilson, W.B., Ray, S.M., Aldrich, D.V. (1975). *Gymnodinium breve*: population growth and development of toxicity in cultures. In: LoCicero, V.R. (ed.) Proceedings of the First International Conference on Toxic Dinoflagellates. Mass. Sci. Tech. Found., Wakefield, MA, pp. 127–141.

Yang, H., Weisberg, R.H., Black, B. 1996. Climatological wind responses of the three-dimensional west Florida continental shelf circulation. *J. Geophys. Res.* (in press)

Bloom Dynamics and Ecophysiology of the *Gymnodinium mikimotoi* Species Complex

P. Gentien
IFREMER, Centre de Brest, DEL/Ec. Pel., B.P.70 , 29280 Plouzané, France

1.Introduction

Known in the literature under different names (*G. aureolum, G.* cf. *aureolum, Gymnodinium nagasakiense, G. mikimotoi*), *G. mikimotoi* Adachi & Fukuyo is one of the most common red tide dinoflagellates proliferating in the Eastern North Atlantic regions and around Japan. Blooms of this species are commonly associated with marine fauna kills, but unlike certain other toxic species, it does not seem to be the cause of human intoxications. Mild symptoms in humans have been reported, following contact with sea water (Mahoney *et al.* 1990), an observation which has not been confirmed. Ecological and economic consequences of *G.* cf. *nagasakiense* red tides are often dramatic (Partensky *et al.* 1991a). Toxic effects are restricted to a direct action on primary consumers (herbivorous fish and shellfish) and other species present on site during the blooms. Recently, in 1995, along the French Atlantic coast, mortality of an estimated 800-900 tons of *Mytilus edulis,* among other species, was observed to coincide with an exceptional bloom. The density of *G.* cf. *nagasakiense* cells must usually reach a few million cells per liter for any measurable effect on marine fauna.

In this chapter, the bloom dynamics of *G.* cf. *nagasakiense* are reviewed in relation to its physiological properties. The contours of the niche of this dinoflagellate, delineated by its ecophysiology, are outlined, and an application to the understanding of a large bloom on the French Atlantic Coast is presented. Toxicity is discussed only with respect to its effects on potential competitors and grazers. A summary of the important questions still pending concludes this chapter.

2.Species of concern

Gyrodinium aureolum Hulburt was first described by Hulburt (1957) from samples collected in a coastal lagoon close to Woods Hole (Massachusetts, USA). This name was subsequently used by Braarud and Heimdal (1970) to describe a dinoflagellate responsible for a "red tide" in 1966 along Norwegian coasts and has been used since in North Sea waters (Tangen 1977).

This European species is morphologically similar to the Japanese species, *Gymnodinium nagasakiense* Takayama & Adachi, which has been responsible for important red tides. Optical and electron microscopical observations confirmed a close similarity between the two species. Both present a similar apical groove, and cytological observations show that both European and Japanese taxa have a large nucleus, elongated along the ante-posterior axis, contrary to *G. aureolum* (US species) in which the nucleus is spherical. Despite their close morphological similarities, some differences remain between the Japanese and European taxa, particularly in terms of DNA content, physiology (subpopulation differentiation) and behavior (vertical migration) which cast some doubts on their conspecificity. Partensky *et al.* (1988) suggested to use of the name *Gymnodinium* cf. *nagasakiense* for the European variant.

Gymnodinium nagasakiense has an junior synonym, *Gymnodinium mikimotoi* Miyake et Kominami ex Oda. (Takayama and Matsuoka 1991) and possibly '*Gymnodinium* type

NATO ASI Series, Vol. G 41
Physiological Ecology of Harmful Algal Blooms
Edited by D. M. Anderson, A. D. Cembella and G. M. Hallegraeff

65' (Takayama 1981), as well. The species are at the theoretical limit between *Gyrodinium* and *Gymnodinium*. The main discriminating criterion between these species groups is the ratio of the cingulum displacement to the total length of the cell, but as this ratio is reported to be greater than 20% in *Gyrodinium* (Kofoid and Swezy 1921), and varying between 17 and 27% in *G.* cf. *nagasakiense*, this distinction appears arbitrary. The two genera intergrade and those species on the boundary could be combined (Partensky *et al.* 1988). Six monoclonal antibodies prepared against *G.* cf. *aureolum* (strain Iroise, France) cross-react only with Japanese and Scandinavian strains of *Gymnodinium mikimotoi*, *G.* (cf.) *nagasakiense*, *Gyrodinium* (cf.) *aureolum*. *G. aureolum* was reported in the St. Lawrence gulf (Quebec) in 1993 (Blasco *et al.* 1996), the first report of this species in this area. Immunological tagging methods indicated that the species found in the Gulf was phenotypically identical to the species found in Northern Europe.

Since close relationships between these species have been noted and due to the fact that the deleterious effects of the blooms are very similar, this chapter presents results concerning the five following species: *Gymnodinium mikimotoi*, *G.* (cf.) *nagasakiense* and *Gyrodinium* (cf.) *aureolum*. *Gymnodinium mikimotoi* will be used as a generic alias for all these species.

3.Geographical distribution

Gymnodinium nagasakiense blooms have been reported around west Japan (Matsuoka *et al.* 1989), where the species occurs in water temperatures ranging from 13°C to 31°C and salinities comprised between 15 to 35 p.s.u.. It has also been reported in Hong-Kong waters (Wong 1989). Red tides have occurred in the Southern Atlantic (Negri *et al.* 1992) and along the East coast of USA (Chang and Carpenter 1985; Mahoney *et al.* 1990). In 1993, blooms were reported for the first time along the entire coast of Quebec, Canada and in offshore waters of the Gulf of St. Lawrence (Blasco *et al.* 1996), where temperature and salinity ranges were between 4-18°C and 17-29 p.s.u. respectively.

Gymnodinium cf. *nagasakiense* has been investigated by many workers on European coasts (for a map, see Partensky *et al.* 1991a). Occurrences have been reported from Senja Island, Norway (70°N) to the Galician coast of Spain. Throughout this range, this dinoflagellate species is essentially neritic. It does not occur in mixed waters like the English Channel or the Celtic sea, but is often associated with tidal fronts on the stratified side. In European waters, red tides occur generally in the 30-36 p.s.u. range, extending to 25 p.s.u. in the Kattegat, and to even lower salinities (9 p.s.u.) at times. Videau (unpubl. results) compared three strains of the western English channel, having the same immunological response (Plymouth, Brest Bay and Ushant front). Optimal growth (0.4 day^{-1}) was reached at 20°C and 100 μmole $m^{-2}s^{-1}$. In contrast with the other strains, the Ushant front strain, from the furthest offshore, was still able to maintain its cell density at 10°C.

The range of environmental conditions in which this species has been reported confirm the opinion of Yamaguchi and Honjo (1989) that *G. nagasakiense* is a eurythermal and euryhaline organism. These physiological features allow it to endure winter conditions as motile cells, which in turn act as the seed population for the summer red tide.

4- Oceanographic observations

There are three basic situations in which the concentration of *G. nagasakiense* cells may reach red tide proportions: 1- semi-confined water bodies; 2- buoyant plumes and estuaries; 3- tidal fronts. Yanagi *et al.* (1992) examined the physical conditions associated with the occurrence of a red tide of *Gymnodinium mikimotoi* in the Seto inland sea. They reported that red tides are favoured by low solar radiation and temperature and high rainfall in early summer. Yamaguchi (1994) included these observations into a more general scheme. Due to increased supply nutrients and establishment of stratification, the population starts to build up from the local initial seed population of overwintering cells. It reaches division rates of 1.0 day^{-1}, when temperature and salinity reach 25°C and 25 p.s.u. respectively at irradiances superior to 110 μmole m^{-2}s^{-1}. Concentration of cells at the surface due to wind related water movement and tidal flow leads to the occurrence of red tides. A numerical simulation (Yanagi *et al.* 1995) taking into account parametrization of the physiology of this species reproduced quite accurately the observed situation. These authors found that ammonium night uptake and hydrological accumulation of cells play essential roles in the bloom development.

Lindahl (1993) examined the role of hydrodynamical processes in the triggering of *G. aureolum* development in the Kattegat and Skagerrak. Low saline water from the Baltic sea, is transported northward as a buoyant coastal current through the Kattegat and along the south coast of Norway. During this transport, nutrient-rich deep waters are advected into the surface layer. Nutrient-rich waters are present in the central Skagerrak. Three different types of blooms can be distinguished in this area: 1-surface blooms which occur in the surface waters above the pycnocline; 2-shallow subsurface blooms at 10-20m depth very close to the pycnocline; 3-deep subsurface blooms (30-50 m depth) which occur offshore of the coastal currents and are in connection with the presence of Atlantic waters.

Blooms of *G. aureolum* occur at 20-50 m depth in the Skagerrak. The deep water containing such a high density population can be transported landwards and up to the surface in the coastal area during westerly or northerly winds. Lindahl (1993) considered this phenomenon to be particularly significant in relation to the occurrence of red tides in coastal areas. The Skagerrak *G. aureolum* blooms subsequently follow the Norwegian coastal current at the base of the pycnocline and can cause nuisance events as far north as the Trondheim area (Dahl and Tangen 1993).

Gyrodinium aureolum is a species which frequently dominates the phytoplankton communities in frontal regions. Raine *et al.* (1993) proposed the following scenario for coastal waters around the west coast of Ireland : 1-Species succession with increasing water stratification to dinoflagellate dominance in summer, 2-Full establishment (Δ T>5°C) of the seasonal pycnocline by mid-July and 3-Development of *G. aureolum* during July.

Once a *G. aureolum* population is established in open coastal water frontal regions, at least two different physical processes can transport the populations shorewards. Some wind-regimes may result in a relaxation of upwelling and a subsequent incursion of water into coastal bays from open coastal regions. Another mechanism could be a wind-driven exchange in the bays, forcing intrusion of warm surface water containing large populations of *G. aureolum* (Raine *et al.* 1993).

The theme common to the majority of these observations is that *G. aureolum* is often found in or near the pycnocline layer during some stage of population development (the main exception seems to be when the population develops in shallow, weakly stratified semi-confined water bodies).

5- Vertical distribution and migration

This topic is critical to the understanding of bloom development, since the ability of a dinoflagellate to migrate through the water column could give it an *a priori* advantage over other algae which depend on turbulence, buoyancy variations and cell ornementations, to counteract sedimentation and remain in a favorable environment for growth. Commonly observed *in vitro*, migration has been seldom reported *in situ*.

Koizumi *et* al. (1996) provided the first observations of diurnal migrations of *G. mikimotoi*. At the two stations monitored, stratification was weak (1°C m^{-1};0.12 psu m^{-1}). Maximum density depth varied from 0-1 m at noon to 20 m at midnight. The authors estimated both upward and downward migration velocities *ca.* 2.2 m h^{-1}. This speed is very high in comparison with swimming speeds reported for other dinoflagellates (Kamykowsky 1995). At the shallowest station (20 m depth), the population reached the bottom layer. The vertical migration of this species may be an ecologically important feature in shallow waters since the cells can utilize nutrients and growth-promoting substances present in the mud interstitial water (Hirayama and Numaguchi 1972).

In the Carmans estuary, Chang and Carpenter (1985) observed a complex interaction between the migration pattern and the tidal circulation which reduces the amount of cells being flushed out of the estuary, leading to bloom formation. They suggested that vertical migration may be a response to salinity as well as to light intensity.

Experiments in *in situ* enclosures of 13-15 meters depth (Dahl and Brockmann 1985, 1989) have shown either no diurnal vertical migration or limited migrations within 2m in the pycnocline layer. In the latter case, the gradient in thc pycnocline was 4 p.s.u. and 2.5 °C at five meters depth. A bloom of *G.* cf. *aureolum* in a shallow stratified ecosystem (Birrien *et al.* 1991) remained confined at the pycnocline level during its development which lasted 2 months.

The offshore Ushant thermal front delimits two water masses: a coastal water mass relatively cold and vertically mixed and an offshore temperature stratified water column (17 - 11 °C) (Fig. 2). In the latter zone, as long as the temperature difference between the two mixed overlying water bodies (ΔT) was lower than 1.2°C, *G.* cf. *aureolum* was located in the upper layers, at high light levels, even when the water was nutrient depleted. When ΔT exceeded 3°C, the chlorophyll maximum occurred in the pycnocline at low light levels (<5% incident light). This chlorophyll maximum was due to *G.* cf. *aureolum* in July or August (Arzul *et al.* 1993). The maximum cell density depth varied in the pycnocline depth range (5m at 20m depth), the deeper being at night.

Bjoernsen and Nielsen (1991) studied the distribution of *G. aureolum* in a pycnocline in the Kattegat with a high resolution sampler. They observed a strong heterogeneity of the dinoflagellate population at the decimeter scale and supposed that *G. aureolum* formed at that time a more or less coherent "magic carpet" in the pycnocline of the Kattegat. They supposed that the inhibition of potential predators could be an important factor in the maintenance of a high phytoplankton biomass in the pycnocline. However, this is a consequence rather than an explanation.

To summarize, in the case of mixed or slightly stratified water columns, *G. mikimotoi* is observed to migrate up and down with a range of up to 15 m. When stratification is greater, the population exhibits a non-migrating maximum in the pycnocline. It is very difficult to estimate from the available data, the lowest density gradient through which migration still occurs. Sharp pycnoclines are associated with high shear between water bodies and, if phototropism modulated by cell quota were the only driving force of population movement, cells escaping from the transition zone would be flushed out; observations of high density layers could be partly the result of this selection. Even if fine layering can result from purely physical processes (Franks 1995), the persistence of populations in narrow layers at such small scale suggests that other factors, such as chemotropism or higher survival rates in some low turbulent energy layers, may be involved in the maintainence of high density populations within these layers. It seems that under certain hydrological and meteorological conditions, these boundary layer waters may exhibit limited residual movement, allowing the population to develop with limited dispersion.

In French Atlantic coastal waters, light penetrating to 20m depth is between 5 and 10 % of incident light. To take advantage of the limited physical dispersion in these layers, *G.* cf. *nagasakiense* must, therefore, be able to efficiently utilize low light energies and to grow in subdued light.

6.Light utilization

Dense concentrations of *G. aureolum* are found both in surface waters and in the vicinity of the pycnocline. The species grows actively in a large range of light regimes and its physiological properties should reflect adaptative capacities.

Richardson and Kullenberg (1987) report that photosynthesis per cell by cells previously exposed to low light (10 μmol m^{-2} s^{-1}) is higher at non-saturating light regimes than for cells preconditioned in high light (150 μmol m^{-2} s^{-1}). However, light saturation occcurs at approximately the same photon flux density for both high- and low-light adapted cells (around 200 μmol m^{-2} s^{-1}, which is higher than for many dinoflagellates). The P-I curves reported by for a Western Channel population do not differ significantly from reports by Richardson and Kullenberg (1987). Garcia and Purdie (1994) report also that cultures adapted to high lights (235-380 μmol m^{-2} s^{-1}) had considerable reduced rates of maximum photosynthesis and lower efficiencies per cell than cultures grown at lower irradiances (35-120 μmol m^{-2} s^{-1}). Compensation irradiance was lower (21- 29 μmol m^{-2} s^{-1}) for cultures adapted to low light levels (<120 μmol m^{-2} s^{-1}) than for high light adapted cultures (>235 μmol m^{-2} s^{-1}) for which the compensation exceeded 50 μmol m^{-2} s^{-1}.

No inhibition is observed in *G. aureolum* exposed to high light. This statement is confirmed by Dixon and Holligan (1989) in the case of a natural population for irradiances up to 1200 μmol m^{-2} s^{-1}.

Light saturated P_{max} range were in the range 0.7-2.3 mgC mg^{-1} chl$a^{-1}h^{-1}$ in the Channel and 0.7 to 1.5 in Norwegian waters. Maximum rates of photosynthesis are likely to occur only in near-surface waters. The fact that low-light-adapted *G. aureolum* cells apparently retain the ability to photosynthetize maximally at high light make this species particularly well suited to exploit dynamic light environments such as those occuring in frontal regions. Populations from the Western English channel composed of more than 90% *G.* cf. *aureolum* presented photosynthetic quotients varying between 1.2 and 3.7 mol O_2 mol^{-1} CO_2 (Garcia and Purdie 1994). The average maximum photosynthetic rate normalized per chlorophyll unit was 0.123 μmol C μg^{-1} chla^{-1} h^{-1}.

According to Yamaguchi and Honjo (1989), irradiances greater than 10 μmol m^{-2} s^{-1} supported growth. Saturation in terms of growth was observed at 110 μmol m^{-2} s^{-1}. The relation between the growth rate and irradiance was hyperbolic with a half saturation constant for the growth rate of 54 μmol m^{-2} s^{-1} and a maximum growth rate of 1.2 div. day^{-1}. Nielsen (1992) reported that, for a 18h day-length, the maximum growth rate of *G. aureolum* (0.4 day^{-1}) was at 150 μmol m^{-2} s^{-1}, whereas for higher irradiances the growth rate decreased quite significantly. Cellular content of chlorophyll *a* and chlorophyll *a*/ Carbon ratio also decrease with both increasing irradiance and day length. The saturation irradiance for photosynthesis of *G. aureolum* is higher than for many other dinoflagellate species (Partensky and Sournia 1986). The difference in light requirements for growth and chl*a*-normalized photosynthesis is due to the decrease in the cellular chl*a*/C ratio with increasing irradiance (Nielsen 1992). Garcia and Purdie (1994) reported that under saturation of growth which occurs *ca.* 100 μmol m^{-2} s^{-1} , decreases in growth rates and cell size, and a relative increase in carbon specific respiration rates could be observed.

The pigment composition and bio-optical characteristics of *G. aureolum* are very plastic with respect to adaptation to the growth-light regime (Johnsen and Sakshaug 1993).

A light regime such as those reported from pycnocline levels (1-5% incident light) could support net growth of the population, albeit at non-saturating rates (Richardson and Kullenberg 1987). If a pycnocline population was then brought to the surface, such results suggest that the population could immediately capitalize on the change in the light regime. *G.* cf. *nagasakiense* is particularly well adapted to low light environment, but can also benefit from high irradiances.

7.Population growth

7.1.Growth rate and cell cycle

Lindahl (1983) gives 0.3-1.0 day^{-1} on the Swedish west coast. Dahl and Brockmann (1985) reported 0.3 day^{-1} in *in situ* experimental bags and 0.5 day^{-1} for the same strain in culture. Iizuka (1979) found growth rates of 1 day^{-1} typical of *Gymnodinium* type-'65 under conditions of very low nutrients in Omura Bay. Johnsen and Sakshaug (1993) reported 0.33 day^{-1} at 170 μmol m^{-2} s^{-1} .

Chang and Carpenter (1985) found a highest rate of 0.90 day^{-1}. Dahl and Brockmann (1985) reported in enclosures rates of 0.27 day^{-1}. Maximum growth rates reported in the literature range from 1 to 0.3 day^{-1}, which is quite a wide for the presumed same species grown in cultures. Subpopulation differentiation could explain such a variability.

Gymnodinium cf. *nagasakiense* populations differentiate into two size related subpopulations, in contrast with *G. nagasakiense* for which the different sizes do not seem to co-occur. Implications of such a feature are not yet clear in terms of population dynamics.

G. cf. *nagasakiense* in culture or in the field forms a subpopulation of reduced size (apical length :16-27 μm) coexisting sometimes with larger cells (26-37 μm). Size distributions during a bloom in the bay of Brest are represented on Fig.1. Apical length tended to be reduced towards the end of the bloom (Gentien and Arzul 1991). In cultures, the two subpopulations can be distinguished in terms of Coulter counter cell volume, chlorophyll and proteins (Partensky and Vaulot 1989). Small cells are usually elongated while

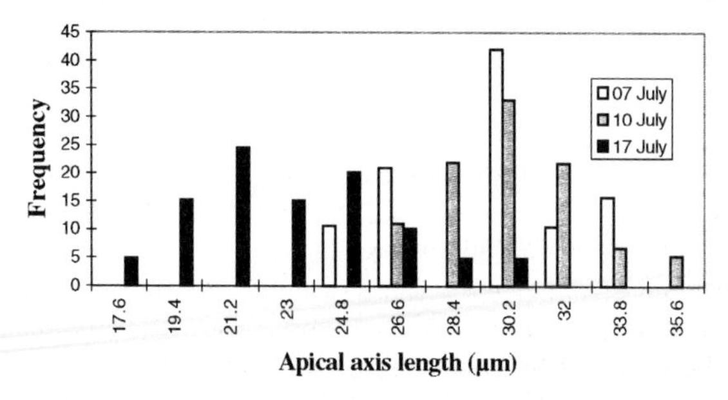

Fig. 1 : Evolution of the apical axis length (Brest Bay)

large cells are more spherical. Both forms were able to divide asexually. Cytofluorometric analyses of the two sub-populations show that both subgroups present cells in G_1, S and G_2+M phases of the cell cycle and that the two cycles are independent. Small and large cells present the same taxonomic characters, including DNA content. The shape and density of the nucleus does, however, vary: in small cells nucleoplasm is highly condensed and chromosoms are very closely bundled, while in large cells chromosomes are in loose parallel bundles. Small cells are produced by budding of the left epicone of a large cell. Separation of small and large strains by flow cytometry and subsequent culturing revealed that small cells could produce large cells by simple enlargement corresponding to
the stationary phase (Partensky and Vaulot 1989). In contrast, regeneration of small cells from large forms took a variable number of generations, ranging from 1 to more than 50. This indicates that the formation of small cells is probably not continuous, but appears sporadically.

Parameters inducing such a differentiation are not yet defined. It does not seem to be related to a N or P depletion (Partensky and Vaulot 1989). Division of both forms was inhibited in the absence of phosphorus, but a sharp increase in the volume of small cells resulting in the production of large ones was observed. Partensky and Vaulot (1989) found that population growth was slowed when ammonium was the sole source of nitrogen, as a result of a significantly decreased growth rate of the small cells. Authors concluded that small cells preferentially use nitrate while the large cells can use either ammonium or nitrate. Under standard conditions, small cells divided once a day whereas large ones divided at 0.6 day^{-1}. On the contrary, Yamaguchi (1994) found that N-limitation induced the production of small cells, while P-limitation tended to produce large cells in *G. nagasakiense*, but did not observe the co-occurrence of the two types (Yamaguchi 1992). He reported that growth rates of *G. nagasakiense* were not significantly different when supplied with the same concentration of ammonium or nitrate. On an urea substrate, the maximum growth rate was about 5 times lower than with nitrate.

Even if a modulation of cell-size by turbulence cannot be excluded in the field, in the case of *G.* cf. *nagasakiense*, obviously, this cell-size differentiation has some other origin. The reduced size of the small subpopulation allows an optimization of photon capture and nutrient uptake (Raven 1986), such that they may take optimal advantage of the conditions prevailing during blooms. In contrast, the large form appears to be more adapted for survival under non-bloom conditions. This biological mechanism could account for a large part of the success of *G.* cf. *nagasakiense* in the field. However, it does not seem to be specific of *G. nagasakiense*: it has been observed in other dinoflagellates.

Subpopulation differentiation could be an explanation to the large variation observed in maximum growth rates. The factors controlling the production of small cells are unknown.

7.2.Role of polyamines in the cell cycle

Recently, Videau *et al.* (1996) reported on a possible link between the decaying spring diatom bloom and *G.* cf. *nagasakiense* bloom through the effect of putrescine. Putrescine is synthetized by decarboxylation of ornithine and arginine. Spermidine (Spd) and spermine are synthetized from Put under the action of polyamine-synthetases. Transformation of putrescine into spermidine and spermine is important in cell division processes. Variations of the ratio Put/Spd follows the cell cycle. Cell cycle studies have been conducted on batch cultures with two subpopulations. The subpopulations do not divide at the same time : small cells divide 1h into the dark period ,whereas large-size cells divide 9 to 10h into the dark period (Partensky *et al.* 1991b). Two hours before each of these peaks in frequency division, an increase in the ratio Put/Spd has been measured. This increase is related to putrescine synthesis which is synchronous with DNA synthesis.

Polyamines stimulate growth in *G.* cf. *nagasakiense,* when added to the growth medium (Videau *et al.* 1996). An increase of 30% in growth rate has been measured *in vitro* for very low concentrations (0.1 - 5 μM putrescine). These levels are a 1000 times less than for other cells. Since the final culture yield is not different from blanks, putrescine cannot be considered as a complementary nitrogen source. Active concentrations of putrescine (0.1 μM) correspond to the concentrations measured on Ushant front at the pycnocline level where diatom decay, previously to the *G.* cf. *nagasakiense* bloom.

These preliminary results indicate that decay of the previous diatom bloom rich in ornithine seems to favour *G.* cf. *nagasakiense* bloom development, not only in providing nitrogen from organic matter but also in elevating polyamines concentrations in sufficient quantities to stimulate elevated dinoflagellate growth rate.

7.3.Nutrients

Gyrodinium aureolum blooms in nutrient-depleted waters are often able to accumulate nitrogen to the extent that cellular N per unit volume of water exceeds the maximum winter concentration of nitrate-N (Holligan *et al.* 1984). These authors related these observations to vertical migration of the dinoflagellate. However, they specified that no vertical migrations have been observed in the western Channel. This fact led them to suppose that the cells could be moving up and down over time scales larger than a day.

Several bloom-forming dinoflagellates have the ability to take up and assimilate nitrate in the dark. This ability confers on them a competitive advantage over diatoms in stratified waters. Paasche *et al.* (1984) have shown that *G. aureolum* was incapable of dark nitrate uptake when nitrogen-sufficient but that it was induced by a period of nitrogen deprivation.

Nielsen (1992) reported mean cellular C, N and P contents of 408, 74 and 39 pg cell^{-1} respectively at 400 μmol m^{-2} s^{-1}. Adaptation to low irradiances results in cells enriched in C and N, but not total P. (Nielsen 1992).

The values of K_S reported by Yamaguchi (1994) for different N and P substrates are smaller than for other dinoflagellates. This indicates that *G. nagasakiense* requires very low concentrations of nutrients for growth, which is a characteristic allowing it to grow more abundantly than its competitors in the natural environment. Hirayama *et al.* (1989) have reported the induction of high alkaline phosphatase activity in *G. nagasakiense* when inorganic phosphorus was depleted in the culture medium. *G. aureolum* capacity in storing phosphorus is large compared to other dinoflagellates (Nielsen and Toenseth 1991)

Paasche et al. (1984) reported that *G. aureolum* did not take up nitrate in the dark when nitrogen sufficient, and that a 24-h starvation induced a capacity for taking up nitrate in the dark. A strong coupling between nitrate assimilation and photosynthetic carbon as milation seemed to occur on the base of the close similarity of the light saturation curves of $^{15}NO_3^-$ and $^{14}CO_2$. These features of *G. aureolum* physiology makes it unlikely that surface blooms of this dinoflagellate would benefit from N-NO_3 incorporation at night time down in the pycnocline. However, the low light levels encountered in the pycnocline are sufficient to satisfy the growth requirements in 20-30m deep layers. Ammonium and nitrate uptake rates have been measured by Le Corre et al. (1993) on the Ushant tidal front (Fig. 2). *G.* cf. *aureolum* developing in stratified waters (Station S) contributed a substantial fraction of the phytoplankton productivity in a short time. Rapid build-up of N biomass requires that the inorganic N is either in sufficient quantity at the beginning of the bloom or that it is supplied in adequate quantities by either physical or recycling processes.

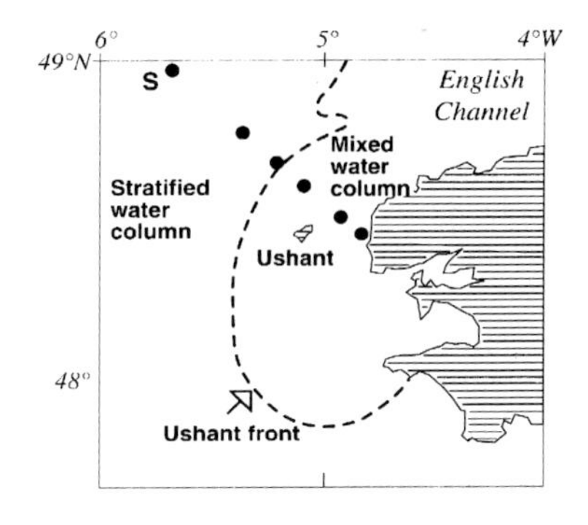

Fig. 2 : Position of Ushant front

The former situation is not possible since the previous diatom bloom consumed inorganic N. Le Corre *et al.* (1993) measured the nitrogen uptake and regeneration rates. Ammonium uptake rates were high (3.75 mmol m^{-2} h^{-1}) while nitrate uptake rate was only 0.4 mmol m^{-2} h^{-1}. Ammonium regeneration (3.4 mmol m^{-2} h^{-1}) was also very high compared to the annual average of a coastal homogeneous system (0.2 mmole m^{-2} h^{-1}), (Maguer 1995). Uptake and regeneration profiles resembled chlorophyll *a* (data not shown) and particulate organic profiles (Fig. 3).

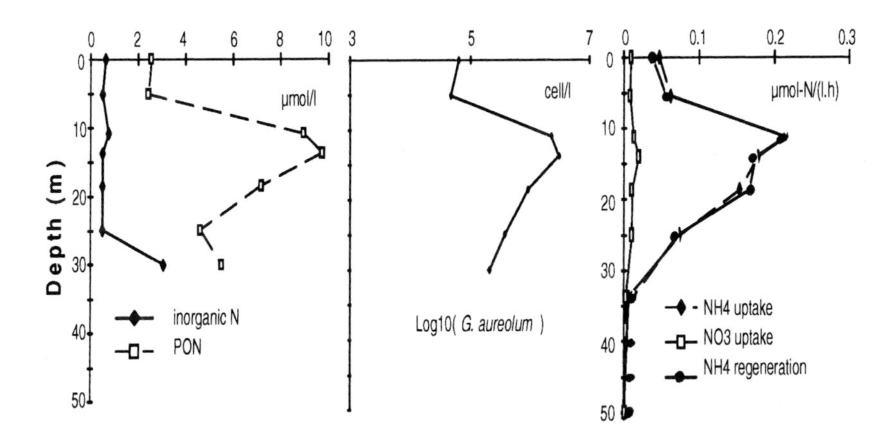

Fig. 3 : Vertical profiles in the stratified zone of the Ushant front

Microheterotrophs under 30 µm accounted for all the ammonium regeneration which provided *ca.* 90% of the N required by *G.* cf. *aureolum* bloom. The N necessary for such fluxes of regeneration is mainly composed of detritus from the previous diatom bloom accumulated in a density barrier. It has also been observed (Gentien, pers.obs.) that some detritic matter from land origin (debris of higher plants) was present in the pycnocline layers. The mass balance of N is a strong evidence demonstrating that in this case the *G.* cf. *aureolum* bloom can be maintained almost exclusively by *in situ* N remineralization. Nitrate flux through the pycnocline has been estimated (Holligan et al., 1984) to be two orders of magnitude lower than the mesured N uptake rate. Calculated ammonium flux was insignificant compared to the regeneration rate.

Le Corre *et al.* (1993) concluded that the *G.* cf. *aureolum* blooms mechanisms in frontal stratified systems differ from those in estuaries and near shore which derive either as riverborne or biodegradable PON that is brought into the system occasionnaly (Birrien *et al.*, 1991). Blasco *et al.* (1996) report similar results in the Gulf of St Lawrence: principal component analysis indicate that *G. aureolum* favors environments with high nitrogen recycling activity. Nishimura (1982) reported that dissolved organic matter from fish farms promoted the growth of *Gymnodinium* type-'65.

Apart from nitrogen and phosphorus, other limiting nutrients cannot be excluded : Ishimaru *et al.* (1989) reported that selenium was the most effective trace element promoting *G. nagasakiense* growth in the form of selenite (SeO_3^{2-}). *G. nagasakiense* reached 0.79 $div.day^{-1}$ and presented a K_S of 0.075nM. The maximum cell yield in cultures was limited by Se at concentrations inferior to 2nM. It appears that the cellular requirement of Se is 4.4 10^{-17} $mole.cell^{-1}$. Koike *et al.* (1993) have found a strong relationship between Se(IV) and the cell concentration of *G. nagasakiense* which required *in situ* 2.9 10^{-17} mole $Se.cell^{-1}$. These results suggest that selenium may play an important role in red tides of *G. nagasakiense* but sources of selenium are still uncertain and need to be clarified.

A large part of the seston produced in the upper mixed layer accumulate on sharp density interfaces. In highly stratified water columns, *G.* cf. *nagasakiense*, being able to grow in subdued light environments, can benefit of nutrient recycling in the pycnocline layers, without vertical migration. It is likely that the dinoflagellate population grows faster because of growth factors produced by the decay of organic matter.

8. Control by external factors

8.1.Allelopathy

Gymnodinium cf. *nagasakiense*, possessing the ability to move and accumulate in specific layers of the water column, can profit from its ability to use low levels of light and of its efficiency in nutrient assimilation. Using a restricted niche restricts the number of potential competitors. Allelopathy further restricts the contours of the niche.

G. cf. *nagasakiense* has been shown to produce exotoxins which are detrimental to the growth of other algae (Gentien and Arzul 1990). Cells of these other species exhibited an apparent volume reduction and a large proportion was not viable, due to induced membrane permeability judged by a viability test (Gentien 1985).

On the Ushant front, on vertical profiles, the *G.* cf. *nagasakiense* maximum density corresponded to a minimum in diatoms (Arzul *et al.* 1993). The minimum *G.* cf. *nagasakiense* cell density for a reduction in diatom growth rate was *ca.* 10^4 $cell.l^{-1}$. The inhibitory power of *G.* cf. *nagasakiense* probably plays a role in the population development before any harmful effects may be observed on marine animals.

Yasumoto *et al.* (1990) reported the presence of 1-acyl-3-digalactosylglycerol and octadecapentaenoic acid in *G. aureolum* cell extracts. The strain from the Brest Bay produces the all-*cis 3,6,9,12,15* octadecapentaenoic acid (Parrish *et al.* 1993), the major unsaturated fatty acid at 18°C. It can reach 34% of total fatty acids at 18°C and 35 $\mu E.m^{-2}s^{-1}$. At the level of 1µM, it inhibits totally the growth of *Chaetoceros gracile*. Like many other ichthyotoxic species, *G.* cf. *nagasakiense* produces oxygen radicals (unpublished results). The high sensitivity of this acid and the acyl glycerols to oxidation indicates that these compounds may be tracers of the toxic effect, but does not exclude other compounds such as volatile sesquiterpenes. Epicubenol, α-cadinol and cubenol, identified from both red tides and cultures of *G. nagasakiense*, presented an algal cell-destroying activity (Kajiwara *et al.* 1992).

This ability of destroying or repressing the growth of other species further restricts the number of potential competitors.

8.2. Grazing

Grazing is an important process controlling phytoplankton biomass. Grazer populations are very diverse and grazing pressure are case dependent. Lee and Hirayama (1992) reported that *N. scintillans* clearance rate of *G. nagasakiense* was in the range 1 to 7 10^{-3} ml. *Noctiluca*$^{-1}$ h^{-1}. Tangen (unpublished result) has observed that the heterotrophic dinoflagellate *Kofoidinium velelloides* has been the cause of a *G. aureolum* bloom decline. Recently, Nakamura *et al.* (1996) reported that a large increase in the population of *Favella ehrenbergii* (10 ml^{-1}) was the cause of a bloom termination, this tintinnid consumed *G. nagasakiense* and the heterotrophic dinoflagellates in a few days.

Frequency and amplitude of movements of cephalothoracic appendages in calanoid copepods determine the speed of the feeding current which controls directly the encounter rate between the algae and the grazer. A modification of the technique developed by Gill and Poulet (1986) has been used by the author and S. Poulet to study the effect of the dinoflagellate on *Calanus helgolandicus* (unpublished results). Simultaneous video recording allowed to measure the ingestion rate. Males and females were presented a *G.* cf. *nagasakiense* suspension. Toxic effect was observed on females only in the case they ingested some algae: they died 30 to 60 minutes following ingestion. In all cases, females got tangled into a mucus web which limited progressively the feeding current even when the appendages were beating normally. This effect has seldom been seen with males since they do not create a feeding current. In the case of females, activity initially decreased, followed by a peak of activity. This period corresponded to specialized movements to disantangle the appendages. Subsequently, activity decreased to a complete stop corresponding to fibrilations of small amplitude.

From these observations, one can conclude that population control by copepod grazing is probably not very efficient. It is not known if dinoflagellate proliferations are a mortality factor for copepods, or if copepods can avoid dinoflagellate accumulation zones. It would be of the greatest interest to determine if a specialized grazer has developed strategies dealing with mucus. This approach can only be conducted in field studies.

8.3. Autotoxicity and sensitivity to agitation

Increased stability of the water column due to stratification and calm weather favours red tides, while storm events terminate them. (for instance, see Iizuka *et al.* 1989).

Cell membranes of *G.* cf. *nagasakiense* are somehow protected from, or more likely present a low sensitivity to, its own toxicity. The effect of one cell on another depends on the mean distance between them. This mean distance decreases with increasing densities or with agitation. In cultures, agitation by bubbling causes a cell density decrease and a production of small globules of cytoplasm of *ca.* 2μm. Filtered cells of *G.* cf. *nagasakiense* do not yield any viable inoculum. Cells forced to collide under the microscope stick together and lyse in a few minutes.

In cultures, few hours into the light phase, most of the algae have migrated to the top few centimeters. Patterns of bioconvection start to develop as observed for other flagellates. Such aggregation patterns, fingers-like incursions, appear to be in some part biotic but not social. As described by Harashima *et al.* (1988), the essence of this phenomenon is that gravity acts not against the water or microorganisms separately but against a mixture of them. The energy source of this phenomenon is the active transport of buoyancy given internally by the upward swimming of microorganisms. These patterns create as a consequence an increased vertical diffusion of dissolved compounds released by the cells, including possible exotoxins.

Culture conditions *G.* cf. *nagasakiense* are optimal when agitation is very limited. An equilibrium between the rate of exotoxin production, the volume under the surface layer and the inactivation of the toxin occurs which allows a normal growth. By drawing off very carefully the bottom layer in the culture flask, it is possible to concentrate an initial cell density up to 6 times. In the two hours following, the cell density in the concentrate is down to a limit number of 3 to 4 10^7 cell.l^{-1} (6 observations). If this density represents the equilibrium between the flux of exotoxin and its degradation, the average free volume required for survival of a 25-μm cell is around 250 μm in diameter. Therefore, in collision models, the radius of the particle to be taken into account should be comprised between the radius of the particle and the radius of the free volume required by each cell ($25\mu m<d<250\mu m$). Encounter rate of two particles is a function of size, of concentrations of particles and of the physical environment. Let us suppose that two cells encountering lyse. Differential sedimentation does not occur in this case because of cell lysis. The encounter rate varies with d^3 ,at a given shear and concentration , in the formulation used by Jackson (1990). It can be seen that an increase by a factor 10 of the effective diameter of the particle corresponds to 1000-times higher sensitivity to shear.

For a deep unstratified water column, the mortality term would be proportional to the square of the wind speed above the surface. Many red tides, including *G.* cf. *nagasakiense*'s, are characterized by high production rates of extracellular polysaccharides of high molecular weight (Jenkinson 1993). According to Moum and Lueck (1985), the shear rate resulting from the energy dissipation can be estimated as $\gamma = \sqrt{\varepsilon / 7.5 \nu}$, where ε is the energy dissipation rate and ν the kinematic viscosity. At a constant ε, the shear rate decreases when the kinematic viscosity increases. Recently, Jenkinson and Biddanda (1996) reported that some samples taken in accumulation microlayers of *Phaeocystis* sp. were 6.5 to 14 times (depending on estimation method) more viscoelastic than bulk-phase samples. Changes in rheological properties due to biological production lead to a decrease of the shear seen by the cells and thus, increase the maximum shear permitting net growth of the population. Some phytoplankters may have evolved the ability to use slime secretion to decouple mixing parameters and to change its physicochemical environment (Jenkinson and Wyatt 1992). It does not appear that shear effect on collision rates could be totally suppressed. Autoxicity of *G.* cf. n*agasakiense* seems to depend on a balance between the toxins and mucus production rates, one counteracting the other.

8.4. Termination of the bloom

Some other mechanisms due to viral or bacterial infestation may be responsible for bloom decay. Doucette (*this volume*) reports on studies associating changes in bacterial flora and development and decline of harmful algal blooms. Fukami *et al.* (1991, 1992) have reported that bacteria isolated during a *G. nagasakiense* red tide could either stimulate or inhibit dinoflgellate growth depending on the stage of the bloom. At times up to and including peak algal concentrations augmented the growth of *G. nagasakiense* cultures, while bacteria obtained during bloom decay produced a strong negative effect on algal growth. These authors isolated from the field a *Flavobacterium* species that exhibited strong inhibitory effects on the dinoflagellate's growth.

In general, bacterial populations seem to vary over the course of bloom formation and decay, apparently resulting from the selective stimulation of bacteria belonging to particular groups (Doucette, *this volume*). In terms of bloom dynamics, it is important to decide if these changes are inducing the fate of the bloom or if they are a consequence.

9. Bloom inoculation (The Bay of Biscay case)

The origin of the inoculum is rarely mentioned in toxic algal bloom events, apart from accidental input of new species from ballast tank waters or shellfish transfer. *G.* cf. *nagasakiense* is present all the year round and does not appear to form cysts. Vegetative cells in the Bay of Brest (France) have been observed in January and February at a density of 5-10 cell l^{-1}. In Gokasho Bay (Japan), Honjo *et al.* (1991) presented a relationship between winter water temperature and the timing of summer red tides. The higher the mean winter temperature, the earlier *G. nagasakiense* reaches a cell density of 1000 cell l^{-1} (corr. coeff. = -0.98). In this location, this prediction allows the implementation of countermeasures well in advance of bloom development. However, this scheme may not be usable in other locations.

Geographical extension of the toxic effects due to *G.* cf. *nagasakiense* along the French Atlantic coast varies from year to year. In most cases, the bloom remains located in the waters around the Western part of Brittany, close to the Ushant front and adjacent areas. This was the case in 1987, a year during which the bloom did not extend further south than the Loire estuary. On the contrary, in 1995, the bloom was noticed as far south as the Arcachon basin: it was the first occurrence of this dinoflagellate since 1984 (when the monitoring was first implemented) south of the Loire estuary. The time lag between the northern and southern maximum was of the order of a fortnight. Hydrodynamics cannot explain this rapid spreading of an already established population.

In order to seek conditions favoring a bloom with such an extent, a 3-D hydrometeorological model of the Bay of Biscay (developed by Pascal Lazure, IFREMER) was employed. This model of the continental shelf allows the study of movements and mixing of water masses at time scales between a day and a year. Positions of water particles are calculated every 15 min. The model extends to the 150-m isobath. Its horizontal mesh size is 5 km and 10 layers are computed on the vertical. Boundary conditions are derived from a large scale 2-D model of the Eastern North-Atlantic.

The basic hypothesis concerning *G.* cf. *nagasakiense,* was that this species, during its early and active development phases, is mainly located in the pycnocline when it exists. The second hypothesis, based on observations from REPHY (French Phytoplankton Monitoring Network) showing that the bloom in normal years occurs in Western Brittany, is that the population originates from this area. Water particles were, therefore, released in the pycnocline along a release transect from the South-West Brittany coast to the 100-m isobath. Particles are released at different dates and their trajectories simulated up to July, the 15th for the two years 1987 and 1995.

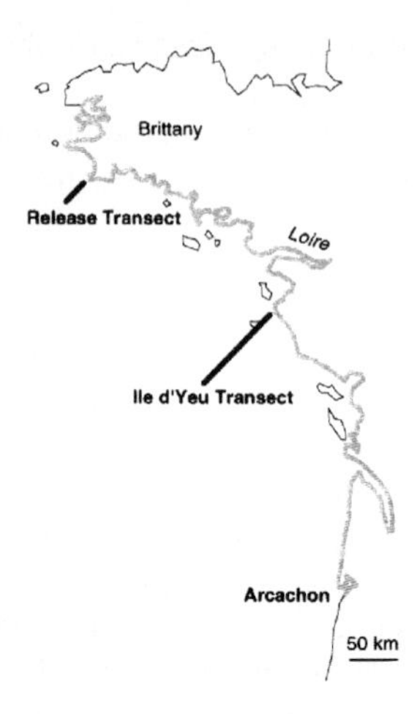

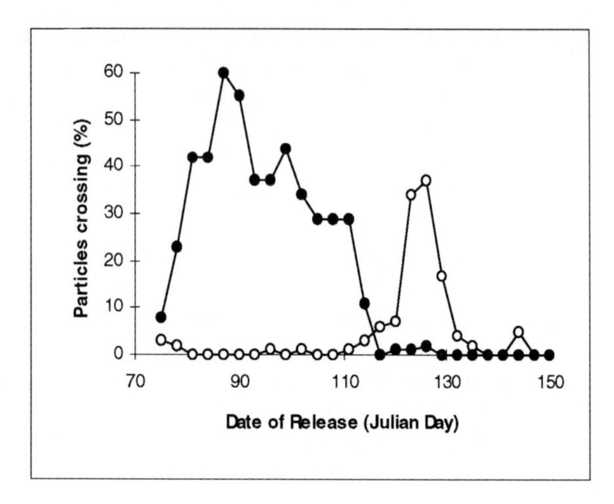

Fig. 4 : left : Map of the French Atlantic coast showing the location of the transects and the extent of the toxic event (shaded coast line)
right : Yield of particles released at different dates crossing the Ile d'Yeu transect before the 15th of July. (open circles : 1987 ; closed circles : 1995)

On the whole, transportation south was more important and reached further south in 1995 than in 1987 (trajectories not shown). The yields of trajectories crossing the Ile d'Yeu transect southward for particles released are presented above. In 1987, percentage of particles released every 3 days from the 17th of March to the 15th of May, crossing the Ile d'Yeu transect before the 15th of July was 5.6% , while the same percentage for the same period was 23% in 1995. It clearly shows that, in April 1995, as a result of meteorological conditions (sustained North Easterly winds and high river discharges) transportation towards the South was of greater magnitude relative to 1987. This was the only period during which such a discrepancy between the two years could be found. If the inoculation area is effectively restricted to the Western Brittany area, the conditions for a large bloom to occur in July 95 on the whole continental shelf of the Bay of Biscay were set at the end of May. A complementary observation of the importance of stratification is that the integrated surface area of high anomalies of potential energy is much higher in April-May 1995 than in 1987 (data not shown), thus being *a priori* more favourable to *G.* cf. *nagasakiense* growth.

10.Conclusions and perspectives

The group of species considered here is eurythermal and euryhaline. In weakly stratified water columns, it migrates from surface layers during the day down to 10 m-depth at night. In presence of a sharp pycnocline, no migration pattern is observed: the maximum density is located in the pycnocline. The size class differenciation reflects probably a adaptation for growth to different conditions. The pigment composition and bio-optical characteristics of *G. mikimotoi* being very plastic with respect to adaptation to the growth-light regime allows it to benefit from low and high levels of light. In the pycnocline layers which are often nutriclines, the population thrives on ammonia from regeneration of seston and benefits from growth factors. The exotoxins and mucus production play an allelopathic role and act as grazing repressing agents. Exotoxins are responsible to the high sensitivity of these species to agitation and wind events. However, it seems that this species is able to modify its physical environment in order to mitigate the effect of its toxin.

Despite the complexity of the problem, operational advices can be drawn without a comprehensive knowledge of the processes involved in the population development as shown by the modelling example which derives from a first order approach. This model predicting only a potential risk needs to be complemented.

The biological quality of the pycnocline water properties should be complemented by net *in situ* growth rate modulated by temperature and salinity in order to produce a prediction of the geographical extension of the bloom. However, a more precise prediction of the cell densities will require a detailed formulation of the biology, and of the physical-biological interactions. Mortality due to grazing requires to search for potential species-specific associations. Understanding of the toxic effect of these events requires a more detailed investigation of the environmental conditions inducing toxicity. It is, however, likely that a strain specific study be required in this case. Finally, it would be of interest to investigate if a functional grouping of ichtyotoxic species is realistic.

Acknowledgements

The reported work of the French teams since 1988 have been partly supported by the French National Program on Harmful Algal Blooms (P.N.E.A.T.). I am grateful to P. Lazure for the simulation of particles trajectories. Time series if toxic events along the French Atlantic coast were provided by the IFREMER monitoring network REPHY.

References

Arzul G., Erard-Le Denn E, Videau C., Jegou A.M., Gentien P. (1993). Diatom growth repressing factors during an offshore bloom of *Gyrodinium* cf. *aureolum*. in "Toxic Phytoplankton Blooms in the Sea" Smayda T.J. and Shimizu Y. Eds., Elsevier Science Publishers: 719-724.

Arzul G., Gentien P., Crassous M.P. (1994). A haemolytic test to assay toxins excreted by the marine dinoflagellate *Gyrodinium* cf. *aureolum*. Water Res., 28, 4:961-965.

Birrien J.L.,. Wafar M.V.M., Le Corre P., Riso R. (1991).Nutrients and primary production in a shallow stratified ecosystem in the Iroise Sea. J. Plankton Res., 13, 721-742.

Bjoernsen P.K.,Nielsen T.G., (1991). Decimeter scale heterogeneity in the plankton during a pycnocline bloom of *Gyrodinium aureolum*. Mar. Ecol. Prog. Ser.,P,vol. 73, no. 2-3: 263-267.

Blasco D., Berard-Therriault L., Levasseur M., Vrieling E.G. (1996). Temporal and spatial distribution of the ichyotoxic dinoflagellate *Gyrodinium aureolum* Hulburt in the St. Lawrence, Canada. J. Plankton Res. (*in press*).

Braarud T., Heimdal B.R. (1970). Brown water on the Norwegian coast in autumn 1966. Nytt Mag. Bot. 17 (2): 91-97.

Chang J., Carpenter E.J. (1985). Blooms of the dinoflagellate *Gyrodinium aureolum* in a Long Island estuary: Box model analysis of bloom maintenance. Mar. Biol.: 89, 83-93.

Dahl E., Brockmann H. (1985).The growth of *Gyrodinium aureolum* Hulburt in *in situ* experimental bags in "Toxic dinoflagellates" Anderson D. M., White A.W. and Baden D.G. (eds.), Elsevier Science Publishing Company, pp. 233-238.

Dahl E., Brockmann H. (1989). Does *Gyrodinium aureolum* Hulburt perform diurnal vertical migrations?. Red tides: Biology, Environmental Science and Toxicology, Okaichi T., Anderson D.M., Nemoto T. (eds.), Elsevier Science Publishing, pp. 225-228.

Dahl E., Tangen K. (1993). 25 years experience with *Gyrodinium aureolum* in Norwegian waters. in " Toxic Phytoplankton in the Sea", Smayda and Shimizu (eds.) , Elsevier Science Publishers B.V.: 15-21.

Dixon G.K., Holligan P.M.(1989). Studies on the growth and nitrogen assimilation of the bloom dinoflagellate *Gyrodinium aureolum* Hulburt. J. Plankton Res. 11, 1:105-118.

Franks P.J.S. (1995). Thin layers of phytoplankton: a model of formation by near-inertial wave shear. Deep-Sea Res. I, 42, 1, 75-91.

Fukami K., Nishijima T., Murata H., Doi S., Hata Y. (1991) .Distribution of bacteria influential on the development and decay of *Gymnodinium nagasakiense* red tide and their effects on algal growth. Nippon Suisan Gakkaishi, 57, 2321-2326.

Fukami K., Yuzawa A., Nishijima T., Hata Y. (1992). Isolation and properties of a bacterium inhibiting the growth of *Gymnodinium nagasakiense*. Nippon Suisan Gakkaishi 58,6: 1073-1077.

Garcia V.M.T., Purdie D.A.(1994). Primary production studies during a *Gyrodinium* cf. *aureolum* (Dinophyceae) bloom in the Western English Channel. Mar. Biol. 119:297-305.

Gentien P. (1985). A method for evaluating phytoplankton viability by induced fluorochromasia. in "Progress in flow cytometry", Becton-Dickinson Ed.:151-164.

Gentien P., Arzul G. (1990). Exotoxin production by *Gyrodinium* cf. *aureolum* (Dinophyceae). J. mar. biol. Ass. U.K., 70:571-581.

Gentien P., Arzul G. (1991). Some physiological features of the development of *Gyrodinium* cf. *aureolum* in " Recent approaches on red tides " J.S. Park and H.G. Kim (eds.).National Fisheries Research and Development Agency Publications, Pusan, South Korea, pp 73-84.

Gill C.W., Poulet S.A. (1986). Utilization of a computerized micro-impedance system for studying the activity of copepod appendages. J. Exp. Mar. Biol. Ecol. 101, 1-2:193-198.

Harashima A., Watanabe M., Fujishiro I. (1988). Evolution of bioconvection patterns in a culture of motile flagellates. Phys. Fluids, 31,4:764-775.

Hirayama K., Numaguchi K. (1972). Growth of *Gymnodinium* Type-'65, causative organism of red tide in Omura Bay, in medium supplied with bottom mud extract. Bull. Plankton Soc. Jpn., 19, 13-21.

Hirayama K., Doma T., Hamamura N., Muramatsu T., (1989). Role of alkaline phosphatase activity in the growth of red tide organisms. in " Red Tides: Biology, Environmental Science and Toxicology", Okaichi T., Anderson D. M., Nemoto T., (eds.): 317-320.

Hulburt E.M. (1957). The taxonomy of unarmored Dinophyceae of shallow embayments on Cape Cod, Massachusetts. Biol. Bull., Mar. Biol. Lab. Woods Hole 112(2) : 196-219.

Holligan P.M.,Williams P.J. leB., Purdie D., Harris R.P. (1984).Photosynthesis, respiration and nitrogen supply of plankton populations in stratified, frontal and tidally mixed shelf waters. .Mar. Ecol. Prog. Ser., 17, 201-213.

Honjo T., Yamaguchi M., Nakamura O., Yamamoto S., Ouchi A., Ohwada K. (1991) . A relationship between winter temperature and the timing of summer *Gymnodinium nagasakiense* red tides in Gokasho Bay. Nippon Suisan Gakkaishi, 57(9):1679-1682.

Iizuka S. (1979). Maximum growth rate of natural population of *Gymnodinium* red tide. in Taylor D.L., Seliger H.H., (eds.). Toxic Dinoflagellate Blooms. Elsevier, New York: 111-114.

Iizuka S., Sugiyama H., Hirayama K. (1989). Population growth of *Gymnodinium nagasakiense* red tide in Omura Bay. in " Red Tides: Biology, Environmental Science and Toxicology", Okaichi T., Anderson D.M., Nemoto T. , (eds.).: 269-272.

Ishimaru T., Takeuchi T., Fukuyo Y., Kodama M. (1989). The selenium requirement of *Gymnodinium nagasakiense*. in " Red Tides: Biology, Environmental Science and Toxicology", Okaichi T., Anderson D.M., Nemoto T., (eds.): 357-360.

Jackson G.A. (1990). A model of formation of marine algal flocs by physical coagulationprocesses. Deep-Sea Res.,37,8:1197-1211.

Jenkinson I. R. (1993). Viscosity and elasticity of *Gyrodinium* cf. *aureolum* and *Noctiluca scintillans* exudates, in relation to mortality of fish and damping of turbulence. in "Toxic Phytoplankton Blooms in the Sea", Smayda T.J. and Shimizu Y., (eds.), Elsevier : 757-762.

Jenkinson I.R., Wyatt T., (1992). Selection and control of Deborah numbers in Plankto ecology. J. Plankton Res.,14, 12, 1697-1721.

Jenkinson I.R., Biddanda B.A., (1996). Bulk-phase visoelastic properties of seawater: relationship with plankton components. J.Plankton Res., 17, 12, 2251-2274.

Johnsen G., Sakshaug E. (1993).Bio-optical characteristics and photoadaptative responses in the toxic and bloom-forming dinoflagellates *Gyrodinium aureolum, Gymnodinium galatheanum* and two strains of *Prorocentrum minimum*. J. Phycol., 29, 627-642.

Kajiwara T., Ochi S., Kodama K., Matsui K., Hatakana A., Fujimura T., Ikeda T. (1992). Cell-destroying sesquiterpenoids from red tide of *Gymnodinium nagasakiense*. Phytochemistry, 31,3:783-785.

Kamykowsky D. (1995). Trajectories of autotrophic marine dinoflagellates. J. Phycol. 31: 200-208.

Kofoid C.A., Swezy O. (1921). The Free-living Unarmored Dinoflagellata. Mem. Univ. Calif. 5:563.

Koike Y., Nakaguchi Y., Hiraki K., Takeuchi T. (1993). Species and concentrations of Selenium and nutrients in Tanabe Bay during red tide due to *Gymnodinium nagasakiense*. J. Oceanogr.,*49,* 6:641-656.

Koizumi Y., Uchida T., Honjo T. (1996). Diurnal vertical migration of *Gymnodinium mikimotoi* during a red tide in Hoketsu Bay, Japan. J. Plankton Res., 18, n°2: 289-294.

Lee J.K., Hirayama K. (1992). The food removal rate by *Noctiluca scintillans* feeding on *Tetraselmis tetrathelle* and *Gymnodinium nagasakiense* . Bull. Fac. Fish. Nagasaki Univ. Chodai Suikenpo., 71:69-175.

Le Corre P.,L'Helguen S., Wafar M. (1993).Nitrogen source for uptake by *Gyrodinium* cf. *aureolum* in a tidal front.Limnol. Oceanogr., 38, 2, 446-451.

Lindahl O. (1983). On the development of a *Gyrodinium aureolum* bloom occurrence on the Swedish west coast in 1982. Mar. Biol. 77:143-150.

Lindahl, O. (1993). Hydrodynamical processes: A trigger and source for flagellate blooms along the Skagerrak coasts?. in " Toxic Phytoplankton in the Sea", Smayda T. and Shimizu Y. (eds.) , Elsevier Science Publishers B.V. :775-782.

Maguer J.F. (1995).Absorption et régénération de l'azote dans les écosystèmes côtiers - Relations avec le régime de mélange vertical des masses d'eaux - Cas du domaine homogène peu profond de la Manche Occidentale. Thèse, Université de Bretagne Occidentale, Brest, France.

Mahoney J.R., Olsen P.,Cohn M. (1990). Blooms of a dinoflagellate *Gyrodinium* cf. *aureolum* in New Jersey coastal waters and their occurrence and effects worldwide. J. coast. Res. 6,1:121-135.

Matsuoka K., Iizuka S., Takayama H., Honjo T., Fukuyo Y., Ishimaru T. (1989). in " Red Tides: Biology, Environmental Science and Toxicology", Okaichi T., Anderson D.M., Nemoto T., Eds.:101-104.

Moum J.N., Lueck R.G. (1985). Causes and implications of noise in oceanic dissipation measurements. Deep-Sea Res. 32:379-392.

Nakamura Y., Suzuki S., Hiromi J. (1996). Development and collapse of a *Gymnodinium mikimotoi* red tide in the Seto inland sea. Aquatic Microb. Ecol., 10:131-137.

Negri R.M., Carreto J.I., Benavides H.R., Akselman R., Lutz V.A.(1992). An unusual bloom of *Gyrodinium* cf. *aureolum* in the Argentine sea: Community structure and conditioning factors. J. Plankton Res.,14,2:261-269.

Nielsen M.V. (1992). Irradiance and daylength effects on growth and chemical composition of *Gyrodinium aureolum* Hulburt in culture. J. Plankton Res., 14, 6:811-820.

Nielsen M.V. and Toenseth C.P. (1991). Temperature and salinity effect on the growth and chemical composition of *Gyrodinium aureolum* Hulburt in culture. J. Plankton Res., 13:389-398.

Nishimura A. (1982). Effects of organic matter produced in fish farms on the growth of red tide algae *Gymnodinium* type-'65 and *Chattonella antiqua*. Bull. Plankton Soc. Japan, 29,1:1-7.
Ouchi A., Aida S., Uchida T., Honjo T. (1994). Sexual reproduction of a red tide dinoflagellate *Gymnodinium mikimotoi*.Fish. Sci., 60, 1: 125-126.
Paasche E., Bryceson I., Tangen K. (1984).Interspecific variation in dark nitrogen uptake by dinoflagellates. J. Phycol., 20, 394-401.
Parrish C.C., Bodennec G., Sebedio J.-L., Gentien P. (1993). Intra- and extracellular lipids in cultures of the toxic dinoflagellate, *Gyrodinium aureolum*. Phytochemistry, 32, 2:291-295.
Partensky F., Sournia A. (1986). Le dinoflagellé *Gyrodinium* cf. *aureolum* dans le plancton de l'Atlantique nord : identification, écologie, toxicité. Cryptogam. Algol. 7:251-275.
Partensky F., Vaulot D., Couté A., Sournia A.(1988). Morphological and nuclear analysis of the bloom-forming dinoflagellates *Gyrodinium* cf. *aureolum* and *Gymnodinium nagasakiense*. J. Phycol. 25(4):741-750.
Partensky F., Vaulot D. (1989). Cell size differentiation of the bloom-forming dinoflagellate *Gymnodinium* cf. *nagasakiense*. J. Phycol. 25(4) : 741-750.
Partensky F., Gentien P., Sournia A. (1991a). *Gymnodinium* cf. *nagasakiense* = *Gyrodinium* cf. *aureolum* (Dinophycées) in "Le Phytoplancton nuisible des côtes de France : de la biologie à la prévention" Sournia A. *et al.* (eds.). IFREMER Publications, Brest :63-82.
Partensky F., D. Vaulot and C. Videau (1991b). Growth and cell cycle of two closely related red tide-forming dinoflagellates: *Gymnodinium nagasakiense* and *G.* cf. *nagasakiense*. J. Phycol. 27, 733-742.
Raine R., Joyce B., Richard J., Pazos Y., Moloney M., Jones K.J., Patching J.W. (1993). The development of a bloom of the dinoflagellate *Gyrodinium aureolum* (Hulburt) on the south-west Irish coast. ICES J. mar. sci. 50:461-469.
Raven J.A. (1986).Plasticity in algae.in Jennings, D.H. & Trewavas A.J. (eds.) Plasticity in Plants. The Company of Biologists Ltd., Cambridge, pp. 347-372.
Richardson K., Beardall J., Raven J.A. (1983). Adaptation of unicellular algae to irradiance: an analysis of strategies. New Phytol., 93:157-191.
Richardson K., Kullenberg G. (1987). Physical and biological interactions leading to plankton blooms: A review of *Gyrodinium aureolum* blooms in Scandinavian waters.Rapp. P.-v. Réun. Cons. int. Explor. Mer, 187: 19-26.
Takayama H. (1981). Observations on two species of *Gymnodinium* with scanning electron microscopy. Bull. Plankton Soc. Jap. 28(2) : 121 -129.
Takayama H., Matsuoka K. (1991). A reassessment of the specific characters of *Gymnodinium mikimotoi* Miyake et Kominami ex Oda and *Gymnodinium nagasakiense* Takayama et Adachi.Bull. Plankton Soc. Jpn. 38: 53-68.
Tangen K. (1977). Blooms of *Gyrodinium aureolum* (Dinophyceae) in North European waters, accompanied by mortality in marine organisms. Sarsia, 63, 2, 123-133.
Thomas W.H., Vernet M., Gibson C.H. (1995). Effects of small-scale turbulence on photosynthesis, pigmentation, cell division and cell size in the marine dinoflagellate *Gonyaulax polyedra*. J. Phycol., 31, 1, pp 50-59.
Videau Ch., Hourmant A., Gentien P., Caroff J. (1996) Putrescine: a possible link between the spring diatom bloom and *Gymnodinium* cf. *nagasakiense* blooms (submitted).
Vrieling E.G., Peperzak L., Gieskes W.W.C., Veenhuis M. (1994). Detection of the ichthyotoxic dinoflagellate *Gyrodinium* (cf.) *aureolum* and morphologically related *Gymnodinium* species using monoclonal antibodies : a specific immunological tool.Mar. Ecol. Progress Ser. 103: 165-174.
Wong P.S. (1989) The occurrence and distribution of red tides in Hong Kong -Applications in red tide management. in " Red Tides: Biology, Environmental Science and Toxicology", Okaichi T., Anderson D.M., Nemoto T., (eds.):125-128.

Yamaguchi M., Honjo T. (1989).Effects of temperature, salinity and irradiance on the growth of the noxious red tide flagellate *Gymnodinium nagasakiense* (Dinophyceae). Nippon Suisan Gakkaishi, 55(11), 2029-2036.

Yamaguchi M. (1992). DNA synthesis and the cell cycle in the noxious red-tide dinoflagellate *Gymnodinium nagasakiense*. Mar. Biol., 112:191-198.

Yamaguchi M. (1994) Physiological ecology of the red tide dinoflagellate *Gymnodinium nagasakiense* (Dinophyceae)-Mechanism of the red tide occurrence and its prediction. Bull. Nansei Natl. Fish. Res. Inst., n°27:251-394.

Yanagi T., Asai Y., Koizumi Y. (1992). Physical conditions for red tide outbreak of *Gymnodinium mikimotoi* . Suisan Kaiyo Kenkyu 56, 2:107-112.

Yanagi T., Yamamoto T., Koizumi Y., Ikeda T., Kamizono M., Tamori H. (1995). A numerical simulation of red tide formation. J. Mar. Systems, 6:269-285.

Yasumoto T., Underdal B., Aune T., Hormazabal V., Skulberg O.M., Oshima Y. (1990). Screening for hemolytic and ichthyotoxic components of *Chrysochromulina polylepis* and *Gyrodinium aureolum* from Norwegian coastal waters. in "Toxic Marine Phytoplankton" Granèli *et al.* Eds., Elsevier Science Publishers: 436-440.

Physiological Ecology of *Pfiesteria piscicida* with General Comments on "Ambush-Predator" Dinoflagellates

J.M. Burkholder,[1] H.B. Glasgow Jr.,[1] and A.J. Lewitus[2]

[1] Department of Botany, Box 7612, North Carolina State University, Raleigh, NC 27695-7612 U.S.A.; and

[2] Belle-Baruch Institute for Marine Biology and Coastal Research, University of South Carolina, P.O. Box 1630, Georgetown, SC 29442 U.S.A.

1. Introduction

The discovery in the early 1980s of a small estuarine organism, *Katodinium* [*Gymnodinium*] *fungiforme* (Anisimova) Loeblich III, represented the first known report of free-living dinoflagellates with "ambush-predator" behavior. Spero and Moree (1981) described a small free-living dinoflagellate of unknown origin that, previously undetected, swarmed up from benthic stages to rip apart a wounded protozoan ciliate, devour it in a "feeding frenzy," and then rapidly settle back to the bottom of the culture vessel. This characteristic behavior has been documented for four species. All were accidentally detected as cryptic culture contaminants of unknown origin from brackish waters. *K. fungiforme*, cosmopolitan in occurrence, is a prey generalist that targets certain species of protozoan ciliates and/or microalgae (Spero and Moree 1981). *Crypthecodinium* [*Gymnodinium*] sp. (Elbrächter, pers. comm.; strain thus far known only from the Mediterranean Sea) is a prey specialist that targets only one species of the rhodophyte, *Porphyridium* (Ucko *et al.* 1989). Available information indicates that the haploid, haplobiontic life cycles of these two dinoflagellates may be comparatively simple; the representative *K. fungiforme* has dominant zoospores that may either encyst or form gametes, which fuse to create planozygotes that may encyst before yielding zoospores (Spero and Moree 1981).

In contrast, three other recognized taxa of toxic ambush-predator dinoflagellates, found thus far on the western Atlantic and Gulf Coasts, are prey generalists with directed toxicity toward an array of targeted finfish and shellfish. The better known species, *Pfiesteria piscicida* Steidinger & Burkholder, is also capable of consuming diverse other prey including bacteria, small algae, microfauna, and mammalian tissues, usually when fish are not available but sometimes while killing both finfish and shellfish (Burkholder and Glasgow 1995, Burkholder *et al.* 1995, Steidinger *et al.* 1996a). Aside from their toxic and prey generalist features, *P. piscicida* and at least two other toxic *Pfiesteria*-like species also apparently differ from other known ambush-predator dinoflagellates in having complex life cycles that involve rapid trans-formations among flagellated, amoeboid, and encysted stages within a 90-fold size range (Burkholder and Glasgow 1995, 1997; Landsberg *et al.* 1995).

Ambush-predator dinoflagellates cross unarmored, weakly armored, and moder-ately armored taxa lines and broad geographic regions, with prey spanning all trophic levels. Small cryptic zoospores (diameter 5-12 µm) dominate the planktonic or free-swimming forms in their life cycles. All stages are heterotrophs, sometimes with limited mixotrophic capability. The zoospores feed primarily by myzocytosis

NATO ASI Series, Vol. G 41
Physiological Ecology of Harmful Algal Blooms
Edited by D. M. Anderson, A. D. Cembella and G. M. Hallegraeff

(Schnepf and Elbrächter 1992), although they can also capture prey via phagotrophy or velum construction. The known species show strong chemosensory interactions with prey, and transformations in their life cycles are highly influenced by prey type and availability. Prey detection stimulates often-rapid excystment and swarming behavior in surrounding the prey, generally enhanced when prey are wounded, senescent, or otherwise weakened.

Although ambush-predator behavior, and the dinoflagellates that manifest it, may be common and is geographically widespread, field detection would be virtually impossible without prior understanding based on cultures, and unless the prey could be readily discerned (e.g., a fish kill rather than microflora or microfauna death). Likely for those reasons, ambush-predator dinoflagellates were unknown or unrecognized until the early 1980s. The recognized taxa have been tracked to natural habitat, but only one has been detected attacking its prey in a natural setting. Additional species, both toxic and nontoxic, probably remain to be discovered. Here we review known information about this group of cryptic organisms that exhibits complex, directed behavior toward targeted prey spanning all levels of estuarine food webs. The known behavior and physiological ecology of ambush-predator dinoflagellates will be summarized, and future research directions will be identified. Emphasis will be placed on *P. piscicida*, for which the most information is available.

2. Nontoxic Ambush-Predator Dinoflagellates

Katodinium fungiforme has been described as an herbivore, carnivore, and scavenger, with transformations in its life cycle highly dependent on feeding (Spero and Moree 1981). In this trait, and in other behavioral and morphological features, it resembles *P. piscicida*. *K. fungiforme* can be cultured indefinitely on the small green flagellate *Dunaliella salina*, which it consumes through myzocytosis (Schnepf and Elbrächter 1992). Its preference for small flagellates without cell walls has been demonstrated in feeding trials with potential prey as 11 other algal species having cell walls. Chemosensory mediation is inferred in prey acquisition by *K. fungiforme* because healthy ciliates, nematodes, polychaete larvae, and brine shrimp fail to elicit any response from the dinoflagellate, even after prolonged periods (days). Using microprobes to wound microfauna such as the large protozoan ciliate, *Condylostoma magnum* (length 600 - 1,000 μm), Spero and Moree (1981) described an immediate response (within 10 - 15 sec) by *K. fungiforme* zoospores. They swarmed in "dynamic aggregations" of hundreds of cells that attached to the ciliate surface with their peduncles and suctioned the prey's cytoplasmic contents. Cytoplasmic materials were pulled from the prey by groups of 1-5 zoospores, which enhanced leakage of fluids through the small wounds and led to attraction of additional zoospores.

Large wounded ciliates are consumed alive by swarms of *K. fungiforme* zoospores in less than 30-minute periods (Spero and Moree 1981). This feeding activity is accompanied by a 20-fold increase in cell size as the food vacuole swells. A similar response has been shown by zoospores of this species when they detect senescent or dying animals that would innately be more structurally "leaky." Both zoospores and planozygotes actively feed by myzocytosis. In food-depleted cultures, swimming cells rapidly disappear by forming resistant, colorless, thick-walled, long-term resting cysts that strongly adhere to the culture vessel walls. Asexual cyst germination is triggered by increased prey availability. Asexual reproduction is initiated by swimming zoospores or, more commonly,after groups of zoospores (10-60) settle out of the water in a gelatinous mass. Each zoospore forms a nonmotile division cell (zoosporangium or asexual division cyst that sloughs off its flagella, becomes rounded and smooth, and divides to produce two small zoospores (dimensions ca. 9 x 7 μm). Isogamous gametes resemble newly formed zoospores, but are smaller (diameter 5-7 μm); planozygotes expand to the maximum species dimensions, then enter a hypnocyst stage via a similar (benthic) process as for vegetative division of zoospores. Controls on hypnocyst germination are not yet known.

Spero and Moree (1981) compared grazing activity by small, cryptic zoospores of *K. fungiforme* (which can ingest, on average, 13 algal prey/day) to that by copepods. The authors effectively argued in support of the premise that the lower range of zoospore bloom densities observed in culture (10^5 cells/mL) could accomplish comparable prey consumption as average field densities of common copepods. Noting the estuarine characteristics of this dinoflagellate, the authors hypothesized that it is cosmopolitan in distribution, but usually missed or overlooked because of its size and potentially poor preservation in most fixatives. They also hypothesized that its grazing activity in field conditions is substantial.

Spero (1982) noted that early authors who described dinoflagellates ingesting prey through a tentacle (1881 report), a pseudopod (1891-1892 reports), or a cytoplasmic extension (1952 report) probably were describing peduncles -- cryptic organelles that could easily be overlooked due to the speed with which prey capture and ingestion occur. The second known non-toxic ambush-predator dinoflagellate, *Crypthecodinium* sp. (but note: possibly a species of *Gymnodinium*; M. Elbrächter, pers. comm., 1997), was described as remarkably difficult to characterize with respect to feeding activity. Simon *et al.* (1991) related numerous attempts to use various technologies to film feeding by this small, colorless dinoflagellate, without success. They described zoospores that made their presence known only when a certain species of the small unicellular red alga, *Porphyridium*, was added to culture vessels. The dinoflagellates swarmed up from a benthic habit, and approached the algae. There the observations would end in frustration; the researchers would observe colorless zoospores at one moment, and zoospores with swollen reddish food vacuoles at the next. The dinoflagellate showed preference for feeding in darkness, which further complicated assessment of the feeding mechanism. Through use of specialized high-speed videotaping under minimal light, they finally caught its extremely rapid peduncle in action, attaching to and myzocytizing the algal prey.

Crypthecodinium sp. (also called *C. cohnii* (Seligo) Chatton [Ucko *et al.* 1994] and a *C. cohnii*-like species [Ucko *et al.* in press] as well as *Gymnodinium* sp. [Simon et al. 1991]) was formerly believed to be highly specialized in prey preference with respect to ambush-predator behavior. Recently, however, it has been observed to consume microbial prey other than *Porphyridium* (M. Elbrächter, pers. comm., 1997). When approaching *Porphyridium* sp., it alters its swimming from a slightly helical track with broad turns to more narrow turns, and apparently uses physical contact with the tip of its longitudinal flagellum to test prey desirability (Ucko *et al.* in press). The zoospore then ceases swimming and extends a feeding tube (peduncle, ca. half the cell diameter), attaches to and penetrates the prey, and suctions the contents, leaving the outer cell envelope. In a manner analogous to the protozoan prey of *K. fungiforme*, wounded *Porphyridium* cells become highly attractive to other zoospores, which swarm about the prey and also attach to feed.

This dinoflagellate has been observed to "disappear" by forming cryptic benthic cysts when *Porphyridium* sp. is absent, and it has not responded similarly to other *Porphyridium* strains; Ucko *et al.* 1989). Its life cycle is being characterized, with vegetative cysts (that divide and produce zoospores or gametes upon excystment),

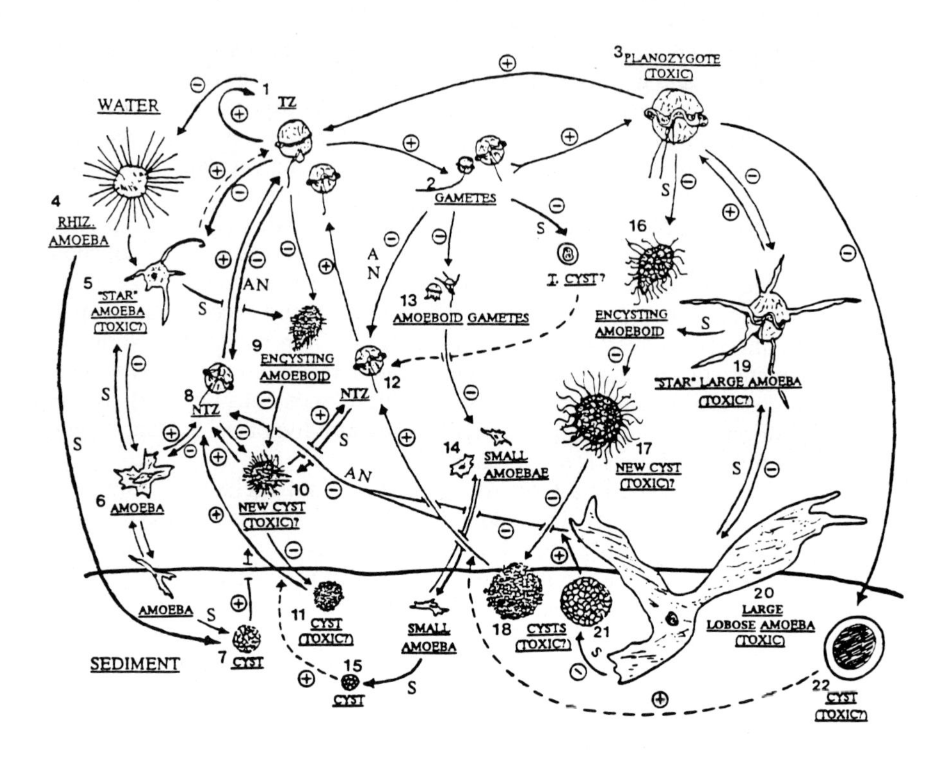

Figure 1. Schematic of the complex life cycle of *Pfiesteria piscicida* with (+) or without (-) live finfish (F = presence of certain flagellated algal prey such as *Cryptomonas* sp., *Dunaliella tertiolecta* Butcher or *Isochrysis galbana* Green; NU = nutrient enrichment; S = environmental stress such as sudden shift in temperature or salinity, or physical disturbance). Solid lines = verified transformations; dashed lines = hypothesized pathways. Nontoxic zoospores (NTZs, stages #8, #12) are produced by various amoeboid and cyst stages, by gametes (stage #2) through reversion or loss of sexual activity, and by toxic zoospores (TZs, stage #1) through cessation of toxin production. They could be considered an additional numbered stage, as well, since they also can be produced by lobose amoebae (stages #6, #20) that, in turn, transformed from TZs or planozygotes (stage #3). NTZs are similar to "-" anisogamous gametes and TZs in appearance, but are of intermediate size; this stage is considered a nontoxic immediate precursor to the most lethal form (TZs). Cysts include stages with (a) reticulate covering (#7, #15, #21 -- from amoebae), (b) scaled covering (± bracts: #10, #11, #16, #17, #18 -- from TZs and planozygotes), and hyaline covering (#22, with interior large swollen, darkened cell -- presumed hypnocyst). Gametes have also been observed to rapidly form what appear to be temporary cysts (T.CYST, with obvious mucous covering; see Taylor 1987), which may settle out of the water column; if these cysts yield viable cells, they likely produce NTZs [reverted gametes with loss of sexuality]). Several other stages are known, but with uncertain position in the life cycle (e.g., two heliozoan amoebae, and a small ephemeral flagellated stage with consistent tear-drop shape that has appeared in abundance on several occasions immediately after fish death (from Burkholder and Glasgow 1997).

isogametes, anisogametes, nuclear cyclosis, planozygotes, and zygotic cysts verified as well as zoospores (trophonts; Ucko *et al.* in press). In experiments to examine chemosensory attraction to *Porphyridium* sp., Ucko and coworkers have found that the dinoflagellates contain enzymes which degrade the red algal cell wall (Simon *et al.* 1992, 1993), but the attraction does not appear to be linked to polysaccharide-degrading activity (Ucko *et al.* 1994). Induction of these enzymes, nonetheless, enables the dinoflagellates to peel the algal cell wall, facilitating access to the intra-cellular materials.

3. Complex prey generalists: trophic controls on toxicity

The known toxic *Pfiesteria*-like species (Burkholder *et al.* 1992; Landsberg *et al.* 1995; Steidinger *et al.* 1996a,b; Burkholder and Glasgow 1997) appear to resemble one another closely in cell structure, life cycles, and behavior. Zoospores demonstrate toxic or apparently toxic behavior (*P. piscicida* and the other toxic *Pfiesteria*-like species, respectively) toward live fish. They also have complex life cycles with filose and lobose amoebae, cysts of various structures, and a wide size range among stages. Little is yet known about a second, apparently subtropical *Pfiesteria* species (Landsberg et al. 1995). *P. piscicida* has been found as far north as we have surveyed (inland bay system, Delaware, U.S.A.), and in the Chesapeake and Albemarle-Pamlico Estuaries as well as other mid-Atlantic and southeastern U.S. areas (Burkholder *et al.* 1995, Lewitus *et al.* 1995, Burkholder and Glasgow 1997). A second apparently toxic *Pfiesteria*-like species also occurs at least from the Chesapeake to the St. Johns Estuary, Florida, USA (Burkholder and Glasgow 1997, K. Steidinger pers. comm.). Information is needed about the distribution of *P. piscicida* and *Pfiesteria*-like species in other geographic regions. Given the paucity of data on the other species, the remainder of this discussion will focus on *P. piscicida.*

3.1. Complex life cycles

Detection of unknown substances secreted or excreted by finfish stimulate zoospores of *P. piscicida* to emerge from benthic cysts or to transform from amoeboid stages (Burkholder and Glasgow 1995; Fig. 1). As the chemosensory stimulation increases, zoospores develop the capacity for toxin production. The TZs retain similar morphology as NTZs, but produce exotoxins that narcotizes finfish, slough their epidermis, and cause formation of open ulcerative lesions (Burkholder *et al.* 1995, Noga *et al.* 1996). The dinoflagellates consume bits of epidermal tissue and blood cells from affected fish, while also engulfing other available phytoplankton. In addition, they produce gametes that complete sexual fusion in the presence of dying fish. The gametes are both osmotrophic and phagotrophic (Glasgow and Burkholder 1993). Upon fish death, TZs and planozygotes form mostly nontoxic amoeboid stages, or encyst and descend back to the sediments. Alternatively, in the absence of live fish, gametes and TZs can revert to nontoxic, asexual zoospores that can remain active in nutrient-enriched waters by consuming bacterial, algal, and microfaunal prey (Burkholder and Glasgow 1995, 1997; Lewitus *et al.* 1995b).

In maintaining an array of active benthic amoebae as well as planktonic flagellates, toxic *Pfiesteria*-like predatory dinoflagellates differ markedly from red tide dinoflagellates and the bloom-forming "hidden flora" that appear in high abundance to cause massive fish death (Smayda 1992; Hallegraeff 1993). By contrast, lethal stages of *Pfiesteria*-like species often represent a minor component of the plankton community during toxic outbreaks, and rarely are detected by water discoloration except for a brownish surface foam that sometimes preceeds a fish kill (Burkholder *et al.* 1995, Glasgow *et al.* 1995). Relatively low cell densities, nonetheless, are highly lethal to fish. *P. piscicida* has 22 stages with known placement in its life cycle (Burkholder and Glasgow 1995), and at least 3 additional stages (e.g., a saccate form and another small flagellated form) with uncertain placement. We are continuing to work toward resolving the role of these additional stages, and to clarify ploidy among stages.

3.2. Physical controls and field behavior

Toxic *Pfiesteria*-like species are estuarine organisms with wide temperature and salinity tolerance. Their toxic outbreaks have been documented and experimentally verified using bioassays with test fish at salinities ranging from freshwater (0 psu, with $\geq$ 4 mg calcium hardness/L) to full-strength seawater (35 psu; optimum 15 psu), and across temperature gradients from 10-33°C (optimum > 26°C; Burkholder *et al.* 1995). Under colder conditions (< 15°C; work thus far with *P. piscicida*), toxic lobose amoebae display similar behavior toward live fish as TZs, which represent the dominant toxic stage at warmer temperatures (Burkholder and Glasgow 1995, 1997). In the field and in large-scale aquaculture facilities (e.g., hybrid striped bass in freshwater ponds), fish kills related to *P. piscicida* have occurred from 0 to 35 psu, supporting the laboratory information. Freshwater aquaculture kills have involved water with high calcium content from the coastal Castle Hayne aquifer. Kills have also been documented at sea, ca. 6 km from shore; these events involved schools of Atlantic menhaden during late autumn, and likely occurred because the dinoflagellates were transported with migrating schools down-estuary and out to sea.

Available light appears to impose little direct influence on *P. piscicida*'s toxic activity. Replicated aquarium bioassays with live fish and TZs yielded comparable numbers of dead fish across an imposed light gradient (neutral density screens; 12:12 L:D at 2, 50, 75, 100, 150, 200, and 250 μEinst/m^2/sec; Burkholder *et al.* 1995). Further, toxic activity has been demonstrated whenever live fish were added over 24-hour periods (Burkholder and Glasgow 1997). In near-darkness (ca. 2 μEinst/m^2/sec for 1-2 minutes/day while adding fresh prey), both zoospores and amoebae can survive for long periods (weeks to months) when given bacterial, algal, or fish prey, with amoebae generally predominating (Burkholder *et al.* 1995).

In laboratory conditions we have noted that TZs tend to encyst prior to fish death when aquaria become anoxic, but under field conditions *P. piscicida* and other *Pfiesteria*-like species have been found swarming in hypoxic to anoxic waters during in-progress fish kills (e.g., > 2 x 10^3 TZs/mL; Goose Creek, Neuse River Estuary, July 1995; Burkholder and Glasgow 1997). Although these dinoflagellates likely are able to more easily attack fish that previously were affected by low oxygen or other stressors, recent stress or physiological weakness of the prey is not a requisite for attack -- *Pfiesteria*-like species also have been implicated as the primary cauative agents of major estuarine fish kills during mid-spring before bottom-water hypoxia was widely established, and in late fall 1 to 3 months after oxygen stress was encountered (Burkholder *et al.* 1992, 1995; Burkholder and Glasgow 1997).

These field observations have been supported by laboratory tests in which relatively dilute cultures of *P. piscicida* (ca. 1,000 TZs/mL) that have been fed live fish repeatedly have proven lethal (often in < 20 minutes) to healthy animals representing 18 native and exotic finfish species in more than 2,000 confirming bioassays. In the field, *Pfiesteria*-like species respond poorly to water-column agitation from storm events, and often gather over the bottom when storms are in progress (Burkholder and Glasgow 1997). Like other toxic dinoflagellates (e.g., Heil *et al.* 1993, Flynn and Flynn 1995), in laboratory conditions they respond poorly to gentle mixing, pressurized filtering, or rapid pouring of culture water. Such perturbations usually lead to cessation of toxin production (indicated by lack of fish death) and transformations from flagellated to amoeboid and encysted stages (H. Glasgow unpubl. data). Most frequently, *Pfiesteria*-like species kill fish in poorly flushed estuaries and small tributaries, areas which would afford relatively long periods for chemical communication between the predators and prey (Burkholder and Glasgow 1997).

3.3. Complex nutrition: stimulation by nutrients

Laboratory experiments with TZs and NTZs of *P. piscicida* have demonstrated both apparent direct stimulation by nutrient enrichment, and indirect stimulation mediated through photosynthetic algal prey. TZs from fish-killing cultures, but without live fish, revert to NTZs that have shown ephemeral stimulation (several days) by inorganic phosphate (P_i) enrichment (Glasgow *et al.* 1995). TZs and gametes readily consume dissolved organic nitrogen and carbon as amino acids (cellular-level resolution provided by track light microscope-autoradiography [Burkholder 1992], from 30 minute-assays with ^{14}C-protein hydrolysate; Amersham, Chicago, IL; Glasgow and Burkholder 1993, Burkholder and Glasgow 1997). NTZs without fish for 10 days exhibited stimulation by both P_i and glycerophosphate (P_o; 100 or 500 μg $PO_4^{-3}P/L$ for either P source used in f/2 medium [McLachlan 1979] with 15 psu-salinity Instant Ocean water, ultrafiltered through 0.22 μm-porosity Millipore filters; Burkholder and Glasgow 1997). These effects were significantly enhanced in the presence of algal prey (*Dunaliella tertiolecta* previously grown in f/2 medium with 1% P_i or P_o depending on the treatment; *Pfiesteria* NTZ : algal prey ratio 1:5 initially, with prey added daily). Other algal contaminants (small chrysophytes, blue-green algae) and bacteria were not significantly correlated with NTZ densities except in the P_o treatments without *Dunaliella*. Amoebae also were stimulated under P enrichments, but were more variable in response. The data suggested that, while NTZs could be stimulated by P_i and P_o enrichments, significantly higher growth occurred in the presence of both nutrients and [certain] algal prey.

These bioassay results are supported by field observations, wherein we have found significantly higher levels of total phosphorus, and significantly higher abundances of both *Pfiesteria*-like species and smaller (potential) algal prey near sewage outfall sites in several estuaries (these three variables were positively correlated; $p < 0.01$ to 0.05; e.g., Burkholder and Glasgow 1997). Other field work in a eutrophic mesohaline estuary has demonstrated significant positive correlations between *Pfiesteria*-like zoospore densities and phytoplankton biomass as chlorophyll *a* during late winter/spring seasons characterized by high nutrient loading from precipitation and runoff, and large phytoplankton blooms (Fensin 1997, Burkholder and Glasgow 1997). The phenomenon of kleptochloroplastidy, first described in *P. piscicida* by Steidinger *et al.* (1995), likely plays a role in zoospore survival under field conditions. Zoospores have been documented to retain the still-functional chloroplasts of algal prey for weeks to months (Lewitus *et al.* 1995b), and apparently derive supplemental nutrition from this "adopted" mode of nutrition.

Two experiments were completed to further examine the response of NTZs to N and P enrichments. It has not yet been possible to grow *Pfiesteria*-like species in high abundance without algal or animal prey; hence, as in previously described

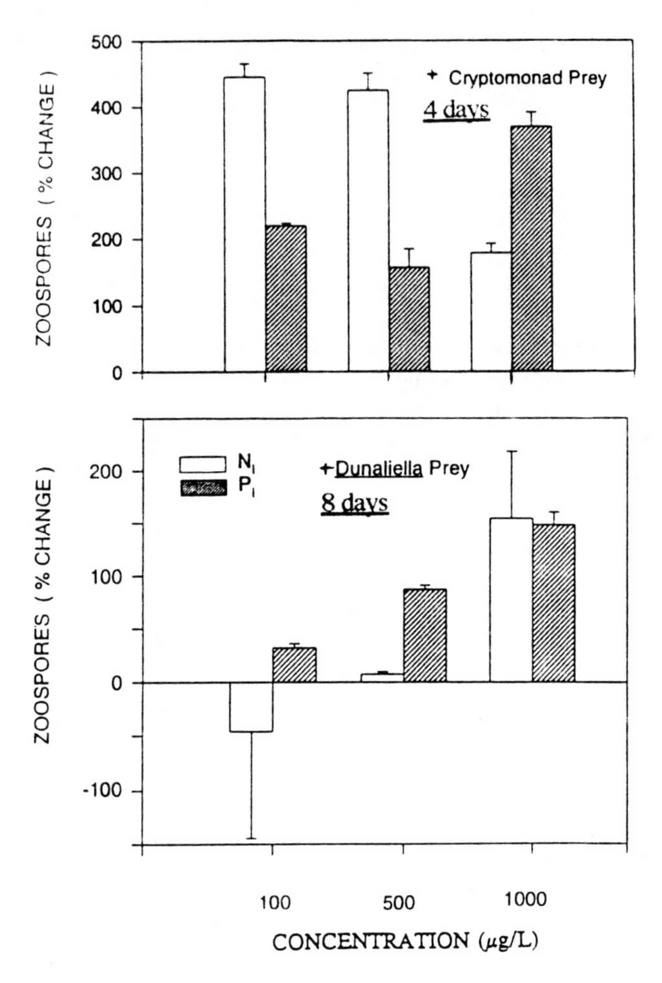

Figure 2. Nutrient experiment I, showing stimulation of *P. piscicida* NTZs by P_i + *Cryptomonas* and *Dunaliella* prey after 4 and 8 days, respectively. Also note general stimulation of NTZs by N_i enrichment as $NaNO_3$, in the presence of either algal prey. Similar tests with NH_4^+N yielded no response. Data are plotted as the percent change from initial NTZ population densities. NTZ growth in controls without nutrient additions was negligible (means ± 1 SE, n = 3).

12-mL well plate batch cultures of ultrafiltered 15 psu-seawater with imposed nutrient gradients (0, 50, 100, 250, and 500 µg P_i, N_i, or P_o from Na_2HPO_4, $NaNO_3$, or glycerophosphate, respectively). Algae aside from cryptomonads were negligible, except for occasional small filaments of *Lyngbya* sp. (< 10^3/ mL); initial bacterial densities were also low (< 10^5 cells/mL). Controls consisted of the various dinoflagellate stocks without cryptomonads, without N or P enrichment, or without both algal prey and N or P.

After 4 days enrichment with P_i generally had stimulated growth, and there was a trend of increasing growth with increasing [P_i] (Fig. 5). Zoospore abundances were

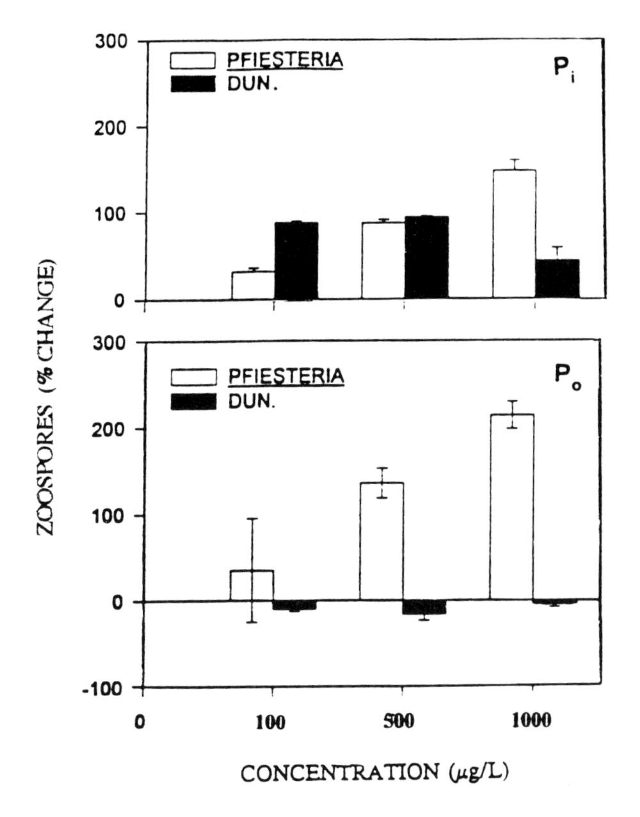

Figure 3. Nutrient experiment I, showing stimulation of *Pfiesteria piscicida* NTZs and the algal prey species *Dunaliella*, after 8 days in P_i enrichment (upper panel); and stimulation of NTZs after 8 days in P_o enrichment (data given as in Fig. 2).

significantly correlated with $[P_i]$ in the DINO1 treatment with history of potential reliance on kleptochloroplasts ($P < 0.05$, $r^2 = 0.76$), and also in the DINO3 treatment (nutrient-poor history; $P < 0.02$, $r^2 = 0.96$), but not in the previously nutrient-enriched *P. piscicida* cultures (DINO2). In the DINO2 treatment, but not in the other *P. piscicida* cultures, NTZ densities were positively correlated with cryptomonad abundances ($P < 0.05$, $r^2 = 0.76$). In the cultures with DINO1, cryptomonad numbers were significantly negatively correlated with glycerophosphate level (Fig. 5). There were no significant correlations between any of the nutritional history dinoflagellate cultures and $[P_o]$, although in the (previously nutrient-replete) DINO2 treatments cryptomonad growth was positively correlated with $[P_o]$. DINO1 responded poorly to $[P_o]$, but appeared to closely track the abundance of its cryptomonad prey. *P. piscicida*'s response to the N_i enrichment gradient generally was positive but variable for DINO1 and DINO3. The previously nutrient-replete culture, DINO2, did not increase relative to the controls without N_i additions, as expected. There was no correlation between DINO3 NTZ abundance and algal prey densities under N_i enrichment, where cryptomonad growth was most variable.

Collectively, our experiments thus far indicate that the nutritional history and prey type / availability are important in controlling *P. piscicida*'s response to nutri-

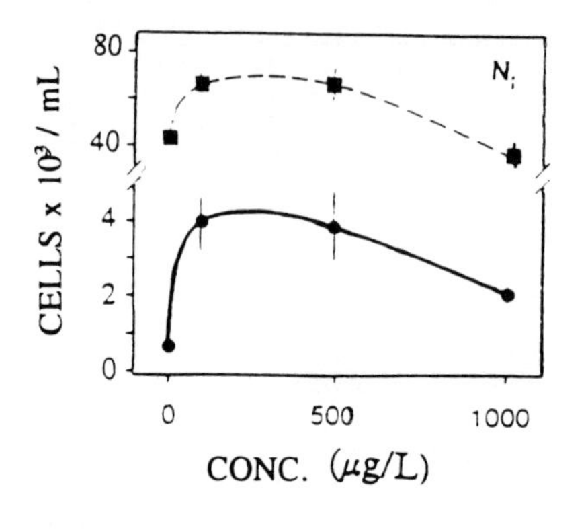

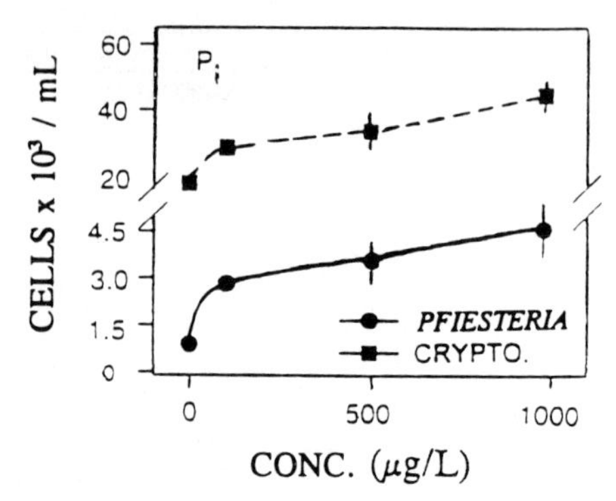

Figure 4. Nutrient experiment I, showing the response of *P. piscicida* NTZs to inorganic N (nitrate, N_i), inorganic phosphorus (phosphate, P_i), and organic phosphorus (glycerophosphate, P_o) when NTZs also were fed algal prey (*Cryptomonas* sp.) for 8 days. Enrichment with "plant" nutrients P_i or N_i, NTZs "tracked" prey densities. However, NTZs appeared to be directly stimulated by the P_o source, irrespective of algal prey abundances (means ± 1 SE, n = 3; Burkholder and Glasgow 1997).

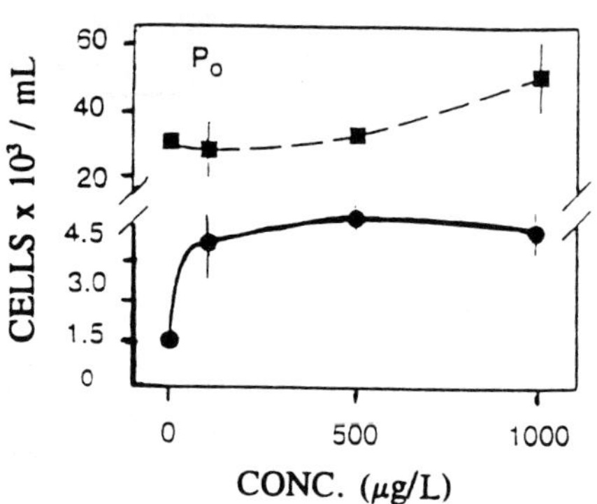

ent enrichment. *P. piscicida* can take up dissolved organic nutrients (for example, N_o; Glasgow and Burkholder 1993), and can be indirectly stimulated by both Ni and P_i through stimulation of algal prey. Experiments are in progress to gain additional insights about the nutritional ecology of *Pfiesteria*-like dinoflagellates, including their stim-ulation by fish secreta/excreta. In planned trials with NTZs we intend to (i) determine P, N, and C budgets for NTZs of different nutritional histories, across inorganic and organic P and N gradients (P_o uptake to be verified using track micro-autoradiography); (ii) examine the influence of bacteria as competitors for P_o and N_o; (iii) assess the role of various forms of dissolved organic carbon on growth of NTZs and TZs (also with aid of autoradiography); and (iv) examine their response to nutrient enrichments within the natural community framework.

In the field we have found significant correlations between *Pfiesteria*-like NTZ abundances and areas characterized by chronically nutrient-enriched conditions (Burkholder and Glasgow 1997). Efforts are underway in our laboratory to model their seasonal relationships with algal biomass and abundance of known algal prey species (e.g., cryptomonads, certain dinoflagellates and flagellated chrysophytes) under field conditions, using an extensive field dataset supported by laboratory and field experiments. It is important to remember, when approaching the nutritional ecology of these dinoflagellates, that they are not simply photosynthetic phytoplankters -- rather, they are flagellated or amoeboid (protist) animals with complex nutrition that extends well beyond mixotrophy with chloroplasts "borrowed" from algal prey (Lewitus *et al.* 1995b). Lewitus and coworkers are currently examining their microalgal prey preferences in detail, and the role of kleptochloroplasts in their nutrition.

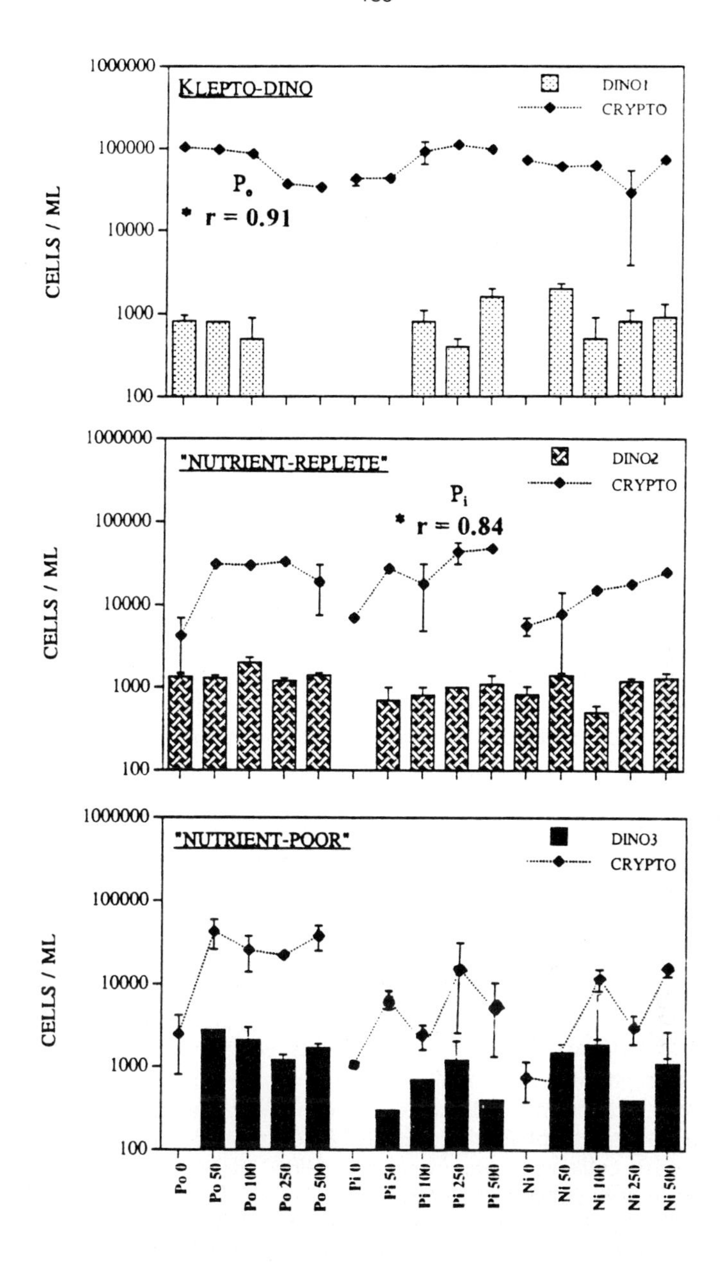

Figure 5. Nutrient experiment II, showing effects of nutritional history on growth of *P. piscicida* NTZs (black bars; data are plotted in comparison to NTZ response with cryptomonad prey densities, and probability values indicate a significant correlation between *Pfiesteria* and prey abundances (means $\pm$ 1 SE, n = 3). Note that under these conditions, NTZs did not exhibit growth under nutrient enrichment alone, except for ephemeral increases ($\leq$ 2 days) with P_O.

experiments, the dinoflagellate cultures were neither axenic nor unialgal. NTZs of *P. piscicida* were selected for study because they could be examined without confounding effects of fish secreta, and because they can provide an inoculum for TZs (Burkholder and Glasgow 1995). The objective of experiment I was to examine the response of NTZs (without live fish for 1.5 years; initial densities at ca. 3 x 10^2 cells/mL) to inorganic and organic N and P, (0, 100, 500, and 1,000 µg/L) +/- algal prey. Prey species included either *Cryptomonas* sp. (WH2423, dimensions 8 x 5 µm; treatment 1) or *Dunaliella tertiolecta* (Pamlico isolate, length 4 µm), each initially added at a density of 1.5 x 10^5 cells/mL (22°C, 12:12 L:D with available light at 50 µEinst/m^2/sec). The P_i (Na_2HPO_4), P_o (glycerophosphate), N_i ($NaNO_3$), and N_o (protein hydrolysate, in similar mixture as radiolabeled substrate from Amersham, Inc. described previously) treatments were imposed in 15 psu-ultrafiltered Instant Ocean water, using f/2 medium prepared without P (for treatments with P_i, P_o) or N (for treatments with N_i, N_o). Nutrient treatments were prepared with/without addition of nutrient-replete *Cryptomonas* prey. Controls consisted of *P. piscicida* in ultrafiltered Instant Ocean f/2 media with algal prey minus P, or lacking both algal prey and P (P treatments); and with algal prey minus N, or lacking both algal prey and N (N treatments; for further details see Burkholder and Glasgow 1997).

After 4 days there was clear stimulation of NTZs by P_i + *Cryptomonas* prey (Fig 2). NTZ growth was also significantly higher under N_i enrichment as $NaNO_3$, in the presence of either algal prey (Fig. 3). After 8 days, there was a general stimulatory effect of both P_i and N_i on NTZ growth with *Dunaliella*, but not with *Cryptomonas* (Fig. 4). *P. piscicida* reponded favorably to increasing P_o in cultures with either prey (Figs. 3, 4). Zoospore population abundance tracked cryptomonad prey densities in the presence of P_i and N_i, but stimulation apparently occurred independently of prey concentration in the P_o treatments (Burkholder et al. 1996; Fig. 4). We inferred direct stimulation by P_o since neither zoospore nor cryptomonad densities were significantly correlated with bacterial numbers. NTZs apparently were outcompeted by bacteria and the cryptomonad for N_o under these experimental conditions, with highest bacterial numbers in cultures containing *Dunaliella* (2.3 - 2.8 x 10^9 cells/mL at $\geq$ 100 µg N_o/L, versus < 0.5 x 10^7 cells/mL in all treatments with cryptomonads). In at least three 60-min periods of viewing per day, cryptomonads were observed to actively phagocytize bacteria in the N_o treatments, whereas occasional to negligible bacterial phagotrophy was noted in *P. piscicida* cultures with *Dunaliella*.

The objective of experiment II was to examine the role of nutritional "history" in population-level growth of *P. piscicida* NTZs (without fish for 6 months; initial densities 1.5 x 10^2 cells/ mL), previously maintained for 17 days (22°C, 12:12 L:D with available light at 50 µEinst/ m^2/sec) as:

(i) DINO1 ("klepto-dino") -- plant-like mode with kleptochloroplasts (f/2 media in 15 psu, ultrafiltered seawater made with Millipore Q water, but containing negligible algal prey [< 2 cells/mL];
(ii) DINO2 ("nutrient-replete") -- previously given high densities of nutrient-rich cryptomonad prey (1.5 x 10^3 cells/mL, grown in f/2 media); and
(iii) DINO3 ("nutrient-poor") -- previously given high densities of N- and P-depauperate cryptomonad prey for 17 days (1.5 x 10^3 cells/mL, grown in 1% f/2 media).

Prey offered to *P. piscicida* TZs during the experiment consisted of *Cryptomonas* sp. (WH2423); other experimental conditions were similar to those used during preacclimation. The dinoflagellates, $\pm$ additional cryptomonads, were maintained in

3.4. Predation: From Microflora to Higher Trophic Levels

P. piscicida's flagellated and amoeboid stages have been observed to be voracious predators on both microflora and microfauna including bacteria, an array of algae (both diatoms and small flagellates, although certain flagellates are preferred; Burkholder and Glasgow 1995, Lewitus *et al.* 1995b), protozoan ciliates, rotifers, and the tissues of finfish, shellfish, and mammals (Burkholder and Glasgow 1995, 1997). NTZs can be maintained indefinitely on algal prey, but become more sporadic in toxin production after ca. 1 year under laboratory conditions. Like *K. fungiforme*, *P. piscicida* consumes protozoan ciliates under certain conditions such as wounding, or sometimes after the ciliates have fed upon TZs for an extended period (hours; Burkholder *et al.* 1992).

When *P. piscicida* zoospores or planozygotes encounter algal or small protozoan ciliate prey of interest, they alter their swimming path from a gentle, random helical pattern to short, abrupt motions about the targeted cell. Within 5-19 seconds, each cell extends a peduncle and suctions the prey contents in a manner similar to that described for *Crypthecodinium* sp. When attacking larger organisms such as rotifers, both zoospores and lobose amoebae spin a velum about the prey which seems to slow and then immobilize it so that it can be more readily attacked and consumed via myzocytosis (H. Glasgow and J. Burkholder, unpubl. data). Mallin *et al.* (1995) observed that the rotifer *Brachionus* could consume TZs for 7 days before fecundity was adversely affected, and that the copepod *Acartia tonsa* could consume TZs for 3 days without apparent ill effects. Interactions between *Pfiesteria*-like species and other microfauna in estuarine microbial food webs are being further assessed. Springer, Shumway and coworkers also have begun to examine interactions between these dinoflagellates and higher trophic levels including shellfish and finfish species.

P. piscicida and at least one other toxic *Pfiesteria*-like species have been implicated as primary causative agents in many major kills affecting 10^3 to 10^9 finfish and shellfish annually, during what has come to be called "fish kill season" in the Albemarle-Pamlico Estuarine System (Burkholder *et al.* 1995, Burkholder and Glasgow 1997). For a 3- to 4-month period during mid - late summer, millions of fish have been affected in most years. The Albemarle-Pamlico is the second largest estuary on the U.S. mainland, and provides about 50% of the total surface area used as nursery grounds by fish from Maine to Florida (Epperly and Ross 1986, Steel 1991). Aside from lethality, these dinoflagellates also have been implicated as major causative agents of fish epizootics such as ulcerative skin and shell diseases (Burkholder *et al.* 1995, Noga *et al.* 1996, Burkholder and Glasgow 1997; Table 1). These events are believed to represent an apparent chronic sublethal effect of the dinoflagellates' toxins, which has been demonstrated to be a causal factor of the disease known as ulcerative "mycosis" in Atlantic menhaden (Noga *et al.* 1996). *P. piscicida*'s toxins are poorly characterized, but are known to include both lipophilic and wter-soluble components that can be aerosolized (Glasgow *et al.* 1995, Burkholder and Glasgow 1997). Ulcerative diseases linked to *Pfiesteria*-like species often extend through warmer seasons (Miller *et al.* 1990, Noga 1993). Up to 98% of all fish in large estuaries can be affected for weeks to months in spring - fall (Burkholder *et al.* 1995), with death usually occurring 1-2 weeks after noticeable body or shell damage. There is a critical need to identify these substances, and to develop *in vitro* assays of cytotoxicity (for example, using fish cell lines).

Zoospores and amoebae of *P. piscicida* and other *Pfiesteria*-like species show directed chemosensory attraction to fresh finfish tissues (e.g., epidermis, blood cells; motion analysis with computerized imagery in the presence vs. absence of prey tissues; Burkholder and Glasgow 1997) as well as fresh shellfish tissues (e.g., gill, siphon, spleen; Springer *et al.* 1996). Other tissues from both finfish and shellfish are consumed, as well. Like some other microheterotrophs, zoospores and amoebae also show strong attraction to mammalian red and white blood cells (e.g.,11 $\pm$ 3 red corpuscles zoospore^{-1} hour^{-1} in short-term feeding trials, using blood from either fish or humans). But after 4 days, *P. piscicida* stages have not maintained growth on mammalian blood -- the dinoflagellates have lysed or encysted without supplemental fish or other food resources (Glasgow, Ph.D. thesis, in preparation).

4. Summary

Little is known about the ecology of ambush-predator dinoflagellates, thus far a small group of species that can repeatedly swarm up from benthic stages, rapidly attack, kill and consume prey, and then descend back to the sediments or to other benthic habits. These cryptic dinoflagellates, thus far, are estuarine with temperate to subtropical distribution. They, and additional species, likely are widely distributed; for example, a fourth *Pfiesteria*-like species, with toxicity not yet confirmed, recently was found in the Chesapeake Bay drainage (Burkholder and Glasgow 1997, K. Steidinger pers. comm.). These dinoflagellates are heterotrophs, but in at least one case photosynthesis from kleptochloroplasts is used to supplement nutrition. They are also prey generalists that consume a wide array of organisms including bacteria, algae, microfauna, invertebrates, and vertebrates; but certain prey induce strong chemosensory attraction manifested in the rapid predation response. Among the recognized species, *Katodinium fungiforme* and *Crypthecodinium* [*Gymnodinium*] sp. are nontoxic and have relatively simple life cycles as presently understood. By contrast, *P. piscicida* and other toxic *Pfiesteria*-like species thus far are known to have complex life cycles including an array of flagellated and amoeboid stages, some of which are toxic to fish and mammals. The best known, *P. piscicida*, is euryhaline and eurythermal. *Pfiesteria*-like species have been implicated as causative agents of major estuarine fish kills and epizootics. They may also represent significant estuarine microbial predators.

Pfiesteria piscicida has shown stimulation by inorganic and organic forms of nitrogen and phosphorus. Both field and laboratory experimental evidence indicates that inorganic N and P can stimulate *Pfiesteria*-like zoospore production as an indirect effect, mediated through algal prey. Additional information is needed on organic nutritional controls on stage transformations and feeding in ambush-predator dinoflagellates. Given its warm-optimal character and its apparent preference for nutrient-enriched conditions, *P. piscicida* would be expected to expand its range and its activity under warming trends in global climate change (Epstein *et al.* 1993), and with established, pervasive human population growth and water quality degradation to coastal areas (Smayda 1992, Hallegraeff 1993, Adler *et al.* 1993).

5. Acknowledgments

Funding support for this work was provided by the National Science Foundation (grant OCE9403920), the U.S. EPA, the National Sea Grant Biotechnology Program, the North Carolina Agricultural Research Service, and the Department of Botany at North Carolina State University. We thank E. Hannon, J. Manning, M. Mallin, S. Shumway, and J. Springer for counsel on the manuscript.

6. References

Adler, R. W., Landman, J. C., and Cameron, D. M. (1993) The Clean Water Act 20 years later. National Resources Defense Council. Island Press, Washington, DC.

Burkholder, J. M. (1992) Phytoplankton and episodic suspended suspended sediment loading: phosphate partitioning and mechanisms for survival. *Limnol. Oceanogr.* 37:974-988.

Burkholder, J. M., and Glasgow Jr., H. B. (1995) Interactions of a toxic estuarine dinoflagellate with microbial predators and prey. *Arch. Protistenkd.* 145:177-188.

Burkholder, J. M., and Glasgow Jr., H. B. (1997) *Pfiesteria piscicida* and other toxic *Pfiesteria*-like dinoflagellates: behavior, impacts, and environmental controls. *Limnol. Oceanogr.* (in press).

Burkholder, J. M., and Glasgow Jr., H. B., and Hobbs, C. W. (1995) Fish kills linked to a toxic ambush-predator dino-...flagellate: Distribution and environmental conditions. *Mar. Ecol. Prog. Ser.* 124: ...43-61.

Burkholder, J. M., Glasgow Jr., H. B., and Lewitus, A. J. (1996) Response of the toxic estuarine dinoflagellate, *Pfiesteria piscicida*, to N and P from organic and inorganic sources (abstract). *Proceedings, Annual Meeting of the American Society of Limnology & Oceanography*, San Diego (Feb.).

Burkholder, J. M., Noga, E. J., Hobbs, C. W., Glasgow Jr., H. B., and Smith, S. A. (1992) New "phantom" dinoflagellate is the causative agent of major estuarine fish kills. *Nature* 358:407-410; *Nature* 360:768.

Epperly, S.P. and Ross, S.W. (1986) Characterization of the North Carolina Pamlico-Albemarle estuarine complex. National Oceanic & Atmospheric Administration Technical Memorandum NMFS-SEFC-175. NOAA, Washington, DC.

Epstein, P. R., Ford, T. E., and Colwell, R. R. (1994) Marine ecosystems. *Lancet*, special issue - Health and Climate Change, p. 14-17.

Fensin, E. E. 1997. Population dynamics of *Pfiesteria*-like dinoflagellates, and environmental controls in the mesohaline Neuse Estuary, 1994-1996. Master of Science thesis. North Carolina State University, Raleigh.

Flynn, K. J., and Flynn, K. (1995) Dinoflagellate physiology: nutrient stress and toxicity. Pages 541-550 *in* Lassus, P., Arzul, G., Erard, E., Gentien, P., and Marcaillou, C., editors. Harmful marine phytoplankton blooms. Lavoisier Intercept, Ltd., Paris.

Glasgow , H.B. Jr. and Burkholder, J.M. (1993) Comparative saprotrophy by flagellated and amoeboid stages of an ichthyotoxic estuarine dinoflagellate. Page 87 ...*in* Abstracts, Sixth International Conference on Toxic Marine Phytoplankton. Ministre de Affaires Etrangeres, Nantes.

Glasgow , H.B. Jr. and Burkholder, J.M., J. M., Schmechel, D. E.,Tester, P. A., and Rublee, P. A. (1995) Insidious effects of a toxic dinoflagellate on fish survival and human health. *J. Toxicol. Environ. Health* 46:101-122.

Hallegraeff, G. M. (1993) A review of harmful algal blooms and their apparent global increase. *Phycologia* 32:79-99.

Heil, C. A., Maranda, L., and Shimizu, Y. (1993) Mucus-associated dinoflagellates: large-scale culturing and estimation of growth rate. Pages 501-506 *in* Smayda, T. J. and Shimizu, Y., editors. Toxic phytoplankton blooms in the sea. Elsevier, New York.

Landsberg, J. H., Steidinger, K. A., and Blakesley, B. A. (1995) Fish killing dinoflagellates in a tropical marine aquarium. Pages 65-70 *in* Lassus, P., Arzul, G., Erard, E., Gentien, P., and Marcaillou, C., editors. Harmful marine algal blooms. Lavoisier Intercept, Ltd., Paris.

Lewitus, A. J., Jesien, R. V., Kana, T. M., Burkholder, J. M., and Glasgow, H. B. Jr. (1995a) Discovery of the "phantom" dinoflagellate in Chesapeake Bay. *Estuaries* 18:373-378.

Lewitus, A. J., Glasgow, H. B. Jr., and Burkholder, J. M. (1995b) Kleptochloroplasts in the toxic estuarine dinoflagellate, *Pfiesteria piscicida. Proceedings, Biennial Meeting of the Estuarine Research Federation*, Houston (abstract).

Mallin, M. A., Burkholder, J. M., Larsen, L. M., and Glasgow , H. B. Jr. (1995) Response of two zooplankton grazers to an ichthyotoxic estuarine dinoflagellate. *J. Plankt. Res.* 17:351-363.

McLachlan, J. (1979) Growth media - marine. Pages 25-51 *in* Stein, J. R., editor. Handbook of phycological methods - culture methods and growth measurements. Cambridge University Press, Boston.

Miller, K. H., Camp, J., Bland, R. W., Hawkins, J. H. III, Tyndall, Cl L., and Adams, B. L. (1990) Pamlico Environmental Response Team report (June - December 1988). North Carolina Department of Environment, Health & Natural Resources, Wilmington.

Noga, E. J. (1993) Fungal diseases of marine and estuarine fishes. Pages 85-100 *in* Couch, J.A. and Fournie, J. W., editors. Pathobiology of marine and estuarine organisms. CRC Press, Boca Raton.

Noga, E. J., Khoo, L., Stevens, J. B., Fan, Z., and Burkholder, J. M. (1996) Novel toxic dinoflagellate causes epidemic disease in estuarine fish. *Mar. Pollut. Bull.* 32: 219-224.

Schnepf, E., and Elbrächter, M. 1992. Nutritional strategies in dinoflagellates - a review with emphasis on cell biological aspects. *Europ. J. Protistol.* 28:3-24.

Simon, B. S., Geresh, S., and (Malis) Arad, S. (1991) Characterization of biodegradation products of the cell-wall polysaccharide of the red microalga *Porphyridium* sp. using enzymatic activity of its predator, the dinoflagellate *Gymnodinium* sp. *Proceedings, Joint Meeting of the Fifth International Phycological Congress and the Phycological Society of America*, Durham, North Carolina (abstract).

Simon, B. S., Geresh, S., and (Malis) Arad, S. (1992) Degradation of the cell wall polysaccharide of *Porphyridium* sp. (Rhodophyta) by means of enzymatic activity of its predator, *Gymnodinium* sp. (Pyrrhophyta). *J. Phycol.* 28:460-465.

Simon, B. S., Geresh, S., and (Malis) Arad, S. (1993) Polysaccharide-degrading activities extracted during growth of a dinoflagellate that preys on *Porphyridium* sp. *Plant Physiol. Biochem.* 31:387-393.

Smayda, T. J. (1990) Novel and nuisance phytoplankton blooms in the sea: Evidence for a global epidemic. Pages 29-40 *in* Graneli, E., Sundstrom, B., Edler, L., and Anderson, D. M., editors. Toxic marine phytoplankton. Proceedings, *Fourth International Conference on Toxic Marine Phytoplankton* (June 1989, Lund, Sweden). Elsevier Publishers, New York.

Smayda, T. J. (1992) Global epidemic of noxious phytoplankton blooms and food chain consequences in large ecosystems. Pages 275-307 *in* Sherman, K., Alexander, L. M., and Gold, B. D., editors. Food chains, models and management of large marine ecosystems. Westview Press, San Francisco.

Smith, S. A., E. J. Noga, and R. A Bullis (1988) Mortality in *Tilapia aurea* due to a toxic dinoflagellate bloom (abstract). Pages 167-168 *in* Proceedings, Third International Colloquiem of Pathology in Marine Aquaculture, Gloucester.

Spero, H. J. (1982) Phagotrophy in *Gymnodinium fungiforme* (Pyrrhophyta): The peduncle as an organelle of ingestion. *J. Phycol.* 18:356-360.

Spero, H. J., and Moree, M. (1981) Phagotrophic feeding and its importance in the life cycle of the holozoic dinoflagellate, *Gymnodinium fungiforme. J. Phycol.*17:43-51

Springer, J. J., J. M. Burkholder, H. B. Glasgow Jr., and S. E. Shumway. Impacts of the toxic dinoflagellate, *Pfiesteria piscicida*, on the scallop *Argopecten irradians* L. (abstract). Proceedings, Annual Meeting of the American Shellfish Association, Baltimore.

Steel, J. (editor). (1991) Status and trends report of the Albemarle-Pamlico Estuarine Study. Albemarle-Pamlico Estuarine Study - United States Environmental Protection Agency National Estuaries Program and North Carolina Department of Environment, Health & Natural Resources, Raleigh.

Steidinger, K. A., Truby, E. W., Garrett, J. K., and Burkholder, J. M. (1995) The morphology and cytology of a newly discovered toxic dinoflagellate. Pages 83-87 *in* Lassus, P., Arzul, G., Erard, E., Gentien, P., and Marcaillou, C., editors. Harmful marine phytoplankton blooms. Lavoisier Intercept, Ltd., Paris.

Steidinger, K. A., Burkholder, J. M., Glasgow Jr., H. B., Truby, E. W., Garrett, J. K., Noga, E. J., and Smith, S. A. (1996a) *Pfiesteria piscicida* gen. et sp. nov. (Pfiesteriaceae, fam. nov.), a new toxic dinoflagellate genus and species with a complex life cycle and behavior. *J. Phycol.* **32:**157-164.

Steidinger, K. A., Landsberg, J. H., Truby, E. W., and Blakesley, B. A. (1996b) The use of scanning electron microscopy in identifying small "gymnodinioid" dinoflagellates. *Nova Hedwigia* 112:415-422.

Ucko, M., Cohen, E., Gordin, H., and Arad, S. (1989) Relationship between the unicellular alga *Porphyridium* sp. and its predator, the dinoflagellate *Gymnodinium* sp. *Appl. Env. Microbiol.* 52:2990-2994.

Ucko, M., Geresh, S., Simon-Berkovitch, B., and Arad (Malis), S. (1994) Predation by a dinoflagellate on a red microalga with a cell wall modified by sulfate and nitrate starvation. *Mar. Ecol. Prog. Ser.* 194:293-298.

Ucko, M., Elbrächter,and Schnepf, E. (1997) A *Crypthecodinium cohnii*-like dinoflagellate feeding myzocytotically on the unicellular red alga *Porphyridium* sp. *Europ. J. Phycol.* (in press).

Bloom Dynamics and Physiology of *Prymnesium* and *Chrysochromulina*

B. Edvardsen and E. Paasche

Section for Marine Botany, Department of Biology, University of Oslo, P.O. Box 1069, N-0316 Oslo, Norway

1. Introduction

The two flagellate genera *Chrysochromulina* and *Prymnesium* both belong to the class Prymnesiophyceae within the division Haptophyta (Jordan and Green 1994). The Haptophyta, which include other important bloom-forming genera such as *Phaeocystis* and *Emiliania*, are of a predominantly marine distribution. While *Chrysochromulina* species form a regular component of the marine plankton (though with a handful of representatives in freshwater), most *Prymnesium* records are from inshore localities and brackish-water lakes and ponds. Harmful blooms of *Chrysochromulina* seem to be exceptional events, while *Prymnesium* blooms are recurrent in many parts of the world. Both types of bloom may result in large fish kills causing great economic losses.

The aim of this chapter is to review blooms of *Prymnesium* and *Chrysochromulina* and to interpret them as far as is feasible in terms of experimental information on the ecophysiology of these algae. Information on the morphology and biology of these flagellates is included since it may help elucidating their ecological strategy.

2. Morphology, phylogeny, and life histories of the *Prymnesium-Chrysochromulina* group

The morphology of Haptophyta was reviewed in Green and Leadbeater (1994). Members of *Chrysochromulina* and *Prymnesium* are in the size range 3-35 µm in cell length. All species are photosynthetic, but particle uptake (possibly indicating mixotrophic tendencies) has been demonstrated in both genera. They usually possess two chloroplasts, two smooth flagella and always a third appendage called haptonema. The haptonema can adhere to suitable substrates, and in some *Chrysochromulina* species it has been shown to be involved in particle capture or handling (Kawachi *et al.* 1991; Kawachi and Inouye 1995). The cells are covered by organic unmineralised scales. The fine structure of these scales, which usually can be seen only in the electron microscope, is considered to be species specific.

At present, about 50 species of *Chrysochromulina* and 10 species of *Prymnesium* are described (Jordan and Green 1994). *Chrysochromulina* species typically have two or more elaborate types of scales and a long, often coiling haptonema, whereas *Prymnesium* species usually have simple plate scales and always have a short, non-coiling haptonema.

A molecular phylogeny of the Prymnesiales is presently under investigation (Medlin *et al.*, in prep.). Analysis of the nuclear-encoded ssu rRNA gene indicates that *Chrysochromulina* is not a natural group and can be divided into two clades. Some species (*C. polylepis*, *C. kappa*, and *C. hirta*) appear more closely related to *Prymnesium* species than to other *Chrysochromulina* species and comprise one clade.

NATO ASI Series, Vol. G 41
Physiological Ecology of Harmful Algal Blooms
Edited by D. M. Anderson, A. D. Cembella and G. M. Hallegraeff

All other *Chrysochromulina* species sequenced to date fall into the other clade. A revision of the taxonomy of the two genera may be needed.

Within the Haptophyta, life cycles with alternating morphologically distinct generations are frequent. Yet sexual reproduction, which is commonly found in many other algal groups, has been established only in a small number of cases (see review by Billard 1994).

Generally, haptophyte life histories embrace an alternation between motile and non-motile stages. In *C. polylepis*, however, two motile cell types have been described (Edvardsen and Paasche 1992) that may be stages in a haplo-diploid life cycle (Edvardsen and Vaulot 1996). The significance of this in *Chrysochromulina* bloom dynamics remains to be investigated.

Two of the *Prymnesium* species, *P. parvum* and *P. patelliferum*, appear morphologically identical except for differences in scale morphology that can be discerned only in the electron microscope (see Larsen *et al.* 1993). At present it seems that they are genetically extremely close (Larsen and Medlin, submitted), to the extent that one may ask if they could be stages in the life-cycle of one and the same species.

Resting cysts have been reported in *Prymnesium* (Green *et al.* 1982; Wang and Wang 1992). There are no reports of cysts from *Chrysochromulina*, but amoeboid or non-motile walled cells have been seen in cultures of a number of species (e.g. Parke *et al.* 1956). The significance (if any) of such non-flagellate forms in nature remains obscure.

3. *Prymnesium*

3.1. Toxic species of *Prymnesium*

Prymnesium parvum (under the provisional name of "Chrysomonadine van Workum") was first connected with fish kills in the Workum See in Holland in 1920 (Liebert and Deerns 1920). It has since caused similar problems in numerous other brackish-water localities. Some other species of *Prymnesium* associated with fish kills are listed in Table 1. Blooms reported to be of *P. parvum* could have been of *Prymnesium patelliferum*, since identification was usually made by light microscope (see above) and *P. patelliferum* was not recognised as a separate taxon until 1982. *Prymnesium saltans* was claimed as the causative organism of fish kills in Germany in 1990 (Kell and Noack 1991). However, Moestrup (1994) noted that the description of the alga forming this bloom agreed more closely with *P. parvum*. Wang and Wang (1992) reported that it was *P. saltans*, and not *P. parvum*, that caused fish kills in the TianJin area, China (Table 1). The bloom of *Prymnesium calathiferum* in New Zealand in 1983 (Chang 1985) was remarkable because it was the first instance of a harmful *Prymnesium* bloom in a fully marine environment. In the laboratory, *P. parvum*, *P. patelliferum*, and *P. calathiferum* have proved consistently toxic according to various tests (Table 1). The toxic potential of the additional *Prymnesium* species remains unknown, but all are suspected to be toxic (Moestrup and Larsen 1992).

P. parvum produces toxins with ichthyotoxic, cytotoxic and antibacterial activity (see reviews by Shilo 1971; Hashimoto 1979; and also Igarashi *et al.* 1996). The toxins, production of which is promoted by phosphorus deficiency, affect mainly gill-breathing animals, such as fish, tadpoles, and molluscs (Shilo 1971).

Table 1. Toxicity data on species of *Prymnesium* (modified from Moestrup 1994)

Species	Toxicity in nature	Toxicity in culture	Habitat	Selected references
P. calathiferum	Toxic to fish and shellfish (?)	Toxic to fish	Marine	1
P. parvum	Toxic to fish, tadpoles, molluscs	Toxic to fish and tadpoles	Marine and brackish water	2; see Table 2
		Toxic to *Artemia*		3; 4
P. patelliferum	Toxic to fish	Toxic to *Artemia*	Marine and brackish water	3; 4; 5; see Table 2
P. saltans	Toxic to fish	-	Brackish water	6

References cited: (1) Chang 1985; (2) Shilo 1971; (3) Larsen *et al.* 1993; (4) Meldahl *et al.* 1994; (5) Valkanov 1964, reference in Moestrup 1994; (6) Wang and Wang 1992.

3.2. Occurrence and distribution of *Prymnesium* species associated with toxic events

P. parvum/ *P. patelliferum* have formed harmful blooms in brackish water localities in Europe, the Middle East, Ukraine, China and USA (Table 2) and Australia (W. Hosja pers. comm., quoted in Hallegraeff 1992).

In the marine domain, there are several non-bloom records of *P. parvum*/ *P. patelliferum* from coastal or low-salinity inshore localities in Europe, Canada, USA, South Africa (references in Throndsen 1969; Green *et al.* 1982), Japan (e.g. Throndsen 1983), and one from the open sea (Throndsen 1983). *P. calathiferum* from New Zealand is a truly marine species. Other *Prymnesium* species have been observed on a few occasions only. Data are lacking on the occurrence of *Prymnesium* spp. in the tropics and in polar waters. Due to their small size and fixation damage, *Prymnesium* cells have probably been overlooked in many routine investigations.

3.3. Dynamics of *Prymnesium* blooms

With the exception of the marine *P. calathiferum* "bloom" in New Zealand (Chang 1985), *Prymnesium* blooms have been restricted to waters of low salinities, between 1-12‰ (Table 2). Most blooms have occurred in shallow lakes, ponds or lagoons of limited area. Exceptions to this rule are the recurrent *P. parvum*/*P. patelliferum* blooms in western Norway (see below). Usually *Prymnesium* blooms develop in the warm season at water temperatures above 10° C, but toxic *P. parvum* blooms have also been recorded at a much lower temperature (5° C, Krashnoshchek and Abramovich 1971). Fish kills have usually occurred only at very high algal concentrations (>50-100$\cdot 10^6$ cells$\cdot L^{-1}$, sometimes even >10^9 cells$\cdot L^{-1}$; Table 2). Considerable amounts of nutrients such as nitrogen and phosphorus are needed to build up *Prymnesium* populations of this size. Many of the affected waters are clearly eutrophic due to cultivation of fish, discharge of sewage, or runoff from agricultural land.

Since 1989, blooms consisting of a mixture of *P. parvum* and *P. patelliferum*, together with a number of other plankton algal species, have taken place yearly in the

Table 2. Information on recorded harmful blooms of *Prymnesium* species. br = brackish, nd = no data

Species Country	Locality	Year / Month	Salinity (‰)	Max. cell conc. (10^6 cells·L^{-1})	Reference
P. parvum					
China	Ponds in Shandong	1988-1990 / 5-11	2-11	44	1
Denmark	1) Ketting Nor	1938 / 9	[a]8	1200	2
	2) Selsø lake	1939 / 9-10	[a]4	655	
	3) Pond in Jutland	1991 / 6	nd	146	3
Finland	Dragsfjärd archipelago	1990 / 6	6	55	4
Germany	1) Waterneverstorfer Lakes	1909, 1914, 1920, 1932 / nd	br	18	5
	2) Pond in Fehmarn	1975 / 9	4.8	400	6
	3) Pond in Schleswig-Holstein	1978/ 5	3	10	7
	[b]4) Rügen	1990 / 4	2-6	300	8
Great Britain	1) Fenland Drainage System	1962, 1963, 1964 / 5-8	[a]1-2	1600	9
	2) Norfolk Broads	1969, 1970, 1973, 1975, 1977 / 3-9	[a]1-6	800	10
Holland	1) Workum lake	1920 / 3	[a]2.7	500	11
	2) Botshol, Utrecht	1990 / 11-12	[a]2	50	12
Israel	Ponds	1945-1946 / 11-2	br	800	13
Norway	Ryfylke fjords	Yearly 1989-1995 / 7-8	4-25	1-3	14; 15
Spain	Lagoon, Ebro delta	1977 / 11-12	[a]11.6	200	16
Sweden	Coastal lake, Kyrkfjärden	1991, 1992 / 5-6	br	64	17
Ukraine	1) Yuzhnyy fish farm	1969 / 11-12	4.4	665	18;
	2) Kazankov fish farm	1971 / 11	4.8	720	19
USA	Pecos River, Texas	1985, 1986, 1988 / 10-12	br	150	20; 21
P. patelliferum					
[c]Bulgaria	1) Varna lakes	1959, 1963 / 7-9	1-12	150	22;
	2) Burgas lake	1964 / 3	br	897	23
Norway	[d]see above				
P. calathiferum					
New Zealand	Bream Bay	1983 / summer	34-35	<<1	24
P. saltans					
China	Ponds in TianJin area	1983-1986 / nd	2-5	nd	25

[a]Salinity was estimated from chlorinity as S‰ = 0.03+1.805Cl‰.

[b]The causative organism was originally identified as *P. saltans* but Moestrup (1994) noted that the description was more close to that of *P. parvum*.

[c]The causative organism was originally identified as *P. parvum*, but Green *et al.* (1982) noted that the TEM-graphs in Valkanov (1964) probably showed *P. patelliferum*

[d]The blooms contained a mixture of *P. parvum* and *P. patelliferum* ([e]Eikrem and Throndsen 1993).

[e]References in Moestrup 1994.

period July-August in the Ryfylke fjords on the west coast of Norway. The bloom in 1989 caused death to 750 tons of farmed fish. These blooms are initiated in the inner end of a large marine fjord system, dominated by a marked brackish surface layer caused by high freshwater input and restricted water exchange. The 1989 bloom developed in this layer at salinities of 4-5‰, at temperatures reaching 18.5° C due to local warming by insolation (e.g. Kaartvedt *et al.* 1991). Initial low exchange rates resulted in low advective loss of algae. When water circulation in the fjord was accelerated by freshwater discharge from a hydroelectric power plant, *Prymnesium* cells became dispersed farther out in the fjord, giving rise to fish mortality at salinities up to 27‰. The Ryfylke fjords are not very eutrophic and *Prymnesium* cell concentrations in open water were surprisingly low ($1\text{-}3 \cdot 10^6$ cells L^{-1}), considering the harm done. Dense accumulations of *Prymnesium* cells were, however, sometimes seen on benthic algae and other substrates. Due to the large component in the surface layer of unpolluted freshwater with a high nitrate:phosphate ratio, the algal production was suspected to be limited by phosphorus. This may have promoted the toxicity of the *Prymnesium* cells (see Section 3.1).

3.4. Autecology of *Prymnesium*

Because of the perennial threat of *P. parvum* to fish stocks, growth conditions for this species have been examined repeatedly. In spite of its predominantly brackish-water distribution, *P. parvum* is extremely euryhaline. Brand (1984) reported growth at both 5 ‰ and 45 ‰ salinity, and McLaughlin (1958) observed growth in artificial media at NaCl concentrations from 1 to 100 ‰, provided sufficient magnesium and calcium was present. Generally, the highest growth rates have been achieved in the salinity range of 10-20 ‰ (e.g. Padilla 1970; Brand 1984; Larsen *et al.* 1993). Blooms of *P. parvum* at temperatures from 5-30° C indicate that the temperature tolerance is very wide in this species. The growth rates of cultures were higher at 26° C than at 20° C (Larsen *et al.* 1993). *P. patelliferum* was similarly found to be euryhaline and temperature tolerant (Larsen *et al.* 1993).

4. *Chrysochromulina*

4.1. Toxic species of *Chrysochromulina*

Prior to the *C. polylepis* bloom in 1988, the genus was not thought to present any kind of menace to the marine environment. In laboratory tests, 14 species have been shown to be non-toxic to fish and/or larvae of the crustacean *Artemia* (references in Moestrup 1994; Edvardsen 1993). Jebram (1980), however, found that old cultures of *C. kappa*, *C. brevifilum*, *C. strobilus* , as well as *C. polylepis,* were toxic to the bryozoan *Electra pilosa* when given as the only food source.

Footnotes from Table 2 (continued).

References cited: (1) Yang *et al.* 1993; (2) [e]Otterstrøm and Steemann Nielsen 1940; (3) [e]Bach and Jacobsen 1991; (4) [e]Lindholm and Virtanen 1992; (5) [e]Lenz 1933; (6) [e]Hickel 1976; (7) Dietrich and Hesse 1990; (8) Kell and Noack 1991; (9) Farrow 1969; (10) [e]Holdway *et al.* 1978; (11) Liebert and Deerns 1920; (12) Rip *et al.* 1992; (13) [e]Reich and Aschner 1947; (14) Kaartvedt *et al.* 1991; (15) Aure and Rey 1992; (16); Comin and Ferrer 1978; (17) Holmquist and Willén 1993; (18) Krasnoshchek and Abramovich 1971; (19) Krasnoshchek *et al.* 1972; (20) [e]James and de la Cruz 1989; (21) Rhodes and Hubbs 1992; (22) [e]Petrova 1966; (23) [e]Valkanov 1964; (24) Chang 1985; (25) Wang and Wang 1992.

Although absence of overt toxicity is the usual case in *Chrysochromulina*, the noxious blooms of *C. breviturrita* in lakes (Canada and USA, 1978-80), *C. parva* in a lake (Denmark, 1991), *C. birgeri* (Canada, 1996) in brackish water, and *C. polylepis* and *C. leadbeateri* in marine waters (Scandinavia, 1988 and 1991, respectively), revealed the potential toxicity in this genus (Table 3). The sometimes dramatic effects in nature have been difficult to reproduce in the laboratory, however. Water samples of *C. leadbeateri* isolated from the fish-killing bloom in 1991 were clearly toxic to larvae of *Artemia* and to nerve cell preparations, but axenic cultures established from the bloom were non-toxic to *Artemia* larvae, erythrocytes and nerve cell preparations (Edvardsen 1993; Meldahl *et al.* 1994). *C. polylepis* in 1988 killed a wide range of animals and plants in nature, but its effects on various test organisms in the laboratory, using cultures isolated from the bloom, were not dramatic (e.g. Dahl *et al.* 1989; Carlsson *et al.* 1990; Tobiesen 1991). However, strains of *C. polylepis* remain toxic to *Artemia* larvae even after a number of years in culture (Edvardsen 1993). This is the only species in the genus that has been shown so far to have strains that are toxic in culture at all times. Yet, in 1994 and 1995 *C. polylepis* occurred in bloom concentrations in the same area as the 1988 bloom without causing visible harm to wild biota (Table 3). This, and the toxicity in a bloom situation of *C. leadbeateri* which later proved non-toxic in culture, indicates that expression of the toxin in *Chrysochromulina* is highly variable even within the same species. It was demonstrated that phosphorus deficiency promotes the hemolytic activity and the toxicity to *Artemia* larvae (Edvardsen *et al.* 1990; Edvardsen 1993). Still, we are very far from understanding how toxicity in this genus is triggered by innate control mechanisms and environmental variables.

The *C. polylepis* toxin or toxins have effects similar to those in *Prymnesium*, and exhibit ichthyotoxic and cytotoxic activity (Underdahl *et al.* 1989; Edvardsen *et al.* 1990; Yasumoto *et al.* 1990). The harmful effects of *C. polylepis* during the bloom in 1988 were apparently due to the production of substances able to increase cell membrane permeability and disturb ion balance in cells of a wide range of organisms (Skjoldal and Dundas 1991). Unlike some dinoflagellate toxins that act specifically on specialised biological structures such as nerve systems, the toxins in *Prymnesium* and *Chrysochromulina* have a generalised membrane action and thus may affect organisms ranging from protozoa to fish. In consequence of this, it is legitimate to ask whether the potential for toxin production provides these haptophyte flagellates with a selective advantage in the form of a chemical defence against grazing (Dahl *et al.* 1989; Nejstgaard *et al.* 1995).

4.2. Occurrence and distribution of *Chrysochromulina* species associated with harmful events

All *Chrysochromulina* species described to date, except four from freshwater lakes, are marine (some of them also having been recorded from brackish water). Most species are known only from a handful of localities, since an examination in TEM is necessary to confirm their identity (Thomsen *et al.* 1994). The majority of records are from coastal waters, but an increasing amount of distributional data indicates that most species are ubiquitous in the world's ocean (see reviews by Estep and MacIntyre 1989; Marchant and Thomsen 1994; Thomsen *et al.* 1994). Most often many *Chrysochromulina* species co-occur in the same water mass, and normally in low cell concentrations (10^3-10^5 cells·L^{-1}).

Chrysochromulina polylepis was recorded on 19 different occasions in the English Channel in the period 1955-1959 in all months except November and December

Table 3. Information on recorded *Chrysochromulina* blooms in fresh, brackish and marine waters. nd = no data

Country	Locality	Habitat	Year	Month	Species	Max. cell conc. (10^6 cells·L^{-1})	Effect	Selected reference
Canada	5 lakes in Ontario and New Hampshire (USA)	Freshwater	1978, 1979, 1980	June-Sept.	*C. breviturrita*	9	Obnoxious odours, Death of tadpoles	1
Canada	Nova Scotia,	Brackish	1996	March	*C. birgeri*	10-12	Fish mortality	2
Denmark	Lake in Zealand	Freshwater	1991	May-June	[a]*C. parva*	600	Fish mortality	3
Denmark/ Germany	Kattegat, Belt Sea, Kieler Bay	Marine	1992	April-May	*C. hirta* *C. spinifera* *C. brevifilum* *C. ericina,* a.o.	45	Fish mortality	4
Finland	Tvärminne archipelago	Brackish, under the ice	1974	March	*C. birgeri*	nd	No mortality reported	5; 6
Norway/ Sweden/	Skagerrak and Kattegat	Marine	1) 1988	May-June	*C. polylepis*	50-100	1) High mortality of a wide range of organisms	1) see text
Denmark			2) 1989	May-June	*C. polylepis*	1-2	2) No mortality reported	2) 7
Norway	Lofoten archipelago	Marine	1991	May-June	*C. leadbeateri*	>3	Fish mortality	8, 9
Norway	Skagerrak	Marine	1994	May-June	*C. polylepis* and *C.* spp.	10	No harmful effects, water toxic to *Artemia*	10
Norway	Skagerrak	Marine	1995	May-June	*C. polylepis* and *C.* spp.	50	Minor fish mortality?, water toxic to *Artemia*	10
Sweden	Kyrkfjärden, coastal lake	Brackish	1991, 1992	May-July	*C. parva* and *Prymnesium parvum*	28	Fish mortality	11

[a]*C. parva* has formed several other non-harmful blooms in fresh water (references in Kristiansen 1971).
References cited: (1) Nicholls *et al.* 1982, reference in Moestrup 1994; (2) Claire Carver and Wenche Eikrem, pers. comm.; (3) Hansen *et al.* 1994; (4) Hansen *et al.* 1995; (5) Hällfors and Niemi 1974; (6) Hällfors and Thomsen 1979; (7) Tangen 1989; (8) Rey and Aure 1991; (9) Aune *et al.* 1992; (10) Einar Dahl, pers comm. and own obs.; (11) Holmquist and Willén 1993.

(Manton and Parke 1962). The same species was observed on a few occasions in Scandinavian waters before 1988. Since the bloom in 1988, *C. polylepis* was especially looked for, and was found to be common in the summer nanoplankton in Skagerrak (Wenche Eikrem pers. comm.; own observations). *C. polylepis* has also been recorded from brackish water in the Baltic Sea (Hajdu *et al.* 1996). Single cells have been observed from Australian waters (P.L. Beech and D. Hill, pers. comm., quoted in Moestrup and Larsen 1992). *C. polylepis* has also been isolated from the North Pacific (46'N, 140'W, Neil Price, pers. comm.). This suggests that the species has a world-wide distribution.

Normally, *Chrysochromulina* species occur in low concentrations, but there are an increasing number of recorded exceptions. In Table 3 are listed recorded blooms of *Chrysochromulina* in the sea, in brackish water and in freshwater lakes.

The first reported mass occurrence of a *Chrysochromulina* species was a bloom of *C. birgeri* that formed under the ice in the Tvärminne archipelago in Finland. No harmful effects were reported, perhaps because there were no farmed fish in the area. *C. birgeri* also formed a bloom in March 1996 in Bras d´ Or Lakes, Nova Scotia in Canada, again in brackish, cold water (Table 3). Farmed rainbow trout in the area died. The bloom of *C. leadbeateri* in Northern Norway in 1991 similarly affected mainly farmed fish. Since the harmful *C. polylepis* bloom in 1988, described below, Skagerrak and Kattegat have been monitored for *Chrysochromulina* species, and local blooms were registered in 1989, 1992, 1994 and 1995. The bloom in 1992 in Danish and German waters consisting of several *Chrysochromulina* species was suspected to cause fish mortality, and the bloom off the Southern coast of Norway in 1995 may have caused fish mortality in one farm. In addition to these blooms in marine or brackish waters, noxious blooms of the freshwater species *C. parva* and *C. breviturrita* have been reported.

4.3. Dynamics of *Chrysochromulina* blooms

Although *Chrysochromulina* species occur world-wide, harmful blooms have, up to now, been reported from coastal waters adjacent to the North Atlantic only, such as in Canada, Denmark, Finland, Norway and Sweden (Table 3). All the *Chrysochromulina* blooms in marine waters have developed from April to June, after the regular spring bloom of diatoms. Cell concentrations reportedly causing death to fish have usually been between $3\text{-}50 \cdot 10^6$ cells$\cdot L^{-1}$. The chlorophyll standing stocks recorded during the *C. polylepis* bloom in 1988 were not very high (40-80 mg m^{-2}), corresponding only to about 10-20% of the previous diatom spring bloom (Dahl *et al.* 1989).

The *C. polylepis* bloom in May-June 1988 was exceptional in its wide extension and in its high toxicity. The bloom has been described in numerous reports (e.g., Dahl *et al.* 1989; Skjoldal and Dundas 1991; Granéli *et al.* 1993). In the final stages it covered an area of approximately 75 000 km^2 including Kattegat and most of Skagerrak (Fig. 1; Granéli *et al.* 1993). In Skagerrak, but not in Kattegat, the bloom caused death to a wide range of marine organisms such as farmed and wild fish, molluscs, echinoderms, ascidians, cnidarians, sponges and red algae (e.g. Skjoldal and Dundas 1991). It was restricted to a well-defined, strongly stratified water mass influenced by an unusually strong outflow of brackish water from the Baltic Sea. Sunny weather reinforced the stratification, while at the same time permitting algal growth in the pycnocline at 10-15 m depth. Nutrient concentrations just before the bloom were generally low above the pycnocline; in particular, silicate was exhausted by the preceding diatom spring bloom, which may have favoured growth of flagellates. The general eutrophication of Danish coastal waters by runoff from land may have been a prerequisite for the high cell concentrations attained in Kattegat. Elevated concentrations of nitrate, probably originating from the southern North Sea, were encountered below the pycnocline in Skagerrak, and it was suggested that phosphate limitation could have contributed to the remarkable toxicity of the bloom in this area (Dahl *et al.* 1989; Skjoldal and Dundas 1991).

Initially other phytoplankton species co-occurred with *C. polylepis*, but it became completely dominating as the bloom developed. In Kattegat the *C. polylepis* populations were concentrated in a narrow band in the pycnocline where they reached cell concentrations close to 10^8 cells·L^{-1} (Kaas *et al.* 1991). Few other autotrophic algae or potential grazers were present in this layer, and bacterial production was extremely low (Nielsen *et al.* 1990). In Skagerrak the lack of grazers or algal competitors was equally striking (Dahl. *et al.* 1989). Here, *C. polylepis* cell concentrations were much lower, and the total biomass in the water column was no greater than what could have been produced in the absence of an input of anthropogenic nitrogen (Dahl. *et al.* 1989).

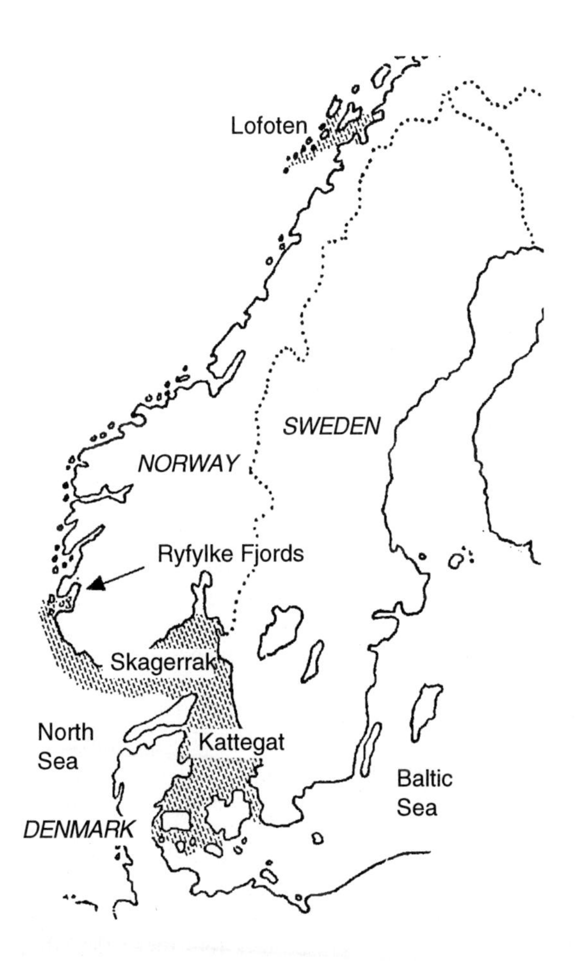

Fig. 1. Map of Scandinavia showing the maximum extent of the blooms (hatched area) of *Chrysochromulina polylepis* in the Kattegat and Skagerrak area in 1988, of *Chrysochromulina leadbeateri* in the Lofoten archipelago in 1991, and the location of Ryfylke fjords where *Prymnesium parvum / P. patelliferum* bloomed yearly in the period 1989-1995.

In May 1991 a bloom dominated by *Chrysochromulina leadbeateri* caused death in fish-farms in the Lofoten archipelago, Northern Norway (Fig. 1; Rey and Aure 1991; Aune *et al.* 1992). This bloom had some features in common with the 1988 *C. polylepis* bloom: it developed, following the decline of the diatom spring bloom, in a surface layer stabilised by admixed freshwater runoff from land, during a period of calm and sunny weather. There were important differences, however: there is no anthropogenic eutrophication in the area; nitrate:phosphate ratios did not deviate from normal; the bloom was not monospecific but contained other *Chrysochromulina* species as well; and no visible damage was done to natural biota.

A bloom of several *Chrysochromulina* species, including *C. polylepis*, occurred off the Norwegian south coast in May and June 1994 (Einar Dahl, pers. comm.). As in 1988, the event was preceded by freshwater outflow, resulting in stable water masses; and warm and sunny weather stimulated algal growth. As in 1988, orthophosphate was almost depleted while some nitrate was left. Up to 10^7 *Chrysochromulina* cells$\cdot L^{-1}$ were recorded locally, but no harmful effects were reported. A similar *C. polylepis* bloom took place in the same area in 1995. A very local fish kill was tentatively ascribed to this bloom (Einar Dahl, pers. comm.).

Normal environmental variability in Scandinavian coastal waters and elsewhere may be such that growth of *Chrysochromulina* species is favoured at times, leading to modest blooms that mostly fail to attract human attention. The *C. polylepis* bloom in 1988 stands out as an exception. Possible reasons why it took place are discussed in Section 4.5.

4.4. Autecology of *Chrysochromulina*

The maximum growth rate of *C. polylepis* is about 1.3 div. day^{-1}, which is of the same order as in many other plankton flagellates. It grows well at temperatures between 9-22° C, and the optimal salinity is about 25‰ although it will grow at lower and higher salinities (Edvardsen and Paasche 1992; Rhodes *et al.* 1994; Throndsen *et al.* 1995). *C. polylepis* thus appears to be eurythermal and euryhaline, adapted to variable conditions as is typically the case in coastal plankton algae. Information on the growth physiology of other marine *Chrysochromulina* species is limited to a few studies (Pintner and Provasoli 1968; Rhodes *et al.* 1994; Rhodes and Burke 1996), and again, nothing extraordinary is indicated.

Several *Chrysochromulina* species were shown to have an unusually high demand for selenium (Pintner and Provasoli 1968; Wehr and Brown 1985; Dahl *et al.* 1989; Rhodes *et al.* 1994). The bloom of *C. polylepis* in 1988 could have been promoted by high selenite concentrations in the outflowing Baltic Current water (Dahl *et al.* 1989). Other workers hypothesised that cobalt in runoff from land had stimulated the growth of *C. polylepis* and altered the species composition of the phytoplankton (Granéli and Haraldsson 1993). Since there are no available data on the selenium or cobalt concentrations in Scandinavian coastal water during the spring of 1988, the role of micronutrients in the success of *C. polylepis* during the bloom remains unknown.

The ability of many *Chrysochromulina* species, including *C. polylepis*, to ingest particles (see review by Jones *et al.* 1994) suggests but does not prove that their growth can be supported by mixotrophy. Jones *et al.* (1995) found that ingestion of prey provided an alternative source of carbon or energy in *C. brevifilum* under light-limiting conditions. Nygaard and Tobiesen (1993) on the other hand suggested that ingestion of bacteria by *C. polylepis* could be a source of phosphate during phosphorus limitation. A mixotrophic behaviour would imply that these flagellates may function both as primary producers and as consumers in the pelagic food web.

Mixotrophy is not obligatory, and *C. polylepis*, for example, can be grown in bacteria-free culture without difficulty. The significance, if any, of mixotrophy in *Chrysochromulina* blooms is unknown.

4.5. Ecological strategies of *Chrysochromulina*

Chrysochromulina species are a normal component of mature communities consisting of auto- and heterotrophic members of the nano- and picoplankton. In such communities, an inherent high growth rate is not expected to be a prerequisite for success. An ability to localise nutrients in short demand, as well as an efficient defence against grazers, may be more important. *Chrysochromulina* species can swim and many species show phototaxis. Several species are able to ingest particles that may provide an alternate source of carbon and/or nutrient under light-limiting or nutrient-limiting conditions. These features may be of advantage during mesotrophic-oligotrophic summer conditions, such as are typical of the stratified waters in Skagerrak.

Populations of nano- and picoplankton, including *Chrysochromulina* species, in mature communities are generally thought to be regulated by grazing protozoans and other microzooplankton. Blooms may result from an imbalance in the grazer-prey relationship. *C. polylepis* was avoided by potential grazers in the 1988 bloom (Nielsen *et al.* 1990), and potentially competing algal species were nearly absent (Dahl *et al.* 1989). Reduced grazing and increased availability of nutrients due to reduced competition could then have resulted in its complete dominance.

Following this reasoning, one may conclude that a relaxation of grazer control was an important aspect of the initiation of the bloom in 1988, as suggested by Dahl *et al.* (1989). However, if toxicity at the level seen in 1988 were a normal feature of *C. polylepis* one would expect toxic blooms to be more frequent, seeing that this species is present in Skagerrak at all times. The extraordinary harm done by *C. polylepis* in 1988 may be interpreted as an abnormally strong expression of a toxicity that is normally either latent or elicited to a limited extent by conditions in the pelagic ecosystem, and which at most comes into play merely as a grazing defence. The exceptional bloom in 1988 may be explained by a combination of meteorological and hydrographical conditions, on increased toxicity of *C. polylepis* due to phosphorus limitation with an attendant reduction of grazing, and possibly other factors of which we are not aware.

5. Conclusions

Although *Prymnesium* and *Chrysochromulina* are closely related morphologically, the two genera have little in common as harmful bloom formers except that they both kill farmed fish, and in exceptional cases wild fish and other marine animals as well.

Most *Prymnesium* blooms have been recorded from low-salinity lakes, lagoons or ponds, limited in area, with high concentrations of nutrients combined with fairly high water temperatures. *Prymnesium* species may bloom in almost any low-salinity, nutrient-rich area in the temperate region of both the Northern and Southern hemisphere.

The *Chrysochromulina* blooms reported from Scandinavian coastal areas have developed in stratified water influenced by high fresh water runoff, with nutrient levels low above the pycnocline and higher below, and during a calm and sunny period. *Chrysochromulina* blooms are natural events that may develop in unpolluted water at low levels of dissolved nitrogen and phosphorus and with normal N:P ratios.

Provided the responsible species has a potential for toxicity, a toxic bloom may result even in an undisturbed environment. This is well illustrated by the *C. leadbeateri* event in Northern Norway in 1991. In eutrophicated areas, toxicity may be accentuated by high N:P ratios caused by transport of nitrate from land, as seems to have been the case in the *C. polylepis* bloom in Skagerrak in 1988.

6. Acknowledgements

We gratefully thank Dr. Hua-Shi Gong for translating papers written in Chinese. B. Edvardsen was funded by the Norwegian Research Council.

7. References

Aune, T., Skulberg, O. M., Underdal, B. (1992). A toxic phytoflagellate bloom of *Chrysochromulina* cf. *leadbeateri* in coastal waters in the north of Norway, May-June 1991. *Ambio* 21: 471-474.

Aure, J., Rey, F. (1992). Oceanographic conditions in the Sandsfjord system, western Norway, after a bloom of the toxic prymnesiophyte *Prymnesium parvum* Carter in August 1990. *Sarsia* 76: 247-254.

Billard, C. (1994). Life cycles. In: Green, J.C., Leadbeater, B.S.C. (eds.), *The Haptophyte Algae*. Systematics Association Special Volume No. 51, Oxford University Press, New York, pp. 167-186.

Brand, L.E. (1984). The salinity tolerance of forty-six marine phytoplankton isolates. *Estuar. Coast. Shelf Sci.* 18: 543-556.

Carlsson, P., Granéli, E., Olsson, P. (1990). Grazer elimination through poisoning: one of the mechanisms behind *Chrysochromulina polylepis* blooms? In: Granéli, E., Sundström, B., Edler, L., Anderson, D.M. (eds.) *Toxic Marine Phytoplankton*, Elsevier, New York, p. 116-122.

Chang, F. H. (1985). Preliminary toxicity test of *Prymnesium calathiferum* n. sp. isolated from New Zealand. In: Anderson, D.M., White, A.W., Baden, D.G. (eds.) *Toxic Dinoflagellates*. Elsevier, New York, pp. 109-112.

Comin, F.A, Ferrer, X. (1978). Desarrollo masivo del fitoflagelado *Prymnesium parvum* Carter (Haptophyceae) en una laguna costera del delta del Ebro. *Oecologia Aquatica* 3: 207-210.

Dahl, E., Lindahl, O., Paasche, E., Throndsen, J. (1989). The *Chrysochromulina polylepis* bloom in Scandinavian waters during spring 1988. In: Cosper, E.M., Bricelj, M., Carpenter, E.J. (eds.) *Novel Phytoplankton Blooms: Causes and Impacts of Recurrent Brown Tides and Other Unusual Blooms*. Springer, Berlin, pp. 383-405.

Dietrich, W., Hesse, K.-J. (1990). Local fish kill in a pond at the German North Sea coast associated with a mass development of *Prymnesium* sp. *Meeresforsch.* 33: 104-106.

Edvardsen, B. (1993). Toxicity of *Chrysochromulina* species (Prymnesiophyceae) to the brine shrimp, *Artemia salina*. In: Smayda,T.J., Shimizu, Y. (eds.), *Toxic Phytoplankton Blooms in the Sea*. Elsevier, Amsterdam, pp. 681-686.

Edvardsen, B., Moy, F., Paasche, E. (1990). Hemolytic activity in extracts of *Chrysochromulina polylepis* grown at different levels of selenite and phosphate. In: Granéli, E., Sundström, B., Edler, L. Anderson, D.M. (eds.), *Toxic Marine Phytoplankton*. Elsevier, New York, pp. 284-289.

Edvardsen, B., Paasche, E. (1992). Two motile stages of *Chrysochromulina polylepis* (Prymnesiophyceae): morphology, growth and toxicity. *J. Phycol.* 28: 104-114.

Edvardsen, B., Vaulot, D. (1996). Ploidy analysis of the two motile forms of *Chrysochromulina polylepis* (Prymnesiophyceae). *J. Phycol.* 32: 94-102.

Estep, K.W., MacIntyre, F. (1989). Taxonomy, life cycle, distribution and dasmotrophy of *Chrysochromulina*: a theory accounting for scales, haptonema, muciferous bodies and toxicity. *Mar. Ecol. Prog. Ser.* 57: 11-21.

Farrow, G.A. (1969). Note on the association of *Prymnesium* with fish mortalities. *Water Res.* 3: 375-379.

Granéli, E., Haraldsson, C. (1993). Can increased leaching of trace metals from acidified areas influence phytoplankton growth in coastal waters? *Ambio* 22: 308-311.

Granéli, E., Paasche, E., Maestrini, S.Y. (1993). Three years after the *Chrysochromulina polylepis* bloom in Scandinavian waters in 1988: some conclusions of recent research and monitoring. In: Smayda, T.J., Shimizu, Y. (eds.), *Toxic Phytoplankton Blooms in the Sea.* Elsevier, Amsterdam, pp. 23-32.

Green, J.C., Hibberd, D.J., Pienaar, R.N. (1982). The taxonomy of *Prymnesium* (Prymnesiophyceae) including a description of a new cosmopolitan species, *P.patellifera* sp. nov., and further observations on *P. parvum* N. Carter. *Br. Phycol. J.* 17: 363-382.

Green, J.C., Leadbeater, B.S.C. (1994). *The Haptophyte Algae.* Systematics Association Special Volume No. 51, Oxford University Press, New York, 446 pp.

Hajdu, S., Larsson, U., Moestrup, Ø. (1996). Seasonal dynamics of *Chrysochromulina* species (Prymnesiophyceae) in a coastal area and a nutrient-enriched inlet of the northern Baltic proper. *Botanica Marina* 39: 281-295.

Hallegraeff, G.M. (1992). Harmful algal blooms in the Australian region. *Mar. Pollution Bull.* 25: 5-8.

Hansen, L.R., Kristiansen, J., Rasmussen, J.V. (1994). Potential toxicity of the freshwater *Chrysochromulina* species *C. parva* (Prymnesiophyceae). *Hydrobiologia* 287: 157-159.

Hansen, P.J., Nielsen, T.G., Kaas, H. (1995). Distribution and growth of protists and mesozooplankton during a bloom of *Chrysochromulina* spp. (Prymnesiophyceae, Prymnesiales). *Phycologia* 34: 409-416.

Hashimoto, Y. (1979). *Marine Toxins and Other Bioactive Marine Metabolites.* Japan Scientific Societies Press, Tokyo, 369 pp.

Holmquist, E., Willén, T. (1993). Fish mortality caused by *Prymnesium parvum. Vatten* 49: 110-115.

Hällfors, G., Niemi, Å. (1974). A *Chrysochromulina* (Haptophyceae) bloom under the ice in the Tvärminne archipelago, southern coast of Finland. *Memo. Soc. Fauna Flora Fenn.* 50: 89-104.

Hällfors, G., Thomsen, H.A. (1979). Further observations on *Chrysochromulina birgeri* (Prymnesiophyceae) from the Tvärminne archipelago, SW coast of Finland. *Acta Bot. Fenn.* 110: 41-46.

Igarashi, T., Satake, M., Yasumoto, T. (1996). Prymnesin-2: a potent ichthyotoxic and hemolytic glycoside isolated from the red tide alga *Prymnesium parvum. J. Am. Chem. Soc.* 118: 479-480.

Jebram, D. (1980). Prospection for a sufficient nutrition for the cosmopolitic marine bryozoan *Electra pilosa* (Linnaeus). *Zool. Jb. Syst.* 107: 368-390.

Jones, H.L.J., Leadbeater, B.S.C., Green, J.C. (1994). Mixotrophy in haptophytes. In: Green, J.C. , Leadbeater, B.S.C. (eds.), *The Haptophyte Algae.* Systematics Association Special Volume No. 51, Oxford University Press, New York, pp. 247-263.

Jones, H.L.J., Durjun, P., Leadbeater, B.S.C., Green, J.C. (1995). The relationship between photoacclimation and phagotrophy with respect to chlorophyll *a*, carbon and nitrogen content, and cell size of *Chrysochromulina brevifilum* (Prymnesiophyceae). *Phycologia* 34: 128-134.

Jordan, R.W., Green, J.C. (1994). A check-list of the extant Haptophyta of the world. *J. Mar. Biol. Ass. U.K.*, 74: 149-174.

Kaartvedt, S., Johnsen, T.M., Aksnes, D.L., Lie, U., Svendsen, H. (1991). Occurrence of the toxic phytoflagellate *Prymnesium parvum* and associated fish mortality in a Norwegian fjord system. *Can. J. Fish. Aquat. Sci.* 48: 2316-2323.

Kaas, H., Larsen, J., Møhlenberg, F., Richardson, K. (1991). The *Chrysochromulina polylepis* bloom in the Kattegat (Scandinavia) May-June 1988. Distribution, primary production and nutrient dynamics in the late stage of the bloom. *Mar. Ecol. Prog. Ser.* 79: 151-161.

Kawachi, M., Inouye, I. (1995). Functional roles of the haptonema and the spine scales in the feeding process of *Chrysochromulina spinifera* (Fournier) Pienaar et Norris (Haptophyta = Prymnesiophyta). *Phycologia* 34: 193-200.

Kawachi, M., Inouye, I., Maeda, O., Chihara, M. (1991). The haptonema as a food-capturing device: observations on *Chrysochromulina hirta* (Prymnesiophyceae). *Phycologia* 30: 563-573.

Kell, V.V., Noack, B. (1991). Fischsterben durch *Prymnesium saltans* Massart im Kleinen Jasmunder Bodden (Rügen) im April 1990. *J. Appl. Ichthyol.* 7: 187-192.

Krasnoshchek, G.P., Abramovich, L.S. (1971). Mass development of *Prymnesium parvam* Cart. in fish-breeding ponds. *Hydrobiol. J.* 7: 54-55.

Krasnoshchek, G.P., Abramovich, L.S., Shemchuk, V.R. (1972). A new case of massive development of *Prymnesium parvum* (Cart.). *Hydrobiol. J.* 8: 80.

Kristiansen, J. (1971). A Danish find of *Chrysochromulina parva* (Haptophyceae). *Bot. Tidskr.* 66: 33-37.

Larsen, A., Eikrem, W., Paasche, E. (1993). Growth and toxicity in *Prymnesium patelliferum* (Prymnesiophyceae) isolated from Norwegian waters. *Can. J. Bot.* 71: 1357-1362.

Liebert, F., Deerns, W. M. (1920). Onderzoek naar de oorzaak van een vischsterfte in den polder Workumer-Nieuwland, nabij Workum. *Rijksintituut voor Visscherijonderzoek. Verhandelingen en rapporten*, pp. 81-93.

Manton, I., Parke, M. (1962). Preliminary observations on scales and their mode of origin in *Chrysochromulina polylepis* sp. nov. *J. Mar. Biol. Ass. U.K.* 42: 565-578.

Marchant, H.J., Thomsen, H.A. (1994). Haptophytes in polar waters. In: Green, J.C., Leadbeater, B.S.C. (eds.), *The Haptophyte Algae*. Systematics Association Special Volume No. 51, Oxford University Press, New York, pp. 209-228.

McLaughlin, J.J.A. (1958). Euryhaline chrysomonads: nutrition and toxigenesis in *Prymnesium parvum*, with notes on *Isochrysis galbana* and *Monochrysis lutheri*. *J. Protozool.* 5: 75-81.

Meldahl, A.-S., Edvardsen, B., Fonnum, F. (1994). Toxicity of four potentially ichthyotoxic marine phytoflagellates determined by four different test methods. *J.Toxicol. Environ. Health* 42: 289-301.

Moestrup, Ø. (1994). Economic aspects:´blooms´, nuisance species, and toxins. In: Green, J.C., Leadbeater, B.S.C. (eds.), *The Haptophyte Algae*. Systematics Association Special Volume No. 51, Oxford University Press, New York, pp. 265-285.

Moestrup, Ø., Larsen, J. (1992). *Potentially Toxic Phytoplankton. 1. Haptophyceae (Prymnesiophyceae).* ICES identification leaflets for plankton No. 179, ICES, Copenhagen, 11 pp.

Nejstgaard, J.C., Båmstedt, U., Bagøien, E., Solberg, P.T. (1995). Algal constraints on copepod grazing. Growth state, toxicity, cell size, and season as regulating factors. *ICES J. Mar. Sci.* 52: 347-357.

Nielsen, T.G., Kiørboe, T., Bjørnsen, P.K. (1990). Effects of a *Chrysochromulina polylepis* subsurface bloom on the planktonic community. *Mar. Ecol. Prog. Ser.* 62: 21-35.

Nygaard, K., Tobiesen, A. (1993). Bacterivory in algae: a survival strategy during nutrient limitation. *Limnol. Oceanogr.* 38: 273-279.

Padilla, G.M. (1970). Growth and toxigenesis of the chrysomonad *Prymnesium parvum* as a function of salinity. *J. Protozool.* 17: 456-462.

Parke, M., Manton, I., Clarke, B. (1956). Studies on marine flagellates. III. Three further species of *Chrysochromulina. J. Mar. Biol. Ass. U.K* 35: 387-414.

Pintner, I.J., Provasoli, L. (1968). Heterotrophy in subdued light of 3 *Chrysochromulina* species. *Bull. Misaki Mar. Biol. Inst.* 12: 25-31.

Rey, F., Aure, J. (1991). The *Chrysochromulina leadbeateri* bloom in the Vestfjord, north Norway, May-June 1991. Environmental conditions and possible causes. *Fisken og Havet* 3: 13-32.

Rhodes, K., Hubbs, C. (1992). Recovery of Pecos River fishes from a red tide fish kill. *The Southwestern Naturalist* 37: 178-187.

Rhodes, L., Burke, B. (1996). Morphology and growth characteristics of *Chrysochromulina* species (Haptophyceae = Prymnesiophyceae) isolated from New Zealand coastal waters. *New Zealand J. Mar. Freshw. Res.* 30: 91-103.

Rhodes, L.L., O´Kelly, C.J., Hall, J.A. (1994). Comparison of growth characteristics of New Zealand isolates of the prymnesiophytes *Chrysochromulina quadrikonta* and *C. camella* with those of the ichthyotoxic species *C. polylepis. J. Plankton Res.* 16: 69-82.

Rip, W. J., Everards, K., Houwers, A. (1992). Restoration of Botshol (The Netherlands) by reduction of external nutrient load: the effects on physico-chemical conditions, plankton and sessile diatoms. *Hydrobiol. Bull.* 25: 275-286.

Shilo, M. (1971). Toxins of Chrysophyceae. In: *Microbial Toxins Volume VII. Algal and Fungal Toxins*. Kadis, S., Ciegler, A., Ajl, S.J. (eds.) Academic Press, New York, pp. 67-103.

Skjoldal, H. R., Dundas, I. (1991). *The Chrysochromulina polylepis Bloom in the Skagerrak and the Kattegat in May-June 1988: Environmental Conditions, Possible Causes, and Effects.* ICES Coop. Res. Rep. No. 175, 59 pp.

Tangen, K. (1989). Algal blooms in Norway in 1989. *Red Tide Newsletter* 2: 2-3.

Thomsen, H.A., Buck, K.R., Chavez, F.P. (1994). Haptophytes as components of marine phytoplankton. In: Green, J.C. , Leadbeater, B.S.C. (eds.), *The Haptophyte Algae*. Systematics Association Special Volume No. 51, Oxford University Press, New York, pp. 187-208.

Throndsen, J. (1969). Flagellates of Norwegian coastal waters. *Nytt Mag. Bot.* 16: 161-216.

Throndsen, J. (1983). Ultra- and nanoplankton flagellates from coastal waters of southern Honshu and Kyushu, Japan (including some results from the western part of the Kuroshio off Honshu). In: Chihara, M., Irie, H. (eds.), *Working Party on Taxonomy in the Akashiwo Mondai Kenkyukai.*, Gakujutsu Tosho Printing, Tokyo, 62 pp.

Throndsen, J., Larsen, J., Moestrup, Ø. (1995). Toxic algae: toxicity of *Chrysochromulina* with new ultrastructural information on *C. polylepis*. In: Wiessner, W., Schnepf, E., Starr, R.C. (eds.), *Algae, Environment and Human Affairs*. Biopress Ltd., Bristol, pp. 201-222.

Tobiesen, A. (1991). Growth rates of *Heterophrys marina* (Heliozoa) on *Chrysochromulina polylepis* (Prymnesiophyceae). *Ophelia* 33: 205-212.

Underdal, B., Skulberg, O.M., Dahl, E., Aune, T. (1989). Disastrous bloom of *Chrysochromulina polylepis* (Prymnesiophyceae) in Norwegian coastal waters 1988 - Mortality in marine biota. *Ambio* 18: 265-270.

Wang, Y., Wang, Y. (1992). Biology and classification of *Prymnesium saltans*. *Acta Hydrobiol. Sin.* 16: 193-199.

Wehr, J.D., Brown, L.M. (1985). Selenium requirement of a bloom-forming planktonic alga from softwater and acidified lakes. *Can. J. Fish. Aquat. Sci.* 42: 1783-1788.

Yang, X., Chen, J., Wang, D., Wei, Y., Wei, Y., Li, W. (1993) The prevention and control of *Prymnesium parvum* caused the disease of fish in ponds of saline and alkaline soil. *J. Fish. China.* 17: 319-324.

Yasumoto, T., Underdal, B., Aune, T., Hormazabal, V., Skulberg, O.M., Oshima, Y. (1990). Screening for hemolytic and ichthyotoxic components of *Chrysochromulina polylepis* and *Gyrodinium aureolum* from Norwegian coastal waters. In: Granéli, E., Sundström, B., Edler, L. Anderson, D.M. (eds.), *Toxic Marine Phytoplankton*. Elsevier, New York, pp. 436-440.

Autecology of the Marine Haptophyte *Phaeocystis sp.*

C. Lancelot[1], M.D. Keller[2], V. Rousseau[1], W. O. Smith, Jr[3]. and S. Mathot[3]

[1]Groupe de Microbiologie des Milieux Aquatiques, Université Libre de Bruxelles, Campus de la Plaine CP 221, Bd du Triomphe, B-1050 Bruxelles, Belgium.

[2]Bigelow Laboratory for Ocean Sciences, W. Boothbay Harbor, ME 04575, USA.

[3]Department of Ecology and Evolutionary Biology, University of Tennessee, Knoxville, TN 37996, USA

1. *Phaeocystis* physiological ecology

1.1. Life forms

The eurythermal and euryhaline genus *Phaeocystis* is one of the most widespread marine haptophytes, with most species sharing the ability to produce nearly monospecific blooms in many environments. Its unusual heteromorphic life cycle, which alternates between gelatinous colonies and different types of free-living cells (vegetative non-motile, vegetative flagellate and microzoospores), sets it apart from other members of the class (Fig.9 in Rousseau *et al.* 1994). The colonies - composed of thousands of cells embedded in a mucilaginous matrix - occasionally reach several mm in diameter. Individual cells, 3-10 µm in diameter, are distributed within the gel matrix of the colonies, which vary in form from little (20 µm) to large (1 mm) homogeneous spheres and to large ill-formed colonies invaded by bacteria and protists. This variety in colony form appears to be largely a function of life stage.

The dominance of one form over the other in natural environments has dramatic consequences for planktonic and benthic ecosystem structure and functioning (Lancelot and Rousseau 1994; Weisse *et al.* 1994) and can have severe environmental (Lancelot *et al.* 1987) and biogeochemical (Wassmann 1994) consequences. Free-living cells are heavily grazed by protozoa (Weisse *et al.* 1994) and stimulate the development of an active microbial food-web (Lancelot 1995) which retains most *Phaeocystis*-derived material in the surface waters ('regeneration-based food chain'). However a linear 'export'-food chain, with mesozooplankton grazing on protozoa (Hansen and van Boekel 1991; Bautista *et al.* 1992; van Boekel *et al.* 1992) may also develop. The trophic and geochemical role of the colonial form is more complex and depends on the colony size (Weisse 1983), the microbial colonization of senescent *Phaeocystis* colonies (Estep *et al.* 1990) and the feeding behavior and life strategy of indigenous mesozooplankton (Weisse *et al.* 1994; Wassmann 1994). The large size of colonies lowers the risk of being eaten because of the considerable time-lag in the response of large herbivores. However, the presence of overwintering meso- and meta- zooplankton in deep water

NATO ASI Series, Vol. G 41
Physiological Ecology of Harmful Algal Blooms
Edited by D. M. Anderson, A. D. Cembella and G. M. Hallegraeff

environments can result in sustained grazing. In shallow, turbid environments, colony grazing is limited due to the prevalence of immature mesozooplankton (Weisse *et al.* 1994). In addition, the gel properties of the *Phaeocystis* mucilaginous matrix (Lancelot and Rousseau 1994) combined with low aggregation properties compared to those of diatoms (Riebesell 1993) tend to maintain healthy *Phaeocystis* colonies in surface waters. *Phaeocystis* supply to the deep ocean and the benthos thus relies on the capacity of colonies to resist microbial degradation and sedimentation i.e., a compromise between the environmental characteristics and the intrinsic features of *Phaeocystis* (colony size and density) determined by the gel properties and the colonization by bacteria and protists.

Phaeocystis colonies, if present in sufficient density, are a nuisance occurrence. The mucopolysaccharide matrix of the colonies is extremely viscous and odorous, clogs nets and upon colony death, either sinks or breaks down into an organic foam. Impressive banks of this foam are regularly observed on North Sea beaches (Lancelot *et al.* 1987 and references therein). There are also many reports of fish avoiding areas of *Phaeocystis* blooms (e.g. Hurley 1982 and references therein) and of deleterious effects on shellfish (Moestrup, 1994 and references therein). In addition, there is one recent report of fish mortalities associated with *Phaeocystis*, with a substantial crop of farmed salmon lost in 1992 in Norway during a bloom period (Tangen pers. comm. in Moestrup 1994). Furthermore *Phaeocystis* is a major planktonic source of the atmospherically-important gases, dimethyl sulfide (e.g. Barnard *et al.* 1984) and methyl bromide (Saemundsdottir and Matrai, 1997).

1.2 Biogeographical distribution

1.2.1 Species number and distribution

There has been considerable confusion in the literature regarding the number of valid *Phaeocystis* species due to the lack of taxonomic criteria (see review by Sournia 1988). Recent molecular data indicate the existence of at least three colony-producing species : *P. pouchetii, P. globosa,* and *P. antarctica* (Medlin *et al.* 1994), in addition to the distinctive, *P. scrobiculata*, which has only been observed in the single cell phase (Moestrup 1979). The latter is ultrastructurally different from the colony-forming *Phaeocystis* species, especially in the occurrence of a nine ray pattern in its filaments and the fine structure of its scales. Aside from resolution at the molecular level, the only criteria for separation of the three colonial species are the original diagnostic features, colony shape and geographic distribution (Baumann *et al.* 1994). In both *P. globosa* and *P. antarctica,* individual cells are uniformly distributed around the periphery of the colony, whereas in *P. pouchetii*, the cells are grouped in clusters, usually of four cells, in lobes of the colony. *Phaeocystis antarctica* is present only in Antarctic coastal waters, whereas *P. globosa* is present in more temperate waters, with a growth temperature optimum of 15°C. The temperature range of *Phaeocystis pouchetii* is intermediate, being present in boreal and cold, temperate waters (Bauman *et al.* 1994).

1.2.2. Life forms distribution: the importance of inorganic nitrogen sources

Strain-related morphological and physiological characteristics appear to be of little significance with respect to the autecology and dynamics of *Phaeocystis* blooms. The shared ability to form large gelatinous colonies, demonstrated for all strains except *P. scrobiculata* constitutes the key ecological factor. Conditions prevailing for the existence of free-living and colonial *Phaeocystis* forms are thus examined irrespective of species.

Despite intensive research efforts, factors controlling the occurrence and dominance of *Phaeocystis* life forms in natural environments, and in particular the transition from the free-living to the colonial stage are not fully understood (Rousseau *et al.* 1994). The nutrient status, in particular phosphate limitation, is now believed to be a major factor driving colony formation from free-living cells (Veldhuis and Admiraal 1987). Furthermore, the dominant form of inorganic nitrogen is likely an important clue for understanding the dominance and the biogeographical distribution of the colonial stage. Experiments performed with cultures of *Phaeocystis* (Riegman *et al.* 1992) demonstrate that free-living cells outcompete colonial forms in ammonium- and phosphate-limited conditions whereas colonies dominate in nitrate-replete cultures. This suggests that free-living *Phaeocystis* cells would be prevalent in environments which rely on regenerated nitrogen and that colonial forms would rely on nitrate supply and thus would be associated with new production. The geographical distribution of free-living cells and colonies supports this hypothesis. Solitary cells are cosmopolitan in distribution, and are an important component of the haptophycean assemblage which dominates oceanic nanophytoplankton in many areas (e.g. Thomsen *et al.* 1994). They are also a seasonal dominant in some relatively pristine coastal areas including the Gulf of Maine (Keller and Haugen 1996) and the Gulf of Alaska (Booth *et al.* 1982). The abundances recorded in these areas however are up to an order of magnitude lower (e.g., *ca.* 2 10^6 cells l^{-1} in the Gulf of Maine, Keller and Haugen 1996) than the bloom concentrations typical of more eutrophic environments. The biomass of free-living cells of *Phaeocystis* is presumably kept in check by protozoan grazing pressure which in turn regenerates ammonium and phosphate.

Massive blooms of *Phaeocystis* colonies have been observed in turbulent, nutrient-rich environments at all latitudes. These dense, near-monospecific blooms regularly occur in spring, in nitrate-rich temperate and polar areas of the world ocean. In the North Atlantic, colonial *Phaeocystis* blooms have been recorded in such physically contrasting areas as temperate estuaries (e.g. Roger and Lockwood 1990), coastal bays (e.g. Jones and Haq 1963; Verity *et al.* 1988), the tidally-mixed continental coastal waters of the North Sea (e.g. Lancelot *et al.* 1987); the Norwegian (e.g. Egge and Asknes 1992) and Danish coastal waters (Rieman unpublished) and most Norwegian fjords (e.g. Sakshaug 1972; Eilertsen *et al.* 1981). In boreal and austral polar waters, *Phaeocystis* blooms have been recorded at the receding ice-edge of the Barents Sea (e.g. Rey and Loeng 1985; Wassmann *et al.* 1990), Greenland Sea (e.g. Smith *et al.* 1991), Islandic waters (e.g. Stefanson and Olafsson, 1991), Bering Sea (e.g. Barnard *et al.* 1984), Ross Sea (e.g. El-Sayed *et al.* 1983; Palmisano *et al.* 1986); Weddell Sea (e.g. Buck and Garrison 1983), Prydz Bay (e.g. Davidson and

Marchant 1992) and Bransfield Strait (e.g. Bodungen *et al.* 1986). Scattered colonies of *Phaeocystis* were also recorded in the permanently ice-free portion of the Barents Sea (Rey and Loeng 1985; Wassmann *et al.* 1990).

In all of these areas, the colonial form largely dominates. Its rapid development is sustained by new sources of nitrate of natural (winter deep convection) or anthropogenic (coastal areas under the influence of river discharge) origin as showed by the positive relationship between the maximum Chl.*a* concentration reached by colonies in each *Phaeocystis*-dominated environment and the nitrate reduction observed during the bloom (Fig.1).

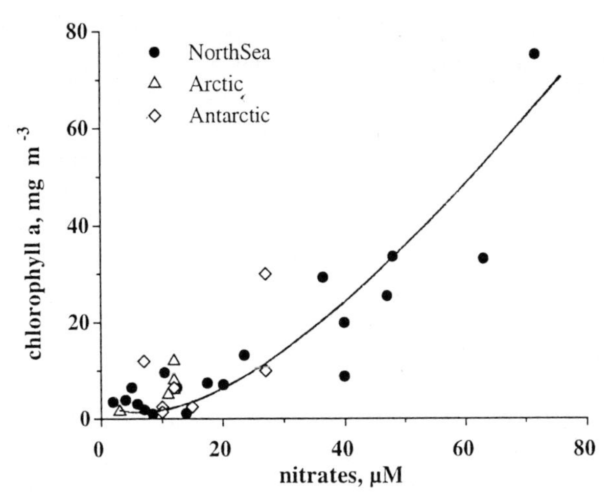

Figure 1: Empirical relationship between maximum *Phaeocystis*-Chl.*a* and nitrate reduction. Data from Rey and Loeng 1985; Wassmann *et al.* 1990; Vernet, 1991; Smith unpublished; Palmisano *et al.* 1986; El-Sayed *et al.* 1988; Rousseau *et al.* unpublished

Accordingly elevated f_{NO3} ratios (the ratio of nitrate uptake to the total inorganic nitrogen uptake rate) have been measured in the Greenland Sea (mean: 0.56, range: 0.09-0.9; Smith 1993) as well as in the continental coastal waters of the North Sea (mean: 0.62, range: 0.5 -0.8; Lancelot *et al.* 1986). In both areas, f_{NO3} decreased from 0.8-0.9 at the beginning of the *Phaeocystis* bloom to 0.4-0.5 at its decline.

1.3. Ecophysiology of *Phaeocystis* colonies

The above characteristics place *Phaeocystis* colonies at an ecological position similar to the spring diatom population which often blooms at the same time. However, the position of the maximum development of *Phaeocystis* colonies with respect to that of diatoms in the spring phytoplankton succession varies among systems and between years. Reasons for this variation - resource versus predator based competition - are not well understood, although the unique ability of *Phaeocystis* to form colonies is a common element of most hypotheses.

The gel-forming exopolysaccharides of the colonial matrix, may enable *Phaeocystis* colonies to outcompete other phytoplankters in turbulent waters by increasing buoyancy and retention in surface waters and by avoiding consumption by indigenous mesozooplankton due to their large size (Lancelot and Rousseau 1994). In addition, the palatability of *Phaeocystis* colonies is still questionable, as large metazooplankton will feed selectively on diatoms when offered a choice between *Phaeocystis* colonies and diatoms (Verity and Smayda 1989). Some diatoms (e.g. *Chaetoceros socialis*, a species which often co-dominates with *Phaeocystis*) have developed similar adaptive mechanisms to resist sinking and grazing.

Table I : Photosynthetic characteristics - photosynthetic capacity K_{max} and light adaptation parameter I_K - of *Phaeocystis* and spring diatoms

	K_{max} mgC mgChl.*a*$^{-1}$ h^{-1}	I_K µmol m^{-2} s^{-1}	Reference
BOREAL POLAR WATERS:			
Phaeocystis cells	0.9-1.5	4-29	Matrai *et al.* 1995
Phaeocystis colonies	0.8-4.2	16-57	Matrai *et al.* 1995
	5.3-13.3	32-140	Cota *et al.* 1994 Verity *et al.* 1991
Diatoms	0.8-1.2	14-104	Cota *et al.* 1994
	0.6-7.5	9-48	Matrai *et al.* 1995
TEMPERATE WATERS:			
Phaeocystis cells	0.8-3	10-120	Lancelot, Mathot 1987
Phaeocystis colonies	2-14	120-180	Lancelot, Mathot 1987 Lancelot unpublished
	5-15	91-144	Colijn 1983
	3-8.5	250	Verity *et al.* 1988
Diatoms	1.6-4	125-236	Lancelot, Mathot 1987 Lancelot unpublished
ANTARCTIC WATERS:			
Phaeocystis colonies	3.5-8.1	47-144	Palmisano *et al.* 1986

Springtime populations and cultures of *Phaeocystis* colonies and diatoms display similar photosynthetic properties (Table I) suggesting that both taxonomic groups are able to adapt their photophysiology to the low light conditions prevailing in early spring. The superior photosynthetic efficiency of *Phaeocystis* free-living cells at lower light levels (Table I) may indirectly promote the prevalence of *Phaeocystis* colonies by seeding the water column with large numbers of cells for colony initiation. Furthermore, the considerable flexibility of *Phaeocystis* colonies to adapt their photosynthetic characteristics to ambient light conditions, as evidenced in the Southern Ocean (Palmisano *et al.* 1986) and in the continental coastal waters of the North Sea (Lancelot unpublished) offers an alternate explanation for the competitive success of *Phaeocystis* colonies in turbulent and turbid systems. In the Ross Sea, *Phaeocystis* colonies associated with sea-ice doubled their photosynthetic efficiency by lowering the typical value of the light adaptation parameter I_k (Platt *et al.* 1980) from about 100 to 50 µmol m^{-2} s^{-1} when drifting underneath the ice from well-illuminated ice-free waters (Palmisano *et al.* 1986). Similarly in the near-shore continental coastal waters of the North Sea, I_k values for *Phaeocystis* exponentially decrease from 250 to 10 μ mol $m^{-2} s^{-1}$ along the SW-NW turbidity gradient (Lancelot, unpublished).

Finally, the energy (Lancelot and Mathot 1985), phosphate (Veldhuis and Admiraal 1987) and trace element (Davidson and Marchant 1987) storage capacity of the colonial matrix may also impart a competitive advantage to *Phaeocystis* colonies over diatoms when energy-costly nitrate is the dominant nitrogen source and/or ambient trace elements concentrations are depleted.

Keeping in mind the peculiar physiology of *Phaeocystis*, we now present the autecology of *Phaeocystis* colony blooms in key areas. Particular attention is given to the diatom-*Phaeocystis* colony succession and to physical and chemical conditions initiating, maintaining and limiting blooms of *Phaeocystis* colonies. For this comparison, we have not included zooplankton grazing pressure, although we acknowledge that in some systems and at certain times, it may be critical.

2. Auto-ecology of *Phaeocystis* colony blooms : case studies

2.1. Boreal and Austral polar waters

Blooms of *Phaeocystis* colonies reaching 6-8 mg Chl.*a* m^{-3} (about 7 10^6 cells per liter) are regularly observed in subarctic and arctic shelf seas at the beginning of the vernal period (April-May). In these sea-ice associated areas, *Phaeocystis* colonies are often associated with the retreating ice-edge. Bloom development is triggered by the stability induced in the upper 15-40 m surface waters by ice melt. An exception, however, is the Atlantic Current region where *Phaeocystis* colonies flourish later in the season, after water column stabilization due to surface heating (Rey and Loeng 1985; Vernet 1991; Wassmann *et al.* 1990).

The sequence of phytoplankton succession at the receding ice-edge is similar for diverse sea-ice environments. In most areas, *Phaeocystis* colonies appear first, contributing up to 95% of cell density (Vernet 1991). Populations peak in late April-early May and are distributed homogeneously in the upper mixed layer at depths of 15-40 m (e.g. Rey and Loeng 1985). The biomass maximum is limited by the winter stock of nitrate (about 12 μM). Winter silicate levels of 5-6 μM remain unutilized during this period (Smith *et al.* 1991; Stefansson and Olafson 1991). Depending on the degree of turbulence, *Phaeocystis* colonies accumulate at the pycnocline, reaching biomass levels of 12 mg Chl.*a* m^{-3} (Vernet, 1991) or densities up to 27 10^6 cells l^{-1} (Thingstad and Martunissen 1991). These deep maxima tend to be very transient however and often sediment abruptly (Wassmann 1994). Remineralization typically is completed within the aphotic water column and little *Phaeocystis*-derived material reaches the bottom (Wassmann *et al.* 1990).

Low concentrations of *Chaetoceros socialis* and *Pseudo-nitzschia delicatissima* are present at the time of the *Phaeocystis* bloom, but the main diatom population, composed of *Chaetoceros* spp., *Thalasiossira* sp., *P. delicatissima* and *N. cylindrus,* develops later, as *Phaeocystis* declines, reaching maximum density at the depth of the nutricline (Rey and Loeng 1985). Occasionally, the diatom community blooms before *Phaeocystis,* as low silicate concentrations (~ 1 μM) have been observed at the time of *Phaeocystis* blooms (Wassmann *et al.* 1990). Reasons for this reverse succession are not known.

In the Southern Ocean, blooms of *Phaeocystis* colonies have been recorded in waters influenced by a receding ice-edge as well. *Phaeocystis* blooms are particularly well documented in the Ross Sea where they dominate in the extreme southern (El-Sayed *et al.* 1983) and southeastern areas (e.g. Palmisano *et al.* 1986). The southwestern region of the Ross Sea is characterized by diatom-dominated blooms, which are also enhanced by ice melt (Smith and Nelson 1985). Melting ice not only increases water column stability but also supplies a significant inoculum of viable sea-ice diatoms into the water column. The sea-ice community in the southwestern Ross Sea is composed largely of diatoms (Smith and Nelson 1985).

The seasonal pattern in phytoplankton biomass in the Ross Sea follows the retreating ice edge which is driven in the southeastern region by catabatic winds and by solar heating at the northern ice-edge. Phytoplankton community succession at the ice-edge from south to north shows a shift from a *Phaeocystis*-dominated to a diatom-dominated population (Fig. 2). The reasons for this change are not known, although changes in the degree of vertical stability may be important. The observed ice-edge diatom bloom in the southwestern Ross Sea typically coincides with the formation and persistence of a sharp halocline and pycnocline at a depth of 20-30 m (Smith and Nelson 1985). High intensity winds, typical in the area adjacent to Ross Island, prevent the establishment of vertical stability. We suggest that *Phaeocystis* colonies, with the buoyancy attributable to the mucilaginous matrix, are better able to maintain themselves in the surface waters than diatoms. We also believe that *Phaeocystis* is better able to adapt to the lower ambient light conditions associated with deep vertical mixing.

Maximum recorded *Phaeocystis* biomass in the Southern Ocean is 12 mgChl.*a* m^{-3} (30 10^6 cells l^{-1}; Palmisano *et al.* 1986). Although ca. 31 μM of nitrate has been consistently measured in this area, it has not been fully utilized during the bloom period. Based on this nitrate level, the observed biomass is *ca.* 50% lower than might be expected (30 mg Chl.*a* m^{-3} or 60 10^6 cells l^{-1}; Fig.1). Nitrate concentrations of 10 to 19 μM have been measured in the water column at the peak of the *Phaeocystis* bloom (e.g. Palmisano *et al.* 1986). Light and/or iron limitation have been suggested to explain this paradigm (de Baar *et al.* 1997). The expected 30 mg Chl.*a* m^{-3} level of biomass has been observed in the spring beneath the annual sea ice in Prydz Bay (Davidson and Marchant 1987). In this area, as in the Ross Sea (Palmisano *et al.* 1986), high densities of *Phaeocystis* colonies were found beneath the sea ice, advected from ice-free surface waters where their growth was initiated. Populations appear to be maintained in this dim environment both from the buoyant properties of the *Phaeocystis* matrix, which keep the colonies just beneath the sea ice, and the rapid photoadaptative capability of *Phaeocystis* to the variable light environment (Palmisano *et al.* 1986). The exceptionally high concentrations of *Phaeocystis* recorded beneath the ice in Prydz Bay (Davidson and Marchant 1992) may have been sustained by additional iron present in the ice. Iron released in this way during sea ice melt has been observed in the Atlantic sector of the Southern Ocean (de Baar *et al.* 1997)

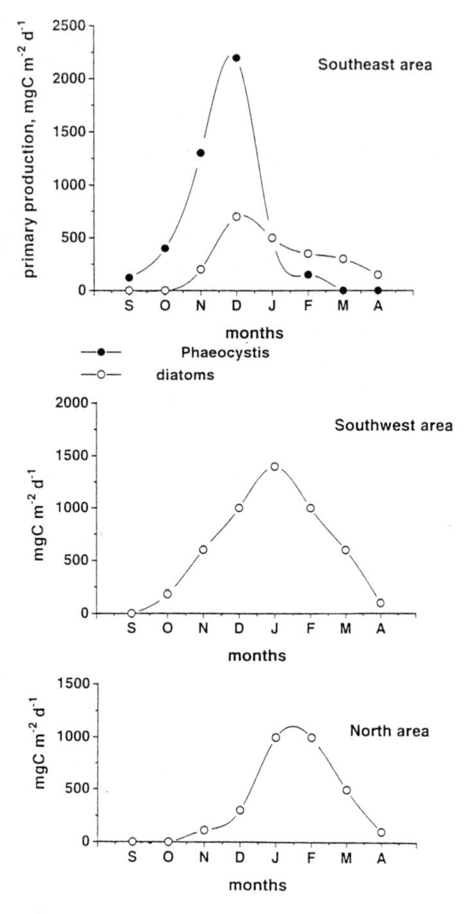

Figure 2: *Phaeocystis*- and diatom- daily growth in the Ross Sea

2.2. North Atlantic coastal waters under the influence of riverine inputs

Large blooms of *Phaeocystis* colonies are recurrent phenomena in the fjords and coastal environments of Norway, occurring in early spring, between early March and late May, depending on the latitude. In these systems, the spring bloom utilizes winter stocks of nutrients while later spring-summer blooms rely on the riverine supply of nutrients enriched in snow melt from the mountains. Phytoplankton seasonal succession is particularly well documented in Trondheimfjord (1963-1966; Sakshaug 1972) and Balsfjorden (1977-1978; Eilertsen *et al.* 1981), respectively at 64°N and 69°N. In both fjords, the spring bloom is initiated by increasing light levels, rather than water column stability, which occurs later in the spring as freshwater inputs increase (Sakshaug 1972; Eilertsen *et al.* 1981). The composition of the spring bloom exhibits considerable interannual variability, sometimes it is dominated by diatoms, other years by *Phaeocystis,* or by co-occurrence (Fig.3). In years when *Phaeocystis* blooms are co-incident with diatoms, the main diatom is typically *Chaetoceros socialis.*

The amplitude and extent of the colonial *Phaeocystis* bloom also show significant interannual variation. In 1977, *Phaeocystis* cell numbers ranged from 0.5 and 2 10^6 cells l^{-1} in Balsfjorden throughout the spring and summer, from March to September. In contrast, the 1978 *Phaeocystis* bloom was short in duration (April), but very intense, with maxima up to an order of magnitude higher than observed in 1977 (Fig.3). Furthermore, the relative abundance of diatoms and *Phaeocystis* in the spring bloom period was quite different, with diatoms dominant in 1977 and *Phaeocystis* in 1978. The reasons for this are not known, although freshwater inputs may contribute to water column and nutrient conditions which may favor one form over the other.

Similar *Chaetoceros-Phaeocystis* successions are typical in the Trondheimfjord as well, although *Phaeocystis* maxima typically persist only for one month (March-April; Sakshaug 1972). As in the northern fjord, large interannual variation occurs with cell densities ranging from 1 to 8 10^6 cells l^{-1}. Higher levels have been observed in areas under river influence (Sakshaug 1972).

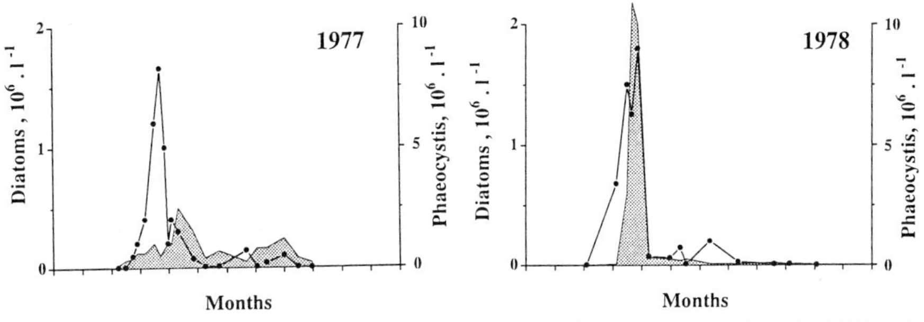

Figure 3: Diatom-*Phaeocystis* (gray area) colonies succession in the Balmsfjord in 1977 and 1978 (redrawn from Eilertsen *et al.* 1981).

Massive blooms of *Phaeocystis* colonies, with cell numbers up to 10^8 cells l^{-1}, are observed every spring in the continental coastal waters of the North Sea, which receives the discharge of seven major west-European rivers. The fluvial basins, characterized by high population densities and intense industrial and farming activities, have introduced new and unbalanced sources of nutrients into coastal waters (Lancelot 1995and references therein). The general N, P, Si enrichment of the coastal area is characterized by winter concentrations an order of magnitude higher than those in adjacent Atlantic waters (Lancelot 1995). Qualitative changes in the nutrient ratios supplied by the freshwater sources have resulted in an excess of nitrate with respect to silicate, which has implications for the growth of coastal diatoms (Lancelot 1995).

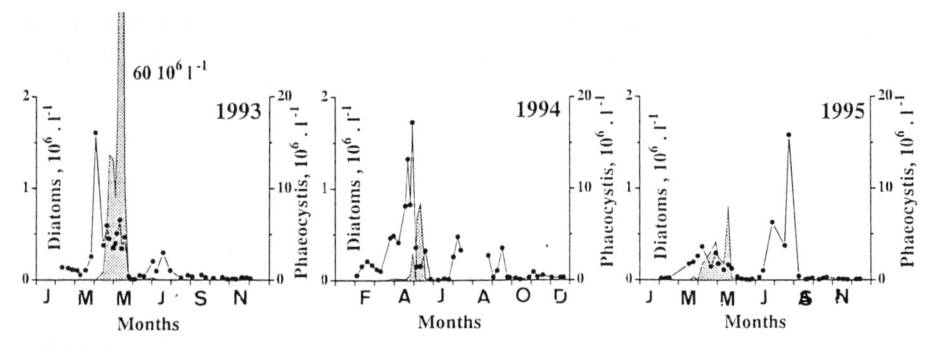

Figure 4: Diatom-*Phaeocystis* (grey area) colonies succession in the continental coastal waters of the North Sea (station 330). Rousseau *et al.* unpublished

Since 1988, the spring phytoplankton community at station 330 (N 51°26.05, E 02°9.08) located in Belgian coastal waters, has been monitored extensively. The general phytoplankton succession is similar to that characterizing Norwegian fjords. Diatoms initiate the vernal bloom in early spring (February-March) (Fig.4). *Phaeocystis* colonies appear somewhat later. Large interannual variations in *Phaeocystis* biomass are also evident in this system, with *Phaeocystis* cell maxima varying over two orders of magnitude (Fig.4). The early-spring diatom community is composed of small, neritic species including *Thalassiosira nordenskoldii*, *T. rotula*, *Asterionella glacialis*, *Thalassionema nitzschoïdes*, *Plagiogramma brockmanii*, and *Skeletonema costatum*. This diatom community, although typical of the North Sea, is not observed in the Norwegian fjords; its growth is controlled by the winter concentration of silicate (Rousseau *et al.* unpublished). *Phaeocystis* colonies appear as the early spring diatom community declines. The later diatom community composed of *Chaetoceros* spp. and *Schroederella* sp., both of which require relatively low levels of silicate, appear at the same time. As the *Phaeocystis* bloom develops, additional diatoms, *Cerataulina* sp. and *Rhizosolenia* spp., mainly *R. delicatula,* become abundant as well. The fluctuations in this community of larger diatoms and *Phaeocystis* colonies (Fig.4) appear to result from the competition for nitrate, suggesting that both occupy the same ecological niche. The extreme differences in the abundance of diatoms and *Phaeocystis* colonies recorded in 1993 and 1994, with *Phaeocystis* dominating in 1993 and diatoms in 1994 (Fig.4) has been correlated to differences in late winter meteorological conditions prevailing during 1993 (cold and dry) and 1994 (temperate, high rainfall). Rousseau *et al.* (unpublished) show evidence that rainfall - strength, frequency and duration - controls the amount and the relative contributions of nitrate and silicate of freshwater origin, with high silicate associated with high rainfall. When silicate is available, the *Rhizosolenia* sp. and *Cerataulina* sp. diatom community may be able to outcompete *Phaeocystis* colonies.

It appears that under non-limiting concentrations of silicate and nitrate, diatoms outcompete *Phaeocystis* colonies in temperate North Atlantic coastal waters. This contrasts with the succession from *Phaeocystis* colonies to diatoms observed in boreal and austral polar waters. Based on its apparently superior photoadaptative properties, *Phaeocystis* should have an advantage in all systems in early spring. However, temperature-dependent growth experiments performed on diatom and

Phaeocystis communities sampled throughout the winter-spring in the coastal North Sea (Fig.5) demonstrate that the early spring diatom community grows better than *Phaeocystis* at the temperatures typical of early spring (5-8°C).

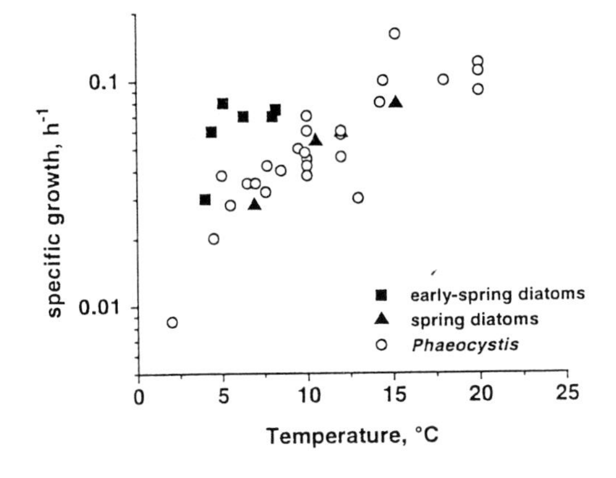

Figure 5: Relationship between temperature and specific growth of diatoms and *Phaeocystis* colonies. (Lancelot *et al.*, unpublished data)

3. Conclusions and perspectives

There are many questions which remain about the physiological ecology of the genus *Phaeocystis.* The success of *Phaeocystis* in marine systems has been attributed to its ability to form large gelatinous colonies during its life cycle (Lancelot and Rousseau, 1994). These colonies are functionally similar to the large, chain-forming or colonial diatoms that occupy the same spring bloom niche in turbulent, tidally- or seasonally-mixed water columns. In most environments, the magnitude of the *Phaeocystis* colony bloom appears to be regulated by nitrate availability. An exception is the Southern Ocean where iron shortage may prevent optimal utilization of the high nitrate resources. An analysis of the diatom-colonial *Phaeocystis* succession in contrasting *Phaeocystis*-dominated ecosystems demonstrates that, while there are large interannual and spatial variations, there appear to be consistent differences between ecosystem types. In polar waters, *Phaeocystis* precedes the main diatom bloom in most areas; in temperate waters, the reverse is true. *Chaetoceros socialis* emerges as a common co-dominant with *Phaeocystis* colonies in each *Phaeocystis*-dominated ecosystem, typically blooming slightly before but never achieving biomass comparable to *Phaeocystis* colonies.

The efficient and adaptable photophysiology of *Phaeocystis*, combined with superior buoyant properties imparted by the colonial matrix, make *Phaeocystis* extremely competitive in turbulent environments or under low light conditions. These conditions are typical in polar waters where *Phaeocystis* colonies initiate the vernal season. Diatoms appear to dominate early spring blooms in these waters only when vertical stratification occurs (Smith and Nelson, 1985). This scenario is not typical of Northeast Atlantic coastal waters and bays where the spring bloom is initiated by a diatom community composed of small neretic species, which also do well at low light levels but grow better than *Phaeocystis* colonies at the temperatures of 5-8°C prevailing in early spring. While a comparison of regional differences in bloom formation and succession is helpful, the differences observed between regions may be simply a result of biogeography, i.e. the occurrence of different species. These observed differences reinforce the need to establish the systematics of this genus in a comprehensive way. In polar environments, the dominant species appear to be *Phaeocystis pouchetii* and *P. antarctica*, in Arctic and Antarctic coastal waters respectively. In Northeast Atlantic temperate waters (North Sea), the species appears to be *P. globosa*. In Northwest Atlantic temperate waters (Gulf of Maine, Narragansett Bay), the species appears to be *P. pouchetii*. The temperature tolerances of these species are different enough to account for at least some of the reported contradictions in autecology from different environments. Thus, it is not unreasonable to conclude that *Phaeocystis* precedes diatoms in polar waters due to its superior photophysiology, while in temperate environments, where the less eurythermal *P. globosa* dominates, diatoms have the edge in early spring. Although what we know about the physiology of these algae is consistent with the empirical observations of diatom-*Phaeocystis* colony succession, our knowledge is incomplete. We need to develop appropriate mechanistic models which consider the unique physiological characteristics of diatoms and of *Phaeocystis*. Comparative laboratory experiments of the physiology of the different species of *Phaeocystis* are also required.

Little can be said about the autecology of the solitary cells of *Phaeocystis. Phaeocystis* colonies are rare in regions with a permanently stratified water column. In these areas, solitary cells of *Phaeocystis*, which appear to be more competitive at low nutrient concentrations, are more prevalent. Changes in trophic function and structure result from these alterations in life stage. With the flagellate stage, an entirely different community develops, based on a microbial food web, with regeneration of nutrients and carbon in surface waters. Understanding the life cycle of *Phaeocystis* is critical to understanding the ecosystems where it occurs. How do the cells overwinter? Is there a benthic stage? Is the flagellate stage linked to sexuality? These are all questions which need attention. The role of other trophic levels in bloom dynamics also warrants further investigation. Do microzooplankton play a critical role by controlling the single cell phase and is bloom development ultimately controlled by metazooplankton grazing upon microzooplankton? Is there any basis for allelopathy in these blooms and if not, why are bacteria and protozoa not associated with healthy colonies? Finally, is there any basis for the observations that *Phaeocystis* is toxic? If so, can this toxicity be induced or are the toxic effects associated with anoxia or hypoxia, due to the viscous character of the mucilage ?

To date efforts have concentrated on understanding the physico-chemical conditions enhancing the exponential development of *Phaeocystis* colonies, rather than the fate of *Phaeocystis* colonies and *Phaeocystis* bloom termination. Grazing and sedimentation in particular appear to rely on the presence (deep environments) or absence (shallow water environments) of overwintering meso- and meta-zooplankton (Weisse *et al.*, 1994) and on the water column characteristics (Wassmann, 1994). The sudden termination of *Phaeocystis* blooms in all *Phaeocystis*-dominated systems highlights the need for additional investigations on the relationship between turbulence and the formation of *Phaeocystis*-derived aggregates. Although *Phaeocystis* colonies do not apparently readily aggregate (Riebesell, 1993), changes in water column vertical structure, especially if driven by salinity change, may be important. The gelling properties of the colony matrix, which equilibrate colony density to that of seawater, may cause *Phaeocystis* colonies to rise in response to salinity increases or sink when salinity decreases. Further investigations should focus on the interaction of physics and *Phaeocystis* colonies at different stages of their development, particularly in frontal structures such as river plumes and the receding ice edge in polar systems.

The unique heteromorphic life cycle of *Phaeocystis* imparts versatility and adaptive abilities to this genus that are not shared by other co-occurring phytoplankters. The colonial life stage is an obvious and important factor in the structure and function of coastal ecosystems, but the ubiquity of the less-well studied solitary flagellate stage may make it an important contributor to oligotrophic environments as well. *Phaeocystis* is present and important as a primary producer in almost every oceanic environment. Its occurrence as a nuisance species is directly linked to eutrophication, and as such, is one of the few species where such a clear, causal relationship is apparent.

4. Acknowledgments

This review is a contribution to the Belgian Impulse Programme 'Marine Sciences' (contract MS/11/070) and to the EC projects on 'Biogeochemistry of *Phaeocystis* colonies and their derived aggregates' (contract EV5V-CT94-0511,DG12-SOL) and on 'Comparative Analysis of food webs based on flow networks - COMWEB (contract MAS3-CT96-0052-DG12-DTEE). Funding for M. Keller was from EPA grant (R819515010). Funding for W.O. Smith was from NSF (OPP-9317587). Partial funding to S. Mathot was provided by a NATO fellowship.

5. References

Barnard, W.R., Andreae, M.O., Iverson, R.L. (1984). Dimethylsulfide and *Phaeocystis pouchetii* in southeastern Bering Sea. *Cont. Shelf Res.* 3: 103-113.

Bauman, M.E.M, Lancelot, C., Brandini, F.P., Sakshaug, E., John, D.M. (1994). The taxonomic identity of the cosmopolitan prymnesiophyte *Phaeocystis* : a morphological and ecophysiological approach. *J. Mar.Syst.* 5: 5-22.

Bautista, B., Harris, R.P., Tranter, P.R.G., Harbour, D. (1992). In situ copepod feeding and grazing rates during a spring bloom dominated by *Phaeocystis* sp. in the English Channel. *J. Plankton Res.* 14: 691-703.

Bodungen, B., Smetacek, V. , Tilzer, M.M., Zeitzschel, B. (1986). Primary production and sedimentation during spring in the Antarctic Peninsula region. *Deep Sea Res.* 33: 177-194.

Booth, B.C., Lewin, J., Norris, R.E. (1982). Nanoplankton species dominant in the subarctic Pacific in May and June 1978. *Deep Sea Res.* 29: 185-200.

Buck, K. R., Garrison, D.L. (1983). Protists from the ice-edge region of the Weddell Sea. *Deep-Sea Res.* 30: 1261-1277.

Colijn, F., (1983). Primary production in the EMS-Dollard estuary. Ph.D.Thesis. University of Groningen (The Netherlands)

Cota, G.F, Smith, W.O.Jr, Mitchell, B.G. (1994). Photosynthesis of *Phaeocystis* in the Greenland Sea. *Limnol. Oceanog.* 39: 948-953.

Davidson, A.T., Marchant, H.J. (1987). Binding of manganese by Antarctic *Phaeocystis pouchetii* and the role of bacteria in its release. *Mar. Biol.* 95: 481-487.

Davidson, A.T., Marchant, H.J. (1992). Protist interactions and carbon dynamics of a *Phaeocystis* dominated bloom at an Antarctic coastal site. *Polar Biol.* 12: 387-395.

de Baar, H.J.W., van Leeuwe, M.A., Scharek, R., Goeyens, L., Koeve, W., Bakker, , K.M.J., Fritsche, P. (1997). Iron availability may affect the nitrate/phosphate ratio (A.C. Redfield) in the Antarctic Ocean. *Deep-Sea Res.* (in press).

Egge, J.K., Aksnes, D.L. (1992). Silicate as regulating nutrient in phytoplankton competition. *Mar. Ecol. Prog. Ser.* 83: 281-289.

Eilertsen, H.C., Schei, B., Taasen, J.P. (1981). Investigations on the plankton community of Balsfjorden, Northern Norway. The phytoplankton 1976-1978. Abundance, species composition and succession. *Sarsia* 66: 129-141.

El-Sayed, S.Z., Biggs, D.L., Holm-Hansen, O. (1983). Phytoplankton standing crop, primary productivity and near-surface nitrogenous nutrient fields in the Ross Sea, Antarctica. *Deep-Sea Res.* 30: 871-886.

Estep, K.W., Nejstgaard, J.C., Skjoldal, R.H., Rey, F. (1990). Predation by copepods upon natural populations of *Phaeocystis pouchetii* as a function of the physiological state of the prey. *Mar. Ecol. Prog. Ser.* 67: 235-249.

Hansen, F.C., van Boekel, W.H.M. (1991). Grazing pressure of the calanoid copepod *Temora longicornis* on a *Phaeocystis* dominated spring bloom in a Dutch tidal inlet. *Mar. Ecol. Prog. Ser.* 78: 123-129.

Hurley, D.E. (1982). The "Nelson Slime". Observations on past occurrences. *N.Z. Oceanogr. Inst. Oceanogr. Summary*. 20: 1-11.

Jones, P.G.W., Haq, S.M. (1963). The distribution of *Phaeocystis* in the Eastern Irish Sea. *Cons. Perm. Int. Explor. Mer* 28: 8-20.

Keller, M.D., Haugen, E.M. (1996). Abundance and distribution of *Phaeocystis* sp. in the Gulf of Maine, U.S.A.: Spring bloom dynamics and bloom initiation. *EOS Transactions* 76(3): 48.

Lancelot, C., Mathot, S. (1985). Biochemical fractionation of primary production by phytoplankton in Belgian coastal waters during short- and long term incubations with 14C-bicarbonate. II. *Phaeocystis pouchetii* colonial population. *Mar. Biol.* 86: 227-323.

Lancelot, C., Mathot, S., Owens, N.J.P. (1986). Modelling protein synthesis, a step to an accurate estimate of net primary production: *Phaeocystis pouchetii* colonies in Belgian coastal waters. *Mar. Ecol. Progr. Ser.* 32: 193-202.

Lancelot, C., Mathot, S. (1987). Dynamics of a *Phaeocystis*-dominated spring bloom in Belgian coastal waters. I. Phyplankton activities and related parameters. *Mar. Ecol. Progr. Ser.* 37: 239-248.

Lancelot, C., Billen, G., Sournia, A., Weisse, T., Colijn, F., Veldhuis, M., Davies, A., Wassman, P. (1987). *Phaeocystis* blooms and nutrient enrichment in the continental coastal zones of the North Sea. *Ambio* 16: 38-46.

Lancelot, C., Rousseau, V. (1994). Ecology of *Phaeocystis* ecosystems: The key role of colony forms. In: Green, J., Leadbeater, B.S.C. (eds.) *The Haptophyte Algae*. Oxford Science Publications. pp. 227-245.

Lancelot, C. (1995). The mucilage phenomenon in the continental coastal waters of the North Sea . *The Science of the Total Environment* 165: 83-102.

Matrai, P.A., Vernet, M., Hood, R., Jennings, A., Brody, E., Saemundsdottir, S. (1995). Light dependence of carbon and sulphur production by polar clones of the genus *Phaeocystis*. *Mar. Biol.*, 124: 157-167.

Medlin, L.K., M. Lange, M.E.M., Baumann. (1994). Genetic differentiation among three colony-forming species of *Phaeocystis*: further evidence for the phylogeny of the Prymnesiophyta. *Phycologia* 33: 199-212.

Moestrup, O. (1979). Identification by electron microscopy of marine nanoplankton from New Zealand, including the description of four new species. New Zea. J. Bot. 17: 61-95.

Moestrup, O. (1994). Economic aspects: blooms, nuisance species, and toxins. In: Green, J., B.S.C. Leadbeater, (eds.) *The Haptophyte Algae*. Oxford Science Publications.. pp. 265-85

Palmisano, A.C., Soo Hoo, J.B., Soo Hoo S.L., Kottmeier, S.T., Craft, L.L., Sullivan, C. W. (1986). Photoadaptation in *Phaeocystis pouchetii* advected beneath annual sea ice in Mc Murdo Sound, Antarctica. *J. Plankton Res.* 8: 891-906.

Platt, T., Gallegos, C.L., Harrison, W.G. (1980). Photoinhibition of photosy,nthesis in natural assemblages of marine phytoplankton. *J. Mar. Res.* 38: 687-701.

Putt, M., Miceli, G., Stoecker, D. K. (1994). Association of bacteria with *Phaeocystis* sp. in Mc Murdo Sound, Antarctica. *Mar. Ecol. Prog. Ser.* 105: 179-189.

Rey, F., Loeng, H. (1985). The influence of ice and hydrographic conditions on the development of phytoplankton in the Barents Sea. In: Gray, J.S., Christiansen, M.E. *Marine Biology of polar regions and effects of stress on marine organisms.* Wiley, Chichester, pp. 49-63.

Riebesell, U. (1993). Aggregation of *Phaeocystis* during phytoplankton spring blooms in the southern North Sea. *Mar. Ecol. Prog. Ser.* 96: 281-289.

Riegman, R., Noordeloos, A., Cadée, G.C. (1992). *Phaeocystis* blooms of the continental zones of the North Sea. *Mar. Biol.* 112: 479-484.

Rogers, S.I., Lockwood, S. J. (1990). Observations on coastal fish fauna during a spring bloom of *Phaeocystis pouchetii* in the Eastern Irish Sea. *J. Mar. Biol. Ass. U.K.* 70: 249-253

Rousseau, V., Vaulot, D., Casotti, R., Cariou, V., Lenz, J. Gunkel, J., Baumann, M. (1994). The life cycle of *Phaeocystis* (Prymnesiophyceae): evidence and hypothesis. *J. Mar. Syst* 5: 23-39.

Saemundsdottir, S., P.A. Matrai (1997). Biological production of the volatile gas methyl bromide by marine phytoplankton. *Limnol. Oceanogr.* in press.

Sakshaug, E. (1972). Phytoplankton investigations in Trondheimsfjord, 1963-1966. *K. norske Vidensk. Selsk.Skr.* 1: 1-56.

Smith, W.O., Nelson, D.M. (1985). Phytoplankton bloom produced by a receding ice edge in the Ross Sea: spatial coherence with the density field. *Science* 227: 163-166.

Smith, W.O. Jr. (1993). Nitrogen uptake and new production in the Greenland Sea: the Spring *Phaeocystis* bloom. *J. Geophys. Res.* 98: 4681-4688.

Smith, W.O., Codispoti, L. A., Nelson, D.M., Manley, T., Buskey, E.J., Niebauer, H.J., Cota, G.F. (1991). Importance of *Phaeocystis* blooms in the high-latitude ocean carbon cycle. *Nature* 352: 514-516.

Sournia, A. (1988). *Phaeocystis* (Prymnesiophyceae): How many species? *Nova Hedwigia* 47, 211-7.

Stefansson, U., Olafsson, J. (1991). Nutrients and fertility of Icelandic waters. *J. Rit Fisk.* 12: 1-56.

Thingstad, F., Martunissen, I. (1991). Are bacteria active in the cold pelagic ecosystem of the Barents Sea?. *Polar Res.* 10: 255-266.

Thomsen, H.A., K.R. Buck, Chavez F.P. (1994). Haptophytes as components of marine phytoplankton. . In: Green, J., B.S.C. Leadbeater, (eds.) *The Haptophyte Algae*. Oxford Science Publications. pp. 187-208,

van Boekel, W.H.M., Hansen, F.C., Riegman, R., Bak, R.P.M. (1992). Lysis-induced decline of a *Phaeocystis* spring bloom and coupling with the microbial foodweb. *Mar. Ecol. Prog. Ser.*, 81: 269-276.

Veldhuis, M.J.W., Admiraal, W. (1987). Influence of phosphate depletion on the growth and colony formation of *Phaeocystis pouchetii* (Hariot) Lagerheim; *Mar. Biol.* 95: 47-54.

Verity, P.G., Villareal, T.A., Smayda, T.J. (1988). Ecological investigations of blooms of colonial *Phaeocystis pouchetii*. I. Abundance, biochemical composition and metabolic rates. *J. Plankton Res.* 10: 219-248.

Verity, P.G., Smayda T.J. (1989). Nutritional value of *Phaeocystis pouchetii* (Prymnesiophyceae) and other phytoplankton for *Acartia* spp. (Copepoda): ingestion, egg production and growth of nauplii. *Mar. Biol.* 100, 161-71.

Verity, P., G. Smayda, T., Sakshaug, E. (1991). Photosynthesis, excretion and growth rates of *Phaeocystis* colonies and solitary cells. *Polar Res.* 10: 117-128.

Vernet, M. (1991). Phytoplankton dynamics in the Barents Sea estimated from chlorophyll budget models. *Polar Res.* 10: 129-145.

Wassmann, P. Vernet, M., Mitchell, B.G., Rey, F. (1990). Mass sedimentation of *Phaeocystis pouchetii* in the Barents Sea. *Mar. Ecol. Prog. Ser.* 66 : 183-195.

Wassmann, P. (1994). Significance of sedimentation for the termination of *Phaeocystis* blooms. *J. Mar. Syst* 5: 81-100.

Weisse, T. (1983). Feeding of calanoid copepods in relation to Phaeocysts *pouchetii* in the German Wadden Sea off Sylt. *Mar. Biol.* 74: 87-94.

Weisse, T., Tande, K., Verity, P., Hansen, F., Gieskes, W. (1994). The trophic significance of *Phaeocystis* blooms. *J. Mar. Syst* 5: 67-79.

Genetic Variation in Harmful Algal Bloom Species: An Evolutionary Ecology Approach

Jane C. Gallagher

Biology Dept., City College of New York, Convent Ave. at 138th St., New York, NY 10031

1. Introduction

A cornerstone of the traditional ecological approach to biodiversity is to assume that there is a critical association between "species" and some critical function in ecosystems. This assumption that species "do" something has been so central to ecology that Mayr has proposed that the biological species concept be modified to be: "...A species is a reproductive community of populations (reproductively isolated from others) that occupies a specific niche in nature...." (Mayr 1982, p. 273). Although the biological species concept is based on interbreeding units, in practice species are commonly defined on morphological grounds (morphospecies). Therefore, it is an implicit assumption in ecology that there is a critical relationship between form and function in most organisms. A corollary of this assumption is that patterns of variation in morphology at the species level are good predictors of patterns of variation in other traits that determine the ecological role of species in natural environments. Even though these assumptions have been challenged for terrestrial organisms through empirical and theoretical investigations (see Otte and Endler 1989; and Eldredge 1992, for reviews), these assumptions still form the primary framework for the design of experiments in phytoplankton ecology.

The purpose of this chapter is to review the evidence that extensive genetic variation exists in all phytoplankton species that have been examined, and to show that morphology is very often only weakly correlated with other traits, such as toxin production and physiological parameters at lower phylogenetic levels, e.g. genera and species. For those organisms from which we have sufficient data, it is evident that complex patterns of inheritance may prevent the simple sorting of taxa into hierarchical groups based on small numbers of characters or genes. A recurrent theme that emerges from a review of all of the data for both harmful and benign phytoplankton is that biogeographic and/or temporal populations within species, or groups of closely related morphospecies located in a single geographic area appear to be the bloom-forming units rather than classically defined morphospecies.

2. Patterns of genetic differentiation in HAB species

2.1 Dinoflagellates

The most extensive data for toxic bloom forming organisms has been amassed by Anderson and co-workers in the U.S., Cembella and co-workers in Canada and by Ishida, Sako and co-workers in Japan on the *Alexandrium tamarensis /catenella*

NATO ASI Series, Vol. G 41
Physiological Ecology of Harmful Algal Blooms
Edited by D. M. Anderson, A. D. Cembella and G. M. Hallegraeff

species complex (Cembella et al. 1988). The group of species in the complex form blooms worldwide (see elsewhere in this volume). The taxonomic and genetic relationships of these organisms has long been a subject of controversy (Taylor 1985; Balech 1985; Cembella *et al.* 1986; 1987; 1988; Hayhome *et al.* 1989; Sako *et al.* 1990; Scholin *et al.* 1994a; 1995). In the North Atlantic, *A. tamarense* and *A. fundyense* form extensive red tide blooms in closely spaced, but disjunct areas (Anderson *et al.* 1994). The disjunct nature of the blooms suggests the presence of local populations in bloom areas, but this conclusion is not supported by the morphology. Isolates from diin the North Atlantic have different signatures for toxins and bioluminescence that are not correlated with their classical species definitions. When the toxin signatures are analyzed by multivariate clustering procedures and correlated with morphology, bioluminescence and place of isolation (Fig. 1), it is apparent that separate populations can be identified on the basis of toxin signature, but not morphology or bioluminescence. These geographically separate populations do not form a simple north-south gradient, as would be expected by populations founded by a simple recent north to south dilution effect, but appear to be remnants of older endemic populations (Anderson *et al.* 1994). The separation of the Cape Cod and Gulf of Maine populations is also supported by the presence of different patterns of biological rhythms in these two groups (Anderson and Keafer 1987; Anderson *et al.* 1994). Cembella et al. (1986; 1988) also demon-

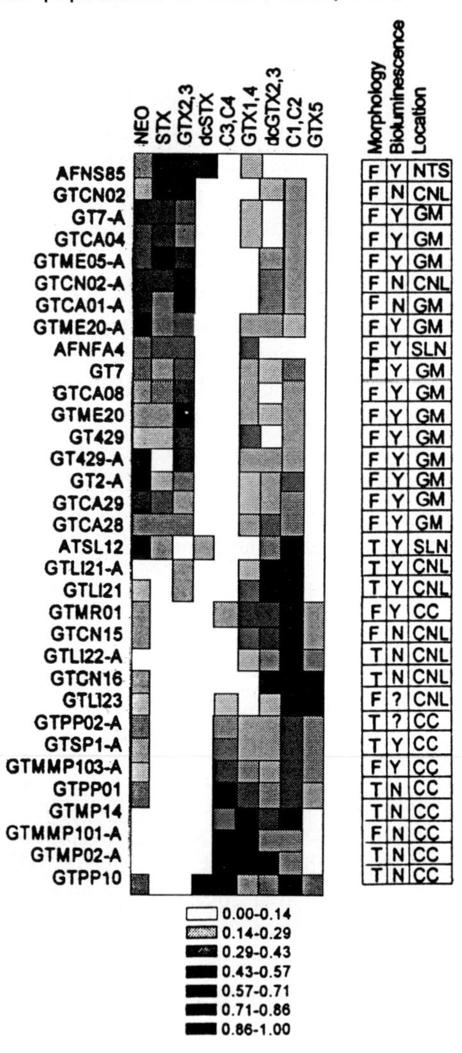

Fig 1: Multivarigate clustering matrix for toxin composition. The intensity of shading in each box indicates relative abundance for each isolate. T= *tamarense* morphology, F= *fundyense* morphology, CC= Cape Cod, GM= Gulf of Maine, SLN= St. Lawrence, NTS= Nantucket shoals (from Anderson et al. 1994).

strated the complexity of the genetic relationships in the Pacific populations using allozyme electrophoresis. Similar to the Atlantic populations, the Pacific ones do not clearly segregate on the basis of morphology. In the waters near British Columbia, populations with *"catenella"* or *"tamarensis"* morphotypes are separated at the extremes of a gradient. Intermediate locations have intermediate types, i.e. short chains of isodiametrical cells (Cembella *et al.* 1986). Morphology can also be plastic in culture, with *"catenella"* changing to *"tamarensis"* types.

The complexity of the relationships between toxicity and morphospecies on a global basis is clearly illustrated by the phylogenetic tree based on large subunit (lsu) rRNA constructed by Scholin *et al.* (1995) (Scholin, this volume). This tree clearly shows the presence of distinct geographical groups of dinoflagellates based on sequences that are not correlated with morphology. Some of these groups are toxic whereas others, such as those sampled in Western Europe to date, are non-toxic. The discordance of some of the geographic groups has been attributed to dispersals of the originally separate local populations by human activities and natural ocurring dispersal and vicariant processes (Scholin *et al.* 1995; Adachi *et al.* 1996). *Alexandrium tamarense* is also one of the few dinoflagellates where enough isolates have been examined for temperature and salinity optima (Brand 1980). These data also demonstrate that genetic variation exists at even among individuals isolated from the same pond for these ecologically important characteristics. Similar patterns of complexity are shown in eastern Canadian populations of *Alexandrium excavatum sensu* Balech (Cembella and Destombe, 1995). Toxin profiles and 18s rDNA sequences indicate population level differences exist between the Gulf of St. Lawrence and Nova Scotia, but that populations from the latter region are more heterogeneous.

Alexandrium is one of the few dinoflagellate genera where the life cycle can be controlled reliably in culture. Various crosses of strains of *A. tamarense* and *A. fundyense* have been performed (Anderson *et al.* 1994; Sako *et al.* 1992; and elsewhere in this volume). These experiments show that the pattern of inheritance of characters is complex. In a cross between *A. tamarense* and *A. fundyense* isolates from the Atlantic, complete sexual compatibility has been observed (Table 1), hence they are not "good" biological species in this instance. However, other crosses between these species performed by Japanese workers do indicate partial reproductive isolation (Sako *et al.* 1990). In the compatible cross (Table 1), inheritance of a ventral pore, which is the distinguishing feature between these two species, does not show a simple Mendelian pattern of inheritance in the haploid progeny. These data indicate that the genetic basis of this feature may involve more than one gene and/or be influenced by cytoplasmic factors. Other crossing experiments (Ishida, in this volume) between isolates with different toxin profiles show that the progeny may be undergoing recombination. Although these experiments are few in number due to their difficulty, they also demonstrate that isolates with extremely divergent traits can retain sexual compatibility. The one attempt to cross *A. tamarense* with *A. fundyense* by Japanese workers involved cultures from different parts of Scholin *et al.*'s (1995) lsu rRNA tree. These isolates different in their allozyme profiles by 50%, yet they formed zygotes (Sako *et al.* 1990). Although the zygotes did not germinate, thus showing post-zygotic isolation mechanisms, it is remarkable that organisms this divergent were able to retain any ability to mate.

Destombe and Cembella (1990) also delineated mating relationships among isolates of the morphospecies *Alexandrium excavatum sensu* Balech (=*A. tamerense).* This is the most complete study of sexual compatibility ever conducted in an HAB organism (Fig. 2). It very clearly demonstrates that varying degrees of compatibility exist within isolates of a single morphospecies and that both prezygotic and postzygotic isolating mechanisms are functional. This study confirmed earlier reports of heterothally in another HAB dinoflagellate, *Gymnodinium catenatum* (Blackburn *et al.* 1989). The morphological aspects of mating in dinoflagellates are reviewed in Gao *et al.* (1989).

Table 1. Characteristics of progeny from mating between *A. tamarense* (GTSP1) and *A. fundyense* (GT7). Isolates grouped together are from the same cyst (Biolum.=bioluminescent)(modified from Anderson *et al.* (1994). n/a=not available

Isolate	Biolum.	Mating type	Ventral pore	Comments
Parents				
GT7	Y	+	-	3/59 cells w/ small pore
GTSP1	Y	-	+	normal pore
Progeny				
17C8A	Y	+	+	38/40 cells w/ normal pore
17C8B	n/a	-	-	3/40 cells w/ small pore
17C8C	Y	-	-	
17C8D	Y	+	-	1/40 cells w/ small pore
17D5A	Y	-	-	
17D5B	n/a	-	-	
17D5C	Y	+	-	1/40 cells w/ small pore
17D5D	Y	+	-	1/40 cells w/ small pore
17D5E	Y	+	-	3/40 cells w/ small pore
17D5F	Y	+	-	2/40 cells w/ small pore
17C3A	Y	n/a	-	
17C3B	n/a	n/a	-	
17C3C	Y	n/a	-	
17C4A	Y	n/a	-	
17C4B	n/a	n/a	-	
17C4C	Y	n/a	-	
17C5A	n/a	n/a	-	
17C5B	n/a	n/a	-	
17C5C	n/a	n/a	-	
17C1A	N	n/a	n/a	
17C1C	Y	n/a	n/a	

The data on the *Alexandrium tamarense* group clearly demonstrates that local geographic populations are the functional units in blooms, not morphospecies. They also demonstrate that crosses between morphospecies are possible and that the inheritance of traits is complex. The retention of sexual compatibility in some crosses and isolation in others is common during the development of sexual isolation. However, the ability of isolates to cross in spite of large genetic distances is likely to result in the confounding of the relationship between traits as these geographic populations are dispersed throughout the world's oceans. This ability may mean that new genetic combinations may evolve as populations, dispersed by anthropogenic activity, mate with endemic ones. The lack of correlation of morphology and other properties is most likely due to the process of lineage sorting during the evolution of this group (see below and Scholin *et al.* 1995).

The *A. tamarense* group is not the only lineage in this genus to show complex distributions of traits. Mackenzie *et al.* (1996) have also discovered that *A. ostenfeldii* in New Zealand also forms disjunct local populations separated by small geographic distances. Isolates from these populations have significantly different toxin profiles. The isolate from Wellington Harbor was revealed to be non-toxic. These data indicate that intraspecific genetic variation for toxin production is widespread in *Alexandrium* and is probably a general feature of the group. At the current time, we lack sufficient information about the photosynthetic properties, motility patterns and nutrient requirements of members of this genus to make statements about the relationship of these parameters to morphology in this group. However, all of the data on the toxic species of *Alexandrium* strongly indicate that local populations seem to form blooms and that these local populations may include morphologically identical units or may include members of different morphospecies.

Other types of dinoflagellates that have been examined for intraspecific genetic variation include *Prorocentrum micans* (Braarud, 1951; Brand, 1980;1981) and *Crypthecodinium cohnii* (Beam and Himes, 1987). *Prorocentrum micans* isolates vary in their responses to temperature and salinity. Braarud (1951), in the first demonstration of infraspecific physiological variation, showed that isolates from Norway have lower temperature optima for growth than those from the Caribbean. Brand (1980; 1981) demonstrated that all populations of *P. micans* were genetically diverse, but that the pattern of diversity varied with location. The Georges Bank population

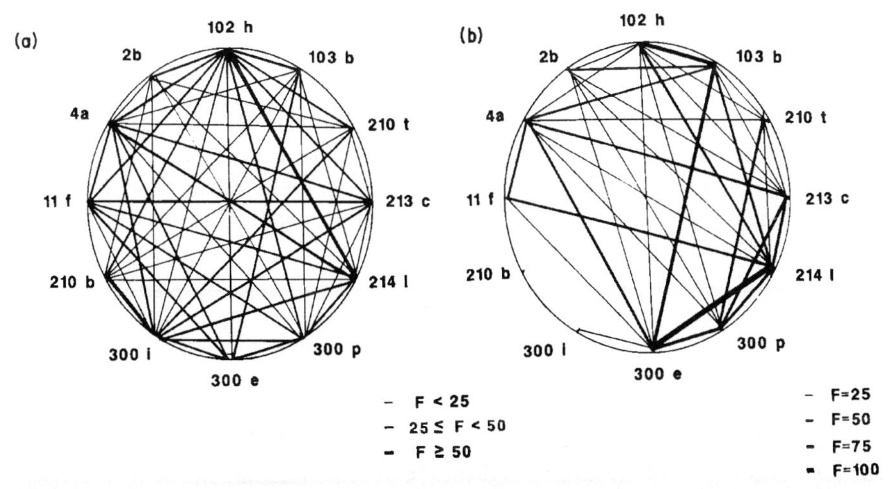

Fig. 2. a. Mating compatibility and b. zygote germination of isolates of A. excavatum (Destombe and Cembella 1990)

had significantly less diversity than that from the Sargasso Sea. These data indicate that selection for genotypes with adaptive traits might be occurring on George's Bank. Brand's data seem to suggest that genotypes within species and/or local populations may be the relevant ecological units in *P. micans* blooms.

Crypthecodinium has been observed to follow a different pattern of genetic divergence in the benthic environment (Beam and Himes, 1987). For this species complex, the morphology indicates one species, *C. cohnii*, whereas allozymes, rRNA and breeding studies demonstrate the existence of many reproductively isolated, genetically divergent "cryptic" species. Twenty-one of these are endemic sibling species with narrow distributions, while seven are global and cosmopolitan. It is not clear how these cryptic biological species relate to ecology because this is a benthic dinoflagellate that does not form extensive blooms.

2.2. Other harmful species and comparisons with non-toxic bloom species

Although the few toxic species of dinoflagellates are the best characterized and probably the most economically important toxic organisms, there are several other groups that contain harmful members. Of the non-dinoflagellate taxa, the prymnesiophyte *Chrysochromulina polylepis* and the toxic diatom *Pseudo-nitzschia multiseries* have received the most attention. The prymnesiophyte *Phaeocystis* spp. secretes acrylic acid and forms extensive blooms (Guillard and Hellebust, 1971). Because these blooms can inhibit the growth of other species and has been implicated in net and beach fouling and anoxia, *Phaeocystis* must also be considered harmful. However, non-toxic species, such as *Emiliania huxleyi* (Prymnesiophyta) and *Skeletonema costatum* (Bacillariophyta) also form extensive blooms. These non-toxic forms can be used to illuminate some of the basic processes that many be occurring in the harmful species that are less well characterized.

In 1988, *C. polylepis* formed an extensive bloom in Scandinavian waters that caused extensive economic damage to commercial fisheries (Dahl *et al.* 1988; Estep and McIntyre, 1989; Dundas *et al.* 1989). This organism bloomed only in 1988 and has not reocurred. The most notable feature from an evolutionary ecology perspective is that this species was never reported to be toxic prior to this time. Furthermore, most of its sister taxa are non-toxic. Some recent data on the morphology of toxic and non-toxic isolates indicates that small morphological differences may exist among them. It is not clear at this time whether they represent different species (see Edvardsen and Paasche, this volume). At this point the paucity of published data on the organism precludes definitive analysis. However, the fact that a local population in what appears to be a widely distributed species shows unique ecological dynamics suggests that the population genetics of the organism should be further examined.

The toxic diatom *Pseudo-nitzschia multiseries* and its non-toxic sister species *P. pungens* are sympatric in Atlantic, Pacific and on the Gulf coasts (Bates, *et al.* 1989; Hasle, 1995; Vrieling *et al.* 1996;Bates *et al.*, this volume). These species are very similar morphologically. However, they differ in 18s rDNA sequences to a small extent (Douglas *et al.* 1994) and to a greater extent in internal transcribed spacer (ITS) regions (Manhart *et al.* 1995) and lsu sequences (Miller and Scholin, 1997; Scholin *et al.* 1995). They would thus appear to be distinct species with a high degree of concordance between gene sequences and morphology. However, there is considerable reason for concern that the issue of toxicity in these forms has not been finally resolved. Isolates of *P. multiseries* vary in toxicity and morphology, and commonly become less toxic with time in culture (Bates, this volume) At present, it is unclear whether this variation has a genetic or an ecophenotypic base. However, the prevalence of the observation of clonal variation indicates that the phenomenon is general enough to warrant further investigation to see if local geographic ecotypes exist.

Another difficulty is that *P. australis* also makes domoic acid (Garrison, *et al.* 1992) and *P. delicatula* may as well (Scholin *et al.* 1995). However, these species are not sister taxa to *P. multiseries,* but are more distantly related, with non-toxic species intervening toxic ones in the lsu rDNA tree (Scholin *et al.* 1995). It is unclear whether the non-toxic species lack the genes for toxin production, which might be an indication of lineage sorting or horizontal gene flow, or if the genes are present but have been silenced. The latter result would indicate true character reversal.

The number of species and ecotypes in the cosmopolitan genus *Phaeocystis* has been the subject of controversy for many years (Sournia, 1988). The amount of acrylic acid and dimethyl sulfide (DMS) secreted and the temperature optima for growth of isolates varies. The data on the properties of various isolates, their characteristics and their distributions have been the subject of several recent reviews and investigations (Baumann *et al.* 1994; Medlin *et al.* 1996; Vaulot *et al.* 1994; Reigman and Van Boekel, 1996). Unfortunately, after a great deal of effort, it is still not clear how many species vs. how many ecotypes or local populations of *Phaeocystis* exist. Baumann *et al.* (1994) recommend the recognition of four species based on colony formation, ultrastructure and physiology: *P. globosa, P. pouchetii, P. scrobiculata* and an undescribed Antarctic type that was later named *P. antarctica* (Medlin *et al.* 1996). Sequences of 18s rRNA showed small differences between isolates ascribed to *P. pouchetii*, *P. globosa* and *P. antarctica* (Medlin *et al.* 1996). However, the taxonomy is still confused (see Keller et al., this volume).

The main morphological features that distinguish *Phaeocystis* species are colony shape, excretion of star-like features and habitat. However, some of these features are only distinguishable in the colony form and others in the flagellate form (Vaulot *et al.* 1994). The massive study by Vaulot *et al.* (1994) also examined ploidy and pigment composition as well as morphology. Using the combined data, they recognized six groups of isolates. One of these, they referred to *P. globosa* and another to *P. antarctica.* However, they also discovered that the isolate used as type material for *P. pouchetii* was probably referable to *P. globosa.* As in the investigations of *Alexandrium*, *Skeletonema* and *Emiliania*, they found that isolates tended to cluster on the basis of geographic location. Vaulot *et al.* (1994) strongly recommend that all characters must be considered together to understand the properties of species. A later review (Riegman and van Boekel, 1996) documents at least five different ecotypes present within the most widespread species, *P. globosa*, based on physiological variation. An additional cause for confusion appears to be the existence of different mating types in some populations whose compatibility is unknown (Vaulot *et al.* 1994). It is unclear whether any of the physiologically distinct variants can interbreed. The patterns of genetic diversity in *Phaeocystis* are unclear, although there seems to be clear evidence for the existence of biogeographic groups of uncertain taxonomic rank.

The non-toxic coccolithophore species, *Emiliania huxleyi* and *Gephyrocapsa oceanica*, are among the most abundant Prymnesiophytes. *Emiliania huxleyi* can form extensive, dense blooms that contribute to global carbon balance in subpolar waters (McIntyre *et al.* 1967; Pond and Harris, 1996). *Emiliania huxleyi* tends to be more common in colder waters and can tolerate temperatures from 1 to 31° C in the laboratory and *Gephyrocapsa oceanica* is more abundant in warmer waters and tolerates 19 to 31° C (Brand, 1982). Both species tend to show greater increase in number in response to macronutrient enrichment compared to open ocean coccolithophores (Brand, 1994). However, temperature tolerance of single isolates alone does not account for the distribution of coccolithophore species. Brand (1980; 1982) demonstrated that the growth rates of *Emiliania huxleyi* isolates vary significantly in response to temperature and salinity conditions. Isolates from colder water in the Gulf of Maine tolerate low temperatures better than those from more Southern areas. Coastal isolates also tolerate lower salinities better than oceanic isolates (Brand, 1982). *Gephyrocapsa oceanica* isolates also show temperature and salinity variations in comparisons of populations located on the neritic and oceanic sides of the Gulf Stream (Brand, 1982). The relationship between the physiological variation observed by Brand (1982) and small morphological differences in coccolith structure (Young and Westbroek, 1991) have yet to be elucidated because observations were made with different isolates. *Emiliania huxleyi* and *Gephyrocapsa oceanica* isolates also cluster on the basis of lipid signatures between neritic and open ocean regions, but the two species were very similar to each other within regions. Different isolates from the same locality were as different as strains originating from widely separate ocean basins (Conte *et al.* 1995). Again, as in other studies of other organisms, geographic distribution seems to yield related groups rather than morphology, at least among those described as *Emiliania huxleyi*.

One of the most complete data sets available on phytoplankton is that for *Skeletonema costatum* (Gallagher, 1980, 1982, Stabile *et al.* 1995) and related species. This species forms massive blooms in coastal areas throughout temperate and subtropical regions. Although this species is considered a nuisance alga only in Japan where its blooms affect the quality of the *Porphyra* harvest (Smayda, pers. comm.), it is relevant to discussions of toxic species because it serves as a model system for phytoplankton population biology. The morphology, ecology and genetics of members of the genus *Skeletonema* are well known (Hasle, 1973; Smayda, 1973; Gallagher, 1980; 1982;1994; Stabile, 1994) and provide insights to processes operating in less well characterized toxic forms. Characters are highly incongruent even when analyzed by identical methods.

A cladistic analysis of the genus *Skeletonema* reveals that it is morphologically distinct from its sister taxa, *Detonula* and *Lauderia* (Gallagher, in prep.) consistent with the earlier interpretations of Hasle (1973). The most abundant species in *Skeletonema, S. costatum,* forms distinct seasonal populations that genetically introgress (Gallagher, 1980; 1982; 1994). A chloroplast DNA (cpDNA) tree shows that genotypes of *S. costatum* are related to each other on a thermal gradient and that *D. confervacea*, a cold water species, is more closely related to the cold water *S. costa-*

tum isolates than it is to *D. pumila*, a warm water species (Stabile, 1994; Gallagher, in prep.). This pattern is also shown in a tree based on rRNA ITS data (Duplessis *et al.*, in prep.). Furthermore a comparison of the allozymes, ITS data and cpDNA types reveal that chloroplasts are sorted into different nuclear backgrounds (Stabile, 1994). This type of character conflict is similar to that shown by Scholin *et al.* (1995) for *Alexandrium*. These patterns are probably due to lineage sorting and/or hybridization (see below). The *Skeletonema* molecular data shows a greater correlation between molecular characters and average environmental temperature than it does with classical morphology.

These examples show that both HAB and non-HAB organisms have complex patterns of genetic divergence and evolution. It is likely that these problems are common to all phytoplankton and that HAB organisms do not have special evolutionary properties other than production of toxins or other characteristics that have nuisance value to humans.

3. Genetic processes

How can incongruent patterns for different characters evolve? There are basically three processes that can commonly be invoked to account for the lack of congruence in data sets. These are: 1. genetic introgression through sexual reproduction, 2. heterogeneity in rates of evolution in different genes followed by lineage sorting, and 3. horizontal gene transfer. The key to understanding these processes is to remember that many features of a cell are determined by combinations of many different genes, whereas scientists often study a single gene or feature at a time. Also, natural selection and random genetic drift will influence the abundance of genotypes in natural environments.

A phylogenetic tree based on detailed analysis of single genes, such as that which codes for 18s (ssu) rRNA, reflects the history of the gene, and is called a "gene tree." The relationship of the evolution of single genes to that of entire organisms that have thousands of genes (the true "species" tree) is difficult to determine. Conflict among trees derived from different genes for the same set of organisms is common (see Hulsenbeck *et al.* 1996 for a review). However, there is no clear consensus on what to do with the data in case of conflict (Hulsenbeck et al. 1996). In theory, over millions of years of evolution various gene trees and the species trees should converge (Pamilo and Nei, 1988). Therefore, higher level phylogenetic relationships such as various "trees of life" derived from single genes are more likely to be reasonably accurate than those for differences among organisms at lower phylogenetic levels, e.g. populations, species and genera. At these lower levels, it is also important to consider how single genes are inherited. For example, chloroplast DNA is inherited uniparentally without recombination. Therefore, although many single genes are located in the chloroplast, they are all genetically linked and must be conceptually treated as a single gene at the species level.

We know very little about the genetic basis of morphology in phytoplankton. It is commonly assumed that these are dependent on combinations of genes. However, the number of genes, their patterns of linkage and epistasis is unknown for any morphological feature in any member of the phytoplankton.

3.1. Sexual reproduction

Reproduction and recombination have the effect of creating non-congruence. The data from Destombe and Cembella (1990), Ishida *et al.* (in this volume) and Anderson *et al.* (1994) clearly demonstrate that genetic compatibility in dinoflagellates can be very complicated and that the patterns of inheritance of characters such as toxin profiles and morphology are complex. Toxin profiles can clearly be affected by recombination independently of morphology. While no one has done the all the definitive experiments, the preliminary data available show very little indication that morphology and toxicity are linked. Therefore, all combinations of toxin profiles and morphology should be possible. In fact, it appears that virtually all recombinants have been found in natural populations (Anderson *et al.* 1994; Cembella *et al.* 1988) and sexual reproduction among different genotypes is occurring in at least some populations of toxic dinoflagellates. The development of sexual isolation between species can be either rapid or gradual (Otte and Endler, 1989) and can proceed at different rates in different parts of species ranges (Mayr 1942; 1982). Therefore failure to mate in one part of a species range e.g. Japan for *Alexandrium* does not always predict responses elsewhere, e.g. Canada (Destombe and Cembella 1990). Also, there is no consensus on what constitutes a “species” in organisms that are predominantly asexual (Otte and Endler, 1989). The biological species concept is not universally accepted. Therefore, barriers to mating and zygote viability may be considered by some workers to be infraspecific or to indicate separate species.

3.2. Rates of evolution and lineage sorting

It is clear that genes evolve at different rates (e.g. Ayala *et al.* 1996) and organisms evolve with different trajectories. This rate heterogeneity is the basis for criticism of the biological clock model of evolution (Avise 1994; Ayala *et al.* 1996). When genes evolve different alleles prior to taxa developing complete sexual isolation, these alleles can wind up in random associations with each other. When phylogeneticists use DNA methods, such as sequencing, they are usually tracing the evolution of single genes. Morphologists examining very small scale characters may also be tracing the phylogeny of features determined by very few genes. Each feature can yield conflicting phylogenies. Lineage sorting is the process that accounts for "monophyletic radiation" scenario for *Alexandrium* (see Scholin, this volume). In lineage sorting, genes evolve different alleles before sexual isolation between organisms is complete. During the speciation process, these different lineages of genes are randomly associated in the developing species. This process results in the phenomenon where individual genes in different species can be more closely related to each other than most of the rest of the DNA in the organisms. In Scholin’s monophyletic radiation scenario, the genes controlling morphology and toxicity diverged before various populations were dispersed around the globe.

Lineage sorting is also likely to account for the patterns of conflict between the morphological and molecular data for *Skeletonema* and *Detonula* (Figs. 3 and 4). Several genera in the Thalassiosiraceae evolved warm and cold water species pairs or populations. Often these forms were described as different species, such as *Lauderia annulata* vs. *Lauderia borealis* and *D. confervacea* vs. *D. pumila* based on morphology. Others such as *S. costatum*, never developed enough morphological divergence to be considered different species (Hasle, 1973). However, in terms of

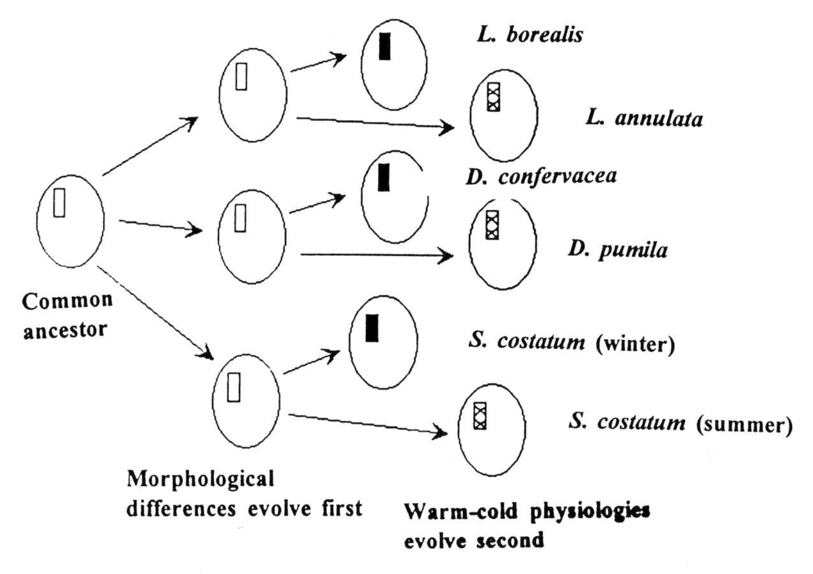

Fig 3: Hypothetical phylogeny if morphological difference evolve first. Circles represent cell walls and the boxes represent linked groups of genes. Filled boxes are genes linked to cold-water adaptation, hatched boxes are genes linked to warm-water adaptation.

physiological characteristics, both the warm water and the cold water species/populations appear similar within their respective group. If the morphology diverged first, all of these characters would have had to develop twice (Fig. 3). However, if the basic warm water and cold water adaptations evolved first and then the morphological differences evolved along with sexual isolation, each set of traits only has to evolve once (Fig. 4).

Lineage sorting only requires that populations be polymorphic for genetic traits. The results of numerous allozyme studies (a method that examines several genes simultaneously) of phytoplankton show that all populations of all taxa examined to date are polymorphic (Gallagher 1986; Wood 1988). This process has the potential to be an extremely important force in phytoplankton evolution for HAB as well as non-HAB organisms.

3.3. Horizontal gene flow

Horizontal gene flow occurs when genes are transferred between genetically isolated lineages by a vector, such as a virus, parasite, or plasmid. We know very little about these processes in HAB species. However, virus infections have been reported as being important in ending blooms of *Chrysochromulina* and *Emiliania* (Suttle and

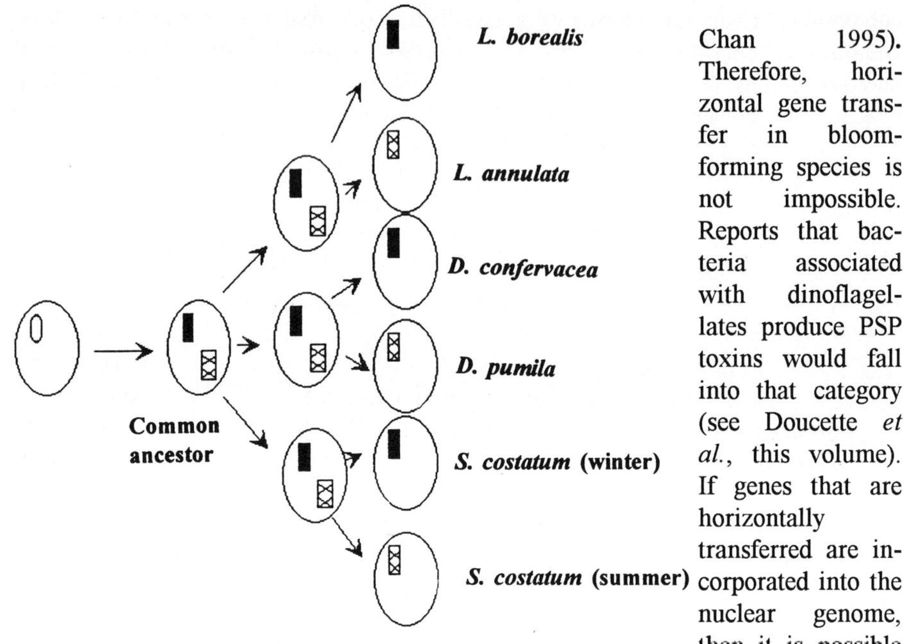

Fig 4: Hypothetical phylogeny if warm-cold adaptive genes evolve first. Circles represent morphologies and boxes represent groups of linked genes. Filled boxes are genes linked to cold-adapted genes, hatched boxes indicate linkages to warm-adapted genes.

Chan 1995). Therefore, horizontal gene transfer in bloom-forming species is not impossible. Reports that bacteria associated with dinoflagellates produce PSP toxins would fall into that category (see Doucette *et al.*, this volume). If genes that are horizontally transferred are incorporated into the nuclear genome, then it is possible that these genes could be transferred vertically into the progeny of the host. This type of genetic pattern would be expected when traits, such as toxicity, appear more or less randomly throughout the tree. This might be the case in the production of domoic acid in an unusual variety of taxa. Clearly, this possibility requires more study.

4. Species concepts and their relationship to marine phytoplankton

There are a variety of species concepts currently proposed by taxonomists (see Otte and Endler, 1989, for a review). The most popular ones are those that rely on sexual isolation (biological species concept) or monophyly (phylogenetic species concept). Morphological data have been used to support both of these theoretical models. However, each of these concepts is best regarded as a model of evolution rather than an absolute standard.

The biological species concept was originally developed by Mayr (1942) to describe "...actually or potentially interbreeding..." groups of organisms. The problem with the biological species concept as originally constituted is the use of the term "potentially" because it implies that laboratory breeding experiments can be used to determine species boundaries. Unfortunately, most laboratory settings seriously perturb any prezygotic isolating mechanisms that may be operating in field populations. The most extreme example of this is found in orchids (Grant 1971; Gill 1989). Six percent of plant species are in the orchid family and hybrids are rare in field populations, yet virtually all of them are interfertile when in the lab. Apparently orchids rarely develop postzygotic isolating mechanisms. If the older version of the biological species concept is applied strictly to orchids, hundreds of species that are clearly morphologically different would have to be synonomized. Clearly this solution is deemed unacceptable. Similarly, it is questionable whether the nitrogen deprivation that is normally used to induce cyst formation in HAB species, such as *Alexandrium,* is a natural way to determine breeding affinities (e.g. Cembella and Destombe 1990). Another set of problems with the old biological species concept arise when conflicting results from mating experiments are obtained in lab experiments using the same organism because sexual isolation may develop with different rates in different parts of a species range. Problems such as these forced Mayr (1982) to redefine his biological species concept as stated in the beginning of this chapter.

The phylogenetic species concept has been proposed in several variants (see Otte and Endler, 1989, for review). The goal is to identify groups with distinct ancestor descendant relationships. These groups can putatively be identified by the presence of a unique character or a unique combination of morphological or molecular characters. This concept has been attacked on the grounds that it will result in the description of unmanageable numbers of new species that are sexually compatible as well as the practical grounds that it is often extremely difficult to apply (e.g. Martin 1996).

Morphological similarity is used to design breeding experiments. Morphological characters are used to describe monophyletic units. In spite of the current emphasis on molecular data, morphology will continue to be used as a standard because it is the only dataset that can be collected relatively quickly and completely in a large variety of organisms. The great problem in systematics is conflict among data sets. When morphology and molecular phylogenies conflict, some investigators conclude that the morphological data set is erroneous because they judge morphological characters to be more likely to be subject to convergent evolution. However, this is not always the case. As in the case of lineage sorting, there are important evolutionary processes that can result in character conflict without convergence. There are many instances where molecular data are ambiguous and trees are actually made more robust by the addition of morphological data (see Hulsenbeck *et al.* 1996) for a recent review on "total evidence theory". Therefore, it is inappropriate *a priori* to discount morphological data as being inherently more faulty than molecular data.

The application of any of these concepts to HAB organisms not only have all the same difficulties as they do in higher organisms, their capacities for asexual growth, complicated patterns of selfing and outbreeding, and their high capability for dispersal will complicate interpretations of their patterns of evolution. Although this poses formidable problems for nomenclature and taxonomy using any of the available species concepts, it need not be insurmountable for ecologists provided they become aware of the lack of coupling between taxonomy practices and ecological roles. In virtually every HAB and non-HAB organism examined populations and/or groups of closely related species in different geographic regions appear to occupy niches, as defined by forming blooms during specified times of year. If phytoplankton ecologists focus on local ecosystems rather than global processes, it may be possible to interpret events.

4.0. Summary

- Morphology is often uncoupled from molecular and morphological characters in HAB organisms and other phytoplankton.
- The pattern between biochemical, physiological and/or molecular characters in HAB's often indicates that local geographic populations and/or groups of closely related morphospecies appear to occupy ecological niches.
- Detection of patterns requires large sample sizes and should include closely related species. The lack of infraspecific variation is more often due to poor sampling than to biological reality.
- No species concept is going to solve all of the taxonomic problems for HAB organisms. It is incumbent on ecologists to recognize that morphology may not indicate ecological roles.
- The distribution of characters, such as toxicity, on phylogenetic trees and the complexity of patterns of mating and the inheritance of features emphasize the need for more studies of the genetics of HAB species.

5. Acknowledgements

This work was supported in part by NSF grants DEB-9119405 and OCE-9217454.

6. References

Adachi, M., Sako, Y. Ishida, Y. 1996. Population analysis of dinoflagellate *Alexandrium* species using sequences of the 5.8s ribosomal DNA and internal transcribed spacer regions. *J. Phycol.* 32:424-432.

Anderson, D. M., Keafer, B. A.. 1987. An endogenous annual clock in the toxic marine dinoflagellate *Gonyaulax tamarensis. Nature* 325:616-617.

Anderson, D. M., Kulis, D. M., Doucette, G. J., Gallagher, J. C., Balech, E.. 1994. Biogeography of toxic dinoflagellates in the genus *Alexandrium* from the northeastern United States and Canada. *Mar. Biol.* 120:467-478.

Avise, J. C. 1994. *Molecular Markers, Natural History and Evolution.* Chapman and Hall., New York. 511 pp.

Ayala, F. J., Barrio, E., Kwiatowski, J. 1996. Molecular clock or erratic evolution? A tale of two genes. *Proc. Natl. Acad. Sci. USA* 93:11729-11734.

Balech, E. 1985. The genus *Alexandrium* or *Gonyaulax* of the *tamarensis* group. In D. M. Anderson, A. W. White and D. G. Baden (eds.), *Toxic Dinoflagellates*, Elsevier, New York, pp. 33-38..

Bates, S. S., Bird, C. J., Boyd, R. K., de Freitas, A. S. W., Foxall, R. M. W. Gilgan. 1989. Pennate diatom *Nitzschia pungens* as the primary source of domoic acid, a toxin in shellfish from eastern Prince Edward Island, Canada. *Can. J. Fish. Aquatic Sci.* 46:1203-1215.

Baumann, M. E. M., Lancelot, C., Brandini, F. P., Sakshaug, E., John, D. M.. 1994. The taxonomic identity of the cosmopolitan prymnesiophyte *Phaeocystis:* a morphological and ecophysiological approach. *J. mar. Syst.* 5:5-22.

Beam, C. A., Himes, M. 1987. Electrophoretic characterization of members of the *Crypthecodinium cohnii* (Dinophyceae) species complex. *J. Protozool.* 34:204-217.

Blackburn, S. I., Hallegraeff, G. M. Bolch, C. J. 1989. Vegetative reproduction and sexual life cycle of the toxic dinoflagellate *Gymnodinium catenatum* from Tasmania, Australia. *J. Phycol.* 25:577-590.
Braarud, T. 1951. Salinity as an ecological factor in marine phytoplankton. *Physiol. Plant.* 4:28-34.
Brand, L. E. 1980. Genetic variability and differentiation in niche components of marine phytoplankton species. Ph.D. Dissertation, Woods Hole Oceanographic Inst.-M.I.T. Joint Program.
Brand, L. E. 1981. Genetic variability in reproduction rates in marine phytoplankton populations. *Evolution* 35:1117-1127.
Brand, L. E. 1982. Genetic variability and spatial patterns of genetic differentiation in the reproductive rates of the marine coccolithophores *Emiliania huxleyi* and *Gephyrocapsa oceanica. Limnol. Oceanogr.* 27:236-245.
Brand, L. E. 1994. Physiological ecology of marine coccolithophores. In A. Winter and W. G. Sieser (eds.), *Coccolithophores*, Cambridge U. Press, New York, pp. 39-49..
Cembella, A. D., Destombe, C. in press. Genetic differentiation among *Alexandrium* populations from Eastern Canada. *In* Yarumoto, T., Oshima, Y. and Fukuyo, Y. (Eds.). *Harmful and Toxic Alagal Blooms,* IOC-UNESCO, Paris, pp. 447-450.
Cembella, A. D., Sullivan, J. J., Boyer, G. L., Taylor, F. J. R. Anderson, R. J. 1987. Variation in paralytic shellfish txoin composition within the *Protogonyaulax tamarensis/catenella* species complex: red tide dinoflagellates. *Biochem. Syst. Ecol.* 15:171-186.
Cembella, A. D., Taylor, F. J. R. 1986. Electrophoretic variability within the *Protogonyaulax tamarensis/catenella* species complex: pyridine linked dehydrogenases. *Biochem. Syst. Ecol.* 14:311-321.
Cembella, A. D., Taylor, F. J. R. Therriault, J.-C. 1988. Cladistic analysis of electrophoretic variants within the toxic dinoflagellate genus *Protogonyaulax tamarensis/catenella* species complex: red tide dinoflagellates. *Bot. Mar.* 31:39-51.
Coleman, A. W. 1977. Sexual and genetic isolation in the cosmopolitan algal species *Pandorina morum. Amer. J. Bot.* 64:361-368.
Conte, M. H., Thompson, A., Eglinton, G. Green, J. C. 1995. Lipid biomarker diversity in the coccolithophorid *Emiliania huxleyi* (Prymnesiophyceae) and the related species *Gephyrocapsa oceanica. J. Phycol.* 31:272-282.
Dahl, E., Lindahl, O., Paasche, E. Throndsen, J.. 1989. The *Chrysochromulina polylepis* bloom in Scandanavian waters during spring 1988. In E. M. Cosper, V. M. Bricelj and E. J. Carpenter (eds.), *Novel Phytoplankton Blooms.*, Springer-Verlag, New York, pp. 383-405..
Destombe, C., Cembella, A. 1990. Mating-type determination, gametic recognition and reproductive success in *Alexandrium excavatum* (Gonyaulacales, Dinophyta), a toxic red-tide dinoflagellate. *Phycologia* 29:316-325.
Douglas, D. J., Landry, D. Douglas, S. E. 1994. Genetic relatedness of toxic and nontoxic isolates of the marine pennate diatom *Pseudonitzschia* (Bacillariophyceae): Phylogenetic analysis of 18s rRNA sequences. *Natural Toxins* 2:166-174.

Dundas, I., Johannessen, O. M., Berge, G. Heimdal, B. 1989. Toxic algal bloom in Scandinavian waters, May-June 1988. *Oceanography* 2:9-14.

Eldredge, N. 1992. *Systematics, Ecology and the Biodiversity Crisis*. Columbia Univ. Press, New York.

Estep, K. W., MacIntyre, F. 1989. Taxonomy, life cycle, distribution and dasmotrophy of *Chrysochromulina:* a theory accounting for scales, haptonema, muciferous bodies and toxicity. *Mar. Ecol. Prog. Ser.* 57:11-21.

Gallagher, J. C. 1980. Population genetics of *Skeletonema costatum* in Narragansett Bay. *J. Phycol.* 16:464-474.

Gallagher, J. C. 1982. Physiological variation and electrophoretic banding patterns of genetically different seasonal populations of *Skeletonema costatum* (Bacillariophyceae). *J. Phycol.* 18:148-162.

Gallagher, J. C. 1986. Population genetics of microalgae. In W. R. Barclay and R. P. McIntosh (eds.), *Algal Biomass Technologies: An Interdisciplinary Perspective*, Nova Hedwigia Beih. 83, pp. 6-14.

Gallagher, J. C. 1994. Genetic structure of microalgal populations. I. Problems associated with the use of strains as terminal taxa. *Mem. Cal. Acad. Sci.* 17:69-86.

Gao, X., Dodge, J. D. Lewis, J. 1989. Gamete mating and fusion in the marine dinoflagellate *Scrippsiella* sp. *Phycologia* 28:342-351.

Garrison, D. L., Conrad, S. M., Eilers, P. P. Waldron, E. M. 1992. Confirmation of domoic acid production by *Pseudonitzschia australis* (Bacillariophyceae) in culture. *J. Phycol.* 28:604-607.

Gill, D. 1989. Fruiting failure, pollinator inefficiency and speciation in orchids. In D. Otte and J. A. Endler (eds.), *Speciation and Its Consequences*, Sinauer Associates, Sunderland, MA. pp. 458-482.

Grant, V. 1971. *Plant Speciation*. Columbia University Press, New York. 452 pp.

Guillard, R. R. L., Hellebust, J. A. 1971. Growth and the production of extracellular substances by two strains of *Phaeocystis pouchettii*. *J. Phycol.* 7:330-338.

Hasle, G. R. 1973. Morphology and taxonomy of *Skeletonema costatum* (Bacillariophyceae). *Norw. J. Bot.* 20:109-137.

Hasle, G. R. 1995. *Pseudo-nitzschia pungens* and *P. multiseries* (Bacillariophyceae): nomenclatural history, morphology and distribution. *J. Phycol.* 31:428-435.

Hayhome, B. A., Anderson, D. M., Kulis, D. M., Whitten, D. J. 1989. Variation among congeneric dinoflagellates from the northeastern United States and Canada. I. Enzyme electrophoresis. *Mar. Biol.* 101:427-435.

Hulsenbeck, J. P., Bull, J. J., Cunningham, C. W. 1996. Combining data in phylogenetic analysis. *Trends Ecol. Evol.* 11:5-11.

Mackenzie, L., White, D., Oshima, Y., Kapa, J. 1996. The resting cyst and toxicity of *Alexandrium ostenfeldii* (Dinophyceae) in New Zealand. *Phycologia* 35:148-155.

Manhart, J. R., Fryxell, G. A., Villac, M. C., Segura, L. Y. 1995. *Pseudo-nitzschia pungens* and *P. multiseries* (Bacillariophyceae) nuclear ribosomal DNA's and species differences. *J. Phycol.* 31:421-427.

Martin, G. 1996. Birds in double trouble. *Nature* 380:666-667.

Mayr, E. 1942. *Systematics and the Origin of Species*. Columbia U. Press, New York. 452 pp.

Mayr, E. 1982. *The Growth of Biological Thought*. Harvard Univ. Press, Cambridge, MA. 582 pp

McIntyre, A., Be, A. W. H., Roche, M. B. 1967. Modern coccolithophorids of the Atlantic Ocean. I. Placolith and cyrtoliths. *Deep-sea Res.* 14:561-597.

Medlin, L. K., Lange, M., Baumann, M. E. M. 1996. Genetic differentiation among three colony-forming species of *Phaeocystis:* further evidence for the phylogeny of the Prymnesiophyta. *Phycologia* 33:199-207.

Miller, P.E., Scholin, C.A. 1997. Identification of *Pseudo-nitzschia* (Bacillariophyceae) using species-specific lsu rRNA targeted fluorescent probes. *J. Phycol.* (in press).

Otte, D., Endler, J. A. E. 1989. *Speciation and its Consequences*. Sinauer, Sunderland, MA. 464 pp.

Pamilo, C., Nei, M.. 1988. Relationships between gene trees and species trees. *Mol. Biol. Evol.* 5:568-583.

Pond, D. W., Harris, R. P. 1996. The lipid composition of the coccolithophore *Emiliania huxleyi* and its possible ecophysiological significance. *J. mar. biol. Ass. U.K.* 76:579-594.

Riegman, R., Van Boekel, W. 1996. The ecophysiology of *Phaeocystis globosa*: A review. *J. Sea Res.* 35:235-242.

Sako, Y., Kim, C. H., Ishida, Y. 1992. Mendelian inheritance of paralytic shellfish poisoning toxin in the marine dinoflagellate *Alexandrium catella. Biosci. Biotech. Biochem.* 56:692-694.

Sako, Y., Kim, C. H., Ninomiya, H., Adachi, M., Ichida, Y. 1990. Isozyme and cross analysis of matine populations in the *Alexandrium catenella/tamarense* species complex. In E. Graneli, B. Sundstrom, L. Edler and D. M. Anderson (eds.), *Toxic Marine Phytoplankton*, Elsevier, New York, pp. 320-323..

Scholin, C. A., Buck, K. R., Britschgi, T., Cangelosi, G., Chavez, F. P. 1996a. Identification of *Pseudo-nitzschia australis* (Bacillariophyceae) using rRNA-targeted probes in whole cell and sandwich hybridization formats. *Phycologica*, in press.

Scholin, C. A., Miller, P., Buck, K., Chavez, F. 1996b. Detection and quantification of *Pseudo-nitzschia australis* in cultured and natural populations using lsu rRNA-targeted probes. *Limnology and Oceanography*, in press.

Scholin, C. A., Hallegraeff, G. M., Anderson, D. M. 1995. Molecular evolution of the *Alexandrium tamarense* 'species complex' (Dinophyceae): Dispersal in the North American and West Pacific regions. *Phycologia* 34:472-485.

Scholin, C. A., Villac, M. C., Buck, K. R., Krupp, J. M., Powers, D. A., Fryxell, G. A., F. Chavez, P. 1994. Ribosomal DNA sequences discriminate among toxic and non-toxic *Pseudonitzschia* species. *Natural Toxins* 2:152-165.

Smayda, T. J. 1973. The growth of *Skeletonema costatum* during a winter-spring bloom in Narragansett Bay, R.I. *Norw. J. Bot.* 20:219-247.

Sournia, A. 1988. *Phaeocystis* (Prymnesiophyceae): How many species? *Nova Hedwigia* 47:211-217.

Stabile, J. E. 1994. Molecular Evolution in Natural Populations of *Skeletonema costatum:* Restriction Mapping and Analysis of the Chloroplast Genome. Ph.D. thesis, City University of New York. 254 pp.

Stabile, J. E., Gallagher, J. C., Wurtzel, E. T.. 1995. Colinearity of chloroplast genomes in divergent ecotypes of the marine diatom *Skeletonema costatum* (Bacillariophyceae). *J. Phycol.* 31:795-800.

Suttle, C. A., Chan, A. M.. 1995. Viruses infecting the marine Prymnesiophyte *Chrysochromulina* spp.: Isolation, preliminary characterization and natural abundance. *Mar. Ecol. Prog. Ser.* 118:275-282.

Taylor, F. J. R. 1985. The taxonomy and relationships of red tide dinoflagellates. In D. M. Anderson, A. M. White and D. G. Baden (eds.), *Toxic Dinoflagellates*, Elsevier, New York. pp. 11-26.

Vaulot, D., Birrien, J.-L., Marie, D., Casotti, R., Veldhuis, M. J. W., Kraay, G. W., Chretinnot-Dinet, M. J. 1994. Morphology, ploidy, pigment composition and genome size of cultured strains of *Phaeocystis* (Prymnesiophyceae). *J. Phycol.* 30:1022-1035.

Vrieling, E. G., Koeman, R. P. T., Scholin, C. A., Scheerman, P., Peperzak, L., Veenhuis, M., Gieskes, W. C. 1997. Identification of a domoic acid-producing *Pseudo-nitzschia* species (Bacillariphyceae) in the Dutch Wadden Sea with electron microscopy and molecular probes. *Eur. J. Phycol.* 31:333-340.

Wood, A. M. 1988. Molecular biology, single cell analysis and quantitative genetics: new evolutionary genetic approaches in phytoplankton ecology. In C. M. Yentsch, F. C. Mague and P. K. Horan (eds.), *Immunochemical Approaches to Estuarine, Coastal and Oceanographic Questions*, pp. 41-71. Elsevier, New York.

Young, J. R., Westbroek, P.. 1991. Genotypic variation in the coccolithophorid species *Emiliania huxleyi*. *Mar. Micropaleontol.* 18:5-23.

Bloom Dynamics and Ecophysiology of *Dinophysis* spp.

Serge Y. Maestrini

Centre de Recherche en Ecologie Marine et Aquaculture de L'Houmeau (CNRS-IFREMER), B.P. 5, 17137 L'Houmeau, France.

1. Introduction

The dinoflagellate genus *Dinophysis* has been known since 1840, when Ehrenberg described several species. About 200 species have been described in the literature; on the basis of these descriptions, the genus is said to be cosmopolitan (Sournia 1986). For a long time, research pertaining to species of that genus was mostly descriptive. Their ecology and ecophysiology remained poorly understood, mostly because of failure to culture any of these species and also their relative scarcity.

Research on *Dinophysis* species increased greatly after they were linked to a new type of shellfish poisoning named "Diarrhetic Shellfish Poisoning" (DSP) (Yasumoto *et al.* 1978, 1980; Avaria 1979). Accordingly, attention has been focused on the few toxin-producing species in the genus. These species are: *D. acuminata, D. acuta, D. caudata, D. fortii, D. norvegica, D. rotundata, D. sacculus, D. skagii* and *D. tripos*. The effects of diarrhetic toxins and/or the occurrence of *Dinophysis* blooms appear to have recently increased both in space and time (Anderson 1989; Smayda 1990; Belin 1993; Hallegraeff 1993). Therefore, research has focused on mechanisms allowing the species to adapt to changes in environmental conditions and increase in cell concentration in natural assemblages. This review examines whether significant progress has been made in understanding the ecophysiology and bloom dynamics of potentially-toxic *Dinophysis* species.

2. Bloom dynamics

Dinophysis blooms seldom are of sufficient density to discolor the water. Moreover, except for two cases (Subba Rao *et al.* 1993; Dahl *et al.*, 1996), *Dinophysis* spp. populations have never been reported to be dominant in dinoflagellate summer assemblages; the maximum contribution was 39% of that of the total dinoflagellate community (Jacques and Sournia, 1978-79). Hereafter "*Dinophysis* bloom" refers to cell densities of at least several thousands per liter. For some reviewed work, values are given for a single species; for others, they refer to an assemblage of several species.

In the Risør area of the southern coast of Norway, Dahl *et al.* (1995) recorded *Dinophysis* spp. concentration from the end of May to the end of September 1987. In July, the water column was stratified. Cell densities of *D. acuminata* and *D.*

NATO ASI Series, Vol. G 41
Physiological Ecology of Harmful Algal Blooms
Edited by D. M. Anderson, A. D. Cembella and
G. M. Hallegraeff

norvegica peaked in late June ($50x10^3$ cells l^{-1} ; Fig. 1). In mid July, both species were present at rather high densities from the surface to 20 m depth. Decline of these two concurrent blooms was very different. *D. acuminata* had become scarce by late July, before the thermal stratification broke down. In contrast, the subsurface population of *D. norvegica* was high until the end of the survey, in late Sept. *D. acuta* was also present, but was only recorded in the upper 10 m layer, from mid Aug. to late Sept. Routine sampling in the upper layer (0 - 3 m) and counting of *Dinophysis spp.* carried out in this area from 1986 to 1995 have shown that blooms occur in May-July for *D. acuminata*, from Aug. to autumn for *D. acuta*, while *D. norvegica* might bloom in all season (Dahl, personnal communication). Moreover, in an area nearby, reddish surface water due to the occurrence of just over $25x10^6$ *Dinophysis* cells l^{-1} was observedon late Sept., in which *D. norvegica* accounted for over 90% (Dahl *et al.,* 1996).

This work provided a time series of *Dinophysis* spp. counts showing that dense populations may develop in the upper part of the water column, whereas many other studies report highest concentrations at the pycnocline. It also showed that environmental conditions which lead to blooms are different according to the different species of the genus. Unfortunately, the study did not indicate whether the high concentrations were due to *in situ* growth or to physical accumulation. Wind conditions and organism behavior can easily account for rapid increases in cell concentrationn (Franks 1992, 1995).

Daily counts of *Dinophysis* spp. in 1987 and 1988 in Antifer Harbour, Normandy, France, have shown that cell concentrations $>10x10^3$ cells l^{-1} occurred in both years from early July to mid Aug. (Lassus *et al.* 1990a). Lassus *et al.* (1993) also described the bloom dynamics of *Dinophysis* species in that area for 1989. Inside the harbour, *Dinophysis* concentrations, mainly *D. sacculus*, *D.* cf. *acuminata* and *D. skagii*, increased in ten days from a few hundred cells l^{-1} to $165x10^3$ cells l^{-1}. Outside the harbour, values ranging from 50-160 $x10^3$ cells l^{-1} were recorded in mid Aug., whereas only a few thousand cells per litre were present in early and late Aug. Within the bloom period, the water column was not thermally stratified, and there was a marked drop in salinity and an increase in nitrate. Such conditions probably originated from an input of water from the nearby River Seine due to a strong southerly wind during the three days preceeding the peak of *Dinophysis* concentration. These authors concluded that the critical factor triggering the increase in cell concentration in that location was the presence of winds from the south and the southwest over several days. They also hypothesized that dense populations result from three complementary mechanisms: (i) cell division in the Seine Bay, (ii) migration upward and concentration in the surface layers subjected to wind action, (iii) confinement of *Dinophysis*-rich water in the harbour-locked area by a southwesterly wind.

Although, this work is of importance in clearly showing the role of hydrodynamics in building dense patchy populations of *Dinophysis* spp., several important aspects

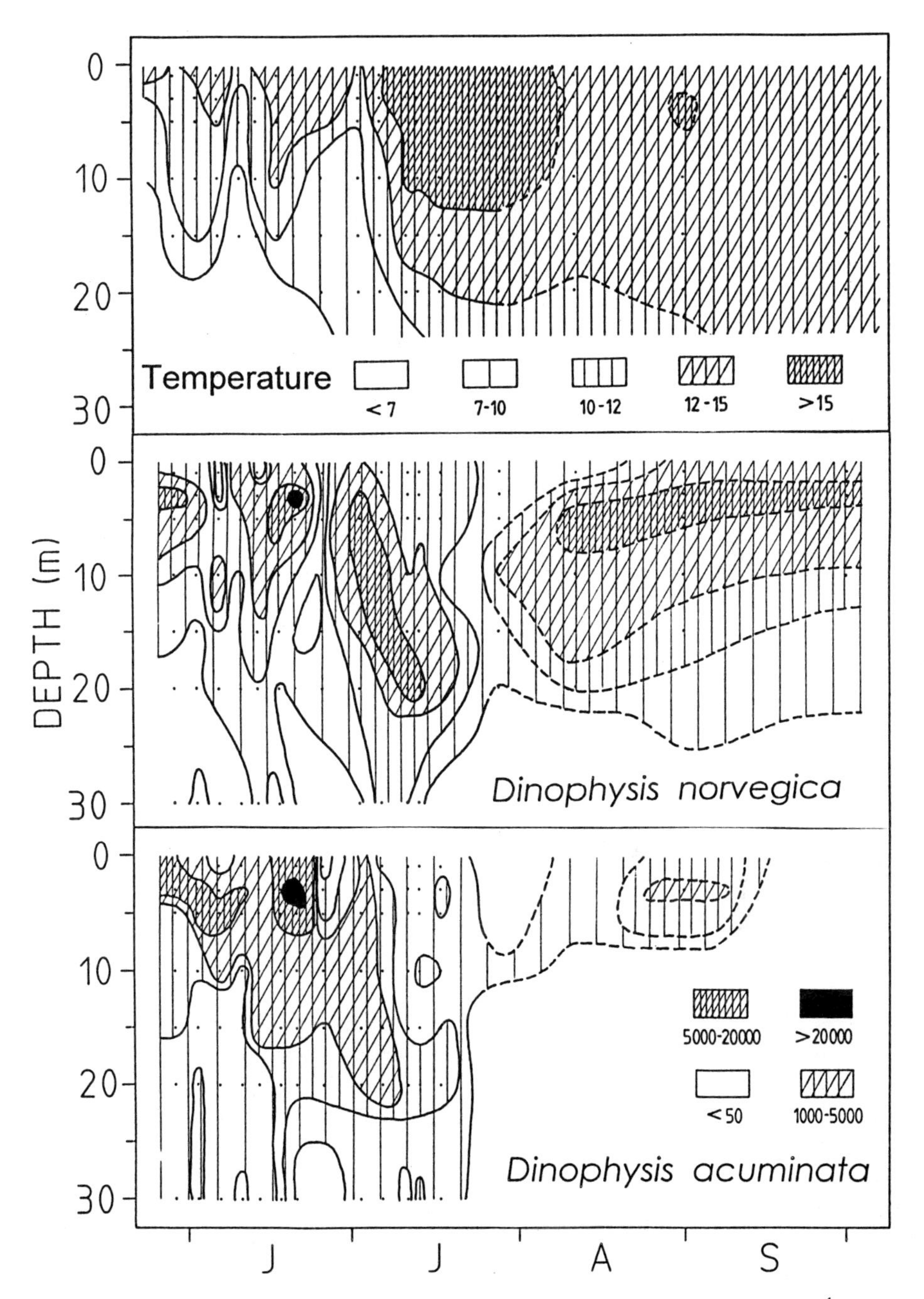

Figure 1. Temperature (°C) and cell concentration (cells l^{-1}) of *Dinophysis acuminata* and *Dinophysis norvegica*, from June to September 1987, in the Risør area, southern coast of Norway (simplified and redrawn from Dahl *et al*. 1995)

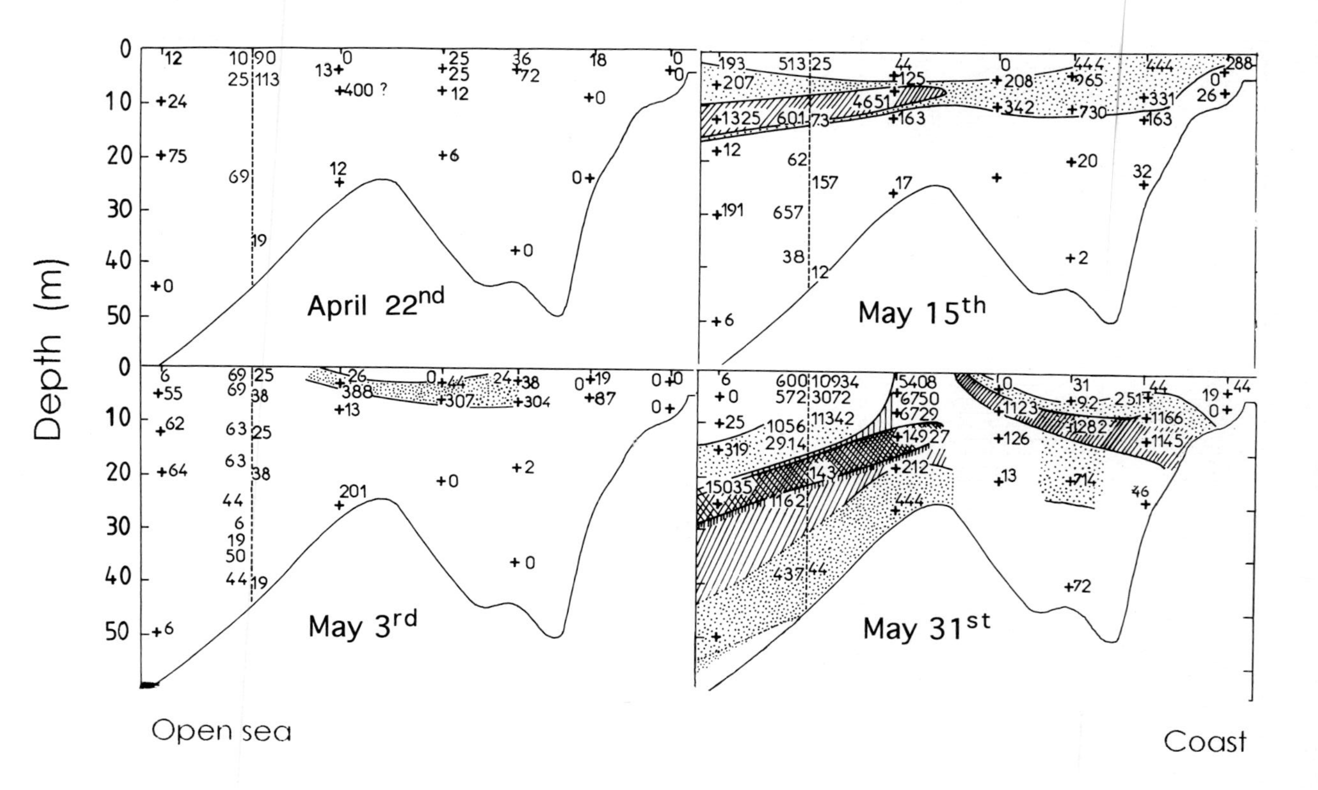

Figure 2. Temperature (°C) and *Dinophysis* spp. cell concentration (cells l^{-1}), from late April to late May 1990, in La Rochelle area, French Atlantic coast (simplified and redrawn from Delmas *et al*. 1992)

remain unclear. Where did the cells divide? What nutrients did they take up to sustain growth? What were the environmental conditions which allowed this growth?

From mid-April to mid-June 1990, Delmas *et al.* (1992) followed the continuous increase in cell concentrationn of the *Dinophysis* spp. population through the water column in the vicinity of La Rochelle, on the French Atlantic coast. Eight stations located from the inshore nutrient-rich water to the offshore nutrient-poor water (50 m depth) were visited once a week. In the offshore area, the water column stratified in late April, and between mid-May and mid-June a thermocline was established. In contrast, temperature stratification did not develop inshore. In April, there were few *Dinophysis* spp. present offshore and none inshore. In late May, up to $15x10^3$ cells l^{-1} were recorded in the 10-15 m layer offshore, whereas inshore waters contained only a few cells l^{-1} (Fig. 2). In late May, a short period of wind partly modified the vertical structure, and *Dinophysis* spp. concentration fell, increasing again when marked stratification was again established. *Dinophysis* spp. were more abundant in the "thermocline layer", scarce below this layer, and absent near the bottom (Fig. 3). No relationship between *Dinophysis* spp. growth and availability of dissolved inorganic nutrients was found. It was concluded that a significant thermocline (Δt = 5°C) and stable stratification of water column (for at least one week) are the critical conditions required in that area for *Dinophysis* to reach 10^3 cells l^{-1} or higher.

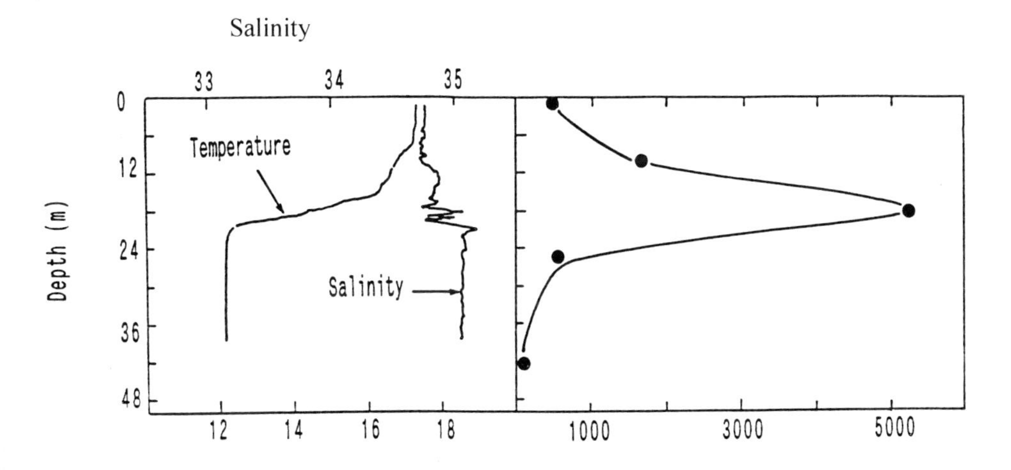

Temperature (°C) *Dinophysis* (Cell.l^{-1})

Figure 3. Typical vertical distribution of temperature (°C), salinity and *Dinophysis* spp. cell concentration (cells l^{-1}), in late spring 1990, in La Rochelle area, French Atlantic coast (simplified and redrawn from Delmas *et al.* 1993).

In Ria de Vigo and Ria de Pontevedra, Galicia, Spain, Reguera *et al.* (1990, 1995) recorded the vertical distribution of *D. acuta* and companion species from early July to late Nov 1989. At the station where *D. acuta* was the most persistent, in July and Aug., a strong thermal gradient occurred between 0 and 20 m. At the beginning of Sept., an upwelling event caused the breakdown of the stratification and a decrease in temperature; the water column became almost isothermal. Renewed stratification occurred in mid-Sept. Then coastal upwelling made the water column isothermal again in mid-October, and these conditions, totally without stratification, continued until the end of the survey. In early July, the upper layer phytoplankton assemblage was dominated by diatoms, whereas in the 10 - 20 m layer flagellates dominated. The most abundant *Dinophysis* species was *D. acuminata; D. acuta* was also present. In August the abundance of *D. acuminata* diminished, while that of *D. acuta* increased and peaked in late Aug. Peak concentration of *D. acuta* coincided with a bloom of *Gyrodinium* sp. Then, while diatoms became dominant again, *Gyrodinium* sp. and *D. acuta* declined. Nevertheless, a new burst of *D. acuta* population (17 x 10^3 cells l^{-1}) was observed in late Sept. when stratification had resumed for a week. When autumn wind and rain resumed and the water column had become isothermal, diatoms practically disappeared and large dinoflagellates dominated. *D. caudata* and *D. tripos* which had been almost absent before appeared at this time in concentrations of a few hundreds cells per litre.

The Aug. peak of *D. acuta* occurred in a community similar to the "red tide assemblage" which was observed blooming in the same ria, in 1990, at the bottom of the thermocline and the top of nutricline layer (Figueiras *et al.* 1994). The Nov peak coincided with cessation of upwelling and a period of downwelling. During this period, diatoms practically disappeared because the downward movement only allowed motile species to remain in the upper unstratified layer. It was thus inferred that the dense dinoflagellate assemblages resulted rather from concentration than from growth (Fraga *et al.* 1988; Figueiras *et al.* 1994).

Similar trends were observed in 1990 in Ria de Pontevedra (Reguera *et al.* 1993). *D. acuminata* occurred from early July to Oct; peak concentrations were recorded in late July and mid-Aug. *D. acuta* was detected in mid-July; then it rapidly grew until mid-Aug., and the peak cell concentration lasted until early Oct.

This work is critically important in showing that two cell maxima of *D. acuta* could occur during two distinct hydrographic regimes. According to the authors, dense populations during stratification of the water column resulted from *in situ* growth in the thermocline layer, while those observed during downwelling originated from physical accumulation of cells from surface populations of the adjacent shelf. This study also showed that *D. acuminata* and *D. acuta* occupied two different water layers.

During the summer of 1994, a bloom of *D. acuminata* was observed in the Southern Bight, Netherlands (Peperzak *et al.*, 1996). Before late July, water temperature was < 19°C, stratification of the water column was weak, and *D. acuminata* was below detection limit. Then, there was a heat wave; in early Aug., that increased surface temperature significantly, while surface salinity decreased and the water column became stratified. Inorganic nitrogen and phosphate concentrations were very low. *D. acuminata* concentration increased rapidly and peaked at $5x10^3$ cells l^{-1} at a station located 10 km offshore, but still in the river plume. Termination of the bloom started in mid-Aug. when water temperature decreased below 20°C, despite increased availability of inorganic nutrients. At the peak of the bloom, cell densities were correlated to density stratification, but not to salinity stratification.

These results add to the evidence that inorganic nutrient replenishment is not a direct actor promoting growth of photosynthetic *Dinophysis* species; they also stress the critical importance of increases in temperature and water column stratification.

At Wedge Point, Queen Charlotte Sound, New Zealand, during a chronic DSP problem, from mid-July 1994 to mid-December 1995, *Dinophysis acuta* occurred below detection limits or was present at a few tens of cells per litre during winter. When stratification of the water column established, cell concentrations of *D. acuta* increased to about 10^4 cells l^{-1} (Fig. 4); most cells accumulated in the thermocline layer.

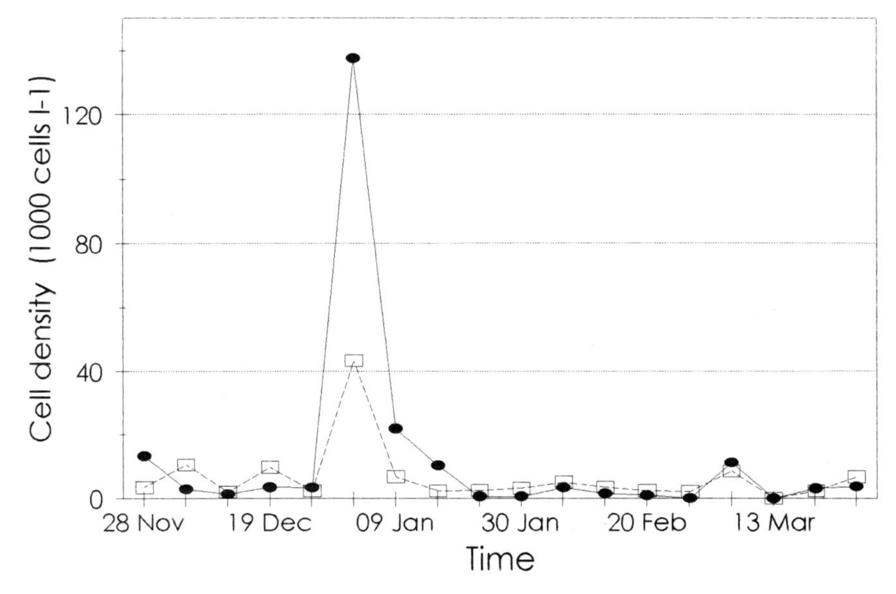

Figure 4. Cell concentration (10^3 cells l^{-1}) of *Dinophysis acuta*, from November 1994 to March 1995, at Wedge Point, Queen Charlotte Sound, New Zealand (drawn from unpublished data of L. MacKenzie; courtesy of the author)

On basis of these few studies, characteristics of *Dinophysis* bloom dynamics can be summarized as follows: (i) in winter, species were absent (truly absent or present below detectable level), or present at the concentration of a few cells per litre; (ii) although a few dense populations have been observed in early autumn, maximum cell densities typically occur in late spring and summer; (iii) most peak cell densities appeared as a bulge, even when samples were taken twice a week over several months; (iv) except for two cases, *Dinophysis* spp. thrived among a dinoflagellate "red tide assemblage" which they never dominated; (v) altogether, the total cell number of companion species during blooms of *Dinophysis* spp. was greater than that of the *Dinophysis* by at least two orders of magnitude; (vi) although few cases of dense populations occuring in the upper layer of a isothermal water column have been reported, most peak densities were related to the pycnocline of a stratified water column; (vi) no relationship between formation of dense populations and uptake of inorganic-nutrients has been established. Some research dedicated to other aspects of *Dinophysis* ecophysiology has provided complementary information of importance in understanding bloom dynamics: all bloom species but one are potentially photosynthetic (see section 4.3).

The most common feature in *Dinophysis* blooms appears to be the formation of a dense patchy population non-randomly distributed through the water column. Therefore, to understand the ecophysiological capabilities of *Dinophysis* spp. and how environmental conditions evolve during a bloom episode is roughly similar, *at this stage*, to answering: Where, when and how are these dense patchy populations generated?

3. Environmental factors affecting bloom dynamics

3.1 Temperature

Lassus and Marcaillou-Le Baut (1991) have summarized data on the areas and times in which potentially-toxic species have been recorded. All dense populations of *Dinophysis* spp. have occurred during warm periods. In contrast, the presence of active cells in winter is poorly documented: only a few cells or tens of cells per litre were recorded by Dahl and Yndestad (1985), MacKenzie (1992), Selina (1993), Maranda (1995) and Sidari *et al.* (1995).

The re-appearance of a species following its disappearance during unfavorable environmental conditions is a critical problem in phytoplankton ecology. Dormant stages are an important phase in the life cycle of many algal species, since they allow survival under conditions which would destroy the corresponding active vegetative stages. Accordingly, when *Dinophysis* spp. are absent from samples collected in winter, it is of critical importance to determine whether cells were too scarce to be detected by usual sampling protocols, or whether they were present in the form of dormant benthic cysts. Unfortunately, most field studies have covered only summer periods as part of a DSP monitoring program (Egmond *et al.* 1993). A significant change in sampling strategy is highly desirable for future surveys, which should involve time series samples taken from winter to the period of usual appearance of the *Dinophysis* species.

A question arising from the possible existence of dormant cysts is: at what temperature does excystment occur? At what temperature do cells start to grow, is also a relevent question if *Dinophysis* spp. remain active in the pelagic in winter. So far there is information only for *Dinophysis fortii*: cells not detected in winter have been observed as soon as the surface temperature exceeded 8°C (Ozaka, 1985), and Yoshimatsu *et al.* (1983) have shown that a significant increase in cell concentration was associated with temperature values of 13-22°C.

Many other workers have reported similar temperatures occuring during blooms of several potentially-toxic *Dinophysis* species, but the role of temperature has been more closely related to increased stratification of the water column than to a direct effect on the behaviour of *Dinophysis* spp. Hence, the life cycle of members of the genus is poorly understood.

3.2 Water column stratification and vertical distribution

3.2.1 Mixing or stability

In addition to the findings of Lassus *et al.* (1993) in Antifer Harbour and that of Reguera *et al.* (1995) in Galician rivers (see section 2), there are only a few reports indicating that a peak concentration of *Dinophysis* spp. occurred in conditions when the structure of the water column was unstable, or in a clear mixing regime. Sidari *et al.* (1995) first observed in the Gulf of Trieste, northern Adriatic, that the *Dinophysis* population (9 species, mostly *D. fortii* and *D. caudata*) was maximal at 7 or 9 m depth until mid-Sept.; then, most *Dinophysis* spp. were present in the surface water where their cell concentration peaked in late Sept.; in October and Nov, their concentration greatly diminished. These results could indicate that *Dinophysis* spp. populations had grown in two hydrodynamical regimes; unfortunately, no indication of the water column structure was recorded. In Oliveri-Tindari Bay, Tyrrhenian Sea, Giacobbe *et al.* (1995) also observed that maximal concentration of *D. sacculus* occurred in April, in surface water, while the water column was homogeneous both in temperature and salinity, and while inorganic nutrient concentrations were low. This suggests that there is a possibility that *Dinophysis* spp. population can grow rapidly in the absence of water column stratification. However the evidence is poor, because the studied area was really shallow (3 m depth). In contrast, members of the genus *Dinophysis* have been frequently observed to be concentrated at specific layers of the water column.

3.2.2 Vertical distribution

Norris and Berner (1970) observed that in the Gulf of Mexico *D. doryphora, D. hastata, D. pusilla* and *D. schuetti* were present from the surface to at least 110 m; while *D. paulseni* appeared to be present only at depths greater than 50 m and *D. uracantha* at depths greater than 100 m. In Mutsu Bay (Yokohama), Japan, Ozaka (1985) observed that *D. fortii* in April-Aug. were mostly concentrated in the 20-30 m layer, while they remained scarce in the layer above. A similar summer distribution was also reported for the coast of the Ibaraki Prefecture (Iwasaki and Kusano, 1985; Iwasaki, 1986).

More recently, Taylor *et al.* (1994) observed the successional trend of *D. acuminata* and *D. fortii* in a fjord of British Columbia, Canada, from May 1988 to Sept. 1990. It was clear that peak values coincided with the highest thermal stratification of the water column. In Big Glory Bay, New Zealand, MacKenzie (1991, 1992) recorded the cell concentration and relative number of couplets of *D. acuta* and *D. dens*. Throughout the sampling period (22 h), *D. acuta* and *D. dens* had a stratified vertical distribution: during the night, maximum cell concentration was located at 15 m depth; during the day and at dusk, it was situated at 10 m depth; peak value was recorded at $19^{h}00$. No indication was found of the causes of that change. In Bedford Basin, eastern Canada, Subba Rao *et al.* (1993) found that most dinoflagellates and diatoms present in the natural assemblage gathered at 10 m depth; *D. norvegica* constituted 59 to 88% of the standing stock. At 10 m depth, cell concentration of *D. norvegica* fell in the range $150\text{-}500 \times 10^{3}$ cells l^{-1}, whereas in the 15-25 m layer it remained $< 100 \times 10^{3}$ cells l^{-1}. In the 0-5 m upper layer, *D. norvegica* were scarce. Most of these these observations have been made during the course of research focused on topics other than bloom dynamics; hence, authors usually did not discuss the causes of the observed distribution.

During further research (Marcaillou-Le Baut *et al.*, 1993) on the bloom reported by Delmas *et al.* (1992; see section 2), a drogue positioned in the thermocline was followed for 10 days, at a time (late spring) when a *Dinophysis* spp. concentration of several thousands of cells per litre was likely to be found. Changes in *Dinophysis* spp. concentration showed two distinct phases: during the first seven days, the number of *Dinophysis* spp. in the subsurface layer did not exceed 100 cells.l^{-1}, while in the thermocline their number ranged between 100 and 1000 cells.l^{-1}. During the following four days, *Dinophysis* spp. concentration increased sharply (~ 3000 cell-l^{-1}) both in the thermocline and subsurface layers. No reason for such change and distribution has been suggested. During the course of the same program, Gentien *et al.* (1995) used an *in situ* particle-size analyser which allows direct determination of particle distribution spectra in 30 size classes ranging from 0.7 to 400 μm equivalent diameter. They showed the presence of a sharp thermocline between 18 and 20 m; particle load in the thermocline layer was 4-fold higher than in the 18-m above and 20-m below (Fig. 5); this increase was not related to any increase in chlorophyll *a* fluorescence, however. Accumulated particles 32-96 μm in diameter were dominant within that layer. Furthermore, species *Pyrocystis* greatly dominated the phytoplankton assemblage in the discontinuity layer of the genus *Dinophysis, Dissodium*, and(20 - 20.5 m). In contrast, no *Dinophysis spp.* were found in the layers 18-m above and 20-m below the thermocline. Profiles recorded in the Po River plume

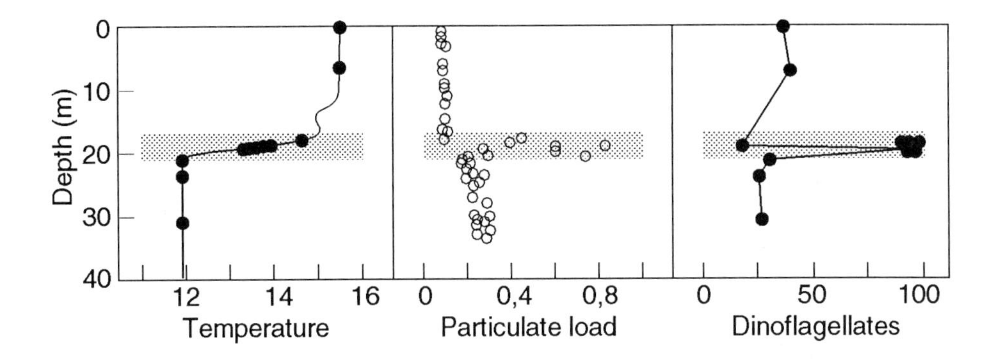

Figure 5. Vertical distribution of temperature (°C), particulate load (relative unit) and fractional cell concentration of dinoflagellates (%) off the « Pertuis d' Antioche », France (simplified and redrawn from Gentien *et al.*, 1995)

with the same particle analyser also demonstrated that there was a 5-fold particle accumulation in the thermocline layer. Most of these particles were over 96 μm; dinoflagellates including *Dinophysis spp.* were also present, however. In contrast, *Dinophysis spp.* were lacking in assemblages of the layers above and below the thermocline, although few other dinoflagellates such as *Ceratium fusus* and *Prorocentrum micans* were present there (Gentien *et al.* 1995). Furthermore, a peak value of 68×10^3 cells l^{-1} has been observed in the Bay of Seine, in a 0.5 m thick layer of the pycnocline, whereas the layers above and below both contained less than 10^3 cells l^{-1} (Gentien, pers. comm.).

3.2.3 Diel changes in the vertical distribution

Reports of active vertical migration for *Dinophysis* species are few. In addition, some of the published data are inconclusive.

The first research on *Dinophysis* vertical migration was carried out in a mesocosm, in order to avoid artefacts due to lateral dispersion. Brockman *et al.* (1977) isolated a body of water off Helgoland in a plastic tank (4-meter deep), and observed the algal species distribution for 28 days. No diel migration was observed. *Dinophysis acuminata* showed only changes in the vertical distribution: up to the 13th day, it occurred predominantly at the lower depth; from the 14th day onwards it reached a maximum at a depth of 1 m. Lassus *et al.* (1990b) used a 1.3 m-diameter polyethylene cylinder, with which they isolated water in Antifer Harbour, Normandy.

An improvement from the work of Brockmann *et al.* (1977) was the increase in the height of the cylinder (30 m). Results showed that changes in the *D. acuminat*a population occurred only within the 0-6 m upper layer; the center of concentration of the population was about at 3 m depth in the afternoon (18^h00), at 5.5 m depth about midnight, and again at 4 m in the morning (06^h00). A 2°C temperature gradient became established between 1 and 7 m depth. The authors infered that *D. acuminata* actively accumulated in the warm surface layer above the thermocline, and they calculated the active migration to be 1.1 m h^{-1}, a value similar to those published for other dinoflagellates. However, such experiments suffer from a potential "wall effect"; *Dinophysis* spp. have been reported to actively migrate upwards in bottles and to concentrate in dense populations, without any clear experimentally-proved relationship with stratification, phototropism or nutritional conditions (Maestrini *et al.* 1995). Nevertheless, in both these mesocosm experiments, the range of vertical migration was modest (a few meters), while aggregations of *Dinophysis spp.* cells in nature have been assumed to result from migration over several tens of meters.

In Vilaine Bay, French Atlantic coast, Durand Clément *et al.* (1988) observed that during the night only 24% of *Dinophysis* cells population were located in the 3 m upper layer. During the day, cells present in that layer represented 62% of the whole population in the water column. They therefore assumed that *Dinophysis spp.* migrated upwards at sunrise as a positive phototactic response, and then downwards before midday, as a negative response to saturating light intensity. This work suffers from no attention being paid to horizontal movements due to tidal currents, which are critical in the studied area, since the tide height is important there (> 5 m) and the water height at sampling stations ranged between 8 and 15 m.

In Ria de Pontevedra, Reguera *et al.* (1993) recorded the peak concentration of *D. acuminata* concentration in the 0-5 upper layer, and that of *D. acuta* between 10 and 15 m (see section 2). The vertical distribution of DSP in mussels suspended on ropes was quite homogeneous, which led the authors to assume that *Dinophysis* cells had migrated vertically through the water column. In another Galician ria nearby, the Ria de Vigo, Villarino *et al.* (1995) reported that *D. acuminata* showed a marked diel migration, in association with three red-tide forming species; the dinoflagellates *Ceratium furca* and *Scrippsiella trochoidea* and the flagellate *Eutreptiella* sp. One chloroplastic ciliate, *Mesodinium rubrum* also displayed the same pattern. At midday, these five organisms were near the surface: 92% of the cells present in the whole water column were gathered in the 0-1.25 m layer. Then they descended to about 10 m depth; during the night they were absent from the upper layer. In the following late morning, 56% of the five species were located at the surface, with only 28% at 5 m depth.

However, a few studies have concluded that *Dinophysis* spp. cells do not migrate within the water column. According to Ozaka (1985), Igarashi (1986) and Iwasaki (1986), *D. fortii* does not vertically migrate in the different Japanese embayments where they worked. Observation of *D. norvegica* made in summer, in the Gotland Deep, Baltic Sea, over three years, by Carpenter *et al.* (1995), mainly indicated that maximal abundance (40 - 150×10^3 cells l^{-1}) was found in the thermocline layer, while cell densities in the upper mixed layer were always low. However, some results recorded at two stations in 1992 conflicted with this: surface water temperature was about 17°C and was isothermal to about 20 m, below which there was a sharp thermocline; nevertheless, the maximum concentration of *D. norvegica* (~ 25 and 45×10^3 cells l^{-1}) was recorded at 15 m and 10 m depths, respectively. Carpenter *et al.* (1995) also concluded that no diel vertical migration occurred during the time of their investigation; absence of data along with a sufficient time series might have offset diel changes in the vertical distribution, however.

4. Ecophysiological mechanisms affecting bloom dynamics

It is generally clear that at the time of greatest cell densities, cells of *Dinophysis* can be concentrated in a layer of water which represents only a small fraction of the water column. Mechanisms responsible for such observed vertical distributions of *Dinophysis* cells could fundamentally be one of a combination of the following: (i) sinking of senescent cells and accumulation because the pycnocline acts as a barrier; (ii) active vertical migration; (iii) better growth due to decrease of stress from lower turbulence; (iv) better growth due to the accumulation of detrital material at the pycnocline which provides nutrients which meet *Dinophysis* spp.'s nutritional requirements; and (v) reduction or absence of grazing.

4.1 Sinking, active vertical migration

Sinking rate is an important ecological factor in understanding the succession of species. At present, there is no information on the sinking capability of *Dinophysis* spp., however.

4.2 Metabolic activity versus turbulence

One explanation for the presence of a dense populations in a small portion of the water column could be cell division. Since most high cell densities have been recorded under conditions of stratification (see section 3.2.2), one can assume that cell division in *Dinophysis* might be better in the pycnocline layer where turbulence is significantly reduced. Unfortunately, no *Dinophysis* species has been studied in the literature dedicated to the role of turbulence. Hence, once again, any opinion on whether or not these organisms divide better in deep calm layers could only be speculative.

4.3 Nutrition

Several of the main potentially-toxic *Dinophysis* species have been proved to be photosynthetic (Granéli *et al.* 1993, 1995; Subba Rao and Pan 1993; Berland *et al.* 1994). Notwithstanding, there is at present no available data on potential uptake of inorganic nutrients in any *Dinophysis* species. In contrast, Delmas *et al.* (1992) did not find any relationship between *Dinophysis* spp. growth and dissolved inorganic nutrient concentrations and concluded that riverborn nutrients from agricultural and domestic origins do not promote growth of the local *Dinophysis* spp. population in inshore waters. Similarly, peak cell values of *D.* cf. *acuminata* reported by Reguera *et al.* (1993) in the Galician Rià de Pontevedra were recorded during a period of low nutrients (*i.e.* < 2 μM N-NO_3), as did those of Giacobbe *et al.* (1995) in a shallow embayment of the Tyrrhenian Sea. That no relationship was found between the biomass of *Dinophysis spp.* and nutrient concentration conflicts with previous assumptions about the nutritional requirements of *Dinophysis* spp. (Menesguen *et al.* 1990). Moreover, in the Seine plume, peak cell densities of *Dinophysis* cf. *acuminata* have been observed to parallel increased nitrate concentration (Lassus *et al.* 1991). As always, correlations or lack thereof between ambient nutrient concentrations and cell abundance can be misleading, as nutrient measurements indicate what is remaining, not what was available earlier that could have promoted growth.

Recent findings suggest that most *Dinophysis* are mixotrophic. From electron-microscope observations, Hallegraeff and Lucas (1988) concluded that Australian coastal bloom-forming species such as *Dinophysis acuminata* and *D. fortii* are truly photosynthetic dinoflagellates, whereas oceanic species such as *D. hastata* and *D. exigua* are predominantly heterotrophic; these species showed ingested algal particles in food vacuoles. Similarly, Jacobson and Andersen (1994) observed that in the West Boothbay-Harbor area, 6%of *D. acuminata* and 36% of *D. norvegica* contained food vacuoles. Vacuoles were filled with transparent globules matching the appearance of the colourless species *D. rotundata*, which in turn feeds upon ciliates (Hansen 1991). Ishimaru *et al.* (1988) reported having obtained a culture of *D. fortii* and *D. acuminata* fed with the cryptomonad *Plagioselmis* sp.; they obtained 22 cells from a single one in three weeks. Although this was not yet a culture in the full sense, this result was important because it supported the possibility of phagotrophy. In addition, Granéli *et al.* (1993, 1995) who measured the light and dark carbon uptake rates of single isolated cells of several autotrophic *Ceratium*, and *Dinophysis acuminata*, *D. acuta* and *D. norvegica* incubated *in situ* in natural assemblages reported that specific biomass rates of carbon uptake in darkness for the *Dinophysis* species were significantly higher than for *Ceratium* species. Although *Dinophysis* species did not systematically show lower carbon fixation in light compared to *Ceratium* species, their positive dark carbon fixation suggested a mixotrophic mode of nutrition.

The absence of clear laboratory-based knowledge on nutrition of *Dinophysis* spp. has left field scientists with no indication about relevant variables to focus on. Up until now, current analyses carried out in *Dinophysis* research have mostly been those used for autotrophic phytoplankon; confused and/or conflicting results have been obtained. Since there is an assumption that mixotrophy is likely to be a common mode of nutrition in many *Dinophysis* species, attention should be directed towards their role as predators.

On the French Atlantic coast, D. Delmas, A. Herbland and S. Maestrini (unpublished data relating to Delmas *et al.*, 1992) have observed a reciprocal decrease in bacterial cell concentration when concentrations of *D. acuminata*, *D. acuta*, *D. rotundata* and *D. sacculus* increased between April to June 1990. From samples collected monthly between April - Sept. 1993 in Oliveri-Tindari Bay, Tyrrhenian Sea, Giacobbe *et al.* (1995) observed that at times of maximal concentration of *D. sacculus*, in April, concentrations of inorganic nutrients were reduced. Albeit, no obvious relationship was found between the two organisms, the authors reported that the concentration of synechoccoid cyanobacteria was maximal during the same period. They also observed that disappearance of *D. sacculus* in summer coincided with a sharp decrease in cyanobacteria concentration. In the Gulf of Riga, Balode and Purina (1996) have observed another relationship between a *Dinophysis* species and the presence of cyanobacteria: in late July, in surface water, there was an intense bloom of cyanobacteria (*Aphanizomenon flos-aquae* and *Nodularia spumigena*); *D. acuminata* was present at a concentration of 5×10^3 cells l^{-1}, representing 1% of the total phytoplankton biomass. One week later, *D. acuminata* concentration had increased to 67×10^3 cells l^{-1}, and its biomass represented 28% of the total algal biomass, while the standing stock and biomass of cyanobacteria greatly decreased. No direct trophic relationship between *D. acuminata* and the filamentous cyanobacteria *A. flos-aquae* and *N. spumigena* is likely. Nevertheless, dissolved organic substances released by these cyanobacteria might have promoted the growth of *D. acuminata* and/or some other unicellular cyanobacteria not taken into account may have been ingested as food. However, since the *Dinophysis* community usually represents a small fraction of the whole phytoplankton assemblage, any correlation between their density concentration and one environmental factor could be misleading.

Altogether, there is at present no evidence to support or to challenge the working hypothesis that there is better availability of dissolved nutrients and/or particulate food, in the layer of maximum concentration for the most abundant photosynthetic *Dinophysis* species.

4.4 Repelling the grazers: allelopathy and/or the "termite strategy"

Population strategies for life vary depending on the respective importance of growth rate and avoidance of losses. There is no available estimation of μ_{max} for any

Dinophysis species in order to discuss to which strategy they belong to. Nevertheless, calculations made from changes in natural assemblages (Table 1) showed that their generation time is likely to fall in the range of 2-3 days, thus indicating that they cannot compensate for the usual losses that a palatable species has to undergo from grazing. Therefore, avoiding losses from grazing is likely to be a critical capability of blooming members of the genus *Dinophysis*. This assumption being accepted, two distinct mechanisms, which are frequently confused, may protect *Dinophysis* spp. against grazers: allelopathy and delayed poisoning.

4.4.1 Allelopathy

Experiments involving *Dinophysis* species have been rare. In Turner and Anderson's (1983) experimental assemblages, *D. acuminata* was one of the dominant species; nevertheless, it was essentially ungrazed by the copedod *Acartia hudsonica* and the larvae of the polychaete *Polydora* sp. Carlsson *et al.* (1995) have studied the grazing of several copepods on *D. acuminata* in a naturally concentrated (40 - 70 μm) phytoplankton assemblage dominated by *D. acuminata, Ceratium fusus, C. furca* and *Leptocylindrus danicus.* Copepod nauplii were also present. From 3 to 5 copepods or copepodites of *Acartia clausi, Isias clavipes* and *Centropages typicus* were incubated together with the phytoplankton. During the first 24 hours, *D. acuminata* was lightly grazed by *I. clavipes* and *C. typicus;* it represented only 5 - 10%, of their total ingested carbon. Then they stopped feeding, and both copepods thrived well. In contrast, *A. clausi* did not avoid *D. acuminata* which represented 30% of its ingested carbon in one day, and most individuals died. In control conditions, with a diatom mixture as food, individuals of *A. clausi* remained healthy, with the females producing eggs. The authors concluded that okadaïc acid produced by *D. acuminata* was responsible for poisoning *A. clausi*, and they assumed that this substance is a potential grazer repellent in natural assemblages.

The results of Turner and Anderson (1983) and those of Carlsson *et al.* (1995) did not provide any evidence for the existence of a true allelopathic mechanism; i.e. the production of an inimical substance acting at some distance from the producing cell. At present, there is no indication whether or not okadaic acid and other *Dinophysis* toxins can be excreted by living *Dinophysis* cells. The question of their possible role as direct grazer repellents is therefore still fully open.

4.4.2 Poisoning the grazers: the "termite strategy"

Carlsson *et al.* (1995) have suggested the possible presence of another defense mechanism in *D. acuminata*: the delayed effect of the sequestered toxin. In order for toxicity to evolve as a defense mechanism, it would appear that the individual which produces intracellular toxic substances should not be eaten and killed; otherwise there is no apparent benefit for it. Nevertheless, it could be evolutionary beneficial for one population to loose some of its individuals in order to decrease the grazing pressure on other cells. In other words, it may be that *Dinophysis* populations could significantly diminish losses from grazing by following the "termite strategy": some individuals are sacrificed in order to destroy or to deter predators and thus to allow the main population to escape capture and death. To what extent such mechanisms could contribute to the dense populations reported in the recent literature is of course unclear, once again.

Table 1. Estimated net *in situ* division rates (div. day^{-1}) for species of the genus *Dinophysis*

Species	Div. rate	Area	Reference
Dinophysis spp.	0.25	French Atlantic coast (near La Rochelle)	Delmas *et al.* (1993)
D. acuminata *D. acuta* *D. norvegica*	0.52-0.73 0.36-0.45 0.25-.038	Skagerrak, Sweden, and French Atlantic coast (near La Rochelle)	Granéli *et al.*, (1995)
D. acuta	0.33	Bantry Bay, Ireland	McMahon *et al.* (1995)
D. acuminata	0.35	Antifer Harbour, northward from the Seine estuary, France	Berland *et al.*, (1994)
D. acuminata	0.78 0.96	Long Island Sound, U.S.A.	Chang and Carpenter, (1991)

5. Conclusion

Since they contribute a negligible fraction to the global carbon flux, species of the genus *Dinophysis* would have been of interest to only a few taxonomists, were it not that DSP and subsequent economic losses have been related to their presence in coastal phytoplankton assemblages, worldwide. Despite the recent great increase in research, little is known about how they thrive or survive in the pelagic. It is especially regrettable that we do not know (i) whether they have several life stages and how they face winter adverse conditions, (ii) exactly what they take from the sea in order to live and to grow, and (iii) to what extent dense vertically-patchy populations result from active or passive concentration or from growth, or from 1a mixture of both mechanisms. Although it is likely that many species are mixotrophic, whether they are mostly preyed upon and thus are top-down limited, or whether they are mostly predators and thus bottom-up limited (Verity and Smetacek 1996) is yet another open question. Short-duration occurrence and scarcity, absence of cultured strains, and sampling strategies devised mostly for DSP monitoring purposes have limited current knowledge. Some of these limitations on research were, and will remain, outside our control. Others can be managed, and should lead to significant advances.

6. Acknowledgements

I warmly thank authors who have provided unpublished data and manuscripts in press, colleagues who helped me spot and/or collect published material, and to Dr. Ian Jenkinson and Cathy Caudwell (ACRO, La Roche Canillac) for improving the English manuscript. French research reported here was supported by the "Programme National Efflorescences Algales Toxiques"

7. References

Anderson D.M. (1989). Toxic algal blooms and red tides: a global perspective. In: Okaichi, Anderson D.M., Nemoto (eds.) *Red tides: Biology, Environmental Science, and Toxicology*, Elsevier Sci. Publish., New York, pp. 11-16.

Arenovski A.L., Lim E.L., Caron D.A. (1995). Mixotrophic nanoplankton in oligotrophic surface waters of the Sargasso Sea may employ phagotrophy to obtain major nutrients. *J. Plankt. Res.* 17: 801-820.

Avaria S. (1979). Red tides off the coast of Chile. In: Taylor D.L., Seliger H.H. (eds.) *Toxic Dinoflagellate Blooms*, Elsevier Sci. Publish., New York, pp. 161-164.

Balode M., Purina I., (1996). Harmful phytoplankton in the Gulf of Riga (the Baltic Sea). In: Yasumoto T., Oshima Y., Fukuyo Y. (eds.) *Harmful and Toxic Algal Blooms*, IOC/UNESCO, Paris, pp. 69-72.

Belin C., (1993). Distribution of *Dinophysis* spp. and *Alexandrium minutum* along French coasts since 1984 and their DSP and PSP toxicity levels . In: Smayda T.J. and Shimizu Y. (eds.) *Toxic Phytoplankton Blooms in the Sea,* Elsevier Sci. Publish., New York, pp. 469-474.

Berland B.R., Maestrini S.Y., Bechemin C., Legrand C. (1994). Photosynthetic capacity of the toxic dinoflagellates *Dinophysis* cf. *acuminata* and *Dinophysis acuta. La Mer*, Tokyo 32: 107-117.

Bonin D.J., Maestrini S.Y., Leftley J.W. (1981). Some processes and physical factors that affect the ability of individual species of algae to compete for nutrient partition. In: Platt T. (ed.) *Physiological Bases of Phytoplankton Ecology, Can Bull. Fish. Aquat. Sci.* 210: 292-309.

Boraas M.E., Estep K.W., Johnson P.W., Sieburth J. McN. (1988). Phagotrophic phototrophs: the ecological significance of mixotrophy. *J. Protozool.* 35: 249-252.

Brockmann U.H., Eberlein K., Hosumbek P., Trageser H., Maier-Reimer E., Schöne H.K., Junge H.D. (1977). The development of a natural plankton population in an outdoor tank with nutrient-poor sea water. I. Phytoplankton succession. *Mar. Biol.* 43: 1-17.

Carlsson P., Granéli E., Finenko G., Maestrini S.Y. (1995). Copepod grazing on a phytoplankton community containing the toxic dinoflagellate *Dinophysis acuminata. J. Plankt. Res.* 17: 1925-1938.

Carpenter E.J., Janson S., Boje R., Pollehne F., Chang J. (1995). The dinoflagellate

Dinophysis norvegica: biological and ecological observations in the Baltic Sea. *Eur. J. Phycol.* 30: 1-9.

Chang J., Carpenter E.J. (1991). Species-specific phytoplankton growth rates via diel DNA synthesis cycles. V. Application to natural populations in Long Island Sound. *Mar. Ecol. Progr. Ser.* 78: 115-122.

Dahl E., Yndestad M., (1985). Diarrhetic Shellfish Poisoning (DSP) in Norway in the autumn 1984 related to occurrence of *Dinophysis* spp.. In: Anderson D.M., White A.W., Baden D.G. (eds.) *Toxic Dinoflagellates*, Elsevier Sci. Publish., New York, pp. 495-500.

Dahl E., Rogstad A., Aune T., Hormazabal V., Underdal B., (1995). Toxicity of mussels related to occurrence of *Dinophysis* species. In: Lassus P., Arzul G., Erard-Le Denn E., Gentien P., Marcaillou-Le Baut C. (eds.) *Harmful Marine Algal Blooms,* Lavoisier, Paris, pp. 783-788.

Dahl E., Aune T., Aase B., (1996). Reddish water due to mass occurrence of *Dinophysis* spp. In: Yasumoto T., Oshima Y., Fukuyo Y. (eds.) *Harmful and Toxic Algal Blooms*, IOC/UNESCO, Paris, pp. 265-267.

Delmas D., Herbland A., Maestrini S.Y. (1992). Environmental conditions which lead to increase in cell density of the toxic dinoflagellates *Dinophysis* spp. in nutrient-rich and nutrient-poor waters of the French Atlantic coast. *Mar. Ecol. Prog. Ser.* 89: 53-61.

Delmas D., Herbland A., Maestrini S.Y. (1993). Do *Dinophysis* spp. come from the "open sea" along the French Atlantic coast?. In: Smayda T.J., Shimizu Y (eds.) *Toxic Phytoplankton Blooms in the Sea,* Elsevier Sci. Publ., New York, pp.489-494.

Durand Clément M., Clément J.C., Moreau A., Jeanne N., Puiseux-Dao S. (1988). New ecological and ultrastructural data on the dinoflagellate *Dinophysis* sp. from the French coast. *Mar. Biol.* 97: 37-44.

Egmond van H.P., Aune T., Lassus P., Speijers G.P.A., Waldock M. (1993). Paralytic and diarrhetic shellfish poisons: occurrence in Europe, toxicity, analysis and regulation. *Jour. nat. Toxins* 2: 41-83.

Ehrenberg C.G. (1840). Uber jetzt wirklich noch zahlreich lebende thier-arten der kreideformation der Erde. *Verh. preuss. Akad. Wiss. (Berl.)* 1839: 152-159.

Figueiras F.G., Jones K.J., Mosquera A.M., Alvarez-Salgado A.M., Edwards A., MacDougall N. (1994). Red tide assemblage formation in an estuarine upwelling ecosystem: Rio de Vigo. *J. Plankt. Res.* 16: 857-878.

Fraga S., Anderson D.M., Bravo I., Reguera B., Steidinger K.A., Yentsch C.M. (1988). Influence of upwelling relaxation on dinoflagellates and shellfish toxicity in Ria de Vigo, Spain. *Estuar. Coast. Shelf. Sci.* 27: 349-361.

Franks P.J.S. (1992). Sink or swim: accumulation of biomass at fronts. *Mar. Ecol. Prog. Ser.* 82: 1-12.

Franks P.J.S. (1995). Thin layers of phytoplankton: a model of formation by near-inertial wave shear. *Deep-Sea Res.* 42: 75-91.

Gentien P., Lunven M., Lehaître M., Duvent J.L. (1995). *In-situ* depth profiling of particles sizes. *Deep-Sea Res*. 42: 1297-1312.

Giacobbe M.G., Oliva F., La Ferla R., Puglisi A., Crisafi E., Maimone G. (1995). Potentially toxic dinoflagellates in Mediterranean waters (Sicily) and related hydrobiological conditions. *Aquat. Microb. Ecol.* 9: 63-68.

Granéli E., Anderson D.M., Maestrini S.Y., Paasche E., (1993). Light and dark carbon fixation by the marine dinoflagellate genera *Dinophysis* and *Ceratium.* In: Li W.K.W., Maestrini S.Y. (eds.) *Measurement of primary production from the molecular to the global scale*, ICES, Copenhagen, pp. 274.

Granéli E., Anderson D.M., Carlsson P., Finenko G., Maestrini S.Y., Sampayo M.A. de M., Smayda T.J. (1995). Nutrition, growth rate and sensibility to grazing for the dinoflagellates *Dinophysis acuminata, D. acuta* and *D. norvegica. La Mer*, Tokyo 33: 149-156.

Hallegraeff G.M. (1993). A review of harmful algal blooms and their apparent global increase. *Phycologia* 32: 79-99.

Hallegraeff G.M., Lucas I.A.N. (1988). The marine dinoflagellate genus *Dinophysis* (Dinophyceae): photosynthetic, neritic and non-photosynthetic, oceanic species. *Phycologia* 27: 25-42.

Hansen P.J. (1991). *Dinophysis* - a planktonic dinoflagellate genus which can act both as a prey and a predator of a ciliate. *Mar. Ecol. Prog. Ser.* 69: 201-204.

Igarashi T. (1986). Occurrence of *Dinophysis fortii*, a dinoflagellate responsible for diarrhetic shellfish poisoning at Kesennuma Bay. *Bull. Tohoku Reg. Fish. Res. Lab.* 18: 137-111.

Ishimaru T., Inoue H., Fukuyo Y., Ogata T., Kodama M. (1988). Cultures of *Dinophysis fortii* and *D. acuminata* with the Cryptomonad, *Plagioselmis* sp. In: Aibara K., Kumagai S., Ohtsubo K., Yoshizawa T. (eds.) *"Mycotoxins and Phycotoxins"*, Jap. Ass. Mycotoxicol., Tokyo, pp. 19-20.

Iwasaki J. (1986). The mechanism of mass occurrence of *Dinophysis fortii* along the coast of Ibaraki prefecture. *Bull. Tohoku Reg. Fish. Res. Lab.* 48: 125-136.

Iwasaki J., Kusano K. (1985). Kashima Nada (in Japanese) . In: Fukuyo Y. (ed.) *Toxic Dinoflagellates - Implication in Shellfish Poisoning. Bull. Jap. Assoc. Sci. Fish., 56*, pp. 82-97.

Jacobson D.M., Andersen R.A. (1994). The discovery of mixotrophy in photosynthetic species of *Dinophysis* (Dinophyceae): light and electron microscopial observations of food vacuoles in *Dinophysis acuminata, D. norvegica* and two heterotrphic dinophysoid dinoflagellates. *Phycologia* 33: 97-110.

Jacques G., Sournia A. (1978/1979). Les "eaux rouges" dues au phytoplankton en Méditerranée. *Vie et Milieu* 28-29: 175-187.

Lassus P, Maggi P, Proniewski F, Truquet P, Bardouil M (1990a). Le maximum saisonnier de *Dinophysis cf acuminata* à Antifer (Normandie, France). Ifremer/DERO-90-03-MR, Nantes, pp. 1-16

Lassus P., Proniewski F., Pigeon C., Veret L., Le Dean L., Bardouil M., Truquet P. (1990b). The diurnal vertical migrations of *Dinophysis acuminata* in an outdoor tank at Antifer (Normandy, France). *Aquat. Living Res.* 3: 143-145.

Lassus P., Herbland A., Le Baut C. (1991). *Dinophysis* blooms and toxic effects along the French coast. *World Aquacult.* 22: 49-54.

Lassus P., Marcaillou-Le Baut C. (1991). Le genre *Dinophysis* (Dinophycées). In: Sournia A., Belin C., Berland B., Erard-Le Denn E., Gentien P., Grzebyk D., Marcaillou-Le Baut C., Lassus P., Partensky F. (eds.) *Le phytoplancton nuisible des Côtes de France - De la biologie à la prévention,* Ifremer-SDP, Plouzané, pp. 11-61.

Lassus P., Proniewski F., Maggi P., Truquet P., Bardouil M., (1993). Wind-induced toxic blooms of *Dinophysis* cf. *acuminata* in the Antifer area (France). In: Smayda T.J., Shimizu (eds.) *Toxic Phytoplankton Blooms in the Sea",* Elsevier Sci. Publish., New York, pp. 519-523.

MacKenzie L. (1991). Toxic and noxious phytoplankton in Big Glory Bay, Stewart Island, New Zealand. *J. Appl. Phycol.* 3: 19-34.

MacKenzie L. (1992). Does *Dinophysis* (Dinophyceae) have a sexual life cycle? *J. Phycol.* 28: 399-406.

Maestrini S.Y., Berland B.R., Grzebyk D., Spano A.-M. (1995). *Dinophysis* spp. cells concentrated from nature for experimental purposes, using size fractionation and reverse migration. *Aquat. Microb. Ecol.* 9: 177-182.

Maranda L., (1995). Population studies of *Dinophysis* spp. in a northern temperate coastal embayment. In: Lassus P., Arzul G., Erard-Le Denn E., Gentien P., Marcaillou-Le Baut C. (eds.) *Harmful Marine Algal Blooms,* Lavoisier, Paris, pp. 609-613.

Marcaillou-Lebaut C., Delmas D., Herbland A., Gentien P., Maestrini S.Y., Lassus P., Masselin P. (1993). Cinétique de contamination de moules *Mytilus edulis* exposées à une population naturelle de Dinoflagellés *Dinophysis* spp. *C.R. Acad. Sci. Paris, Sci. Vie* 316: 1274-1276.

McMahon T., Nixon E., Silke J. (1995). Ireland: extended "toxic season" in 1994. *Harmful Algae News* 10/11: 6.

Menesguen A., Lassus P., de Cremoux R., Boutibonnes L. (1990). Modelling *Dinophysis* blooms: a first approach. In: Granéli E., Sundström B., Edler L., Anderson D.M. (eds.) *Toxic Marine Phytoplankton*, Elsevier Sci. Publish; New York, pp. 195-214.

Neuer S., Cowles T.J. (1995). Comparative size-specific grazing rates in field populations of ciliates and dinoflagellates. *Mar. Ecol. Prog. Ser.* 125: 259-267.

Norris E.R., Berner L.D. Jr. (1970). Thecal morphology of selected species of *Dinophysis* (Dinoflagellata) from the Gulf of Mexico. *Contrib. Mar. Sci.* 15: 145-192.

Ozaka Y. (1985). Mutsu Bay (in Japanese). In: Fukuyo Y. (ed.) *Toxic Dinoflagellates - Implication in Shellfish Poisoning. Bull. Jap.. Assoc.. Sci.. Fish.,* 56: 59-70.

Peperzak L., Snoeijer G.J., Dijkema R., Gieskes W.W.C., Joordens J., Peeters J.C.H.,

Schol C., Vrieling E.G., Zevenboom W. (1996). Development of a *Dinophysis acuminata* bloom in the River Rine Plume (North Sea). In: Yasumoto T., Oshima Y., Fukuyo Y. (eds.) *Harmful and Toxic Algal Blooms*, IOC/UNESCO, Paris, pp. 273-276.

Reguera B, Bravo I, Fraga S. (1990). Distribution of *Dinophysis acuta* at the time of a DSP outbreak in the rias of Pontevedra and Vigo (Galicia, NW Spain), ICES 1990, C.M. 1990/L: pp. 14-12.

Reguera B., Bravo I., Fraga S. (1995). Autoecology and some life history stages of *Dinophysis acuta* Ehrenberg. *J. Plankt. Res.* 17: 999-1015.

Reguera B., Bravo I., Marcaillou-Le Baut C., Masselin P., Fernandez M.L., Miguez A., Martinez A., (1993). Monitoring of *Dinophysis* spp. and vertical distribution of okadaic acid on mussel rafts in Ria de Pontevedra (NW Spain). In: Smayda T.J., Shimizu Y. (eds.) *Toxic Phytoplankton Blooms in the Sea*, Elsevier Sci. Publish., Amsterdam, pp. 553-558.

Schnepf E., Elbrächter M. (1992). Nutritional strategies in Dinoflagellates. A review with emphasis on cell biological aspects. *Europ. J. Protistol.* 28: 3-24.

Selina M.S. (1993). Distribution of potential toxic dinoflagellates *Dinophysis* (Dinophyta) in Vostok Bay, Sea of Japon. *Russ. J. Mar. Biol.* 19: 299-302..

Sidari L., Cok S., Cabrini M., Tubaro A., Honsell G. (1995). Temporal distribution of toxic phytoplankton in the Gulf of Trieste (northern Adriatic Sea) in 1991 and 1992. In: Lassus P., Arzul G., Erard-Le Denn E., Gentien P., Marcaillou-Le Baut C. (eds.) *Harmful Marine Algal Blooms,* Lavoisier, Paris, pp. 231-236.

Smayda T.J., (1990). Novel and nuisance phytoplankton blooms in the sea : evidence for a global epidemic. In: Granéli E., Sundström B., Edler L., Anderson D.M. (eds.) *Toxic Marine Phytoplankton*, Elsevier Sci. Publish., New York, pp. 29-40.

Sournia A. (1986). *Atlas du phytoplancton marin. Vol. I : Introduction, Cyanophycées, Dictyochophycées, Dinophycées et Raphidophycées*, Edit. du CNRS, Paris, 219 p.

Subba Rao D.V., Pan Y. (1993). Photosynthetic characteristics of *Dinophysis norvegica* Claparede and Lachmann, a red-tide dinoflagellate. *J. Plankt. Res.* 15: 965-976.

Subba Rao D.V., Pan Y., Zitko V., Bugden G., MacKeigan K. (1993). Diarrhetic shellfish poisoning (DSP) associated with a subsurface bloom of *Dinophysis norvegica* in Bedford Basin, eastern Canada. *Mar. Ecol. Prog. Ser.* 97: 117-126.

Taylor F.J.R., Haigh R., Sutherland T.F. (1994). Phytoplankton ecology of Sechelt Inlet, a fjord system on the British Columbia coast. II. Potentially harmful species. *Mar. Ecol. Prog. Ser.* 103: 151-164.

Turner J.T., Anderson D.M. (1983). Zooplankton grazing during dinoflagellate blooms in a Cape Cod embayment, with observations of predation upon tintinnids by copepods. *P.S.Z.N.I.: Mar. Ecol. Prog. Ser.* 4: 359-374.

Verity P.G., Smetacek V. (1996). Organism life cycles, predation, and the structure of marine pelagic ecosystems. *Mar. Ecol. Prog. Ser.* 130: 277-293.

Villarino M.L., Figueiras F.G., Jones K.J., Alvarez-Salgado X.A., Richard J., Edwards

A. (1995). Evidence of *in situ* diel vertical migration of a red-tide microplankton species in Ria de Vigo (NW Spain). *Mar. Biol.* 123: 607-617.
Yasumoto T., Oshima Y., Yamaguchi M. 1978). Occurrence of a new type of shellfish poisoning in the Tohoku District. *Bull. Jap. Soc. Scient. Fish.* 44: 1249-1255.
Yasumoto T., Oshima Y., Sugawara W., Fukuyo Y., Oguri H., Igarashi T., Fujita N. (1980). Identification of *Dinophysis fortii* as the causative organism of diarrhetic shellfish poisoning. *Bull. Jap. Soc. Scient. Fish.* 46: 1405-1411.
Yoshimatsu S., Ono C., Ohkawa T. (1983). Occurrence of *Dinophysis fortii* Pavillard (Dinophyceae), Kagawa Prefectural Region, Inland Sea of Seto (in Japanese). *Sci. Rep. Kagawa Prefec. Fish. Exp. Sta.* 20: 15-21.

Bloom Dynamics and Physiology of Domoic-Acid-Producing *Pseudo-nitzschia* Species

Stephen S. Bates[1], David L. Garrison[2], and Rita A. Horner[3]

[1]Fisheries and Oceans Canada, Gulf Fisheries Centre, P.O. Box 5030, Moncton, NB, Canada E1C 9B6
[2]Institute of Marine Sciences, University of California Santa Cruz, CA, USA 95064
[3]School of Oceanography, Box 357940, University of Washington, Seattle, WA, USA 98195-9740

1 Introduction

In late November to December 1987, Canadians awoke to headlines such as "Mystery Toxin Taints Mussels", "Killer Shellfish Algae Found in River", and "Fatal Mussel Toxin Found". After an unprecedented 104 h of detective work, the neurotoxin domoic acid (DA) was identified as the contaminating agent in blue mussels (*Mytilus edulis*) from Cardigan Bay, eastern Prince Edward Island (PEI), Canada (Wright *et al.* 1989). This toxin caused at least 107 illnesses, killed at least three elderly people (Perl *et al.* 1990; Todd 1993), and temporarily devastated the molluscan shellfish aquaculture industry (Addison and Stewart 1989; Wessells *et al.* 1995). Symptoms included abdominal cramps, vomiting, and neurologic responses involving disorientation and memory loss that could persist indefinitely. Due to the latter, the term Amnesic Shellfish Poisoning (ASP) was given to this clinical syndrome. Domoic Acid Poisoning (DAP) is also sometimes used for this illness since shellfish are not always a vector.

The identification of DA as a toxin was at first treated with skepticism. This water-soluble tricarboxylic amino acid with a molecular weight of 311 (Fig. 1) was known as a folk medicine to treat intestinal pinworm infestations in young children in Japan (Takemoto and Daigo 1958). However, an order of magnitude greater dose was consumed in the eastern Canadian toxic episode, and those most affected were elderly or infirm (Perl *et al.* 1990). As an analog of glutamate, an excitatory neurotransmitter, DA binds to the kainate type of glutamate receptors, but with a binding capacity three times greater and 20 times more powerful than kainic acid (Teitelbaum *et al.* 1990). In the presence of endogenous glutamate, DA causes massive depolarization of the neurons, with a subsequent increase in cellular Ca^{2+}, leading to neuronal swelling and death. These nerve cells, located in the hippocampus, are associated with memory retention, hence the memory loss characteristic of ASP.

This overview focuses on the distribution and bloom dynamics of pennate diatoms of the genus *Pseudo-nitzschia*, some (but not all) of which produce DA. A review of the physiology of DA production, based on laboratory studies, is given by Bates (this volume). Laboratory research helps to interpret field studies of *Pseudo-nitzschia* blooms, but there is still a paucity of field data. Nevertheless, data sets are becoming available from eastern Canada, western North America, and the Gulf of Mexico. This permits a comparison of commonalities and differences among these events, in order to understand mechanisms of bloom formation, and eventually, to be able to predict their occurrence. Indeed, information gathered to date has enabled monitoring programs that now protect consumers of molluscan shellfish in an increasing number of countries.

NATO ASI Series, Vol. G 41
Physiological Ecology of Harmful Algal Blooms
Edited by D. M. Anderson, A. D. Cembella and G. M. Hallegraeff

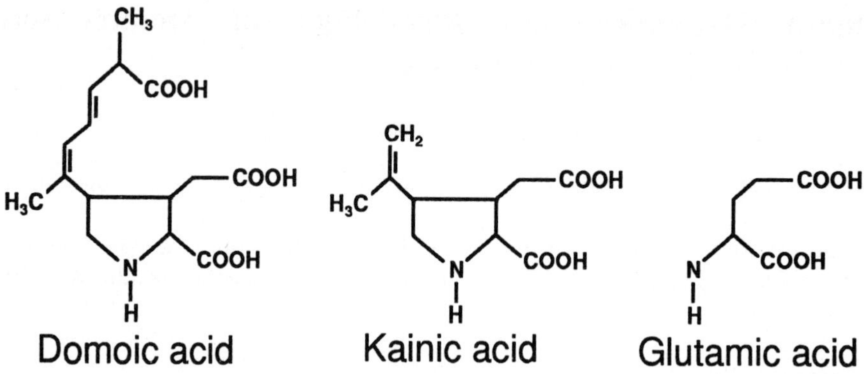

Domoic acid Kainic acid Glutamic acid

Fig. 1. Structure of domoic acid and its analogues.

1.1 Domoic-Acid-Producing Algae

Domoic acid was first isolated from the rhodophycean macroalga *Chondria armata* in Japan (e.g., Maeda *et al.* 1986), and is named after the Japanese word for this seaweed, "domoi". It was later found in the rhodophytes *Alsidium corallinum*, from the east coast of Sicily (Impellizzeri *et al.* 1975), and *Chondria baileyana*, from southern Nova Scotia and PEI, Canada (Laycock *et al.* 1989). It has since also been found in *Amansia glomerata*, *Digenea simplex*, and *Vidalia obtusiloba*, all belonging to the family Rhodomelaecae (Sato *et al.* 1996). The source of DA in the 1987 episode on PEI was the pennate diatom, *Pseudo-nitzschia multiseries* (Subba Rao *et al.* 1988a; Bates *et al.* 1989). This was the first known instance of a diatom producing a phycotoxin. The nomenclature of the genus *Pseudo-nitzschia*, once in a state of flux, has now been clarified: the DA-producing diatom was originally reported as *Nitzschia pungens* forma *multiseries*, then as *Pseudonitzschia pungens* f. *multiseries*, *Pseudo-nitzschia pungens* f. *multiseries*, and finally as *Pseudo-nitzschia multiseries* (Hasle 1965; 1994; Hasle *et al.* 1996). Based on morphological, physiological, and genetic features, it has been raised to the rank of species, to distinguish it from the nominate form, *P. pungens* f. *pungens*, now named *P. pungens* (Hasle 1995). One of the major distinctions between diatoms of the genera *Nitzschia* and *Pseudo-nitzschia* is that the latter form stepped colonies, i.e., chains of cells with overlapping tips, when seen in girdle view (Hasle 1994).

The 1987 DA episode generated an awareness of this toxin and resulted in its discovery in several other locations and phytoplankton species around the world (Table 1). To date, at least eight species of diatoms have been shown to produce DA in culture (Table 2). Curiously, all are pennate diatoms, and all but one belong to the genus *Pseudo-nitzschia*. It would not be surprising, however, if this toxin were to be found in numerous other microorganisms, including bacteria and other phytoplankton genera. Indeed, the source of DA that contaminated sea scallops on Georges/Browns Bank, Nova Scotia in 1995, cultured blue mussels in Newfoundland in 1994, and razor clams and Dungeness crabs in Oregon, Washington, and British Columbia since 1991, has not been determined. The taxonomic identity of other possible producers, such as *P. turgidula* (Rhodes *et al.* 1996), must be confirmed.

Until recently, it was believed that *P. pungens* was non-toxic, as shown in reports from Atlantic Canada (e.g., Smith *et al.* 1990b; Bates *et al.* 1993b), the Atlantic

Table 1. Locations around the world where species of *Pseudo-nitzschia* have been shown to produce domoic acid.

Geographic Area	Species Name	Reference
Prince Edward Island, Canada	*P. multiseries*	Bates *et al.* 1989; Smith *et al.* 1990a; b
Massachusetts Bay, MA, USA		Villareal *et al.* 1994
Narragansett Bay, RI, USA		Hargraves *et al.* 1993
Galveston Bay, TX, USA		Fryxell *et al.* 1990; Reap 1991; Dickey *et al.* 1992; Villac 1996
Monterey Bay, CA, USA		Villac *et al.* 1993b; Villac 1996
Willapa Bay, WA, USA		Sayce and Horner 1996
Hood Canal, WA, USA		Horner *et al.* 1996
Dutch Wadden Sea		Vrieling *et al.* 1996
Ofunato Bay, Japan		Kotaki *et al.* 1996
Jinhae Bay, Korea		Lee and Baik 1995
Bay of Fundy, NB, Canada	*P. pseudodelicatissima*	Martin *et al.* 1990; 1993
Prince Edward Island, Canada	*P. delicatissima*	Smith *et al.* 1990b
Denmark	*P. seriata*	Lundholm *et al.* 1994
Galicia, Spain	*P. australis*	Míguez *et al.* 1996
Monterey Bay, CA, USA		Buck *et al.* 1992; Fritz *et al.* 1992; Garrison *et al.* 1992; Villac *et al.* 1993b; Walz *et al.* 1994
Ilwaco, WA, USA		Villac *et al.* 1993b
Coos Bay, OR, USA		Villac *et al.* 1993b
New Zealand		Rhodes *et al.* 1996
Marlborough Sd., New Zealand	*P. pungens*	Rhodes *et al.* 1996
Bay of Plenty, New Zealand	*P. fraudulenta*	Rhodes *et al.* 1997

Table 2. Diatoms shown to produce domoic acid in culture, although some clones are non-toxic.

Species Name	Reference
Amphora coffeaeformis	Shimizu *et al.* 1989; Maranda *et al.* 1990
Pseudo-nitzschia multiseries	Subba Rao *et al.* 1988a; 1990; Bates *et al.* 1989; 1991; 1993b; 1995; 1996; Fryxell *et al.* 1990; Reap 1991; Dickey *et al.* 1992; Douglas and Bates 1992; Wohlgeschaffen *et al.* 1992; Douglas *et al.* 1993; Lewis *et al.* 1993; Villac *et al.* 1993a; b; Villareal *et al.* 1994; Wang *et al.* 1993; Hargraves *et al.* 1993; Whyte *et al.* 1995a; b; Pan *et al.* 1996a; b; c; Vrieling *et al.* 1996; Villac 1996
P. pseudodelicatissima	Martin *et al.* 1990
P. delicatissima	Smith *et al.* 1990b; Rhodes *et al.* 1997; Lundholm *et al.* 1997
P. australis	Garrison *et al.* 1992; Villac *et al.* 1993b; Rhodes *et al.* 1996
P. seriata	Lundholm *et al.* 1994
P. fraudulenta	Rhodes *et al.* 1997
P. pungens	Rhodes *et al.* 1996; G.J. Doucette; R.A. Horner; C.A. Scholin; V.L. Trainer; J.N.C. White, pers. commun.

coast of the USA (Villareal *et al.* 1994; Wang *et al.* 1993), the Gulf of Mexico (Villac *et al.* 1993b), Monterey Bay (Villac *et* al. 1993b), Europe (Lundholm *et al.* 1994; Vrieling *et al.* 1996), and New Zealand (Mackenzie *et al.* 1993). However, there is now a published report of DA production by a *P. pungens* isolate from one location in New Zealand, although isolates from other sites were not toxic (Rhodes *et al.* 1996). Several additional cases of toxic *P. pungens* have now been reported: four clones from Monterey Bay, CA, collected during two separate years (G.J. Doucette and C.A. Scholin, unpubl.); one isolate from Hood Canal, WA, and another from open coastal waters of WA (R.A. Horner and V.L. Trainer, unpubl.). Although the cellular DA levels are low in these examples (generally < 0.1 pg $cell^{-1}$) relative to *P. multiseries* (cf. Table 3), certain molluscan shellfish may still become contaminated over time.

A similar contradictory finding regarding toxicity was reported for *P. seriata* (Lundholm *et al.* 1994), which was previously shown to be non-toxic in culture (Bates *et al.* 1989). Likewise, there are reports of non-toxic strains of *P. multiseries* (Villareal *et al.* 1994), *P. pseudodelicatissima* (Reap 1991; Villareal *et al.* 1994; Lundholm *et al.* 1994; Hallegraeff 1994; Walz *et al.* 1994), *P. australis* (Villac *et al.* 1993b), and *P. delicatissima* (Villac *et al.* 1993b; Lundholm *et al.* 1994). In contrast to Shimizu *et al.* (1989) and Maranda *et al.* (1990), *Amphora coffeaeformis* (strain CCMP127; Bates *et al.* 1989), and another *A. coffeaeformis* strain isolated from Esquimalt Lagoon on southern Vancouver Island, Canada, failed to produce DA (L.M. Brown; R.F. Addison, unpubl.).

These findings suggest a) genetic variability among strains of the same *Pseudo-nitzschia* and *Amphora* species from different geographic locations; b) differences in factors controlling DA production (e.g., nutrients, bacteria); c) an incomplete study of growth conditions conducive for toxin production; d) mis-identification of the organism, given that species of the genus *Pseudo-nitzschia* are difficult to distinguish morph-ologically; or e) mis-interpretation of analytical results. To avoid the latter, it is essential that the identity of the toxin be confirmed, e.g., by tandem mass spectrometry, in reporting any new producer of DA. The analytical technique used for measuring cellular DA should be sensitive, and its limit of detection should be reported. A rigorous examination of growth conditions (see Bates, this volume) must also be carried out before the ability for toxin production can be ruled out with confidence. Finally, it is essential that the species identity be confirmed by a taxonomic authority. Many countries are beginning to expand their monitoring programs, at great expense, to include *Pseudo-nitzschia* spp. and DA. It is therefore critical to be certain about the taxonomic identity of any new toxigenic species and about the chemical identity DA.

The cosmopolitan distribution of toxigenic *Pseudo-nitzschia* species along our coasts (Hasle 1965; Hasle *et al.* 1996) is cause for concern about the safety of the natural and aquaculture harvest of molluscan shellfish that may accumulate DA (Addison and Stewart 1989). Food chain transfer has been demonstrated, as pelicans (*Pelecanus occidentalis*) and cormorants (*Phalacrocorax penicillatus*), and their food source, anchovies (*Engraulis mordax*), have been affected (Fritz *et al.* 1992; Work *et al.* 1993; Wekell *et al.* 1994; McGinness *et al.* 1995), as have Dungeness crabs (*Cancer magister*) and razor clams (*Siliqua patula*) (Horner *et al.* 1993; Wekell *et al.* 1994). Indeed, the number of countries that report problems with this diatom genus is growing by the year (Table 1). An understanding of factors that control the physiology of DA production by *Pseudo-nitzschia* species (see Bates, this volume) and its bloom dynamics is therefore essential.

One of the more difficult tasks with field populations is to be able to discriminate among the various *Pseudo-nitzschia* species, which are morphologically similar and tend to be confused with one another (Hasle *et al.* 1996). Aside from its ability to produce DA, at one time *P. multiseries* could only be distinguished from the non-toxic *P. pungens* by the number of rows of poroids on the silicon frustule, as seen by electron microscopy (e.g., Villac *et al.* 1993a). The two species can now be discriminated by an immunofluorescence assay (Bates *et al.* 1993a), lectin-binding assays (Fritz 1992; K.E. Pauley, unpubl.), and by differences in nucleic acid sequences (Scholin *et al.* 1994; 1996; Douglas *et al.* 1994; Manhart *et al.* 1995; Miller and Scholin, 1996; Scholin, this volume). Because of the species' morphological similarity, DNA probes developed against *P. australis*, *P. delicatissima*, *P. fraudulenta*, *P. americana*, and *P. heimii* (Scholin *et al.* 1994; 1996) are invaluable for identification purposes. Molecular probes for identifying *P. seriata* are thus far still lacking. An alternative approach for determining if one species is different from or the same as another is by carrying out clonal breeding experiments in culture (see Section 4).

2 Case Histories of *Pseudo-nitzschia* Blooms

Pseudo-nitzschia blooms differ from, e.g., certain dinoflagellate blooms, in that high concentrations (at least 100,000 cells L^{-1}) of the DA-producing species must be present in order for shellfish to become contaminated to a level that closes harvesting (20 μg DA g^{-1} wet weight of tissue). A complicating factor is that the morphologically-similar *P. pungens* is generally non-toxic, thus cell numbers alone will not indicate the certainty of DA intoxication; hence the need for methods to distinguish the species (see above). There are no reports of *Pseudo-nitzschia* blooms causing problems independent of their toxicity. With the exception of the dense bloom of *P. multiseries* in 1987 in PEI, there appears to be nothing unusual about the occurrence of *Pseudo-nitzschia* blooms.

2.1 Prince Edward Island, Canada

The first-reported site of a DA-producing *P. multiseries* bloom was in estuaries of Cardigan Bay, eastern PEI (Fig. 2), where cell concentrations and DA levels in mussels reached record levels in 1987 (Table 3) (Bates *et al.* 1989). Meteorological conditions

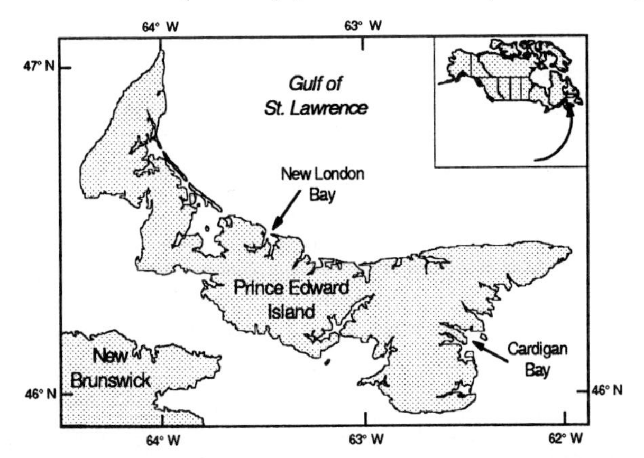

Fig. 2. Location of *P. multiseries* and *P. pungens* blooms in eastern and northern Prince Edward Island, Canada.

Table 3. *Pseudo-nitzschia multiseries* plus *P. pungens* cell concentrations (cells L^{-1}) and domoic acid in mussels ($\mu g\ g^{-1}$) in Cardigan Bay, PEI; harvesting closure occurs at 20 $\mu g\ g^{-1}$.

Year	*Pseudo-nitzschia*	Domoic Acid	Conditions
1987	15,000,000	790	Dry summer, rainy autumn; Calm weather, light S winds
1988	1,200,000	280	Dry summer, early autumn rains; Strong SE gale
1989	460,000	16	Wet summer, no autumn runoff
1990	132,000	0.6	N winds disperse incipient bloom of *P. multiseries*
1991	354,000	not detected	Violent storms, then calm
1992	48,000	not detected	No unusual weather
1993	92,000	not detected	Mostly *P. pungens*
1994	79,000	not detected	Mostly *P. pungens*
1995	14,000	not detected	Mostly *P. pungens*

apparently contributed to the bloom formation: a prolonged dry period in summer, followed by an unusually rainy autumn (Smith 1993). This may have provided nutrients, via river runoff, to drive the bloom. Similar, but less severe meteorological conditions followed in the autumn of 1988 (Smith *et al.* 1990a). The *Pseudo-nitzschia* bloom that year was therefore less intense, but still resulted in closure of mussel harvesting due to elevated levels of DA. A 9-day lag between the peak of DA in the algae and in the mussels provided an early warning of an impending health hazard (Fig. 3A). Early in the bloom, cellular DA appeared to accumulate when the *Pseudo-nitzschia* concentration stopped increasing (Fig. 3B), consistent with laboratory studies linking toxin production to the stationary phase of growth (see Bates, this volume). Increases in cell concentration corresponded with pulses of nitrate from rivers, following rain events. The bloom was likely terminated by the decrease in light and temperature in December, as illustrated in a generalized schematic (Fig. 4).

In 1989, there was no dry summer, nor was there any substantial runoff during the autumn; the *Pseudo-nitzschia* bloom was again smaller than in 1987 and 1988. Cells of *P. pungens* were present in early August, with no detectable DA (Fig. 5) (Smith *et al.* 1990b). This was followed by a small bloom of *P. pungens* plus *P. multiseries* in late August to September, with the appearance of DA. The major autumn bloom then began in mid-October, with *P. multiseries* reaching 100% of the *Pseudo-nitzschia* spp. population in late November, when mussels rapidly accumulated DA. In 1990, non-toxic *P. pungens* bloomed in October, but an incipient *P. multiseries* bloom was dispersed seaward by a series of violent, primarily northerly storms (Smith 1993). Since then, there have been only minor *Pseudo-nitzschia* blooms, composed mainly of *P. pungens*, and no detectable levels of DA in Cardigan Bay mussels (Table 3). In contrast, mussel harvesting was closed due to DA in New London Bay (Fig. 2), in October 1991 and 1994. The dynamics of these latter blooms have not been studied.

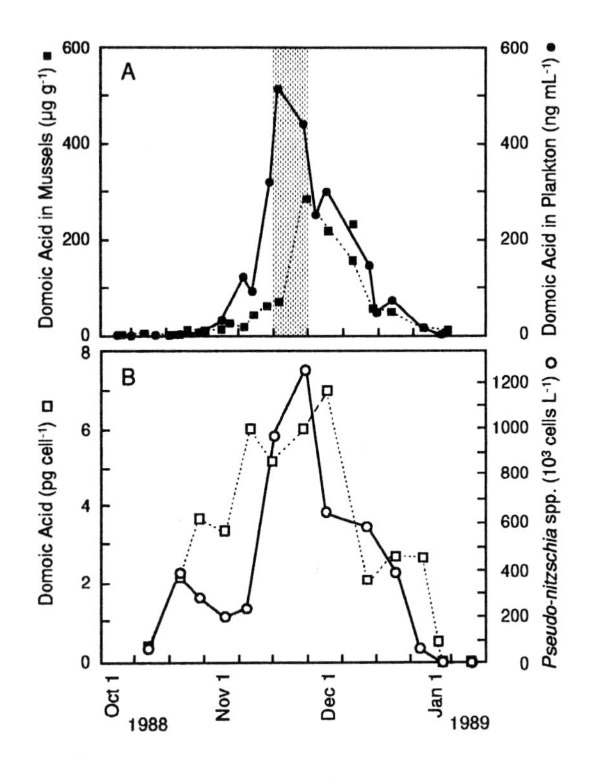

Fig. 3. Cardigan Bay, Prince Edward Island, Canada. A) Domoic acid concentration in mussels and phytoplankton, and time lag (shaded area) between peaks of the two; B) cellular domoic acid and *Pseudo-nitzschia* spp. concentration. Redrawn from Smith *et al.* 1990a.

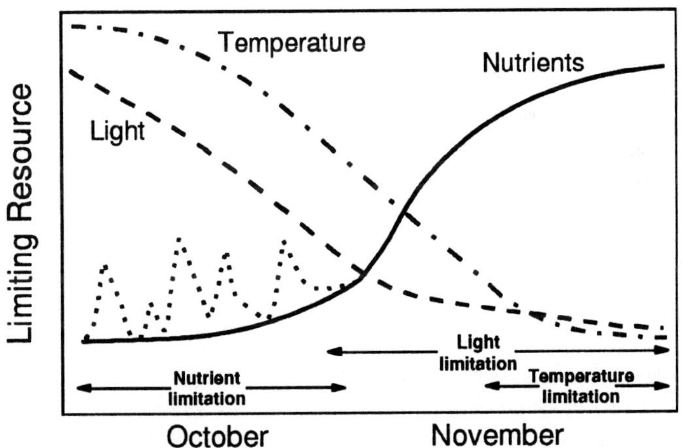

Fig. 4. Schematic of changes in temperature, light and nutrients, and periods of limitation that may affect the decline of *P. multiseries* blooms in Prince Edward Island, Canada. The dotted line represents pulses of nutrient input from rivers or sediments.

It is clear that the large interannual variability and general decline in bloom intensity are the most obvious characteristics of the Cardigan Bay blooms. The time series is still too short to unequivocally determine the cause of the decline in intensity since the 1987 outbreak. However, several hypotheses may be advanced: decreases in nutrient loading from rivers; an increase in predation by cultured mussels; cycles in the sexual reproduction of *P. multiseries*; changes in bacterial populations affecting diatom growth; and attack by parasites. For the latter, parasitic oomycetes and/or chytrids have infected *P. multiseries* and *P. pungens* cells in eastern PEI (K.E. Pauley, unpubl.; L.A. Hanic, pers. commun.; Elbrachter and Schnepf this volume). Their importance in regulating bloom dynamics has yet to be determined, but their presence is suggestive of one cause of the bloom decline since 1987. In support of this, Horner *et al.* (1996) observed a fungal parasite in some *Pseudo-nitzschia* spp. cells during the decline of a bloom in coastal Washington, USA. Hasle *et al.* (1996) reported an unexplained decrease in *P. multiseries* abundance in the Skagerrak between 1991 and 1993, and also raised the possibility of control by parasitic fungi.

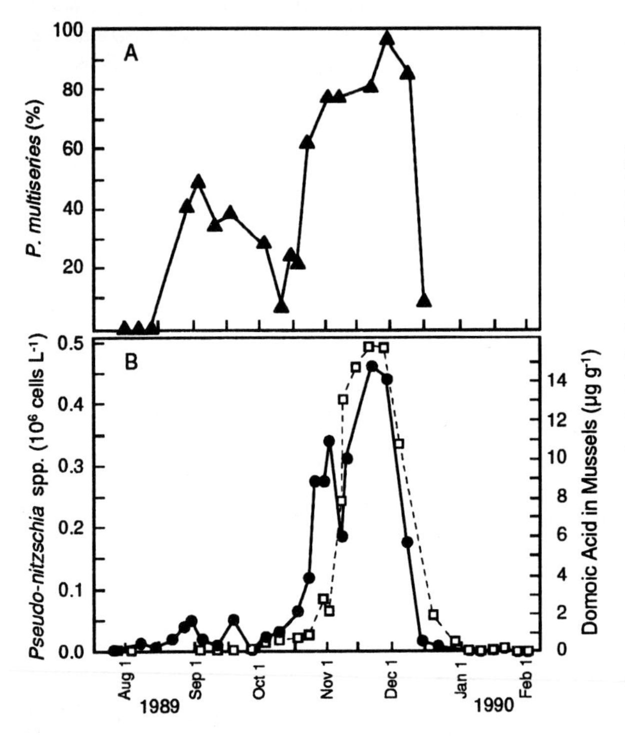

Fig. 5. Cardigan Bay, Prince Edward Island, Canada. A) Percentage of *Pseudo-nitzschia multiseries* in *P. pungens* plus *P. multiseries* populations (determined using scanning electron microscopy); B) concentration of *P. pungens* plus *P. multiseries* cells in the seawater (●), and domoic acid in mussels (0). Redrawn from Smith *et al.* 1990b.

2.2 Bay of Fundy, Eastern Canada

Because of the 1987 ASP event in PEI, monitoring for *Pseudo-nitzschia* and DA was initiated in the southwest Bay of Fundy in 1988. Regions of the Bay of Fundy are commonly closed due to PSP toxins, but no shellfish toxicity had previously been associated with diatom blooms. In late July, 1988, monitoring revealed the presence of DA in blue mussels and soft-shell clams (*Mya arenaria*) in Passamaquoddy Bay, and harvesting was closed in late August; the highest DA concentration was 74 µg g^{-1}. The dominant phytoplankter at the time was *Pseudo-nitzschia pseudodelicatissima*, which produced DA in laboratory culture (Martin *et al.* 1990). It must be noted, however, that strains of *P. pseudodelicatissima* from other parts of the world have failed to produce DA in culture (see Section 1.1). Likewise, no DA was detected in at least two predominantly *P. pseudodelicatissima* blooms in coastal California (Walz *et al.* 1994).

Although *P. pseudodelicatissima* has historically been present in the Bay of Fundy, no toxicity had previously been associated with it. This may be because shellfish harvesting is often prohibited anyway, due to PSP toxins, during the summer months when DA may also be present. Subsequent surveys have shown that this diatom occurs throughout the year, but with a small bloom generally in June, followed by a larger bloom in late August to early September (Fig. 6) (Martin *et al.* 1993). Shellfish harvesting was again closed due to DA from this diatom in September, 1995. The late summer appearance of *P. pseudodelicatissima* links it to elevated water temperatures, although neither its temperature preference, nor the conditions conducive for growth and toxin production, have been studied in culture. Cells are distributed throughout the water column (Fig. 6), consistent with the vigorous tidal mixing in the area, and making them available to contaminate benthic shellfish.

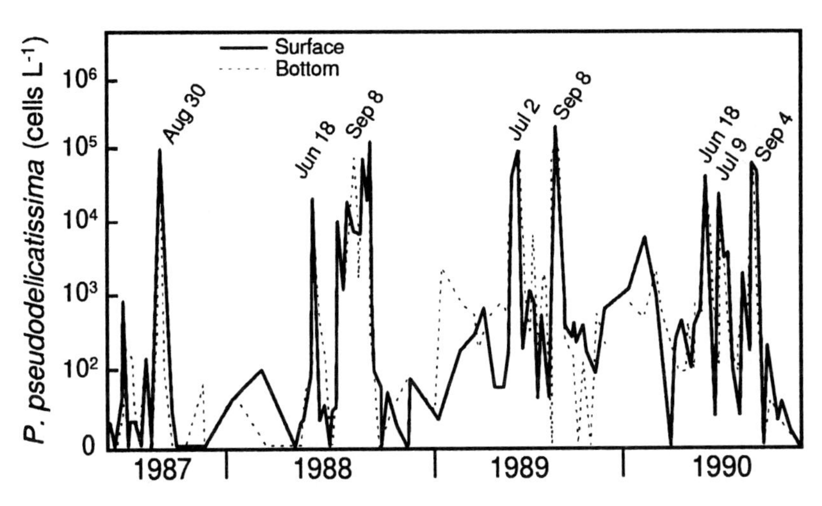

Fig. 6. Concentration of *Pseudo-nitzschia pseudodelicatissima*; Passamaquoddy Bay, southwest Bay of Fundy, Canada. Redrawn from Martin *et al.* 1993.

2.3 Monterey Bay, California, USA

Until 1991, eastern Canada was the only region documented to have problems associated with toxigenic *Pseudo-nitzschia* species. Then, in September, 1991, unusual neurological symptoms and deaths of more than 100 brown pelicans and Brandt's cormorants led to the discovery of DA in Monterey Bay, CA (Work *et al.* 1993). The DA was traced to the birds' food source, northern anchovies, which in turn had fed on the diatom *Pseudo-nitzschia australis* (Buck *et al.* 1992; Fritz *et al.* 1992). This diatom produced DA in culture (Garrison *et al.* 1992; Villac *et al.* 1993b; Bates, this volume).

The Monterey Bay event was significantly different from the Cardigan Bay, PEI outbreak, not only because of the other *Pseudo-nitzschia* species involved, and birds rather than humans were poisoned, but also because the vector was northern anchovies, not shellfish; hence the use of the term DA Poisoning (DAP). The highest concentrations of DA, up to 2,300 μg g^{-1} wet weight, were found in the viscera of the anchovies (Loscutoff 1992), along with the *P. australis* cells; these fish continue to be contaminated with DA (McGinness *et al.* 1995). Further, DA has been found in some grazing zooplankton (Buck *et al.* 1992; Haywood and Silver 1994). With the exception of the seabirds, there are no effects or impacts of DA on the pelagic food web. There is evidence, based on historical accounts of seabird behavior and mortalities, and on archived plankton samples, that massive blooms of *P. australis* have occurred since at least the 1920's (Garrison *et al.* 1992; Villac *et al.* 1993b; Lange *et al.* 1994). It is likely, however, that the species had been erroneously reported as *Nitzschia seriata*, because at the time *P. australis* was believed to be limited to the southern hemisphere.

Pseudo-nitzschia species and DA were monitored in Monterey Bay between 1991 and 1994. At the peak of the 1991 toxic event, DA levels in coastal waters reached >10 $\mu g\ L^{-1}$ and abundances of *P. australis* were >10^6 cells L^{-1} (Walz *et al.* 1994). Since the massive bloom in 1991, DA has been detected in both autumn and spring plankton assemblages, but with concentrations usually <1 $\mu g\ L^{-1}$ and cell densities 1-2 orders of magnitude lower (Fig. 7). In contrast to the eastern Canadian situation, blooms have often been simultaneously comprised of up to four potentially-toxic species, i.e., *P. australis*, *P. multiseries*, *P. pseudodelicatissima*, and *P. pungens*. Based on cell volume dominance by *P. australis*, however, most of the DA measured in plankton assemblages was assumed to have been contributed by this species (Fig. 7) (Walz *et al.* 1994). Production of DA has been demonstrated for California isolates of *P. australis* and *P. multiseries* (Garrison *et al.* 1992; Villac *et al.* 1993b; Villac 1996), and recently for four clones of *P. pungens* (see Section 1.1), but not for *P. pseudodelicatissima* (Villac *et al.* 1993b; Walz *et al.* 1994; Villac 1996).

During 1991-1994, blooms of *P. australis* were most common and persisted the longest during late summer to autumn, when the hydrography is usually characterized by warm sea-surface temperatures, thermal stratification, and low nutrient concentrations (Buck *et al.* 1992; Walz *et al.* 1994), conditions associated with the end

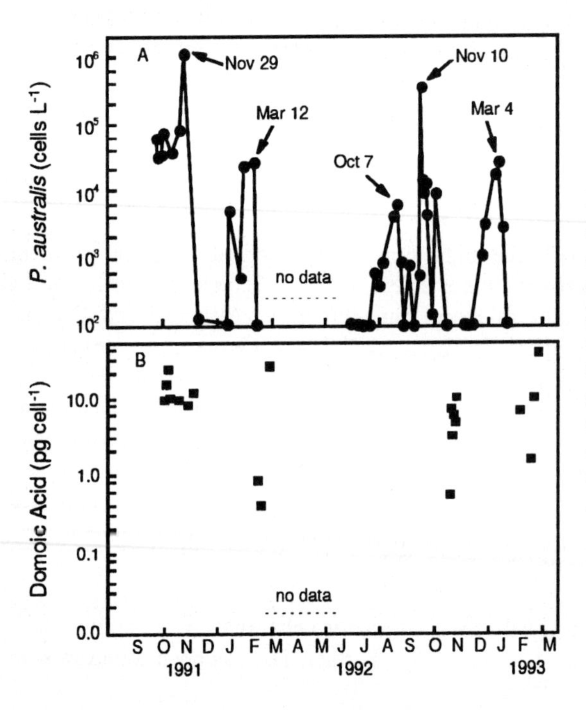

Fig. 7. A) Concentration of *P. australis* in near-surface waters (0 or 5 m); B) cellular domoic acid in *P. australis* calculated from species abundance and domoic acid in net-collected samples, assuming that the domoic acid is apportioned according to biovolume and cell abundance; Monterey Bay, CA, USA. Redrawn from Walz *et al.* 1994.

of the upwelling season. Blooms also occurred during the spring upwelling season, but they were less well-developed and generally shorter in duration (Walz *et al.* 1994). In contrast, the *P. australis* blooms in southern California were most common in the late spring to early summer months, not in the autumn (Lange *et al.* 1994). They were associated with intrusions of cool, high-nutrient waters, probably related to upwelling events. In spite of cell concentrations equivalent to those reached in Monterey Bay, no major toxic episodes were reported. It must be noted, however, that no toxin measurements were made in the phytoplankton and that nearshore bivalves are not good indicators of toxicity in offshore anchovies, and vice versa (Langlois *et al.* 1993).

Reports (e.g., Malone 1971; Garrison 1979) suggest that diatom blooms in the central California region are mostly a coastal (i.e., continental shelf) phenomenon, and this is also likely the case with *Pseudo-nitzschia* spp. blooms. Moreover, the well-developed blooms in Monterey Bay and along the continental shelf to the north appear to be associated with a local mesoscale circulation pattern that develops during periods of active upwelling and results in water being retained near the coast (Graham *et al.* 1992). Coastal populations, however, are frequently carried offshore in jets or filaments of coastal upwelling water and then may find their way back onshore in the complexity of eddies and meanders that occur with upwelling episodes followed by relaxations (e.g., Garrison 1981). Although the California Current is a continuous system extending from approximately Vancouver Island to northern Baja California, there are pronounced latitudinal variations in physical and biological dynamics (Horner *et al.* 1997). These variations may account for the differing descriptions of *Pseudo-nitzschia* bloom dynamics reported in southern California (Lange *et al.* 1994), central California, and the Oregon/Washington region.

2.4 Oregon and Washington, USA

The September, 1991 DA episode in Monterey Bay prompted an examination of shellfish to the north. By late October and November, DA was found in razor clams (*Siliqua patula*) and Dungeness crabs (*Cancer magister*) on the Oregon and Washington coasts. Several people in Washington suffered mild symptoms after eating razor clams, but the illnesses were mild and short-lived and DA poisoning was not confirmed. Moreover, neither the source(s) nor pathway(s) of DA to the razor clams and Dungeness crabs have been identified. No phytoplankton samples were available from offshore, and samples collected from ocean beaches since 1991 usually contain few cells of *Pseudo-nitzschia* spp. (Horner and Postel 1993; R. Horner, unpubl.). However, several potentially-toxic *Pseudo-nitzschia* species (i.e., *P. multiseries*, *P. australis*, *P. pungens*, and *P. pseudodelicatissima*) are found in local waters. The 1991 open coast event occurred after a record hot, dry period lasting 45 d followed by rain (Horner and Postel 1993), conditions similar to eastern PEI in 1987. It is possible that similar oceanographic conditions, including unusually warm weather associated with El Niño, and phytoplankton assemblages existed along the US west coast, with the DA source(s) part of these widespread populations (Taylor and Horner 1994). However, levels of DA within razor clams vary considerably along the Washington coast. All clam beaches were not affected to the same extent and there is considerable clam-to-clam variability in DA content on a single beach (J.C. Wekell, pers. commun.).

In coastal embayments, i.e., Willapa Bay, *Pseudo-nitzschia* abundances show some correlation with weather, but usually do not occur until after a salinity near 29 ‰ is reached (Sayce and Horner 1996), suggesting the intrusion of high-salinity ocean water. Regardless of the source of cells, oysters within Willapa Bay have not contained DA.

Pseudo-nitzschia pungens, *P. multiseries*, *P. pseudodelicatissima*, and *P. australis* have been identified in inland marine waters of western Washington, i.e., Puget Sound and Hood Canal, since the summer of 1990. They often persist together in some combination at low concentrations (ca. 1-5 x 10^3 cells L^{-1}) for several months each year and not just in the autumn. When increases (blooms) occur, sometimes after warm dry periods followed by rain, they are usually short-lived and concentrations rarely exceed 10^5 cells L^{-1} (Horner and Postel 1993). An exception was a bloom of *P. pungens*, *P. multiseries* and *P. australis* in Hood Canal in the autumn of 1994 that lasted from mid-October to mid-December with cell concentrations near 5 x 10^5 L^{-1}. Mussels used as the sentinel organism for toxins by the Washington Department of Health contained about 10 µg DA g^{-1}, the first time DA was found at relatively high levels in these inland waters (Horner *et al.* 1996).

The seed stock for *Pseudo-nitzschia* spp. along the open Pacific coasts and coastal embayments of Oregon and Washington appears to be from offshore, although few offshore samples are available to confirm this hypothesis. In contrast with the Gulf of Mexico (see Section 2.5), no *Pseudo-nitzschia* cells have been found at depth or near the bottom in samples collected over the continental shelf and nearshore, but sampling has not been extensive (R. Horner, J. Postel, M. Ross, unpubl.). In inland waters of Puget Sound, *Pseudo-nitzschia* populations apparently originate *in situ* (R. Horner, unpubl.), as is the case in eastern PEI (J.C. Smith, pers. commun.). As noted above, several *Pseudo-nitzschia* species co-exist within one water mass along the US west coast. Within Coos Bay, OR, species shifts are prevalent on short-term, seasonal, and annual time scales (G.A. Fryxell, pers. commun.). During one season in Coos Bay, for example, *P. multiseries* was replaced by *P. australis* in the course of less than three weeks, perhaps reflecting subtle differences in physiological preferences.

2.5 Other Global Locations

Awareness of DA has prompted the search for DA and DA-producing algal species in other parts of the world. Not surprisingly, DA-contaminated Dungeness crabs and several species of molluscan shellfish, including razor clams, were found on the coast of British Columbia (BC), western Canada, since 1992 (Forbes and Chiang 1994). As on the Pacific coast of the USA, the responsible organism(s) has not yet been determined, but *P. multiseries*, *P. australis*, *P. pungens*, *P. pseudodelicatissima*, *P. seriata*, and *Amphora coffeaeformis* are present (Forbes and Denman 1991; Taylor *et al.* 1994; Taylor and Haigh 1996). In Sechelt Inlet, BC, *P. pungens* bloomed in the summer and autumn, similar to eastern PEI, but the autumn abundance of *P. multiseries* was not great (Taylor *et al.* 1994). Abundant populations of *Pseudo-nitzschia* spp. also occur in Barkley Sound, an open coastal embayment on western Vancouver Island (Taylor and Haigh 1996). Populations of *P. pungens* and/or *P. multiseries* (a distinction was not made) persisted from June to September; *P. australis* was present mainly during May and June. Domoic acid appeared in *Mytilus californianus* in September, only after a rise in the concentrations of both *P. multiseries* (and/or *P. pungens*) and *P. delicatissima*. The strong prevailing onshore winds during the summer suggested that populations of *P. australis* and *P. multiseries* (and/or *P. pungens*) were advected from offshore. Large differences in species abundance between the two years of the study indicate that a longer time series is required to ascertain trends. The question of a possible link between *Pseudo-nitzschia* populations along the continental coast could not be answered, but it was suggested that local variations in physical regimes may lead to blooms in disparate locations.

Most of the earlier *Pseudo-nitzschia* blooms and DA-producing species have been reported in colder, northern waters (Table 1). However, because this genus is cosmopolitan (Hasle 1994), it is not surprising to also find blooms in the more southern latitudes. For example, an as yet unconfirmed species of *Pseudo-nitzschia,* along with DA, was found in the stomachs of sardines and mackerel along the southern tip of the Baja Peninsula, Mexico, in early 1996 (Ochoa *et al.* 1996). As in Monterey Bay, this was accompanied by the deaths of more than 100 brown pelicans and other seabirds. Elsewhere, strains of DA-producing *P. multiseries* were isolated from Galveston Bay, TX, northwest Gulf of Mexico (Fryxell *et al.* 1990; Dickey *et al.* 1992). In the northern Gulf of Mexico, annual blooms of *Pseudo-nitzschia* spp. occur during the spring and sometimes in the autumn (Fig. 8) (Dortch *et al.* 1997). Potentially-toxic *P. multiseries* (present in 60% of the samples), *P. pseudodelicatissima* (in 80% of the samples), and *P. delicatissima* are found in the northern Gulf, although no DA contamination of molluscan shellfish has yet been reported. Elsewhere, several *Pseudo-nitzschia* spp., including the dominant *P. fraudulenta*, *P. pungens*, and *P. pseudodelicatissima,* are found in Australian waters, but none have produced DA in culture (Hallegraeff 1994).

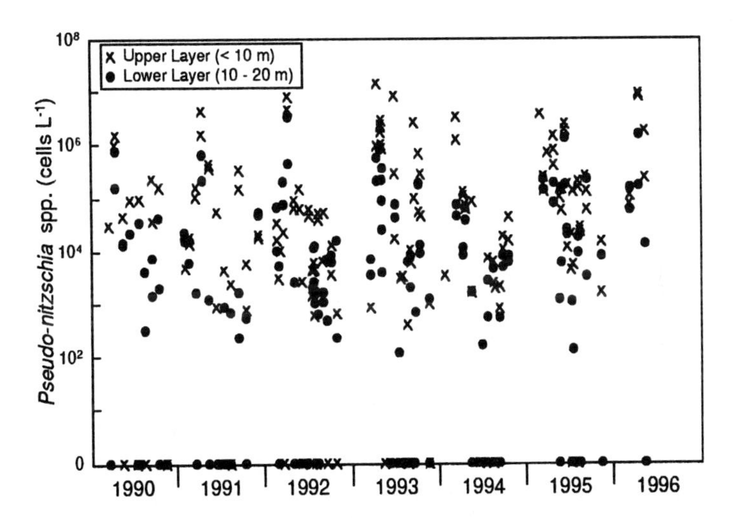

Fig. 8. Temporal variation in *Pseudo-nitzschia* spp. abundance; shelf site, northern Gulf of Mexico. Redrawn from Dortch *et al.* 1997, with updates from Q. Dortch.

3 Factors Influencing *Pseudo-nitzschia* Bloom Dynamics

3.1 Temperature and Irradiance

One of the more curious aspects of the *P. multiseries* blooms in PEI is that they have occurred during the autumn, when both temperature and irradiance levels rapidly decrease (Fig. 4). Furthermore, *P. multiseries* appears to have an affinity for cold waters elsewhere. For example, its population increased during winter and spring in Galveston Bay, TX, following cold fronts (Fryxell *et al.* 1990; Dickey *et al.* 1992; Villac *et al.* 1993a), and during the winter on the Pacific coast of Mexico (Ochoa *et al.* 1996). It must be remembered, however, that winter-spring temperatures in southern latitudes do not necessarily bracket those in the north. In Europe, *P. multiseries* was recorded in the Oslofjord, Norway, during the autumn and early winter at temperatures from 14° to about 2°C (Hasle 1965), and during the autumn and early winter in the Skagerrak (Hasle *et al.* 1996). Blooms of *P. seriata* occurred in the autumn in Galicia, Spain (Míguez *et al.* 1996). In Japan, toxic *P. multiseries* and DA in bivalves were observed from October to February (Kotaki *et al.* 1996). Perhaps these species have unique physiological adaptations that enable them to survive and to outcompete other algae during the cold seasons. Exceptions to this pattern may be the summer blooms of *P. australis* in Monterey Bay (14° - 16°C) when upwelling relaxes (Walz *et al.* 1994), *P. pungens* in British Columbia and Washington (Taylor and Haigh 1996; R. Horner, pers. observ.), and *P. pseudodelicatissima* in the Bay of Fundy (Martin *et al.* 1993).

Pseudo-nitzschia multiseries, the only species studied thus far, has a marked temperature dependence for growth, photosynthesis (P_{max} and I), and DA production (Lewis *et al.* 1993; Pan *et al.* 1993; Bates, this volume). The Q_{10} for growth was 1.8 to 2.8 between 5° - 15°C, and 2.1 between 10° - 20°C, which is within the range of other algae. Optimal growth and photosynthesis occurred in the temperature range of 15° - 25°C, well above the ambient autumn to winter temperatures (13° to -1°C) at which *P. multiseries* typically blooms. The broad range of temperature tolerance (5° to 25°C) was confirmed for *P. multiseries* isolated from northern Nova Scotia, Canada (Seguel 1991); the upper tolerance limit was not determined. It is evident that factors other than temperature must have initiated bloom development. These studies suggest that *P. multiseries* has no particular physiological adaptations that would enable it to outcompete other species at low temperatures. The broad temperature optima are consistent with its world-wide distribution at temperatures ranging from 2° to 28°C (Hasle 1965; Hasle *et al.* 1996), and may also indicate genetically-different strains (see Section 4). Indeed, *P. multiseries* was found under the ice in January, in eastern PEI at temperatures of -1.5°C (Bates *et al.* 1989), and during the summer in Galveston Bay, TX, at temperatures up to 30°C (Reap 1991; Dickey *et al.* 1992).

One consistent finding, perhaps related to temperature, is that there is a species succession from the usually non-toxic *P. pungens* to the toxic *P. multiseries* in several locations in North America and elsewhere. During the 1989 field season in northern and eastern PEI, *P. pungens* appeared in early August when no DA was detected in the mussels (Smith *et al.* 1990b). This was followed by a decline in the proportion of *P. pungens* and an increase in the proportion of *P. multiseries* to nearly 100% in late November, when the mussels quickly became toxic (Fig. 5). The two species did co-occur, although at varying proportions, during September to October. Seasonality between the two species is also found in Galveston Bay, TX (Fryxell *et al.* 1990; Reap 1991; Dickey *et al.* 1992). Although *P. multiseries* is resident all year, it dominates *P. pungens* from winter through spring, decreasing in abundance during the summer and into the autumn. It was especially abundant during cold "blue norther" storms in February (Reap 1991). In Sechelt Inlet, British Columbia, Canada, the summer populations of *Pseudo-nitzschia* spp. were predominantly *P. pungens* (Taylor *et al.* 1994). This species was also found in the autumn, as was *P. multiseries*, but at a lower concentration. Occurrences of *P. multiseries* and *P. pungens* usually coincided in Monterey Bay, CA, during the autumn, but with *P. pungens* being numerically less abundant (Walz *et al.* 1994). In the Skagerrak (Hasle *et al.* 1996), there was no evidence of a seasonal succession as in North America. If there is a temperature dependence to its seasonal succession, however, this cannot be supported by other physiological measurements. For example, when grown at 18°C, the only growth temperature studied, *P. pungens* gave a maximal nitrate reductase activity at a lower temperature (15°C) than did *P. multiseries* (23°C) (L. Rivard, unpubl.). It may be that the degree of water column mixing, rather than temperature, is important. Vigorous mixing may be required for *P. multiseries*, while the summer appearance of *P. pungens* was associated with more stratified waters (Fryxell *et al.* 1990; Taylor *et al.* 1994).

The other characteristic peculiar to the autumn *P. multiseries* blooms in eastern PEI is that they occur when irradiance levels are rapidly decreasing, both in duration and intensity (Fig. 4). In spite of this property, relatively few laboratory, and no field studies have been devoted to the effects of irradiance on growth and DA production. Photosynthesis vs irradiance (P-I) characteristics of *P. multiseries* (Pan *et al.* 1991; 1996d) do not indicate any extraordinary adaptations for growth at low irradiance levels. Cellular chlorophyll *a* levels and I^B (the initial slope of the P-I curve, a measure of the efficiency of photon absorption) in stationary-phase cultures do increase at low growth irradiance, but other phytoplankton also show this type of photoadaptation. The irradiance level necessary to saturate photosynthesis (I_m = 200 - 600 μmol photons m^{-2} s^{-1}) is substantially greater than that reported above for saturation of growth. Another study (Bates and Léger, unpubl.) gives a lower range (I_m = 100 - 200 μmol photons m^{-2} s^{-1}). Again, these irradiance levels are generally greater than those found in the growth habitat of *P. multiseries* in the autumn (e.g., 200 μmol photons m^{-2} s^{-1} at 2 m; Pan *et al.* 1996d). No P-I values are reported for monospecific blooms of *P. multiseries* in the field. Laboratory experiments, however, demonstrate that *P. multiseries* can outcompete other phytoplankton species at low irradiance levels when grown with a relatively short light:dark cycle of 8:16 h (Sommer 1994).

3.2 Nutrients

Based on laboratory studies, nutrient conditions conducive for DA production include limitation by silicate or phosphate, and an excess of nitrogen (see Bates, this volume). This is sometimes, but not always, seen in natural situations due to their complexity and our inability to quantify nutrient fluxes adequately. A low silicate concentration (0.62 μM) was found during the original 1987 Cardigan Bay bloom (Subba Rao *et al.* 1988b), consistent with optimum conditions for DA production. Nitrate and ammonium values, however, were also low (0.89 μM and 0.75 μM, respectively), contrary to the N requirement for DA production. The highest DA values did occur several days later, when nitrate increased and phosphate fell to a low concentration (0.04 μM), suggesting that limiting P concentrations may also have triggered DA production. The following autumn, a major *P. multiseries* bloom was preceded by a prolonged, intense bloom of the diatom *Skeletonema costatum* (Smith *et al.* 1990a). This, plus the presence of the *P. multiseries* bloom itself, may have depleted the silicate and phosphate concentrations in the water column, creating conditions conducive for toxin production. Much could be learned from following the seasonal succession of these diatoms. The 1988 blooms of *P. multiseries* in Cardigan Bay, PEI, were associated with pulses of nitrate, either from river runoff after rain events or resuspended sediments after wind events (Smith *et al.* 1990a). Increases in cellular DA appeared to roughly coincide with periods of nitrate influx, and preceded an increase in cell number (Fig. 3B).

In the northern Gulf of Mexico, *Pseudo-nitzschia* spp. abundance was negatively related to nitrate, ammonium, and silicate concentrations (Dortch *et al.* 1997). Walz *et al.* (1994) observed the highest concentrations of DA in a *P. australis*-dominated bloom in Monterey Bay, CA, in late summer and autumn, during post-upwelling conditions when nitrate concentrations were low (0.2 - 3.0 μM). Regenerated nutrients may have supported growth and DA production under these low-nitrogen conditions.

The riverine or sediment input of nutrients could lead to an enrichment of N and P relative to Si. This may occur in Cardigan Bay, PEI, as well as in the northern Gulf of Mexico, where spring blooms of *Pseudo-nitzschia* spp. are associated with peaks in outflow from the Mississippi River (Dortch *et al.* 1997). The substantial increase in N inputs relative to Si due to eutrophication have decreased the Si/N ratio by a factor of four, thus favoring *Pseudo-nitzschia* blooms and the possibility of DA production in that part of the Gulf of Mexico. This problem is compounded by an apparent ability of *P. multiseries* to grow and outcompete other phytoplankton at low concentrations of silicate relative to nitrate (Sommer 1994). Indeed, Smayda (1990) hypothesized that altered nutrient ratios as a result of changing patterns of anthropogenic inputs of N and P into the sea may favor the growth of certain harmful algal species.

Coastal environments where toxigenic blooms occur are particularly rich in ammonium and organic forms of N. Only minimal work, however, has been carried out to study growth and DA production in relation to these forms of N. Growth rates of *P. multiseries* with nitrite, glutamine and urea are similar to those with nitrate (Hillebrand and Sommer 1996); unfortunately, DA production was not measured. Ammonium prevented the growth of *P. multiseries* at a concentration of 880 μM and decreased the stationary phase cell number at 220 - 440 μM, as well as enhanced DA production (Bates *et al.* 1993b). A similar growth-inhibiting effect of ammonium, but starting at 20 μM, was found by Hillebrand and Sommer (1996). These deleterious concentrations of ammonium are lower than those found for most other diatoms. This is curious, given that *P. multiseries* grows in waters where ammonium levels may be high due to remineralization, agricultural runoff, and excretion from wild and cultured mussels.

3.3 Salinity

One commonality of *Pseudo-nitzschia* spp. is that they are coastal or estuarine. Not surprisingly, therefore, an isolate of *P. multiseries* from Pomquet Harbour, Nova Scotia, Canada, showed a broad salinity range (30 - 45 ‰) for optimum growth (Fig. 9); cells did not grow below 9 Y (Seguel 1991; Jackson *et al.* 1992). Interestingly, the salinity optimum is higher than the range found in estuaries (5 - 29 ‰). The salinity range for optimum growth of *P. multiseries* is higher than that of *P. pungens*, but this does not explain their seasonal succession (see Section 3.1). Isolates of *P. multiseries* from Galveston Bay, TX, grew successfully at salinities of 13 - 34 ‰ (Reap 1991), and a mixture of *Pseudo-nitzschia* spp. was found over a broad salinity range (0 - 36 ‰) near the Mississippi River (Dortch *et al.* 1997), also showing that they are very halotolerant.

The physical regime of low salinity associated with peak abundances of *P. australis* in Monterey Bay, CA, led Buck *et al.* (1992) to conclude that this species has different environmental constraints compared to those of *P. multiseries* from eastern Canada. An examination of their salinity plots, however, reveals that while peaks of *P. australis* coincide with periods of lower salinity, the range shown (33.2 - 33.8 ‰) is, in fact, higher than that encountered in eastern Canadian embayments.

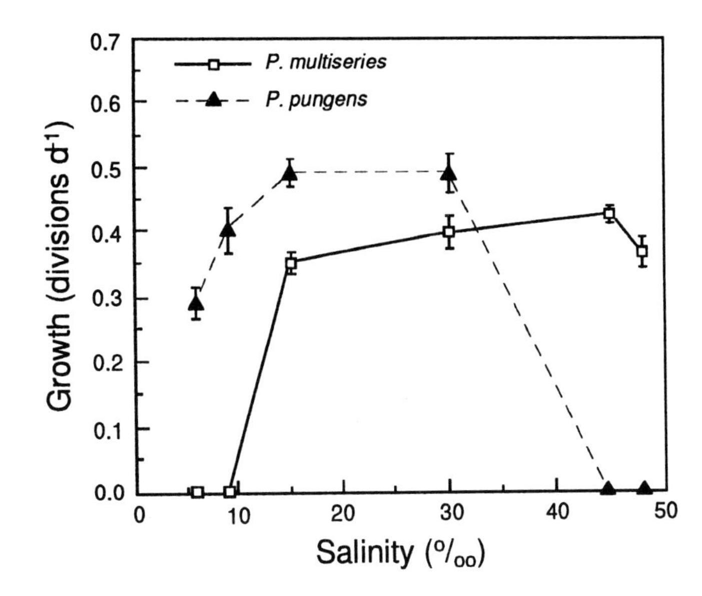

Fig. 9. Growth of *Pseudo-nitzschia multiseries* and *P. pungens*, isolated from northern Nova Scotia, Canada, as a function of salinity. Redrawn from Jackson *et al.* 1992.

Thus far, none of the chemical or physiological characteristics, including maximum division rates (Table 4), is necessarily unique to this genus. This leads one to conclude that *Pseudo-nitzschia* species will bloom when there are a sufficient number of such cells initially present, and the conditions generally bracket the cells' rather wide physiological tolerances.

Table 4. Division rates (d^{-1}) of *Pseudo-nitzschia multiseries* in culture at different growth temperatures (°C) and irradiance levels (μmol photons m^{-2} s^{-1}).

Temperature	Light Intensity	Division Rate	Reference
10	45	0.24	Bates *et al.* 1991
10	145	0.44	Bates *et al.* 1991
10-12	105*	0.56	Pan *et al.* 1991
10-12	1100*	0.74	Pan *et al.* 1991
15	70-80	1.07	Reap 1991
15	100	0.30	Jackson *et al.* 1992
5	180	0.25	Lewis *et al.* 1993
20	180	0.65	Lewis *et al.* 1993
0	350-440	0.21	Pan *et al.* 1993
15-25	350-40	1.20	Pan *et al.* 1993
10-12	1100	0.32	Subba Rao *et al.* 1995
10	53*	0.21	Pan *et al.* 1996d
10	250*	0.21	Pan *et al.* 1996d
10	410*	0.85	Pan *et al.* 1996d
10	815*	0.55	Pan *et al.* 1996d
10	1100*	0.80	Pan *et al.* 1996d

*Continuous irradiance　　*P. australis*: 0.56 d^{-1} at 15°C (Garrison *et al.* 1992)

4. General Biology of *Pseudo-nitzschia* Species

Because they are diatoms, *Pseudo-nitzschia* species decrease in cell size as a result of normal vegetative division. However, until recently it was not known why they did not undergo sexual reproduction in culture to revert to the original, large cell dimensions. It has now been demonstrated that *P. multiseries*, *P. pungens*, and *P. pseudodelicatissima* are dioecious, so that sexual reproduction can be induced by mixing "male" and "female" clones of the appropriate minimal cell size (N. Davidovich and S. Bates, unpubl.). The following has been observed: pairing of parent cells (allogamy); production of four morphologically isogamous and non-flagellated gametes per gametangial pair; fusion of gametes to form zygotes, revealing physiological anisogamy; enlargement of auxospores; and formation of long initial cells. These observations of allogamous reproduction are consistent with those reported for other pennate diatoms, and differ significantly from those in the report of Subba Rao *et al.* (1991), which we regard as erroneous for reasons also given in Fryxell *et al.* (1991) and Rosowski *et al.* (1992).

An understanding of the life history of *P. multiseries* may provide some insight into the reasons for the yearly variability in bloom intensity and toxicity of this species in eastern PEI, Canada. Knowledge of the minimal cell size suitable for sexual reproduction and of the rate of decrease in cell size allows one to predict when sexual reproduction can take place. A dense bloom in a given year would favor sexual reproduction by increasing the chance for the cells to pair while in the water column. This would lead to an increase in the proportion of large cells, hypothesized to be more toxic and to have a faster division rate, ultimately allowing the species to persist into the following seasons. It likely takes several years for cells to decrease to the suitable size, thus resulting in a long-term periodicity in bloom intensity and toxicity. Measurements of cell size-frequency distributions in the field over a period of several years would increase our understanding of *Pseudo-nitzschia* spp. bloom dynamics.

Another unknown is why *P. multiseries* cells bloom only during the autumn in eastern Canada. Few, if any, cells are present in the water column during the spring and early summer. To where, and in what form, do these cells disappear? There is evidence that the autumn blooms originate within the bays and estuaries in eastern PEI, not from the immediate coastal waters (J.C. Smith, pers. commun.). In this case, the cells may overwinter in the sediments, perhaps as "resting cells" that are morphologically indistinguishable from vegetative cells, but with condensed chloroplasts. Sedimentation of *Pseudo-nitzschia* cells can be significant (Fryxell *et al.* 1990; Villac 1996; Dortch *et al.* 1997), so the sediments may be one repository for dormant cells. Surprisingly, however, there is thus far no evidence of an abundance of *Pseudo-nitzschia* cells in the sediments, at least in PEI in late autumn (S. Bates, pers. observ.), or in the California Bight (C. Lange, unpubl.). Cells of *P. pungens* were nearly always found in the upper 20 m off the northwestern WA coast in July, 1996 (R. Horner, pers. observ.). Thus, it is difficult to generalize about *Pseudo-nitzschia* cells sinking to the sediments at all locations. Because of their tendency to sink out of the euphotic zone, one might think that *Pseudo-nitzschia* species would have a requirement for a well-mixed water column. This is the case in the turbulent bays of PEI. Similarly, in the northern Gulf of Mexico, the abundance of a mixture of *Pseudo-nitzschia* species was negatively related to delta σ_t (difference in σ_t from surface to bottom), an approximation of water column stability (Dortch *et al.* 1997). However, *P. multiseries* (and/or *P. pungens*) was found under stratified conditions in Sechelt Inlet, British Columbia (Taylor *et al.* 1994).

When interpreting laboratory experiments in relation to field studies, it is important to consider that *Pseudo-nitzschia* spp. cells tend to become deformed after relatively short periods in culture (Subba Rao and Wohlgeschaffen 1990; Subba Rao *et al.* 1991; Reap 1991; Garrison *et al.* 1992; Hillebrand and Sommer 1996), perhaps when they decrease to <50% of their original cell length (Villac 1996). These variants are described variously as being "beaked", "sickle-shaped", or "lobed" (Subba Rao and Wohlgeschaffen 1990). Teratologous *P. multiseries* cells were reported at low frequency (<1%) in eastern PEI (Subba Rao and Wohlgeschaffen 1990), southern Nova Scotia, Canada (Y. Pan, pers. commun.), and in Monterey Bay, USA (C.A. Scholin, pers. commun.). However, others have not seen this type of cell in nature (J.C. Smith; K.E. Pauley; P.E. Hargraves, pers. commun.; R. Horner, unpubl.). *Pseudo-nitzschia pungens* cells with 1-3 "swellings" were described from eutrophic coastal waters of Japan (Takano and Kikuchi 1985). Once they appear, deformities are passed onto succeeding generations because they become part of the frustule "template". Cultures of *Pseudo-nitzschia* spp. may also form stacked colonies (similar to *Fragilariopsis* spp.), rather than "stepped' chains; thus, the chain morphology may not be a good generic character. The deformities may occur during stationary phase, perhaps as a result of extracellular autoinhibiting metabolites or of Si limitation. Villac (1996) hypothesized that the dense nucleus, located at the mid-section of the long cell, may produce an undulation of the valve mantle in Si-limited cells. Deformities are not part of the sexual life cycle as once proposed by Subba Rao *et al.* (1991). Their existence, however, reminds us that if the morphology of cells can change under unnatural culture conditions, other physiological and chemical characteristics may also be aberrant.

5 Conclusions

- Toxigenic *Pseudo-nitzschia* species are cosmopolitan in coastal waters. However, different species or mixtures of species generally dominate in different regions of the world (*P. multiseries* in eastern PEI, Canada; *P. pseudodelicatissima* in the Bay of Fundy, Canada; co-existence of *P. multiseries*, *P. australis*, and *P. pseudodelicatissima* along the west coast of North America; *P. multiseries* and *P. delicatissima* in Holland; *P. seriata* in Denmark; *P. australis* in Spain; *P. australis* and *P. delicatissima* in New Zealand).
- Thus far, the same *Pseudo-nitzschia* species may be toxic in one part of the world but not in another (*P. pseudodelicatissima* in the Bay of Fundy, Canada; *P. delicatissima* originally in eastern Canada, but now also in Denmark and New Zealand; *P. seriata* in Denmark). Although still rare, there is also the enigma of toxic strains of *P. pungens* (so far only in the Pacific Ocean) and non-toxic strains of *P. multiseries*. The most sensitive analytical techniques must be used to confirm the presence or absence of toxicity.
- The life cycle of *Pseudo-nitzschia* species in relation to bloom dynamics is not understood, nor is the location of the seed beds (sediments, water column) and overwintering populations.
- Blooms of *P. multiseries* generally occur during the colder seasons (autumn to spring); blooms of *P. pungens*, *P. pseudodelicatissima*, and *P. australis* tend to occur during the warmer seasons. No unusual physiological adaptations to irradiance or temperature have yet been found to account for their presence at those times of the year.
- A better understanding is required of the effects of mesoscale features and short-term events on bloom dynamics, and of triggers of DA production in natural populations.

6 Acknowledgments

We thank G. Doucette, C. Scholin, J.C. Smith, and V. Trainer for providing unpublished information; Q. Dortch, S. Fraga, and L. Rhodes for copies of manuscripts "in press"; and Q. Dortch, J.L. Martin, C. Scholin, and J.N.C. Whyte for comments.

7 References

Addison, R.F., Stewart, J.E. (1989). Domoic acid and the eastern Canadian molluscan shellfish industry. *Aquaculture* 77: 263-269.

Bates, S.S., Bird, C.J., de Freitas, A.S.W., Foxall, R., Gilgan, M., Hanic, L.A., Johnson, G.R., McCulloch, A.W., Odense, P., Pocklington, R., Quilliam, M.A., Sim, P.G., Smith, J.C., Subba Rao, D.V., Todd, E.C.D., Walter, J.A., Wright, J.L.C. (1989). Pennate diatom *Nitzschia pungens* as the primary source of domoic acid, a toxin in shellfish from eastern Prince Edward Island, Canada. *Can. J. Fish. Aquat. Sci.* 46: 1203-1215.

Bates, S.S., de Freitas, A.S.W., Milley, J.E., Pocklington, R., Quilliam, M.A., Smith, J.C., Worms, J. (1991). Controls on domoic acid production by the diatom *Nitzschia pungens* f. *multiseries* in culture: nutrients and irradiance. *Can. J. Fish. Aquat. Sci.* 48: 1136-1144.

Bates, S.S., Léger, C., Keafer, B.A., Anderson, D.M. (1993a). Discrimination between domoic-acid-producing and nontoxic forms of the diatom *Pseudonitzschia pungens* using immunofluorescence. *Mar. Ecol. Prog. Ser.* 100: 185-195.

Bates, S.S., Worms, J., Smith, J.C. (1993b). Effects of ammonium and nitrate on domoic acid production by *Nitzschia pungens* in batch culture. *Can. J. Fish. Aquat. Sci.* 50: 1248-1254.

Bates, S.S., Douglas, D.J., Doucette, G.J., Léger, C. (1995). Enhancement of domoic acid production by reintroducing bacteria to axenic cultures of the diatom *Pseudo-nitzschia multiseries*. *Nat. Toxins* 3: 428-435.

Bates, S.S., Léger, C., Smith, K.M. (1996). Domoic acid production by the diatom *Pseudo-nitzschia multiseries* as a function of division rate in silicate-limited chemostat culture. In: Yasumoto, T., Oshima, Y, Fukuyo, Y. (eds.) *Harmful and Toxic Algal Blooms*, Intergov. Oceanogr. Comm., UNESCO, Paris, pp. 163-166.

Buck, K.R., Uttal-Cooke, L., Pilskaln, C.H., Roelke, D.L., Villac, M.C., Fryxell, G.A., Cifuentes, L., Chavez, F.P. (1992). Autecology of the diatom *Pseudonitzschia australis* Frenguelli, a domoic acid producer, from Monterey Bay, California. *Mar. Ecol. Prog. Ser.* 84: 293-302.

Dickey, R.W., Fryxell, G.A., Grande, H.R., Roelke, D. (1992). Detection of the marine toxins okadaic acid and domoic acid in shellfish and phytoplankton in the Gulf of Mexico. *Toxicon* 30: 355-359.

Dortch, Q, Robichaux, R., Pool, S., Milsted, D, Mire, G., Rabalais, N.N., Soniat, T.M., Fryxell, G.A., Turner, R.E., Parsons, M.L. (1997). Abundance and vertical flux of *Pseudo-nitzschia* in the northern Gulf of Mexico. *Mar. Ecol. Prog. Ser.* 146:249-264.

Douglas, D.J., Bates, S.S., Bourque, L.A., Selvin, R.C. (1993). Domoic acid production by axenic and non-axenic cultures of the pennate diatom *Nitzschia pungens* f. *multiseries*. In: Smayda, T.J., Shimizu, Y. (eds.) *Toxic Phytoplankton Blooms in the Sea*, Elsevier, Amsterdam, pp. 595-600.

Douglas, D.J., Landry, D., Douglas, S.E. (1994). Genetic relatedness of toxic and nontoxic isolates of the marine pennate diatom *Pseudonitzschia* (Bacillariophyceae): phylogenetic analysis of 18S rRNA sequences. *Nat. Toxins* 2: 166-174.

Forbes, J.R., Chiang, R. (1994). Geographical and temporal variability of domoic acid in samples collected for seafood inspection in British Columbia, 1992-1994. In: Forbes, J.R. (ed.) *Proceedings of the Fourth Canadian Workshop on Harmful Marine Algae. Can. Tech. Rep. Fish. Aquat. Sci.* 2016, pp. 11-12 (abstract).

Forbes, J.R., Denman, K.L. (1991). Distribution of *Nitzschia pungens* in coastal waters of British Columbia. *Can. J. Fish. Aquat. Sci.* 48: 960-967.

Fritz, L. (1992). The use of cellular probes in studying marine phytoplankton. *Korean J. Phycol.* 7: 319-324.

Fritz, L., Quilliam, M.A., Wright, J.L.C., Beale, A., Work, T.M. (1992). An outbreak of domoic acid poisoning attributed to the pennate diatom *Pseudonitzschia australis*. *J. Phycol.* 28: 439-442.

Fryxell, G.A., Reap, M.E., Valencic, D.L. (1990). *Nitzschia pungens* Grunow f. *multiseries* Hasle: observations of a known neurotoxic diatom. *Beih. Nova Hedwigia* 100: 171-188.

Fryxell, G.A., Garza, S.A., Roelke, D.L. (1991). Auxospore formation in an Antarctic clone of *Nitzschia subcurvata* Hasle. *Diatom Res.* 6: 235-245.

Garrison, D.L. (1979). Monterey Bay Phytoplankton. I. Seasonal cycles of phytoplankton assemblages. *J. Plankton Res.* 1: 241-256.

Garrison, D.L. (1981). Monterey Bay phytoplankton. II. Resting spore cycles in coastal diatom populations. *J. Plankton Res.* 3: 137-156.

Garrison, D.L., Conrad, S.M., Eilers, P.P., Waldron, E.M. (1992). Confirmation of domoic acid production by *Pseudonitzschia australis* (Bacillariophyceae) cultures. *J. Phycol.* 28: 604-607.

Graham, W.H., Field, J.G., Potts, D.C. (1992). Persistent upwelling shadows and their influence on zooplankton distributions. *Mar. Biol.* 114: 561-570.

Hallegraeff, G.M. (1994). Species of the diatom genus *Pseudonitzschia* in Australian waters. *Bot. Mar.* 37: 397-411.

Hargraves, P.E., Zhang, J., Wang, R., Shimizu, Y. (1993). Growth characteristics of the diatoms *Pseudonitzschia pungens* and *P. fraudulenta* exposed to ultraviolet radiation. *Hydrobiol.* 269: 207-212.

Hasle, G.R. (1965). *Nitzschia* and *Fragilariopsis* species studied in the light and electron microscopes. II. The group *Pseudonitzschia*. *Skr. Norske Vidensk-Akad. I. Mat.-Nat. Kl. Ny Serie* 18: 1-45.

Hasle, G.R. (1994). *Pseudo-nitzschia* as a genus distinct from *Nitzschia* (Bacillariophyceae). *J. Phycol.* 30: 1036-1039.

Hasle, G.R. (1995). *Pseudo-nitzschia pungens* and *P. multiseries* (Bacillariophyceae): nomenclatural history, morphology, and distribution. *J. Phycol.* 31: 428-435.

Hasle, G.R., Lange, C.B., Syvertsen, E.E. (1996). A review of *Pseudo-nitzschia*, with special reference to the Skagerrak, North Atlantic, and adjacent waters. *Helgol. Meeresunters.* 50: 131-175.

Haywood, G.J., Silver, M.W. (1994). The concentration of domoic acid by zooplankton. *EOS* 75: 89 (abstract).

Hillebrand, H., Sommer, U. (1996). Nitrogenous nutrition of the potentially toxic diatom *Pseudonitzschia pungens* f. *multiseries* Hasle. *J. Plankton Res.* 18: 295-301.

Horner, R.A., Postel, J.R. 1993. Toxic diatoms in western Washington waters (U.S. west coast). *Hydrobiol.* 269: 197-205.

Horner, R.A., Kusske, M.B., Moynihan, B.P., Skinner, R.N., Wekell, J.C. (1993). Retention of domoic acid by Pacific razor clams, *Siliqua patula* (Dixon, 1789): preliminary study. *J. Shellfish Res.* 12 451-456.

Horner, R.A., Hanson, L., Hatfield, C.L., Newton, J.A. (1996). Domoic acid in Hood Canal, Washington, USA. In: Yasumoto, T., Oshima, Y, Fukuyo, Y. (eds.) *Harmful and Toxic Algal Blooms*, Intergov. Oceanogr. Comm., UNESCO, Paris, pp. 127-129.

Horner, R.A., Garrison, D.L., Plumley, F.G. (1997). Harmful algal blooms and red tide problems on the U.S. west coast. *Limnol. Oceanogr.* (in press).

Impellizzeri, G., Mangiafico, S., Oriente, G., Piattelli, M., Sciuto, S., Fattorusso, E., Magno, S., Santacroce, C., Sica, D. (1975). Amino acids and low-molecular-weight carbohydrates of some marine red algae. *Phytochem.* 14: 1549-1557.

Jackson, A.E., Ayer, S.W., Laycock, M.V. (1992). The effect of salinity on growth and amino acid composition in the marine diatom *Nitzschia pungens*. *Can. J. Bot.* 70: 2198-2201.

Kotaki, Y., Koike, K., Ogata, T., Sato, S., Fukuyo, Y., Kodama, M. (1996). Domoic acid production by an isolate of *Pseudonitzschia multiseries*, a possible cause for the toxin detected in bivalves in Ofunato Bay, Japan. In: Yasumoto, T., Oshima, Y, Fukuyo, Y. (eds.) *Harmful and Toxic Algal Blooms*, Intergov. Oceanogr. Comm., UNESCO, Paris, pp. 151-154.

Lange, C.B., Reid, F.M.H., Vernet, M. (1994). Temporal distribution of the potentially toxic diatom *Pseudonitzschia australis* at a coastal site in Southern California. *Mar. Ecol. Prog. Ser.* 104: 309-312.

Langlois, G.W., Kizer, L.W., Hansgen, K.H., Howell, R., Loscutoff, S.M. (1993). A note on domoic acid in California coastal molluscs and crabs. *J. Shellfish Res.* 12: 467-468.

Laycock, M.V., de Freitas, A.S.W., Wright, J.L.C. (1989). Glutamate agonists from marine algae. *J. Appl. Phycol.* 1: 113-122.

Lee, J.H., Baik, J.H. (1995). Domoic acid-producing *Pseudonitzschia pungens* f. *multiseries* from Jinhae Bay, Korea: morphology, growth and population dynamics. *Proceed. 7th Internat. Conf. Toxic Phytoplankton*, Sendai, Japan p. 96 (abstract).

Lewis, N.I., Bates, S.S., McLachlan, J.L., Smith, J.C. (1993). Temperature effects on growth, domoic acid production, and morphology of the diatom *Nitzschia pungens* f. *multiseries*. In: Smayda, T.J., Shimizu, Y. (eds.) *Toxic Phytoplankton Blooms in the Sea*, Elsevier, Amsterdam, pp. 601-606.

Loscutoff, S. (1992). The west coast experience - overview. In: *Proceedings Domoic Acid Workshop, February 6-8, 1992, San Pedro, California.* U.S. FDA.

Lundholm, N., Skov, J., Pocklington, R., Moestrup, Ø. (1994). Domoic acid, the toxic amino acid responsible for amnesic shellfish poisoning, now in *Pseudonitzschia seriata* (Bacillariophyceae) in Europe. *Phycologia* 33: 475-478.

Lundholm, N., Skov, J., Pocklington, R., Moestrup, Ø. (1997). Studies on the marine planktonic diatom *Pseudo-nitzschia.* II. Autecology of *P. pseudodelicatissima* (Hasle) Hasle based on isolates from Danish coastal waters. *Phycologia* (in press).

Maeda, M., Kodama, T., Tanaka, T., Yoshizumi, H., Takemoto, T., Nomoto, K., Fujita, T. (1986). Structures of isodomoic acids A, B and C, novel insecticidal amino acids from the red alga *Chondria armata*. *Chem. Pharm. Bull.* 34: 4892-4895.

Malone, T.C. (1971). The relative importance of nannoplankton and netplankton as primary producers in the California Current System. *Fish. Bull.* 69: 799-820.
Manhart, J.R., Fryxell, G.A., Villac, C., Segura, L.Y. (1995). *Pseudo-nitzschia pungens* and *P. multiseries* (Bacillariophyceae): nuclear ribosomal DNAs and species differences. *J. Phycol.* 31: 421-427.
Maranda, L., Wang, R., Musauda, K., Shimizu, Y. (1990). Investigation of the source of domoic acid in mussels. In: Granéli, E., Sundström, B., Edler, L., Anderson, D.M. (eds.) *Toxic Marine Phytoplankton*, Elsevier, New York, pp. 300-304.
Martin, J.L., Haya, K., Burridge, L.E., Wildish, D.J. (1990). *Nitzschia pseudodelicatissima* - a source of domoic acid in the Bay of Fundy, eastern Canada. *Mar. Ecol. Prog. Ser.* 67: 177-182.
Martin, J.L., Haya, K., Wildish, D.J. (1993). Distribution and domoic acid content of *Nitzschia pseudodelicatissima* in the Bay of Fundy. In: Smayda, T.J., Shimizu, Y. (eds.) *Toxic Phytoplankton Blooms in the Sea*, Elsevier, Amsterdam, pp. 613-618.
McGinness, K.L., Fryxell, G.A., McEachran, J.D. (1995). *Pseudonitzschia* species found in digestive tracts of northern anchovies (*Engraulis mordax*). *Can. J. Zool.* 73: 642-647.
Míguez, Á., Fernández, M.L., Fraga, S. (1996). First detection of domoic acid in Galicia (NW of Spain). In: Yasumoto, T., Oshima, Y, Fukuyo, Y. (eds.) *Harmful and Toxic Algal Blooms*, Intergov. Oceanogr. Comm., UNESCO, Paris, pp. 143-145.
Miller, P.E., Scholin, C.A. (1996). Identification of cultured *Pseudo-nitzschia* (Bacillariophyceae) using species-specific LSU rRNA-targeted fluorescent probes. *J. Phycol.* 32: 646-655.
Ochoa, J.L., Sierra-Beltrán, A., Cruz-Villacorta, A., Sánchez-Paz, A., Núñez-Vázquez, E. (1996). Recent observations of HABs and toxic episodes - Mexico. *Harmful Algae News* No. 14, pp. 4.
Pan, Y., Subba Rao, D.V., Warnock, R.E. (1991). Photosynthesis and growth of *Nitzschia pungens* f. *multiseries* Hasle, a neurotoxin producing diatom. *J. Exp. Mar. Biol. Ecol.* 154: 77-96.
Pan, Y., Subba Rao, D.V., Mann, K.H., Li, W.K.W., Warnock, R.E. (1993). Temperature dependence of growth and carbon assimilation in *Nitzschia pungens* f. *multiseries*, the causative diatom of domoic acid poisoning. In: Smayda, T.J., Shimizu, Y. (eds.) *Toxic Phytoplankton Blooms in the Sea*, Elsevier, Amsterdam, pp. 619-624.
Pan, Y., Subba Rao, D.V., Mann, K.H., Brown, R.G., Pocklington, R. (1996a). Effects of silicate limitation on production of domoic acid, a neurotoxin, by the diatom *Pseudo-nitzschia multiseries*. I. Batch culture studies. *Mar. Ecol. Prog. Ser.* 131: 225-233.
Pan, Y., Subba Rao, D.V., Mann, K.H., Brown, R.G., Pocklington, R. (1996b). Effects of silicate limitation on production of domoic acid, a neurotoxin, by the diatom *Pseudo-nitzschia multiseries*. II. Continuous culture studies. *Mar. Ecol. Prog. Ser.* 131: 235-243.
Pan, Y., Subba Rao, D.V., Mann, K.H. (1996c). Changes in domoic acid production and cellular chemical composition of the toxigenic diatom *Pseudo-nitzschia multiseries* under phosphate limitation. *J. Phycol.* 32: 371-381.

Pan, Y., Subba Rao, D.V., Mann, K.H. (1996d). Acclimation to low light intensity in photosynthesis and growth of *Pseudo-nitzschia multiseries* Hasle, a neurotoxigenic diatom. *J. Plankton Res.* 18: 1427-1438.

Perl, T.M., Bédard, L., Kosatsky, T., Hockin, J.C., Todd, E., Remis, R.S. (1990). An outbreak of toxic encephalopathy caused by eating mussels contaminated with domoic acid. *New England J. Med.* 322: 1775-1780.

Reap, M.E. (1991). *Nitzschia pungens* Grunow f. *multiseries* Hasle: growth phases and toxicity of clonal cultures isolated from Galveston Bay, Texas. M.Sc. Thesis, Texas A&M Univ., College Station, TX, 78 p.

Rhodes, L., White, D., Syhre, M., Atkinson, M. (1996). *Pseudo-nitzschia* species isolated from New Zealand coastal waters: domoic acid production *in vivo* and links with shellfish toxicity. In: Yasumoto, T., Oshima, Y, Fukuyo, Y. (eds.) *Harmful and Toxic Algal Blooms*, Intergov. Oceanogr. Comm., UNESCO, Paris, pp. 155-158.

Rhodes, L., Scholin, C., Garthwaite, I., Haywood, A. (1997). New records of domoic acid (DA) producing *Pseudo-nitzschia* species deduced by concurrent use of whole cell DNA probe-based and DA immunochemical assays. In: Reguera, B. *et al.* (eds.) *VIII International Conference on Harmful Algae, Vigo, Spain, June 1997* (in press).

Rosowski, J.R., Johnson, L.M., Mann, D.G. (1992). On the report of gametogenesis, oogamy, and uniflagellated sperm in the pennate diatom *Nitzschia pungens* (1991. J. Phycol. 27: 21-26). *J. Phycol.* 28: 570-574.

Sato, M., Nakano, T., Takeuchi, M., Kanno, N., Nagahisa, E., Sato. Y. (1996). Distribution of neuroexcitatory amino acids in marine algae. *Phytochem.* 42: 1595-1597.

Sayce, K., Horner, R.A. (1996). *Pseudo-nitzschia* spp. in Willapa Bay, Washington, 1992 and 1993. In: Yasumoto, T., Oshima, Y, Fukuyo, Y. (eds.) *Harmful and Toxic Algal Blooms*, Intergov. Oceanogr. Comm., UNESCO, Paris, pp. 131-134.

Scholin, C.A., Villac, M.C., Buck, K. R., Krupp, K.M., Powers, D.A., Fryxell, G.A., Chavez, F.P. (1994). Ribosomal DNA sequences discriminate among toxic and non-toxic *Pseudonitzschia* species. *Nat. Toxins* 2: 152-165.

Scholin, C.A., Buck, K.R., Britschgi, T., Cangelosi, G., Chavez, F.P. (1996). Identification of *Pseudo-nitzschia australis* (Bacillariophyceae) using rRNA-targeted probes in whole cell and sandwich hybridization formats. *Phycologia* 35: 190-197.

Seguel, M.R. (1991). Interactive effects of temperature-light and temperature-salinity on growth of five phytoplanktonic species isolated from a shallow-water embayment of Nova Scotia. M.Sc. Thesis, Acadia Univ., Wolfville, Nova Scotia, Canada, 218 p.

Shimizu, Y., Gupta, S., Masuda, K., Maranda, L., Walker, C.K., Wang, R. (1989). Dinoflagellate and other microalgal toxins: chemistry and biochemistry. *Pure Appl. Chem.* 61: 513-516.

Smayda, T.J. (1990). Novel and nuisance phytoplankton blooms in the sea: evidence for a global epidemic. In: Granéli, E., Sundström, B., Edler, L., Anderson, D.M. (eds.) *Toxic Marine Phytoplankton*, Elsevier, New York, pp. 29-40.

Smith, J.C. (1993). Toxicity and *Pseudonitzschia pungens* in Prince Edward Island, 1987-1992. *Harmful Algae News* No. 6, pp. 1 and 8.

Smith, J.C., Cormier, R., Worms, J., Bird, C.J., Quilliam, M.A., Pocklington, R., Angus, R., Hanic, L. (1990a). Toxic blooms of the domoic acid containing diatom *Nitzschia pungens* in the Cardigan River, Prince Edward Island. In: Granéli, E., Sundström, B.,

Edler, L., Anderson, D.M. (eds.) *Toxic Marine Phytoplankton*, Elsevier, New York, pp. 227-232.

Smith, J.C., Odense, P., Angus, R., Bates, S.S., Bird, C.J., Cormier, P., de Freitas, A.S.W., Léger, C., O'Neil, D., Pauley, P., Worms, J. (1990b). Variation in domoic acid levels in *Nitzschia* species: implications for monitoring programs. *Bull. Aquacult. Assoc. Can.* 90-4: 27-31.

Sommer, U. (1994). Are marine diatoms favoured by high Si:N ratios? *Mar. Ecol. Prog. Ser.* 115: 309-315.

Subba Rao, D.V., Quilliam, M.A., Pocklington, R. (1988a). Domoic acid - a neurotoxic amino acid produced by the marine diatom *Nitzschia pungens* in culture. *Can. J. Fish. Aquat. Sci.* 45: 2076-2079.

Subba Rao, D.V., Dickie, P.M., P. Vass. (1988b). Toxic phytoplankton blooms in the eastern Canadian Atlantic embayments. *Comm. Meeting Internat. Cons. Explor. Sea*, C.M. ICES 1988/L:28 1-16.

Subba Rao, D.V., Wohlgeschaffen, G. (1990). Morphological variants of *Nitzschia pungens* Grunow f. *multiseries* Hasle. *Bot. Mar.* 33: 545-550.

Subba Rao, D.V., Partensky, F., Wohlgeschaffen, G., Li, W.K.W. (1991). Flow cytometry and microscopy of gametogenesis in *Nitzschia pungens*, a toxic, bloom-forming, marine diatom. *J. Phycol.* 27: 21-26.

Subba Rao, D.V., Pan, Y., Smith, S.J. (1995). Allelopathy between *Rhizosolenia alata* (Brightwell) and the toxigenic *Pseudonitzschia pungens* f. *multiseries* (Hasle). In: Lassus, P., Arzul, G., Erard, E., Gentien, P., Marcaillou-Le Baut, C. (eds.) *Harmful Marine Algal Blooms*, Lavoisier, Paris, pp. 681-686.

Takano, H., Kikuchi, K. (1985). Anomalous cells of *Nitzschia pungens* Grunow found in eutrophic marine waters. *Diatom* 1: 18-20.

Takemoto, T., Daigo, K. (1958). Constituents of *Chondria armata*. *Chem. Pharmaceutical Bull.* 6: 578-580.

Taylor, F.J.R., Haigh, R. (1996). Spatial and temporal distributions of microplankton during the summers of 1992-1993 in Barkley Sound, British Columbia, with emphasis on harmful species. *Can. J. Fish. Aquat. Sci.* 53: 2310-2322.

Taylor, F.J.R., Horner, R.A. (1994). Red tides and other problems with harmful algal blooms in the Pacific northwest coastal waters. In: Wilson, R.C.H., Beamish, R.J., Aitkens, F., Bell J. (eds.). *Review of the Marine Environment and Biota of Strait of Georgia, Puget Sound and Juan de Fuca Strait. Can. Tech. Rep. Fish. Aquat. Sci.* 1948, pp. 175-185.

Taylor, F.J.R., Haigh, R., Sutherland, T.F. (1994). Phytoplankton ecology of Sechelt Inlet, a fjord system on the British Columbia coast. II. Potentially harmful species. *Mar. Ecol. Prog. Ser.* 103: 151-164.

Teitelbaum, J.S., Zatorre, R.J., Carpenter, S., Gendron, D., Evans, A.C., Gjedde, A., Cashman, N.R. (1990). Neurologic sequelae of domoic acid intoxication due to the ingestion of contaminated mussels. *New England J. Med.* 322: 1781-1787.

Todd, E.C.D. (1993). Domoic acid and amnesic shellfish poisoning - a review. *J. Food Protection* 56: 69-83.

Villac, M.C. 1996. Synecology of the genus *Pseudo-nitzschia* H. Peragallo from Monterey Bay, California, U.S.A. Ph.D. Thesis, Texas A&M Univ., 258 p.

Villac, M.C., Roelke, D.L., Villareal, T.A., Fryxell, G.A. (1993a). Comparison of two domoic acid-producing diatoms: a review. *Hydrobiol.* 269: 213-224.

Villac, M.C., Roelke, D.L., Chavez, F.P., Cifuentes, L.A., Fryxell, G.A. (1993b). *Pseudonitzschia australis* Frenguelli and related species from the west coast of the U.S.A.: occurrence and domoic acid production. *J. Shellfish Res.* 12: 457-465.

Villareal, T.A., Roelke, D.L., Fryxell, G.A. (1994). Occurrence of the toxic diatom *Nitzschia pungens* f. *multiseries* in Massachusetts Bay, Massachusetts, U.S.A. *Mar. Environ. Res.* 37: 417-423.

Vrieling. E.G., Koeman, R.P.T., Scholin, C.A., Scheerman, P., Peperzak, L., Veenhuis, M., Gieskes, W.W.C. (1996). Identification of a domoic acid-producing *Pseudo-nitzschia* species (Bacillariophyceae) in the Dutch Wadden Sea with electron microscopy and molecular probes. *European J. Phycol.* 31: 333-340.

Walz, P.M., Garrison, D.L., Graham, W.M., Cattey, M.A., Tjeerdema, R.S., Silver, M.W. (1994). Domoic acid-producing diatom blooms in Monterey Bay, California: 1991-1993. *Nat. Toxins* 2: 271-279.

Wang, R., Maranda, L., Hargraves, P.E., Shimizu, Y. (1993). Chemical variation of *Nitzschia pungens* as demonstrated by the co-occurrence of domoic acid and bacillariolides. In: Smayda, T.J., Shimizu, Y. (eds.) *Toxic Phytoplankton Blooms in the Sea*, Elsevier, Amsterdam, pp. 607-612.

Wekell, J.C., Gauglitz, E.J. Jr., Barnett, H.J., Hatfield, C.L., Eklund, M. (1994). The occurrence of domoic acid in razor clams (*Siliqua patula*), Dungeness crab (*Cancer magister*), and anchovies (*Engraulis mordax*). *J. Shellfish Res.* 13: 587-593.

Wessells, C.R., Miller, C.J., Brooks, P.M. (1995). Toxic algae contamination and demand for shellfish: a case study of demand for mussels in Montreal. *Mar. Resources Economics* 10: 143-159.

Work, T.M., Barr, B., Beale, A.M., Fritz, L., Quilliam, M.A., Wright, J.L.C. (1993). Epidemiology of domoic acid poisoning in brown pelicans (*Pelecanus occidentalis*) and Brandt's cormorants (*Phalacrocorax penicillatus*) in California. *J. Zoo Wildlife Medicine* 24: 54-62.

Wright, J.L.C., Boyd, R.K., de Freitas, A.S.W., Falk, M., Foxall, R.A., Jamieson, W.D., Laycock, M.V., McCulloch, A.W., McInnes, A.G., Odense, P., Pathak, V.P., Quilliam, M.A., Ragan, M.A., Sim, P.G., Thibault, P., Walter, J.A., Gilgan, M., Richard, D.J.A., Dewar, D. (1989). Identification of domoic acid, a neuroexcitatory amino acid, in toxic mussels from eastern Prince Edward Island. *Can. J. Chem.* 67: 481-490.

Community Dynamics and Physiology of Epiphytic/Benthic Dinoflagellates Associated with Ciguatera

Donald R. Tindall[1] and Steve L. Morton[2]

[1] Department of Plant Biology, Southern Illinois University, Carbondale, IL 62901, USA

[2] Center for Culture of Marine Phytoplankton, Bigelow Laboratory for Ocean Sciences, West Boothbay Harbor, ME 04575, USA

1. Introduction

Ciguatera or ciguatera fish poisoning (CFP) is a perplexing, multifaceted human syndrome which is often debilitating and sometimes fatal. It results from consuming toxic but presumably wholesome finfish from coral reef and inshore habitats in subtropical and tropical regions of the world (see Anderson and Lobel 1987; Withers 1988). Due to their frequent occurrence in reef fish, ciguatoxin and maitotoxin have been generally regarded as the primary toxins responsible for CFP. However, the coral reef food web contains several potent toxins which originate in an assemblage of epiphytic and benthic dinoflagellates including species of *Gambierdiscus, Ostreopsis, Prorocentrum* and *Amphidinium*. *Gambierdiscus* toxins include maitotoxin and analogs of ciguatoxin (Murata *et al.* 1989, 1992). *Ostreopsis* toxins include lipid soluble and water soluble entities the latter of which is an analog of palytoxin (Tosteson *et al.* 1989; Tindall *et al.* 1990; Usami *et al.* 1995). *Prorocentrum* toxins include okadaic acid and methylokadaic acid which are confirmed diarrhetic shellfish poisons (Murakami *et al.* 1982, Dickey *et al.* 1990), and a water soluble fast-acting toxin (Tindall *et al.* 1989; Aikman *et al.* 1993) recently characterized as prorocentrolide B (Hu *et al.* 1996). *Amphidinium* toxins have been characterized as hemolysins. Hemolytic activity also has been reported for extracts of most of the toxic species comprising this assemblage (Yasumoto *et al.* 1987). Thus, unlike other algae-induced illnesses, CFP may involve a number of unrelated toxins which originate in an inconspicuous assemblage of epiphytic species rather than one or a group of related toxins from a single bloom-forming species.

Although CFP is of common occurrence at certain locations, outbreaks are generally sporadic and unpredictable. Furthermore, abundance of toxic dinoflagellates, especially *G. toxicus*, does not consistently correlate with levels of fish toxicity or human intoxications (Bagnis *et al.* 1985a; Lewis *et al.* 1987). Clearly, the presence of such a vast array of toxic species and toxins in a relatively confined ecosystem and the complex nature of the ciguatera syndrome warrant detailed study of the entire assemblage of epiphytic dinoflagellates. It will not be possible to understand the ecology of one species without also understanding that of its cohabitants.

2. Epiphytic and Benthic Dinoflagellate Communities

Prior to 1990 communities of ciguatera-associated dinoflagellates could be described in terms of a few prominent species that typically occurred in close association with macroalgae (epiphytic) and sometimes occupied various bottom materials including coral rubble, sand, and detritus (benthic). These communities usually included *G. toxicus, O. lenticularis* or *O. siamensis* (occasionally *O. heptagona* or *O. ovata*), *C.*

NATO ASI Series, Vol. G 41
Physiological Ecology of Harmful Algal Blooms
Edited by D. M. Anderson, A. D. Cembella and G. M. Hallegraeff

monotis, P. lima, P. concavum, P. mexicanum, P. emarginatum, A. carterae and *A. klebsii* (Yasumoto *et al.* 1980; Fukuyo 1981; Steidinger 1983; Tindall *et al.* 1984; Carlson and Tindall 1985; Bomber *et al.* 1988a, 1989). Quod *et al.* (1995) provided the first confirmation of a more-or-less typical epiphytic dinoflagellate community in the Indian Ocean. The most notable exception to the norm was the regular occurrence of a new toxic species of *Ostreopsis* (Quod 1994). Faust (1990-1995) and Faust and Morton (1995) discovered a more diverse benthic dinoflagellate community in the barrier reef system of Belize. These studies culminated in the description of fifteen new taxa (including species of *Prorocentrum, Gambierdiscus, Ostreopsis* and *Coolia*). A list of the prominent species of dinoflagellates known to occur in epiphytic, benthic, and associated planktonic communities in ciguatera-endemic regions is included in Table 1. Individual species may be found in a wide range of specific habitats even within relatively small reef, reef flat, or mangrove systems. However, most tend to show a preference for one habitat type.

The most widely studied specific habitat type supporting ciguatera-associated dinoflagellates is sessile macroalgae. Some species are loosely or firmly affixed to surfaces of macroalgae by a coating of mucus of variable consistency, whereas, other species appear to be more mobile within the interstices of macroalgal thalli. As a result of this variable affixation, most species are quite vulnerable to the physical pressures of the environment, especially turbulence. Thus, reproducible quantitative analyses of epiphyte communities are extremely difficult, if not impossible, in energetic environments.

Species of macroalgae reported to host significant numbers of epiphytic dinoflagellates are numerous and include members of the phyla Rhodophyta, Phaeophyta, Chlorophyta, and Cyanophyta. Although it is not uncommon for individual species of dinoflagellates to demonstrate an apparent preference for a particular macroalgal species (or phylum) at a single site, there is little hint of consistency between sites. However, there is some evidence suggesting that colonization favors three dimensional, flexible, high surface area algae (Taylor 1985, Lobel *et al.* 1988, Bomber 1989). For example, macroalgae most frequently reported to host significant numbers of epiphytic dinoflagellates include species of *Spiridia, Acanthophora, Heterosiphonia, Digenia, Laurencia, Dictyota, Rosenvingea, Turbinaria, Sargassum, Padina, Chaetomorpha, Cladophora* and *Penicillus*.

The highest density reported for *G. toxicus* was from the coralline red alga *Jania* (over 5.0 x 10^5 cells^{-1} gram fresh weight macroalga) in a Gambier Island reef (Yasumoto *et al.* 1980). Carlson and Tindall (1985) reported a maximum density of *G. toxicus* associated with the green alga *Chaetomorpha* (7.5 x 10^4 cells^{-1} g f.w. alga) in a protected cove in the British Virgin Islands. The highest maximum number of *O. lenticularis* was found in association with *Dictyota* (2.35 x 10^5 cells^{-1} g f.w. alga) at Laurel Reef, Puerto Rico (Ballantine *et al.* 1985). Also, significant numbers of cells of *G. toxicus* and *Ostreopsis* spp. are known to occur in association with the seagrass, *Thalassia testudinum* (Ballantine *et al.* 1985). Maximum numbers for each of eight other epiphytic species reported by Carlson and Tindall (1985) ranged from 1.48 x 10^3 (*P. emarginatum*) to 1.5 x 10^6 (*P. mexicanum*)$^{-1}$ g f.w. macroalgae. These and most exceptionally high numbers reported were only occasionally observed during extended surveys of the same site. More common maximum densities for epiphytic species range from about 100 to 10,000 cells^{-1} g f.w. of macroalgae.

Benthic communities which form in back reefs and shallow reef flats are usually short-lived due to increases in turbulence. However, in low energy systems such as protected coves and mangrove island lagoons benthic communities are more stable and quite productive (Carlson and Tindall 1985; Faust 1996). Regardless, as researchers begin to recognize the recently described benthic species throughout their geographic ranges, the perception of their preferred habitat type may change. For example, both *P. hoffmannianum* and *P. belizeanum* were described from attached and floating detritus but are now known to be abundant on macroalgae in Atlantic and Pacific regions (Tindall and Morton unpubl.).

Also included in Table 1 are planktonic species that form extensive blooms in close association with epiphytic and benthic communities in certain low energy environments. Planktonic species such as *S. subsalsa, G. grindleyi* and *G. polyedra* regularly comprise associated blooms but often are equally abundant on macroalgae and bottom substrates. They display marked diurnal migration patterns that result in considerable variation in their apparent preferred habitat types similar to that described for *Peridinium quinquecorne* in the Philippines (Horstmann 1980).

Table 1. Species of dinoflagellates reported to form closely associated communities in ciguatera-endemic regions of the world and their apparent preferred habitat type(s).

Species	Habitat type
Gambierdiscus toxicus Adachi & Fukuyo	epiphytic
G. belizeanus Faust	benthic & epiphytic
Ostreopsis siamensis Schmidt	epiphytic
O. lenticularis Fukuyo	epiphytic
O. ovata Fukuyo	epiphytic
O. heptagona Norris, Bomber & Balech	epiphytic
O. mascarenensis Quod	epiphytic
O. labens Faust and Morton	benthic & epiphytic
Coolia monotis Meniuer	epiphytic
C. tropicalis Faust	benthic
Prorocentrum lima (Ehrenberg) Dodge	epiphytic
P. arenarium Faust	benthic
P. emarginatum Fukuyo	epiphytic
P. sculptile Faust	benthic
P. maculosum Faust	benthic
P. foraminosum Faust	benthic
P. concavum Fukuyo	epiphytic
P. hoffmannianum Faust	epiphytic
P. belizeanum Faust	epiphytic
P. sabulosum Faust	benthic
P. ruetzlerianum Faust	benthic
P. elegans Faust	planktonic
P. formosum Faust	planktonic
P. micans Ehrenberg	planktonic
P. carribaeum Faust	planktonic
P. mexicanum Tafall (also see *P. rhathymum*)	epiphytic
Amphidinium carterae Hulburt	epiphytic
A. klebsii Kofoid and Swezy	epiphytic
Scrippsiella subsalsa (Ostenfeld) Steidinger & Balech	plankt., benth. & epiphtic
S. trochoidea (Stein) Loeblich	planktonic
Gonyaulax grindleyi Reinecke	planktonic & epiphytic
G. polyedra Stein	planktonic & epiphytic
Protoperidinium quinquecorne (Abe) Balech	planktonic
Ceratium hircus Schroder	planktonic
Cochlodinium polykirkoides Margalef	planktonic
Gymnodinium sanguineum Hirasaka	planktonic
Gyrodinium fissum (Levandor) Kofoid and Swezy	planktonic

3. Geographic Distribution

Gambierdiscus toxicus (as *Diplopsalis* sp.) was the first dinoflagellate species from coral reef habitats found to produce ciguatoxin- and maitotoxin-like compounds (Yasumoto *et al.* 1977; Bagnis *et al.* 1977). This distinction established *G. toxicus* as the preeminent ciguatoxigenic organism. As a result, published records of its occurrence far outnumber those of associated species. Clearly, the geographic range of *G. toxicus* is circumtropical-subtropical (Fig. 1). With the exception of established populations in the Bermuda Islands (about 32°N), the known distribution of perennial populations of *G. toxicus* is between 28°N and 28°S. Ephemeral populations have been reported in southern Japan (about 32°N) and northern New Zealand (about 35°S) (Hara *et al.* 1982; Chang 1995). *Gambierdiscus belizeanus* is confirmed only from South Water Cay and Carrie Bow Cay, Belize (16°49'N) (Faust 1995).

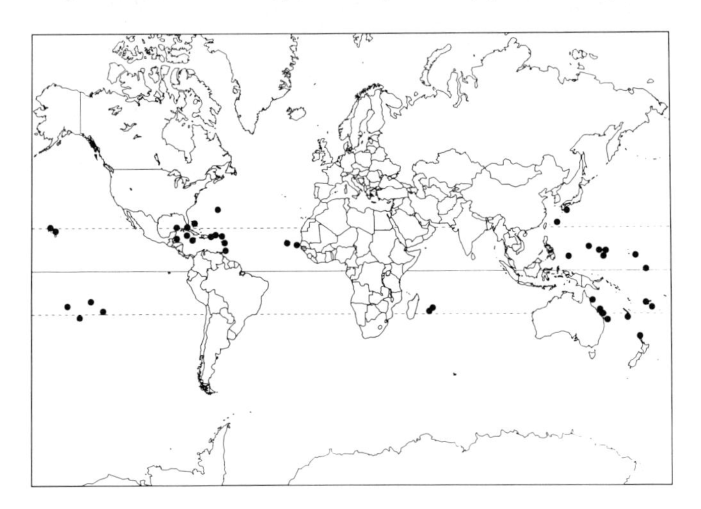

Figure 1. A summary of the known distribution of *Gambierdiscus toxicus*.

Species of the Ostreopsidaceae are represented to some extent in most established epiphytic/benthic dinoflagellate communities in all ciguatera-endemic regions of the world. The geographic ranges of *O. siamensis, O. lenticularis* and *O. ovata* appear to be more or less identical to that of *G. toxicus*, with two notable exceptions: *O. siamensis* in the Mediterranean Sea (Taylor 1979) and *O. ovata* in the Tyrrhenian Sea (Tognetto *et al.* 1995). *Ostreopsis heptagona* is known only from the Florida Keys and Bahamas (Norris *et al.* 1985). *Ostreopsis labens* was reported from a limited number of sites in the Caribbean Sea, Indian Ocean, and Pacific Ocean (about 25°30'N to 21°10'S); whereas, *O. mascarenensis* was reported from the same sites in the Caribbean Sea and the Indian Ocean (16°49'N to 21°10'S) (Quod 1994; Faust and Morton 1995; Faust *et al.* 1996). *Coolia tropicalis* was reported from the same sites as those reported for *O. labens*, and one additional site in Puerto Rico (25°30'N to 17°56'S) (Faust 1995). *Coolia monotis* is ubiquitous in tropical and subtropical regions and its range extends well into temperate regions.

Species of *Prorocentrum* are well represented in most epiphytic and benthic dinoflagellate communities. The distribution of the more frequently reported species (*P. lima, P. concavum, P. mexicanum,* and *P. emarginatum*) is adequately represented by that depicted for *G. toxicus* in Figure 1. However, *P. lima* is also widely distributed in temperate regions. Faust (1990-1994) described eleven new species of *Prorocentrum* from a variety of habitats in the reef system of Belize. At least three of these new species, namely *P. hoffmannianum, P. belizeanum* and *P. elegans*, are wide-spread in the Atlantic and Pacific Oceans.

Species of *Amphidinium* also are significant components of ciguatera-associated dinoflagellate communities. The two species most often reported are *A. klebsii* and *A. carterae*. Morphotypes representing these two species commonly occur in abundance in both epiphytic and benthic communities but are rarely found in associated dinoflagellate blooms. Based on their reported distribution, both species occur most commonly in subtropical and tropical waters.

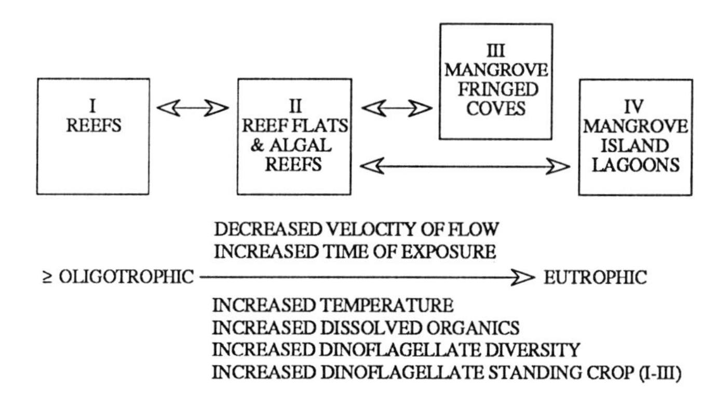

Figure 2. Four subsystems of the ecosystem supporting epiphytic dinoflagellate communities. Arrows between subsystems indicate transfer of dinoflagellates within the ecosystem. Some changes in the nature of the water and dinoflagellate communities that occur between subsystems are noted.

4. The Ecosystem

The physical boundaries of the ecosystem supporting these epiphytic/benthic dinoflagellates extend from beach (or shore) to ledge or outer forereef (allowing for depth limitations) throughout tropical and subtropical regions of the world. Of course, not all areas within these boundaries possess the essential abiotic and biotic prerequisites for development of epiphytic or benthic dinoflagellate communities. We suggest that this ecosystem consists of a series of four intergrading but relatively distinct subsystems (Fig. 2), hereafter referred to as systems. These systems are distinguished to a large extent by their respective hydrodynamic characteristics. Changes in the physical and chemical nature of the water as it transverses these systems are noted in Figure 2. There are discernible patterns to the distribution, structure, and density of macroalgal-associated dinoflagellate communities which relate directly to these intergrading physical and chemical factors. Also, it is apparent that members of these communities are regularly transported shoreward and seaward via rafting on detached macroalgae and floating detritus (Besada 1982; Bomber 1988a; Faust 1995). Movement and concentration of dinoflagellate cells may also occur by way of outflow channels (Bomber 1989). These transport mechanisms in addition to the tendency of certain species to sink probably account for the presence of *G. toxicus* and *P. lima* in deep forereef sites such as those reported by de Silva (1982).

Although there are several published reports on the distribution and abundance of *G. toxicus* and one or two associated species, complete descriptions and analyses of entire epiphytic dinoflagellate communities are limited. Thus, the basic blueprint for the proposed model is a year-long study of the entire epiphyte community conducted at seven stations in the Virgin Islands (Carlson 1984; Tindall *et al.* 1984; Carlson and Tindall 1985). However, general descriptions of type I and type II systems incorporate observations from world-wide locations.

4.1. System I

Type I systems include high energy barrier, patch, and fringing reefs and associated channels. The most distinctive feature of this system is its nearly constant, moderate to strong flow of water (including surf, swells, and currents). Due to the sustained movement of water, physical and chemical conditions are generally characteristic of those of the open ocean (oligotrophic). Depths at which extensive epiphytic communities develop range from 0.5 to 4 meters. Macrophytes that support dinoflagellate communities consist of moderate to dense patches of macroalgae (*e.g. Turbinaria, Sargassum, Padina, Jania, Galaxaura*) on dead coral and/or small to expansive stands of mixed macroalgae and seagrasses in the backreef flats and crest zones. During periods of unusual calm epiphytic communities in type I systems may temporarily take on the appearance of communities typical of type II systems.

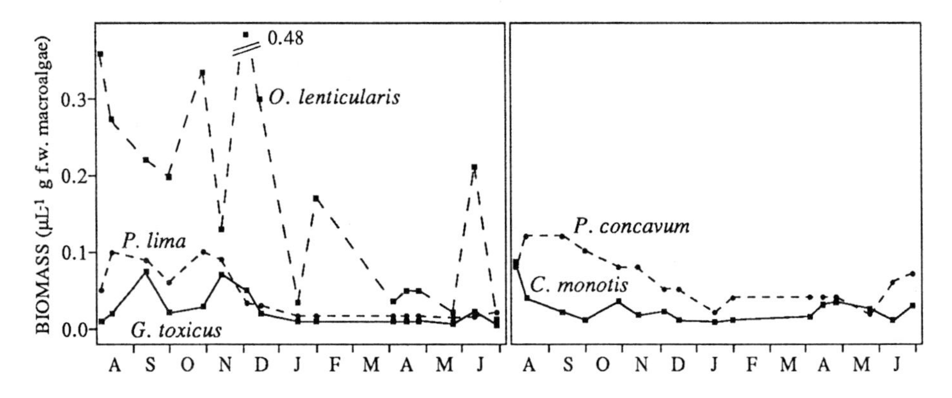

Figure 3. Dynamics of 5 species of dinoflagellates on macroalgae at West End, Anegada during August 1980-June 1981. Data from Carlson (1984) and Carlson and Tindall (1985).

Two type I systems were examined in the Virgin Islands. These included northerly exposed, high energy fringing reefs at Anegada and Necker Island (Carlson and Tindall 1985). The total numbers of species of dinoflagellates recorded at these stations were 16 and 14, respectively. The dynamics (as biomass) of the five most prominent species at the Anegada station over the course of one year is shown in Figure 3. Although density of all species appeared to correlate negatively with temperature at this station (Carlson and Tindall 1985), it was most negatively influenced by increased wave action characteristic of the cooler winter season. Lowest densities for all species were observed during January through May which was a period of heavy ground swells, so severe that the Anegada station was not accessible during February and March (Carlson 1984). However, collections at the more accessible Necker Island station revealed strong peaks of *O. lenticularis* in mid February and early April.

Table 2. Summary of mean density and standing crop biomass of dinoflagellates on macroalgae at 7 stations in the British and United States Virgin Islands during June 1980 through August 1981. Stations are arranged according to system type (data from Carlson and Tindall 1985).

System	Station	Annual mean cell density $(\times 10^3)^{-1}$ g f.w. macroalgae	D	Annual Mean Biomass (μL^{-1} g f.w. macroalgae)
I	Necker Island	7.5 ± 4.2	14	0.28
I	Anegada	9.0 ± 6.3	16	0.29
II	Mattie Point	31.8 ± 14.6	23	0.43
III	Water Creek	33.9 ± 25.1	26	0.90
III	South Creek	121.4 ± 252.4	31	2.10
III	Salt Creek	111.3 ± 114.9	29	1.36
III	Biras Creek	146.6 ± 277.6	28	1.67

D = diversity represented by the total number of species of dinoflagellates at each station

The similarity in mean density and biomass of epiphytic dinoflagellates observed at Anegada and Necker Island is shown in Table 2. Likewise, there was consistency in the structure of the communities at these stations (Fig. 4). *Ostreopsis lenticularis* contributed 63-66% and *G. toxicus* contributed 8-11% of the mean standing crop biomass at these stations. Much of the remaining biomass was a product of *P. lima, P. concavum* and *A. carterae* (19-23%). Although mean cell density and biomass were relatively low, all of these species appeared to be well-adapted to maintain their presence on surfaces of macroalgae despite a nearly constant high degree of turbulence. Nevertheless, standing crop estimations of macroalgal-associated dinoflagellate communities in high energy systems will seldom reflect real production rates. It is reasonable to assume that much biomass is regularly distributed to other habitats within and outside the reef system. The frequent occurrence of epiphytic species on inorganic bottom materials at Anegada and Necker Island (Carlson and Tindall 1985) and the presence of *G. toxicus* and *P. lima* at deep forereef sites in the Caribbean (de Silva 1982) support this assumption.

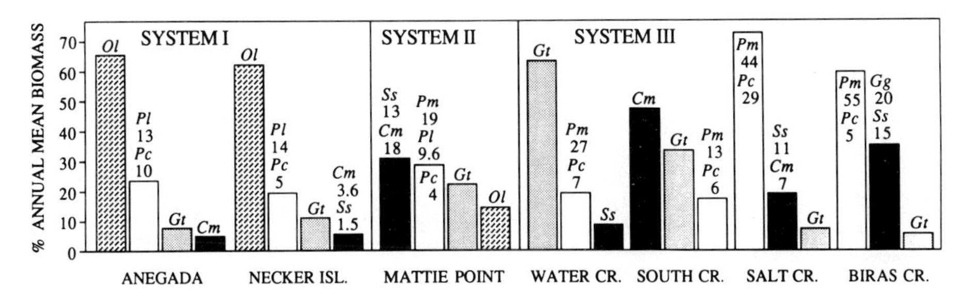

Figure 4. Epiphytic dinoflagellate community structure represented by percent annual mean biomass of predominant species at 7 stations (3 systems) in the British and United States Virgin Islands.

4.2. System II

Type II systems include permanently submerged flats, algal reefs, and tide pools extending seaward from shore to ledge or backreef, and shallow flats extending into lagoons and channels. Protection afforded by reef or land mass, established macroalgal/seagrass communities, absence of substantial freshwater inflow, and minimal tide flux may be prerequisites to extensive epiphytic dinoflagellate community development in this system. Maximum depth for significant community development is about 4 m. Although several authors have reported a minimum depth of 0.5 m for sustained growth of epiphytic dinoflagellates, we have observed dense communities at 2-3 cm at low tide in seagrass-macroalgal-coral flats in the Virgin Islands. Type II systems are usually characterized by moderate to dense stands of macroalgae (*e.g. Dictyota, Achanthophora, Laurencia, Spiridia, Pinicillus*, *Halimeda*) and sparse to dense meadows of seagrasses. Type II systems often display dense to scattered patches of live (or dead) coral.

Constant moderate flow associated with tide fluxes characterizes type II systems. As a result of the extended duration of exposure of the same water to sunlight and substrates the physical and chemical features of the water are affected somewhat by local conditions. Thus, type II waters may frequently display mesotrophic characteristics. More eutrophic conditions may characterize flats adjoining islands that serve as rookeries and flats that receive substantial domestic sewage. Of course, many type II systems are subject to periodic surges, at which times conditions may be more oligotrophic. Exposed shoreline flats that are regularly subjected to high velocities of flow should be classified as type I systems.

The only type II system examined by Carlson and Tindall (1985) was Mattie Point. This station is located in the South Sound, Virgin Gorda. South Sound is a large, easterly exposed bay enclosed by an extensive bay mount barrier reef. Mattie Point is a shoreline coral-rock formation situated 200 meters behind the barrier reef. The bottom, which consists of dead coral pavement and moderately dark sand, slopes from shore to about 4 m deep.

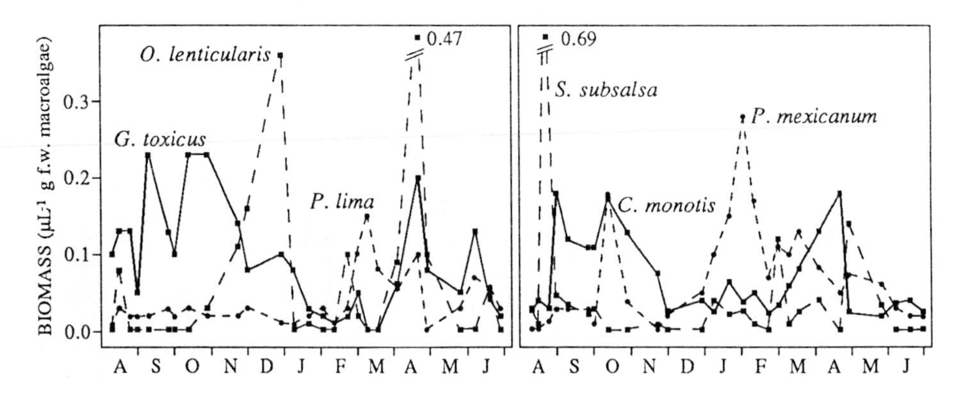

Figure 5. Dynamics of 6 species of dinoflagellates on macroalgae at Mattie Point during August 1980-June 1981. Data from Carlson (1984) and Carlson and Tindall (1985).

Twenty-six species of dinoflagellates were recorded at Mattie Point, of which six represented significant biomass over the course of one year (Fig. 5). The responses of individual species to variable conditions at Mattie Point resulted in a very dynamic community. However, annual mean cell density at this station was over three times that recorded at the Anegada station. Due to the numerical dominance of small species such as *A. carterae, P. mexicanum,* and *C. monotis* mean biomass production was only 1.5 times that recorded at Anegada (Table 2). *Prorocentrum lima*, *P concavum*, and *S. subsalsa* were also represented by consistently high numbers. Numbers of *O. lenticularis* and *G. toxicus* were relatively low, but due to their large cell size they added significantly to the total biomass. One of the most interesting features of the epiphyte community at Mattie Point was the balance observed in annual mean biomass of its most predominant species. Six species produced from about 10 to 20% of the annual mean biomass (Fig. 4). Although considerably less energetic than the reef environment, shoreline flats are exposed to continuous (variable) flow which will take its toll on the epiphytic community. Also, such flats are subject to immigration from other sites.

4.3. System III

Our description of the type III system is based on observations at four stations in the Virgin Islands (Carlson and Tindall 1985). This system includes protected mangrove-fringed coves of larger enbayments of high islands. Based upon the observations of Yasumoto *et al.* (1980) and Taylor and Gustavson (1985) and results of culture studies, we exclude from this system those mangrove lagoons or other enbayments that regularly receive moderate to high volumes of fresh water. Frequent salinities below 28 psu will probably preclude permanent epiphytic dinoflagellate community development. Type III systems experience only limited water movement, usually a gentle rise and fall associated with tide flux (slow flushing rate). Due to near stagnation, physical and chemical qualities of the water and bottom materials are affected directly by the organic load from the mangrove stand and by local geologic structures. Also, this standing water is subject to greater extremes in temperature, salinity, and other features that are influenced by light intensity and atmospheric conditions. Depths at which epiphytic dinoflagellate communities develop range from about 0.2 to 3 m. The bottoms are usually gray-brown in color consisting of carbonate sand-silt and mud. The substrate a few mm beneath the sand-silt surface layer is often distinctly organic and anoxic. Generally, conditions can be described as eutrophic. The stations examined displayed sporadic or permanent dinoflagellate blooms which restricted macroalgal growth to clear water along the shores. The macrophyte assemblage often included an abundance of filamentous chlorophytes such as *Chaetomorpha* and *Cladophora* and assorted attached cyanophytes. Diversity and abundance of other macroalgae and seagrasses increased with decreased phytoplankton growth. Diatom growth was common on bottom substrates.

The four type III stations examined in the Virgin Islands included Water Creek, St. John; Salt Creek, Salt Island; and South Creek and Biras Creek, Virgin Gorda (Carlson and Tindall 1985). Our selected example of system III is South Creek, a protected, mangrove-fringed cove about 400 meters south of the Mattie Point station in South Sound, Virgin Gorda. The most pronounced biological features of this station was its permanent dinoflagellate bloom which consisted of expansive dark red-brown patches comprised of *Cochlodinium polykirkoides*, *Scrippsiella trochoidea, Gymnodinium sanguineum, Gonyaulax polyedra*, and other lesser abundant species. The lighter margins of the bloom contained the same species and significant numbers of *Gyrodinium fissum* .

Although this station displayed greater dinoflagellate diversity (31 spp.) than the type I and II stations described, only four epiphytic species were regularly observed in significant numbers (Fig. 6, *P. concavum* not shown). Notably absent from macroalgae were *O. lenticularis, P. lima* and *A. carterae*. Notwithstanding the absence of these species, South Creek was the most productive station with four species displaying a combined mean biomass of 2.10 μL^{-1} g f.w. macroalgae (Table 2). This unusually high mean biomass was due to occasional extraordinarily high numbers of *C. monotis* and fairly consistent high numbers of *G. toxicus, P. mexicanum*, and *P. concavum*. The highest mean density (and biomass) recorded for *G. toxicus* during the Virgin Island study was at South Creek. This species accounted for 34 % of the annual mean biomass (Fig. 4). Although *G. toxicus* contributed 64 % of the mean biomass at Water Creek, its mean cell density was somewhat less than that observed at South Creek. The four type III stations consistently displayed higher mean standing crop biomass than that observed at type I and II stations (Table 2). Generally, conditions characterizing the four stations were conducive to growth of several species (*e.g. P. mexicanum, C. monotis, S. subsalsa, G. grindleyi, P. concavum,* and *G. toxicus*) but were unfavorable for growth of *O. lenticularis* and *P. lima*. Between-station differences in community structure appeared to be a result of individual species responses to the nutritional, physical, and biotic features unique to each station.

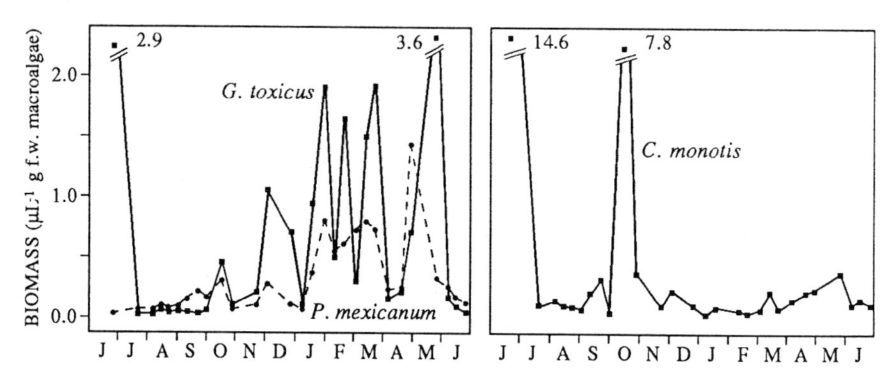

Figure 6. Dynamics of 3 species of dinoflagellates on macroalgae at South Creek during June 1980-June 1981. Data from Carlson (1984) and Carlson and Tindall (1985).

4.4. System IV

We tentatively characterize the type IV system as mangrove island lagoons such as that described by Faust (1995, 1996). Characteristics of this system are similar to those of system III. Both systems are normally characterized by the consequences of a very slow flushing rate. However, significant unique features of mangrove island lagoons, which we infer from the work of Faust, include a much heavier organic load from extensive mangrove forests, limited macroalgal development, extensive "living" detrital mat formation, and a very diverse benthos-associated dinoflagellate community. However, it is conceivable that the mangrove island lagoon is a large-scale version of a microenvironment commonly manifested in type III systems. Since studies of the dinoflagellate community in mangrove island lagoons have been of a short-term nature, we do not have a basis on which to compare its dynamics with that of Virgin Island communities.

5. Carrying Capacity

Based on the observations discussed above, we concur with the suggestion of Lobel *et al.* (1988) that individual macroalgae have a "carrying capacity," but it is not a static entity. The carrying capacity of a species of macroalgae varies depending on prevailing conditions (especially velocity of flow). Each species of macroalgae has a characteristic surface area (or space) which once occupied, can support no additional cells. Under conditions of high turbulence, space available for association is limited to the surface layer. However, as flow decreases, space available for association is extended to include the medium surrounding the branches of the macroalga. Under stagnant conditions emigration ceases and the dinoflagellates multiply to fill all spaces within the macroalgal canopy. The concept of a variable carrying capacity as it applies to communities of macroalgae situated along the longitudinal axis of the systems gradient is illustrated in Figure 7. Although the single most important factor regulating carrying capacity is velocity of flow, as system III is approached along the systems gradient, carrying capacity is increasingly affected by other physical and chemical factors (temperature, salinity, gasses, inorganic nutrients, and organic compounds). Thus, the unique nature of each type III system will govern carrying capacity to some extent (as well as community structure and dynamics). The result is consistently high but variable standing crop biomass among type III systems.

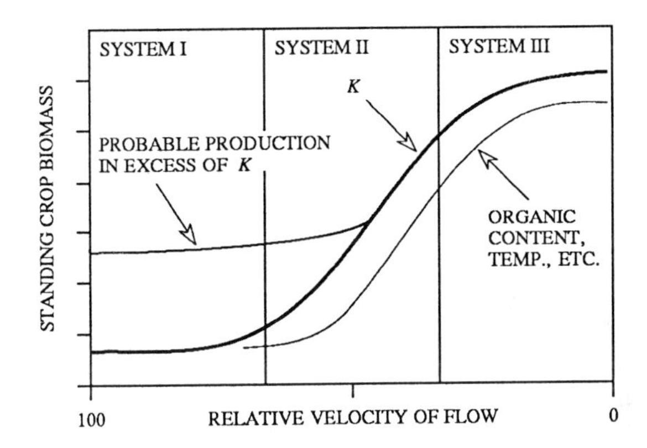

Figure 7. The carrying capacity (*K*) of macroalgae in the intergrading systems of the epiphytic dinoflagellate ecosystem. There is a progressive increase in *K* and variations in other factors that accompany decrease in velocity of flow as the systems gradient is transversed. Also shown is the probable routine production of biomass in excess of normal *K* in high energy segments of the gradient.

The biomass produced in system I which is in excess (surplus) of carrying capacity is continuously distributed to all parts of the reef and adjacent areas. That component representing excess production may also represent increased carrying capacity (and standing crop biomass) during extended periods of calm in energetic systems. Also, we suggest that a gradient of diminishing velocity of flow exists on the vertical axis in high energy systems. Thus, there will be a corresponding change (zonation) in carrying capacity and community structure with increasing depth. The latter, in addition to redistribution probably account for the high concentrations of epiphytic dinoflagellates observed on macroalgae and in detritus in deeper pockets of backreefs and forereefs.

Figure 8 is a conceptual model showing the dynamic nature of macroalgal carrying capacity along the longitudinal and lateral axes of the ecosystem. Inclusion of changes that occur on the vertical axis in energetic segments would provide a relatively complete model of carrying capacity as it relates to velocity of flow. We conclude that this model accounts for (or could predict) variations in mean biomass production at points along the systems gradient. It is also useful for explaining or predicting some aspects of dinoflagellate community dynamics in each systems. Also, there is an intergrading but definable mean dinoflagellate community structure that parallels the systems (and carrying capacity) gradient. Predictability of community structure and biomass production allows for generalizations regarding probable toxins (and potency) in fish at selected sites within the ecosystem.

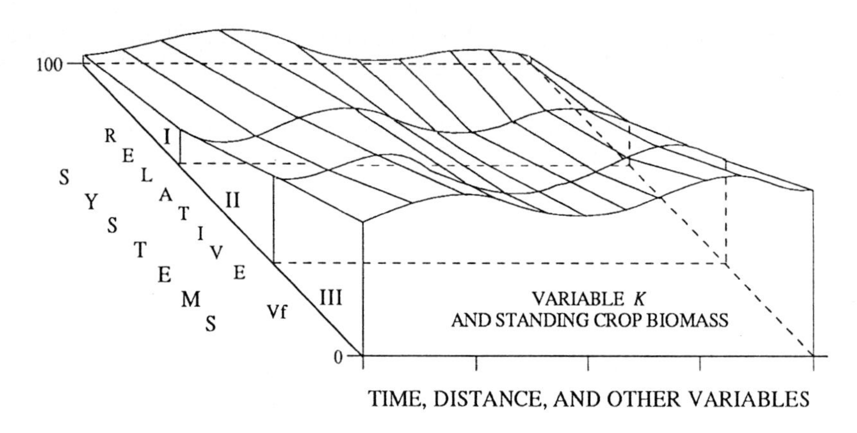

Figure 8. The dynamic carrying capacity (*K*) of the macroalgal canopy. *K* increases with decreases in velocity of flow (Vf) along longitudinal and lateral (and vertical) axes. As *K* increases it is increasingly altered by other abiotic and biotic factors including temperature, salinity, carbon dioxide, organics, nutrients, phytoplankton etc. Epiphytic dinoflagellate populations (and communities) adjust accordingly.

6. Growth in Culture

Most of the more prominent species of epiphytic/benthic dinoflagellate from ciguatera-endemic regions have been isolated and cultured. Clearly, under controlled laboratory conditions some amount of enrichment of natural seawater is required in order to induce growth rates and biomass production comparable to that observed in nature. Media which have been utilized to achieve relatively good biomass and toxin production for these species were reviewed by Bomber and Aikman (1989). These media were similar in that they consisted of natural seawater enriched with a limited number of nitrogen (KNO_3, $NaNO_3$ and/or NH_4Cl) and phosphorus (K_2HPO_4, NaH_2PO_4, $NaH_2PO_4{\cdot}H_2O$, or Na_2 glycero$\cdot PO_4$) sources, chelators (EDTA, soil extract, or algal extract), and essential micronutrients. Although vitamins are frequently added to culture media, amounts greater than those occurring in natural seawater do not appear to be essential for growth of some species (Yasumoto *et al.* 1980; Durand-Clement 1987).

Members of the epiphytic/benthic community appear to be photoautotrophic but auxotrophy under some circumstances should not be ruled out. Furthermore, Faust and Morton (1995) have provided evidence of heterotrophy in species of *Ostreopsis.*

Growth of some species in certain seawater-based media has been stimulated by the addition of macroalgal extracts (Withers 1981; Carlson *et al.* 1984; Tosteson *et al.* 1989; Bomber *et al.* 1989). However, the same species have displayed comparable growth rates in K medium (high EDTA) without the addition of extracts (Bomber *et al.* 1988b, 1989; Morton *et al.* 1992). Although algal extracts appear to provide beneficial chelators which enhance growth in culture, it is too soon to conclude that chelation is the only nutrition-related function of macroalgae in the natural environment.

6.1. Nitrogen and Phosphorus

Epiphytic/benthic dinoflagellates do not appear to be unique in their requirements for the two major limiting macronutrients, nitrogen and phosphorus. They readily utilize nitrate and ammonium with some showing an apparent preference for the latter (Bomber and Aikman 1989; Aikman *et al.* 1993). All epiphytic species appear to grow well on either inorganic phosphorus (H-PO_4) or organic phosphorus (glycero PO_4). Six geographically distinct clones of *G. toxicus* acclimated to the same environmental conditions demonstrated between-clone variations but no significant within-clone variation in growth rates when subjected to N:P ratios ranging from 5:1 to 50:1 (Sperr and Deucette 1996). Aikman *et al.* (1993) reported rapid depletion of NH_4 and HPO_4 from the medium during early log phase growth (3 and 7 days, respectively) of *P. hoffmannianum.* Little depletion of NO_3 was noted throughout the complete growth cycle. Total intracellular phosphorus rose from an initial level of 3 pg^{-1} cell to 12 pg^{-1} cell at day 3, at which time it began a progressive decline to less than 3 pg at day 11 and remained at about 2 pg^{-1} cell throughout the remainder of the growth cycle (log growth extended through day 22). These results suggest high efficiency in HPO4 uptake, effective use of intracellular phosphorus reserves under external phosphorus-limited conditions, and that phosphorus is likely to be a limiting factor in some natural environments. Interestingly, the pattern of H-PO4 (and NH_4) uptake and utilization of intracellular phosphorus closely followed the pattern of toxin (okadaic acid and fast-acting toxin) synthesis in *P. hoffmannianum* (see Aikman 1992). Initial declines followed by rapid increases in toxin content are characteristic of most *Gambierdiscus* and *Prorocentrum* growth cycles that we have examined. These observations suggest that toxin synthesis in a variety of species might be linked to intracellular phosphorus utilization, especially when phosphorus is limited in the medium. Aspects of phosphorus uptake and utilization and toxin synthesis also closely parallel certain aspects of pigment synthesis during the growth cycles of *G. toxicus* (Tindall unpubl.) and *P. hoffmannianum* (Aikman *et al.* 1993).

6.2. Salinity

Yasumoto *et al.* (1980) examined growth of *G. toxicus* at five salinity levels ranging from about 27 to 42 psu. They observed little or no growth at 27 and 42 psu, moderate growth at about 30 and 39 psu, and best growth at 34 psu. Bomber *et al.* (1988b) reported optimal growth of *G. toxicus* at 32 psu and a considerable reduction in growth rates at 25 and 40 psu. Morton *et al.* (1992) reported salinity optima of 30 psu for *P. lima* and *P. concavum*; 33 psu for *A. klebsii*, *C. monotis*, *O. siamensis*, and *O. heptagona*; 30-33 psu for *G. toxicus*; and 36 psu for *P. mexicanum*. Minimal growth occurred at salinities of 25 and 40 psu. Morton *et al.* (1994) reported moderate growth at 28 and 40 psu and optimal growth at 34 psu for *P. hoffmannianum.* Thus, growth rates for the community as a whole in the natural environment are probably highest at salinities between 30-34 psu and are limited at salinities less than 28 and greater than 38 psu. This conclusion is supported by the positive correlation between dinoflagellate biomass production and annual rainfall observed in the Virgin Islands (Carlson and Tindall 1985). Also, the intolerance of *G. toxicus* to salinity below 28 psu explains its absence in areas receiving continuous or periodic heavy surface runoff as reported by Yasumoto *et al.* (1980) and Taylor (1985).

6.3. Irradiance

Guillard and Keller (1984) concluded that dinoflagellates generally grow well in any light source at an irradiance of about 10% of full daylight. The epiphytic/benthic species comply with this generality, but 10% of full daylight is near the upper limit for optimal growth. A comparison of light regimes used for culture of epiphytic species of *Gambierdiscus*, *Ostreopsis*, *Coolia, Prorocentrum* and *Amphidinium* reveals some discrepancies regarding quantities of light required for optimal growth. However, these discrepancies appear to relate to light source. Cells grown under daylight or cool-white fluorescent tubes displayed optimal growth at about 3-4% of full daylight, whereas those grown under Vita-Lite tubes (Duro-Test Corp.) displayed optimal growth at about 8-10% of daylight. Vita-Lite tubes simulate the full color and ultraviolet spectrum of sunlight more closely than do other general purpose fluorescent tubes.

Bomber *et al.* (1988b) observed highest growth rates for *G. toxicus* at 11% of full sunlight (Vita-Lite) within the blue range of the spectrum but found growth efficiency to drop (under blue light) when intensities decreased. They concluded that optimal depth for growth of *G. toxicus* in clear tropical waters was 1-4 m. Thus, we may conclude that *G. toxicus* and probably other epiphytic dinoflagellates struggle to maintain a position in the environment where there is a balance between appropriate light quality and acceptable intensity. The location of certain macroalgae in the natural environment may represent that position where light quality acceptable to epiphytic dinoflagellates is generally maintained. The branches of those macroalgae provide a substrate on which the dinoflagellates can affix and spread out for maximum exposure or be utilized as a shield (filter) from the damaging effects of periodic high intensities (see Ballantine *et al.* 1988; Heil *et al.* 1993). Also, the close proximity of inorganic substrates and detrital layers afford immediate access to even greater protection from high light intensities. The observation of Carlson and Tindall (1985) that maximum community development was on macroalgae growing at 1-3 m in substrates of low reflectivity supports this hypothesis. Highly reflective substrates would counter the positive shading effects of the macroalgal canopy.

6.4. Temperature

Most epiphytic/benthic species, with the exception of clones of *P. lima* and *C. monotis* isolated from temperate regions (Jackson *et al.* 1993; Silva and Faust 1995) and possibly natural populations of some tropical species inhabiting sites at northern or southern extremes of their geographic ranges (see Gillespie *et al.* 1985), appear to have rather narrow ranges of tolerance with regard to temperature. Generally, isolates from subtropical or tropical locations display sustained growth (above 0.1 division day^{-1}) in culture between 22 and 31° C. However, competitive growth rates are only achieved near optimum temperature for each species. Optimal temperatures for this group as a whole (in culture) range from about 25 to 29° C (Bomber *et al.* 1988b; Morton *et al.* 1992). However, Heil *et al.* (1993) reported good growth rates for *P. hoffmannianum* and *P. mexicanum* from the Florida Keys cultured at 20° C and Zhou and Fritz (1993) reported growth of *P. maculosum* from the Virgin Islands at 20° C.

Apparent discrepancies regarding precise temperature optima for any one or all isolates of a species in culture or in the field may be attributed to genetic adaptations to local climatic conditions. Jackson *et al.* (1993) observed significant growth of *P. lima* (isolated form the coast of Nova Scotia) in cultures maintained at 10, 15 and 20° C, and best growth at the highest temperature tested (25° C). They concluded that since optimum growth occurred at 25° C they were culturing a "warm-water" isolate. However, since this isolate grew quite well at 15 and 20° C, it also demonstrated a degree of adaptation to temperate conditions. Regardless, temperature limitations probably restrict the distribution of populations of most epiphytic species and influence their periodicity throughout subtropical and tropical regions.

Although temperatures below 20° C may limit growth of most tropical epiphytic species, vegetative cells or cysts of some of these species may remain viable at lower temperatures for long periods of time. The ability of quiescent cells of tropical species to tolerate low temperatures might result in seasonal growth of some species in more temperate regions. Clearly, some populations of *C. monotis* and *P. lima* are well-adapted to survive in cold temperate waters. Other examples of such a phenomenon might be represented by the occurrence of ephemeral populations of *O. ovata* in the Tyrrhenian Sea and *G. toxicus* in southern Japan and northern New Zealand.

7. Summary and Conclusions

Growth studies conducted thus far confirm that isolates of most species of ciguatera-associated dinoflagellates display competitive growth rates under nearly identical controlled environmental conditions. This explains their distribution and common association but does not provide clear-cut answers to questions pertaining to periodicity, competition, epiphyte-host interaction, colonization and other aspects of community dynamics. Of course, we suggest that community structure and dynamics at a given site are dictated to a large extent by degree of water movement. The effects of other physical and chemical factors increase coincident with diminishing water movement.

High energy type I systems select for a rather precise hierarchy of species capable of withstanding the pressures of flowing water (*e.g. O. lenticularis, P. lima, A. carterae,* and to a lesser extent *G. toxicus* and *P. concavum*). Also, some species such as *O. lenticularis* may exhibit a preference or requirement for the more oligotrophic conditions that characterize these systems. Due to low carrying capacity, standing crop estimations of the epiphytic community in these systems seldom reflect real production. Biomass in excess of carrying capacity is regularly distributed throughout and adjacent to the reef system. Published reports suggest that community structure in type I systems might vary, however few studies have included the entire epiphyte community. Ballantine *et al.* (1988) described persistent co-dominant populations of *O. lenticularis* and *G. toxicus* at Laurel Reef, Puerto Rico. Also, exceptionally high numbers of *G. toxicus* were reported at reef sites in the Gambier Islands (Yasumoto *et al.* 1980). However, Bagnis *et al.* (1985b) described long-term dynamics of populations of *O. lenticularis, P. lima* and *G. toxicus* on fringing and barrier reefs in Tahiti similar to that described for type I stations in the Virgin Islands. We suggest that differences in community structure within and between high energy reefs will relate to position of the community in the horizontal and vertical gradients of velocity of flow which characterize type I systems.

In many respects (including carrying capacity) the type II system is intermediate between type I and type III systems. The type II system represents a moderately energetic environment in which several epiphytic species (*e.g. C. monotis, P.*

mexicanum, G. toxicus, O. lenticularis, P. lima, S. subsalsa and *P. concavum*) maintain a dynamic but relatively strong and balanced presence. However, variations in community structure and macroalgal carrying capacity in type II systems will be governed by local hydrodynamic features. Bomber *et al.* (1989) examined epiphyte communities in a wide range of flats and algal reefs in the Florida Keys. Some of their stations displayed community structures similar to those observed in the Virgin Islands but others were dominated by *G. toxicus* followed by *Ostreopsis* spp., *P. lima, P. mexicanum* and *P. concavum.* Furthermore, type II systems are often positioned in the first line of exposure to anthropogenic influences and their community structures and biomass production will respond accordingly. Although standing crop estimations are probably more indicative of real production rates, emigration and immigration probably characterize most type II systems.

Although type III systems are consistently highly productive, carrying capacity and community structure vary in response to the unique physical, chemical, and biological characteristics of each site. Observations made in the Virgin Islands revealed *Gambierdiscus toxicus* to dominate macroalgae in some type III environments but to be nearly excluded by species such as *P. mexicanum, C. monotis, S. subsalsa,* and *G. grindleyi* in others. These diminutive species appear to challenge for position in the epiphytic communities of all systems with competitive success increasing with decreasing velocity of flow (increased carrying capacity). *Ostreopsis lenticularis* and *P. lima* were noticeably absent on macroalgae in the type III systems examined. Possible explanations for their absence include the lack of water movement to facilitate association with macroalgae, competitive exclusion from macroalgae, and/or inhibition by chemical or physical factors characteristic of type III systems. The possibility that these two species might occupy the water column or inorganic substrates in type III systems should not be overlooked. For example, there are several reports of *Ostreopsis* spp. (and other epiphytic species) dwelling in the water column and on various bottom materials (Bomber 1989; Tognetto *et al.* 1995; Faust 1996).

Yasumoto *et al.* (1980), Ballantine *et al.* (1988) and Lobel *et al.* (1989) described enormous variations in numbers of cells of *G. toxicus* on adjacent macroalgae within their respective collecting sites. Such variations are common to most sites supporting abundant epiphytic dinoflagellates. Although there is not a single explanation for this within-site variation, we believe that it fits into the general scheme of system mechanics. Epiphytic species are individually or collectively subjected to the pressures of water movement and the result is redistribution of cells on a small or large scale. Also, each species opportunistically and aggressively seeks to fill its segment of the many overlapping niches in this very dynamic ecosystem, resulting in within-site variability. Bomber (1989) suggested that the presence of *Prorocentrum* spp., *C. monotis* and *O. siamensis* in the water column was a consequence of vertical migration which facilitated redistribution and concentration of cells. Niche separation (Bomber *et al.* 1989) which could be a consequence of allelopathy might also contribute to variability. Compounds produced by several species of epiphytic dinoflagellates are known to be allelopathic to a variety of potential competitor organisms such as fungi, phytoflagellates, and diatoms (Carlson 1984; Durand *et al.* 1985; Dickey *et al.* 1990; Bomber 1991; Nagai *et al.* 1992).

Due to this characteristic within-site variability, the most important consideration when examining epiphytic communities is appropriate sampling. We concur with Lobel *et al.* (1988) that numbers of samples analyzed in most past studies have been inadequate for determining precise causes of community dynamics. Also, it is clear that single spot samples will seldom yield reliable results regarding epiphyte communities in any area or region. However, we believe that the total numbers of samples analyzed during long term studies have provided good characterizations of populations and communities in a variety of environments. Since the dynamics of an individual species is directly affected by other members of the community, it is most essential that new field studies include an assessment of all associated species.

8. References

Aikman, K. E. (1992). The physiology, biochemistry and toxicity of the marine dinoflagellate *Prorocentrum hoffmannianum* Faust (Pyrrophyta), and its role in seafood-borne human diseases. Ph.D. Dissertation, Southern Illinois University, Carbondale. 124 p.

Aikman, K. E., Tindall, D. R., Morton, S. L. (1993). Physiology and potency of the dinoflagellate *Prorocentrum hoffmannianum* during one complete growth cycle. In: Smayda, T. J, Shimizu, Y. (eds.) Toxic Phytoplankton Blooms in the Sea. Elsevier, Amsterdam. p. 463-468.

Anderson, D. M., Lobel, P. S. (1987). The continuing enigma of ciguatera. Biol. Bull. 172:89-107.

Bagnis, R. A., Chanteau, S., Yasumoto, T. (1977). Mise en evidence d'un dinoflagellate responsible en puissance de la ciguatera. Rev. Int. Oceanogr. Med. 45-46:29-34.

Bagnis, R. A., Bennett, M., Barsinas, M., Chebret, M., Jacquet, G., Lechat, I., Mitermite, Y., Perolat, P. H., Rohgeras, S. (1985a). Epidemiologie de la ciguatera en Polynesie Francaise de 1960-1984. Proc. 5th Int. Coral Reef Cong., Tahiti. 4:475-482.

Bagnis, R. A., Bennett, J. Prieur, C., LeGrand, A. M. (1985b). The dynamics of three benthic dinoflagellates and the toxicity of ciguateric surgeonfish in French Polynesia. In: Anderson, D. M., White, A. W., Baden, D. G. (eds.) Toxic Dinoflagellates. Elsevier, New York. p. 177-182.

Ballantine, D. L., Bardales, A. T., Tosteson, T. R., Durst, H. D. (1985). Seasonal abundance of *Gambierdiscus toxicus* and *Ostreopsis* sp. in coastal waters of southwest Puerto Rico. Proc. 5th Int. Coral Reef Cong., Tahiti. 4:417-422.

Ballantine, D. L., Tosteson, T. R., Bardales, A. T. (1988). Population dynamics and toxicity of natural populations of benthic dinoflagellates in southwestern Puerto Rico. J. Exp. Mar. Biol. Ecol. 119:201-212.

Besada, E. G., Loeblich, L. A., Loeblich III, A. R. (1982). Observations on tropical, benthic dinoflagellates from ciguatera-endemic areas: *Coolia, Gambierdiscus,* and *Ostreopsis*. Bull. Mar. Sci. 32:723-735.

Bomber, J. W. (1991). Toxigenesis in dinoflagellates: genetic and physiological factors. In: Miller, D. M. (ed.) Ciguatera Seafood Toxins. CRC Press, Boca Raton, FL. p. 135-170.

Bomber, J. W., Morton, S. L., Babinchak, J. A., Norris, D. R., Morton, J. G. (1988a). Epiphytic dinoflagellates of drift algae - another toxigenic community in the ciguatera food chain. Bull. Mar. Sci. 43:204-14.

Bomber, J. W., Guillard, R. R. L., Nelson, W. G. (1988b). Roles of temperature, salinity and light in seasonality, growth and toxicity of ciguatera-causing *Gambierdiscus toxicus* Adachi et Fukuyo (Dinophyceae). J. Exp. Mar. Biol. Ecol. 115:53-65.

Bomber, J. W., Aikman, K. E. (1989). The ciguatera dinoflagellates. Biol. Oceanogr. 6:291-311.

Bomber, J. W., Rubio, M. G., Norris, D. R. (1989). Epiphytism of dinoflagellates associated with ciguatera: substrate specificity and nutrition. Phycol. 28:360- 368.

Carlson, R. D. (1984). Distribution, periodicity and culture of benthic/epiphytic dinoflagellates in a ciguatera endemic region of the Caribbean. Ph.D. Dissertation, Southern Illinois University. 308 p.

Carlson, R. D., Morey-Gaines, G., Tindall, D. R., Dickey, R. W. (1984). Ecology of toxic dinoflagellates from the Caribbean Sea: effects of macroalgal extracts on growth in culture. In: Ragelis, E. P. (ed.) Seafood Toxins. Am. Chem. Soc. Symp. Series, Book 262, Washington, D. C. p. 271-287.

Carlson, R. D., Tindall, D. R. (1985). Distribution and periodicity of toxic dinoflagellates in the Virgin Islands. In: Anderson, D. M., White, A. W., Baden, D. G.(eds.) Toxic Dinoflagellates. Elsevier, New York. p. 171-176.

Chang, F. H. (1995). More on marine toxic algae in New Zealand. Water and Atmosphere 4:18-19.

de Silva, D. P. (1982). A comparative survey of the populations of a dinoflagellate *Gambierdiscus toxicus* in the vicinity of St. Thomas, U. S. Virgin Islands. Final report, NOAA contract NA-80-RAA-04083, p. 69.

Dickey, R. W., Bobzin, S. C., Faulkner, D. J., Bencsath, F. A., Andrzejewski, D. (1990). Identification of okadaic acid from a Caribbean dinoflagellate, *Prorocentrum concavum.* Toxicon 28:371-7.

Durand-Clement, M. (1987). Study of production and toxicity of cultured *Gambierdiscus toxicus*. Biol. Bull 172:108-121.

Durand, M., Sguiban, A., Viso, A.-C., Pesando, D. (1985). Production and toxicity of *Gambierdiscus toxicus*. Effects of its toxins (maitotoxin and ciguatoxin) on some marine organisms. Proc. of the 5th Int. Coral Reef Cong. Tahiti. 4:483-487.

Faust, M. A. (1990). Morphological details of six benthic species of *Prorocentrum* (Pyrrophyta) from a mangrove island, Twin Cays, Belize, including two new species. J. Phycol. 26:548-558.

Faust, M. A. (1993a). *Prorocentrum belizeanum, Prorocentrum elegans* and *Prorocentrum caribbaeum*, three new benthic species (Dinophyceae) from a mangrove Island, Twin Cays, Belize. J. Phycol. 29:100-107.

Faust, M. A. (1993b). Three new benthic species of *Prorocentrum* (Dinophyceae) from Twin Cays, Belize: *P. maculosum* sp. nov., *P. foraminosum* sp. nov, *P. formosum* sp. nov Phycologia 32:410-408.

Faust, M. A. (1993c). A further SEM study of marine benthic dinoflagellates from a mangrove island, Twin Cays, Belize, including *Plagiodinium belizeanum* gen. et sp. nov. J. Phycol. 29:826-832.

Faust, M. A. (1994). Three new benthic species of *Prorocentrum* (Dinophyceae) from Carrie Bow Cay, Belize: *P. sabulosum* sp. nov., *P. sculptile* sp. nov., and *P. arenarium* sp. nov. J. Phycol. 30:755-763.

Faust, M. A. (1995). Observations of sand-dwelling toxic dinoflagellates (Dinophyceae) from widely differing sites, including two new species. J. Phycol. 31:996-1003.

Faust, M. A. (1996). Dinoflagellates in a mangrove ecosystem, Twin Cays, Belize. Nova Hedwigia 112:445-458.

Faust, M. A., Morton, S. L. (1995). Morphology and ecology of the marine dinoflagellate *Ostreopsis labens* sp. nov. (Dinophyceae). J. Phycol. 31:456-463.

Faust, M. A., Morton, S. L., Quod, J. P. (1996). Further SEM study of Marine dinoflagellates: the genus *Ostreopsis* (Dinophyceae). J. Phycol. 32:1053-1065.
Fukuyo, Y. (1981). Taxonomical study on benthic dinoflagellates collected in coral reefs. Bull. Jap. Soc. Sci. Fish. 47:967-978.
Gillespie, N., Holmes, M. J., Burke, J. B., Doley, J. (1985). Distribution and periodicity of *Gambierdiscus toxicus* in Queensland, Australia. In: Anderson, D. M., White, A. W., Baden, D. G.(eds.) Toxic Dinoflagellates. Elsevier, New York. p. 183-188.
Guillard, R. R. L., Keller, M. D. (1984). Culturing Dinoflagellates. In: Spector, D. (ed.) Dinoflagellates, Academic Press, New York. p. 391-442.
Hara, Y, Horiguchi, T. (1982). Marine microalgae along the coast of the Izu Peninsula. Memoirs of the National Science Museum 15:99-110.
Heil, C. A., Maranda, L., Shimizu, Y. (1993). Mucus-associated dinoflagellates: large scale culturing and estimation of growth rate. In: Smayda, T. J, Shimizu, Y. (eds.) Toxic Phytoplankton Blooms in the Sea. Elsevier, B.V. p. 501-506.
Horstmann, W. (1980). Observations on the peculiar diurnal migrations of a red tide Dinophyceae in tropical shallow waters. J. Phycol. 16:481-485.
Hu, T., deFreitas, A. S. W., Curtis, J. M., Oshima, Y., Walter, J. A., Wright, J. L. C. (1996). Isolation and structure of Prorocentrolide B, a fast-acting toxin from *Prorocentrum maculosum*. J. Nat. Prod. 59:1010-1014.
Jackson, A. E., Marr, J. C., McLachlan, J. L. (1993). The production of diarrhetic shellfish toxins by an isolate of *Prorocentrum lima* from Nova Scotia, Canada. In: Smayda, T. J, Shimizu, Y. (eds.) Toxic Phytoplankton Blooms in the Sea. Elsevier, B.V. p. 513-518.
Lewis, R. J., Gillespie, N. C., Burke, J. B., Holmes, M. J., Keys, A., Fifoot, A. (1987). Ciguatoxin levels in fish in relation to population densities of *Gambierdiscus toxicus*. In: Marsh. H, Heron, M. L. (ed.) Aust. Mar. Sci. Assoc. Aust. Physical Oceanogr. Joint Conf. pp. 36-42.
Lobel, P. S., Anderson, D. M., Durand-Clement, M. (1988). Assessment of ciguatera dinoflagellate populations: sample variability and algal substrate selection. Biol. Bull. 175:94-101.
Morton, S. L., Norris, D. R., Bomber, J. W. (1992). Effect of temperature, salinity, and light intensity on the growth and seasonality of toxic dinoflagellates associated with ciguatera. J. Exp. Mar. Biol. Ecol. 157:79-90.
Morton, S. L., Bomber, J. W., Tindall, P. M. (1994). Environmental effects on the production of okadaic acid from *Prorocentrum hoffmannianum* Faust I. temperature, light, and salinity. J. Exp. Mar. Biol. Ecol. 178:67-77.
Murakami, Y., Oshima, Y., Yasumoto, T. (1982). Identification of okadaic acid as a toxic component of a marine dinoflagellate *Prorocentrum lima*. Bull. Jap. Soc. Sci. Fish. 48:69-72.
Murata, M., Legrand, A. M., Ishibashi, Yasumoto, T. (1989). Structures of ciguatoxin and its congener. J. Am. Chem. Soc. 111:8929-8931.
Murata, M., Iwashita, T., Yokoyama, A., Sasaki, M., Yasumoto, T. (1992). Partial structures of maitotoxin, the most potent marine toxin from dinoflagellate *Gambierdiscus toxicus*. J. Am. Chem. Soc. 114:6594-6596.
Nagai, H., Torigoe, K., Satake, M., Murata, M., Yasumoto, T. (1992). Gamberic acids: unprecedented potent antifungal substances isolated from cultures of a marine dinoflagellate *Gambierdiscus toxicus*. J. Am. Chem. Soc. 114:1103-1105.
Norris, D. R., Bomber, J. W., Balich. (1985). Benthic dinoflagellates associated with ciguatera from the Florida Keys. I. *Ostreopsis heptagona* sp. nov. In: Anderson,

D. M., White, A. W., Baden, D. G.(eds.) Toxic Dinoflagellates. Elsevier, New York. p. 34-44.

Quod, J. P. (1994). *Ostreopsis mascarenensis* sp. nov. (Dinophyceae), dinoflagellé toxique associé à la ciguatera dans l'Océan Inden. Cryptogamie Algol. 15:243-251.

Quod, J. P., Turquet, J., Diogene, G., Fessard, V. (1995). Screening of extracts of dinoflagellates from coral reefs (Reunion Island, SW Indian Ocean), and their biological activities. In: Lassus, P., Arzul, G., Erard, E., Gentien, P., Marcaillou, C. (eds) Harmful Marine Algal Blooms. Intercept Ltd., Lavoisier. p. 815-820.

Sperr, A. E., Doucette, G. J. (1996). Variation in growth rate and ciguatera toxin production among geographically distinct isolates of *Gambierdiscus toxicus*. In Yasumoto, T., Oshima, T., Fukuyo, Y. (eds.) Harmful and Toxic Algal Blooms. Intergov. Oceanog. Comm. of UNESCO. p. 309-312.

Steidinger, K. A. (1983). A re-evaluation of toxic dinoflagellate biology and ecology. In: Round, F.E, Chapman, D.J. (eds.) Progress in Phycological Research, Vol. 2. Elsevier, New York. p. 147-188.

Taylor, F. J. R. (1979). A description of the benthic dinoflagellate associated with maitotoxin and ciguatoxin, including observations on Hawaiian material. In: Taylor, D. L., Seliger, H. H. (eds.) Toxic Dinoflagellate Blooms. Elsevier, North Holland, New York. p. 71-76.

Taylor, F. J. R. (1985). The distribution of the dinoflagellate *Gambierdiscus toxicus* in eastern Caribbean. In: Salvat, B. (ed.) Proceedings of the Fifth International Coral Reef Congress. Antenne Museum-EPHE. p. 423-428.

Taylor, F. J. R., Gustavson, M. S. (1985). An underwater survey of the organism chiefly responsible for "ciguatera" fish poisoning in the eastern Caribbean region: the benthic dinoflagellate *Gambierdiscus toxicus*. In: Stefanon, A., Flemming, N. J. (eds) Proc. Seventh Internat. Diving Science Symp. Padova, Italy (1983).

Tindall, D. R., Dickey, R. W., Carlson, R. D., Morey-Gaines, G. (1984). Ciguatoxigenic dinoflagellates from the Caribbean. In: Ragelis, E. P. (ed.) Seafood Toxins. Am. Chem. Soc. Symp. Series, Wash., D.C. p. 225-240.

Tindall, D. R., Miller, D. M., Bomber, J.W. (1989). Culture and toxicity of dinoflagellates from ciguatera endemic regions of the world. Ninth World Conf. on Animal, Plant and Microbial Toxins. Toxicon 27:83.

Tindall, D. R., Miller, D. M., Tindall, P. M. (1990). Toxicity of *Ostreopsis lenticularis* from the British and United States Virgin Islands. In: Graneli, E., Sundstrom, B., Edler, L., Anderson, D. M. (eds.) Toxic Marine Phytoplankton. Elsevier, New York. p. 424-429.

Tognetto, L., Bellato, O, Moro, I., Andreoli, C. (1995). Occurrence of *Ostreopsis ovata* (Dinophyceae) in the Tyrrhenian Sea during summer 1994. Botanica Marina 38:291-295.

Tosteson, T. R., Ballantine, D. L., Tosteson, C. G., Hensley, V., Bardales, A. T. (1989). Associated bacterial flora, growth and toxicity of cultured benthic dinoflagellates *Ostreopsis lenticularis* and *Gambierdiscus toxicus*. Appl. Env. Microbiol. 55:137-141.

Usami, M. Satake, M., Ishida, S., Inoue, A., Kan, Y., Yasumoto, T. (1995). Palytoxin analogs from the dinoflagellate *Ostreopsis siamensis*. J. Am. Chem. Soc. 117:5389-5390.

Withers, N. W. (1981). Toxin production, nutrition, and distribution of *Gambierdiscus toxicus* (Hawaiian strain). Proc. 4th Int. Coral Reef Symp., Manilla 2:449-451.

Withers, N. W. (1988). Ciguatera fish toxins and poisoning. In: Tu, A. T. (ed.) Handbook of Natural Toxins: Marine Toxins and Venoms, Vol. 3. Marcel Dekker, New York. p. 31-61.

Yasumoto, T. Nakajima, I., Bagnis, R., Adachi, R. (1977). Finding of a dinoflagellate as a likely culprit of ciguatera. Bull. Jap. Soc. Sci. Fish. 43:1021-1026.

Yasumoto, T., Fujimoto, K., Oshima, Y., Inoue, A., Ochi., T., Adachi, R., Fukuyo, Y. (1980). Ecological and distributional studies on a toxic dinoflagellate responsible for ciguatera. Report to the Jap. Min. of Edu., Tokyo. 50 p.

Yasumoto, T., Raj, U., Bagnis, R. (1984). Seafood poisonings in tropical regions. Lab. of Food Hygiene, Fac. of Agric., Tohoku Univ., Kagoshima, Japan. 74 p.

Yasumoto, Y., Seino, N., Murakami, Y., Murata, M. (1987). Toxins produced by benthic dinoflagellates. Biol. Bull. 172:128-131.

Aspects of *Noctiluca* (Dinophyceae) Population Dynamics

M. Elbrächter [1] and Y.-Z. Qi [2]

[1] Biologische Anstalt Helgoland, Taxonomische Arbeitsgruppe,Wattenmeerstation Sylt, Hafenstr. 43, D-25992 List/Sylt, Federal Republic of Germany
[2] Institute of Hydrobiology, Jinan University, Guangzhou, P.R. China

1. Introduction

The dinoflagellate *Noctiluca* is one of the most common "red tide" organisms. Dense accumulations of cells result from multiple physical and biological processes. Physical forcings such as currents or upwelling, in conjunction with cell behaviour such as vertical migration or buoyancy, combine to accumulate large numbers of cells at fronts, and concentrate low cell concentrations into visible blooms as a result of physical /biological coupling. Red tides of *Noctiluca* are often characterised by sharp fronts, indicating that the cells are directly affected by physical processes. This paper will not deal with the physical processes involved in *Noctiluca* red tide formation but will instead elaborate on some of the biological processes necessary for population development and decline. The effects of these red tides on the food web and fisheries will also be discussed.

2. Description, taxonomy, life cycle

Several species have been described in the genus *Noctiluca* Suriray ex Lamarck but at the moment these all are regarded as conspecific (Kofoid and Swezy 1921). The correct name of the type species is *Noctiluca scintillans* (Macartney) Kofoid although its synonym *Noctiluca miliaris* Lamarck is also in common use (Sournia 1984). The vegetative cells are bladder-shaped, large (up to 2000 µm), and have an oral pouch which is an invagination of the more or less spherical cell. At this pouch are located the tentacle and one emergent flagellum (Lucas 1982), a permanent cytostome, a rod organelle and a projecting tooth (Takayama 1983). The function of the tooth is not well understood. The tentacle beats slowly and repeatedly toward the oral pouch, and the distal end reaches the cytostome. The flagellum is poorly visible whereas the tentacle and its movement are a characteristic feature of this taxon (Fig. 1). The flagellum and the tentacle are not suitable for effective swimming. The vertical movement of *Noctiluca* in the water column is due to active ionic regulation of the specific gravity. According to Kesseler (1966), potassium ions are accumulated about 3.8 fold in the cell sap relative to its concentration in seawater, whereas the concentration of the heavier ions calcium and magnesium are only 0.5 and 0.15 times seawater concentration, respectively. In addition, large amounts of ammonia are concentrated inside the cells.

Noctiluca scintillans has no chloroplasts, and therefore is heterotrophic. In addition, intracellular bacteria are common, at least at some times during its seasonal cycle (Lucas 1982, Kirchner *et al.* 1996). Another symbiotic association known since the last century,

NATO ASI Series, Vol. G 41
Physiological Ecology of Harmful Algal Blooms
Edited by D. M. Anderson, A. D. Cembella and G. M. Hallegraeff

is the symbiosis of *Noctiluca* and a photosynthetic green flagellate. The flagellate, *Pedinomonas noctilucae* (Subr.) Sweeney was originally believed to be euglenoid (Subrahmanyan 1954) but it is now placed among the prasinophytes (Sweeney 1976). The association is not a permanent one, since the flagellates swim inside the protoplasmic vacuoles of *Noctiluca* (Sweeney 1978). When the endosymbiont is present, *Noctiluca* can thrive photosynthetically, apparently without the need for phagotrophic food uptake. In large areas of the Indian and Pacific Ocean, dense blooms of green *Noctiluca* have been recorded, lasting for several months up to more than one year (Anton, Corrales, Usup, pers. comm.). The conditions for such blooms are quite different from those of the phagotrophic red populations. In the former, light and nutrients such as nitrogen and phosphorus may be limiting, whereas the latter may be prey-limited. Since the green *Noctiluca* lives only in warmer waters at temperatures > 25° C , this may be a different taxon compared to the red *Noctiluca* which lives in colder waters . Another taxonomical problem relates to variation in the reported size range of the vegetative cells. In the North Sea most cells are in the range of 400 to 600 µm, and occasionally cells up to 700 µm are found (Hanslik 1987). The same size range was reported by Polishchuk *et al.* (1981) for *Noctiluca* populations of the Black Sea. In contrast, Kofoid and Swezy (1921) report cells with a diameter of up to 2000 µm. In addition, the swarmers depicted by Kofoid and Swezy (1921) differ from those found in the North Sea (Fig. 6). Whereas most populations show a brilliant bioluminescence, there exist several populations of *Noctiluca* which do not bioluminesce (Taylor 1993). Whether this feature is species specific as argued by Hastings (1975) has to be investigated. Clearly, more detailed investigations will be needed to determine whether all the *Noctiluca* populations reported world-wide belong to a single species.

Life cycle: Vegetative cell division is by binary fission as in most free-living dinoflagellates. This may result in equal or unequally sized cells, as demonstrated by Qi and Li (1994). The process of cell division starts with a reorganisation of cell organelles. Tentacle, flagellum, rod organelle and cytostome are resorbed, changing the appearence of the species. During cell division, no food uptake takes place since the cytostome and tentacle are missing. Phased cell division in the German Bight occurs during the dark period, with highest cell numbers in division occurring about midnight. The duration of cell division is temperature dependent (Fig. 7) (Uhlig and Sahling 1995). Mean doubling time of *Noctiluca* in experiments with North Sea strains was dependent on salinity and temperature, see sections 4.1 and 4.2.

Zoospore formation: In addition to binary fission, *Noctiluca* exhibits zoospore formation. The nucleus migrates to the cell surface, dividing several times without cytoplasmic division (Figs 3-5) and eventually, after 11 nuclear divisions, up to 2000 zoospores are formed from one mother cell in less than 24 hours (Uhlig and Sahling 1995). Zoospores have one longitudinal flagellum, while the transverse flagellum is concealed (Höhlfeld and Melkonian 1995). Frequently cell division is not complete after separation from the mother cell, and thus "zoospores" with two, three or even four emergent flagella can be recorded. However, mature zoospores possess only one long flagellum (Fig. 6). The morphology of the zoospores reported by Kofoid and Swezy (1921) is slightly different from that observed from North Sea populations. Nuclear division during vegetative binary fission and zoospore formation can be distinguished morphologically, and this can be used

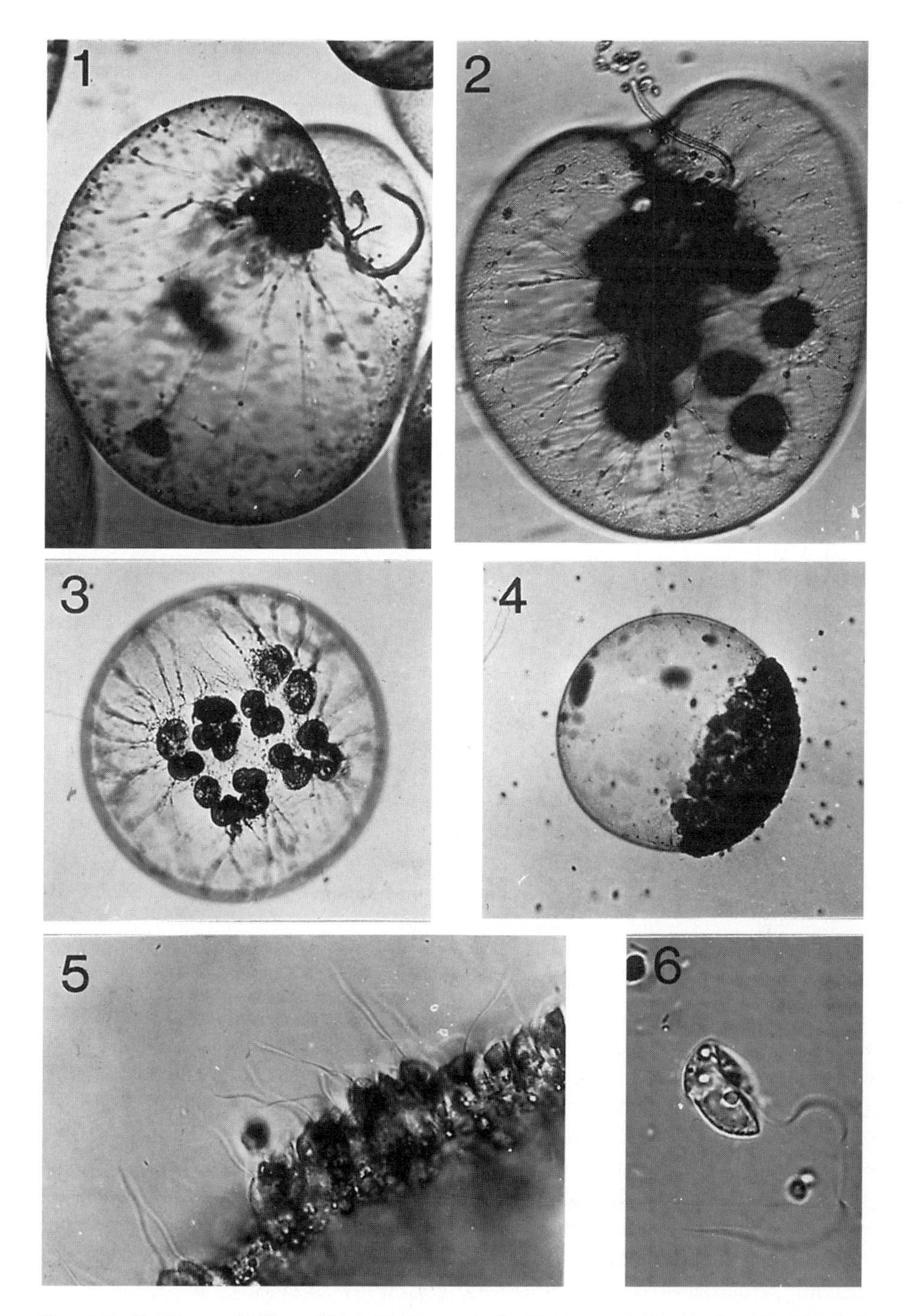

Figs. 1-6: *Noctiluca scintillans*. Fig. 1: Veget[illegible] cell with characteristic tentacle and a few food vacuoles; Fig. 2. Cell with food particles adhering to the tip of the tentacle and many food vacuoles present; Fig. 3. Early stage of zoospore formation; Fig.4. Late stage of zoospore formation, with zoospores formed only at one end of the cell; Fig. 5. The same stage at higher magnification, showing the longitudinal flagella of many zoospores; Fig. 6. Mature zoospore with one longitudinal flagellum and the transverse flagellum concealed. Photos courtsey Dr. Drebes

to determine the percentage of zoospore forming cells of a population. Uhlig and Sahling (1995) found a maximum of 5% of cells in zoospore formation in field samples. Zoospore formation is common in spring and early summer in the North Sea, when the populations are increasing, but rare during the decline of the population. Zoospore formation therefore cannot be responsible for the decline of the *Noctiluca* population (Uhlig and Sahling 1995). The fate of the zoospores is not known. Zingmark (1970) claimed, that they are fusing isogametes and give rise to new large cells. Schnepf and Drebes (1993) suggested that the zoospores are microgametes whereas the macrogametes are identical or very similar to vegetative cells. In field samples, no cells of intermediate size have been found up to now, favoring the latter hypothesis. The claim by Zingmark (1970) that *Noctiluca* is a diploid dinoflagellate is questionable. Resting stages of *Noctiluca* are so far not known.

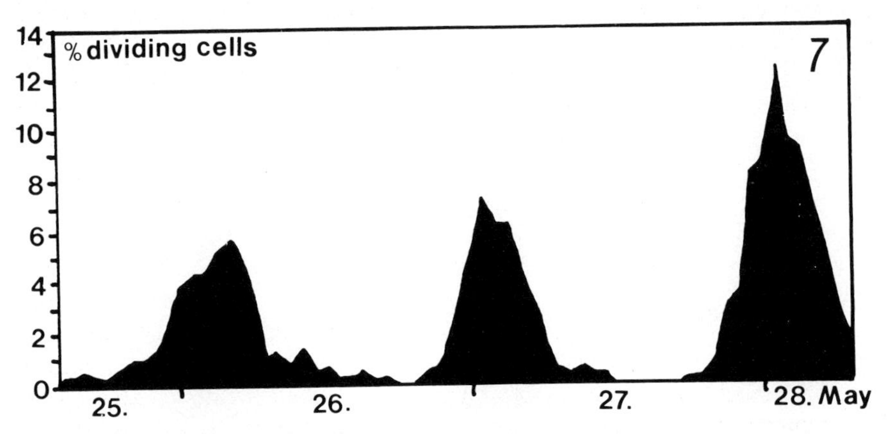

Fig. 7: Percentage of *Noctiluca scintillans* cells in division during a 3-day period. Note the phased cell division occurring around midnight (after Uhlig and Sahling 1995).

3. Geographic distribution

Noctiluca is found in all temperate, subtropical and tropical coastal waters. Brongersma-Sanders (1948) pointed out, that *Noctiluca* red tides are observed in all major upwelling regions. In these regions *Noctiluca* may be observed outside the direct continental shelf region. Taylor (1976) recorded it far offshore from the equatorial Indian Ocean. Water discolorations due to *Noctiluca* have been reported from the east coast of the Atlantic, e.g the west coast of Norway up to 61°N (Tangen 1979); the North Sea (Uhlig and Sahling 1995); the Netherlands (Kat 1979); Spain (Fraga and Sanchez 1979); the Northwest African upwelling region off Mauretania /Senegal (Margalef 1973, Elbrächter, unpubl.); South Africa (Horstmann 1981 ,west coast; and Grindley and Taylor 1971, south coast). Apparently, *Noctiluca* is not common on the northwest coast of the Atlantic Ocean. Marshall and Cohn (1983) did not mention *Noctiluca* in their review of the phytoplankton of northeastern coastal waters of the United States, but Marshall (1976) reports the species from the eastern coast of the USA. It forms blooms in Florida Waters and the Gulf of

Mexico (Buskey 1995) as well as in the Carribean (Ferraz-Reyes et al. 1979) and on the east coast of South America (Odebrecht et al. 1995: Brazil; Mendez 1993 : Uruguay; Balech 1988: Argentina). *Noctiluca* is regarded as a warm water indicator species in subantarctic waters, but it was absent from the Antarctic Ocean . In the Indian Ocean it is found at all coasts e.g. South Africa (Grindley and Heydorn (1970), the Arabian Sea (Saifullah and Chaghtai 1990), both coasts of India (Subrahmanyan 1954, Devassy 1989) and Australia (Brongersma-Sanders 1948, Jeffrey and Carpenter 1974, Hallegraeff, pers. comm.). *Noctiluca* is reported as bloom forming also in coastal zones in the Pacific Ocean. From the far eastern coast of Russia there exist reports from the beginning of the century (Ostroumoff 1924). Konovalova (1989) stated that red tides including those of *Noctiluca* have become more frequent. Kim *et al.* (1993) described dense blooms from Korea and many reports exist from Japanese and Chinese waters including Hong Kong (Kuroda 1995; Qi et al. 1993; Wong 1989). *Noctiluca* forms dense blooms in Vietnam waters (Lam and Hai 1996), the Gulf of Thailand (Suvapepun 1989), in Indonesian Waters (Adnan 1989) and New Guinea (Sweeney 1976). On the west coast of America *Noctiluca* is distributed from Alaska (Lutz and Incze 1979) to British Columbia (Quaile 1969), California (e.g. Kofoid and Swezy 1921, Smayda 1974), Mexico (Altamirano et al. 1996), Peru (Brongersma-Sanders 1948), and in Chilean coastal waters (Lembeye, pers comm.). *Noctiluca* forms blooms also in the Mediterranean Sea (Jacques and Sournia 1978), the Black Sea (Polishchuk et al. 1981) and the Red Sea (Brongersma-Sanders 1948) and can invade the Baltic Sea to the east in the Arkona Sea (Pankow 1990). The distribution of the green *Noctiluca* remains unclear as many reports do not explicitly state, whether blooms are formed by the phagotrophic red or the phototrophic green *Noctiluca*. Nowadays the green *Noctiluca* is apparently restricted to tropical waters of the Pacific and Indian Ocean. Subrahmanyan (1954) described it off Calicut, India, but in waters off Pakistan, Saifullah and Chaghtai (1990) reported on blooms of the red form. Devassy (1989) reported off India on both red and green *Noctiluca* blooms. Adnan (1989) provided a review on green *Noctiluca* blooms in Jakarta Bay, Indonesia and Sweeney (1978) reported it from Northern New Guinea and eastern Borneo in September and October, but during April, May and early June 1975 she found it only very sparsely in the Banda, Celebes and Sulu Seas. Reports of *Noctiluca* blooms from the Chinese coast are all of the red form, as do those occurring in Japan (Fukuyo, pers. comm.). Famous are the spectacular red tides of Hong Kong and it is therefore surprising, that Ostroumoff (1924) described in some detail the occurrence of green *Noctiluca* in the Bay of Vladivostok, Russia. The green *Noctiluca* did not show any bioluminescence whereas red *Noctiluca* cells from other parts of the bay showed bioluminescence. Unfortunately, no associated temperature data are available from Russia.

4. Autecological aspects

4.1 Temperature

In nature, *Noctiluca* has been reported from temperatures below 0°C up to about 30°C. No temperature values are available for the waters in which *Noctiluca* bloomed off Point Barrows, Alaska (Lutz and Incze 1979) but in the North Sea off Sylt this species is found during the whole winter even at temperatures at or below 0°C (Drebes and Elbrächter, unpubl.). The temperature range of the phagotrophic *Noctiluca* is restricted to water temperatures below 25 °C. One needs to be aware, however, that *Noctiluca* cells, thriving below the thermocline at colder temperatures may have been advected into warmer waters. The laboratory experiments of Uhlig and Sahling (1995) clearly show that the optimal temperature for growth (24°C) is near the upper temperature limit of about 25°C for cell division in cultures (Fig. 9). The results from extensive field studies of Huang and Qi (1997) in Dapeng Bay, China, are in agreement with the laboratory studies of the North Sea population where the temperature rarely exceeds 20°C. Lee and Hirayama (1992) also found a temperature optimum at about 23°C but reported reduced growth at 27°C. All cells died at 31 to 32 °C in cultures of their Japanese *Noctiluca* strain. Where temperature-data for *Noctiluca* blooms are provided in the literature, these are below 25°C except for station 185 (26.40°C) of Saifullah and Chaghtai (1990). These authors did not clearly state whether the *Noctiluca* cells were sampled at the surface or by vertical net hauls below the thermocline.

In contrast, the green *Noctiluca* blooms for long periods in waters warmer than 25 °C and are rarely found in colder waters. The temperature range of the green *Noctiluca* still needs to be defined in culture. It also has to be verified whether the host without its endosymbiont has the same upper and lower temperature range as the consortium of *Noctiluca* with *Pedinomonas*. Temperature influences not only the mean doubling time (Fig. 9) but also the time needed for cell division (Fig. 8). At 12°C the time between the first incidence of cell division to the final separation of the daughter cells is about 7 hours, whereas at 24°C the same process lasts only about 3 hours.

4.2 Salinity

Noctiluca is euryhaline, occurring at salinities of about 10 to 37 PSU. In the Baltic Sea its distribution boundary occurs in the Arkona Sea at about 7 to 10 PSU but it is not clear whether the species is actively dividing there or is being introduced by currents from regions (or deeper water layers) with higher salinity. The natural distribution is in accordance with laboratory studies. Lee and Hirayama (1992) reported that for their Japanese strain 14 PSU was the lowest salinity for minimal and continual growth, with optimal growth occurring at about 22 PSU, while a sudden transfer from 34 to 14 PSU was lethal. Uhlig and Sahling (1995), working with cultures isolated from the North Sea, found 10 PSU to be the lowest salinity supporting growth, with optimum growth in the range of 21 to 25 PSU (Fig. 9).

4.3 Light

Light seems to have no influence on the growth rate of the red Noctiluca. In contrast the green *Noctiluca* can divide in culture in the light without the need for additional food , but if cells are kept in the dark, they cannot survive. The exact relationship between the endosymbiont and *Noctiluca* with regard to transmission of chemical compounds and/ or energy is not known. Cell division shows a clear circadian rhythm with a peak at midnight, both for field populations (Fig. 7) as well as for laboratory studies. In the field, there was no influence of tides on the diurnal cell division cycle. Bioluminescence capacity also did not show a circadian rhythm in contrast to reports for most other dinoflagellates. It can be stimulated at all times of the day (Nicols 1958, Buskey et al.

distal end of the tentacle (Fig.2) on which adhesive compounds are thought to be secreted. The second stage begins with a strong flexion of the tentacle towards the cytostome where the accumulated algae are ingested preceeded by vigorous cytoplasmic streaming . At the start of ingestion, the food is carried towards the apex, where the rod organelle is located which may facilitate ingestion of large food items such as copepod eggs (Takayama 1983). Tentacle movement and feeding is apparently controlled by bioelectrical

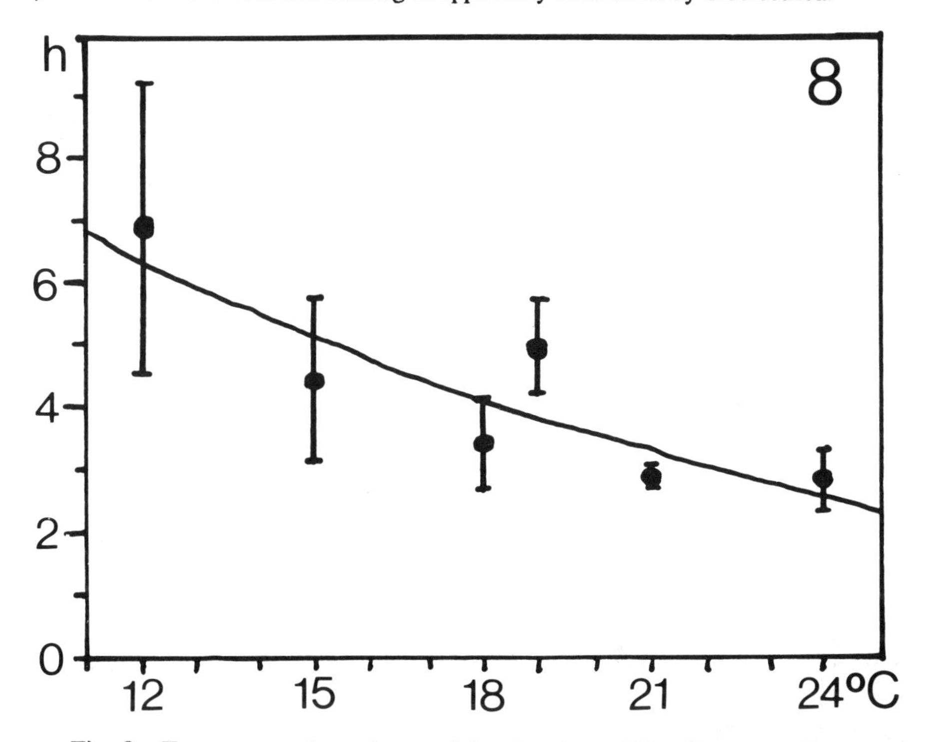

Fig. 8: Temperature dependence of the duration of *Noctiluca scintillans* cell division (after Uhlig and Sahling 1995).

membrane 1992). Uhlig and Sahling (1995) speculated that the sudden decline of North Sea populations in August may be related to an endogenous circannual rhythm triggered by the change of the light-dark cycle. Huang and Qi (1997) found a similar sudden decline of *Noctiluca* populations in May in Dapeng Bay, China. Although Dapeng Bay is at a latitude of 34°N to 36° N, and the North Sea study area at a latitude of about 53°N to 55°N, there is no major difference in the light regimes which may explain the different times of *Noctiluca* population decline. The impact of UV-light on *Noctiluca* is controversial. According to Balch and Haxo (1984), *Noctiluca* exhibits high absorption in the near-ultraviolet region. Although mycosporine-like amino acids, which are regarded as protective UV-absorbing compounds, are found in high concentrations in *Noctiluca*, it is not clear whether these are produced by *Noctiluca* itself or have accumulated by ingestion of phytoplankton (Carretto *et al.* 1996). Nothing is known on the possible differences of UV-toleration of red and green populations.

5. *Noctiluca* in the food web

Noctiluca scintillans is a voracious feeder, ingesting all particles which it can engulf (e.g. Kimor 1979). *Noctiluca* has a permanent cytostome near the tentacle from which faeces are also excreted. The tentacle is used for food uptake and to eliminate faeces from the cell (Uhlig 1972). Feeding on small particles like small phytoplankton occurs into two successive phases. First the food gathering phase, in which flexion and extension of the tentacle is repeated several times to trap the algae. The food organisms adhere to the potential changes (Nawata and Sibaoka 1986; Oami and Naitoh 1989). Whether the failure of food uptake by *Noctiluca* cells in populations just before their decline is induced by the absence of membrane potential controlling ability remains to be investigated. *Noctiluca* does not discriminate between food particles and ingests all particles even if they have no food value, e.g. glass particles, or lethal crude oil droplets after oil spills. The species is able to take up large food items such as eggs of copepods (Kimor 1979). Best growth was obtained if food particles had a size between 5 and 25 µm in diameter. Chen and Qi (1991) found that feeding activity was highest during night time and that

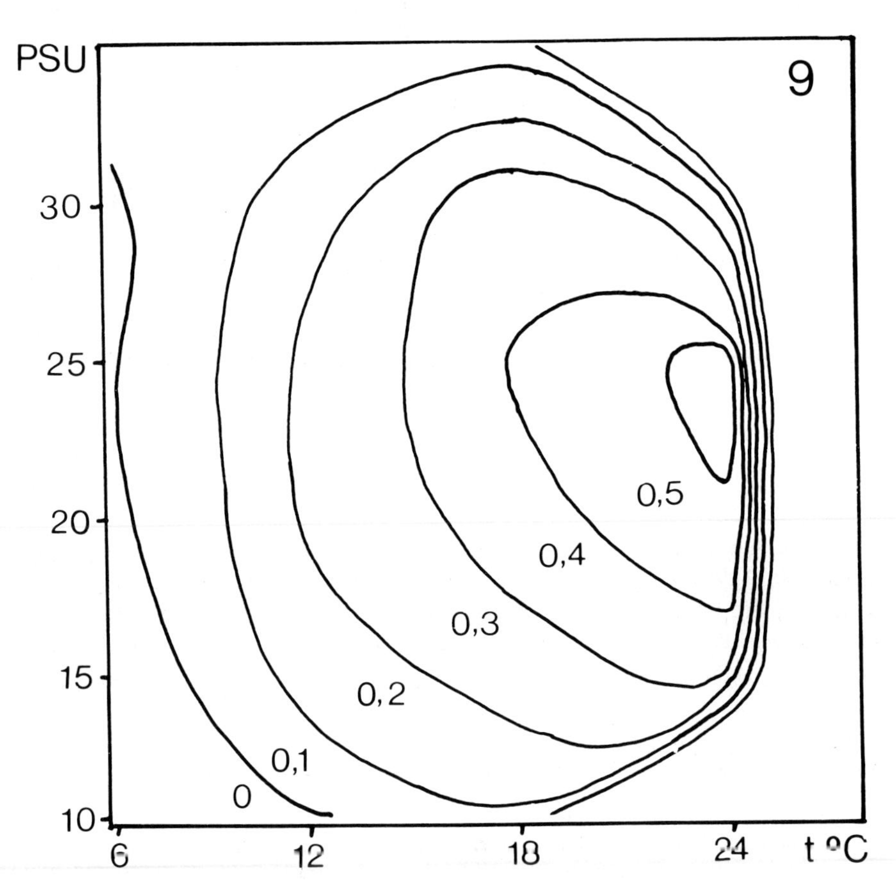

Fig. 9: Temperature and salinity dependence of *Noctiluca scintillans* cell division rate (after Uhlig and Sahling 1995).

most food items are digested in about 4.5 ± 2.5 hours. After digestion, faeces are expelled through the cytostome and transported out off the oral pouch by tentacle movement. Not all phytoplankton do equally support growth and cell division. Hanslik (1987) demonstrated experimentally, that the diatom *Skeletonema costatum* and the chlorophyte *Dunaliella tertiolecta* are good food items; in contrast, the dinoflagellate *Scrippsiella trochoidea* and the yeast *Debaromyces hansenii* did not sustain growth. Feeding stopped if the food concentration was < 1000 µg C dm^{-3}, independent of whether the food promoted growth or not. Buskey (1995) found that *Noctiluca* had its highest growth-rate (0.5 day^{-1}) with the diatom *Thalassiosira* sp. at food concentrations of 0.5 mg C dm^{-3}. Jeong and Shim (1996) found negative growth rates when only *Gymnodinium sanguineum* was offered as food. Holligan *et al.* (1983) reported that *Noctiluca* had ingested up to 50 cells of the ichthyotoxic dinoflagellate *Gyrodinium aureolum*. In North Sea waters, phytoplankton seems to be limiting during long periods, and hence *Noctiluca* must ingest other components of the seston. It is likely that *Noctiluca* exerts significant feeding pressure on the phytoplankton, especially on diatoms. The seasonal variation in chlorophyll *a* concentrations in Dapeng Bay, South China Sea shows a negative correlation with *Noctiluca* populations (Fig. 10; Huang and Qi 1997). There was one exception on March 25th, 1991, when a *Noctiluca* bloom coincided with a bloom of the diatom *Thalassiosira*. Healthy populations of *Noctiluca* appear to feed effectively on phytoplankton (Enomoto 1956; Brockmann et al. 1977; Hanslik 1987; Weisse and Schefflel-Möser 1992; Huang and Qi 1997). In contrast, Schaumann *et al.* (1988) found a positive correlation between a *Noctiluca* bloom and diatom abundance in the North Sea. This phenomenon was attributed to *Noctiluca* cells in the red tide patch, which no longer were feeding but were damaged by starvation. Such cells excrete ammonia while other nutrients are liberated due to cell lysis. *Noctiluca* thus generated nutrient enrichment in the patch which favoured the developement of diatoms such as *Odontella sinensis* (Grev.), *Guinardia flaccida* (Castr.) Perag., *Rhizosolenia shrubsolei* Cleve, and *R. setigera* Brightw. In contrast, the surrounding water contained the dinoflagellate *Ceratium fusus* (Ehrenb.) Duj. as the most common species. The same explanation may apply to the *Thalassiosira* associated with a *Noctiluca* red tide as observed by Huang and Qi (1997). *Noctiluca* is also able to feed effectively on large particles such as eggs of copepods (Sekiguchi and Kato 1976; Kimor 1979,1981; Daan 1987) or fish eggs (Enomoto 1956; Hattori 1962). Although *Noctiluca* ingests up to 50% of the copepod egg production per day, Daan (1987) did not find that population dynamics of copepods were negatively influenced by this predation. A second type of food uptake occurs in *Noctiluca*, namely social feeding with a mucoid web (Omori and Hamner 1982, Elbrächter 1991, Kirchner *et al.* 1996). Many *Noctiluca* cells excrete a mucoid web, the cells then sink together with this web scavenging particles from the water column. Bacteria are effectively collected by this method. Therefore *Noctiluca* has not only an enormous impact on phyto- and zooplankton but also on bacterial populations(Kirchner *et al.* 1996). There is increasing evidence, that intracellular bacteria are common in *Noctiluca*, at least during some parts of its seasonal cycle (Lucas 1982, Kirchner et al. 1996). Whether these bacteria are also used as a carbon or energy source remains to be established. The green *Noctiluca* is phototrophic and apparently can rely on the photosynthesis of the endosymbiotic prasinophyte. The mode of material and/or energy transfer from endosymbiont to host remains unclear. In continous darkness, *Noctiluca* digests its endosymbionts, but whether this is a regular mode of nutrition in natural blooms is not known.

Few data are available how *Noctiluca* is used in the food web. Petipa (1960) reported that the species served as food for the copepod *Calanus helgolandicus*. This shows how complicated the relationships between aquatic organisms can be, since at the same time *Noctiluca* is able to ingest copepod eggs. Malej (1982) reported that the scyphomeduse *Pelagia noctiluca* feeds on *Noctiluca* in the Gulf of Trieste, Adriatic Sea. Taylor and Pollingher (1987) pointed out, that *Noctiluca* plays an important in coastal ecosystems in many parts of the world but that interactions with associated organisms have been largely neglected. This equally applies to the understanding how *Noctiluca* changes its physical and chemical environment. For instance, Grindley and Taylor (1964) reported a higher water-temperature inside a *Noctiluca* patch. This may change growth rates of phytoplankton species serving as food. Jenkinson (1993) demonstrated that *Noctiluca* can also change the viscosity and elasticity of the sea water.

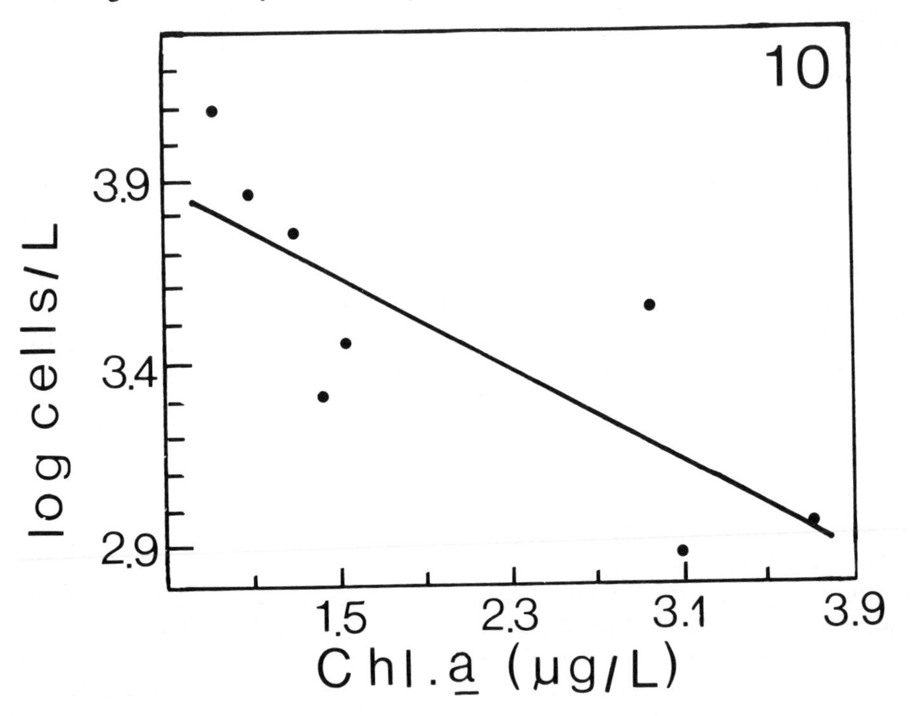

Fig. 10: Relationship between the log cell number of *Noctiluca* and chlorophyll *a* concentration of the waters of Dapeng Bay, China (after Huang and Qi 1997).

6. Population dynamics

High cell numbers of *Noctiluca* distributed over the water column do not necessarily cause a red water discolouration. These spectacular events are only apparent when many cells are concentrated at the surface . This may result by different mechanisms, either physically or biologically driven and some of them may occur at the same time. The vertical distribution of *Noctiluca* cells in the water column is largely dependent on vertical mixing due to turbulence, either by currents, tides or wind. Thus during calm weather, accumulation of large cell numbers resulting in water discolouration may occur one day but may disappear some hours later if wind stress is mixing the water column . Accumulation of large cell numbers may be forced by another physical process that is the formation of fronts (LeFèvre and Grall 1970; Holligan 1979). Indeed , most red tides of *Noctiluca* have very sharp boundaries demonstrating the presence of fronts (Schaumann *et al.* 1988). In Dapeng Bay, Huang and Qi (1997) reported of a 100-fold population increase in 3 days. This increase was due to wind driven accumulation but not increase of cell numbers by cell division. On the other hand, *Noctiluca* can regulate its specific gravity and thus its depth distribution. In early bloom conditions, the species is negatively buoyant and distributed in subsurface waters. At the end of the bloom, *Noctiluca* accumulates ammonia inside the cells resulting in positive buoyancy and accumulation of cells at the sea surface, provided wind forced turbulence is not too strong. Thus the occurrence of water discolouration occurs only, if high cell numbers are present and suitable physical/biological conditions (fronts, calm weather, positive buoyancy) prevail.

6.1. Case study German Bight, North Sea

At Helgoland, a long term study over 25 years has been conducted on the annual occurrence of *Noctiluca* using synoptic ship surveys of the German Bight. This data set provides a unique opportunity for population dynamic studies (Uhlig and Sahling 1995, Uhlig *et al.* 1995).

Pre-bloom conditions: In the North Sea *Noctiluca* is present throughout the entire year, but in winter months only very few cells may be found in net tow samples. In the German Bight increasing cell numbers are found first in shallow waters at the coast. Cell numbers > 100 cells dm^{-3} are found only at water temperatures above 8°C from April /May onwards (Fig. 11). In shallow coastal bays water temperatures may be locally higher and the salinity lower, so that cell division rates may be enhanced. Normally April/May is also the time of the spring phytoplankton bloom (Radach and Bohle-Carbonell 1990) From coastal regions, *Noctiluca* is then spreading into the open German Bight. At this time, in the open German Bight higher cell numbers occur in subsurface layers than in the upper few meters (Uhlig and Sahling 1995, Uhlig et al. 1995), but in coastal waters cell concentrations are higher near the surface. The higher cell numbers are correlated with warmer waters of lower salinity coming with ebb-tides from the Wadden Sea, which is in agreement with the temperature/salinity optimum for cell division rates (Fig 9). Cell numbers increase until June/July showing *in situ* cell division rate of about 15% in coastal waters but only 2% in the open German Bight. During this period < 1% of the population transforms into cells forming zoospores. Zoospore formation comes to a halt in late June, when the population does no longer increase. There is no evidence that the formation of zoospores is causally related to either population increase or decline. Maximal population densities varied from year to year but never occurred if water temperatures were below 15°C. Apparently a prerequisite for bloom formation are water temperatures of 15°C and a high biomass of phytoplankton.

In years with high cell numbers of *Noctiluca* the phytoplankton biomass is not sufficient and other seston components therefore may also be used as food.
Bloom formation : In June cell densities reach their maximum. In each year, the population densities vary and apparently there is a 3-year cycle with 2 years of low population densities but in the third year there is a maximum (Fig.12). During July-August, cell division rates decrease and *Noctiluca* stops feeding. These empty cells with no food accumulate ammonia and rise to the surface if water turbulence is not too high. These cells are no longer able to feed even when isolated and provided with optimal food. Only few cells survive the sudden population decline as in September and October there are actively dividing *Noctiluca* cells, which may lead to an autumn peak.

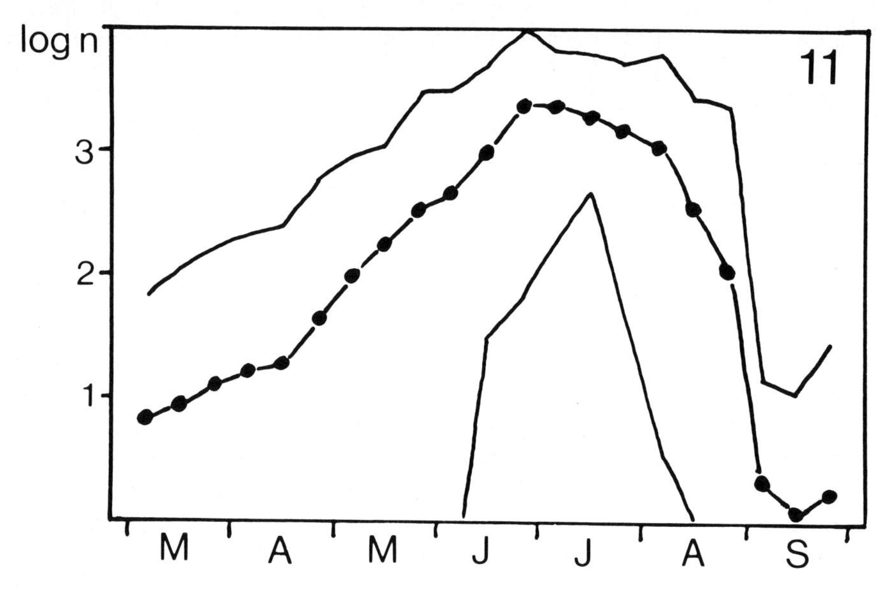

Fig. 11: Annual cycle of the occurrence (log cells per L) of *Noctiluca scintillans* from March to September in the North Sea. Integrated values over 24 years, together with the minimum and maximum values during this period are indicated (after Uhlig and Sahling 1995).

6.2 Case study Dapeng Bay, South China Sea

Huang and Qi (1997) conducted a 3-year study on population dynamics of *Noctiluca* in Dapeng Bay, South China Sea, where the temperature is higher, ranging from 15°C to more than 30°C and salinity ranged from 15.6 to 33.7 PSU. The seasonal occurrence of *Noctiluca* was from January to May/June with peak abundances from March to early May. *Noctiluca* was found at water temperatures between 15.8°C to 28.6°C, but mass development was at temperatures between 19 to 22°C. Culture experiments with cells isolated from Dapeng Bay showed that *Noctiluca* could not survive 26°C.The densities of *Noctiluca* increased with water temperatures up to 20°C and decreased rapidly when water temperatures exceeded 25°C. However, temperature seems not to be the only cause for population decrease. In 1991 temperature was below 23°C when a dramatic decline of *Noctiluca* population densities occurred. There was the same time lag of about 4 weeks

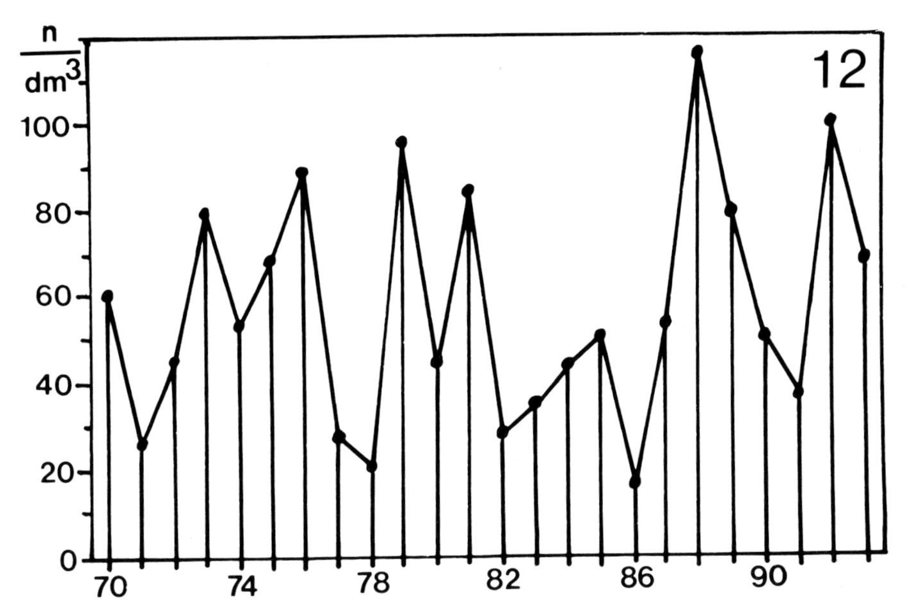

Fig. 12: Mean cell numbers of *Noctiluca* at the Helgoland Reede station, North Sea, during the summer months of 1970 to 1973 (after Uhlig and Sahling 1995).

between peak abundance of *Noctiluca* and the population decline as observed in the North Sea. A clear negative correlation between *Noctiluca* cell numbers and the chlorophyll content of the water column was noted (Fig.10). The species may be food limited and population dynamics showed the same pattern as in the North Sea. As *Noctiluca* does not tolerate high water temperatures, it dies out in Dapeng Bay in summer. Ho (pers. comm.) suggested that *Noctiluca* is generally absent from the waters around Hong Kong and in the Pearl River Estuary from August to October because of high temperatures, and the Kuroshio Current is responsible for introducing the vegetative cells from other areas to the South China Coast. Since the *Noctiluca* blooms first appear in eastern waters of Hong Kong in late December every year (Ho pers. comm.), the vegetative cells are transported by wind or tidal forces to Dapeng Bay at the end of each year.

6.3 Bloom formation and eutrophication

It has repeatedly been discussed that long-term trends in increased frequency of harmful algal blooms are associated with nutrient enrichment of coastal waters and inland seas (Smayda 1990). In South-Asian waters an increase of *Noctiluca* blooms has been reported especially from Japan (Okaichi and Nishio 1976), Hong Kong (Lam and Ho1989), and the South China Sea (Qi et al. 1993; Huang and Qi 1997). In contrast in the German Bight, North Sea, there is no evidence for an increase in population densities nor in bloom occurrence of *Noctiluca* over a period of 25 years (Uhlig and Sahling 1995), although at the same time there was evidence of considerable eutrophication and an increase of phytoplankton biomass (Raddach and Bohle-Carbonelle 1990). As discussed by Uhlig and Sahling (1995) other effects such as short term climatic changes, changes in water currents or trophic structure or other effects may create a non-causal correlation with eutrophication. Only long-term studies over decades at various localities can solve the question whether *Noctiluca* blooms are stimulated by eutrophication.

7. Impact on fisheries and toxic effects

There exist many reports of *Noctiluca* bloom impacts on fisheries but up to now there is no evidence for the production of an ichthyotoxic chemical compound. Ogawa and Nakahara (1979) showed that fish avoid large *Noctiluca* aggregations.The negative correlation between *Noctiluca* and fish was interpreted as competition for space. Subrahmanian (1985) suggested that fish avoid dense *Noctiluca* blooms due to massive slime production which can clogg the gills. Fish kills have been associated with *Noctiluca* mass occurrences by various authors (e.g Brongersma-Sanders (1948), LeFèvre and Grall (1970), Subrahmanian (1985). Fish kills may be caused by different mechanisms. In many cases oxygen depletion is discussed as causative agent for fish kills (Subrahmanyan 1985, Ho and Hodgliss 1992; Xie et al. 1993). *Noctiluca* is a heterotrophic organism, not producing but consuming oxygen. Hanslik (1987) determined in cultures under various conditions respiration rates of 2 to 4 ng oxygen per cell per hour. Large amounts of oxygen are also used by bacteria when *Noctiluca* is decaying. Thus oxygen depletion may cause fish kills in particular in caged fish. Another principle of fish kill is the release of ammonia (Okaichi and Nishio 1976). This is particularly the case at the end of the blooms when *Noctiluca* do no longer feed and accumulate ammonia in their cells before lysing (Schaumann et al. 1988). Fish kills caused by the green *Noctiluca* have also been reported (Adnan 1989). Devassy (1989) reported on adverse effects on fisheries of a green *Noctiluca* bloom extending over 600 km on the coast of Goa, India, which lasted for more than 3 months. In contrast, during earlier red tides of red *Noctiluca* in the same region no adverse effects on fisheries had been reported. Chen and Gu (993) documented a loss of about $ 100 million for prawn mariculture in the Pei Hai Sea, North China Sea. According to Suvapepun (1989) heavy blooms of the green *Noctiluca* deplete oxygen in sheltered bays of the Gulf of Thailand having impacts on shrimp farming.. Although there are each year massive blooms of *Noctiluca* in Hong Kong waters, Wong (1989) considered this species not as a harmful alga. Lam and Hai (1996) stated that *Noctiluca* had no toxic effect on prawns or fish cages in Vietnam waters but they mention that fishermen claimed that *Noctiluca* had adverse effects on fisheries. Uhlig and Sahling (1995) reported no adverse effect of *Noctiluca* blooms in the North Sea.

8. Concluding remarks

Noctiluca is a widespread organism known for more than two centuries and forming blooms in arctic, temperate and tropical regions, causing spectacular water discolourations. There exist many gaps in our knowledge of its taxonomy, biology, population dynamics and impact of blooms on the ecosystem. At least four species have been described which are regarded as conspecific, although there are indications that the green form harbouring endosymbionts and occuring at much higher temperatures than the obligate heterotrophic form may be a separate taxon. A thorough taxonomical reevaluation including classical morphological and modern techniques such as molecular genetic studies is needed. Although *Noctiluca* is a wide-spread and common organism, its life cycle including sexual reproduction and resting stage formation are not known. In particular the biology of the *Noctiluca* with prasinophyte endosymbiont is unknown. Does it actively grow by photosynthesis or is it obligately mixotrophic ? The conditions for bloom formation and for bloom decline seem to be different in different geographical regions and under different hydrographical regimes. A clear relationship between bloom frequency or duration to eutrophication has so far not been established. Only in long-term studies over more than two decades such trends can be separated from short-term variations of various other parameters. For some of the blooms either an impact on fisheries or fish kills have been reported whereas in other regions no such adverse effects have been found. Up to now no toxic compound has been found from *Noctiluca*, but accumulation and release of ammonia can cause harm to fisheries and aquaculture.

9. References

Adnan, Q. (1989). Red tides due to *Noctiluca scintillans* (Maccartney) Ehrenb. and mass mortality of fish in Jakarta Bay. In: Okaichi,T., Anderson, D.M., Nemoto, T. (eds.) *Red Tides. Biology, Environmental Science and Toxicology.* Elsevier Science Publishing B.V. pp. 53-55.

Altamirano, C., Hernandéz-Becerril, D.U., Luna-Soria, R. (1996). Red tides in Mexico, a review. In: Yasumoto, T., Oshima, Y, Fukujo, Y. (eds.) *Harmful and toxic algal blooms.* Interg. Ocean. Comm. UNESCO pp. 45-48.

Balch, W.W., Haxo, F.T. (1984). Spectral properties of *Noctiluca miliaris* Suriray, a heterotrophic dinoflagellate. J. Plankt. Res. 6 : 515-525.

Balech, E. (1988). Los dinoflagellados del Atlantico Sudoccidental. Publ. Esp. Inst.Esp. Oceanogr. 1 : 1-219.

Brockmann, U.H., Eberlein, K., Hosumbeck, P., Trageser, H., Meier-Reimer, E., Schöne, H.K., Junge, H.D. (1977). The development of a natural plankton population in an outdoor tank with nutrient-poor sea water. I. Phytoplankton succession. Mar. Biol., 43 :1-17.

Brongersma-Sanders, M. (1948) The importance of upwelling waters to vertebrate, palaeontology and oil production. Verh. K. ned. Akad. Afd.Natuurk. 2 : 1-112.

Buskey, E.J. (1995). Growth and bioluminescence of *Noctiluca scintillans* on varying algal diets. J. Plankt. Res. 17: 29-40.

Buskey, E.J., Strom, S., Coulter, C. (1992). Bioluminescence of heterotrophic dinoflagellates from Texas coastal waters. J. Exp. Mar. Biol. Ecol. 159 : 37-49.

Cachon, J. (1964). Contribution a l'étude des péridiniens parasites.Cytologie, cycles évolutives. Ann. Sci. Nat. Zool. 6: 1-158.

Carreto, J.I., Benavides, H.R., Carigman, M.O., Negri, R.M., Akselman, R., Cucchi, C.A.D. (1996). Photosynthetic response of natural phytoplankton populations to environmental ultraviolet radiation. In: Yasumoto, T., Oshima, Y, Fukujo, Y. (eds.) *Harmful and toxic algal blooms.* Interg. Ocean. Comm. UNESCO pp. 325-328.

Chen, H., Qi, S. (1991). The feeding and vegetative reproduction diurnal rhythms of *Noctiluca scintillans.* J. Jinan Univ. China 12 : 104-107.

Chen, Y.Q., Gu, X.G. (1993) An ecological study of red tides in the East China Sea. In: Smayda, T.J., Shimizu, Y. (eds.) *Toxic phytoplankton blooms in the sea.* Elsevier Science Publishers B.V. pp. 217-221.

Daan, R. (1987). Impact of egg predation by *Noctiluca miliaris* on the summer development of copepod populations in the southern North Sea. Mar. Ecol. Progr. Ser. 37 : 9-17.

Devassy, V.P. (1989). Red tide discolouration and ist impact on fisheries. In: Okaichi, T., Anderson, D.M., Nemoto, T. (eds.) *Red Tides. Biology, Environmental Science and Toxicology*. Elsevier Science Publishing B.V. pp. 57-60.

Elbrächter, M. (1991). Food uptake mechanisms in phagotrophic dinoflagellates and classification. In: Patterson, D.J., Larsen, J. (eds.) *The biology of free-living heterotrophic flagellates*. Systematic Association Special Volume 45, pp. 303-312.

Enomoto, Y., (1956). On the occurrence of *Noctiluca scintillans* (Macartney) in the waters adjacent to the west coast of Kyushu, with special reference to the possible damage caused to the fish eggs by that plankton. Bull. Jap. Soc. Scient. Fish. 22 : 82-88.

Ferraz-Reyes, G., Ferraz-Reyes, E. Vasques, E. (1979). Toxic dinoflagellate blooms in northeastern Venezuela during 1977. In Taylor, D.L., Seliger, H.H. (eds.) *Toxic Dinoflagellate Blooms*. Elsevier/North Holland, pp.191-194.

Fraga, S., Sanchez, F.J. (1979). A bloom of *Amphidinium* sp. in the Ria de Vigo (N.W. Spain). In: Taylor, D.L.Seliger, H.H. (eds.) *Toxic Dinoflagellate Blooms*. Elsevier/North Holland, pp.165-168.

Grindley, J.R. and Heydorn, A.E.F. (1970). Red water and associated phenomena in St. Lucia. S. Afr. J. Sci. July 1970 : 210-213.

Grindley, J.R., Taylor, F.J.R. (1964). Red water and marine fauna mortality near Cape Town. Trans. R. Soc. S. Afr. 37 : 111-130.

Grindley, J.R. and Taylor, F.J.R. (1971). Factors affecting plankton blooms in False Bay. Trans. R. Soc. S. Afr. 39 : 201-210.

Hanslik, M. (1987). Nahrungsaufnahme und Nahrungsverwertung beim Meeresleuchttierchen *Noctiluca miliaris*. - Diss., Univ. Bonn, 92 pp.

Hattori, S. (1962). Predatory activity of *Noctiluca* on anchovy eggs. Contr.Tokai Reg. Fish. Res. Lab. 16 : 211-220.

Hastings, J.W. (1975). Dinoflagellate bioluminescence: molecular mechanisms and circadian control. In: LoCicero, V.R. (ed.) *Proceedings of the First International Conference on Toxic Dinoflagellate Blooms*. Massachusetts Science Technology Foundation, Wakefield pp.235-248.

Ho, K.C., Hodgliss, I.J. (1992). Severe fishkill in Hong Kong caused by *Noctiluca scintillans*. Red Tide Newslett. 5 : 1-2.

Holligan, P.M. (1979) Dinoflagellate blooms associated with tidal fronts around the British Isles. In: Taylor, D.L.Seliger, H.H. (eds.) *Toxic Dinoflagellate Blooms*. Elsevier/North Holland, pp. 249-256.

Holligan, P.M. Viollier, M., Dupouy, C., Aikens, J. (1983). Satellite studies on the distribution of chlorophyll and dinoflagellate blooms in the western English Channel. Contin. Shelf Res. 2 : 81-96.

Höhfeld, I., Melkonian, M. (1995). Ultrastructure of the flagellar apparatus of *Noctiluca miliaris* Suriray swarmers (Dinophyceae). Phycologia 34: 508-513.

Horstmann, D.A. (1981). Reported red-water outbreaks and their effects on fauna of the west and south coasts of South Africa. Fish. Bull. S. Afr.15: 71-88.

Huang, C., Qi, Y. (1997). The abundance cycle and influence factors on red tide phenomena of *Noctiluca scintillans* (Dinophyceae) in Dapeng Bay, theSouth China Sea. J. Plankt. Res. 19 : 303-318.

Jacques, G., Sournia, A. (1978). Les "Eaux Rouges" dues au phytoplancton en Méditerranée. Vie Milieu 28,sér. AB : 175-187.

Jeffrey, S.W., Carpenter, S.M. (1974). Seasonal succession of phytoplankton at a coastal station off Sydney. Austr. J. mar. Freshwat. Res. 25 : 361-369.

Jenkinson, I.R. (1993). Viscosity and elasticity of *Gyrodinium cf.aureolum* and *Noctiluca scintillans* exudates, in relation to mortality of fish and damping turbulence. In: Smayda, T.J., Shimizu, Y. (eds.) *Toxic phytoplankton blooms in the sea*. Elsevier Science Publishers B.V. pp. 757-762.

Jenkinson, I.R., Wyatt, T. (1995). Does bloom phytoplankton manage the physical oceanographic environment ? In: Lassus,P., Arzul, G., Erard-Le Denn, E., Gentien, P., Marcaillou-Le Baut, C. (eds.) *Harmful Marine Algal Blooms*. Lavoisier, Paris, pp. 603-608.

Jeong, H.J., Shim, J.H. (1996). Interactions between the red-tide dinoflagellate Gymnodinium sanguineum and its microzooplankton grazers. In: Yasumoto, T., Oshima, Y, Fukujo, Y. (eds.) Harmful and toxic algal blooms.Interg. Ocean. Comm. UNESCO pp. 377-380.

Kat, M. (1979). The occurrence of Prorocentrum species and coincidental gastrointestinal illness of mussel consumers. In: Taylor, D.L.Seliger,H.H. (eds.) *Toxic Dinoflagellate Blooms*. Elsevier/North Holland, pp.215-220..

Kesseler, H. (1966). Beitrag zur Kenntnis der chemischen und physikalischen Eigenschaften des Zellsaftes von *Noctiluca miliaris*. Veröff. Inst.Meeresf.Bremerh. Suppl. II : 357-368.

Kim, H.G., Park, J.S., Lee, S.G., An, K.H. (1993). Population cell volume and carbon content in monospecific dinoflagellate blooms. In: Smayda, T.J., Shimizu, Y. (eds.), *Toxic phytoplankton blooms in the sea*. Elsevier Science Publishers, pp.769-773.

Kimor, B. (1979). Predation by *Noctiluca miliaris* Suriray on *Acartia tonsa* Dana eggs in the inshore waters of southern California. Limol. Oceanogr.24: 568-572.

Kimor, B. (1981). The role of phagotrophic dinoflagellates in marine ecosystems. Kieler Meeresforsch. Sonderh. 5 : 164-173.

Kirchner, M., Sahling,, G., Uhlig, G., Gunkel, W., Klings, K.-W. (1996). Does the red tide-forming dinoflagellate *Noctiluca scintillans* feed on bacteria? Sarsia 81:45-55.

Kofoid, C.A., Swezy, O. (1921). The free-living unarmoured dinoflagellates. Mem. Univ. Calif. 5 : 1-564.

Konovalova, G.V. (1989). Phytoplankton blooms and red tides in the far east coastal waters of the USSR. In: Okaichi, T., Anderson, D.M., Nemoto, T. (eds.). *Red tides: Biology, Environmental Science and Toxicology*. Elsevier Science Publishing, pp. 97-100.

Kuroda, K. (1995). *Noctiluca scintillans* (Maccartney) Ehrenberg. In: Fukuyo, Y., Takano, H., Chihara,M., Matsuoka, K. (eds.) *Red tide organisms in Japan.* pp. 78-79.

Lam, N.N., Hai, D.N. (1996). Harmful marine phytoplankton in Vietnam waters. In: Yasumoto, T., Oshima, Y, Fukujo, Y. (eds.) *Harmful and toxic algal blooms.* Interg. Ocean. Comm. UNESCO pp. 45-48

Lam, C., Ho, K.C. (1989). Red tides in Tolo Harbour, Hong Kong. In: Okaichi,T., Anderson, D.M., Nemoto, T. (eds.) *Red Tides. Biology, Environmental Science and Toxicology.* Elsevier Science Publishing B.V pp.49-52.

Lee, J.K., Hirayama, K. (1992). Effects of salinity, food level and temperature on the population growth of *Noctiluca scintillans* (Macartney). Bull. Fac. Fish. Nagasaki Univ. 71 : 163-168.

LeFèvre, J., Grall, J.R. (1970). On the relationships of *Noctiluca* swarming off the western coast of Brittany with hydrobiological features and plankton characteristics of the environment. J. Exp. Mar. Biol. Ecol. 4: 287-306.

Lucas, I.A.N. (1982). Observations on *Noctiluca scintillans* Macartney (Ehrenb.) (Dinophyceae) with notes on an intracellular bacterium. J.Plankton Res. 4 : 401-409.

Lutz, R.A., Incze, L.S. (1979). Impact of toxic dinoflagellate blooms on the North American shellfish industry. In: Taylor, D. L.Seliger, H. H. (eds.) *Toxic Dinoflagellate Blooms.* Elsevier/North Holland, pp. 476-483.

Malej, A. (1982). Unusual occurrence of *Pelagia noctiluca* in the Adriatic. I. Some notes on the biology of *Pelagia noctiluca* (Scyphomedusa) in the Gulf of Triest. Acta Adriat. 23 : 97-102.

Margalef, R. (1973). Fitoplancton marino de la regi6n de afloramiento del NW de África. Invest. Pesq. Suppl. 2 : 65-94.

Marshall, H.G. (1976). Phytoplankton distribution along the eastern coast of the USA. I. Phytoplankton composition. Mar. Biol. 38 : 81-89.

Marshall, H.G., Cohn,M.S. (1983). Distribution and composition of phytoplankton in northeastern coastal waters of the United States. Est.Cost. Shelf Sci. 17 : 119-131.

Mendez, S. (1993). Uruguayan red tide monitoring programme: preliminary results (1990-1991). In: Smayda, T.J., Shimizu, Y. (eds.) *Toxic phytoplankton blooms in the sea.* Elsevier Science Publishers B.V. pp.287-291.

Nawata, T, Sibaoka,T. (1986). Membrane potential controlling the initiation of feeding in the marine dinoflagellate, *Noctiluca.* Zool. Sci. 3 :49-58.

Nicol, J.A. (1958). Observations on luminescence in *Noctiluca.* J. mar.biol. Ass. U.K. 37: 535-549.

Oami, K., Naitoh, Y. (1989). Bioelectric control of effector responses in the marine dinoflagellate, *Noctiluca miliaris.* Zool. Sci. 6 : 833-850.

Odebrecht, C., Röhrig, L., Garcia, V.T., Abreu, P.C. (1995). Shellfish mortality and a red tide event in southern Brazil. In: Lassus,P., Arzul, G., Erard-Le Denn, E., Gentien, P., Marcaillou-Le Baut, C. (eds.) Harmful Marine Algal Blooms. Lavoisier, Paris, pp. 213-218.

Ogawa, Y., Nakahara, T. (1979). Interrelationships between pelagic fishes and plankton in the coastal fishing ground of the southwestern Japan Sea. Mar. Ecol. Progr. Ser. 1:115-122.

Okaichi, T., Nishio, Y., (1976). Identification of ammonia as the toxic principle of red tide of *Noctiluca miliaris*. Bull. Plankt. Soc. Japan. 23 : 25-30.

Omori, M., Hamner, W.M. (1982). Patchy distribution of zooplankton: Behaviour population assessment and sampling problems. Mar. Biol. 72:193-200.

Ostroumoff, A. (1924). *Noctiluca miliaris* in Symbiose mit grünen Algen.Zoologischer Anzeiger 68:193.

Pankow, H. (1990). Ostsee-Algenflora. G. Fischer Verl. Jena, 1-648.

Petipa, T.S. (1960). The role of *Noctiluca miliaris* Sur. in the nutrition of Calanus helgolandicus Claus. Dokl. Akad. Nauk. USSR 132 : 961-963 (in russ.)

Polishchuk, L.N., Kotsegoi, T.P., Trofanchuk, G.M. (1981). Body size and mass of *Noctiluca miliaris* in various regions of the Black Sea.Gidrobiol. Zurn., Kiew 17 : 26-31 (in russ., engl. summ.).

Rebecq, J. (1965). Considérations sur la place des trématodes dans le zooplancton marin. Annls Fac. Sci. Marseille 38: 61-84.

Qi, S., D. Li, D. (1994). Unequal cell division of *Noctiluca scintillans*. Oceanogr. Limnol. Sinica 25: 158-161.

Qi, Y., Zhang, Z., Hong, Y., Lu, S., Zhu, C., Li, Y. (1993). Occurrence of red tides on the coasts of China. In: Smayda, T.J., Shimizu, Y. (eds.) *Toxic phytoplankton blooms in the sea.* Elsevier Science Publishers B.V. pp.43-46.

Quaile, D.B. (1969). Paralytic shellfish poisoning in B.C.. Bull. Fish.Res. Bd Canada 168 :1-68.

Radach, G., Bohle-Carbonell (1990). Strukturuntersuchungen der metereologischen, hydrographischen, Nährstoff- und Phytoplankton-Langzeitreihen in der Deutschen Bucht bei Helgoland. Ber.Biol. Anst. Helgoland 7 : 1-425.

Saifullah, S.M., Chaghtai, F. (1990). Incidence of *Noctiluca scintillans* (Macartney) Ehrenb., blooms along Pakistan's shelf. Pak. J. Bot. 22 : 94-99.

Schaumann, K., Gerdes, D., Hesse, K.J. (1988). Hydrographic and biological characteristics of a *Noctiluca scintillans* red tide in the German Bight, 1984. Meeresforsch. 32: 77-91.

Schnepf, E., Drebes, G. (1993). Anisogamy in the dinoflagellate *Noctiluca* ? Helgolaender Meeresunters. 47: 265-273.

Sekiguchi, H., Kato, T. (1976). Influence of *Noctiluca predation* on *Acartia* population in the Ise Bay, Central Japan. J. oceanogr. Soc. Japan. 32 : 195-198.

Smayda, T.J. (1975). Net phytoplankton and the greater than 20-micron phytoplankton size fraction in upwelling waters off Baja California. Fish.Bull. 73 : 38-50.

Smayda, T.J. (1990). Novel and nuisance phytoplankton blooms in the sea:evidence for a global epidemic. In: Granéli, E., Sundström,B., Edler,L.,Anderson, D.M. (eds.). *Toxic marine phytoplankton*. Elsevier Science Publishing Co. New York, pp. 29-40.

Sournia, A. (1984). Classification et nomenclature de divers dinoflagellés marins (classe de Dinophyceae). Phycologia 23: 345-355.

Subrahmanian, A. (1985). Noxious dinoflagellates in Indian waters. In: Anderson,D.M., White, A.W., Baden, D.G. (eds.) *Toxic dinoflagellates*.Elsevier Publishing Co. New York pp. 525-528.

Subrahmanyan, R. (1954). A new member of the Euglenieae, Protoeuglena noctilucae gen. et spec. nov., occurring in *Noctiluca miliaris* Suriray, causing green discoloration of the sea off Calicut. Proc. Indian Acad. Sciences 39 : 118-127.

Suvapepun S. (1989). Occurrence of red tides in the Gulf of Thailand.In:Okaichi,T., Anderson, D.M., Nemoto, T. (eds.) *Red Tides. Biology, Environmental Science and Toxicology*. Elsevier Science Publishing B.V.pp.41-44.

Sweeney, B. (1976). *Pedinomonas noctilucae* (Prasinophyceae), the flagellate symbiotic in *Noctiluca* (Dinophyceae) in southeast Asia. J. Phycol. 12: 460-464.

Sweeney, B. (1978). Ultrastructure of *Noctiluca miliaris* (Pyrrophyta) with green symbionts. J. Phycol. 14: 116-120.

Takayama, H. (1983). Studies on *Noctiluca scintillans* (Dinophyceae).I. Tentacle and rod organ: their functions. Jap. J. Phycol. 31 :44-50.

Tangen, K. (1979). Dinoflagellate blooms in Norwegian waters. In:Taylor, D.L.Seliger, H.H. (eds.) *Toxic Dinoflagellate Blooms*. Elsevier/North Holland, pp. 179-182.

Taylor, F.J.R. (1976). Dinoflagellates from the international IndianOcean Expedition. Bibliotheca Botanica 132 : 1-234.

Taylor, F.J.R., Pollingher, U. (1987). Ecology of dinoflagellates. In :Taylor, F.J.R. (ed.). *The biology of dinoflagellates*. - Blackwell Sci.Pub., Oxford, pp. 398-529.

Taylor, F.J.R. (1993). The species problem and its impact on harmful phytoplankton studies. In: Smayda, T.J., Shimizu, Y. (eds.) *Toxic phytoplankton blooms in the sea*. Elsevier Science Publishers B.V. pp.81-86.

Uhlig, G. (1972). Entwicklung von *Noctiluca miliaris*. Film C 897, Inst.wiss. Film, Göttingen.

Uhlig, G., Sahling, G. (1982). Rhythms and distributional phenomena in *Noctiluca miliaris*. Annls Inst. océanogr., Paris 58 : 277-284.

Uhlig, G., Sahling, G. (1990). Long-term studies on *Noctiluca scintillans* in the German Bight. Population dynamics and red-tide phenomena 1968-1988.Neth. J. Sea Res. 25: 101-112.

Uhlig, G., Sahling, G. (1995). *Noctiluca scintillans* : Zeitliche Verteilung bei Helgoland und räumliche Verbreitung in der Deutschen Bucht (Langzeitreihen 1970-1993). Ber. Biol. Anst. Helgoland 9 : 1-127.

Uhlig, G., Sahling, G., Hanslik, M. (1995). Zur Populationsdynamik von *Noctiluca scintillans* in der südlichen Deutschen Bucht 1988-1992. - Ber. Biol. Anst. Helgoland 10: 1-32.

Weisse, T., Scheffel-Möser, U. (1992). Growth and grazing loss rates in single-celled *Phaeocystis* sp. (Prymnesiophyceae). Mar. Biol. 106, 153-158.

Wong, P.S. (1989). The occurrence and distribution of red tides in Hong Kong-application in red tide management. In: Okaichi,T., Anderson, D.M., Nemoto, T. (eds.) *Red Tides. Biology, Environmental Science and Toxicology*. Elsevier Science Publishing B.V. pp. 125-128.

Xie, J., Li, J., Lu, S., Cheng, Q., Yang, L. (1993). Features of red tide caused by *Noctiluca scintillans* off Yantian coast within Dapeng Bay. Mar.Sci. Bull. 12:1-6. (in Chinese, English abstract)

Zingmark, R.G. (1970). Sexual reproduction in the dinoflagellate *Noctiluca miliaris* Suriray. J. Phycol. 6:122-126.

Development of Nucleic Acid Probe-Based Diagnostics for Identifying and Enumerating Harmful Algal Bloom Species

Christopher A. Scholin

Monterey Bay Aquarium Research Institute
P.O. Box 628
Moss Landing, CA 95039-0628 USA

1. Introduction

Routine monitoring for harmful algal bloom (HAB) species is severely hampered by the lack of simple, sensitive, rapid diagnostic tests which can be used to identify and quantify particular species as they occur in natural assemblages. New molecular probe technologies designed to meet this challenge are now emerging from a number of laboratories around the world (Anderson 1995). The objective of these techniques is to identify and, in some cases, quantify specific molecules that delineate a particular group of organisms. A "probe" is any one of a number of molecular tools to accomplish this task. The purpose of this contribution is to review nucleic acid probes and their use in HAB research.

1.1 Concepts and Terminology

The phrase "nucleic acid probe" can actually refer to an "oligonucleotide," "polynucleotide," or "riboprobe" (Table 1). The first two consist of DNA, while the latter consists of RNA. Oligonucleotides are relatively small molecules (ca. 15-50 nucleotides) whose sequence is specified by the researcher after reference to a sequence data base. Polynucleotide probes are longer segments of DNA (ca. 100's to 1000's of nucleotides) isolated by any one of a number of molecular cloning techniques, or that are generated *in vitro* enzymatically using the polymerase chain reaction (PCR; Saiki *et al.* 1988). Riboprobes (also referred to as polyribonucleotides) are single-stranded RNA molecules produced *in vitro* by transcription of a DNA template.

"Target," as used here, is a generic term that can refer to a particular group of organisms ("target species") or a specific segment of DNA or RNA ("target sequence"). Target sequences are characterized by two major criteria: specificity and copy number. Specificity concerns the *uniqueness* of the nucleic acid sequence; the degree to which it discriminates target from non-target species, for example. Copy number refers to the amount, or concentration, of target sequences per cell. High copy number targets are desirable because the greater the number of target molecules per cell, the greater a probe's response to that cell and thus the fewer cells required to elicit a detectable reaction.

Probes pair, or hybridize, with unique, complementary target sequences of DNA or RNA to form double stranded helixes (Britton and Davidson 1985; Freifelder 1987). The formation of probe/target pairs is governed by hydrogen bonds between nucleic

NATO ASI Series, Vol. G 41
Physiological Ecology of Harmful Algal Blooms
Edited by D. M. Anderson, A. D. Cembella and G. M. Hallegraeff

Table 1. Types of nucleic acid probes.

Common Name	Typical Size (Nucleotides)	Source	Literature Examples
oligonucleotide	15-50	synthetic	DeLong et al. 1989 Raskin et al. 1994
polynucleotide (DNA)	100's – 1,000's	cloned fragment or PCR product	Ludwig et al. 1994 Ho et al. 1994
riboprobe (RNA)	100's – 1,000's	*in vitro* transcription of a DNA template	Ludwig et al. 1994 Trebesius et al. 1994

acid bases in one strand with their complement in the other (for a review see Freifelder 1987). The fidelity with which a probe interacts with its target is dependent on such factors as the size and sequence of the probe, the propensity of the probe to bind to itself (intra and/or intermolecular hybridization), as well as the three dimensional structure of the target sequence. Physical factors such as temperature, and ionic strength and ionic composition of the hybridization buffer also play an enormous role in modulating the specificity with which a probe binds to or dissociates from nucleic acid templates (Freifelder 1987; Van Ness and Chen 1991). Optimizing a protocol to discriminate target from non-target sequences is often achieved by trial and error after reference to published guidelines (e.g., Ausubel *et al.* 1987; Van Ness and Chen 1991). The utility of any nucleic acid probe depends on the nature of the sequence it binds to and on the manner in which the probe is applied (see below).

2. Probe Labels and Labeling Strategies

A probe must be "labeled" in order to measure its response towards a specific target sequence and the organism it denotes. A label can consist of a radioisotope, fluorescent molecule, or an enzyme that catalyzes colorimetric or chemiluminescent reactions (e.g., Matthews and Kricka 1988; Vaheri *et al.* 1991). "Label" is also equivalent to "reporter," "signal group," or "signal hapten". Non-radioactive reporters are highly desirable because they offer the greatest flexibility for field and laboratory use, and are therefore given special attention here.

Probe/target complexes are ultimately visualized by one of a number of strategies (e.g., Fig. 1). The "primary label" refers to the signal moiety that is physically attached to the probe (fluorochrome, enzyme, biotin, etc.). Primary labels may be used as direct or indirect indicators of the whereabouts of a probe. "Direct labeling" refers to a situation in which the primary label is also the active group that is used to visualize the probe

(Fig. 1a). For example, a probe attached directly to a fluorescent molecule is one that could be applied in a direct labeling scenario. In contrast, "indirect labeling" refers to a situation in which the primary label serves as a scaffold to which a secondary reporter group binds specifically (Fig. 1b). Visualizing probes using such secondary reporters is termed "indirect" because it is the secondary label that carries the active signal moiety (fluorochrome, enzyme, etc.). Although indirect labeling requires more steps to complete, it offers the advantage of incorporating multiple, active signal moieties per single primary probe label. This in turn amplifies the total signal that emanates from a single probe, thus increasing the sensitivity of that probe (e.g., Lim *et al.* 1993). A third type of probe/target visualization strategy is termed "probe capture," or "sandwich hybridization." Here, one probe serves to anchor the target sequence to a solid support, while a second probe bound to a different region of the target molecule functions as the active signaling platform (Fig. 1c). The second probe ("signal probe") may be applied with a direct or indirect labeling as outlined above. The advantage of the sandwich assay is that a successful reaction requires both capture of the target (one unique sequence) *and* binding of the signal probe to that target (a second unique sequence). In this way one can achieve a highly specific diagnostic response because specificity of the assay is exerted at two independent hybridization reactions.

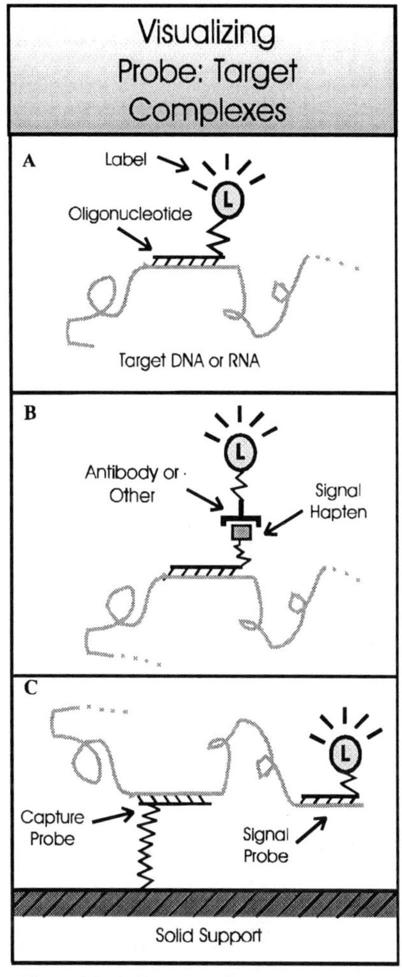

Figure 1.Labeling strategies used to visualize nucleic acid probe/target interaction; a) direct, b) indirect, c) sandwich hybridization.

3. Identification of Species-Specific Target Sequences

Development of a species-specific nucleic acid probe requires three basic areas of research: 1) elucidation of a nucleic acid sequence that serves as a unique signature of the organism in question, 2) synthesis of a probe that recognizes that signature sequence, and 3) empirical tests to determine whether or not the novel probe identifies the target species specifically (e.g., Stahl and Amann 1991, Scholin *et al.* 1994b; Miller and Scholin 1996; Scholin *et al.* 1996). Within a cell there are many nucleic acid targets that have the potential to serve as unique species-specific markers (e.g., Avise 1994). In choosing a nucleic acid target one must consider not only the potential "uniqueness" of that sequence as may be gleaned from literature examples, but also a host of other attributes such as: copy number, methods required to purify and sequence the target, the availability of a pre-existing data base for that class of sequence, and the success other researchers have had when utilizing that target as a species-specific indicator (etc).

At this time very little molecular sequence data are known for HAB species as a whole. Much of what is known stems from sequences of ribosomal RNA (rRNA) genes (rDNA) and the internal transcribed spacer regions (ITS 1 and 2) that separate them (e.g., Adachi *et al.* 1994, 1996a; Douglas *et al.* 1994; Scholin *et al.* 1994a, b; Manhart *et al.* 1995). Ribosomal RNA sequences are not necessarily the best molecules for discriminating between taxa, but they do have a proven record as valuable taxonomic markers and excellent probe targets (Maidak *et al.* 1994). Indeed, many other portions of HAB species' genomes will also harbor sequences useful in delineating target and non-target taxa. Chloroplast DNA is one such region that could reveal interesting genetic patterns among a wide range of HAB species (eg., Boczar *et al.* 1991; Avise *et al.* 1994; van Ham *et al.* 1994).

4. Nucleic Acid Probe Application Strategies

Nucleic acid probes may be applied in whole cell or cell homogenate formats (Fig. 2, Table 2). As a first step, both methods demand concentration of cells in the sample followed by resuspension in a chemically defined buffer. Thereafter, the mechanics of the two methods and analysis of end products differ substantially. The lower limit of detection using either approach will depend on numerous factors that include the volume of water collected and processed, the absolute and relative abundance of the target species, target sequence copy number, the choice of probe label, the labeling strategy, and the equipment used for measuring the resulting signal. The qualitative (presence/absence) versus quantitative (e.g., cells l^{-1}) nature of the result can depend on whether or not the target sequence copy number per cell varies over time (e.g., in response to different physiological states, circadian rhythms, life cycle stage, cellular permeability, etc.), and the extent to which cells are recovered quantitatively from a given sample.

4.1 Whole Cell Hybridization

Whole cell hybridization necessitates that cells remain intact throughout the labeling procedure. Typical end products of this procedure are cells that contain fluorescently labeled oligonucleotides bound to complementary nuclear-encoded rRNA sequences in the cytoplasm (e.g., DeLong *et al.* 1989, Lange *et al.* 1996). Harmful algae labeled in this fashion include *Alexandrium* (Scholin and Anderson *et al.*, unpubl.), *Heterosigma* (Tyrrell *et al.*, unpubl.), *Pfiesteria* (Burkholder, pers. comm), and *Pseudo-nitzschia* (Scholin *et al.* 1996; Miller and Scholin 1996). Adachi *et al.* (1996b) have shown that fluorescently labeled probes targeted towards internal transcribed spacer regions of rRNA are useful tools for identifying particular species of *Alexandrium.* In this case, the probe is localized in the nucleolus, a membrane bound intra-nuclear organelle responsible for rRNA synthesis and processing. Similarly, Miller and Scholin (1996) have shown that it is also possible to label chloroplast rRNA for a suite of *Pseudo-nitzschia* species. Thus, there are multiple target sequences within a cell that may be accessed in a whole cell hybridization format and those targets may occur in any one of a number of intra-cellular locations.

Detection and quantification of fluorescently labeled, intact cells may be accomplished by epifluorescence microscopy or flow cytometry (e.g., Amann 1990; Anderson 1994, Lange *et al.* 1996). A specific advantage of whole cell hybridization is the option of visual identification of labeled cells. That is, one can *see* what cells have retained the probe and make an informed judgment as to whether or not the reaction is a "true positive" or an unfortunate cross reaction. Visualization of labeled cells potentially offers a means of identifying morphologically divergent life cycle stages of the same species (e.g., planktonic form, amoeboid form, resting cyst) as they occur in

Table 2. Comparison of common probe application strategies.

Format	Application Technique	Purification of Nucleic Acid	Primary Type of Probe Used	Literature Examples
Whole Cell	*in situ* hybridization	not required	oligonucleotide; short polynucleotides and riboprobes	DeLong et al. 1989 Amann et al. 1994 Lim et al. 1993 Adachi et al. 1996b Miller & Scholin 1996
Cell Homogenate	"blotting"	generally required	all	Raskin et al. 1994 Thiem et al. 1994 Majiwa et al. 1994
	PCR	generally required	oligonucleotide	Sogin 1990 Miyakawa et al. 1993 Ersek et al. 1994
	RFLP	required	restriction enzyme	Adachi et al. 1994 Johnson & Aust 1994 Scholin & Anderson 1996
	sandwich hybridization	not required	oligonucleotide polynucleotide	Ranki et al. 1983 Van Ness et al. 1991 Scholin et al. 1996, 1997

complex natural assemblages. It is also possible that flow cytometry could be used to concentrate labeled cells from cultured and natural populations.

A difficulty associated with whole cell hybridization is the need for multiple incubation, rinse, and in some instances cell concentration steps. Maintaining constant and specific temperatures throughout the hybridization protocol is also very important since probe/target interaction is influenced substantially by thermal energy (Britton and Davidson 1985). Hybridization buffers must permeabilize cells and enhance accessibility of the target sequence, but not to such an extent that cells become fragile and break during sample processing. Reduction of background autofluorescence and the choice of fluorescent reporter groups and fluorescence filter sets for visualizing labeled cells are also important variables (e.g., Scholin *et al.* 1996). Enzymatic labels offer one means to eliminate fluorescence-based detection schemes (Amann *et al.* 1994), but this possibility remains untested for HAB species. Note that a protocol optimized for labeling one species may not be effective for labeling another, regardless of the specificity of the probe and the target sequence copy number (see below).

Use of fluorescently labeled probes for identification of particular phytoplankton is in its infancy. With respect to HAB species, the vast majority of this work has centered on *Alexandrium* (Anderson 1995; Adachi *et al.* 1996b; Scholin and Anderson *et al.*, unpubl.) and *Pseudonitzschia* (Scholin *et al.* 1996; Miller and Scholin 1996; Scholin *et al.* 1997). In both cases, unique species and strains are revealed by their nuclear-encoded rRNA sequences, and probes targeted towards some of those regions are useful for staining whole cells. The utility of these tools for *routine* analysis of field samples remains an open question, largely because of a lack of field-based trials. However, field studies are planned for 1997 and beyond, and thus the costs/benefits of these assays will be known soon.

A comparison of whole cell hybridization using *A. tamarense/catenella* and *P. australis* as model target species illustrates the wide range of considerations that must be given when one applies the same experimental approach to different organisms. Each group exhibits different types of cell wall structures, and the relative numbers and positions of chloroplasts in the two groups also differs. *Alexandrium* and *Pseudo-nitzschia* can also respond differently to stress induced by cell collection and concentration. *Alexandrium* sometimes form pellicular (temporary) resting cysts rapidly, which may in turn reduce their permeability to nucleic acid probes. *Pseudo-nitzschia*, on the other hand, do not appear to undergo such rapid transformations (Scholin, unpublished). Lastly, "optimal" methods for chemical preservation and autofluorescence reduction also differ, with formalin/methanol treatment being used for *Alexandrium* and a saline/ethanol solution for *Pseudo-nitzschia* (Scholin *et al.* 1996 and unpublished data.; see also Lange *et al.* 1996).

4.2 Analysis of Cell Homogenates

An alternative to whole cell identification is the detection of target molecules in cell homogenates. In this case the initial cell concentrate is disrupted in a solution that liberates targeted molecules. Probes are then applied to purified nucleic acid preparations, or directly to the unpurified lysate (cell homogenate). Necessary developmental steps of identifying target species in sample homogenates include confirming that a positive reaction (e.g., macroscopic color change, Scholin *et al.* 1996) is in fact related

Figure 2. Comparison of whole cell and cell homogenate nucleic acid probe application strategies.

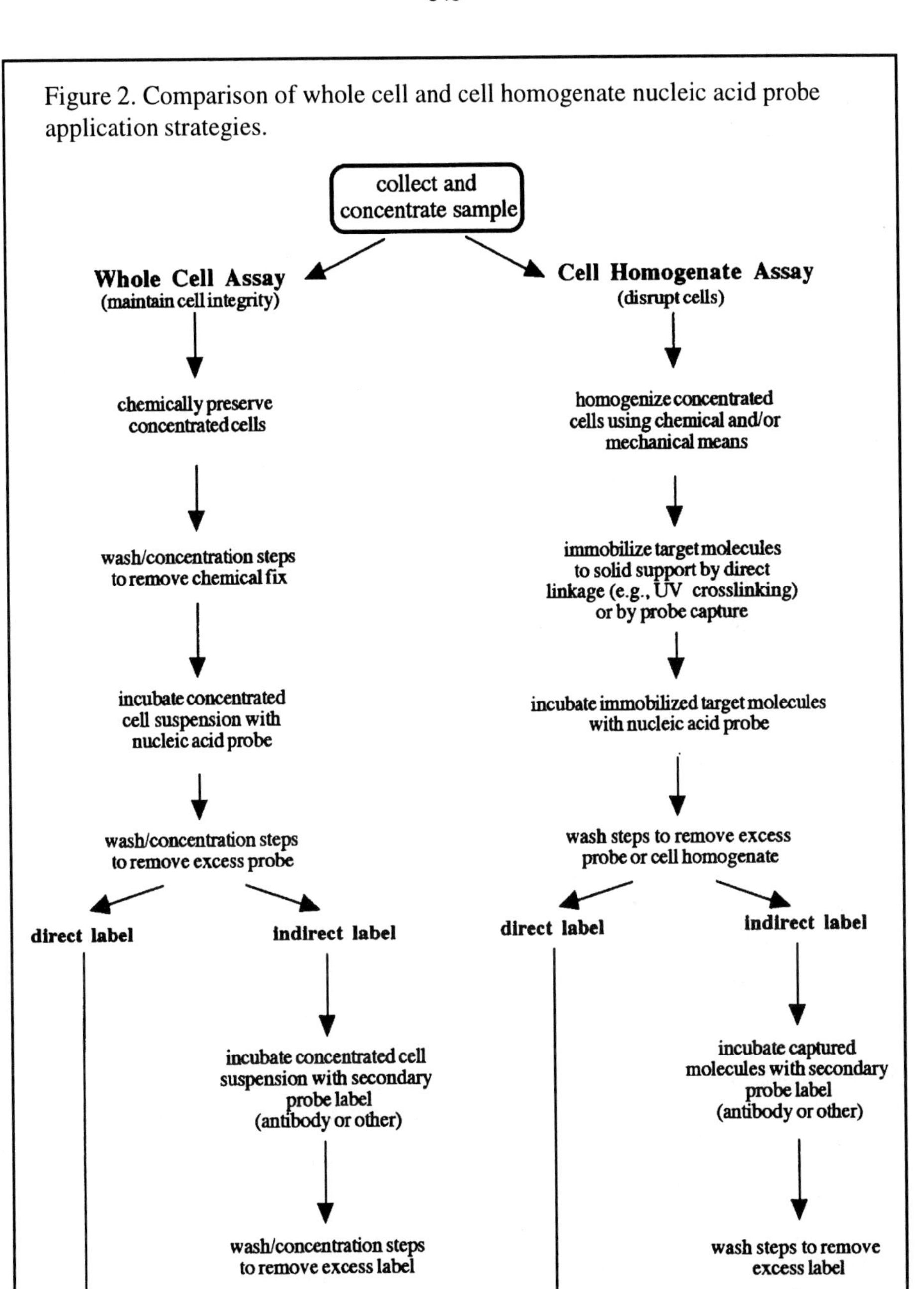

to the intended target species. The latter relationship must be determined empirically by correlating results of the novel diagnostic assay with corresponding observations that could include molecular probes applied in a whole cell format, toxin analyses, and light and electron microscopy (e.g., Scholin et al. 1997).

"Blotting," the polymerase chain reaction (PCR), restriction fragment length polymorphism (RFLP) analysis, and sandwich hybridization are among the most popular techniques used for detecting particular nucleic acid sequences collected from cell homogenates (Table 2). Blotting begins by immobilizing nucleic acids onto a solid support (such a specialized filtration membrane). The position and relative amount of a target sequence on that surface is subsequently visualized by use of a labeled probe (e.g., Ausubel *et al.* 1987). PCR is an enzymatic process whereby minute numbers of target sequences present in a sample are amplified specifically, thus facilitating their detection and manipulation (e.g., Innis 1990). PCR offers one of the most sensitive techniques for seeking minute amounts of target sequences in complex samples. To date, this author knows of only one PCR-based assay for identifying a HAB species, a large-subunit rRNA-targeted assay for *Dinophysis* spp. (Lassus *et al. personal communication*). RFLP analysis employs enzymes that cleave DNA at sequence-specific sites. The resulting fragments are resolved by gel electrophoresis to reveal the sample's pattern of digestion products (e.g., Adachi et al. 1994, Scholin and Anderson 1996). Each of these techniques, either alone or in combination, offer exquisite fidelity in detecting low numbers of target sequences in complex samples (see example references in Table 2). Protocols for each of the cell homogenate-based methods are well-established. The vast majority applied to HAB species have employed nucleic acid purification followed by PCR amplification and sequencing of a defined target sequence and/or RFLP analysis of that fragment. These approaches have focused almost exclusively on nuclear-encoded rRNA and rDNA sequences. In at least one case, however, researchers purified intact chloroplast DNA from isolates of *A. tamarense* and *A. catenella* and then subjected those samples to RFLP analyses (Boczar *et al.* 1991).

One limitation of assays that employ blotting, PCR, sequencing and/or RFLP analysis is that they demand a certain level of specialized training and dedicated laboratory equipment. In addition, all of these methods generally require fastidious nucleic acid purification as a first step after sample collection and cell concentration. Moreover, the PCR/RFLP assays available now (with the exception of that for *Dinophysis*) are designed for cultured material or cells that are concentrated artifically from natural samples. While these assays are quite useful for conducting experiments in the confines of a modern laboratory, they are at present not very portable nor suitable for field application in real-time.

In a step towards circumventing these problems, Scholin *et al.* (1996) proposed that sandwich hybridization could form the basis of a rapid technique to screen large numbers of environmental samples for particular HAB species. The potential advantages of this approach are discussed in detail elsewhere (Scholin *et al.* 1996, 1997). Among those attributes are several worth mentioning here: purification of nucleic acid is not required, requisite reactions are rapid and easy to perform, and the technique is highly amenable to automation. A prototype sandwich hybridization assay for detecting *Pseudo-nitzschia* spp. that embodies these attributes is shown in Fig. 3. Species-specific probes were incorporated into a semi-automated, colorimetric "dipstick" type assay that allows detection of *P. australis*, *P. multiseries*, *P. pseudodelicatissima* and *P. pungens* in a single sample, simultaneously. The extent of color development can reflect the abundance of particular species in a given sample. For example, tests conducted thus far using *P. australis* as a model target species have shown that it is possible to detect and quantify this species collected from natural populations by comparing reaction intensity (color development) against a standard curve derived from cultured material (e.g., Fig. 4). This assay requires less than 1 hour to complete (from live, fresh sample to assay result), with hands-on time being ~10-15 minutes.

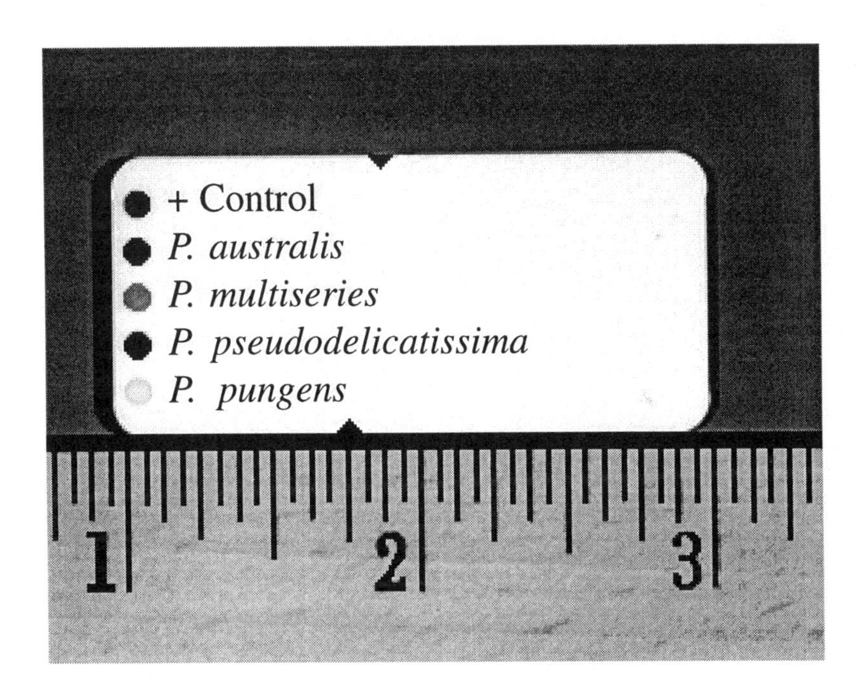

Fig. 3. Plastic analytical card (PAC) used in the semi-automated sandwich hybridization assay for simultaneous detection of multiple Pseudo-nitzschia species in a single sample. Species-specific probe/nylon bead conjugates, in addition to a positive control bead, are pressed into the PAC at known positions. In all cases, the positive control bead should turn blue. Development of blue color of one or more species-specific beads signifies presence of one or more species, and the intensity of that color is a measure of their abundance (Fig. 4). This PAC was exposed to a sample that contained an abundance of *P. australis* and *P. pseudodelicatissima,* a lesser amount of *P. multiseries,* and no *P. pungens* (image scanned from Polaroid photo). Scale = inches.

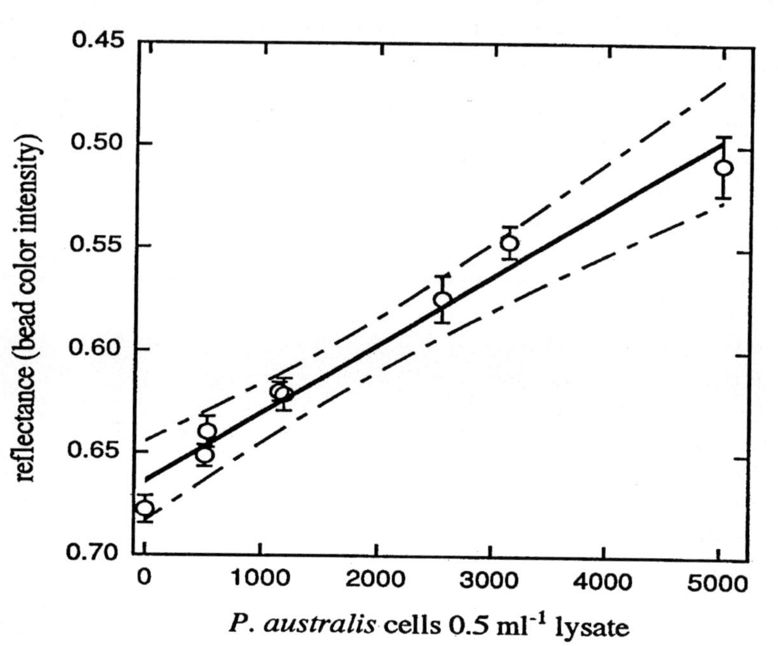

Fig.4. Standard curve showing the response of the semi-automated sandwich hybridization assay for *P. australis* using cell densities ranging from approximately 5×10^2 - 5×10^3 cells 0.5 ml $^{-1}$ lysate. A known number of cells were collected onto a filter, homogenized, and processed as described by Scholin et al. (1997). The resultant PACs (see Fig. 3) were placed in a reflectance scanner to quantify assay results. The latter measurement is based on reflectance of the bead/probe conjugates relative to that of a white surface (internal standard); decreasing reflectance values indicates increasing color intensity. Reflectance readings are the mean values ± SD (n = 4). The bold line shows a first order linear regression (correlation coefficient 0.99); dashed lines represent 99% confidence intervals.

5. Consideration of Genetic Variation Among Targeted Species

Detailed morphological features and life histories are the primary criteria used to distinguish between phytoplankton species. A single group of organisms that share a unique set of such characters — a species — may in fact include multiple genetic variants or "strains." Different strains of the same species may not be recognizable at the micro scopic level, but may nevertheless diverge on a molecular level. A good example of this concerns *A. tamarense* and *A. catenella* (see Scholin, this volume). Although two individuals of the same species may look alike in a macroscopic sense, their associ-

8. References

Adachi, M., Sako, Y., Ishida, Y. (1994). Restriction fragment length polymorphism of ribosomal DNA internal transcribed spacer and 5.8S regions in Japanese *Alexandrium* species (Dinophyceae). *J. Phycol.* 30:857-863.

Adachi, M., Sako, Y., Ishida, Y. (1996a). Analysis of *Alexandrium* (Dinophyceae) species using sequences of the 5.8S ribosomal DNA and internal transcribed spacer regions. *J. Phycol.* 32:424-432.

Adachi, M., Sako, Y., Ishida, Y. (1996b). Cross-reactivity of fluorescent DNA probes to isolates of the genus *Alexandrium* by in situ hybridization. *In:* Oshima, Y. and Fukuyo, Y. [Eds.] *Harmful and Toxic Algal Blooms*, Intergovernmental Oceanographic Commission of UNESCO, Paris. pp 455-458.

Amann, R.I., Binder, B.J., Olson, R.J., Chisholm, S.W., Devereux, R., Stahl, D.A. (1990). Combination of 16S rRNA-targeted oligonucleotide probes with flow cytometry for analyzing mixed microbial populations. *Appl. Environ. Micribiol.* 56: 1919-25.

Amann, R., Zarda,B., Trebesius, K.H., Ludwig, W., Schleifer, K-H. (1994). Rapid identification of micro-organisms using fluorescent rRNA-targeted oligonucleotides. *In*: Spencer, R.C., Wright, E.P., & Newsom, S.W.B. [Eds.] *Rapid Methods and Automation in Mcrobiology and Immunology*. Intercept, Ltd. Andover, U.K. pp. 237-244.

Anderson, D.M. (1994). Identification of harmful algal species using molecular probes: an emerging perspective. *In*: Lasus, P., Arzul, G., Erard, E., Gentien, P., Marcaillou, C. [Eds.] *Harmful Marine Algal Blooms*, Intercept, Ltd., pp 3-13.

Avise J.C. (1994). *Molecular Markers, Natural History and Evolution.* Chapman and Hall, New York. 511 pp.

Ausubel, F.M., Brent, R., Kingston, R.E., Moore, D.D., Siedman, J.G., Smith, J.A., Struhl, K. [Eds.] (1987). *Current Protocols in Molecular Biology.* Wiley Interscience, New York.

Boczar, B.A., Liston, J., Cattolico, R.A. (1991). Characterization of Satellite DAN from three marine dinoflagellates (Dinophyceae): *Glenodinium* sp. and two members of the toxic genus, *Protogonyaulax. Plant Physiol.* 97:613-618.

Britton, R.J., Davidson, E.H. (1985). Hybridization strategy. *In*: Hames, B.D., Higgins, S.J. [Eds.]. *Nucleic Acid Hybridization - A Practical Approach.* IRL Press, Oxford. pp. 3-15.

DeLong, E.F., Wickham, G.S., Pace, N.R. (1989). Phylogenetic stains: ribosomal RNA-based probes for the identification of single cells. *Science (Washington, D.C.).* 243:1360-1363.

Douglas, D.J., Landry, D., Douglas, S.E., (1994). Genetic relatedness of toxic and nontoxic isolates of the marine pennate diatom *Pseudonitzschia* (Bacillariophyceae): phylogenetic analysis of 18S rRNA sequences. *Natl. Toxins* 2:166-174.

Ersek, T., Schoelz, J.E., English, J.T. (1994). PCR amplification of species-specific DNA sequences can distinguish among *Phytophthora* species. 1994. *Appl. Environ. Microbiol.* 60: 2616-2621.

Freifelder, D. (1987). *Molecular Biology.* Jones and Bartlett, Inc. Boston, MA. 834 pp.

Ho, M.S.Y., Conrad, P.A., Conrad, P.J., LeFebvre, R.B., Perez, E., BonDurant, R.H. (1994). Detection of bovine trichomoniasis with a specific DNA probe and PCR amplification system. *J. Clin. Microbiol.* 32: 98-104.

Innis, M.A., Gelfand, D.H.,Sninsky, J.J., White, T.J. [Eds.] (1990). *PCR Protocols: A Guide To Methods And Applications.* Academic Press. San Diego, CA.

Johnston, C.G. and Aust, S.D. (1994). Detection of *Phanerochaete chryosporium* in soil by PCR and restriction enzyme analysis. *Appl. Environ. Microbiol.* 60:2350-2354.

Lange, M., Guillou, L., Valuot, D., Simon, N., Amann, R.I., Ludwig, W., Medlin, L.K. (1996). Indentification of the class *Prymnesiophyceae* and the genus *Phaeocystis* with ribosomal RNA-targeted nucleic acid probes detected by flow cytometry. *J. Phycol.* 32: 858-868.

Lim, E.L., Amaral, L.A., Caron, D.A., DeLong, E.F. (1993). Application of rRNA-based probes for observing marine nanoplankton protists. *Appl. Env. Microbiol.* 59: 1647-1655.

Ludwig, W., Dorn, S., Springer, N., Kirchhof, G., Schleifer, K-H (1994). PCR-based preparation of 23S rRNA-targeted group-specific polynucleotide probes. *Appl. Environ. Microbiol.* 60:3236-3244.

Maidak, B.L., Larsen, N., McCaughey, M.J., Overbeek, R., Olsen, G.J., Fogel, K., Blandy, J., Woese, C.R. (1994). The ribosomal database project. *Nucleic Acids Research* 22:3485-3587.

Majiwa, P.A.O., Thatthi, R., Moloo, S.K., Nyeko, J.H.P., Otieno, L.H., Maloo, S. (1994). Detection of trypanosome infections in the saliva of tsetse flies and buffy-coat samples from antigenaemic but aparasitaemic cattle. *Parasitology.* 108: 313-322.

Manhart, J. R., Fryxell, G.A., Villac, M.C., Segura, L.Y. (1995). *Pseudo-nitzschia pungens* and *P. multiseries* (Bacillariophyceae): nuclear ribosomal DNAs and species differences. *J. Phycol.* 31:421-427.

Matthews, J.A., Krika, L.J. (1988). Analytical strategies for the use of DNA probes. *Analytical Biochemistry* 169:1-25.

Miller, P.E., Scholin, C.A. (1996). Identification of cultured *Pseudo-nitzschia* (Barcillariophyceae) using species-specific LSU rRNA-targeted fluorescent probes. *J. Phycol.* 32:646-655.

Miyakawa, Y., Mabuchi, T., Fukazawa, Y. (1993). New Method for detection of *Candida albicans* in human blood by polymerase chain reaction. *J. Clin. Microbiol.* 31:3344-3347.

Ranki, M., Palva, A., Virtanen, M., Laaksonen, M., Soderlund, H. (1983). Sandwich hybridization as a convenient method for the detection of nucleic acid samples in crude samples. *Gene.* 21:77-85.

Raskin, L., Stromley, J.M., Rittman, B.E., Stahl, D.A. (1994). Group-specific 16S rRNAhybridization probes to describe natural communities of methanogens. *Appl. Environ. Microbiol.* 60:1232-1240.

Saiki, R.K., Gelfand, D.H., Stoffel, S., Scharf, S.J., Higuchi, R., Horn, G.T., Mullis, K.B., Erlich, H.A. (1988). Primer-directed enzymatic amplification of DNA with a thermo stable DNA polymerase. *Science (Washington, D.C.).* 239:487-491.

Scholin, C.A., Anderson, D.M., Sogin, M.L. (1993). The existence of two distinct small-subunit rRNA genes in the toxic dinoglagellate *Alexandrium fundyense. J Phycol.* 29:209-216.

Scholin, C.A., Anderson, D.M. (1994). Identification of species and strain-specific genetic markers for globally distributed *Alexandrium* (Dinophyceae). I. RFLP analysisof SSU rRNA genes. *J. Phycol.* 30: 744-754.

Scholin, C.A.,Herzog, M.,Sogin, M.L., Anderson, D.M. (1994a). Identification of group and strain-specific genetic markers for globally distributed *Alexandrium* (Dinophyceae). II. Sequence analysis of a fragment of the LSU rRNA gene. *J. Phycol.* 30:999-1011.

Scholin, C.A., Villac, M.C., Buck, K.R., Krupp, J.M., Powers, D.A., Fryxell, G.A., Chavez, F.P. (1994b). Ribosomal DNA sequences discriminate among toxic and non-toxic *Pseudonitzschia* species. *Natural Toxins* 2:152-165.

Scholin, C.A., Anderson, D.M. (1996). LSU rDNA-based RFLP assays for discriminating species and strains of *Alexandrium* (Dinophyceae). *J. Phycol.* 32:1022-1035.

Scholin, C.A., Buck, K.R., Britschgi, T., Cangelosi, J., Chavez, F.P. (1996). Identification of *Pseudo-nitzschia australis* (Bacillariophyceae) using rRNA-targeted probes in whole cell and sandwich hybridization formats. *Phycologia* 35:190-197.

Scholin, C.A., Miller, P., Buck, K.R., Chavez, F.P., Harris, P., Haydock, P., Howard, J., Cangelosi, G. (1997). Detection and quantification of *Pseudo-nitzschia australis* in cultured and natural populations using LSU rRNA-targeted probes. *Limnology and Oceanography* (in press).

Sogin, M.L. (1990). Amplification of ribosomal RNA genes for molecular evolution studies. *In*: Innis, M.A., D.H. Gelfand, J.J. Sninsky, & T.J. White [Eds.] *PCR Protocols: A Guide To Methods And Applications.* Academic Press. San Diego, CA. pp 307-314.

Stahl, D.A., Amann, R. (1991). Development and application of nucleic acid probes. *In:* Stackebrandt, E., & M. Goodfellow [Eds.] *Modern Microbiological Methods. Nucleic Acid Techniques in Bacterial Systematics.* pp. 205-248.

Thiem, S.M., Krumme, M.L., Smith, R.L., Tiedje, J. (1994). Use of molecular techniques to evaluate the survival of a microorganism injected into an aquifer. *Appl. Environ. Microbiol.* 60:1059-1067.

Trebesius, K., Amann, R., Ludwig, W., Muhlegger, K., Schleifer, K-H. (1994). Identification of whole fixed bacterial cells with nonradiactive 23S rRNA-targeted polynucleotide probes. *Appl. Environ. Microbiol.* 60: 3228-3253.

Vaheri, A., Tilton, R.C., Balows, A. [Eds.] (1991). *Rapid Methods and Automation in Microbiology and Immunology.* Springer-Verlag, New York. 573 pp.

van Ham, R.C.H.J., Hart, H., Mes, T.H.M., Sandbrink, J.M. (1994). Molecular evolution of noncoding regions of the chloroplast genome in the Crassulaceae and related species. *Curr. Genet.* 25:558-566.

Van Ness, J., Chen, L. (1991). The use of oligonucleotide probes in chaotrope-based hybridization solutions. *Nucleic Acids Res.* 19:5143-5151.

Van Ness, J., Kalbfleisch, S., Petrie, C.R., Reed, M.W., Tabone, J.C., VermeulenM.J. (1991). A versatile solid support system for oligonucleotide probe-based hybridization assays. *Nucleic Acids Res.* 19:3345-3350.

Parasites of Harmful Algae

M. Elbrächter[1] and E. Schnepf[2]

[1]Biologische Anstalt Helgoland, Taxonomische Arbeitsgruppe, Wattenmeerstation Sylt; Hafenstr.43; D-25992 List/Sylt, Federal Republic of Germany
[2]Zellenlehre, Universität Heidelberg, Im Neuenheimer Feld 230, D-69120 Heidelberg, Federal Republic of Germany

1. Introduction

Prokaryotic and eukaryotic microparasites such as viruses, bacteria, fungi and other protists have been recorded from marine and freshwater environments for many decades (Chatton 1920, Canter and Lund 1951, Spencer 1955). However, their relative abundance and ecological significance in pelagic ecosystems have received little attention until recently. In particular, this applies to their possible role in harmful plankton dynamics. Even less is known regarding phagotrophic flagellates which attack larger phytoplankton species like diatoms and dinoflagellates, e.g. as ectoparasites. The occurrence of flagellated parasites on and in phytoplankton cells often has been misinterpretated as life cycle stages, e.g. gametes. This applies to fungal infections as well. Parasites generally do not differentiate between harmful and non-harmful algae, and phytoplankton species not regarded as harmful may be found eventually to produce toxins. Therefore, parasites of diatoms, dinoflagellates, prymnesiophytes and raphidophytes as well as of cyanobacteria are discussed here.

2. The Parasites

2.1. Viruses

Knowledge of lytic viruses and their potential role in causing population level changes in host organisms on time scales of hours and days is only emerging recently (Bratbak *et al.* 1993, Nagasaki et al. 1994, Suttle 1994, Suttle and Chan 1995). Zingone (1995) published a review on this topic, providing a list of 22 phytoplankton species infected by viruses. She concluded that complex virus-algae relationships may represent a key factor in the dynamics of phytoplankton blooms. Virus-like particles (VLPs) have been described more than 20 years in dinoflagellates ago (e.g., Franca 1976, Pienaar 1976, Soyer 1978) but some of these particles may have been confused with cellular structures (Soyer 1978). Viral infections can be recognized only by transmission electron microscopy and apparently only during the late phase of infection (Reisser 1993). This may be the reason why so few reports are published on this topic. The decline of the brown tide caused by the chrysophycean picoplankter *Aureococcus anophagefferens* may be related to viral infection (Sieburth et al. 1988). The prymnesiophyte *Emiliania huxleyi*, a very common coccolithophorid, and species of the genus *Chrysochromulina* have been shown to be infected by viruses (Bratbak et al. 1993, Suttle and Chan 1995). Viruses known to lyse *C. brevifilum* and *C.*

NATO ASI Series, Vol. G 41
Physiological Ecology of Harmful Algal Blooms
Edited by D. M. Anderson, A. D. Cembella and G. M. Hallegraeff

strobilus did not affect the toxic *C. polylepis*, which had caused severe economic losses to Scandinavian fishermen in 1988 (Dahl *et al.* 1989). A further nine species of this genus were not infected by this virus. This shows that viruses may be very host-specific, and do not necessarily infect all species of a genus. However, Cottrell and Curtis (1995a) demonstrated that viruses infecting the prasinophyte *Micromonas pusilla* from different geographic regions showed a genetic diversity. Another group of red tide organisms, the raphidophytes, are also known to be infected by viruses. Nagasaki *et al.* (1994) report on a possible disintegration of a bloom produced by *Heterosigma akashiwo*. The VLPs were located near the nucleus of the cell.

The dynamics of virus infections have been rarely investigated. The first studies were done in freshwater on the *Chlorella*-virus system, (for literature review see Reisser 1995). Cottrell and Curtis (1995b) approached this question by investigating the effect of *Micromonas pusilla* viruses on its host.

Cyanobacteria are known to contain plasmids and viruses. About 3% of *Synechococcus* spp. populations may be lysed daily by viruses (Suttle 1994). The role plasmids and viruses play in toxic cyanobacteria is not well understood (Bose and Carmichael 1990, Suttle and Chan 1994).

2.2. Bacteria

For many years it has been known that certain phytoplankton species harbour endocytic bacteria which may be located either in the cytoplasm or even in the nucleus (e.g. Leedale 1969, Gold and Pollingher 1971, Silva 1978). In most cases the association is regarded as a symbiosis, not as a deleterious infection. For toxic cyanobacteria it has been argued that an intracellular bacterium might be responsible for the decline of the bloom (Gumpert *et al.* 1987). Lysis of bloom-forming cyanobacteria by lytic bacteria is described from freshwater habitats (Steward and Draft 1976). Schnepf *et al.* (1974) reported on a pathogenic bacterium in the freshwater planktonic chlorophyte *Scenedesmus*, and Peterson *et al.* (1993) found associations with freshwater benthic diatoms. Imai *et al.* (1993, 1995) isolated algicidal bacteria, which killed raphidophytes and a dinoflagellate - the red tide flagellates *Chattonella antiqua*, *Heterosigma akashiwo* and *Gymnodinium mikimotoi* - as well as two diatoms. They found two different types of algicidal bacteria. The first type excretes compounds into the water, which are deleterious for the algae. When they added medium from which the bacteria had been filtered to algal cultures, the algae died. The second type does not excrete such compounds but they apparently attack the algal cells directly. When one bacterial cell was added to a thriving *Chattonella antiqua* culture, all algal cells died within 7 days. These processes have so far not been adequately considered for marine phytoplankton blooms. The interactions between bacteria and harmful algae highly varied and will be dealt with in other chapters of this volume (see Doucette *et al.* this volume).

2.3. Dinoflagellates

There are more than 80 dinoflagellate species parasitizing various groups of protists and metazoans. Several parasitic dinoflagellates infect exclusively dinoflagellates including toxic or noxious species, others are restricted to diatom hosts.

a) *Amoebophrya* (Figs. 1a,b) is a well known member of this group. It is an endonuclear or endocytic parasite of Dinophyta, Heliozoa, Radiolaria, and Ciliata, although one species infects chaetognaths and siphonophores (Cachon 1964, Cachon and Cachon 1987). Infection is by biflagellated zoospores (dinospores) of the shape of *Gymnodinium* or by highly modified forms. After attachment to the host cell zoospores penetrate into the cytoplasm, and there they differentiate into a trophont. The trophont soon increases in size by apparently resorptive food uptake. It differentiates into an episoma which sinks progressively in the hyposoma. A cavity, the socalled mastigocoel, forms with a double system of helical structures, one parietal involving the hyposoma, the other axial on the part comprising the episoma. Nuclear divisions are not followed by cytoplasmic divisions. Thus, trophonts of late infections are large, polynucleate (Fig. 1b), and later multiflagellate organisms that occupy most of the host cell. By superficial observations, the parasite may be confused with a hypertrophic nucleus. At maturity, the trophont ruptures through the host's cell wall and transforms into a long, vermiform strongly motile stage that finally divides to produce numerous spores which apparently can infect a new host without maturation phase. A diagram of the life cycle is given by Cachon and Cachon (1987). The vegetative cycle lasts less than 2 days under optimal conditions. To date, about 7 species are recognized, of which only 3 infect dinoflagellates. Whereas *A. grassei* is a hyperparasite of the fish-parasitizing *Oodinium*, and *A. leptodisci* of the heterotrophic *Leptodiscus medusoides*, only *A. ceratii* is known to parasitize several free-living relatives. Little host specificity is postulated (Fritz and Nass 1992). The species is broadly distributed at least in the northern hemisphere, infecting more than 20 host species from at least 19 genera of dinoflagellates (Cachon 1964, Taylor 1968, Elbrächter 1973, Nishitani *et al.* 1985, Fritz and Nass 1992, Coats and Bockstahler 1994). The following dinoflagellate genera with species forming toxic or noxious events have been reported to be infected by *Amoebophrya ceratii*: *Alexandrium*, *Amphidinium*, *Ceratium*, *Dinophysis*, *Gonyaulax*, *Gymnodinium*, *Gyrodinium*, *Prorocentrum*. There is emerging evidence that *Amoebophrya ceratii* is a heterogenic taxon. Cachon (1964) discovered that athecate host-dinoflagellates are infected intracytoplasmatically as is the thecate *Prorocentrum micans*, whereas in other thecate dinoflagellates infestation is intranuclear. It is unlikely that the infection site should be so variable from host to host but constant in a given host. In addition observations on field populations with high infection rates of *Ceratium tripos* and simultaneous occurrence of *Prorocentrum micans* with high cell numbers, but without any infection, does not promote the hypothesis of a broad host spectrum. From these field samples, a subsample with infected *Ceratium* cells inoculated into a *Prorocentrum* culture caused no infection, although in the same experiment uninfected *Ceratium* cells had been infected (Elbrächter, unpubl.). Coats (pers. comm. and in press) achieved similar results in culture experiments. Further investigations are required to prove whether *Amoebophrya* is a collective taxon of different species or host races and whether or not a population from a given host can infect other dinoflagellate species. If host-specificity occurs, *Amoebophrya* might serve as a biological control for red-tides, as first suggested by Taylor (1968) but later dismissed by Nishitani *et al.* (1985).

b) *Coccidinium* includes 4 species (Chatton and Biecheler, 1934,1936) which are poorly known intracytoplasmic or intranuclear parasites of some dinoflagellates, of which *Coolia monotis* is regarded as potentially toxic. The parasite forms multinucleate stages inside the host and eventually produces aberrant dinospores. Cachon and Cachon (1987) related them to the Syndiniales. Their importance is unknown and they are easily overlooked as are most intracellular parasites.

c) *Dubosquella* includes about 8 species, of which one species, *D. melo* parasitizes the noxious-bloom forming dinoflagellate *Noctiluca scintillans* (Cachon 1964). The other species parasitize ciliates. The first stages of the life cycle are very similar to those of *Amoebophrya* but the episoma is covered by a shield - a thickened part of the amphiesma and it is delimited by a fibrous ring, the perinema. Dinospore formation may lead either to the formation of macrospores or microspores.

d) *Paulsenella* (Figs. 2-4) includes 3 species which are exclusively parasites on diatoms. *P. chaetoceratis* (Figs. 2-3) sucks out the contents of *Chaetoceros* cells, including those species that form harmful blooms which damage fish gills with their long setae. The vegetative life cycle is simple: a gymnodinoid athecate dinospore infests on the setae of the host, extends a feeding tube (Fig. 2) of up to 150 μm length through the setae into the cell and then "sucks out" (myzocytosis) the cytoplasm of the cell (Drebes and Schnepf 1988). The driving forces of this process are not known (Schnepf and Elbrächter 1992). After termination of food uptake, a primary cyst wall is formed. It often detaches from the host at this stage. Binary fission inside the primary cyst wall (Fig. 3) is initiated. Both daughter cells form new thick cyst walls, becoming secondary cysts in which binary fission takes place again. Secondary cysts may remain enclosed in the primary cyst wall which later becomes thinner and eventually disappears. Secondary cysts may divide again, forming a tertiary cyst or even a quaternary cyst. The final cysts represent dinosporangia, each giving rise to two dinospores. After a postmaturation phase, the dinospore can infect a new host. Sexual reproduction, indicated by nuclear cyclosis, has been observed. Formation of temporary cysts or resting cysts is known (Drebes and Schnepf 1988). The known host range is *Chaetoceros borealis* and *Ch. decipiens*. Other species of *Paulsenella* have been reported from a variety of diatom genera, see Drebes and Schnepf (1988).

e) Pallium-feeding Peridiniales: Elbrächter (1991) depicted a small unidentified *Protoperidinium* species feeding by pallium-extension on a *Dinophysis fortii* cell. Although this does not present a parasitic interaction, it should be mentioned. Heterotrophic dinoflagellates which exhibit pallium-feeding may have a severe effect on toxic phytoplankton populations. For example, *Dinophysis* or diatoms such as *Pseudo-nitzschia* may be attacked by several species of the *Diplopsalis*-group (Elbrächter unpubl.). This has also been documented in detail by Nakamura et al. (1995) for a *Gymnodinium mikimotoi* bloom.

2.4. Euglenozoa

a) Euglenida: *Rhynchopus coscinodiscivorus* (Figs. 5,6) has recently been described (Schnepf 1994a) as a small (about 10 μm long) heterotrophic flagellate with concealed flagella. It feeds mainly on large diatoms including the occasionally harmful blooms of *Coscinodiscus concinnus* (Figs. 5,6) and *C. wailesii*. The flagellate invades the diatom in the girdle region. The diatom retracts its cytoplasm and the flagellate attacks the protoplasm upon which it feeds. Inside the diatom frustule, *Rhynchopus* multiplies, such that eventually a single frustule may contain 30 to more than 100 *Rhynchopus* cells. Other diatom genera like *Odontella* may also serve as hosts. This parasite has been formerly confused with *Pronoctiluca phaeocysticola* which is a kinetoplastid although it was classified earlier as a dinoflagellate (see below).

b) Kinetoplastida : Cachon (1964, p. 33) reported in a foot-note that he repeatedly found a small bodonid (size 3-4μm long) inside *Noctiluca.* A special paper on this flagellate as announced by Cachon apparently has not been published. Whether this flagellate is a parasite or a commensal organism is not known.

The polykinetoplastid flagellate *Hemistasia phaeocysticola* [synonyms: *Hemistasia klebsii*; *Pronoctiluca phaeocysticola*] (Fig. 7) feeds on a variety of phytoplankton species, preferentially on the large colony-forming and foam-producing noxious prymnesiophyte *Phaeocystis globosa* but it also feeds on large diatoms, e.g. *Coscinodiscus concinnus* and *C. wailesii.* In cultures it is also able to attack *Gonyaulax polyedra* [synonym: *Lingulodinium polyedrum]* (Elbrächter *et al.* 1996). In culture, the diatoms and dinoflagellates are only attacked during stationary phase. Whether this species attacks logarithmically growing cells in the field is not known.

2.5. "Zooflagellates"

Zooflagellates are not a monophyletic taxon but a collection of unicellular, eukaryotic organisms with one or more flagella and without chloroplasts, which cannot be aligned within a natural, well circumscribed monophyletic class of organisms.

a) *Cryothecomonas* has been described from ice biota as a heterotrophic flagellate with two heterodynamic flagella of unequal length (Thomsen *et al.* 1990). In the North Sea, a new member of this genus, *Cryothecomonas aestivalis* (Figs. 8-9) has been discovered which exclusively feeds on the centric diatom *Guinardia delicatula* [synonym : *Rhizosolenia delicatula*] (Drebes *et al.* 1996). It penetrates into the cell and phagocytizes the host cytoplasm by means of a pseudopodium that emerges from the posterior cell pole (Fig. 8). The mature trophont gives rise to 8 to 32 biflagellate swarmers (Fig. 9) which leave the emptied diatom frustule. The vegetative life cycle takes about 18 to 20 hours. During blooms, a large percentage of *Guinardia delicatula* may be infected by *Cryothecomonas.*

b) *Pirsonia* (Figs. 10-11) so far includes 6 species (Schnepf et al. 1990; Kühn *et al.* 1996). They are parasites of various diatom genera and infection of noxious or even toxic diatom genera can be expected. Cells are small (< 15 μm). Two flagella are inserted subapically, the anterior one is 10 to 18 μm long, whereas the posterior flagellum is twice or 3 times as long. The movements of the swimming flagellates are slightly jerky. Infestation of the host by the flagellate is by the posterior end which forms a broad foot, while the flagella are coiled around the apex of the cell. The diatom is invaded by a small pseudopodium which enlarges inside the frustule to become the trophosome (Fig. 10). During feeding on the host, the flagellate outside the host attached to the diatom frustule divides several times, forming multicellular colonies (Fig. 11). After feeding has terminated, swarmers are liberated. They may form resting cysts. The generation time is about 1 day, 8 to more than 30 swarmers may be formed during one generation. The infective phase lasts for about 3 days. Hosts include the noxious bloom forming diatom *Coscinodiscus wailesii.*

2.6. Rhizopda: Amoebidae

Amoeba biddulphiae (Fig. 14) feeds on the diatom *Odontella sinensis* [synonym: *Biddulphia sinensis*] (Zuelzer 1927). Apparently, the parasitic amoeba has been introduced into the North Sea together with its host as no other diatom has been reported to be infected by this amoeba. The amoeba penetrates into its host and digests the protoplasm. Infection rates may be high in field populations (Drebes 1974). Recently, a new species of *Rhizamoeba* has been discovered, which feeds mainly on *Chaetoceros didymus*. It's doubling time is higher than that of the host, as in cultures all diatoms were infected after a few days (Kühn 1995). In fixed plankton samples, amoebae cannot be recognized inside diatoms or any other phytoplankton species, therefore the abundance and possible effect of amoebae on their hosts is unknown.

Pseudaphelidium drebesii (Figs. 12-13) parasitizes the diatom *Thalassiosira punctigera*. This organism is tentatively included in the Rhizopoda by Schweikert and Schnepf (1996). The motile stage consists of a zoospore with a single opisthokont flagellum. It attaches to the host cell, encysts, penetrates the frustule and develops into a plasmodium. This plasmodium phagocytizes portions of host cytoplasm which are included in a single large digestion vacuole (Fig. 12). At the end of the trophic phase the plasmodium is a hollow sphere filling the frustule completely. It cleaves to form amoeboid cells (Fig. 13) which are weakly motile and finally encyst. The cyst releases 4 (or sometimes less) zoospores.

2.7. "Fungi"

Fungi are not a natural assemblage but rather a polyphyletic group of obligate heterotrophic organisms which do not feed by resorption rather than by phagotrophy. In freshwater, fungi have long been known as parasites of diatoms and dinoflagellates (e.g. Canter and Lund 1948, Canter and Heaney 1984, Boltovskoy 1984). Knowledge of marine fungi is mostly restricted to those groups which can be grown on pine pollen. Parasitic fungi of marine phytoplankton are rarely studied. The application of epifluorescence microscopy staining techniques will result in an increase of knowledge. A recent review on the ecology of aquatic fungi in and on algae is given by Van Donk and Bruning (1995).

a) *Lagenisma* belongs to the Oomycota (Drebes 1968) and is probably the best known parasitic "fungus" of marine phytoplankton species (Drebes 1966, Schnepf *et al.* 1978 a,b,c,d,e, Schnepf and Heinzmann 1980, Wetsteyn and Peperzak 1991). It feeds on centric diatoms of the genus *Coscinodiscus*, some of which produce noxious blooms. Before infecting the host, the freshly released zoospores pass through two different cyst stages. The primary zoospores are kidney-shaped and laterally biflagellated. They form a primary cyst with a spiny cyst wall which is left by isomorphic secondary zoospores. The latter form a secondary cyst which is smooth-walled. The secondary cyst germinates and infects a new diatom cell by means of a tube which enters the cell in the cingular region. Then it develops a thallus (Figs. 15). The cytoplasm of the thallus cleaves to form many zoospores (Fig. 16) which, after maturation, are liberated through only one discharge tube. Sexual reproduction is documented (Schnepf *et al.* 1978 b). Sometimes multiinfection is recorded. Infection rates in field samples may exceed 70% of the population.

b) *Ectrogella* belongs to the Saprolegniales and these species are more or less exclusive parasites of diatoms. The species are not very well described, and host specificity is not completely known from experiments. The development of these holocarpic fungi (Figs. 17-19) is similar to that of *Lagenisma*, but the biflagellated zoospores are released by two to several discharge tubes, in contrast to *Lagenisma*, which has only one. Various diatom genera have been observed to be infected by *Ectrogella-* like parasites (Drebes 1974). One would expect that harmful and toxic species may be also infected.

c) *Olpidium* and *Rhizophydium* both belong to the Chytridiales (Figs. 20-27). The zoospores of these parasites have only 1 flagellum. In *Rhizophydium*, the zoospores attach to the cell wall of the host and form a thread-like rhizoid inside the host. Outside the host, the fungus grows and forms a spherical zoosporangium in which the uniflagellated zoospores are formed. Some species which infect *Pseudo-nitzschia* (Figs. 23-27) and *Chaetoceros* species belong to this group of parasites.

d) *Phagomyxa* Karling belongs to the Plasmodiophoromycota. This genus differs from all other "fungi" in that it feeds by phagotrophy, not by resorption. Recently, Schnepf (1994b) discovered a parasite in the marine diatom *Bellerochea malleus* (Figs. 28-31). The parasite infects the host by biflagellated zoospores, the flagella are unequal in length. Inside the diatom, the parasite develops an endocytoplasmic plasmodium and incorporates host cytoplasm into a large central digestion vacuole (Fig. 28) by a form of phagocytosis. Later on, the plasmodium cleaves into several zoosporangia (Fig. 29), each surrounded by a thin wall. Infection rate in field populations may be quite high. The occurrence of such parasites in fixed phytoplankton samples will not be recognized in routine investigation. Whether this or similar parasites are wide-spread has to be investigated in the future.

2.8. Trematoda

According to Rebecq (1965), Pouchet frequently observed an unidentified microscopic trematode clinging to *Noctiluca scintillans* at Concarneau, France. The nature of this association is so far not known but all trematodes are parasites of metazoans.

3. Concluding remarks

In the review, we intend to direct attention to a more or less neglected field of general phytoplankton ecology and red-tide research : parasitism. In terrestrial and marine ecology parasites are a well known factor in population dynamics of many organisms. In dense populations, parasites regulary contribute to their decline. Cultivation of animals or plants always require precautions against parasites, otherwise the stocks may be destroyed by infections.

It is surprising, that so little is known on parasitism of red tide organisms. The mass occurrence of harmful algae should be an ideal target for parasites. Until recently viral infections may have escaped notice due to methodological problems. A sudden decline of a phytoplankton bloom hardly can be correlated to a lysis caused by viral infection. The most simple explanations for a sudden disappearence of a mass occurence of phytoplankton, causing water discoloration at a given geographical location like aquaculture sites, are physical processes such as water mass exchange or vertical mixing. To exclude this possibility, a hydrographical survey has to be done. To prove viral infection, electron microscopy has to be performed on many cells as the presence of viral-like particles have been reported also from healthy populations. The same applies for bacterial infections. Viral infections with lytic effects are typically reported for only nano- or picoplankton species like *Aureococcus*, *Chrysochromulina* and *Emiliania*, but not from diatom and dinoflagellate blooms. Although the latter may produce very high biomasses per volume water, cell numbers are much higher in the blooms of the former species. Are infection chances in diatom and dinoflagellate blooms too low to induce a lytic process?

Bacteria are definitely present in many phytoplankton cells but most of them seem to be symbiotic or at least not responsible for death of the cells (e.g. Schnepf and Feith 1992). Thus infection apparently is not uncommon. Nevertheless, bacterial diseases of red tides have so far not been reported. Are there no phytoplankton-pathogenic bacteria in the sea (although documented in freshwater) ?

Some groups of protista which are obligate parasites such as Apicomplexa (earlier called Sporozoa) and also parasitic ciliates are thus far not reported as parasites in phytoplankton.. Both groups are known as parasites of other protists (for a review see Lauckner 1980).

Is this real or a question of methodology? Phytoplankton ecologists normally fix samples and count cell numbers in order to convert these into biomass or carbon, others only measure chlorophyll or other chemical components. Taxonomists working on preserved samples typically examine only the external morphology as diagnostic features. For cultures, only healthy cells are isolated and if a culture is infected by a parasite, in most cases, it will not survive until the reason for cell death has been recognized. Ultrastructure research on phytoplankton is mostly done on cultured material; if field material is used, it is unlikely that an infected cell will be recorded by chance if infection rate is low. These organisms are recorded only if scientists are specially looking for phytoplankton parasites. At our institute, the systematic search for phytoplankton parasites has enhanced the number of records by a factor of 5 in only two years, compared to the records of parasites made during a decade of weekly alive phytoplankton observations. Several new parasites have been discovered in the last years and there are quite a lot of them so far undescribed. About a dozen parasitic taxa have been observed but their morphology, life cycle and ultrastructure could not be investigated since either the host was not available for experiments or the parasite did not grow under laboratory conditions. The role of parasites in harmful algal dynamics has been underestimated and it is inevitable that further parasites will be described from red tide organisms.

4. References

Boltovskoy, A. (1984). Relacion huesped-parasito entre el quiste de *Peridinium willei* y el oomicete *Aphanomycopsis peridiniella* n. sp.. *Limnobios* 2(8): 635-645.

Bose, S.G., Carmichael, W.W. (1990). Plasmid distribution among unicellular and filamentous toxic cyanobacteria. *J. Appl. Phycol.* 2: 131-136.

Bratbak, G., Egge, G.J., Heldal, M. (1993). Viral mortality of the marine alga *Emiliania huxleyi* (Haptophyceae) and termination of algal blooms. *Mar. Ecol. Prog. Ser.* 93: 39-48.

Cachon, J. (1964). Contribution a l'étude des péridiniens parasites. Cytologie, cycles évolutives. *Ann. Sci. Nat. Zool.* 6: 1-158.

Cachon, J., Cachon, M. (1987). Parasitic dinoflagellates. In : Taylor, F.J.R. (ed.) : *The biology of dinoflagellates*. - Blackwell Sci.Pub., Oxford, pp. 571-610.

Canter, H.M., Heaney, S.I. (1984). Observations on zoosporic fungi of *Ceratium* spp. in lakes of the English Lake District: Importance for phytoplankton population dynamics. *New Phytologist* 97 : 601-612.

Canter, H.M., Lund, J.W.G. (1948). Studies on plankton parasites. I. Fluctuations in the numbers of *Asterionella formosa* Hass. in relation to fungal epidemics. *New Phytologist* 47: 238-261.

Canter, H.M., Lund, J.W.G. (1951). Studies on plankton parasites. III. Examples of the interaction between parasitism and other factors determining the growth of diatoms. *Ann. Bot. N.S.* 15: 359-371.

Chatton, E. (1920). Les péridiniens parasites. Morphologie, reproduction, ethologie. *Arch. Zool. Exp. Gen.* 59: 1-475.

Chatton, E., Biecheler, B. (1934). Les Coccidinidae, dinoflagellés coccidiomorphes parasites de dinoflagellés, et le phylum Phytodinozoa. *Compt. Rend. Hebd. Séances Acad. Sci.* 199 : 252-255.

Chatton, E., Biecheler, B. (1936). Documents nouveaux relatifs aux Coccidinides (dinoflagellés parasites). La sexualité du *Coccidinium mesnili* n. sp. *Compt. Rend. Hebd. Séances Acad. Sci.* 203: 573-576.

Coats, D.W., Bockstahler, K.R. (1994). Occurrence of the parasitic dinoflagellate *Amoebophrya ceratii* in Chesapeake Bay populations of *Gymnodinium sanguineum. J. Euk. Microbiol.* 41: 586-593.

Cottrell, M.T., Curtis, C.A. (1995a). Genetic diversity of algal viruses which lyse the photosynthetic picoflagellate *Micromonas pusilla* (Prasinophyceae). *Appl. Environm. Microbiol.* 61: 3088-3091.

Cottrell, M.T., Curtis, C.A. (1995b). Dynamics of a lytic virus infecting the photosynthetic marine picoflagellate *Micromonas pusilla. Limnol. Oceanogr.* 40: 730-739.

Dahl, E., Lindahl, O., Paasche, E., Throndsen, J. (1989). The *Chrysochromulina polylepis* bloom in Scandinavian waters during spring 1988. In: Cosper *et al* (eds) *Novel phytoplankton blooms*. Springer-Verlag New York, pp. 383-405.

Drebes, G. (1966). Ein parasitischer Phycomycet (Lagenidiales) in *Coscinodiscus. Helgoländer wiss. Meeresunters.* 13: 426-435.

Drebes, G. (1968). *Lagenisma coscinodisci* gen. nov. spec. nov., ein Vertreter der Lagenidiales in der marinen Diatomee *Coscinodiscus*. *Veröff. Inst. Meeresforsch. Bremerhaven, Suppl.* 3: 67-70.

Drebes, G. (1974). *Marines Phytoplankton. Eine Auswahl der Helgoländer Planktonalgen (Diatomeen, Peridineen)*. G. Thieme Verl. Stuttgart 186 pp.

Drebes, G., Kühn, S.F., Gmelch, A., Schnepf, E. (1996). *Cryothecomonas aestivalis* sp. nov., a colourless nanoflagellate feeding on the marine centric diatom *Guinardia delicatula* (Cleve) Hasle. *Helgoländer Meeresunters*. 50:497-515.

Drebes, G., Schnepf, E. (1988). *Paulsenella* Chatton (Dinophyta), ectoparasites of marine diatoms: development and taxonomy. *Helgoländer Meeresunters*. 42: 563-581.

Elbrächter, M. (1973). Population dynamics of *Ceratium* in coastal waters of Kiel Bay. *Oikos* 15 (Suppl.): 43-48.

Elbrächter, M. (1991). Faeces production by dinoflagellates and other small flagellates. *Marine Micobial Food Webs* 5: 189-204.

Elbrächter, M., Schnepf, E., Balzer, I. (1996). *Hemistasia phaeocysticola* (Scherffel) comb. nov., redescription of a free-living, marine, phagotrophic kinetoplastid flagellate. *Arch. Protistenkd.* 147: 125-136.

Franca, S. (1976). On the presence of virus-like particles in the dinoflagellate *Gyrodinium resplendens* (Hulburt). *Protistologica* 12: 425-430.

Fritz, L., Nass, M. (1992). Development of the parasitic dinoflagellate *Amoebophrya ceratii* within host dinoflagellate species. *J. Phycol.* 28: 312-320.

Gold, K., Pollingher, U. (1971). Occurrence of endosymbiotic bacteria in marine dinoflagellates. *J. Phycol.* 7: 264-265.

Gumpert, J., Smarda, J., Hübel, M., Hübel, H. (1987). Ultrastructural studies on the cyanobacterium *Nodularia spumigena* and its epiphytic bacteria of the genus *Seliberia*. *J. Basic Microbiol.* 27: 543-555.

Hasle, G.R., Lange, C.B., Syvertsen, E.E. (1996). A review of *Pseudo-nitzschia*, with special reference to the Skagerrak, North Atlantic, and adjacent waters. *Helgoländer Meeresunters*. 50: 131-175.

Imai, I., Ishida, Y., Hata, Y. (1993). Killing of marine phytoplankton by a gliding bacterium *Cytophaga* sp., isolated from the coastal sea of Japan. *Mar. Biol.* 116: 527-532.

Imai, I., Ishida, Y., Sakaguchi, K., Hata, Y. (1995). Algicidal marine bacteria isolated from northern Hiroshima Bay, Japan. *Fisheries Sci.* 61: 628-636.

Kühn, S. (1995). *Untersuchungen zum Befall von Phytoplankton durch parasitoide Protisten (Nordsee)*. - Diss. Universität Bremen.

Kühn, S., Drebes, G. Schnepf, E. (1996). Five new species of the nanoflagellate *Pirsonia* in the German Bight, North Sea, feeding on planktic diatoms. *Helgoländer Meeresunters*. 50: 205-222.

Lauckner, G. (1980). Diseases of Protozoa. In : O.Kinne (ed.) : *Diseases of marine animals.1. General aspects, Protozoa to Gastropoda*. - J. Wiley & Sons, Chichester, New York, Brisbane, Toronto pp. 75-134.

Leedale, G. F. (1969). Observations on endonuclear bacteria in euglenoid flagellates. *Österr. Bot. Z.* 116: 279-294.

Nagasaki, K., Ando, M., Itakura, S., Imai, I., Y. Ishida, Y. (1994). Viral mortality in the final stage of *Heterosigma akashiwo* (Raphidophyceae) red tide. *J. Plankt. Res.* 16: 1595- 1599.

Nakamura, Y., Suzuki, S.Y. Hiromi, J. (1995). Population dynamics of heterotrophic dinoflagellates during a *Gymnodinium mikimotoi* red tide in the Seto Inland Sea. *Mar. Ecol. Prog. Ser.* 125: 269-277.

Nishitani, L., Erickson, G., Chew, K.K. (1985). Role of the parasitic dinoflagellate *Amoebophrya ceratii* in control of *Gonyaulax catenella* populations. in : Anderson, D.M. et al. (eds) : *Toxic dinoflagellates*.- Elsevier Sci. Publ. Co., Inc. New York, pp. 225-230.

Peterson, C.G.,Dudley, T.L., Hoagland, K.D., Johnson, L.M. (1993). Infection, growth, and community-level consequences of a diatom pathogen in a Sonoran Desert stream. *J. Phycol.* 29: 442-452.

Pienaar, R.N. (1976). Virus-like particles in three species of phytoplankton from San Juan Island, Washington. *Phycologia* 15: 185-190.

Rebecq, J. (1965). Considérations sur la place des trématodes dans le zooplancton marin. *Annls Fac. Sci. Marseille* 38: 61-84.

Reisser, W. (1993). Viruses and virus-like particles of freshwater and marine eukaryotic algae - a review. *Arch. Protistenkd.* 143: 257-265.

Reisser, W. (1995). Phycovirology : aspects and prospect of a new phycological discipline. in: Wiessner, W., Schnepf, E., Starr, R. (eds) : *Algae, environment and human affairs*. Biopress Ltd, Bristol, England, pp. 143-158.

Schnepf, E. (1994a). Light and electron microscopical observations in *Rhynchopus coscinodiscivorus* spec. nov., a colorless, phagotrophic euglenozoon with concealed flagella. *Arch. Protistenkd.* 144: 63-74.

Schnepf, E. (1994b). A *Phagomyxa*-like endoparasite of the centric marine diatom *Bellerochea malleus*: a phagotrophic plasmodiophoromycete. *Botanica Acta* 107: 374-382.

Schnepf, E., Deichgräber, G., Drebes, G. (1978a) : Development and ultrastructure of the marine, parasitic Oomycete, *Lagenisma coscinodisci* Drebes (Lageniales). The infection. *Arch. Microbiol.* 116: 133-139.

Schnepf, E., Deichgräber, G., Drebes, G. (1978b). Development and ultrastructure of the marine, parasitic Oomycete, *Lagenisma coscinodisci* Drebes (Lageniales). Thallus, zoosporangium, mitosis, and meiosis. *Arch. Microbiol.* 116: 141-150.

Schnepf, E., Deichgräber, G., Drebes, G. (1978c). Development and ultrastructure of the marine, parasitic Oomycete, *Lagenisma coscinodisci* Drebes (Lageniales): Formation of the primary zoospores and their release. *Protoplasma* 94: 263-280.

Schnepf, E., Deichgräber, G., Drebes, G.(1978d). Development and ultrastructure of the marine, parasitic Oomycete, *Lagenisma coscinodisci* Drebes (Lageniales): encystment of primary zoospores. *Can. J. Bot.* 56: 1309-1314.

Schnepf, E., Deichgräber, G., Drebes, G. (1978e). Development and ultrastructure of the marine, parasitic Oomycete, *Lagenisma coscinodisci* Drebes (Lageniales): sexual reproduction. *Can. J. Bot.* 56: 1315-1325.

Schnepf, E., Drebes, G., Elbrächter, M. (1990). *Pirsonia guinardiae*, gen. et spec. nov. : a parasitic flagellate on the marine diatom *Guinardia flaccida* with an unusual mode of food uptake. *Helgoländer Meeresunters.* 44: 275-293.

Schnepf, E., Elbrächter, M. (1992). Nutritional strategies in dinoflagellates. A review with emphasis on cell biological aspects. *Europ. J. Protistol.* 28: 3-24.

Schnepf, E., Feith, R. (1992). Experimental studies to modify the number of endocytic bacteria in *Cryptomonas* Strain SAG 2580 (Cryptophyceae) and their lysis by bacteriophages. *Arch. Protistenkd.* 142: 95-100.

Schnepf, E., Hegewald, E., C.J. Soeder, C.J. (1974). Elektronenmikroskopische Untersuchungen an Parasiten aus *Scenedesmus*-Massenkulturen. 4. Bakterien. *Arch. Mikrobiol.* 98: 133-145.

Schnepf, E., Heinzmann, J. (1980). Nuclear movement, tip growth and colchicine effects in *Lagenisma coscinodisci* Drebes (Oomycetes, Lagenidiales). *Biochem. Physiol. Pflanzen* 175: 67-76.

Schnepf, E., Meier, R., Drebes, G. (1988). Stability and deformation of diatom chloroplasts during food uptake of the parasitic dinoflagellate, *Paulsenella* (Dinophyta). *Phycologia* 27: 283-290.

Schweikert, M., Schnepf, E. (1996). *Pseudaphelidium drebesii*, gen. et spec. nov. (incerta sedis), a parasite of the marine centric diatom *Thalassiosira punctigera*. *Arch. Protistenkd.* 147: 11-17.

Sieburth, J. McN., Johnson, P.W., Hargraves, P.E. (1988). Ultrastructure and ecology of *Aureococcus anophagefferens* gen. et sp. nov. (Chrysophyceae) : the dominant picoplankter during a bloom in Narragansett Bay, Rhode Island, summer 1985. *J. Phycol.* 24: 416-425.

Silva, E. S. E. (1978). Endonuclear bacteria in two species of dinoflagellates. *Protistologica* 14: 113-119.

Soyer, M.-O. (1978). Particules de type viral et filaments trichocystoides chez les dinoflagellés. *Protistologica* 14: 53-58.

Spencer, R. (1955). A marine bacteriophage.*Nature* 175: 690-691.

Steward, W.D.P., Daft, M.J. (1976). Algal lysing agents of freshwater habitats. *Soc. appl. Bact. Symp. Ser.* 4: 63-90.

Suttle, C.A. (1994). The significance of viruses to mortality in aquatic microbial communities. *Microb. Ecol.* 28: 237-243.

Suttle, C.A., Chan, A.M. (1994). Dynamics and distribution of cyanophages and their effect on marine *Synechococcus* spp.. - *Appl. Environ. Microbiol.* 60: 3167-3174.

Suttle, C.A. Chan, A.M. (1995). Viruses infecting the marine prymnesiophyte *Chrysochromulina* spp.: isolation, premliminary characterization and natural abundance. *Mar. Ecol. Prog. Ser.* 118: 275-282.

Taylor, F.J.R. (1968). Parasitism of the toxin-producing dinoflagellate *Gonyaulax catenella* by the endoparasitic dinoflagellate *Amoebophrya ceratii*. *J. Fish. Res. Bd. Canada* 25: 2241-2245.

Thomsen, H.A., Buck, K.R., Bolt, P.A., Garrison, D.I. (1990). Fine structure and biology of *Cryothecomonas* gen. nov. (Protista incerta sedis) from ice biota. *Can. J. Zool.* 69: 1048-1070.

Van Donk, E., Bruning, K. (1995). Effects of fungal parasites on planktonic algae and their role of environmental factors in the fungus-algae relationship. in: Wiessner, W., Schnepf, E., Starr, R. (eds) : *Algae, environment and human affairs*. Biopress Ltd, Bristol, England, pp. 223-234.

Wetsteyn, L.P., Peperzak, L. (1991). Field observations in the Oosterschelde (the Netherlands) on *Coscinodiscus concinnus* and *Coscinodiscus granii* (Bacillariophyceae) infected by the marine fungus *Lagenisma coscinodiscus* (Oomycetes). *Hydrobiol. Bull.* 25: 15-21.

Zingone, A. (1995). The role of viruses in the dynamics of phytoplankton blooms. *Giorn. Botanico Italiano* 129: 415-423.

Zuelzer, M. (1927). Über *Amoeba biddulphiae* n. sp., eine in der marinen Diatomee *Biddulphia sinensis* Grev. parasitierende Amöbe. *Arch. Protistenkd.* 57: 247-284.

5. Legend of Figures

Fig. 1a ,b. *Amoebophrya ceratii* in *Ceratium longipes*. a) Early infection stage, b) late tage, short before sporulation; parasite (arrow); nucleus of dinoflagellate (arrowhead). Scale bar. 20 µm. Permanent slide courtesy of W. Coats.

Figs 2-7. Scale bar 20 µm.

Fig. 2 *Pauslenella chaetoceratis* feeding myzocytotically on *Chaetoceros decipiens*. The dinoflagellate is attached to the tip of a seta and drives the feeding tube through the seta into the cell interior (arrowhead). The host cystoplasm H is plasmolyzed. From Schnepf et al. (1988), modified.

Fig. 3 *Paulsenella chaetoceratis* on *Chaetoceros decipiens*. One dividing primary cyst (arrowhead) and secondary cysts.

Fig. 4. *Paulsenella kornmannii* feeding myzocytotically on *Eucampia zodiacus*. Feeding tubes arrowhead.

Fig. 5 *Rhynchopus coscinodiscivorus*, two cells feeding in a *Coscinodiscus concinnus* cell. From Schnepf (1994a).

Fig. 6. *Rhynhopus coscinodiscivorus* in an empty frustule of *Coscinodiscus concinnus*. From Schnepf (1994a).

Fig. 7. *Hemistasia phaeocysticola*, one cell leaving the frustule of *Coscinodiscus concinnus*. From Elbrächter et al. (1996).

Fig. 8-13. Scale bar 20 µm.

Fig. 8. *Cryothecomonas aestivalis*, two flagellates (flagella arrowhead) feeding in a *Guinardia delicatula*. From Drebes *et al.* (1996).

Fig. 9. *Cryothecomonas aestivalis*, many new formed flagellates in a *Guinardia delicatula* frustule. From Drebes *et al.* (1996).

Fig. 10. *Pirsonia guinardiae* feeding on *Guinardia flaccida*. Secondary auxosomes (arrowheads) connected with the trophosome (T), a food vacuole (v) is about to fuse with the trophosome. From Schnepf *et al.* (1990).

Fig. 11. *Pirsonia diadema* on a largely emptied *Coscinodiscus granii* frustule. One just attached flagellate (arrowhead) and a group of developing flagellates and auxosomes connected with a trophosome (T). From Kühn et al. (1996), modified.

Fig. 12. *Pseudaphelidium drebesii* feeding in a *Thalassiosira punctigera* cell. The pseudopodium includes a big digestion vacuole (V). From Schweikert and Schnepf (1996), modified.

Fig. 13. *Pseudaphelidium drebesii*. The pseudopodium has cleaved into numerous globular cells which fill the frustule of the host, *Thalassiosira punctigera*.

Fig. 14. *Amoeba biddulphiae* (arrowhead) feeding in the diatom *Odontella sinensis*. Scale bar 20 µm.

Fig. 15. *Lagenisma coscinodisci*. Thallus in a *Coscinodiscus concinnus* cell. Scale bar. 100µm.

Fig. 16. *Lagenisma coscinodisci*. Thallus, zoosporogenesis. The cytoplasm of the host cell, *Coscinodiscus concinnus* is broken. Scale bar 100 µm.

Fig. 17. *Ectrogella* sp. . young thallus (arrowhead) feeding in *Guinardia delicatula*. Scale bar. 20 µm.

Fig. 18. *Ectrogella* sp. in *Guinardia delicatula.* Thallus with zoosporogenesis. Scale bar. 20 µm.
Fig. 19. *Ectrogella* sp. in a *Guinardia delicatula.* Empty zoosporangium. Scale bar. 20 µm.
Fig. 20. An operculate chytrid, feeding on the diatom *Bellerochea malleus.* Two empty zoosporangia (operculum arrowhead) and a nearly mature zoosporangium (arrow, out of focus). Scale bar 20 µm.
Fig. 21. An operculate chytrid, feeding on the diatom *Cylindrotheca closterium*, multiple infection. Scale bar. 20 µm.
Fig. 22. The same operculate chytrid on *Cylindrotheca closterium* as in Fig. 21. Empty zoosporangium (operculum arrowhead). Scale bar. 20 µm.
Figs 23-27. Unidentified Olpidiaceae, different infection stages of *Pseudo-nitzschia*, raw culture of a field sample. Scale bar 20 µm.
Fig. 23. Just attached zoospore (arrowhead).
Fig. 24. Developing thallus (arrowhead).
Fig. 25. Zoosporogenesis. Discharge tube (arrowhead), residual bodies (arrow).
Fig. 26. Zoosporangium with a few remaining zoospores. Residual body (arrow).
Fig. 27. Empty zoosporangium. Residual bodies (arrow).
Figs 28-31. *Phagomyxa* sp. feeding on *Bellerochea malleus.* . From Schnepf (1994b).
Fig. 28. Plasmodium (arrows) with collapsed host cytoplasm, digested host material within the central vacuole, remnants of host cytoplasm in the form of thin strands (arrowheads). Scale bar. 20 µm
Fig. 29. Cleaving plasmodium and sporangisorus with cleaving zoosporangia. Scale bar. 20 µm
Fig. 30. Newly released zoospores. Scale bar. 10 µm
Fig. 31. Sporangisori with empty zoosporangia and a central residual body. Scale bar. 20 µm

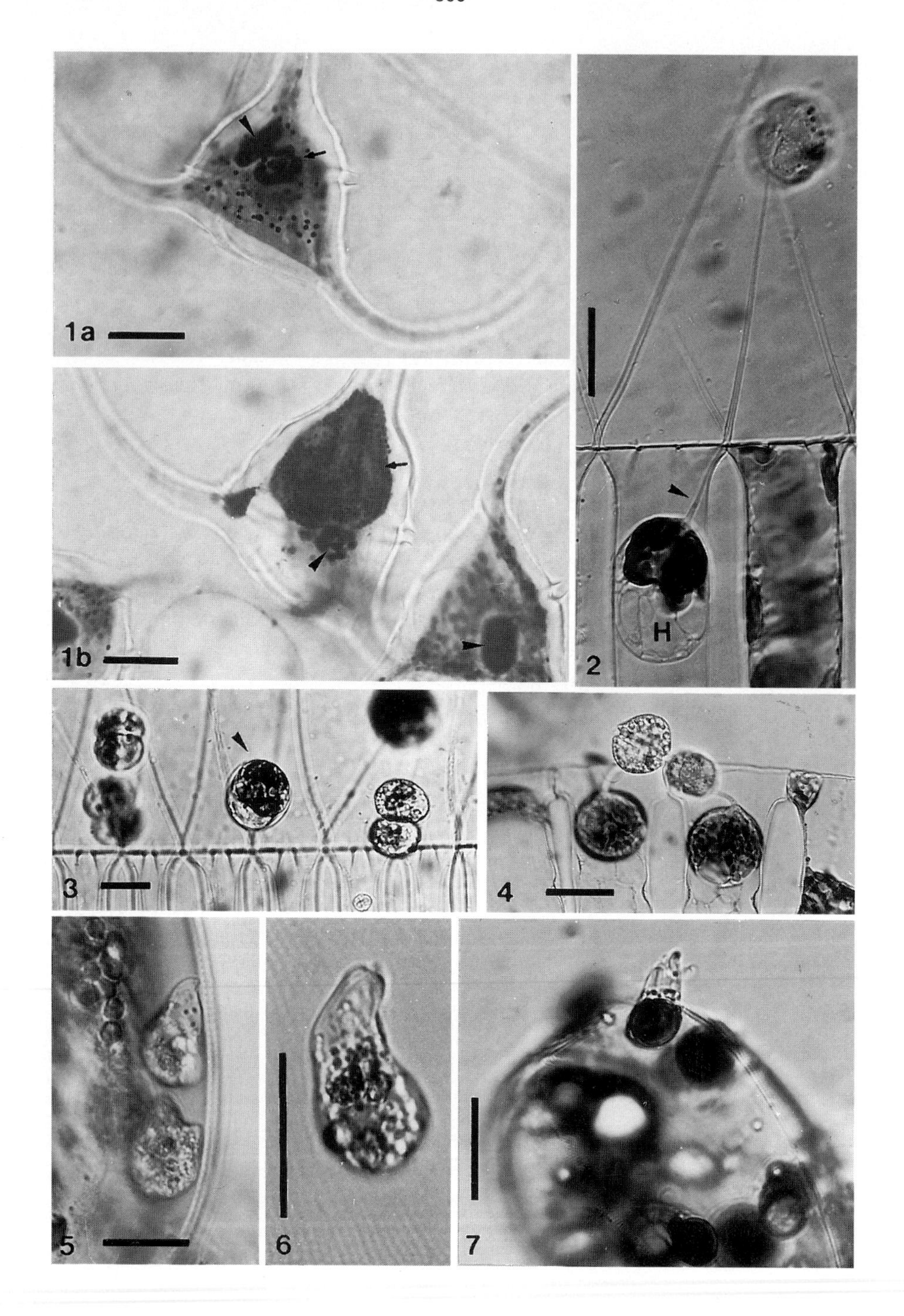
1a
1b
2
H
3
4
5
6
7

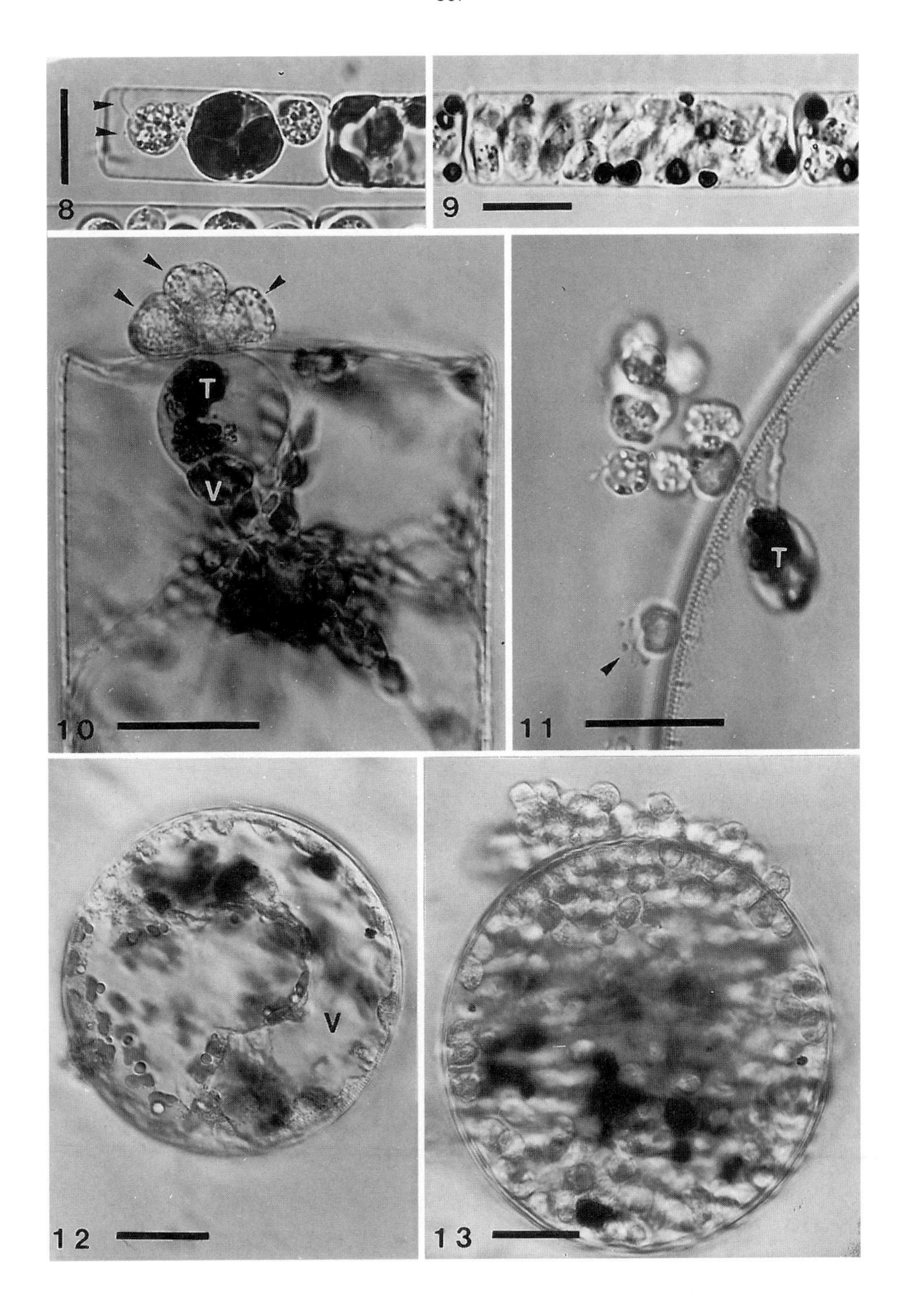
8
9
10
T
V
11
T
12
V
13

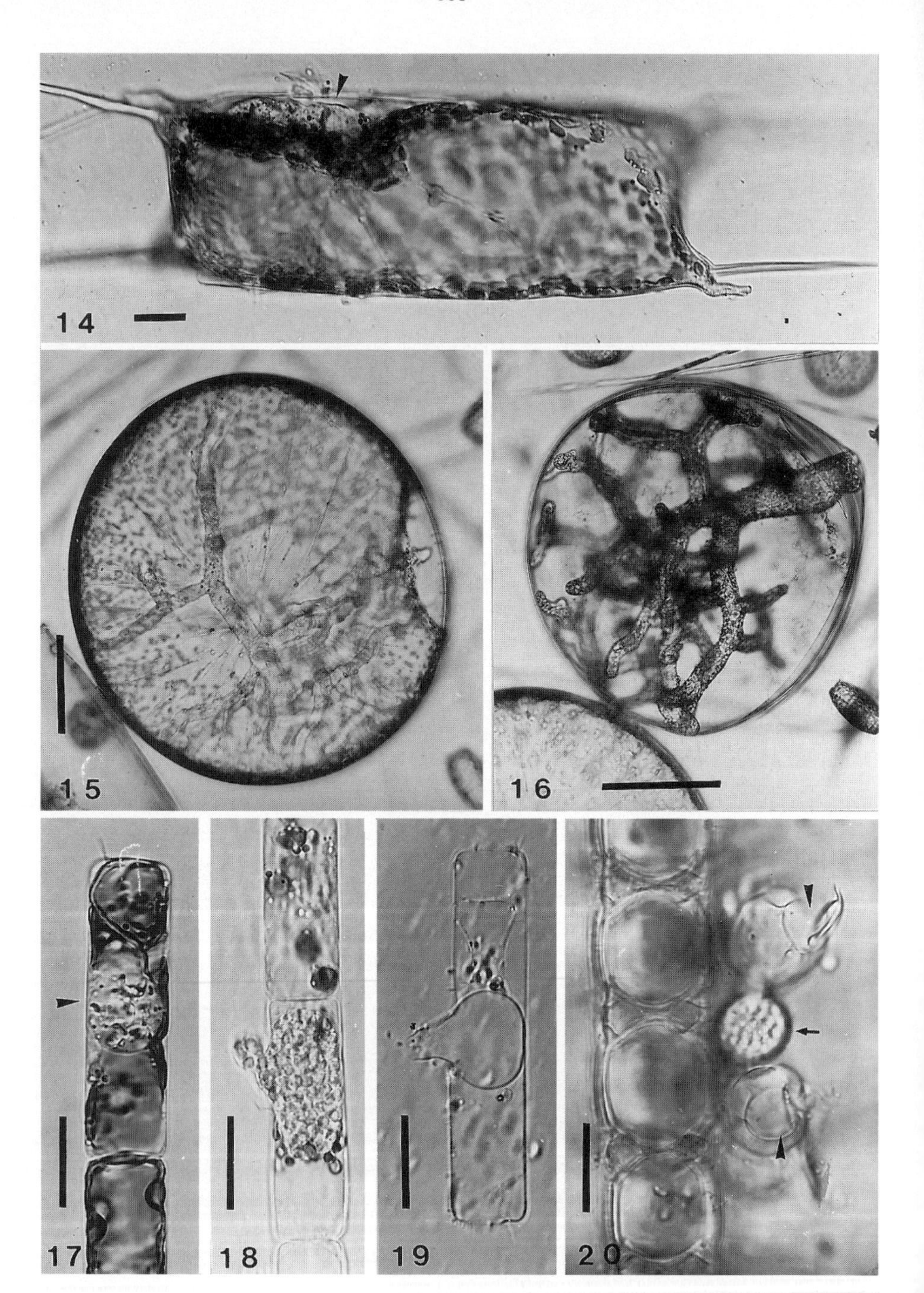
14
15
16
17
18
19
20

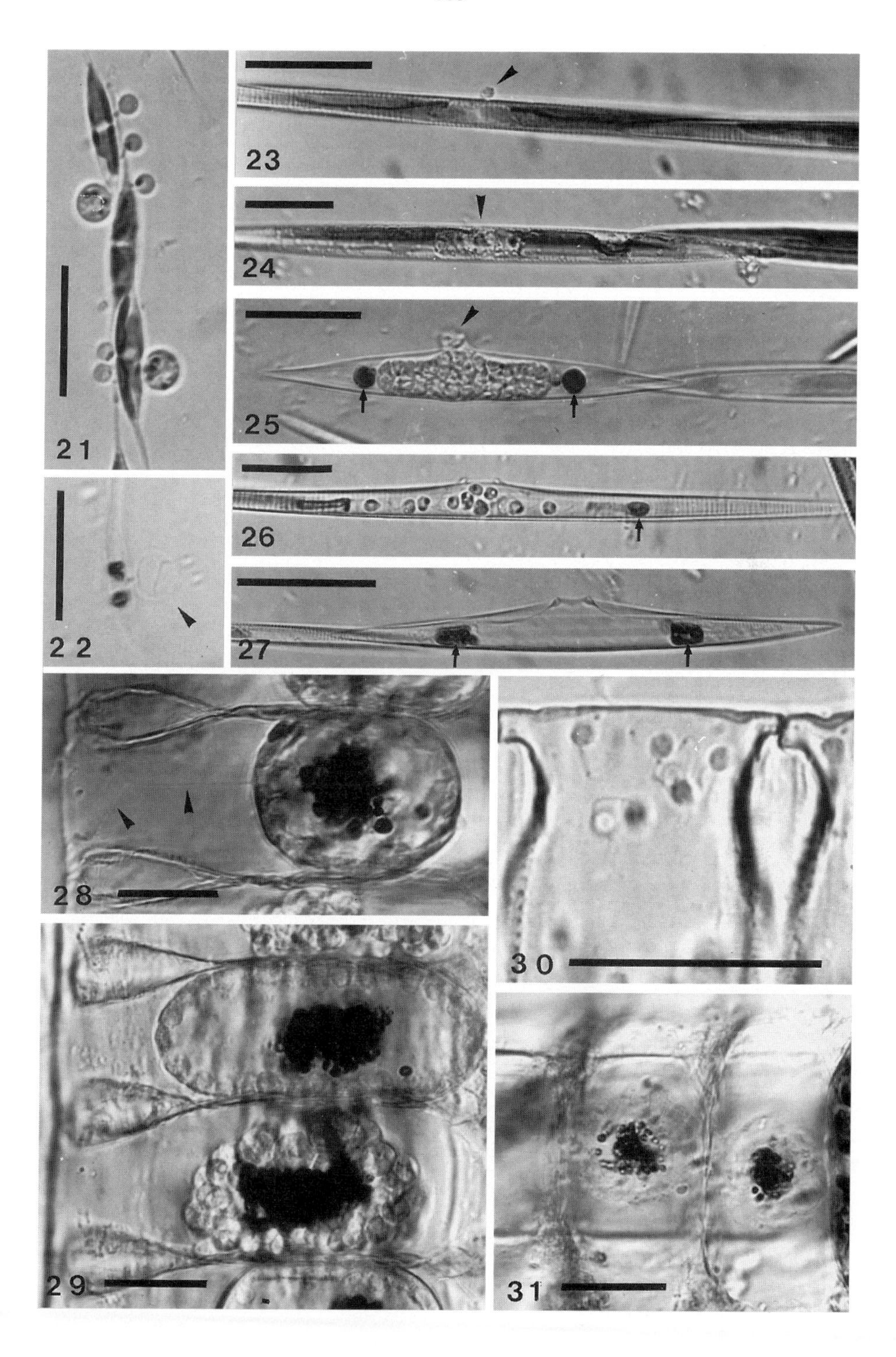
21
22
23
24
25
26
27
28
29
30
31

Concluding Remarks on the Autecology of Harmful Algal Blooms

G.M.Hallegraeff
Department of Plant Science, University of Tasmania, GPO Box 252-55, Hobart, Tasmania 7001, Australia

1. Aim of Bermuda workshop

Most harmful algal blooms are more or less monospecific events, and the autecology of the causative organisms thus becomes crucial in understanding the factors that trigger these phenomena. A primary aim of the Bermuda NATO-ASI workshop was to review and analyze available knowledge which explains why, where and when a particular HAB species blooms under which particular suite of environmental conditions. In turn, this may allow us to predict the impact on HABs of anthropogenic alterations to the environment from sewage outfalls, climate change, coastal engineering and changes in land catchment use. Authors of chapters dealing with the autecology of individual HAB taxa were asked to define the physiological, biochemical and behavioral characteristics ("niche"-defining factors) which cause these species to be successful in Nature, i.e. to cause blooms. The emphasis was on interpreting and reviewing field studies of bloom phenomena in widely different environments in order to look for common principles. In particular, we aimed to uncover gaps in our knowledge and encourage future research efforts to complete the missing information needed to define "niche-hyperspace" of individual HAB taxa. Having previously being involved in conducting an autecology workshop which focused for a whole week on a single HAB species, *Pyrodinium bahamense* (Hallegraeff and Maclean 1989), the broad scope of the Bermuda workshop which covered as many as 16 HAB categories has been ambitious indeed.

2. Functional groups among HAB species

An implicit assumption in ecological studies is that there exists a critical relationship between form and function in organisms, and that life form therefore is a better predictor of fitness than phylogenetic affinities. A corollary of this assumption is that patterns of variation in morphology at the species level are also good predictors of variation in other traits. The Chapter by Gallagher, however, challenges this traditional "morphospecies" approach by indicating that at the lower taxonomic levels morphology of HAB organisms is usually a poor predictor of other traits such as toxicity, bioluminescence and mating compatibility. Similarly, the Chapter by Scholin illustrates that geographic populations (defined on the basis of rRNA sequences) and not morphospecies are the functional units in bloom dynamics of the dinoflagellate species complex *Alexandrium tamarense.* Crossing experiments with *Alexandrium* isolates with different toxin profiles show that the progeny may undergo complex patterns of recombination (Chapter by Ishida *et al.*). All phytoplankton

NATO ASI Series, Vol. G 41
Physiological Ecology of Harmful Algal Blooms
Edited by D. M. Anderson, A. D. Cembella and G. M. Hallegraeff

species tested thus far appear to be genetically polymorphic and detection of genetic patterns requires larger sample sizes than used by higher plant researchers. Furthermore, it is important to work with low resolution, broad spectrum genetic approaches covering large numbers of genes (e.g. allozymes, RFLP's) rather than precise sequencing of a single molecule of DNA. By and large, morphotaxonomy has worked well with HAB species and this traditional species concept continues to provide useful information .

PSP Dinoflagellates

The day of lectures devoted to the PSP-producing dinoflagellates *Alexandrium, Pyrodinium* and *Gymnodinium catenatum* dealt with the most homogeneous grouping of HAB species. Their commonality lies in the absence of a rapid growth strategy and the reliance on benthic resting cysts in life cycle transitions. *Alexandrium* do not usually produce dense biomass blooms that persist throughout the year. Instead, seasonal bloom events appear to be restricted in time by cyst production. The persistence of these cysts through long-term unfavorable conditions allows these dinoflagellates to colonize a wide spectrum of habitats and hydrographic regimes (Chapter by Anderson). The tropical dinoflagellate *Pyrodinium bahamense* (Chapter by Usup and Azanza) represents the most serious HAB problem in the Indo-West Pacific region. This species prefers high salinities (30-35°/psu) and high temperatures. Despite recent progress in culture studies with Malaysian and Philippines strains, factors that trigger *Pyrodinium* bloom events in Nature are still poorly understood. A selenium/ soil extract requirement was identified in culture, which may explain this species' association with rainfall events and land runoff from mangrove areas. The role of vegetative cells as opposed to resting cysts (dormancy period 2.5-3 months) in the seeding of blooms still needs to be elucidated. A more detailed physiological and molecular genetic comparison between tropical Indo-West Pacific and Atlantic populations is still required. Benthic cyst stages of *G. catenatum* (dormancy period of 2 wk) appear not to play a role in seasonal bloom dynamics, but their major function is to sustain this species through long periods when water column conditions are unfavorable for bloom formation A comparison between inshore Tasmanian and offshore Spanish bloom dynamics, although exhibiting some apparently contradictory behavior (association with calm stable weather vs post-upwelling conditions), has elucidated similarities in the important role of temperature and vertical migration (Chapter by Hallegraeff and Fraga).

Fish-killing HAB species

The day of lectures devoted to fish-killing species of raphidophytes *(Chattonella, Heterosigma*), dinoflagellates (*Gymnodinium breve, G. mikimotoi)* and haptophytes (*Chrysochromulina, Prymnesium*) clearly represented an artificial grouping of organisms biased by our human focus on how these species affect aquaculture operations, the only commonality being their similar toxic effects. The fish-killing ambush-predator dinoflagellate *Pfiesteria piscicida* (Chapter by Burkholder *et al.*) but also the gelatinous haptophyte *Phaeocystis* (Chapter by Lancelot *et al.*) represent bloom scenarios which stand on their own. *Pfiesteria* forms a minor component of the warm-temperate to subtropical estuarine plankton and is atypical in maintaining

Table 1. Summary of niche-defining factors of HAB species: responses to the physico-chemical environment

Temperature

Alexandrium (seasonal germination of cysts)
Chattonella , Heterosigma (seasonal trigger for cyst germination)
cyanobacteria (seasonal bloom window)
Gymnodinium breve (demise of blooms)
Gymnodinium catenatum (seasonal bloom window)
Noctiluca (seasonal bloom window)

Salinity

Nodularia spumigena

Inorganic Nutrients

Pseudo-nitzschia (nitrate)
cyanobacteria (phosphate)
Noctiluca (indirect; high prey biomass)
Pfiesteria ? (responds to changing nutrient regime)
Phaeocystis (responds to changing nutrient regime)

Micronutrients (Land Runoff)

Alexandrium (Fe,Cu)
Chattonella , Heterosigma (Mn, Fe stimulates; ?can sense freshwater on top of water column; Cu)
Chrysochromulina (high demand for Se , vit B12; Fe,Cu)
Gymnodinium breve ?(Fe index, riverflow;Mn)
Gymnodinium catenatum (Se?)
Phaeocystis (Fe?)
Pseudo-nitzschia ? (respond to rainfall)
Pyrodinium bahamense (Se, mangrove runoff?)
Trichodesmium (Fe needed for N fixation)

an array of 22 benthic amoebae and planktonic flagellate stages. This heterotrophic dinoflagellate can repeatedly swarm up from benthic stages, attack, kill and consume prey, and then rapidly descend back to the sediments. The success of *Phaeocystis* in marine systems has been attributed to its ability to form large gelatinous colonies during its life cycle. These colonies occupy the same niche in turbulent, tidally or seasonally mixed water columns as colony-forming spring diatom blooms. This haptophyte is one of the few HAB species for which there exists a clear causal relationship between increasing bloom frequency and eutrophication (in the North Sea).

The fish-killers *Heterosigma, Chattonella, Prymnesium, Chrysochromulina* and *G. mikimotoi* have in common the production of high biomass blooms together with the incidence of allelopathic chemicals that play a role in predator avoidance. Raphidophyte blooms of *Heterosigma* are sensitive to temperature for cyst germination, but chemical conditioning of the water by land runoff (Fe?) and other growth promoters (by aquaculture wastes?) determines the outcome of competition with diatoms (Chapter by Smayda). Similarly, the raphidophyte *Chattonella* (Chapter by Imai *et al.*) includes a benthic cyst stage in its life history but again the growth of the germling cells as affected by nutrient conditions and the presence of diatom competitors hold the real key to bloom development. The capacity of vertical migration by *Chattonella* in stratified water columns with a shallow nutricline (i.e. nutrients available only at depth under dim light) provide it with a competitive advantage. While harmful marine blooms of *Chrysochromulina* appear to be exceptional events (in Scandinavia in 1988 and 1991), fish-killing *Prymnesium* bloom events in inshore (low salinity) , eutrophic waters are recurrent in many parts of the world (Chapter by Edvardsen and Paasche). The expression of toxicity by *Chrysochromulina* is variable, can be enhanced by phosphate limitation, and the extraordinary toxic potential of the Skagerrak 1988 event remains largely unexplained. Well-studied blooms in Scandinavia develop in stratified water influenced by freshwater runoff and during calm and sunny periods.

The fish-killing dinoflagellate *Gymnodinium breve* is a K-strategist, adapted to low nutrient, oligotrophic environments. Blooms in the Gulf of Mexico are initiated offshore before being transported into near-shore waters where they cause fish kills, discolored water, human respiratory irritation and occasionally neurotoxic shellfish poisoning in human shellfish consumers (Chapter by Steidinger *et al.*). Taxonomically related dinoflagellate species of the eurythermal and euryhaline *Gymnodinium mikimotoi* species complex (including *Gyrodinium aureolum*, *Gymnodinium nagasakiense*) are associated with marine fauna kills but not normally human intoxications (Chapter by Gentien). Poorly characterized exotoxins and mucus production play an allelopathic role against other algae and also act as zooplankton grazing repressing agents. This species is especially successful in frontal regions and in stratified water columns where it accumulates in the pycnocline (often also a nutricline), thriving on regenerated ammonia and benefiting from polyamine growth factors from decaying diatoms. The question of whether ichthyotoxic, fucoxanthin-containing dinoflagellates comprise a functional group is still open.

Other HAB species

The day dealing with other HAB species, including benthic and planktonic DSP and ciguatera producing dinoflagellates, the heterotrophic *Noctiluca* and toxic diatoms and cyanobacteria (Chapter not included in these Proceedings) served to demonstrate the diversity of life cycle strategies. In the absence of any success in culturing the presumed mixotrophic dinoflagellate *Dinophysis,* it is not clear whether the incidence of occasionally higher biomass in Nature is the result of active growth or passive cell accumulation. Life cycle and overwintering strategies are not understood (Chapter by Maestrini) and the focus of current research efforts on DSP toxin accumulation in shellfish rather than on the causative plankton cell dynamics may be partly to blame. The unusual, large, phagotrophic dinoflagellate *Noctiluca* (Chapter by Elbrachter and Qi) depends upon high prey biomass and optimal water temperatures during the pre-bloom stage, with starved cells coming to the surface and aggregating at fronts during calm weather conditions, and wind mixing terminating blooms. Diatom blooms of the cosmopolitan genus *Pseudo-nitzschia* are common in coastal waters all over the world (Chapter by Bates, Garrison and Horner). Blooms generally occur during colder seasons but no unusual physiological adaptations to low light or low temperatures

Table 2. Summary of niche-defining factors of HAB species (continued): properties of the organisms

Organic Nutrition / Mixotrophy
Chrysochromulina
Dinophysis
Gymnodinium breve (can use vitamins, amino acids)
Gyrodinium aureolum (role of polyamines such as putrescine)
Heterosigma (feeds on bacteria?; responds to aquaculture wastes)
Noctiluca (phagotrophy)
Pfiesteria
Allelopathy / Grazer Avoidance
Chrysochromulina
Dinophysis
Gyrodinium aureolum
Heterosigma
Microcystis
Phaeocystis
Prymnesium
Life History Parameters
Alexandrium
ciguatera dinoflagellates
(in terms of macroalgal substrate,water movement, species associations)
Gymnodinium catenatum
Pfiesteria
Pyrodinium bahamense
Vertical Migration (Sensitivity to Turbulence)
Alexandrium (notably chainforming species)
Chattonella, Heterosigma
Dinophysis? (responding to stability increase)
Gymnodinium breve ?
Gymnodinium catenatum
Gyrodinium aureolum (associated with pycnoclines, fronts)

could be elucidated. Seed populations can derive from both inshore or offshore waters, and the understanding of toxic events of amnesic shellfish poisoning is particularly hindered by the existence of both toxic and non-toxic strains of *P. delicatissima, P. multiseries, P. pseudodelicatissima, P. pungens* and *P. seriata*. The community dynamics of epiphytic/ benthic tropical ciguatera dinoflagellates and their associated macroalgal canopy (Chapter by Tindall and Morton) are dictated to a large extent by the degree of water movement, with other physical and chemical factors such as temperature, salinity, gasses, inorganic and organic nutrients only playing a role with diminishing hydrodynamics. Elbrachter and Schnepf in their Chapter on parasites of harmful algae call for a systematic search for viral, bacterial, dinoflagellate, euglenoid, zooflagellate, amoeboid and fungal parasites of phytoplankton. Their role in phytoplankton dynamics (notably the demise of blooms) of both harmful and non-harmful species is likely to have been seriously underestimated.

3. The Margalef Mandala

Margalef and coworkers (1978, 1979) pioneered efforts to search for unifying principles that defined the niche of "red-tide" phytoplankton, with emphasis on the environmental factors turbulence (intensity of vertical mixing) and nutrients. Neither

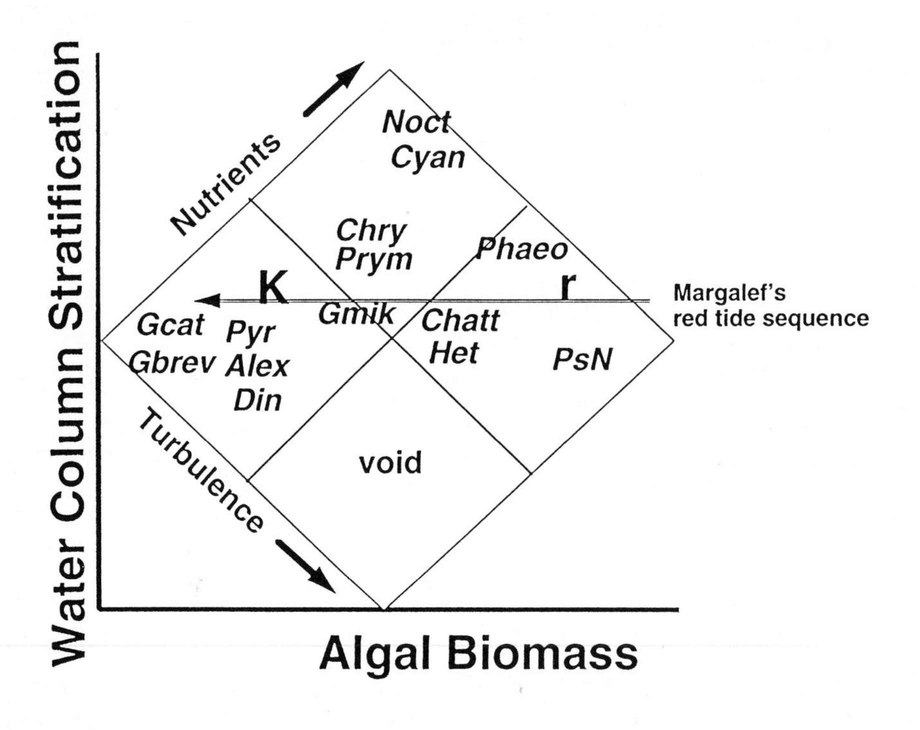

Fig.1. An adaptation of Margalef's Phytoplankton Mandala, which focuses on the environmental factors turbulence and nutrients as major determinants of the ecological space of phytoplankton blooms (inner square), which is superimposed on the plane defined by the coordinates production potential (algal bloom biomass) and the steepness of the environmental gradient (water column stratification). Margalef's original "red tide sequence" is indicated (horizontal arrow), together with the position of the HAB species covered by the Bermuda workshop: Alex=*Alexandrium*; Chatt=*Chattonella*; Chrys= *Chrysochromulina*; Cyan=cyanobacteria; Din= *Dinophysis* ; Gbrev= *Gymnodinium breve*; Gcat= *Gymnodinium catenatum*; Gmik= *Gymnodinium mikimotoi;* Het= *Heterosigma*; Phaeo=*Phaeocystis*; Prym=*Prymnesium*; Noct=*Noctiluca*; PsN= *Pseudo-nitzschia*; Pyr= *Pyrodinium.*

light nor temperature or salinity were considered major determinants of the ecological space of phytoplankton blooms. Most commonly, nutrient availability and turbulence go hand in hand (e.g. during nutrient upwelling) . In the graphic presentation known as Margalef's Mandala (Fig.1) , the nutrient-turbulence plane is superimposed on the plane defined by the coordinates production potential (here interpreted as algal bloom biomass) and the steepness of the environmental gradient (here interpreted as water column stratification). Margalef's "red tide sequence", which reflects dinoflagellate blooms during upwelling relaxation in Spanish rias, is indicated in Fig.1 (arrow), while interactions with zooplankton were not considered in detail in Margalef's original scheme. Superimposing upon Fig.1 the information gleaned during the Bermuda workshop, it becomes abundantly clear that the niche of HAB species is much wider than originally envisaged (see also Smayda and Reynolds 1997). HAB species are not restricted to dinoflagellates, but also include diatoms, prymnesiophytes, raphidophytes and cyanobacteria (see also Sournia 1995). Furthermore, they cover the complete range from r-strategists (e.g. *Pseudo-nitzschia, Chattonella*) whose success is due to their high growth rates (r) and efficient use of nutrients, to K-strategists (e.g. *G. catenatum*) who can achieve high biomass levels by being energy (light) efficient, e.g. by vertical migration. Table 1 summarizes for HAB species niche-defining factors that primarily relate to their responses to the physico-chemical environment (temperature, salinity, nutrients), while Table 2 summarizes factors that reflect properties of the organisms themselves (mixotrophy, allelopathy, life history, vertical migration).

A detailed overview of the niche of depth-regulating phytoplankton by selected HAB flagellates is further provided in the Chapter by Cullen and MacIntyre, while the niche concept as circumscribed by macronutrient and light limitation is further discussed in the Chapter by Riegman. Most HAB species have now been demonstrated to have some capability of mixotrophy /organic nutrient uptake or having a requirement for micronutrients, but the quantitative significance of these factors in contributing to the success of these species in Nature remains largely unknown. Temperature plays a crucial role in the bloom dynamics of the cyst-forming PSP dinoflagellates and raphidophytes, as well as for species such as *G. catenatum, Noctiluca* and many cyanobacteria which have well-defined seasonal temperature windows. However, once these species enter the water column other factors such as nutrients, turbulence and grazing determine the outcome of competition. HAB species show a perplexing diversity of biomass and toxicity patterns, ranging from species such as *Dinophysis* and *Chrysochromulina* which can cause toxicity problems even at very low cell densities to species such as *Phaeocystis* and *Noctiluca* which are basically non-toxic but whose nuisance value derives largely from their high biomass production. The

production of allelochemicals and other grazer deterrent mechanisms appears widespread among HAB species, but it is still not clear to which extent the incidence of an algal bloom reflects breakdown of grazing control. We have a long way to go before we can confidently predict algal blooms at the species level under all environmental scenarios, and we can expect many more "novel and unusual" HAB events to come to light in years to come. However, for selected sites long-term HAB data sets now start to reveal key variables (Tables 1,2) for use in predictive models (see Hallegraeff *et al*, 1995, and references therein), whether they be primary driving factors or environmental correlates (e.g. rainfall reflecting micronutrient input or water column stability) . We hope that the contributions made by the Bermuda workshop will pave the way for consolidated research efforts to fill in the gaps in our knowledge.

References

Hallegraeff, G.M.. , J.L. Maclean (1989). (eds) *Biology, Epidemiology and Management of Pyrodinium bahamense Red Tides.* ICLARM Conference Proceedings 21, 286 pp.

Hallegraeff, G.M., McCausland, M.A., Brown, R.K. (1995). Early warning of toxic dinoflagellate blooms of *Gymnodinium catenatum* in southern Tasmanian waters. *J. Plankton Res.* 17: 1163-1176.

Margalef, R. (1978). Life-forms of phytoplankton as survival alternatives in an unstable environment. *Oceanologia Acta* 1, 493-509

Margalef, R., Estrada, M., Blasco, D. (1979). Functional morphology of organisms involved in red tides, as adapted to decaying turbulence. In: D. Taylor and H.Seliger (eds), *Toxic dinoflagellate blooms.* Elsevier North Holland, pp. 89-94.

Smayda, T.J., Reynolds, C.S. (1997). Principles of species selection and community assembly in the phytoplankton: beyond the Margalef mandala. Abstracts 8th Int. Conf. Harmful Algae, Vigo, Spain, p. 187.

Sournia, A. (1995). Red tide and toxic marine phytoplankton of the world ocean: an inquiry into biodiversity. In: P. Lassus, G. Arzul, E.Erard, P. Gentien and C. Marcaillou (eds), *Harmful Marine Algal Blooms*, pp. 103-112. Lavoisier, Intercept.

Ecophysiological Processes and Mechanisms

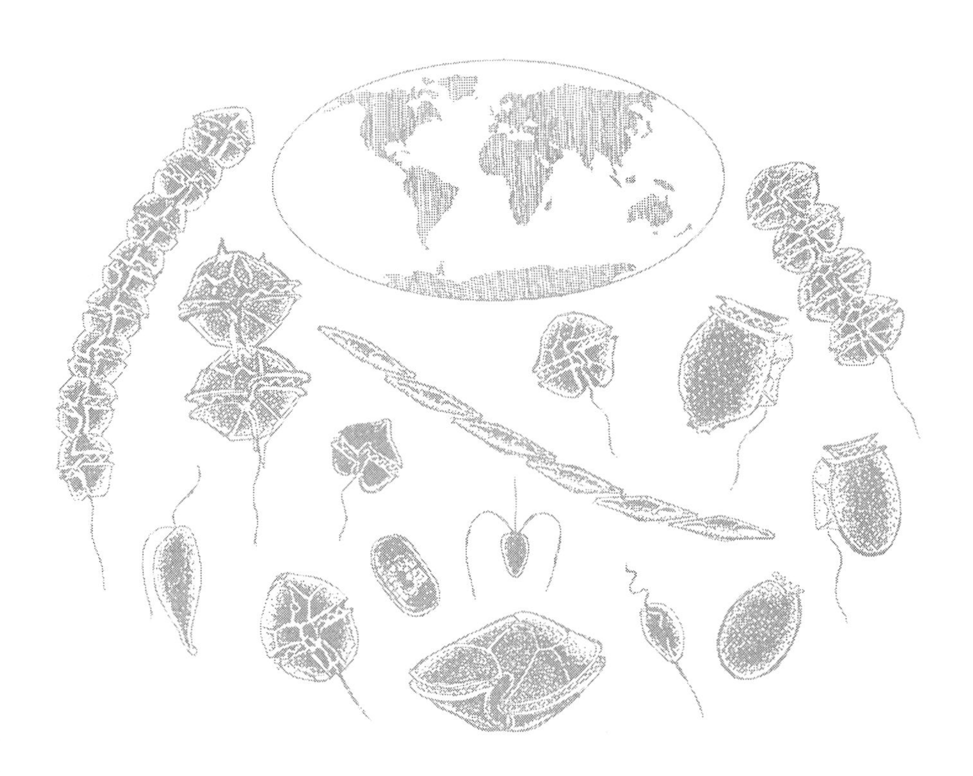

Ecophysiology and Metabolism of Paralytic Shellfish Toxins in Marine Microalgae

Allan D. Cembella

Institute for Marine Biosciences, National Research Council
Halifax, Nova Scotia, Canada B3H 3Z1

1. Introduction

The low molecular weight perhydropurine compounds commonly associated with paralytic shellfish poisoning (PSP), a potent neurotoxic syndrome in vertebrates, are produced primarily by species of marine dinoflagellates and a few cyanobacteria. Recent evidence indicates that these neurotoxins may also be distributed among certain members of the eubacteria.

Considerable progress has been achieved in the structural elucidation of PSP toxins and in their quantitation in a variety of producing and vectoral organisms. Since the structure of the parent compound saxitoxin (STX) was first described, approximately two dozen naturally-occurring derivatives have been found among various organisms (Shimizu 1996). These analogues differ in C-13, N-1-hydroxyl-, 11-hydroxysulfate-, and 21-N-sulfo-substitutions, including epimerization at the C-11 position. Based upon these substitutions, the compounds are now classified as the high potency carbamate toxins (STX; NEO [neosaxitoxin]; GTX 1-4 [gonyautoxins 1,2,3,4]), the low potency N-sulfocarbamoyl group (B1,B2, C1-C4), and decarbamoyl derivatives (dc-GTX1-4, dc-NEO, dc-STX). The decarbamoyl toxins, originally believed to be exclusively a catabolic byproduct of toxin digestion in shellfish, now include recently discovered 13-deoxy-decarbamoyl derivatives (doSTX, doGTX2 and doGTX3), and can occur in low relative abundance in certain strains of marine dinoflagellates (Oshima *et al.* 1993). Toxigenic species typically produce more than a single toxin derivative, but no individual strain is known to synthesize the entire suite of compounds. Unfortunately, examination of the structural relationships among the members of the PSP toxin complex does not provide detailed insight into the mechanisms of toxin biosynthesis and interconversions.

In this review, the rationale for toxin-based taxonomy is provided, as well as an indication of the importance of "molecular logic" in phylogenetic relationships among toxigenic HAB species. Finally, an overview of the basic biosynthetic pathway for toxin production, the regulatory effects of environmental factors versus genetic stability, and the possible eco-evolutionary role for PSP toxins is presented.

2. Phylogenetic origin and biogeographical distribution

The biosystematic and geographical distribution of PSP toxins among organisms considered to be primary toxin sources is unusual. Among the microalgae, the production of PSP toxins is known from divergent genera of free-living planktonic marine dinoflagellates and several closely related genera of cyanobacteria. Among dinoflagellates the production of PSP toxins is restricted to photoautotrophs, and there

NATO ASI Series, Vol. G 41
Physiological Ecology of Harmful Algal Blooms
Edited by D. M. Anderson, A. D. Cembella and
G. M. Hallegraeff

is a propensity for chain-formation in genera of both dinoflagellates and cyanobacteria which include toxic species.

Most of the toxigenic gonyaulacoid dinoflagellate species associated with PSP toxins lie within the genus *Alexandrium*, which was emended and enlarged to include *Protogonyaulax* Taylor, and now comprises ten known toxic species *sensu* Balech (1985). Several of these toxic "species" may be minor morphological variants ("varieties") or ecotypes, e.g., members of the "tamarensis group" (*tamarense/excavatum/fundyense/acatenella/catenella*) or the "minutum group" (*minutum/ibericum/angustitabulatum*). Members of these respective groups are not always clearly differentiated at the molecular DNA level (see Scholin, this volume), nor based upon their toxicity and toxin profiles (Cembella *et al.* 1987; Anderson *et al.* 1994). The distribution of toxigenic *Alexandrium* species is largely within coastal and temperate waters, although toxic populations are found in the Mediterranean Sea, and toxic sub-tropical, tropical, and (possibly) Arctic variants exist.

The heavily armored dinoflagellate *Pyrodinium bahamense* var. *compressum* is responsible for most PSP episodes in the tropical waters of the Indo-Pacific and in Central America (e.g., Guatemala) (see Usup and Corrales, this volume). Significantly, toxigenicity appears to be restricted to the chain-forming var. *compressum* and this may be a constant characteristic; toxicity has never been associated with the non-chain-forming var. *bahamense*, which is common in the Gulf of Mexico and Caribbean. Based upon plate tabulation and structural homology (Fig.1), *Pyrodinium* is considered to be a close phylogenetic relative of *Alexandrium* (Taylor 1987), suggesting that the capacity for PSP toxin production may reflect common descent.

The production of PSP toxins among athecate gymnodinoid dinoflagellates is restricted to a single chain-forming species, *Gymnodinium catenatum*, with a predominately temperate water distribution, including coastal areas of Atlantic Iberia, Mexico, Venezuela, Tasmania, the Philippines and Japan (see Hallegraeff and Fraga, this volume). Early suggestions of a close phylogenetic affinity between *G. catenatum* and *A. catenella* (Morey-Gaines 1982) — based upon arguments of common features such as overall morphology, chain-formation, and tendency for toxin production, now appear to be clearly contradicted by ultrastructural features (Rees and Hallegraeff 1991) and molecular evidence (Zardoya *et al.* 1995). This athecate species is a true, albeit rather divergent gymnodinoid, raising the possibility that the origin of the capacity for PSP toxin production is polyphyletic (Fig.1).

An early report which associated the occurrence of PSP toxins with the calcareous rhodophyte *Jania* sp. was later confirmed (as GTX1,2,3) by more advanced analytical techniques (Oshima *et al.* 1984). This evidence must now be considered as compelling, as efforts were made to clean the tissues to remove potentially toxigenic epiphytic bacteria and/or dinoflagellates.

Both molecular phylogeny (Zardoya *et al.* 1995) and phylogenetic inferences based upon plate homology and ultrastructure (Fig.1) are more suggestive of a divergent (perhaps random) distribution of the expression of PSP toxins than a monophyletic origin. The apparently paradoxical appearance of PSP toxins in prokaryotes, including freshwater cyanobacteria (cited in Shimizu 1996), also supports a polyphyletic origin for PSP toxicity in dinoflagellates, perhaps via multiple endosymbioses. Recent reports of the occurrence of "PSP toxins", or at least Na^+ channel-blocking analogues, in some eubacterial species (Doucette *et al.*, this volume) lends further credence to the

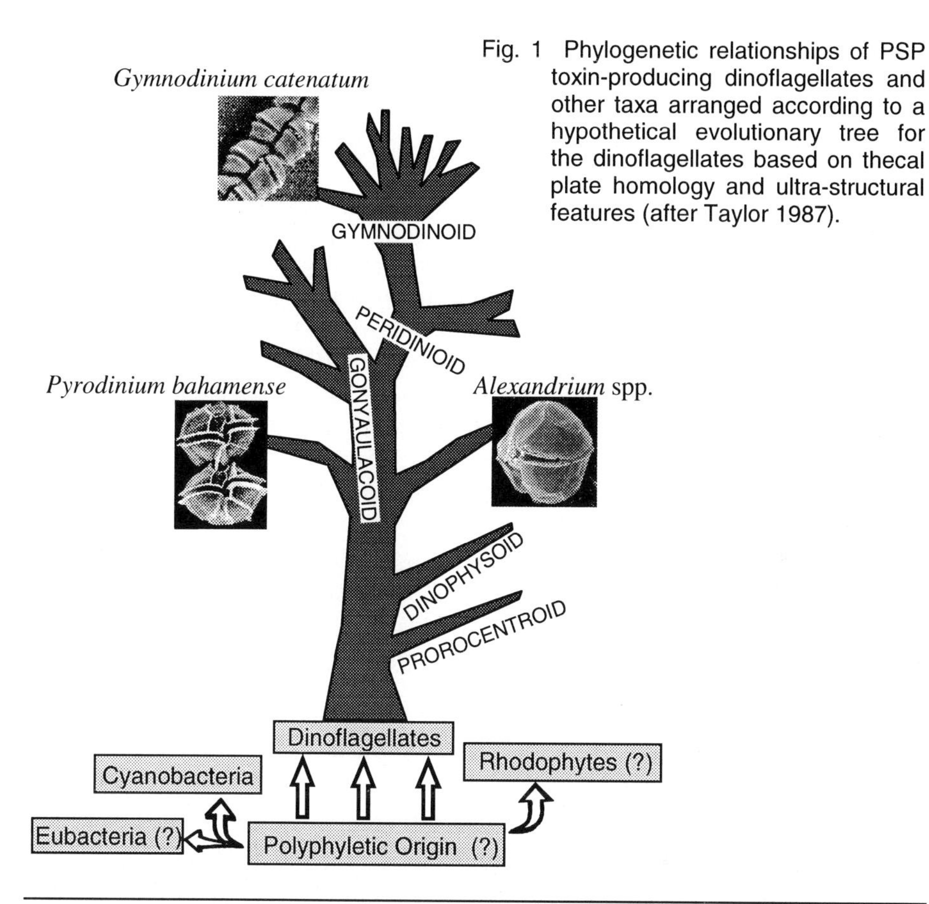

Fig. 1 Phylogenetic relationships of PSP toxin-producing dinoflagellates and other taxa arranged according to a hypothetical evolutionary tree for the dinoflagellates based on thecal plate homology and ultra-structural features (after Taylor 1987).

endosymbiotic origin hypothesis. The frequent failure to detect endocellular bacteria by TEM even in highly toxic dinoflagellate strains may reflect the total incorporation of the basic biosynthetic genes for PSP toxin production into the dinoflagellate genome.

3. The pathway for toxin biosynthesis

The fundamental structural similarity between the perhydropurine saxitoxin (STX) (a guanidine derivative) and common purines of primary metabolism has suggested to some researchers (see Ishida *et al.*, this volume) that PSP toxins may be produced via modifications of the conventional pathways of purine biosynthesis. However, based upon studies of the incorporation of labeled precursors, there is no evidence for a biosynthetic link with purine metabolism. The basic biosynthetic pathway leading to cyclization of the perhydropurine ring to form (STX) was determined more than a decade ago (Shimizu *et al.* 1985). The initial experiments were performed with *Aphanizomenon flos-aquae* but the results were later confirmed to be identical with *Alexandrium*. The precise elements of the pathway have undergone some subsequent revisions (Shimizu 1996) and some steps have yet to be confirmed. For example, Shimizu (1996) contends that the guanidinium group (see **X** on Fig.2) and the 21-N are derived from arginine (ARG), however these may plausibly come from the highly activated intermediate carbamyl phosphate. At present, the three fundamental units in STX biosynthesis are considered to be a high-energy CH_3-group donor (S-adenosyl-methionine: SAM), a two-carbon unit (acetate), and an amino acid precursor (ARG). The exact number of enzymatic steps and sequence of the biosynthetic reactions is still uncertain.

Subsequent modifications to the basic skeleton, e.g., addition of: 1) a hydroxyl group at N1 (=R_1), 2) an O-sulfate at C-11 (R_2, R_3) or 3) a 21-N-sulfo-group, to form the N-sulfocarbamoyl toxins, are presumably mediated by enzymatic reactions (see Ishida et al., this volume). The high ratio (often >10:1) of β:α 11-hydroxysulfated derivatives found in *Alexandrium* spp. and *Gymnodinium catenatum* after careful extraction of cells suggests that only the 11-β-epimers (C2, C4, GTX3, GTX4) are biosynthesized and that the 11-α-epimers (C1, C3, GTX1, GTX2) are formed *in vitro* via spontaneous epimerization (Hall 1982; Oshima *et al.* 1993).

4. Biogeographical and genetic variation in toxicity and toxin composition

There has been considerable debate in the literature regarding the use of cellular toxicity and toxin profiles as chemotaxonomic indicators among populations and species. Some authors contend that the relative toxin composition (''profile") of *Alexandrium* spp. (if not toxin content) is a relatively conservative characteristic within an isolate or natural population, and that this trait can be used to evaluate genetic differentiation among species in diverse geographical regions (Cembella and Taylor, 1985; Cembella *et al.*, 1987; Hall 1982; Anderson *et al.* 1994). Other research has indicated that the high variability of toxin profile which can occur by environmental manipulation of culture conditions may make this an unsuitable trait for intra- and inter-specific characterization (Boczar *et al.* 1988; Anderson *et al.* 1990b). This apparent dichotomy may be more apparent than real because studies which have shown significant mutability in toxin profile have typically found such changes to occur only in advanced senescence (Boczar *et al.* 1988) or under stresses such as chronic nutrient limitation in quasi-steady state conditions (Anderson *et al.* 1990b).

The lack of toxicity has frequently been noted in *Alexandrium*, even among isolates and natural populations of species which are known to contain toxigenic members (e.g., non-toxic *A. tamarense* from the Tamar estuary [Cembella *et al.* 1987]). In retrospect, non-toxic *Alexandrium* isolates reported from New England (e.g., Maranda *et al.* 1985) are usually referable to *A. andersonii*; no non-toxic strains of the *A. tamarense/excavatum/fundyense* species complex are known from this region (D.M. Anderson, pers. comm.). In any case, lack of toxicity among isolates of a species which contains toxigenic members is not limited the *A. tamarense* group, as non-toxic strains are also known for *A. ostenfeldii* (MacKenzie *et al.* 1997) and *Gymnodinium catenatum* (Oshima *et al.* 1993).

It is now dogmatic to note that there is an apparent clinal gradient in increasing cellular toxicity among *Alexandrium* populations from the northeastern coast of Canada and the United States, with a continuous northward increase through the known geographical range of this genus (Maranda *et al.* 1985; Cembella *et al.* 1988; Anderson *et al.* 1994). Re-analysis of previous data sets (ANOVA, with linear regression model) now indicates that this may be a simplistic interpretation (Fig.3), depending upon whether or not non-toxic isolates (which may be *A. andersonii*) are

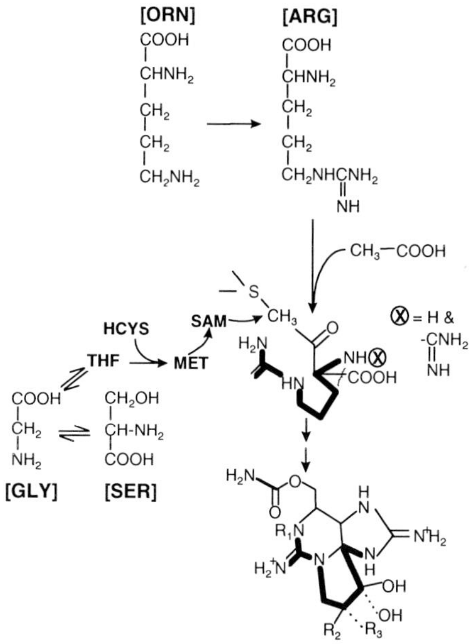

Fig. 2
Putative biosynthetic scheme for PSP toxins based upon the incorporation of labeled precursors into *Alexandrium* and *Anabaena flos-aquae* (modified from Shimizu *et al.* 1985 and Shimizu 1996). R_1=OH; $R_2/R_3=OSO_3^-$; GLY= glycine; SER=serine; ORN=ornithine, ARG= arginine; HCYS=homocysteine; MET=methionine; THF=tetrahyrofolate; SAM=S-adenosyl-methionine.

Heavy lines indicate confirmed origin from ARG.

included in the analysis. When the three data sets from Fig. 3 are combined, the interpretation shows only a weakly positive linear correlation (Pearson product-moment: r=0.39; p<0.05; n=82) of a northward increase in mean cellular toxicity with latitutude along the east coast of North America. Based upon a linear regression model, there is also a weak (but statistically significant) positive relationship between latitude and *Alexandrium* cell toxicity within the Gulf of Maine, including the Bay of Fundy, but not with cell toxin content (Anderson *et al.* 1994; Fig.3). This was attributed to a higher ratio of more potent carbamate derivatives, relative to the low toxicity N-sulfocarbamoyl toxins (C1/C2, B1/B2), in northern populations. A north-south geographical gradient in toxin profile has also been documented for Japan, with higher concentrations of the potent carbamate toxins (NEO, GTX1-4) associated with northern isolates, whereas southern isolates contain primarily the less toxic N-sulfocarbamoyl toxins (B1/ B2, C1/C2, C4) (Kim *et al.* 1993).

However, there is no apparent systematic geographical trend in toxicity among populations from the Atlantic coast of Canada, indicating that there may be a northern threshold of cellular toxicity for the east coast of North America (Cembella and Destombe 1996), or a clustering effect reflecting genetic isolation from the Gulf of Maine populations. Given the prominence of toxins C1/C2 (up to 60% on a molar basis) in highly toxic isolates from populations from the lower estuary and Gulf of St. Lawrence, the suggestion that the high toxicity of northern *Alexandrium* strains is due to a relative paucity of low potency N-sulfocarbamoyl toxins (Anderson *et al.* 1994) does not appear to be valid for this region.

Therefore, inferences regarding cellular toxicity and toxin composition cannot simply be extrapolated by northward range extension from New England. Nevertheless, it does appear that the highest toxicity isolates of the *A. tamarense* group occur in northern waters, whereas weakly toxic isolates are more broadly dispersed. Conversely, it may

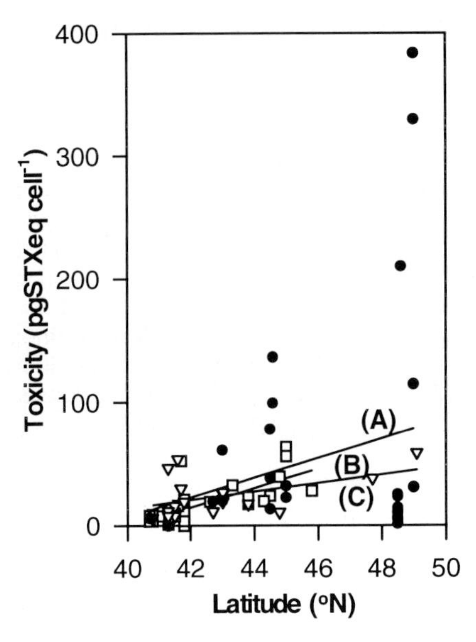

Fig. 3
Relationship between cell toxicity and latitude of origin of *Alexandrium* isolates from the Atlantic coast of Canada and the United States. Data plotted from: (A) filled circles, Cembella and Destombe (1996), plus A. Cembella, unpublished. (r^2=0.06; n=30; p=0.20); (B) open squares, Maranda *et al.* (1985) (r^2=0.50; n=34; p=0.0001); (C) open triangles, Anderson *et al.* (1994) (r^2=0.23; n=18; p=0.04).

Toxicity per cell was calculated from HPLC toxin profile data using appropriate conversion factors for specific toxicity and from bioassays assuming 1 mouse unit (MU) = 0.22 μgSTXeq.

also be accurate to state that the frequency of occurrence of non-toxic populations increases along a southward latitudinal gradient. There is no evidence of increasing complexity in toxin profile among more northern populations which might be related to a shorter permissible growth period or higher genetic diversity.

In any case, it has been repeatedly established that cultured *Alexandrium* cells are typically less toxic than those collected from natural populations in the same region (White 1986; Cembella *et al.* 1988; 1990), and therefore do not yield representative toxicity values. Furthermore, wide temporal variation in toxicity per cell among blooms in the same area is also a common feature in natural ecosystems (White 1986). Re-analysis of the data from Cembella *et al.* (1988) indicates that the acclimated mean toxicity of clonal isolates (n=15) of *A. tamarense* (cf. *excavatum*) from the St. Lawrence estuary, Quebec, was significantly lower than for natural populations (n=9) (ANOVA, p=.0017). It has yet to be determined if this effect is an artifact of "forced growth" in culture (e.g., under high nutrient concentrations), a relaxation of the selective pressure for high toxin production in natural populations, genetic drift in culture expressed as reduced toxin biosynthetic enzyme activity, or higher cell division rates in culture, resulting in reduction of cell toxin quota (Q_t).

Significant regional variation in toxin composition has been found among *Alexandrium* populations and this has provided insights into genetic stability of toxin expression. For example, *Alexandrium* isolates from Alaska were found to contain relatively high amounts of the N-sulfocarbamoyl toxins (C1/C2 and B1/B2) (Hall 1982) and this was also generally true for Pacific coast populations from British Columbia and Washington State (Cembella and Taylor 1985). Among Pacific coast populations, the toxin profile correlated poorly with conventional morphotypic descriptors (non-chain forming isodiametrical cells [tamarensoid] versus chain-forming apically compressed cells [catenelloid]), however contemporaneous isolates from a particular geographical region are tightly clustered (Cembella *et al.* 1987).

Relative to populations from the northeast Pacific, *Alexandrium* populations from the estuary and Gulf of St. Lawrence in Atlantic Canada are conservative in toxin composition, exhibiting less intra-specific and geographical variation (Cembella and Destombe 1996) (Fig. 4 and 5). Representative isolates from eastern coastal Nova Scotian populations tend to be more heterogeneous in their respective toxin profiles and are morphologically distinct from their northern counterparts, suggesting that little recent genetic exchange has occurred.

Analysis of toxin profiles is particularly revealing when applied at the inter-specific and inter-generic levels where large differences in toxin composition are apparent. For example, toxigenic *Alexandrium* species of the "minutum group" *sensu* Balech typically express rather low cellular toxicity, characteristically with the gonyautoxins (GTX1-4) as the dominant analogues — the N-sulfocarbamoyl toxins are usually absent or present in only trace quantities (Cembella *et al.* 1987; Franco *et al.* 1994). However, exceptions do occur among members of the "minutum group", as moderately toxic isolates from New Zealand contained surprisingly high relative levels of NEO and STX (Chang *et al.* 1997).

A wide genetic divergence is apparent between *A. ostenfeldii* populations from diverse geographical regions. In northern Europe, this species is known to be weakly toxic and the toxin profile is dominated by the N-sulfocarbamoyl toxin B2, with lesser amounts of C1/C2 and GTX2/3, whereas New Zealand isolates are considerably more toxic and usually contain STX, GTX3 and often B1 as major components (MacKenzie *et al.* 1997).

In sharp contrast to toxin profiles in *Alexandrium* spp., *Pyrodinium bahamense* var. *compressum* isolates from Palau in the south Pacific are rich in toxins B1/B2, STX, and NEO, with a low percentage of dc-STX (Oshima *et al.* 1984; 1987); 11-hydroxysulfate toxins (GTX1-4) are absent. Other strains isolated from southeast Pacific sites from Borneo (Malaysia), Brunei and the Philippines contain essentially the same suite of toxins (Oshima *et al.* 1990), even under variable growth conditions (Usup *et al.* 1994). Nevertheless, there are also indications of some biogeographical variation in toxin profile within *P. bahamense* — natural populations from Guatemala contain toxins B1, STX, NEO, GTX2/3, and GTX4, but not dc-STX, which characterizes southeast Pacific populations (cited in Usup and Corrales, this volume).

Isolates of *G. catenatum* exhibit an unusual toxin profile, producing mainly N-sulfocarbamoyl toxins (C1-C4, B1/B2), and the carbamate toxins are virtually absent (Oshima *et al.* 1987; 1993). Novel components such as 13-deoxycarbamoyl toxins are found in Tasmanian isolates and decarbamoyl derivatives (dc-STX, dc-GTX2 and dc-GTX3) are common to all known isolates. In spite of the consistent dominance of the N-sulfocarbamoyl toxins in this species, *Gymnodinium catenatum* blooms and cultured isolates from Tasmania can be discriminated from those from Japan and Spain by the high relative amounts of B1 and B2 in the latter, and the absence of toxins C3/C4 in the former.

Relative to the complex PSP toxin profiles often expressed among the dinoflagellates, the number of toxin components produced by certain cyanobacteria tends to be rather restricted. For example, *Anabaena flos-aquae* produces only the carbamate toxins NEO and STX, whereas only GTXs and dc-STX have been found in *Lyngbya wollei* and *Oscillatoria morigeotii. Anabaena circinalis* yields primarily toxins C1/C2, but also GTX2/3 and STX (cited in Shimizu 1996).

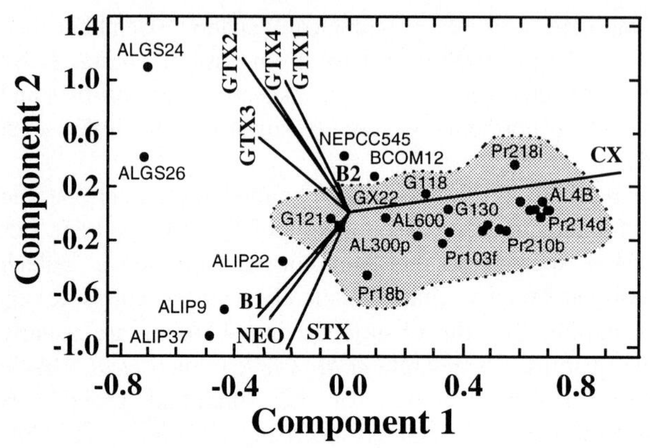

Fig. 4 Biplot of the first two principal components and ordinated scatterplot of the isolates and field populations. This ordination was derived from analysis of standardized toxin profiles after transforming the relative toxin compositional data (% molar) according to the highest ratio of each toxin among all taxa (= 1.0), with relative values for other taxa expressed as a proportion of the maximum. The delimited cluster indicates St. Lawrence populations which group at Euclidean distances < 0.6 (see Fig. 5). (Adapted from Cembella and Destombe 1996).

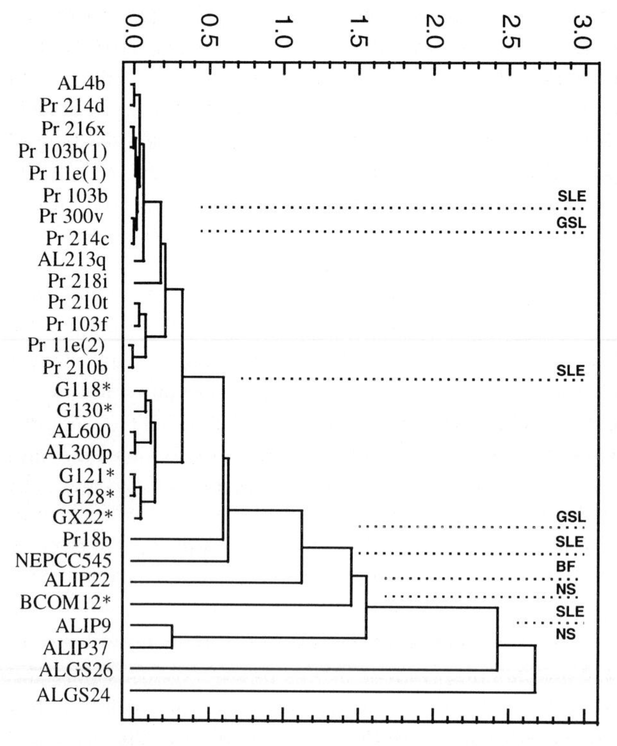

Fig. 5 A centroid-linkage, single cluster dendrogram showing Euclidean distances between isolates and natural populations of *Alexandrium spp.*, based upon ratios of equally weighted toxins.

SLE = St. Lawrence estuary; GSL = Gulf of St. Lawrence; BF = Bay of Fundy; NS = eastern Nova Scotia.
*= *in situ* populations

(modified from Cembella and Destombe 1996).

Research on the toxin composition of isolates and natural *Alexandrium* populations from the lower St. Lawrence estuary, Quebec (Cembella *et al.* 1990; Cembella and Destombe 1996) demonstrated that the toxin profile of cultured vegetative cells in exponential growth phase was indeed typical of rapidly dividing cells in nature. The fact that the spectrum of toxins in cultured isolates generally reflected that found in natural blooms from this region, and that the toxin profile remains qualitatively consistent throughout the culture cycle among alternative life history stages (Cembella *et al.* 1990), suggests that the capacity for toxin production is genetically fixed, and relatively resistant to spontaneous mutation.

Three basic criteria should be adopted for comparisons of toxin profile among isolates: 1) growth conditions must be rigidly standardized (growth medium, light, photoperiod, temperature, nutrients, etc.); 2) cells must be acclimated, i.e. the cultured isolate must have had sufficient time to physiologically adjust to ambient environment through several culture cycles; 3) cells should be harvested in exponential growth, after the same number of cell divisions rather than after a fixed time period. When these conditions are met, genotypic differences in toxin profile are expressed and can be identified.

5. Genetic regulation of toxin biosynthesis

The essential stability of the toxin profile and the differences in toxin composition maintained among toxigenic isolates of all the PSP toxin-producing groups argues forcefully for the dominance of genetic factors in determining and regulating toxin biosynthesis. Recent evidence indicates that the inheritance of toxin composition in strain-crossing experiments involving *Alexandrium* spp. follows a predictable Mendelian segregation pattern — indicative of encoding by nuclear genes — and this biparental inheritance appears to be independent of the mating type locus (see Ishida *et al.*, this volume). The high level of variation in growth rate and toxin production among sub-clones of *Alexandrium tamarense* was interpreted as evidence of involvement of epi-genetic factors in toxin production, e.g., endocellular bacteria (Ogata *et al.* 1987a; see Doucette *et al.*, this volume). The source of this variation is unclear, although at least some may be attributable to stochastic variation in the microenvironement and/or failure to achieve acclimated growth and environmental homogeneity.

In fact, there are no "toxin genes" *per se* as these metabolites are not primary gene products. However, there must be genes which code for at least a dozen enzymes involved in the basic biosynthetic scheme (Fig.2) and more enzymes are undoubtedly involved in the addition or deletion of substituent groups (e.g., N-(amino) sulfo-transferases, O-sulfotransferases, oxidoreductases [hydrolases], carbamoylases, etc.). A number of these biosynthetic enzymes have been identified and purified, including the N-sulfotransferase enzyme responsible for C-11- and N-21-sulfation in *G. catenatum* and *A. catenella*, and an O-sulfotransferase from the former species (discussed in detail by Ishida *et al.*, this volume). Some preliminary data even suggest that the enzymes responsible for toxin interconversions may be linked along a single chromosome (cited in Ishida *et al.*, this volume). This would explain the sharp demarcation between toxigenic and non-toxic strains of the same species. The genetic basis for the stability in toxin profile remains to be determined, although the diversity and variation in toxin composition among populations is circumstantial evidence of a high level of genetic recombination. Particular strains of species may lack one or more genes coding for key enzymes in the assembly of the basic perhydropurine structure or the enzyme(s) may be present in an inactive form (hence no toxicity is expressed). Alternatively, there may be a defect or absence of an enzyme required for the addition of substituent groups (e.g., O-sulfotransferase required for synthesis of the C-11 sulfated carbamates [GTX1-4] or the N-(amino)sulfotransferases coding for the synthesis of the 21-N-sulfocarbamoyl-derivatives [C1-C4; B1/B2]).

Here it is sufficient to underscore the point that the Mendelian segregation of the products of these biosynthetic enzymes does not prove that production of the perhydropurine base is regulated by nuclear genes. There may be only a few highly specific genes associated exclusively with toxin biosynthesis, whereas the remainder of the modifications may be regulated by a various "housekeeping genes" coding for rather broad spectrum enzymes.

6. Cellular toxin dynamics

Extrinsic environmental factors are known to exert profound effects upon the general physiological status of toxigenic species. The toxin cell quota (Q_t) may exhibit substantial variation within a given strain, to the extent that the processes of toxin biosynthesis, sequestration, interconversions, catabolism, and excretion are directly or indirectly linked to intermediary metabolism. Thus Q_t represents an equilibrium between the rate of anabolism and catabolism, leakage/excretion, cell growth and division. Environmental factors, such as irradiance (Ogata *et al.* 1987b; Parkhill and Cembella 1997), temperature (Hall 1982; Anderson *et al.* 1990a,b), salinity (White 1978; Parkhill and Cembella 1997), turbulence (see Estrada and Berdalet, this volume), macronutrients (Hall 1982; Boyer 1987; Anderson *et al.* 1990a,b), and growth promoting substances (GPS) and allelochemical interactions (Ogata *et al.* 1996), have each been shown to affect Q_t, when they were varied as single parameters in cell culture experiments. To a lesser extent, and typically under rather extreme conditions, the PSP toxin composition (molar ratios of the various derivatives) may also be altered (Boczar *et al.* 1988; Anderson *et al.* 1990b). The regulatory pathways underlying these physiological changes are still poorly understood .

6.1. Growth rate and cell physiological status

Most studies on PSP toxin production in microalgae are conducted in batch cultures where the shifting equilibrium of cell density and ambient environmental variables can lead to unbalanced growth, particularly outside of exponential phase. Furthermore, growth may be forced at unrealistically high nutrient loadings (>500 µM inorganic N) and high N:P ratios (often >20:1) used in nutrient media. Among toxigenic microalgae, *Alexandrium* spp. have been most often employed to investigate the endogenous and external factors which regulate toxin synthesis, but with some significant exceptions, the general conclusions on the effects of environmental factors on toxin content also apply to *Pyrodinium bahamense* (Usup *et al.* 1994) and *Gymnodinium catenatum* (Flynn *et al.* 1996b). In batch cultures, cells typically experience various growth stages: lag phase ($\mu \geq 0$, but ≤ 0.1), exponential phase ($\mu \geq 0.2$, but ≤ 1), stationary phase ($\mu = 0$), and senescence/death ($\mu \leq 0$). These nominal values of growth rate (μ) are for *Alexandrium* spp., but they are characteristic of other toxigenic dinoflagellates as well. In *Alexandrium* spp., the cell toxin quota (Q_t) is usually highest in exponential phase and declines as the cells enter stationary phase (Singh *et al.* 1982; Boyer *et al* 1987; Boczar *et al.* 1988), with the exception of P-limitation as the cause for cessation of growth. Phosphorus limitation results in an increased Q_t in *Alexandrium tamarense/fundyense* (Boyer *et al.* 1987; Anderson *et al.* 1990a), and also in *Pyrodinium bahamense* (Usup *et al.* 1994), but not necessarily in *A. minutum* (Flynn *et al.* 1994) or *Gymnodinium catenatum* (Flynn *et al.* 1996b). A key question is whether the decrease in cell toxin content in stationary phase is due to leakage, cell lysis, decrease in the rate of production, increased turnover, or partitioning of toxin into daughter cells via cell division — these processes are not mutually exclusive. The relationship between growth rate and cell toxin content is yet poorly defined; some have argued that there is no definitive relationship between growth rate and cell toxin content (White 1978; Hall 1982; Ogata *et al.* 1987a) whereas others have claimed that toxin production is inversely related to the growth rate (Proctor *et al.* 1975). A multi-variable study of toxin production in *A. tamarense* (Parkhill and Cembella, 1997) indicated that cellular toxicity was independent of exogenous environmental factors (light, salinity, inorganic-N) throughout exponential growth phase (r^2= 0.006 for growth rate - toxicity relationship, Fig. 6), although it varied over the growth stages.

Toxin *production* has frequently been misinterpreted as the *cell toxin quota* (Q_t) (toxin per cell), rather than as the *net rate of toxin production*, i.e., the biosynthetic rate minus the various loss terms. The introduction of the concept of toxin production rate, R_{TOX} (net quantity of toxin produced per cell per unit time) has greatly aided the interpretation of the effects of specific environmental parameters versus growth rate (Anderson *et al.* 1990a). Using this model, Anderson *et al.* (1990a) were not able to determine a universally consistent relationship in *A. fundyense* between R_{TOX} and the specific growth rate (μ) among different treatments, although there was a correlation with individual variables (except for P-limitation). Examination of Fig. 6 indicates that the hypothesized inverse relationship between growth rate and Q_t (Proctor 1975) is not universally valid. The relative stability of toxicity and toxin profile in exponential growth may be due to the fact that minor disequilibrium in R_{TOX} is roughly compensated for by shifts in μ, thus variations in Q_t are often no more than two fold (Anderson *et al.* 1990a).

6.1.1. Diurnal rhythms and cell cycle events

Diurnal and circadian oscillations in many physiological parameters are well known for the microalgae (see Kamykowski *et al.*, this volume). Nevertheless, the effect of exogenous factors on endogenous rhythms has rarely been addressed for toxin production. Diel vertical migrations require photosynthetic reserves and dark uptake of NO_3 imposes an additional bioenergetic demand, since C- and N-assimilation become uncoupled (see Cullen and McIntyre, this volume). Many diel rhythms are entrained by the photoperiod but the response may be species-specific — in non-toxic *A. tamarense* and toxic *Gymnodinium catenatum* the magnitude of the intracellular free amino acid (IFAA) pool is maximum at the end of the dark period (in contrast to other non-toxic species) (Flynn *et al.* 1993). Cells of *A. tamarense* are capable of diel vertical migration (DVM) from N-depleted surface waters to the N-rich nutricline, and thereby regenerate normal C:N and chlorophyll *a* per cell ratios and Q_t values (McIntyre *et al.* 1997). Under severe N-stress, the cells initiate their descent prior to

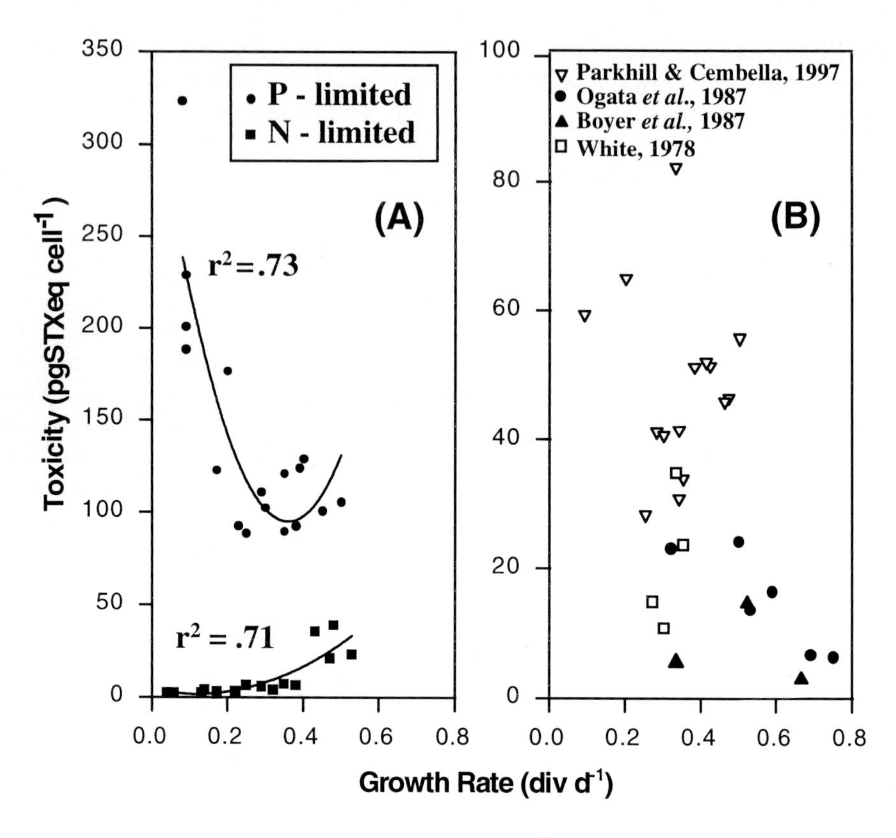

Fig. 6 Plot of cell toxicity versus growth rate (div d^{-1}): A) recalculated from Anderson *et al.* (1990b) for *Alexandrium fundyense* grown in N- and P-limited semi-continuous cultures; B): data plotted from various authors for batch cultures of *Alexandrium* spp. in exponential growth under differing environmental regimes. Toxicity per cell was calculated from HPLC toxin profile data using appropriate conversion factors for specific toxicity and from bioassay data assuming 1 mouse unit (MU) = 0.22 μgSTXeq.

the onset of the dark period and adjust time spent at the nutricline to optimize N-assimilation and toxin regeneration.

In *Alexandrium* spp., toxin production is tightly coupled to particular stages of the cell cycle. Cellular toxin content rises gradually from the latter half of the light period through the middle of the dark period where the sudden decrease is associated with cell division (Kim *et al.* 1993). New evidence from *Alexandrium fundyense*, based upon induced cell synchrony, indicates that toxin synthesis is initiated early in the G1 phase of the cell cycle in response to a light trigger and ceases prior to exiting this phase (Taroncher-Oldenburg *et al.* 1997). In this species, S-phase is discrete, therefore DNA synthesis is discontinuous, and it is not directly linked to toxin synthesis as previously suspected. These authors hypothesize that the relative increase in carbamate toxins GTX3 and STX accompanied by a decline in toxins C2, GTX4 and NEO over time suggests that the latter toxins may be precursors, modulated by the activity of a hydrolase and/or an N-(amino)sulfotransferase for toxin conversions.

6.2. Physical factors

6.2.1. Light effects

The direct and indirect effects of light (photon flux density, PFD) on PSP toxin production are multifaceted and complex. The light harvesting efficiency and photosynthetic capacity are linked to nutritional status in most phytoplankton species, and these parameters are known to decrease by several fold in response to N-depletion in *A. tamarense* (Glibert *et al.* 1988). Toxin biosynthesis imposes an extra demand for photosynthetically-derived carbon skeletons, e.g., for amino acids and acetate, as well as high energy intermediates (e.g., ATP, NADH/NADPH, etc.). The light-dependence of toxin production may be related to the photoassimilation of NO_3 or NH_4 into amino acid precursors. Yet Langdon (1987) found nothing remarkable in the photosynthetic parameters (I_c, α_p), growth rate-irradiance (μ-I) relationship, and carbon-specific growth rates of *A. tamarense*. The bioenergetic cost of toxin production may be below the threshold of detection in these parameters. Nevertheless, the obligatory requirement for photosynthetic activity is clear; growth under extreme low light conditions arrests both growth and toxin synthesis (Ogata *et al.* 1996), perhaps by depressing the activity of the primary photoassimilation enzyme (RuBisCo: ribulose bis-phosphate carboxylase) and/or NO_3-uptake. High chlorophyll levels typically found under low light conditions in toxigenic species (Usup *et al.* 1994; Parkhill and Cembella 1997) may also shunt available cell-N from toxins to pigment biosynthesis. In *A. tamarense,* the increase in cell toxicity in late exponential growth was associated with a decrease in light intensity, but this was clearly coupled to a corresponding decrease in growth rate (Ogata *et al.* 1987b). Re-analysis of these data (Fig. 6B) indicates that there is a high linear correlation between cell toxicity and light- or temperature-dependent growth rate (r^2=0.76), but the dependence is more pronounced for temperature. In *Pyrodinium bahamense* (but not in *Alexandrium* spp.) growth at higher light levels results in an inversion of the NEO/B1 ratio (Usup *et al.* 1994), although low μ is also associated with the lowest light levels, suggesting that N-(amino)sulfotransferase activity may be light-dependent in this species.

6.2.2. Temperature effects

Low temperature is often cited a factor contributing to high PSP toxin "production" and this has even been used as a possible explanation for the apparent high cell toxicity of *Alexandrium* populations at high latitudes (White 1986; Cembella *et al.* 1988). Usually what it meant is that growth at low temperatures favors a high cell toxin quota. Metabolic reactions, particularly those associated with enzymatic activity, are susceptible to the Q_{10} effect, whereby an increase in temperature of 10 °C yields a doubling in the metabolic rate. Thus there may be an indirect temperature-dependent effect on the rate of toxin synthesis and turnover, if μ and toxin production become uncoupled. Most studies of temperature-dependent effects on PSP toxin production have not clearly discriminated the differential effect of temperature on the respective rates of cell division and toxin biosynthesis. For example, in *Alexandrium spp.*, the lower cell yield of toxin at higher temperature is also associated with higher growth rate, but toxin profile is typically constant (Proctor *et al.* 1975; Boyer *et al.* 1985;

Ogata *et al.* 1987b). In contrast, in *Pyrodinium bahamense,* the gradual inversion of the high ratio of toxin B1/NEO with increasing temperature gradient indicates a temperature-dependent process (Usup *et al.* 1994). There is evidence that high Q_t at low temperatures in *Alexandrium* reflects the allocation of more cell-N to toxin synthesis and away from general protein, as this condition favors high intracellular ARG levels (Anderson *et al.* 1990a).

6.2.3. Salinity effects

In principle, salinity could play a regulatory role in toxin biosynthesis by exerting ionic effects on the mechanisms of nutrient uptake — "permeases", and a variety of active transport and facilitated diffusion systems. There could also be an indirect effect on cell toxin yield by interference with osmoregulation via effects on amino acid biosynthesis (e.g., of proline) and cell membrane receptors required for the maintenance of homeostasis and balanced growth. However, most of the toxigenic dinoflagellates are rather euryhaline species and it is unlikely that salinity fluctuations affect either growth rate or toxin per cell in nature. Re-analysis of the data from White (1978) on the effects of salinity-dependent growth on toxin production in *Alexandrium* confirmed that PSP toxin cell quota in exponential growth phase increases with salinity up to a threshold, but the response is non-linear and not directly correlated to the growth rate (r^2=0.43, see Fig. 6). The decrease in toxin per cell observed in senescence appears to be a function of growth stage and not salinity. For *A. fundyense* there is no systematic variation in Q_t associated with salinity gradients either under long term acclimation or in response to short term changes (Anderson *et al.* 1990a). In a high toxicity clone of *A. tamarense*, the highest cell toxicity occurs at the highest salinity-dependent growth rates during exponential phase (Parkhill and Cembella 1997). In *Gymnodinium catenatum*, there is no increase in toxin content in response to decreased salinity (Flynn *et al.* 1996b), whereas in *Pyrodinium bahamense* the enhanced Q_t at the lowest salinity also coincided with the lowest growth rate (Usup *et al.* 1994). In summary, most of these differences can be explained by differences in acclimation time and the steepness and range of the salinity gradient, i.e., whether cells are subjected to salinity-induced osmotic stress.

6.3. Nutritional factors

All of the dinoflagellates and cyanobacteria implicated in PSP toxin production are ultimately dependent upon the uptake and assimilation of low molecular weight dissolved macronutrients from the surrounding milieu for growth and toxin biosynthesis, although certain cyanobacteria have the additional capability of fixing molecular nitrogen (diazotrophy). The PSP toxins are extremely N-rich compounds (*ca.* 30% by weight) and as much as 5-10% of the total cellular-N may be bound as PSP toxins in highly toxic strains of *A. tamarense* (cf. *excavatum*) (calculated from McIntyre *et al.* 1997), *A. minutum* (Flynn *et al.* 1994) and *Gymnodinium catenatum* (Flynn *et al.* 1996b). In contrast, the PSP toxins contains no phosphorus (P), although P-metabolites may be required as key elements of biosynthetic or regulatory enzymes or energy-rich intermediates (e.g., ATP) to drive toxin biosynthesis. Inorganic nitrogen pools in *Alexandrium* spp. and *Gymnodinium catenatum* are small (Flynn *et al.* 1996a,b) and therefore their depletion might be expected to have a rapid short term effect on toxin production. Therefore, it was originally assumed that toxin biosynthesis and Q_t would be particularly sensitive to changes in intracellular N-status and the availability of external N- and C-sources, but much more refractory to modulation by changes in P-nutrient status. Paradoxically, according to the experimental evidence (Anderson *et al.* 1990b), this does not appear to be the case. It is instructive to consider why not — perhaps the higher storage capacity of cells for intracellular-P can mitigate during periods of P-starvation as long as IFAA pools are of sufficient magnitude to supply toxin biosynthesis.

There is no reason to suspect the existence of fundamentally atypical mechanisms of N-uptake and assimilation in PSP toxin-producing species. For all of the toxigenic dinoflagellates, both NO_3 and NH_4 are suitable N-substrates for growth. However, at much higher concentrations (>200 μM) than are typically found in the natural environment, NH_4 may suppress the growth of certain *Alexandrium* strains (A. Cembella, unpubl. data). The bioenergetic cost of NO_3 reduction is not significant for *Alexandrium tamarense* (cf. *excavatum*) (at saturating irradiance), although the higher Q_t for NH_4- versus NO_3-dependent growth might indicate a limitation of toxin biosynthetic rate resulting from rate limitation of NO_3 reduction (Levasseur *et al.* 1995). Toxin composition is essentially unaffected by the N-growth substrate, indicating that toxin interconversions are not subject to feedback effects from unassimilated intracellular pools of inorganic N.

Intracellular reduction of NO_3 to NO_2 is effected by the enzyme nitrate reductase, with subsequent reduction of NO_2 (by nitrite reductase) to NH_4 (Fig.7). Ammonium ion is assimilated into intracellular amino acids by one of two alternative enzymatic pathways: 1) via glutamate dehydrogenase (GDH) by addition of NH_4 to α-ketoglutarate (α-ketoGLU), or 2) via the combined activity of glutamine synthetase (GS) to form glutamine (GLN) from glutamate (GLU), then glutamine-oxoglutarate amino-transferase (GOGAT) which converts GLN plus α-ketoGLU to GLU. The product of these pathways is the primary source of other amino acids for protein biosynthesis. Both pathways may coexist in an organism and may be alternatively favored depending upon the N-nutrient status and bioenergetic considerations.

The degree to which nutrient limitation plays a role in determining Q_t for natural blooms is uncertain. Batch culture experiments on various *Alexandrium* species and strains have shown dramatically different effects of P- versus N-limitation. In members of the *Alexandrium tamarense/catenella* species complex, Q_t typically increases upon entry into stationary phase under P-limitation (beyond what can be accounted for in the increase in cell volume), whereas the quota declines under N-limitation (Boyer *et al.* 1985, 1987; Anderson, 1990a,b). However, this response to P-limitation is not universal; in *Alexandrium minutum* both N- and P-limitation cause a decline in toxin synthesis and Q_t (Flynn *et al.* 1994; 1995). The decline in Q_t when cells enter stationary phase under N-limitation may be linked to remobilization of toxin-bound N into primary metabolism as chlorophyll levels per cell also drop under these conditions, but the contribution is likely to be small. Significantly, maximal toxin synthesis in *A. minutum* occurs following re-addition of N to N-deficient cells ("upshock response"), after which cell toxin quota covaries with free intracellular ARG (Flynn *et al.* 1994; 1995).

The effect of P-limitation on toxin production has been attributed to the increased availability of ARG, perhaps due to reduced demand from competing P-dependent pathways (Anderson *et al.* 1990a,b). In *A. fundyense*, free ARG varies inversely with

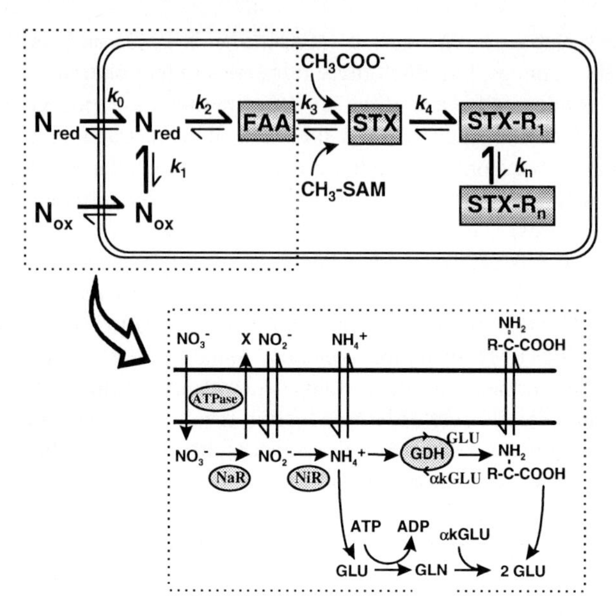

Fig. 7 Regulatory model of PSP toxin production in a hypothetical algal cell showing pathways of N-incorporation.

If $\mu > k_i$, Q_t decreases; if $\mu < k_i$, Q_t increases, whereas if $k_3 \neq k_4 \neq k_n$,a change in toxin ratios occurs. μ = the growth rate; k_i = substrate conversion rate of the rate-limiting enzyme; Q_t = cell toxin quota. FAA = free amino acids; SAM = S-adenosyl-methionine; R = substituent moieties. Other symbols in text.

Q_t (Anderson *et al.* 1990a). But under P-limitation, low levels of ARG indicated high incorporation into toxin synthesis even when cell division was arrested. Since both N- and P-limitation depress amino acid biosynthesis and the size of IFAA pools (particularly ARG) in *A. minutum*, the differential response of *A. fundyense/ catenella/tamarense* versus *A. minutum* to P-limitation may be related to the lack of bioenergetic substrates for cell division rather than amino acid precursors for toxin biosynthesis. There is evidence of a difference in the effect of N- versus P-limitation under quasi-steady state chronic limitation in semi-continuous culture, where the toxin profile was dramatically altered, accompanied by reduced heterogeneity in the toxin composition under the most severe nutrient stress (lowest growth rate) (Anderson *et al.* 1990b). Re-analysis of the data from Anderson *et al.* (1990b) (Fig. 6A) indicates that toxicity per cell increases as a non-linear function of cell division rate under N-limitation ($Y = 4.7 - 44.6X + 187.3X^2$; $r^2 = 0.71$) and decreases non-linearly at low growth rates under P-limitation ($Y = 331.4 - 1313.2X = 1825X^2$; $r^2 = 0.73$).

Amino acid metabolism in *Alexandrium* spp. is somewhat peculiar relative to other algae in that abnormally high ratios of GLN:GLU are attained after N-addition to N-stressed cells (Flynn *et al.* 1996a). However, there is no dramatically discernible pattern in the composition or concentration of particular amino acids (e.g. ARG) among toxic versus non-toxic species or strains which could be attributed to toxin metabolism. It is instructive to note that in toxigenic *G. catenatum* the size of the IFAA pool, ARG content and GLN:GLU remain abnormally low relative to *Alexandrium* spp. and toxin production is only weakly stimulated by N-addition to deprived cells (Flynn *et al.* 1996b). This may indicate that the demand on the IFAA pool for toxin biosynthesis is relatively small.

It is not evident why natural populations of PSP toxin-producing dinoflagellates generally contain a significantly higher Q_t than cultured isolates from the same geographical populations. Levasseur *et al.* (1995) suggested that this may be due to the differential utilization of N-sources — primarily NH_4 in natural ecosystems, whereas NO_3 is most commonly added as the primary substrate in cultures. These nutrient limitation effects on toxin ecophysiology must be considered carefully among different toxigenic organisms, since P is usually considered the primary limiting nutrient in freshwater ecosystems (the habitat of toxigenic *Aphanizomenon flos-aquae* and *Anabaena circinalis*), whereas N is more often growth limiting in coastal marine environments (where blooms of *Alexandrium* spp., *Gymnodinium catenatum* and *Pyrodinium bahamense* var. *compressum* occur).

The effect of organic nutrients on toxin production in phototrophic dinoflagellates has been little studied, but the limited evidence suggests that heterotrophy does not play a major role. The results are difficult to uncouple from the stimulus to growth; cell toxin yields of *Alexandrium* spp. are higher in axenic than bacterized cultures (Singh *et al.* 1982; Ogata *et al.* 1996), but higher growth rates in the presence of bacteria suggest that they may contribute to "toxin dilution" through increasing the rate of cell division. Nevertheless, the depression of cell toxicity in N-limited semi-continuous cultures can be relieved by additions of organic extracts, including live or autoclaved bacteria (Ogata *et al.* 1996). *Alexandrium tamarense* has some ability to take up dissolved free amino acids (DFAA) at environmentally realistic levels (nM) during exponential growth, but this is not apparently linked to increased toxin production even during "force feeding" at high levels (John and Flynn 1997). *Alexandrium tamarense* (cf. *excavatum*) was not able to utilize urea as a growth substrate, indicating the lack of urease activity in this species (Levasseur *et al.* 1995).

There is little information on the specific trace metal requirements for toxin biosynthesis. One study has shown that *Alexandrium tamarense* cells driven into Fe-limitation have only a slightly higher (perhaps not significant) toxin content than control cultures in early stationary phase (Boyer *et al.* 1985). Any effect is probably indirect, through interference with the photosynthetic apparatus, including demands for Fe-S proteins, ferredoxin, and cytochromes.

7. Alternate life history stages

Difficulties in inducing sexuality in PSP toxin producing species and recognizing and segregating the various cell types for independent analysis have greatly impeded the determination of toxin content and composition among alternative life history stages. Early reports of extreme toxicity in *Alexandrium* hypnozygotes (more than an order of magnitude greater than vegetative cells) (Dale *et al.* 1978; Oshima *et al.* 1982), must now be viewed with skepticism. For example, a comparison was made between fractionated bulk sediments rich in *Alexandrium* cysts versus cultured vegetative cells isolated for the same general geographical area of the Gulf of Maine (Dale *et al.* 1978). In addition to the potential artifacts possible in the injection of sediment extracts, cultured cells are typically more toxic than vegetative cells from natural populations. Others have considered cysts to be of approximately equal toxicity to vegetative cells (White and Lewis 1982; White 1986) or even less toxic (Onoue *et al.* 1981). Although some discrepancies may arise through strain-specific differences and variation in cellular metabolic status, in certain studies (Dale *et al.* 1978; White and Lewis 1982; Oshima *et al.* 1982) potentially misleading comparisons have been made between natural benthic cyst populations and vegetative cells from cultured isolates. It has also been assumed that the toxicity of cultures from ecdysed cysts was typical of natural cyst populations (Onoue *et al.* 1981).

A comparison of the toxicity and toxin spectrum of natural populations of motile cells from the water column and benthic hypnozygotes from the lower St. Lawrence estuary, Quebec showed that hypnozygotes contained less toxin on a molar basis than vegetative cells, but the difference in toxicity was much less, given the higher specific toxicity of the toxin components found in hypnozygotes. (Cembella *et al.* 1990). Significantly, hypnozygotes harvested in early spring had a less diverse toxin spectrum and were relatively richer in carbamate derivatives (STX and GTX2), than vegetative cells from cultures or natural populations. The reduced toxin spectrum and relative enrichment in the β-epimer (GTX2) over the α-epimer (GTX3) in hypnozygotes presumably indicates that low rate of metabolism and biosynthetic activity in overwintering cysts. Oshima *et al.* (1992) also found a high ratio of β-: α-epimers in hypnozygotes originating from Japan - but not a reduced toxin profile. The latter observation may be related to the relative age or state of dormancy in these different populations. High levels of ARG found in freshly-formed hypnocysts of *A. catenella* in culture (Lirdwitayaprasit *et al.* 1990) is circumstantial evidence for reduced incorporation into toxins during early quiescence.

Analysis of toxin content in fractionated cells of alternative life history stages from stationary phase cultures of *A. tamarense* (cf. *excavatum*) — small and large vegetative cells, putative gametes, pellicular cysts and a few planozygotes — revealed a relatively homogeneous toxin spectrum (Cembella *et al.* 1990). However, vegetative cells had a higher Q_t (and per unit cell volume) than pellicular cysts and fractions rich in putative gametes and planozygotes. This co-existence of multiple morphological, physiological, and life-history variants with different cell toxin quotas in stationary phase cultures is a significant observation, but it is not known if these results can be extrapolated to field populations.

8. Conclusions on the eco-evolutionary role of PSP toxins

Due to the lack of evidence of their involvement in known pathways of primary or intermediary metabolism, the PSP toxins have been deemed to be secondary metabolites. Non-toxigenic strains have not been shown to spontaneously acquire toxicity and although environmental manipulations can depress or enhance toxin production, it is not usually possible to completely suppress toxicity in strains capable of expressing this trait. Hence the PSP toxins in dinoflagellates and cyanobacteria do not follow the classic pattern of most secondary metabolites in lower organisms (bacteria and fungi) — induction of synthesis only under stress conditions or during a particular phase of the life history.

The eco-evolutionary role of these compounds has engendered considerable speculation but there is little hard evidence. The bacterial endosymbiotic hypothesis is discussed in detail by Doucette *et al.* (this volume). It is sufficient to note here that this hypothesis is consistent with the possible polyphyletic origin of the toxin biosynthetic genes. In any case, regardless of the origin of the biosynthetic genes, either derived via evolution from a common ancestor or as a secondarily acquired characteristic, there must be a selective advantage for retention of this complex and metabolically expensive pathway.

The fact that PSP toxins are N-rich has led to speculation that these compounds serve as N-reserves for remobilization under N-limitation, shunts to reduce NH_4 toxicity in eutrophic environments, or even (rather whimsically) as a mechanism to deposit excess N to the sediments (cited in Wyatt and Jenkinson 1997). These nutrient arguments are difficult to sustain since only a relatively small fraction of cell-N (<10%) is bound in the form of toxins even in the most toxic strains. It is also difficult to conceive why this could not be accomplished in a more bioenergetically efficient fashion by synthesizing or storing less complex low molecular weight forms (free amino acids, inorganic N, urea, small peptides).

This structural affinity with the purines led to early suggestions that PSP toxins may represent the product of a relict or archaic pathway to nucleic acid biosynthesis — dinoflagellates are recognized as a very ancient group of protists, and cyanobacteria are even older. However, the biosynthetic evidence does not support the hypothetical relationship to pathways of purine biosynthesis (Shimizu 1996).

There is a resemblance between the purine guanine, which is a major structural element in the scintillons associated with bioluminescence in dinoflagellates, and the PSP toxins. There are bioluminescent strains within each genus of PSP toxin-producing dinoflagellates. However, there is no apparent correlation between bioluminescence and toxin production or strain potency (Anderson *et al.* 1994), therefore the superficial analogy of PSP toxins with guanine may be coincidental.

The typical dinoflagellate chromosome contains highly condensed chromosomes with a high level of divalent cations (Kearns and Sigee 1979), but no nucleosomes or histones. The immunolocalization of STX within the nucleus in close proximity to the chromosomes (Anderson and Cheng 1988) offers the intriguing possibility that the toxins may serve a primary functional role in chromosome structural organization — in much the same way as polyamines or cations. Unfortunately, this hypothesis begs the question of how this is accomplished in non-toxigenic strains.

The most often touted hypothesis for toxin production is for chemical defence, as kairomones or allelochemical agents against zooplankton predators, such as copepods and tintinnids — after all, these compounds are potent Na^+ channel blockers. To have a selective advantage, this toxic effect would have to be expressed directly against key predators by deterring grazing or reducing reproductive viability. There is some evidence of grazing deterrence in a few species of macrozooplankton (see Turner *et al.*, this volume) and microzooplankton (see Hansen, this volume), but the response is weak (relative to vertebrates) and equivocal or non-existent for many species.

By virtue of their ion-channel blocking activity, PSP toxins could conceivably play a role as channel effectors or ion transporter systems across the cytoplasmic or nuclear membranes. The highly charged quanidinium groups of STX might even serve as surface recognition sites involving in cell mating responses, but there is no evidence for these speculations.

Finally, the most recent suggestion (and one which explains the greatest number of facts) is that PSP toxins may serve as pheromones (Wyatt and Jenkinson 1997). As such, the pattern of release of toxins in senescence via leakage, excretion or cell lysis would make sense in terms of the timing of induction of sexuality during bloom decline. The pheromone hypothesis could also explain why there is such a wide diversity in toxin composition (number of components) and why differences are noted among geographical populations. A complex battery of such compounds determined by a genetically fixed biosynthetic regime might be required to ensure fidelity of mating groups.

Future research should be targeted towards the identification of the genes required for toxin biosynthesis and the means by which they are regulated. Once this task is accomplished this should yield new insights into the functional significance of PSP toxins in the producing organisms.

9. Acknowledgments

The author is grateful to Nancy Lewis, Andrew Bauder, and Youlian Pan, Institute for Marine Biosciences, NRC for assistance in preparing the graphics and for data interpretation. This publication is NRC (Canada) No. 39770.

10. References

Anderson, D.M., Cheng, T.P.O. (1988). Intracellular localization of saxitoxins in the dinoflagellate *Gonyaulax tamarensis*. *J. Phycol.* 24: 17-22.

Anderson, D.M., Kulis, D.M., Sullivan, J.J., Hall, S. (1990b). Toxin composition variations of the dinoflagellate *Alexandrium fundyense*. *Toxicon* 28: 885-893.

Anderson, D.M., Kulis, D.M., Sullivan, J.J., Hall, S., Lee, C. (1990a). Dynamics and physiology of saxitoxin production by the dinoflagellates *Alexandrium* spp. *Mar. Biol.* 104: 511-524.

Anderson, D.M., Kulis, D.M., Doucette, G.J., Gallagher, J.C., Balech, E. (1994). Biogeography of toxic dinoflagellates in the genus *Alexandrium* from the northeastern United States and Canada. *Mar. Biol.* 120: 467-478.

Balech, E. (1985). The genus *Alexandrium* or *Gonyaulax* of the tamarensis group. In: Anderson, D.M., White, A.W., Baden, D.G. (eds.) *Toxic Dinoflagellates,* Elsevier, N.Y., pp. 33-38.

Boczar, B.A., Beitler, M.K., Liston, J., Sullivan, J.J., Cattolico, R.A. (1988). Paralytic shellfish toxins in *Protogonyaulax tamarensis* and *Protogonyaulax catenella* in axenic culture. *Plant. Physiol.* 88: 1285-1290.

Boyer, G.L., Sullivan, J.J., Andersen, R.J., Harrison, P.J., Taylor, F.J.R. (1985). Toxin production in three isolates of *Protogonyaulax* sp. In: Anderson, D.M., White, A.W., Baden, D.G. (eds.) *Toxic Dinoflagellates,* Elsevier, N.Y., pp. 281-286.

Boyer, G.L., Sullivan, J.J., Anderson, R.J., Harrison, P.J., Taylor, F.J.R. (1987). Effects of nutrient limitation on toxin production and composition in the marine dinoflagellate *Protogonyaulax tamarensis*. *Mar. Biol.* 96: 123-128.

Cembella, A.D., Destombe, C. (1996). Genetic differentiation among *Alexandrium* populations from eastern Canada. In: Yasumoto, T., Oshima, Y., Fukuyo, Y. (eds.) *Harmful and Toxic Algal Blooms*, IOC (UNESCO), Paris, pp. 447-450.

Cembella, A.D., Taylor, F.J.R. (1985). Biochemical variability within the *Protogonyaulax tamarensis/catenella* species complex. In: Anderson, D.M., White, A.W., Baden, D.G. (eds.) *Toxic Dinoflagellates,* Elsevier, N.Y., pp. 55-60.

Cembella, A.D., Destombe, C., Turgeon, J. (1990). Toxin composition of alternative life history stages of *Alexandrium excavatum*, as determined by high-performance liquid chromatography. In: Granéli, E., Sundström, B., Edler, L., Anderson, D.M. (eds.) *Toxic Marine Phytoplankton*, Elsevier, N.Y., pp. 333-338.

Cembella, A.D., Therriault, J.C., Béland, P. (1988). Toxicity of cultured isolates and natural populations of *Protogonyaulax tamarensis* from the St. Lawrence Estuary. *J. Shellfish Res.* 7: 611-621.

Cembella, A.D., Sullivan, J.J., Boyer, G.L., Taylor, F.J.R., Andersen, R.J. (1987). Variation in paralytic shellfish toxin composition within the *Protogonyaulax tamarensis/catenella* species complex. *Biochem. Syst. Ecol.* 15: 171-186.

Chang, F.H., Anderson, D.M., Kulis, D.M., Till, D.G. (1997). Toxin production of *Alexandrium minutum* (Dinophyceae) from the Bay of Plenty, New Zealand. *Toxicon* 35: 393-409.

Dale, B., Yentsch, C.M., Hurst, J.W. (1978). Toxicity in resting cysts of the red tide dinoflagellate *Gonyaulax excavatum* from deeper water coastal sediments. *Science* 201: 1223-1224.

Flynn, K., Flynn, K.J., Jones, K.J. (1993). Intracellular amino acids in dinoflagellates; effects of diurnal changes in light and of N-supply. *Mar. Ecol. Prog. Ser.* 100: 245-252.

Flynn, K., Jones, K.J., Flynn, K.J. (1996a). Comparisons among species of *Alexandrium* (Dinophyceae) grown in nitrogen- or phosphorus-limiting batch culture. *Mar. Biol.* 126: 9-18.

Flynn, K., Franco, J., Fernández P., Reguera, B., Zapata, M., Flynn, K.J. (1995). Nitrogen and phosphorus limitation in cultured *Alexandrium minutum* does not promote toxin production. In: Lassus, P., Arzul, G., Erard-Le Denn, E., Gentien, P., Marcaillou-Le Baut, C. (eds.) *Harmful Marine Algal Blooms*, Lavoisier, Paris, pp. 439-444.

Flynn, K.J., Flynn, K., John, E.H., Reguera, B., Reyero, M.I., Franco, J.M. (1996b). Changes in toxins, intracellular and dissolved free amino acids of the toxic dinoflagellate *Gymnodinium catenatum* in response to changes in inorganic nutrients and salinity. *J. Plank. Res.* 18: 2093-2111.

Flynn, K., Franco, J.M., Fernández, P., Reguera, B., Zapata, M., Wood, G., Flynn, K. (1994). Changes in toxin content, biomass and pigments of the dinoflagellate *Alexandrium minutum* during nitrogen refeeding and growth into nitrogen or phosphorus stress. *Mar. Ecol. Prog. Ser.* 111: 99-109.

Franco, J.M., Fernandez, P., Reguera, B. (1994). Toxin profiles of natural populations and cultures of *Alexandrium minutum* Halim from Galician (Spain) coastal waters. *J. App. Phycol.* 6: 275-279.

Glibert, P.M., Kana, T.M., Anderson, D.M. (1988). Photosynthetic response of *Gonyaulax tamarensis* during growth in a natural bloom and in batch culture. *Mar. Ecol. Prog. Ser.* 42: 303-309.

Hall, S. (1982). Toxins and toxicity of *Protogonyaulax* from the Northeast Pacific. Ph.D. Dissertation, Univ. Alaska, Fairbanks. 196 pp.

John, E.H., Flynn, K.J. (1997). Amino acid uptake by *Alexandrium tamarense.* In: *Proceedings of the VIII Int. Conf. on Harmful Algae*, Vigo, Spain, in press.

Kim, C.H., Sako, Y., Ishida, Y. (1993). Comparison of toxin composition between populations of *Alexandrium* spp. from geographical distant areas. *Nipp. Suis. Gakk.* 59: 641-646.

Kearns, L.P., Sigee, D.C. (1979). High levels of transition metals in dinoflagellate chromosomes. *Experientia* 35: 1332-1334.

Langdon, C. (1987). On the cause of interspecific differences in the growth-irradiance relationship of phytoplankton. Part 1. A comparative study of the growth irradiance relationship of three marine phytoplankton species: *Skeletonema costatum*, *Olisthodiscus luteus* and *Gonyaulax tamarensis*. *J. Plank. Res.* 9: 459-482.

Levasseur, M., Gamache, T., St.-Pierre, I., Michaud, S. (1995). Does the cost of NO_3-reduction affect the production of harmful compounds by *Alexandrium excavatum*? In: Lassus, P., Arzul, G., Erard-Le Denn, E., Gentien, P., Marcaillou-Le Baut, C. (eds.) *Harmful Marine Algal Blooms*, Lavoisier, Paris, pp. 463-468.

Lirdwitayaprasit, T., Nishio, S., Montani, S., Okaichi, T. (1990). The biochemical processes in cyst formation in *Alexandrium catenella*. In: Granéli, E., Sundström, B., Edler, L., Anderson, D.M. (eds.) *Toxic Marine Phytoplankton*, Elsevier, N.Y., pp. 294-299.

McIntyre, J.G., Cullen, J.J., Cembella, A.D. (1997). Vertical migration, nutrition and toxicity in the dinoflagellate *Alexandrium tamarense*. *Mar. Ecol. Prog. Ser.* 148: 201-216.

MacKenzie, L., White, D., Oshima, Y., Kapa, J. (1997). The resting cyst and toxicity of *Alexandrium ostenfeldii* (Dinophyceae) in New Zealand. *Phycologia*, in press.

Maranda, L., Anderson, D.M., Shimizu, Y. (1985). Comparison of toxicity between populations of *Gonyaulax tamarensis* of eastern North American waters. *Est. coast. shelf Sci.* 21:401-410.

Morey-Gaines, G. (1982). *Gymnodinium catenatum* Graham (Dinophyceae): morphology and affinities with armoured forms. *Phycologia* 21: 154-163.

Ogata T., Ishimaru, T., Kodama, M. (1987b). Effect of water temperature and light intensity on growth rate and toxicity change in *Protogonyaulax tamarensis*. *Mar. Biol.* 95: 217-220.

Ogata, T., Kodama, M., Ishimaru, T. (1987a). Toxin production in the dinoflagellate *Protogonyaulax tamarensis*. *Toxicon* 25: 923-928.

Ogata, T., Koike, K., Nomura, S., Kodama, M. (1996). Utilization of organic substances for growth and toxin production by *Alexandrium tamarense*. In: Yasumoto, T., Oshima, Y., Fukuyo, Y. (eds.) *Harmful and Toxic Algal Blooms*, IOC (UNESCO), Paris, pp. 343-346.

Onoue, Y., Noguchi, T., Maruyama, J., Ueda, Y., Hashimoto, K., Ikeda, T. (1981). Comparison of PSP compositions between toxic oysters and *Protogonyaulax catenella* from Senzaki Bay, Yamaguchi Prefecture. *Bull. Japan. Soc. Sci. Fish.* 47: 1347-1350.

Oshima, Y., Blackburn, S.I., Hallegraeff, G.M. (1993). Comparative study on paralytic shellfish toxin profiles of the dinoflagellate *Gymnodinium catenatum* from three different countries. *Mar. Biol.* 116: 471-476.

Oshima, Y., Bolch, C.J., Hallegraeff, G.M. (1992). Toxin composition of resting cysts of *Alexandrium tamarense* (Dinophyceae). *Toxicon* 30: 1539-1544.

Oshima, Y., Kotaki, Y., Harada, T., Yasumoto, T. (1984). Paralytic shellfish toxins in tropical waters. In: Ragelis, E. (ed.) *Seafood Toxins,* Am. Chem. Soc., Wash. D.C. pp. 160-170.

Oshima, Y., Singh, H.T., Fukuyo, Y., Yasumoto, T. (1982). Identification and toxicity of the resting cysts of *Protogonyaulax* found in Ofunato Bay. *Bull. Japan. Soc. Sci. Fish.* 48: 1303-1305.

Oshima, Y., Yasumoto, T., Hallegraeff, G., Blackburn, S. (1987). Paralytic shellfish toxins and causative organisms in the tropical Pacific and Tasmanian waters. In:

Progress in Venom and Toxin Research, Univ. of Singapore, Singapore, pp. 423-428.

Oshima, Y., Sugino, K., Itakura, H., Hirota, M., Yasumoto, T. (1990). Comparative studies on paralytic shellfish toxin profile of dinoflagellates and bivalves In: Granéli, E., Sundström, B., Edler, L., Anderson, D.M. (eds.) *Toxic Marine Phytoplankton*, Elsevier, N.Y., pp. 391-396.

Oshima, Y., Itakura, H., Lee, K.C., Yasumoto, T., Blackburn, S., Hallegraeff, G. (1993). Toxin production by the dinoflagellate *Gymnodinium catenatum.* In: Smayda, T.J., Shimuzu, Y. (eds.) *Toxic Phytoplankton Blooms in the Sea*, Elsevier, N.Y., pp. 907-912.

Parkhill, J.P., Cembella, A.D. (1997). The effects of salinity, light and inorganic nitrogen on the growth and toxigenicity of the marine dinoflagellate *Alexandrium tamarense. J. Plank. Res.*, submitted.

Proctor, N.H., Chan, S.L., Trevor, A.J. (1975). Production of saxitoxin by cultures of *Gonyaulax catenella. Toxicon* 13: 1-9.

Rees, A.J.J., Hallegraeff, G.M. (1991). Ultrastructure of the toxic, chain-forming dinoflagellate *Gymnodinium catenatum* (Dinophyceae). *Phycologia* 30: 90-105.

Shimizu, Y. (1996). Microalgal metabolites: a new perspective. *Ann. Rev. Microbiol.* 50: 431-465.

Shimizu, Y., Gupta, S., Norte, M., Hori, A., Genenah, A., Kobayashi, M. (1985). Biosynthesis of paralytic shellfish toxins. In: Anderson, D.M., White, A.W., Baden, D.G. (eds.), *Toxic Dinoflagellates,* Elsevier, N.Y., pp. 271-274.

Singh, H.T., Oshima, Y., Yasumoto, T. (1982). Growth and toxicity of *Protogonyaulax tamarensis* in axenic culture. *Bull. Jap. Soc. Sci. Fish.* 48: 1341-1343.

Taroncher-Oldenburg, G., Kulis, D.M., Anderson, D.M. (1997). Toxin variability during the cell cycle of the dinoflagellate *Alexandrium fundyense. Limnol. Oceanogr.*, in press.

Taylor, F.J.R. (1987). *The Biology of Dinoflagellates.* Blackwell Scientific, Oxford, England. 785 pp.

Usup, G., Kulis, D.M., Anderson, D.M. (1994). Growth and toxin production of the toxic dinoflagellate *Pyrodinium bahamense* var. *compressum* in laboratory cultures. *Nat. Tox.* 2: 254-262.

White, A.W. (1978). Salinity effects on growth and toxin content of *Gonyaulax excavata*, a marine dinoflagellate causing paralytic shellfish poisoning. *J. Phycol.* 14: 475-479.

White, A.W. (1986). High toxin content in the dinoflagellate *Gonyaulax excavata* in nature. *Toxicon* 24: 605-610.

White, A.W., Lewis, C.M. (1982). Resting cysts of the toxic, red tide dinoflagellate *Gonyaulax excavatum* in the Bay of Fundy. *Can. J. Fish. Aquat. Sci.* 39: 1185-1194.

Wyatt, T., Jenkinson, I.R. (1997). Notes on *Alexandrium* population dynamics. *J. Plank. Res.* 19: 551-575.

Zardoya, R., Costas, E., López-Rodas, V., Garrido-Pertierra, A., Bautista, J.M. (1995). Revised dinoflagellate phylogeny inferred from molecular analysis of large-subunit ribosomal RNA gene sequences. *J. Mol. Evol.* 41: 637-645.

Ecophysiology and Metabolism of ASP Toxin Production

Stephen S. Bates

Fisheries and Oceans Canada, Gulf Fisheries Centre, P.O. Box 5030, Moncton, NB, Canada E1C 9B6

1 Introduction

Amnesic Shellfish Poisoning (ASP), caused by the neurotoxin domoic acid (DA), was first discovered as a result of human consumption of contaminated blue mussels (*Mytilus edulis*) from Prince Edward Island (PEI), eastern Canada in late 1987 (see Bates *et al.*, this volume). The source of the toxin was traced to the pennate diatom *Pseudo-nitzschia multiseries*, and this was the first known instance of a diatom producing a neurotoxin. Since then, other DA-producing *Pseudo-nitzschia* species have been discovered in other parts of the world. Their world-wide distribution poses a threat to human health and to the aquaculture industry. There is therefore a need to understand its bloom dynamics (see Bates *et al.*, this volume) and the ecophysiological factors that control *Pseudo-nitzschia* growth and the production of DA.

2 The Physiology of *Pseudo-nitzschia* Species in Culture

In evaluating the physiology of *Pseudo-nitzschia* species in culture, it is essential to consider clonal differences. This aspect is often neglected in studies of other algae, but in the case of *Pseudo-nitzschia* there have been many isolates from different locations and times, so that it is possible to compare clones or strains. The enigma of the presence of non-toxic strains of *P. multiseries*, *P. seriata*, *P. australis*, *P. delicatissma*, and *P. pseudodelicatissima,* and of toxic strains of *P. pungens* is discussed by Bates *et al.* (this volume). However, there are also examples of variability in DA concentration in different clones growing under presumably the same conditions (Bates *et al.* 1989; Reap 1991). This may be accounted for by genetic differences and/or the bacterial composition (see Section 2.1.9). Yet to be examined in detail is the change in cell physiology that may accompany: a) cell deformities that tend to appear after a certain period in culture, b) the decrease in diatom cell size due to vegetative division, or c) sexual reproduction in mixtures of clonal cultures (see Bates *et al.*, this volume).

Of the DA-producing *Pseudo-nitzschia* species, *P. multiseries* has thus far been known the longest, is found in the most locations, and has been studied the most extensively in laboratory culture. The paucity of laboratory studies on other DA-producing *Pseudo-nitzschia* species makes comparisons with *P. multiseries* difficult, with the exception of one study of *P. australis* (Garrison *et al.* 1992) and one of *P. seriata* (Lundholm *et al.* 1994). This has led to certain beliefs, e.g., regarding the timing of DA production in relation to the growth phase in batch culture, that have been generalized from *P. multiseries* to the other species, without prior testing. Clearly, parallel experiments are required with the remaining confirmed, plus any new, DA-producing species. Nevertheless, the numerous studies with *P. multiseries* do allow a comparison among strains isolated from different geographic locations and at different times of the year. As well, certain comparisons may be made with *P. pungens*, once thought to be a different form of the same species (Hasle 1995).

NATO ASI Series, Vol. G 41
Physiological Ecology of Harmful Algal Blooms
Edited by D. M. Anderson, A. D. Cembella and G. M. Hallegraeff

2.1 Factors Influencing Domoic Acid Production

2.1.1 Phase in the Growth Cycle of Batch Culture

After its isolation into culture, the very first experiment with *P. multiseries* showed that it did not produce DA during the exponential growth phase (Bates *et al.* 1989). Although initially disheartening because of the desire to have a toxigenic species for study, toxin production become evident when stationary phase was reached (Fig. 1), in

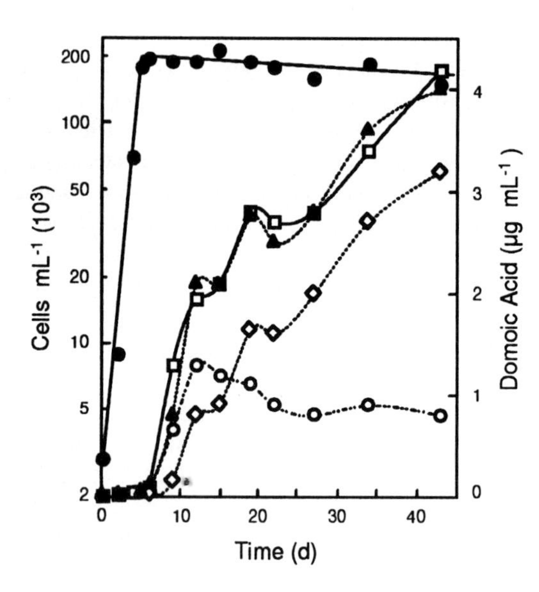

Fig. 1. Growth and domoic acid production by *P. multiseries* in batch culture. Cell number (●); domoic acid in "whole culture" (cells sonicated with growth medium) (■); in cells plus filtrate (▲); in filtrate (◇); and in cells (○). Redrawn from Bates *et al.* 1991.

this case due to silicate limitation. These findings of post-exponential phase production of DA have been confirmed by Subba Rao *et al.* (1990), Reap (1991), Bates *et al.* (1991; 1993; 1995), Wohlgeschaffen *et al.* (1992), Douglas and Bates (1992), Douglas *et al.* (1993), and Lewis *et al.* (1993). If a late-stationary phase culture is used to inoculate a culture flask, these cells are still capable of producing DA during lag phase (Douglas and Bates 1992; Douglas *et al.* 1993). Because the cells already contain DA at the start of the experiment, one observes a decline in DA per cell during exponential growth, as intracellular DA becomes "diluted" into the newly-produced daughter cells (Douglas and Bates 1992; Douglas *et al.* 1993; Pan *et al.* 1996a). Cells in mid-exponential phase, which do not yet produce the toxin, should therefore be used as an inoculum, in order to interpret the dynamics of DA production with less confusion.

The above has led to the conclusion that one condition necessary for DA production by *P. multiseries* is the cessation of cell division (Bates *et al.* 1991). However, this appears to be true only for nitrate as the N source (Bates *et al.* 1993), and only for the mid-exponential phase of batch culture; small amounts of DA are also produced during late-exponential phase (Bates *et al.* 1991; Reap 1991; Pan *et al.* 1996a).

There are at least two interpretations for pre-stationary phase DA production. First, only those cells that have initially stopped dividing may be capable of producing the toxin. The transition from exponential to stationary phase is not always clear-cut, and this period consists of a mixture of physiologically distinct cells. Some cells have stopped dividing, while others have slowed their division due to nutrient limitation in batch culture. Alternatively, this transition period may reflect a general slowing of the division rate of the entire population due to limitation by a specific nutrient or other factor. In this latter interpretation, there is a threshold, but non-zero, division rate at which DA production starts. Pan *et al.* (1996a) referred to a "first stage", with low level DA production occurring during the mid- to late-exponential phase when division rate starts to decline, followed by a "second stage", where the DA production rate is higher and Si becomes depleted. However, the "first stage" is difficult to distinguish in the data presented. The above studies point out that there are at least two different, but perhaps related, triggering mechanisms for DA production: a slowing of division rate, and a depletion of Si or P (see Sections 2.1.2 and 2.1.4). Presently, the exact triggering mechanism(s) for toxin production can only be understood at a descriptive, not a biochemical level (see Section 2.2). Furthermore, studies should look in more detail for the possibility of DA production by cells in the mid-exponential phase. For example, growth of *P. multiseries* in 440 μM of ammonium, a toxic concentration which slows cell division, results in pre-stationary phase production of DA (Bates *et al.* 1993). Other growth conditions should be similarly tested.

Time-course studies in batch culture show that cellular DA reaches a maximum about one week after the beginning of stationary phase, and slowly declines thereafter (Fig. 1). The *P. multiseries* cells then release increasing amounts of DA into the medium as the culture ages during stationary phase (Bates *et al.* 1991; Pan *et al.* 1996a; c). In the case of P-limited cells, at least, this may be caused by improperly formed cell membranes as a result of the reduced supply of P to form the phospholipid bilayer (Pan *et al.* 1996c). However, cell integrity is apparently also compromised in Si-limited cells. The extra effort required to take samples for cellular DA, in addition to filtrate and "whole" (i.e., cells plus filtrate) samples is worth the supplemental information gained (alternatively, the data are obtained "by difference"). It must be noted that cells can leak DA if net tow samples, rather than membrane-filtered samples, are taken in the field for determination of toxin content.

Only one study each of *P. australis* and *P. seriata* has examined toxin production in relation to the growth cycle in batch culture. In two strains of *P. australis* (Fig. 2), DA production began during exponential phase and remained constant or decrease slightly

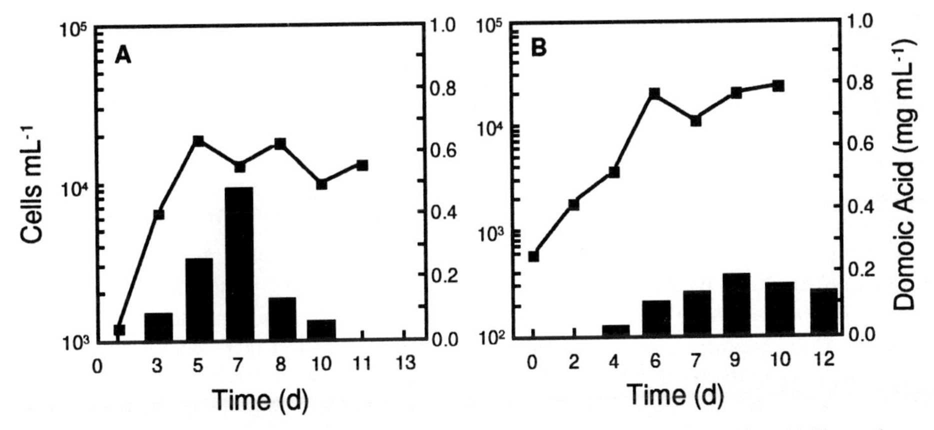

Fig. 2. Cell number (■) and domoic acid concentration (bars) in two strains A) Doma-1 and B) Doma-2 of *P. australis*. Redrawn from Garrison *et al.* 1992.

during stationary phase (Garrison *et al.* 1992), in contrast to *P. multiseries*. For *P. seriata*, traces of DA were found during exponential phase, but most toxin was produced during stationary phase (Fig. 3) (Lundholm *et al.* 1994), as for *P. multiseries*. These examples illustrate the importance of carrying out time-course experiments, in

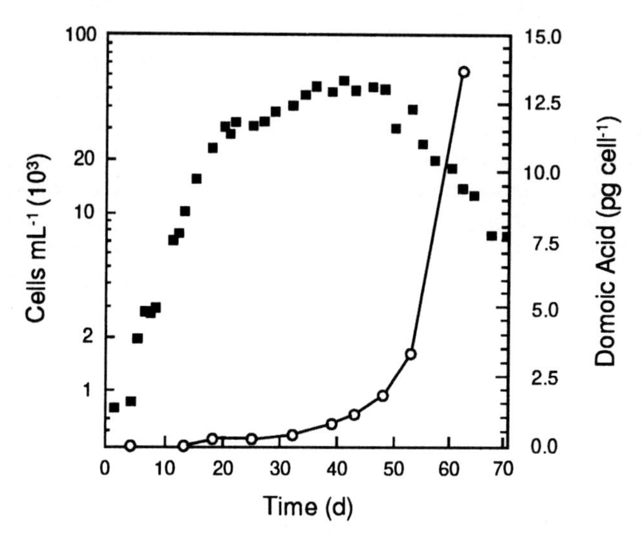

Fig. 3. Cell number (■) and domoic acid in cells plus medium (**O**) of *P. seriata* (strain 1877C) grown at 4°C. Redrawn from Lundholm *et al.* 1994.

which samples are taken regularly (e.g., at least every other day) during both the exponential and stationary phases of growth, in order to adequately describe the kinetics of DA production. This also allows one to determine, under the prevailing conditions, the rate of toxin production and maximum toxin content per cell (Table 1).

2.1.2 Silicon

The growth medium f/2, or one of its derivatives, was used to carry out all of the initial culture studies with *P. multiseries* (see Subba Rao *et al.* 1988). A comparison of the N:P:Si ratio in medium f/2 and in diatoms suggests that Si, rather than N or P, limits the cell yield of diatoms grown in batch culture. This was verified experimentally in Si-addition experiments (Bates *et al.* 1991; Pan *et al.* 1991). Because DA production became pronounced during stationary phase, in this case initiated by Si limitation in medium f/2, Si was suspected of playing a role in controlling DA production by *P. multiseries*. Indeed, decreasing the initial concentration of Si in the growth medium resulted in higher cellular levels of DA at stationary phase (Bates *et al.* 1991). This was more fully explored by Pan *et al.* (1996a), who showed that the production of DA by *P. multiseries* was inversely correlated with the ambient silicate concentration in batch culture. Cells began accumulating DA only when the division rate declined as a result of partial or total depletion of Si.

Presently, one may only hypothesize what role Si may play in toxin production. Diatoms require Si not only for frustule formation but also for DNA synthesis (e.g., Darley and Volcani 1969). Other studies, using flow cytometry, have provided insights

Table 1. Cellular domoic acid (pg $cell^{-1}$) and maximum rate of domoic acid production (pg DA $cell^{-1}$ d^{-1}) by *Pseudo-nitzschia* species in culture, measured at the specified temperature and day.

Species	Temp. (°C)	Day	Cellular DA	Rate	Reference
P. multiseries	10	68	0.8	n.a.[1]	Subba Rao *et al.* 1988
	10	21	21.0	n.a.	Bates *et al.* 1989
	10	42	2.1	2.0	Subba Rao *et al.* 1990
	17	40	9.5	n.a.	Bates *et al.* 1991
	20	28	12.0	n.a.	Reap 1991
	15	15	2.0	0.5	Douglas and Bates 1992
	10-15	11	2.0	n.a.	Wohlgeschaffen *et al.* 1992
	17	43	8.7	n.a.	Bates *et al.* 1993
	5	30	0.17	0.01	Lewis *et al.* 1993
	20	30	1.5	0.17	Lewis *et al.* 1993
	25	30	11.4	0.50	Lewis *et al.* 1993
	15	49	0.1	n.a.	Wang *et al.* 1993
	17	14	4.25	n.a	Whyte *et al.* 1995a
	17	14	5.25	n.a	Whyte *et al.* 1995b
	15	50	0.30	n.a.	Pan *et al.* 1996a
	15	Ch[2]	5.62	1.35	Pan *et al.* 1996b
	15	Ch[2]	2.21	0.26	Pan *et al.* 1996c
	14	19	12.4	n.a.	Bates *et al.* 1996
	16	55	19.0	n.a.	Vrieling *et al.* 1996
	18	15	1.03	n.a.	Lee and Baik 1997
P. australis	15	5-11	12-37	n.a.	Garrison *et al.* 1992
	18	n.a.	2.0	n.a.	Rhodes *et al.* 1966
P. turgidula	18	n.a.	0.03	n.a.	Rhodes *et al.* 1966
P. pungens[3]	18	n.a.	0.47	n.a.	Rhodes *et al.* 1966
P. seriata	4	60	1.0-33.6	n.a.	Lundholm *et al.* 1994
	15	60	0.31-1.6	n.a.	Lundholm *et al.* 1994

[1]n.a. = not available [2]Ch = chemostat culture [3]see Bates *et al.*, this volume

into the links between Si metabolism and the cell cycle, which consists of a period of cell division (D), separated from the period of DNA replication (S) by two gap phases (G1 and G2). Silicon deprivation in diatoms alters the normal progression through the cell cycle by arresting the cells at the G1/S boundary and in G2 and/or M phases (Brzezinski *et al.* 1990). These are also the parts of the cell cycle during which Si transport can take place, enabling them to take advantage of any Si encountered in the environment. As the cells become more Si-limited, the G1 or G2 phase lengthens and their division rate consequently slows down.

Domoic acid production may start during the G2 phase (Laflamme 1993; Pan *et al.* 1996a), or at the end of the G1 phase, just prior to DNA synthesis (Bates and Richard 1996); this needs to be defined more precisely. Cells could be arrested at the end of the G1 phase because DNA synthesis would be prevented by Si deficiency in the growth medium. In batch culture, all of the stationary phase cells would be permanently stopped at, say, the end of the G1 phase, and DA production could proceed maximally. Re-addition of Si permits the cells to continue progressing through the cell cycle, and DA production would slow or cease.

The possible relationship between DA production and DNA synthesis was tested by growing *P. multiseries* in a Si-limited chemostat culture with a 12:12 h light:dark cycle to partially synchronize cell division (Bates and Richard 1996). Cell division occurred during the light period (Fig. 4A), and coincided with the period of lowered DA production (Fig. 4B) and increased Si uptake (Fig. 4C), as expected. The concentration of cellular DNA also generally decreased during the light period, in agreement with the partitioning of the DNA into daughter cells during cell division. Near the end of the light period, therefore, the concentration of Si in the growth medium is lowest, as is the cellular DNA concentration. This may represent the end of the G1 phase of the cell cycle, just prior to DNA synthesis. That same period corresponds to the beginning of DA synthesis. The initial depletion of Si may thus be a trigger for initiating DA synthesis in Si-limited chemostats, as well as well as in batch cultures. The timing of cell division and Si uptake, occurring during the majority of the light period when DA production apparently decreases, is consistent with a competition for energy between primary metabolism (C and Si uptake) and secondary metabolism (DA production).

In contrast to other diatoms, the lipid content of *P. multiseries* decreases, rather than increases, in response to Si deficiency during stationary phase (Parrish *et al.* 1991). This was hypothesized to be related to a decrease in light level due to self-shading. An alternative explanation (Pan *et al.* 1996c) is that shared precursors, such as acetyl-CoA (see Section 2.2), are channeled into DA rather than lipid synthesis at that time. Levels of total, as well as certain individual, fatty acids were positively correlated with cellular DA content (Whyte *et al.* 1995b). Cultures that received the most light also produced the most fatty acids and DA, suggesting a high energy requirement for the synthesis of both of these compounds.

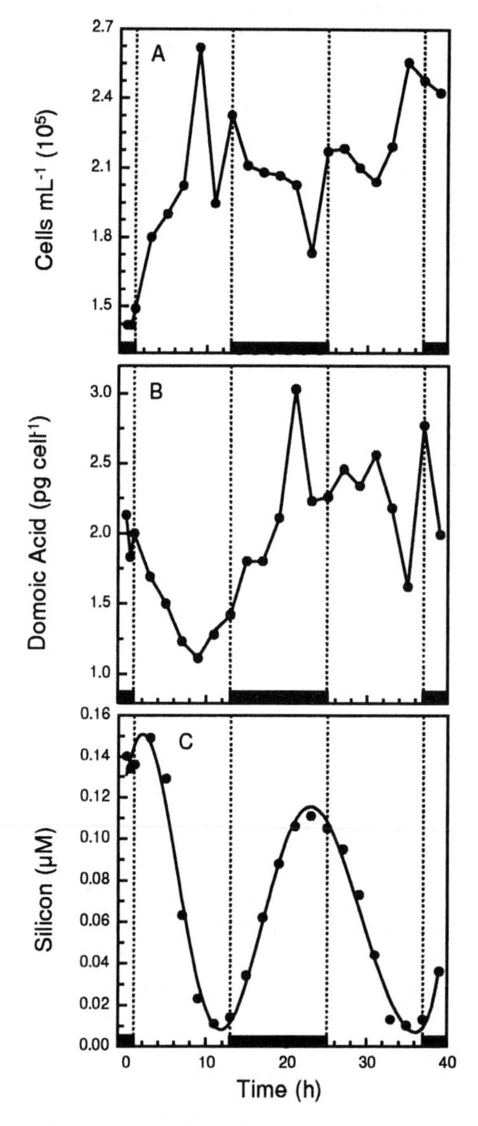

Fig. 4. Growth of *P. multiseries* in a Si-limited chemostat with a 12:12 h light:dark cycle. A) cell number; B) cellular domoic acid; and C) silicon concentration. The dark period is shown by the horizontal bars. Redrawn from Bates and Richard 1996.

2.1.3 Nitrogen

Because DA is an amino acid, N is an essential element for its synthesis. The N requirement was demonstrated when a N-depleted stationary phase batch culture of *P. multiseries* failed to produce DA until supplemented with nitrate and when cell division again ceased (Fig. 5) (Bates *et al.* 1991). Thus far, no other element has been shown to be similarly essential for DA synthesis, but this aspect has not been specifically investigated. No comparable N-limitation/addition experiments have been carried out with any other toxigenic *Pseudo-nitzschia* species. There is also a paucity of studies on growth and DA production using forms of N other than nitrate and ammonium (Bates *et al.* 1993). The low N content of DA (4.5% on a molecular weight basis) argues against its use as a N storage compound (Bates *et al.* 1991).

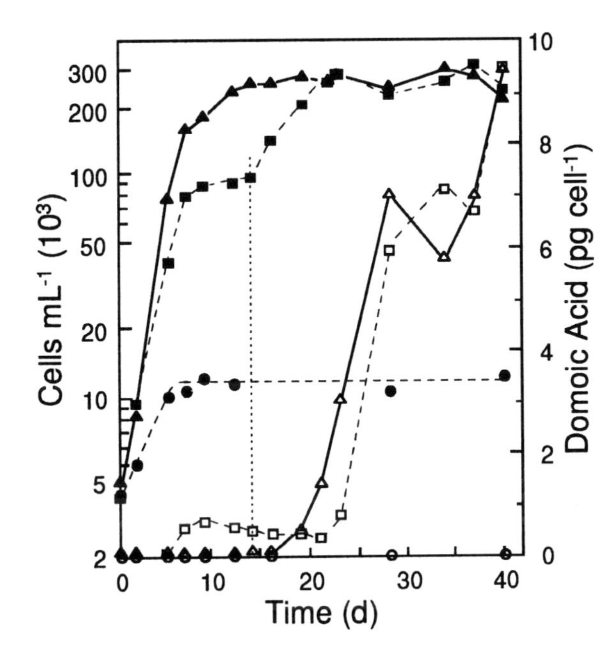

Fig. 5. Effect of nitrate addition on cell number (solid symbols) and cellular domoic acid (open symbols) of *P. multiseries* growing in batch culture. Culture grown with an initial nitrate concentration of 1 mM (triangles); culture grown with an initial nitrate concentration of 0.05 mM and then supplemented with 1 mM nitrate at the time indicated by the dotted line (squares); culture grown in the absence of nitrate (circles). Redrawn from Bates *et al.* 1991.

2.1.4 Phosphorus

Although P is only occasionally a limiting nutrient in coastal marine environments, its potential role in controlling DA production is still of interest. A preliminary experiment showed that *P. multiseries* continued to produce DA in stationary phase batch culture when Si-depleted medium was replaced with one containing no P (Bates *et al.* 1991). This aspect was further studied by using P-limited batch and chemostat cultures (Pan *et al.* 1996c). Production of DA continued in batch culture when the phosphate supply was low (<1 μM) and presumably limiting (Fig. 6). Alkaline phosphatase activity (APA), which increases as a result of phosphate limitation, was positively correlated with DA production. These experiments demonstrate that DA is produced during conditions of P, as well as Si limitation. The mechanism by which P limitation may trigger DA production is presently unknown, except that the cells are placed under "physiological stress" (see Section 2.2). It is not yet known what other macro- or micro-nutrient (e.g., other trace metal) limitations may also trigger DA production.

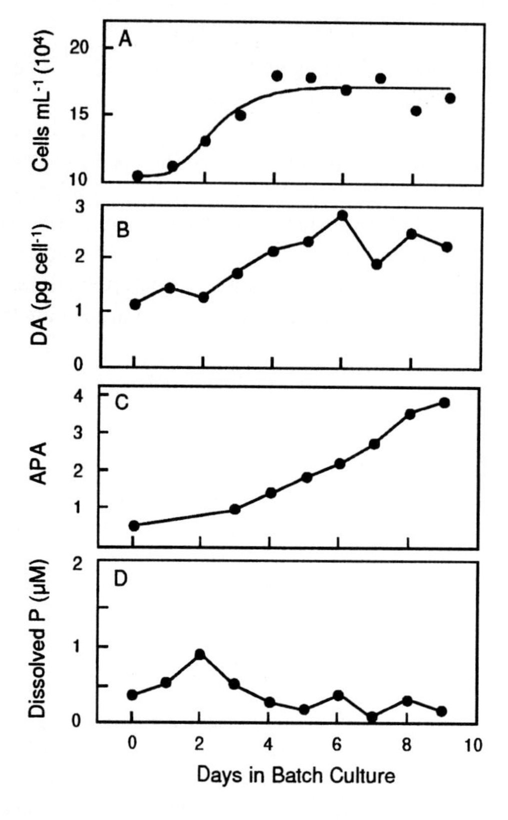

Fig. 6. Batch culture growth of *P. multiseries*, after having turned off the pump of a steady-state P-limited chemostat culture. Variations with time in A) cell number; B) cellular domoic acid (DA); C) alkaline phosphatase activity (APA) (µg P µg [Chl *a*]$^{-1}$ h^{-1}); and D) dissolved reactive phosphate. Redrawn from Pan *et al.* 1996c.

2.1.5 Nutrient Limitation in Continuous Culture

All of the initial studies on defining the conditions for DA production were carried out with *P. multiseries* growing in batch culture, the most straightforward approach at the time. It was also questioned whether this diatom was even capable of producing DA in a continuous culture, where, by definition, cells are continuously dividing. This would be contrary to batch culture studies which showed production predominantly by non-dividing, stationary phase cells. Nevertheless, *P. multiseries* has since been shown to be capable of DA production in Si-limited (Laflamme 1993; Pan 1994; Pan *et al.* 1996b; Bates *et al.* 1996) and P-limited (Pan 1994; Pan *et al.* 1996c) chemostat cultures.

In addition, much valuable information is presently being collected on the effects of nutrients, temperature, and irradiance level on growth and toxin production by growing *P. multiseries* in turbidostat cultures (T. Windust, pers. commun.).

Growth of *P. multiseries* in Si-limited continuous cultures adheres to that predicted by chemostat theory, in that addition of silicate results in a proportional increase in cell number (Laflamme 1993). However, the rate of DA production tended to vary in spite of the fact that the culture was in "steady state", based on cell numbers (Fig. 7); this fluctuation was hypothesized to be due to changes in bacterial numbers in the culture (see Section 2.1.9). The general term "continuous culture" may therefore be more appropriate than "chemostat culture" for this situation, as the latter denotes constancy in chemical composition of the cells as well as the growth medium. Such fluctuations in DA production have not been reported in any other continuous culture study.

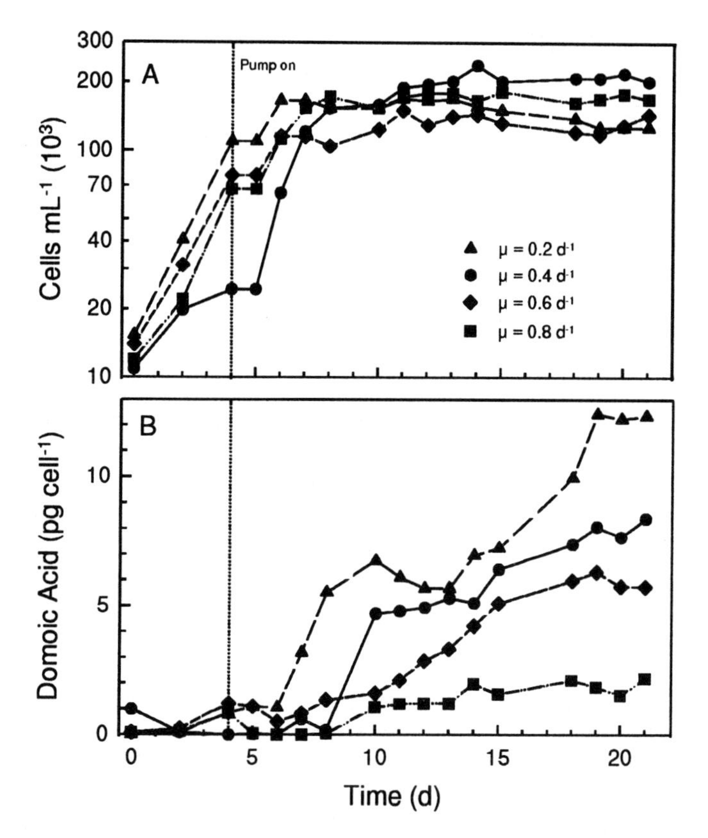

Fig. 7. Growth (A) and cellular domoic acid content (B) of *P. multiseries* growing in Si-limited chemostat cultures at four division rates. Redrawn from Bates *et al.* 1996.

At first it may seem paradoxical that *P. multiseries* would produce DA at all in a culture whose cells are continuously dividing. However, an examination reveals similarities between the late-exponential to stationary phase transition in batch cultures and growth conditions in chemostat cultures. In both cases, cell division rate is slowed, relative to a maximum, as a result of limitation by a specific nutrient. This is the exact condition under which DA is produced in batch, as well as in chemostat culture. Furthermore, there is an inverse linear (Fig. 8A) or curvilinear (Fig. 8B) relationship between division rate and cellular DA content (Bates *et al.* 1996; Pan *et al.* 1996b) or rate of production (Pan *et al.* 1996c) in these Si- or P-limited continuous cultures.

Given the above, one must modify the original conclusion that cessation of cell division is a necessary condition for toxin production (Bates *et al.* 1991). It is more accurate to state that the slowing of cell division due to physiological stress is also conducive to DA production. In spite of a N requirement for DA production in batch culture, one approach that has not yet been explored is the use of N-limited (as opposed to Si- and P-limited) chemostats to investigate toxin production. The goal of these laboratory studies is to help interpret field studies of *P. multiseries* blooms. At present, this has been met with only limited success, due in part to a paucity of field data with which to compare results (see Bates *et al.*, this volume).

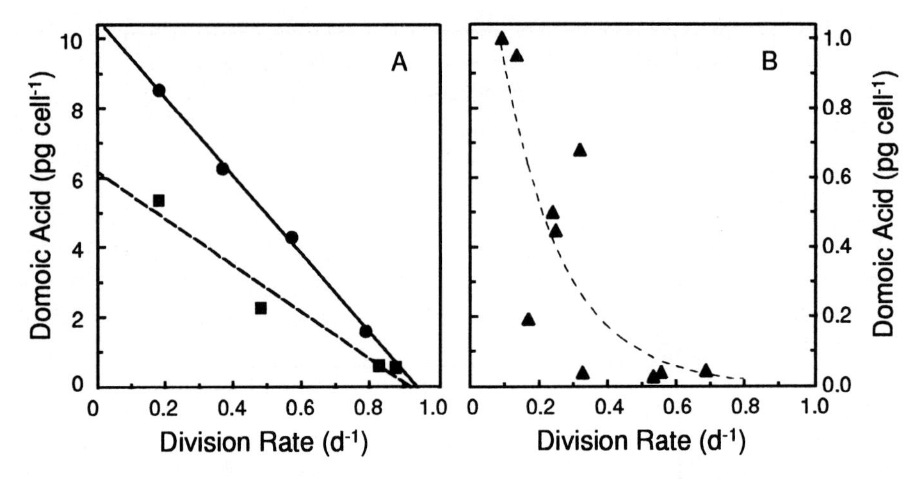

Fig. 8. Relationship between cellular domoic acid content and division rate of *P. multiseries* growing in Si-limited chemostat cultures. A) Examples of two experiments; that shown by (●) is from Fig. 7A (Redrawn from Bates *et al.* 1996); and B) Redrawn from Pan *et al.* 1996b.

2.1.6 Temperature

Laboratory studies show that the division rate and rate of DA production decrease greatly with decreasing temperature, as expected (Fig. 9) (Lewis *et al.* 1993). *Pseudo-nitzschia multiseries* is still capable of growth and DA production when shifted from a growth temperature of 13°C to 0°C, consistent with the finding of toxin-contaminated mussels in ice-covered water (Bates *et al.* 1991; Smith *et al.* 1993). From the above, it may be concluded that low temperature, although a physiological stress, does not boost

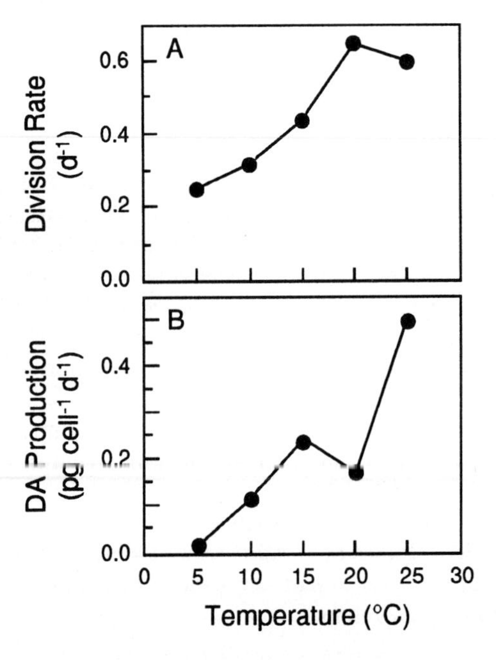

Fig. 9. Relationship between A) cellular division rate of *P. multiseries*, B) rate of domoic acid production, and growth temperature. Redrawn from Lewis *et al.* 1993.

the rate of DA production by *P. multiseries* in the same way that Si or P deprivation does. Cell growth at 25°C and at an irradiance level of 180 µmol photons m^{-2} s^{-1}, however, did increase the cellular DA content relative to growth at 80 µmol photons m^{-2} s^{-1} and at lower temperatures (Lewis *et al.* 1993). These conditions may have imposed a physiological stress or provided an increased supply of photosynthetic energy (see Section 2.2) such that DA production was enhanced.

In contrast to *P. multiseries*, *P. seriata* produced higher levels of cellular DA at 4°C than at 15°C (Lundholm *et al.* 1994). Whether this is due to physiological stress or to real differences in the mechanism of toxin production (*P. seriata* is a cold-water species) must still be determined. Temperature studies are lacking for other toxigenic *Pseudo-nitzschia* species.

2.1.7 Salinity

There are few studies on the effects of salinity on the physiology of *P. multiseries*. An isolate of *P. multiseries* from northern Nova Scotia, Canada, grown at various salinities (6 - 48 ‰), showed a broad salinity range (30 - 45 ‰) for optimum growth (Jackson *et al.* 1992; see Bates *et al.*, this volume). No DA was detected, but this may have been because an amino acid analyzer was used, which is less sensitive than high performance liquid chromatography, and samples were harvested only during early stationary phase.

2.1.8 Irradiance

Relatively few laboratory, and no field studies have been devoted to the effects of light on growth and DA production. Batch (Bates *et al.* 1991) and semi-continuous (Bates and Léger 1992) cultures of *P. multiseries* had lower division rates at low (35 - 45 µmol photons m^{-2} s^{-1}) than at high (90 - 200 µmol photons m^{-2} s^{-1}) irradiance levels, as expected. Division rates, as well as DA production rates, were similar at 80 and 180 µmol photons m^{-2} s^{-1} (Lewis *et al.* 1993). The rate of DA production became lower only when the irradiance level was decreased to 35 µmol photons m^{-2} s^{-1} (Fig. 10) (Bates and Léger 1992). Compared to wide column cultures, an order of magnitude

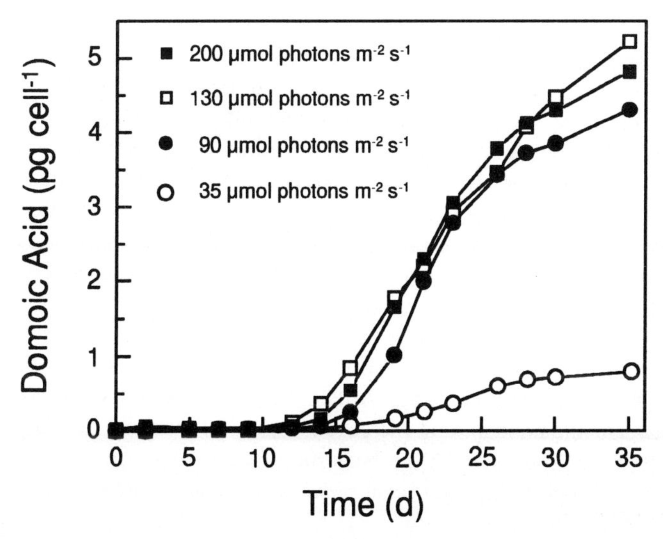

Fig. 10. Production of cellular domoic acid at four irradiance levels by *P. multiseries* growing in batch culture. Stationary phase began on day 13. Redrawn from data presented by Bates and Léger 1992.

greater cellular toxin level was achieved in narrow bag cultures, because *P. multiseries* cells in the latter received more light (Whyte *et al.* 1995b). Low irradiance apparently does not impose the same degree of stress as does Si or P limitation, and therefore does not promote DA production. Toxin production may, in fact, be limited without an adequate supply of light, because there is a photosynthetic energy requirement for DA synthesis (see Section 2.2). These experiments indicate that *P. multiseries* should be grown, conservatively, at 100 µmol photons m^{-2} s^{-1} or greater in order to avoid light limitation of growth or toxin production.

The cycle at which light is delivered may be important for DA production, as was suggested by Sommer (1994) for growth. Except for one experiment (reported in Villac *et al.* 1993), *P. multiseries* is shown to be capable of DA production in constant light, as well as with a light:dark cycle. No experiments have been carried out to determine if a shorter photoperiod, as would be found during the autumn in northern waters, would affect DA production.

Studies are lacking on the effect of irradiance level on other toxigenic *Pseudo-nitzschia* species. Hargraves *et al.* (1993), however, studied the effects of ultraviolet (UV) radiation on *P. multiseries*, *P. pungens*, and *P. fraudulenta*. They found that the growth of *P. multiseries* was unaffected by UV radiation compared to the other two species, which were slightly or significantly inhibited, respectively. Unfortunately, the DA data were ambiguous; UV exposure reduced the cellular toxin content in 35-d old *P. multiseries* cultures, but the effect is not clear in 63-d old cultures. No DA was detected in *P. pungens* or *P. fraudulenta*. The authors conclude that the apparent UV-resistance of *P. multiseries* has implications for the composition of coastal phyto-plankton communities if ambient UV light continues to increase as a result of global ozone depletion.

2.1.9 Bacteria

Because some bacteria play a role in the production of PSP toxins, and others are capable of autonomous toxin production (see Doucette *et al.*, this volume), it was essential to determine if bacteria were involved in DA production by *P. multiseries*. This was not possible until 1991, when the first axenic (bacteria-free) culture of *P. multiseries* was obtained by treatment with antibiotics (Douglas and Bates 1992). Prior to this, all investigations used nonaxenic (bacteria-containing) cultures, so it was not known if bacteria or the diatom produced the toxin. Axenic cultures did produce DA (Douglas and Bates 1992), but only $^1/_{20}$ or less than nonaxenic cultures (Douglas *et al.* 1993). Reintroducing bacteria to axenic cultures (Fig. 11) resulted in an enhancement of DA production by 2 to 115-fold, depending on the *P. multiseries* and bacterial strain used (Bates *et al.* 1995; Osada and Stewart 1997). The bacterial strains were isolated from different *P. multiseries* cultures and were only distantly related to each other. A bacterial strain isolated from a non-toxic *Chaetoceros* sp. culture also enhanced DA production, suggesting that specific bacteria may not be necessary.

Bacteria may be responsible for the variable production of DA observed in continuous cultures, which are in "steady-state" based on the constant diatom cell concentration (Laflamme 1993; see Section 2.1.5). A positive relationship was found, after glucose addition, between the concomitant increase in bacterial numbers and DA production. Similar results were reported with the bacterium *Alteromonas* sp. (Osada and Stewart 1997). Pan (1994) reported a significant positive correlation between DA

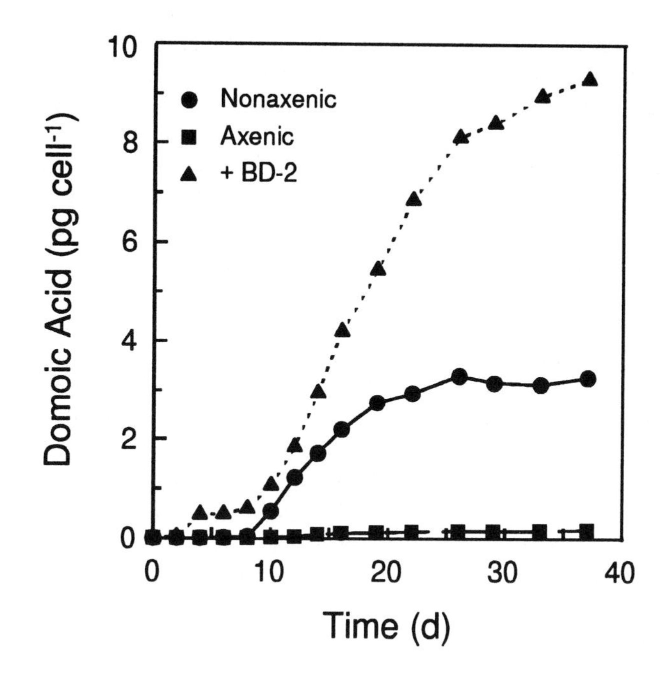

Fig. 11. Production of domoic acid by cultures of *P. multiseries* that are nonaxenic (●), axenic (■), and reintroduced with bacterial strain BD-2 (▲). Redrawn from Bates *et al.* 1995.

production and bacterial abundance in Si-limited chemostat cultures growing at several division rates. The relationship was less clear when the incoming nutrient flow was cut off, effectively creating a batch culture. This points out that, although bacteria and Si or P limitation both enhance DA production, the mechanisms may differ. Bacterial abundance was lowest during active growth of *P. multiseries* in batch culture (Pan 1994), suggesting the production of a bactericidal compound by the diatom, or the absence of production of a compound required for the growth of the bacteria. Studies are required to more thoroughly monitor changes in bacterial numbers in relation to DA production during growth of *P. multiseries* in batch culture.

Thus far there is no evidence of intracellular bacteria (MacPhee *et al.* 1992), nor that isolated extracellular bacteria are capable of autonomous production of DA (G.J. Doucette, pers. commun.; unpubl. data). The mechanism for bacterial enhancement of DA production has not yet been elucidated. Bacteria may produce precursors that are used directly in DA production, or indirectly as "elicitors" of toxin production, or they may regenerate nitrogenous nutrients required for toxin production. Douglas and Bates (unpubl. data) examined the possibility that elicitors associated with the bacteria were responsible for the enhancement. Adding bacterial extracts, sometimes separated into different molecular weight fractions, to axenic *P. multiseries* cultures did not enhance DA production. It was concluded that the enhancement mechanism requires a dynamic interaction between the diatoms and live bacteria. In another study, addition of gluconic acid/gluconolactone to axenic cultures of *P. multiseries* doubled the production of DA (Osada and Stewart 1997). This mixture of compounds was shown to be produced by the bacterium *Alteromonas* sp., a symbiont (Stewart *et al.* 1997), in the presence of glucose. It was suggested that both DA and gluconic acid/gluconolactone are nutrient scavengers, and that *P. multiseries* may counter the effects of gluconic acid/gluconolactone by increasing the production of a counter chelating agent, DA.

Curiously, there is a characteristic decrease in both the viability and toxicity of *P. multiseries* cultures over a period of a year or more (Villac *et al.* 1993; unpubl. data). Of the several possible explanations, including genetic alterations and the lack of sexual reproduction in clonal culture, one may involve changes and/or decreases the bacterial population over time. In one nonaxenic strain (KP-104), for example, cellular DA levels decreased from 4 to 0.01 pg $cell^{-1}$ over a period of two years. Introduction of bacteria (strains BD-1 and PM-1; Bates *et al.* 1995) resulted in a marginal increase in cellular DA to 0.7 pg $cell^{-1}$ during a 23 day incubation period (unpubl. data).

2.2 Biosynthesis of Domoic Acid

Douglas *et al.* (1992) investigated the pathway for DA biosynthesis by pulsing an axenic stationary-phase culture of *P. multiseries* with [^{13}C]-labelled acetate to study the resulting ^{13}C labelling pattern of the intermediate molecules leading to DA, using nuclear magnetic resonance (NMR). They determined that the toxin is biosynthesized from two different precursor units. One unit was shown to be derived from the direct incorporation of acetate into oxaloacetate after multiple rounds of the citric acid cycle, forming an activated derivative, perhaps 3-hydroxyglutamic acid. The other precursor, drawn from a different pool, is thought to be an isoprenoid unit, geranyl pyrophosphate. Assembly of the proline ring portion of the DA molecule is then believed to occur by condensation of the activated glutamate derivative with the isoprenoid chain, consistent with the pathway proposed by Laycock *et al.* (1989). The differences in the level of acetate incorporation into the two key biosynthetic intermediates indicate they are drawn from different acetate pools, though it is also possible that the isoprenoid portion is principally derived from an alternate pathway involving a different primar precursor (J.L.C. Wright, pers. commun.).

All experimental evidence with *P. multiseries* indicates that the initiation of toxin biosynthesis is associated with physiological stress due to limitation by at least Si or P. One index of stress is a lowering of the maximum rate of photosynthesis (P_{max}) during stationary phase in batch culture (Pan *et al.* 1991; 1996d; Bates *et al.* 1993) or accompanying low rates of cell division in continuous cultures (Pan *et al.* 1996b; c). Other indices, which are also correlated with increased DA production, include: decreased cellular content of N, C, and Si (Pan *et al.* 1991; 1996a; Bates *et al.* 1993), and decreased rates of N, Si and P uptake (Fig. 12), as well as various ratios of these elements (Pan *et al.* 1996b; c). In particular, *P. multiseries* appears to have a more conservative range of N:P ratios (6 - 23) than other phytoplankton (Pan *et al.* 1996c).

One hypothesis (cf. Beardall *et al.* 1976) for the triggering of DA production is that there is a switch from C3 carboxylation during exponential growth, when ribulose bisphosphate carboxylase (RuBPCase) activity is prominent, to C4 β-carboxylation, via phosphoenolpyruvate carboxylase (PEPC) or phosphoenolpyruvate carboxykinase (PEPKase), during stationary phase when DA is produced. Each of these carboxylating enzymes is present in both toxic *P. multiseries* and non-toxic *P. pungens* (unpubl. data). However, there was no significant difference between the two diatom species in the activity of these enzymes. Furthermore, the activities of RuBPCase, PEPC, and PEPKase were highest at the end of exponential phase and each declined during stationary phase, suggesting no major switch in pathways for carbon assimilation.

Pan *et al.* (1996b; c) argued that there is a competition for free energy, in the form of adenosine triphosphate (ATP), between DA production and primary metabolism.

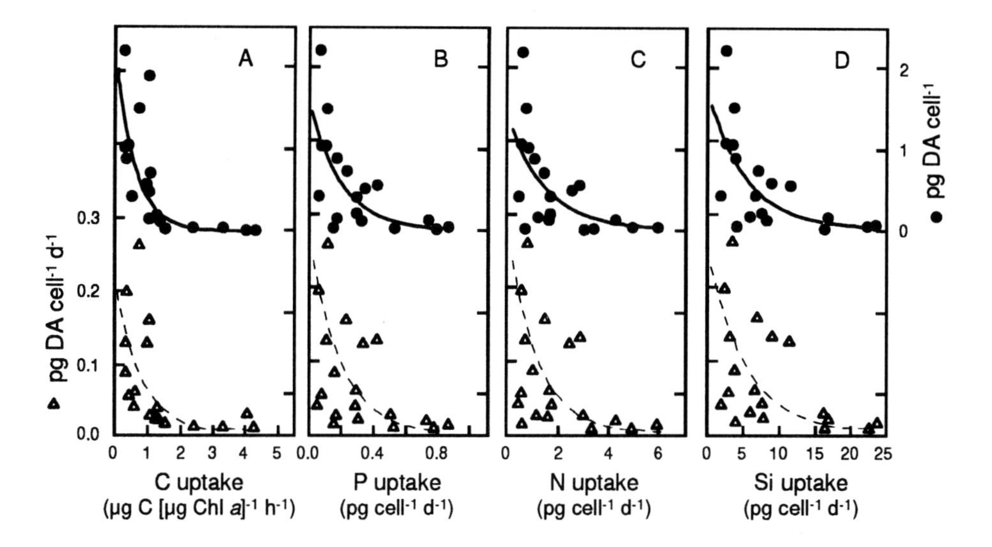

Fig. 12. Relationship between cellular domoic acid (●) and domoic acid production (▲) with uptake of A) carbon, B) phosphate, C) nitrate, and D) silicate by *P. multiseries* growing in a P-limited chemostat. Redrawn from Pan *et al.* 1996c.

During active growth (e.g., exponential phase in batch culture), photosynthetic rates are high, as are rates of N, P, and Si uptake (Fig. 12). This establishes a high demand for metabolic energy and results in less ATP for secondary metabolism. Consequently, DA synthesis is low because it requires at least 5 ATP (Fig. 13) (Pan 1994). When growth slows as a result of Si or P limitation, this free energy can be diverted to toxin production. Addition of Si or P to a culture in stationary phase temporarily reduces DA production until these elements again become limiting (Pan *et al.* 1996a; c). Growth limitation by low irradiance or temperature would not necessarily stimulate toxin production because less photosynthetic energy and reduced enzyme activities, respectively, would become the limiting factor. Because of this energy requirement, there may also be a minimum level of chlorophyll *a* necessary for toxin synthesis.

Pan (1994) proposed that α-oxoglutarate from the citric acid cycle was an intermediate for glutamate, and eventually for DA synthesis (Fig. 13). Werner (1977) reported that Si deprivation resulted, within 1 to 3 hours, in decreased pools of 2-oxoglutarate and glutamate in the diatom *Cyclotella cryptica*. It would be interesting to investigate the relationship between these metabolites and the proposed pathway of DA synthesis. Another avenue to explore is the interaction between the pathways for DA and chlorophyll *a* synthesis, which both draw on a pool of glutamate.

One goal of these biosynthesis studies is to identify enzymes that control DA synthesis and then to generate molecular probes against them. Once obtained, they can be used to probe *Pseudo-nitzschia* species and other genera (including bacteria) for their ability to biosynthesize DA. This would also address the question about the contradictory findings for toxin production among strains of *Pseudo-nitzschia* species (see Bates *et al.*, this volume).

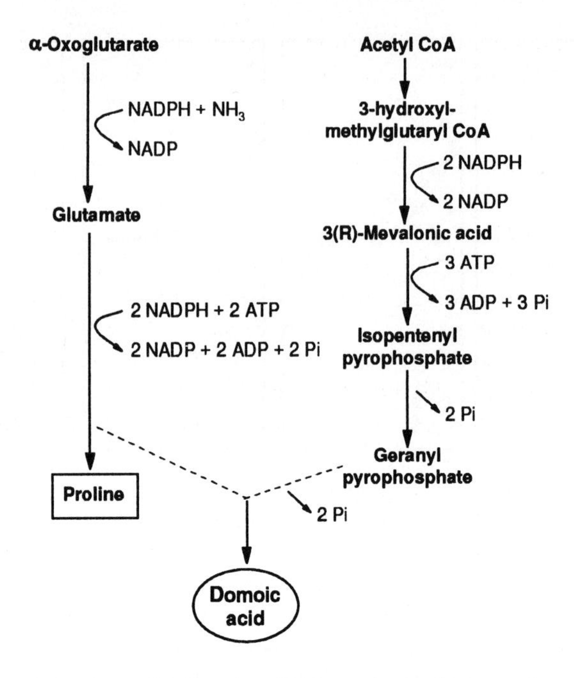

Fig. 13. Potential pathway for domoic acid biosynthesis. Redrawn from Pan 1994.

2.3 Physiological and Ecological Roles of Domoic Acid

In spite of several years of research with toxigenic species of *Pseudo-nitzschia*, the physiological or ecological role of DA has not yet been elucidated. Many such compounds are referred to as "secondary" metabolites, to distinguish them from the "primary" metabolites that include protein, carbohydrates, lipids, and nucleic acids (Luckner 1984). The ability to produce DA is thus far restricted to a relative narrow group of pennate diatoms and red macroalage (Bates *et al.*, this volume), the latter which can also synthesize the neurotoxin, kainic acid (Ramsey *et al.* 1994). Nevertheless, one wonders why this ability is even present in at least these two algal groups. Luckner (1984) argues that the probability that a given secondary metabolic pathway developed independently during evolution in different groups of organisms decreases with increasing complexity of the compound. Although DA is a relatively small molecule, it may still have a complex metabolic pathway.

2.3.1 Excretion Hypothesis

In the case of toxigenic *Pseudo-nitzschia* species, DA is produced only under specific conditions, suggesting that it is not essential for vegetative growth. For example, it is produced during Si or P limitation, when light and C are still in excess. Domoic acid may therefore simply serve as a way of dispensing with "excess" photosynthetic energy when cells are no longer able to grow optimally, analogous to the excretory role of glycolate. This is consistent with the hypothesis that toxin is synthesized when ATP from photophosphorylation is not used for primary metabolism (Pan *et al.* 1996b; c).

Most DA thus produced is excreted from the cells (Fig. 1), as long as they remain in the unfavorable environment (Bates *et al.* 1991; Pan *et al.* 1996c). That DA is also a neurotoxin may or may not be an added benefit to the diatom. The low N content of this amino acid and the presence of a N-sufficient medium make its excretion an inconsequential loss of N (see Section 2.1.3).

2.3.2 Osmolyte Hypothesis

The possibility that DA could serve as an osmolyte in response to increasing salinity was explored by Jackson *et al.* (1992). Although no DA was detected (see Section 2.1.7), the concentration of the amino acid taurine increased significantly when *P. multiseries* was grown in the range of 15 - 48 ‰. Taurine, and also sorbitol (Stewart *et al.* 1997), were hypothesized to be osmoregulators, but their ecological relevance may be minimal if this diatom does not experience the higher salinity values in nature.

2.3.3 Antifeedant Hypothesis

One hypothesis is that DA may act as an antifeedant (see Turner, this volume), thus, for example, enabling the 1987 eastern PEI bloom to intensify and persist for over three months (Bates *et al.* 1989). This was based on the finding that DA, from the red macroalga *Chondria armata*, has a strong insecticidal activity against houseflies and cockroaches, as well as activity against intestinal worms (Maeda *et al.* 1984; 1987). Domoic acid is toxic to the small estuarine copepods *Temora longicornis* and *Pseudocalanus acuspes*, but not to the larger *Calanus glacialis*, and only at relatively high concentrations (LC_{50} at 72 hours of 135 and 38 $\mu g\ mL^{-1}$, respectively) in seawater (Windust 1992). When fed toxic *P. multiseries*, and using non-toxic *P. pungens* as a control, the copepods *T. longicornis* and *C. glacialis* showed no decrease in feeding rate, egg hatching success (only *T. longicornis* was studied), unusual feeding behaviour, nor mortality in short-term (24 h) feeding experiments; mortality was <90% over 9 days. For these copepod species under these conditions, at least, DA does not function as an antifeedant. However, both copepod species retained about 50% of the DA they ingested. Copepods can therefore be a potential vector for the transfer of this neurotoxin through the food chain. Additional studies with the copepods *Temora* sp. and *Acartia* sp. essentially confirmed these results (Turner, this volume).

The lack of toxicity to copepods is surprising, given that DA is capable of depolarizing neuromuscular junctions in crayfish and insects (Maeda *et al.* 1987). Aside from the known adverse affects of DA on mammalian and avian nervous systems, there is some evidence of physiological stress on the Pacific oyster (*Crassostrea gigas*) (Jones *et al.* 1995), spiny scallop (*Chlamys hastata*), and reddish scallop (*C. rubida*) (J.N.C. Whyte, pers. commun.) as a result of feeding on *P. multiseries*. However, DA appears to have no adverse affect on the physiology of anchovies, the California mussel (*Mytilus californianus*) (Jones *et al.* 1995; Whyte *et al.* 1995a), sea scallops (Douglas *et al.* 1997), or blue mussels, some of the known predators of *P. multiseries*. One may question the role of DA as an antifeedant.

2.3.4 Allelopathy Hypothesis

Allelopathy (plant-plant interactions via toxic substances) is one property that is characteristic of certain secondary metabolites (Luckner 1984). Production of a compound that is deleterious to other algae could explain the virtually monospecific bloom of *P. multiseries* in Cardigan Bay, PEI, in 1987 (Bates *et al.* 1989). In bioassays, DA (up to 500 ng mL^{-1}) had no significant influence on the growth of the diatoms *Chaetoceros gracilis* or *Skeletonema costatum* (Windust 1992). Subba Rao *et al.* (1995) tested for allelopathy between the diatoms *Rhizosolenia alata* and *P. multiseries*. They found a more apparent affect on the division rate and maximum cell number of *P. multiseries* due to increasing the proportion of. *R. alata* in the culture mixture than the other way around. The concentration of DA, specifically, was not varied or measured. Competition experiments between pairs of *Pseudo-nitzschia* spp. isolated from Monterey Bay, CA, showed the eventual dominance of one species over another (Villac 1996), but no relationship was made to the presence or absence of DA. Interestingly, however, growth rates were lower (with the exception of a variety of *P. pungens*) when the species were cultured together than when alone, irrespective of their toxicity.

Thus far, therefore, there is no strong evidence that DA mediates allelopathic interactions. Likewise, there were no deleterious effects on 13 species of terrestrial and marine bacteria, indicating that DA is not broadly antibacterial (Windust 1992). Although other secondary products may improve the producer's fitness by acting at specific receptors in competing organisms, this does not seem to be the case with DA.

3 Directions for Further Research

In spite of considerable progress made since the production of DA by *P. multiseries* became a serious topic of research after the 1987 "mussel crisis", there are still many gaps in our knowledge. These are listed below:

- Triggers of DA production, other than Si and P limitation.
- Role of trace metals (especially iron) in mediating DA production and *Pseudo-nitzschia* growth.
- Role of viruses in transferring genetic material that may enable DA production.
- Details of life cycle of *Pseudo-nitzschia* species in relation to DA production.
- Reasons for the decrease in viability and toxicity of *P. multiseries* in culture.
- Complete biosynthetic pathway of DA, including controlling enzymes and genetic control of DA production.
- Ecological and physiological roles of DA.
- Physiological studies of *Pseudo-nitzschia* species other than *P. multiseries*.
- Other producers of DA.

The toxin event of 1987 in eastern PEI was indeed a crisis for the aquaculture mussel industry and for human health. However, it opened up a fruitful area of research that enabled new insights into the physiology and ecology of DA-producing species of *Pseudo-nitzschia*, as well as into the application of DA as pharmacological model for diseases such as Alzheimer's dementia and Huntington's chorea. There are thus practical as well as basic research avenues to pursue.

4 Acknowledgments

I thank G. Doucette, T. Windust, and J.N.C. Whyte for providing me with unpublished information; Y. Pan, L. Rhodes, J.E. Stewart, and E. Vrieling for copies of manuscripts "in press"; and Y. Pan and J.L.C. Wright for commenting on parts of the text.

5 References

Bates, S.S., Léger, C. (1992). Response of *Nitzschia pungens* f. *multiseries* to irradiance: growth and domoic acid production. In: Therriault, J.-C., Levasseur, M. (eds.) *Proceedings of the Third Canadian Workshop on Harmful Marine Algae. Can. Tech. Rep. Fish. Aquat. Sci.* 1893, pp. 9-10 (abstract).

Bates, S.S., Richard, J. (1996). Domoic acid production and cell division by *P. multiseries* in relation to a light:dark cycle in silicate-limited chemostat culture. In: Penney, R. (ed.) *Proceedings of the Fifth Canadian Workshop on Harmful Marine Algae. Can. Tech. Rep. Fish. Aquat. Sci.* 2138, pp. 140-143 (extended abstract).

Bates, S.S., Bird, C.J., de Freitas, A.S.W., Foxall, R., Gilgan, M., Hanic, L.A., Johnson, G.R., McCulloch, A.W., Odense, P., Pocklington, R., Quilliam, M.A., Sim, P.G., Smith, J.C., Subba Rao, D.V., Todd, E.C.D., Walter, J.A., Wright, J.L.C. (1989). Pennate diatom *Nitzschia pungens* as the primary source of domoic acid, a toxin in shellfish from eastern Prince Edward Island, Canada. *Can. J. Fish. Aquat. Sci.* 46: 1203-1215.

Bates, S.S., de Freitas, A.S.W., Milley, J.E., Pocklington, R., Quilliam, M.A., Smith, J.C., Worms, J. (1991). Controls on domoic acid production by the diatom *Nitzschia pungens* f. *multiseries* in culture: nutrients and irradiance. *Can. J. Fish. Aquat. Sci.* 48: 1136-1144.

Bates, S.S., Worms, J., Smith, J.C. (1993). Effects of ammonium and nitrate on domoic acid production by *Nitzschia pungens* in batch culture. *Can. J. Fish. Aquat. Sci.* 50: 1248-1254.

Bates, S.S., Douglas, D.J., Doucette, G.J., Léger, C. (1995). Enhancement of domoic acid production by reintroducing bacteria to axenic cultures of the diatom *Pseudo-nitzschia multiseries*. *Nat. Toxins* 3: 428-435.

Bates, S.S., Léger, C., Smith, K.M. (1996). Domoic acid production by the diatom *Pseudo-nitzschia multiseries* as a function of division rate in silicate-limited chemostat culture. In: Yasumoto, T., Oshima, Y, Fukuyo, Y. (eds.) *Harmful and Toxic Algal Blooms*, Intergov. Oceanogr. Comm., UNESCO, Paris, pp. 163-166.

Beardall, J., Mukerji, D., Glover, H.E., Morris, I. (1976). The path of carbon in photosynthesis by marine phytoplankton. *J. Phycol.* 12: 409-417.

Brzezinski, M.A., Olson, R.J., Chisholm, S.W. (1990). Silicon availability and cell-cycle progression in marine diatoms. *Mar. Ecol. Prog. Ser.* 67: 83-96.

Darley, W.M., Volcani, B.E. (1969). Role of silicon in diatom metabolism. A silicon requirement for deoxyribonucleic acid synthesis in the diatom *Cylindrotheca fusiformis* Reimann and Lewin. *Exp. Cell Res.* 58: 334-342.

Douglas, D.J., Bates, S.S. (1992). Production of domoic acid, a neurotoxic amino acid, by an axenic culture of the marine diatom *Nitzschia pungens* f. *multiseries* Hasle. *Can. J. Fish. Aquat. Sci.* 49: 85-90.

Douglas, D.J., Ramsey, U.P., Walter, J.A., Wright, J.L.C. (1992). Biosynthesis of the neurotoxin domoic acid by the marine diatom *Nitzschia pungens* forma *multiseries*, determined with [13C]-labelled precursors and nuclear magnetic resonance. *J. Chem. Soc. Chem. Commun.* 1992: 714-716.

Douglas, D.J., Bates, S.S., Bourque, L.A., Selvin, R.C. (1993). Domoic acid production by axenic and non-axenic cultures of the pennate diatom *Nitzschia pungens* f. *multiseries*. In: Smayda, T.J., Shimizu, Y. (eds.) *Toxic Phytoplankton Blooms in the Sea*, Elsevier, Amsterdam, pp. 595-600.

Douglas, D.J., Kenchington, E.R., Bird, C.J., Pocklington, R., Bradford, B., Silvert, W. (1997). Accumulation of domoic acid by the sea scallop *Placopecten magellanicus* fed cultured cells of toxic *Pseudo-nitzschia multiseries*. *Can. J. Fish. Aquat. Sci.* (in press).

Garrison, D.L., Conrad, S.M., Eilers, P.P., Waldron, E.M. (1992). Confirmation of domoic acid production by *Pseudonitzschia australis* (Bacillariophyceae) cultures. *J. Phycol.* 28: 604-607.

Hargraves, P.E., Zhang, J., Wang, R., Shimizu, Y. (1993). Growth characteristics of the diatoms *Pseudonitzschia pungens* and *P. fraudulenta* exposed to ultraviolet radiation. *Hydrobiol.* 269: 207-212.

Hasle, G.R. (1995). *Pseudo-nitzschia pungens* and *P. multiseries* (Bacillariophyceae): nomenclatural history, morphology, and distribution. *J. Phycol.* 31: 428-435.

Jackson, A.E., Ayer, S.W., Laycock, M.V. (1992). The effect of salinity on growth and amino acid composition in the marine diatom *Nitzschia pungens*. *Can. J. Bot.* 70: 2198-2201.

Jones, T.O., Whyte, J.N.C., Townsend, L.D., Ginther, N.G., Iwama, G.K. (1995). Effects of domoic acid on haemolymph pH, PCO_2 and PO_2 in the Pacific oyster, *Crassostrea gigas* and the California mussel, *Mytilus californianus*. *Aquat. Toxicol.* 31: 43-55.

Laflamme, M.Y. (1993). Contrôle de la croissance et de la production d'acide domoïque par la diatomée *Pseudonitzschia pungens* forma *multiseries* en cultures continues. M.Sc. Thesis, Univ. of Moncton, Moncton, New Brunswick, Canada, 141 p.

Laycock, M.V., de Freitas, A.S.W., Wright, J.L.C. (1989). Glutamate agonists from marine algae. *J. Appl. Phycol.* 1: 113-122.

Lee, J.H., Baik, J.H. (1997). Neurotoxin-producing *Pseudonitzschia multiseries* (Hasle) Hasle, in the coastal waters of southern Korea. II. Production of domoic acid. *Algae (Korean J. Phycol)*. 12: 31-38.

Lewis, N.I., Bates, S.S., McLachlan, J.L., Smith, J.C. (1993). Temperature effects on growth, domoic acid production, and morphology of the diatom *Nitzschia pungens* f. *multiseries*. In: Smayda, T.J., Shimizu, Y. (eds.) *Toxic Phytoplankton Blooms in the Sea*, Elsevier, Amsterdam, pp. 601-606.

Luckner, M. (1984). *Secondary Metabolism in Microorganisms, Plants, and Animals*. Springer-Verlag, New York, 576 p.

Lundholm, N., Skov, J., Pocklington, R., Moestrup, Ø. (1994). Domoic acid, the toxic amino acid responsible for amnesic shellfish poisoning, now in *Pseudonitzschia seriata* (Bacillariophyceae) in Europe. *Phycologia* 33: 475-478.

MacPhee, D.J., Hanic, L.A., Friesen, D.L., Sims, D.E. (1992). Morphology of the toxin-producing diatom *Nitzschia pungens* Grunow forma *multiseries* Hasle. *Can. J. Fish. Aquat. Sci.* 49: 303-311.

Maeda, M., Kodama, T., Tanaka, T., Ohfune, Y., Nomoto, K., Nishimura, K., Fujita, T. (1984). Insecticidal and neuromuscular activities of domoic acid and its related compounds. *J. Pesticide Sci.* 9: 27-32.

Maeda, M., Kodama, T., Saito, M., Tanaka, T., Yoshizumi, H., Nomoto, K., Fujita, T. (1987). Neuromuscular action of insecticidal domoic acid on the American cockroach. *Pesticide Biochem. Biophysiol.* 28: 85-92.

Osada, M., Stewart, J.E. (1997). Gluconic acid/gluconolactone: physiological influences on domoic acid production by bacteria associated with *Pseudo-nitzschia multiseries*. *Aquat. Microbial Ecol.* (in press).

Pan, Y. (1994). Production of domoic acid, a neurotoxin, by the diatom *Pseudonitzschia pungens* f. *multiseries* Hasle under phosphate and silicate limitation. Ph.D. Thesis, Dalhousie Univ., Halifax, Nova Scotia, Canada, 245 p.

Pan, Y., Subba Rao, D.V., Warnock, R.E. (1991). Photosynthesis and growth of *Nitzschia pungens* f. *multiseries* Hasle, a neurotoxin producing diatom. *J. Exp. Mar. Biol. Ecol.* 154: 77-96.

Pan, Y., Subba Rao, D.V., Mann, K.H., Brown, R.G., Pocklington, R. (1996a). Effects of silicate limitation on production of domoic acid, a neurotoxin, by the diatom *Pseudo-nitzschia multiseries*. I. Batch culture studies. *Mar. Ecol. Prog. Ser.* 131: 225-233.

Pan, Y., Subba Rao, D.V., Mann, K.H., Brown, R.G., Pocklington, R. (1996b). Effects of silicate limitation on production of domoic acid, a neurotoxin, by the diatom *Pseudo-nitzschia multiseries*. II. Continuous culture studies. *Mar. Ecol. Prog. Ser.* 131: 235-243.

Pan, Y., Subba Rao, D.V., Mann, K.H. (1996c). Changes in domoic acid production and cellular chemical composition of the toxigenic diatom *Pseudo-nitzschia multiseries* under phosphate limitation. *J. Phycol.* 32: 371-381.

Pan, Y., Subba Rao, D.V., Mann, K.H. (1996d). Acclimation to low light intensity in photosynthesis and growth of *Pseudo-nitzschia multiseries* Hasle, a neurotoxigenic diatom. *J. Plankton Res.* 18: 1427-1438.

Parrish, C.C., de Freitas, A.S.W., Bodennec, G., MacPherson, E.J., Ackman, R.G. (1991). Lipid composition of the toxic marine diatom, *Nitzschia pungens*. *Phytochem.* 30: 113-116.

Ramsey, U.P., Bird, C.J., Shacklock, P.F., Laycock, M.V., Wright, J.L.C. (1994). Kainic acid and 1'-hydroxykainic acid from Palmariales. *Nat. Toxins* 2: 286-292.

Reap, M.E. (1991). *Nitzschia pungens* Grunow f. *multiseries* Hasle: growth phases and toxicity of clonal cultures isolated from Galveston Bay, Texas. M.Sc. Thesis, Texas A&M Univ., College Station, TX, 78 p.

Rhodes, L., White, D., Syhre, M., Atkinson, M. (1996). *Pseudo-nitzschia* species isolated from New Zealand coastal waters: domoic acid production *in vivo* and links with shellfish toxicity. In: Yasumoto, T., Oshima, Y, Fukuyo, Y. (eds.) *Harmful and Toxic Algal Blooms*, Intergov. Oceanogr. Comm., UNESCO, Paris, pp. 155-158.

Smith, J.C., McLachlan, J.L., Cormier, P.G., Pauley, K.E., Bouchard, N. (1993). Growth and domoic acid production and retention by *Nitzschia pungens* forma *multiseries* at low temperatures. In: Smayda, T.J., Shimizu, Y. (eds.) *Toxic Phytoplankton Blooms in the Sea*, Elsevier, Amsterdam, pp. 631-636.

Sommer, U. (1994). Are marine diatoms favoured by high Si:N ratios? *Mar. Ecol. Prog. Ser.* 115: 309-315.

Stewart, J.E., Marks, L.J., Wood, C.R., Risser, S.M., Gray, S. (1997). Symbiotic relations between bacteria and the domoic acid producing diatom *Pseudo-nitzschia multiseries* and the capacity of these bacteria for gluconic acid/gluconolactone formation. *Aquat. Microbial Ecol.* (in press).

Subba Rao, D.V., Quilliam, M.A., Pocklington, R. (1988). Domoic acid - a neurotoxic amino acid produced by the marine diatom *Nitzschia pungens* in culture. *Can. J. Fish. Aquat. Sci.* 45: 2076-2079.

Subba Rao, D.V., de Freitas, A.S.W., Quilliam, M.A., Pocklington, R., Bates, S.S. (1990). Rates of production of domoic acid, a neurotoxic amino acid in the pennate marine diatom *Nitzschia pungens*. In: Granéli, E., Sundström, B., Edler, L., Anderson, D.M. (eds.) *Toxic Marine Phytoplankton*, Elsevier, New York, pp. 413-417.

Subba Rao, D.V., Pan, Y., Smith, S.J. (1995). Allelopathy between *Rhizosolenia alata* (Brightwell) and the toxigenic *Pseudonitzschia pungens* f. *multiseries* (Hasle). In: Lassus, P., Arzul, G., Erard, E., Gentien, P., Marcaillou-Le Baut, C. (eds.) *Harmful Marine Algal Blooms*, Lavoisier, Paris, pp. 681-686.

Villac, M.C. 1996. Synecology of the genus *Pseudo-nitzschia* H. Peragallo from Monterey Bay, California, U.S.A. Ph.D. Thesis, Texas A&M Univ., 258 p.

Villac, M.C., Roelke, D.L., Chavez, F.P., Cifuentes, L.A., Fryxell, G.A. (1993). *Pseudonitzschia australis* Frenguelli and related species from the west coast of the U.S.A.: occurrence and domoic acid production. *J. Shellfish Res.* 12: 457-465.

Vrieling. E.G., Koeman, R.P.T., Scholin, C.A., Scheerman, P., Peperzak, L., Veenhuis, M., Gieskes, W.W.C. (1996). Identification of a domoic acid-producing *Pseudo-nitzschia* species (Bacillariophyceae) in the Dutch Wadden Sea with electron microscopy and molecular probes. *European J. Phycol.* 31: 333-340.

Wang, R., Maranda, L., Hargraves, P.E., Shimizu, Y. (1993). Chemical variation of *Nitzschia pungens* as demonstrated by the co-occurrence of domoic acid and bacillariolides. In: Smayda, T.J., Shimizu, Y. (eds.) *Toxic Phytoplankton Blooms in the Sea*, Elsevier, Amsterdam, pp. 607-612.

Werner, D. (1977). Silicate metabolism. In: Werner, D. (ed.) *The Biology of Diatoms*, Univ. Calif. Press, Berkeley, pp. 110-149.

Whyte, J.N.C., Ginther, N.G., Townsend, L.D. (1995a). Accumulation and depuration of domoic acid by the mussel, *Mytilus californianus*. In: Lassus, P., Arzul, G., Erard, E., Gentien, P., Marcaillou-Le Baut, C. (eds.) *Harmful Marine Algal Blooms*, Lavoisier, Paris, pp. 351-357.

Whyte, J.N.C., Ginther, N.G., Townsend, L.D. (1995b). Formation of domoic acid and fatty acids in *Pseudonitzschia pungens* f. *multiseries*. *J. Appl. Phycol.* 7: 199-205.

Windust, A. (1992). The response of bacteria, microalgae and zooplankton to the diatom *Nitzschia pungens* f. *multiseries* and its toxic metabolite domoic acid. M.Sc. Thesis, Dalhousie Univ., Halifax, Nova Scotia, Canada, 107 p.

Wohlgeschaffen, G.D., Subba Rao, D.V., Mann, K.H. (1992). Vat incubator with immersion core illumination - a new, inexpensive setup for mass phytoplankton culture. *J. Appl. Phycol.* 4: 25-29.

Ecophysiology and Biosynthesis of Polyether Marine Biotoxins

Jeffrey L.C. Wright and Allan D. Cembella

Institute for Marine Biosciences, National Research Council
Halifax, Nova Scotia, Canada

1. Introduction

Many harmful marine microalgae form blooms in the upper water column of coastal ecosystems or constitute an important component of the epibenthic microflora. Some species express their harmful effects via the production of noxious or toxic secondary metabolites (phycotoxins), which may exhibit potent biological activity against other members of marine food webs and/or human consumers of seafood. These biologically active compounds are considered to be secondary metabolites, as distinct from primary metabolites involved in primary and intermediary metabolism, because they have no known key role in the basic functioning of the producing cell. By default, the category of "secondary metabolites" can include components produced by a conserved and unique biosynthetic pathway, or by relict pathways (hence these could be erstwhile primary metabolites), as well as compounds with a cryptic role in an unconventional metabolic scheme.

Among the several dozen species of harmful marine eukaryotic microalgae known to produce biologically active noxious or toxic substances, flagellates are almost invariably the source (exception: domoic acid from *Pseudonitzschia* spp. and their allies, see Bates, this volume). It is also noteworthy that among the toxigenic flagellate species, dinoflagellates are overwhelmingly represented. Approximately two percent of the extant species of free-living dinoflagellates produce substances which are inimical to the growth, health, viability or survival of other organisms.

The toxicity of biologically active secondary metabolites is a function of many factors, including the dosage, route of administration, time of exposure, chemical lability, and susceptibility of the target organism. The classic view - that such secondary metabolites are either "toxic" or "non-toxic" and that they can be delineated as "water soluble" versus "lipid soluble" based upon solvent partitioning, is not a particularly useful concept in identifying homologous biosynthetic or phylogenetic relationships.

In this review, all marine polyether metabolites are considered, with attention directed primarily towards the biosynthesis and physiology of the major polyether phycotoxins of dinoflagellate origin (brevetoxins, okadaic acid/dinophysistoxins, ciguatoxins/maitotoxins). These phycotoxins are associated with the toxic syndromes known as neurotoxic shellfish poisoning (NSP), diarrhetic shellfish poisoning (DSP), and ciguatera fish poisoning (CFP), respectively. Nevertheless, comparative observations of the mechanism and sequence of biosynthesis and studies of apparently homologous pathways of other polyketide components for which toxicity is absent, unknown, or poorly defined also warrant consideration.

NATO ASI Series, Vol. G 41
Physiological Ecology of Harmful Algal Blooms
Edited by D. M. Anderson, A. D. Cembella and G. M. Hallegraeff

2. Structural Homology and Biosynthesis

2.1. Secondary metabolites

There has been considerable speculation as to the functional significance of secondary metabolites, particularly those with potent biological activity. A variety of hypothetical roles have been ascribed to these secondary metabolites (Nisbet 1992), including their putative function as chemical defense agents, hormonal and ion channel regulators, xenobiotics, and co-factors, and their involvement in sequestration of other metabolites. Biologically active secondary metabolites may also act as multiple effectors in eliciting a variety of cellular responses. For most such metabolites, evidence of functional significance is merely speculative or supported by rather weak (and often indirect) data. For example, in eco-evolutionary terms, it is premature to conclude that a "phycotoxin" acts as a chemical defence mechanism against *in situ* competitors in nature, based upon its high lethality (i.e., low LD_{50}) towards conventional bioassay subjects (e.g., toxin administered intraperitoneally into laboratory mice). Nevertheless, it is difficult to accept that such a complex biosynthesis evolved (or was retained) without functional significance, since biosynthetic pathways require the coordinated interplay of an array of biosynthetic enzymes, regulatory genes and the expenditure of considerable metabolic energy (in the form of ATP, nucleotide co-factors, and other high-energy intermediates) to produce these exotic and often elaborate chemical structures.

Identification of the critical biosynthetic pathways is the crux of secondary metabolism in that it defines a suite of genetic characteristics which are directly linked to the evolutionary history of the organism. For example, although all organisms produce fatty acids (many of them common), they can be assembled by different biosynthetic enzymes and pathways (Simpson 1995). In species *A* the fatty acid synthase might be a single multi-domain complex (Type 1 synthase), whereas in species *B*, fatty acids might be assembled by a complex consisting of several enzymes (Type 2 synthase). Knowledge of the biosynthetic pathways and mechanisms employed provides information on the number and nature of the genes required for biosynthesis and how they interact in the assembly of the molecule. Such information can then be used to determine if any biosynthetic steps or assembly mechanisms are common or if they are unique to the organism. From an evolutionary perspective, biosynthetic genes for a particular secondary metabolite are often clustered on the genome, lending further credence to the idea that the metabolite is designed for a specific purpose and hence serves a useful function for the producing organism.

Among these studies on secondary metabolism in microalgae, the synthesis of biologically active compounds by marine flagellates, particularly dinoflagellates, merit particular scrutiny. Unfortunately, compared with the number of studies on other eukaryotes including higher plants and animals, as well as prokaryotes, detailed investigations of the secondary metabolites produced by dinoflagellates are few, and these have been primarily limited to the isolation and characterization of sterols (Shimizu 1993), novel phycotoxins (Yasumoto and Murata 1993) or cytotoxic macrocycles (Kobayashi and Ishibashi 1993).

2.2. Polyketide biosynthetic pathways in marine dinoflagellates

Upon reviewing the literature of dinoflagellate secondary metabolites (Shimizu 1993; Yasumoto and Murata 1993), a remarkable fact emerges: the overwhelming majority of metabolites (whether toxic or not) are either linear (Fig.1) or macrocyclic polyether compounds (Fig.2) or fused (ladder-frame) polyether structures (Fig.3). Previous work (see Garson 1993; Wright *et al.* 1996), and simple inspection of the structural relationships among biologically active compounds (including polyether-type phycotoxins) from dinoflagellates, suggests that they are derived by the successive addition of acetate units to a growing polyketide chain. This is characteristic of polyketide biosynthesis mediated by the enzyme(s) polyketide synthase (PKS) (Simpson 1995). The fatty acids of primary metabolism are produced in this way, and polyketides are also a common and widely distributed class of secondary metabolites. However, from the few biosynthetic experiments reported from toxigenic dinoflagellates (Garson 1993; Wright *et al.* 1996), some significant differences have been found in the assembly of dinoflagellate polyketides compared with those from other organisms. For example, in the early labeling experiments with dinoflagellates, the patterns of acetate label incorporation into brevetoxins and okadaic acid did not fit exactly with successive additions of intact acetate units as expected. Instead, the sequential addition of units was interrupted by the appearance of single carbons derived from the methyl group of a cleaved acetate unit. This unexpected labeling pattern is apparently unique, and initially this was interpreted as the involvement of acetate units, as well as other biosynthetic intermediates such as succinate and ketoglutarate, in the assembly of the brevetoxins (Chou and Shimizu 1987; Lee *et al.* 1989b) and DSP toxins (Yasumoto and Torigoe 1991).

However, later biosynthetic experiments on DSP toxins from *Prorocentrum lima* (Needham *et al.* 1994, 1995) failed to support this theory. In these experiments, the positions in the molecule labeled from acetate were all equally enriched, indicating that these carbons were all derived from a common pool of acetate without the intervention of other non-acetate precursors. A further peculiarity is that two molecules of glycolate were required in the assembly process - the first time that this has been observed in the biosynthesis of any molecule. Later, a monoxygenase reaction was found to occur during biosynthesis, and another monoxygenase-mediated rearrangement which results in deletion of a single carbon from an intact acetate unit in the backbone chain of the DSP molecule was proposed (Wright *et al.* 1996). The same rearrangement mechanism can be invoked to explain the occurrence of cleaved acetate units in the brevetoxins produced by *Gymnodinium breve* (=*Ptychodiscus brevis*), as well as the macrocyclic lactone amphidinolide J from *Amphidinium* sp. (Kobayashi *et al.* 1995). This deletion mechanism is unique to dinoflagellates, though a modified version has been observed in a few Streptomycete products (e.g., lankacidins). Another unusual feature of PKS-mediated biosynthesis in dinoflagellates is that the pendant methyl groups attached to the polyketide chain are derived from the methyl group of acetate by an aldol-type condensation, rather than the more common methyl donor mechanism involving S-adenosyl methionine (SAM). Although the carbon deletion process and the aldol methylation step have been observed in a limited number of other organisms (all bacteria), the combination of both biosynthetic mechanisms has only been demonstrated in dinoflagellates.

From comparisons of structural homology, there are two apparent distinct groups of polyether compounds (linear and fused) that share many similarities in the initial assembly of the polyketide chain by a conventional process. However, subsequent modification of the chain results in the observed structural differences (Wright *et al.* 1996). Although this proposal is founded only on a limited number of examples, the close chemical similarity between members of each group suggests that these complex biosynthetic pathways probably operate in every case, and are very characteristic of dinoflagellates. Thus it is possible to conclude that many or all dinoflagellates contain active PKS enzymes that ultimately produce characteristic polyether analogues. The structural type and variety (Figs. 1-3) arises through different modes of assembly of the polyketide chain, as well as post-PKS structural modifications.

The occurrence of such biosynthetic mechanisms, which are unique among eukaryotes but previously found in a few bacteria, raises the possibility that some or all of the biosynthetic genes may have originated in prokaryotes. However, it is unlikely that the array of polyether phycotoxins are directly biosynthesized by bacteria associated with dinoflagellates (either free-living or as endosymbionts), as has been suggested for the neurotoxins responsible for paralytic shellfish poisoning (PSP) (see Doucette *et al.*, this volume). A bacterial origin for the polyether phycotoxins implies the lateral transfer of at least the basic biosynthetic genes involved in polyketide biosynthesis to the dinoflagellate genome, with subsequent post-translational modification resulting in the unique array of polyketide-derived polyether compounds. Beyond the circumstantial evidence of structural homology, there is no other support for the hypothesis of evolutionary association of "dinoflagellate" polyketides with bacteria.

2.3. Structural classification and homology among polyethers

2.3.1. Linear and macrocyclic polyethers

The group described as the linear or macrocyclic polyether compounds all contain furan and pyran rings in a chain, sometimes in the form of spiroketal ring systems. The chain also contains pendant methyl and hydroxyl groups. The structures characteristic of the linear polyether type include the DSP toxin complex (okadaic acid [OA], dinophysistoxins [DTXs], plus OA-diol-esters) from *Prorocentrum* and *Dinophysis* spp., the amphidinols (not illustrated) from *Amphidinium klebsii*, which are similar to luteophanol A from an endosymbiotic marine *Amphidinium* sp. (Fig.1). Other polyethers, which have cyclized to form a macrocyclic lactone ring, are illustrated in Fig.2. The general structure of these compounds is very similar to the linear polyethers except that they have self -condensed, presumably under enzymatic control, to produce a macrocyclic lactone. Nonetheless, opening the lactone ring (e.g., as in pectenotoxins [PTXs]) would produce a linear polyether with many common features of DSP toxins. Even larger molecules in the macrocyclic group include the zooxanthellatoxins from an endosymbiotic *Symbiodinium* sp. Collectively, the zooxanthellatoxins (not illustrated) possess a complex array of chemical structures all containing a series of 5- and 6-membered ether rings, linked by a backbone chain often heavily substituted with hydroxyl and methyl groups. A similar structural pattern is found in palytoxin, a product originally isolated from *Palythoa* soft corals, and

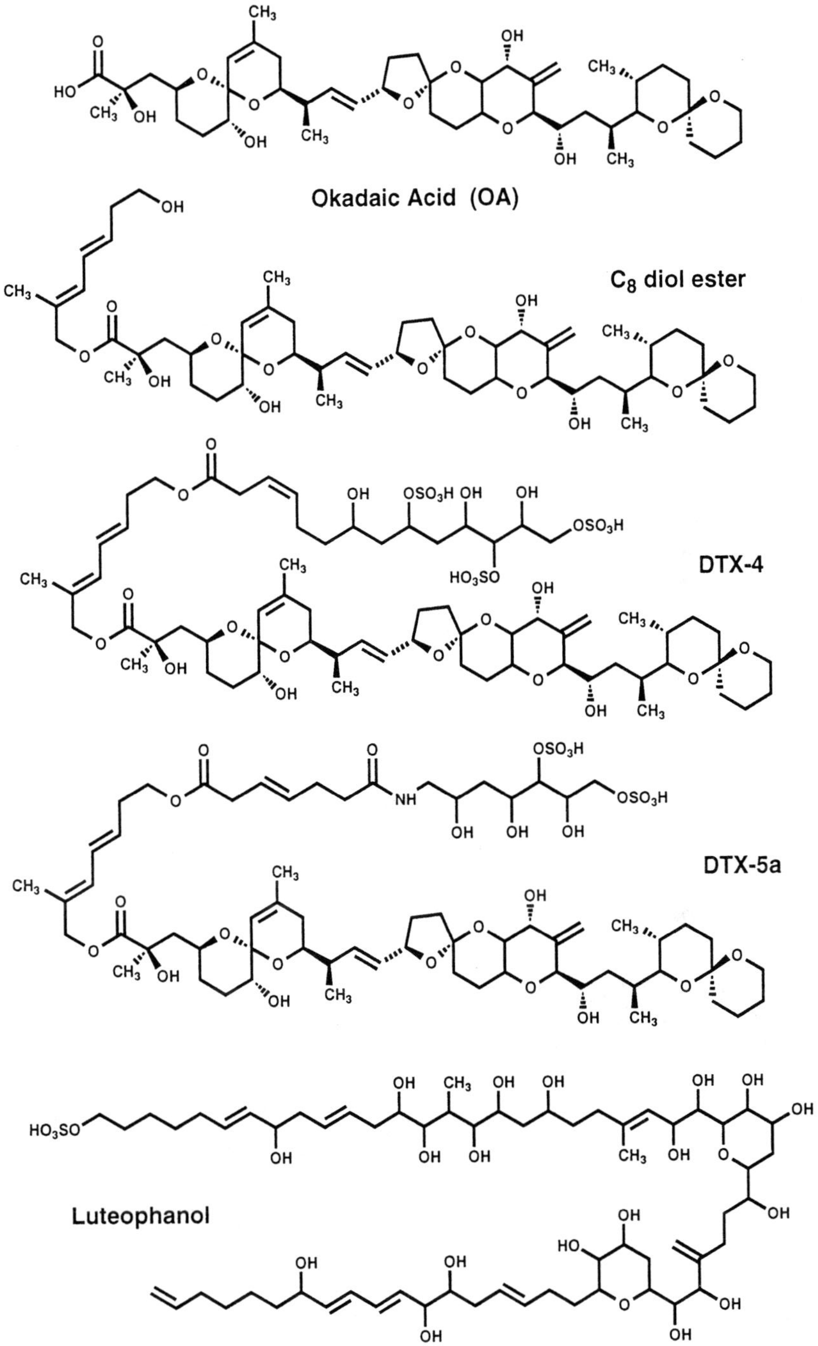

Okadaic Acid (OA)

C_8 diol ester

DTX-4

DTX-5a

Luteophanol

Figure 1. Selected examples of linear polyethers

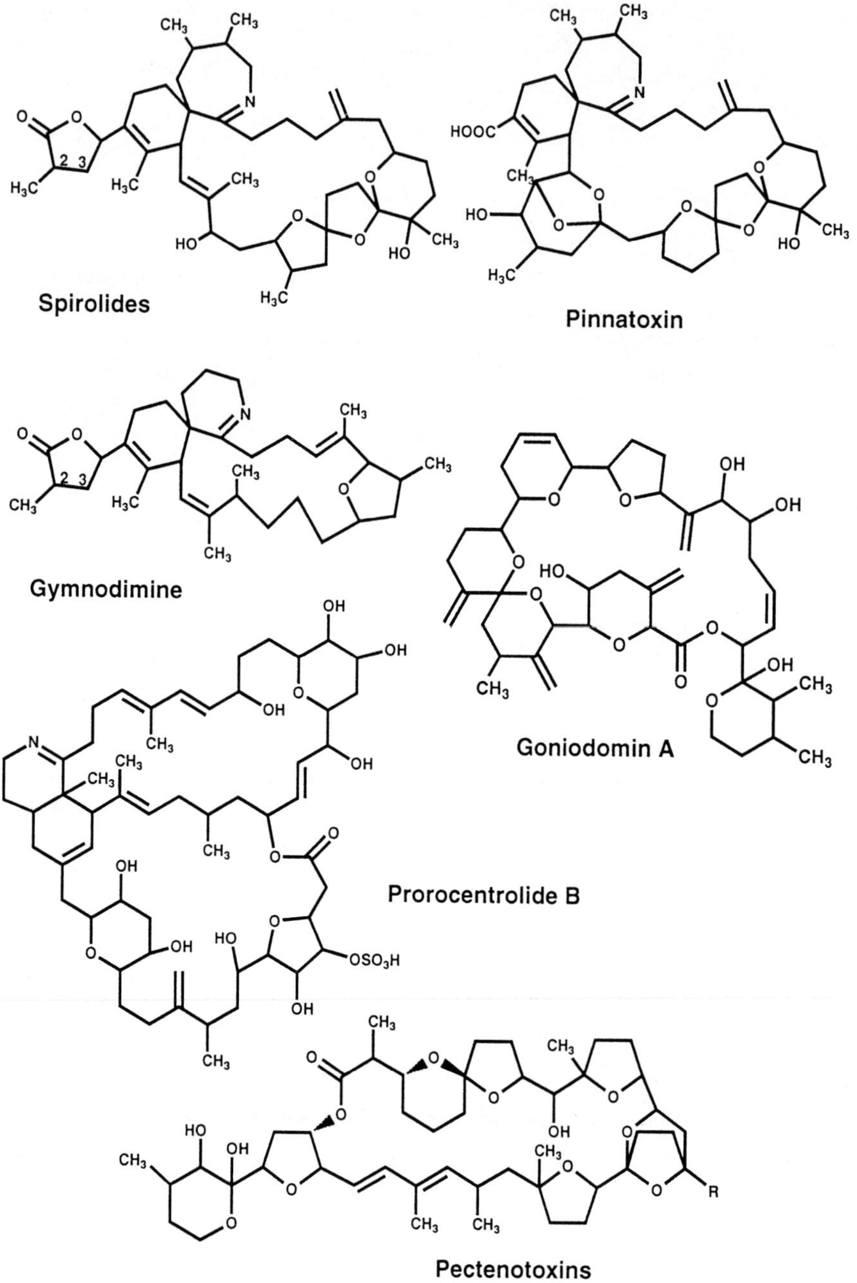

Spirolides

Pinnatoxin

Gymnodimine

Goniodomin A

Prorocentrolide B

Pectenotoxins

Figure 2. Selected examples of macrocyclic polyethers

subsequently found in other diverse organisms including certain species of seaweed, crustaceans, and fish (Yasumoto and Murata 1993) and the benthic dinoflagellate, *Ostreopsis siamensis*. In most cases, these polyethers are unique to eukarytotic microalgae (particularly dinoflagellates) with no exact counterparts in nature, although there are some similarities to the polyether macrocycles in prokaryotes such as Actinomycetes (e.g., tetronasin, monensin) and Myxobacteria (e.g., myxopyronin A). However, comparison of the biosynthetic data for bacterial and dinoflagellate polyethers reveals that they are assembled by different pathways (see Garson 1993 and Robinson 1991).

2.3.2. *Ladder-frame polyethers*

The second major group of polyether metabolites can be described as ladder-frame polyethers, and comparison with the first group (Sec. *2.3.1.*) reveals some remarkable differences. These fused polyethers do not posses an obvious backbone chain - rather they consist of ether rings linked or fused together in a rigid ladder-frame structure. Members of this group, which includes the brevetoxins (GBTX=PbTx), the ciguatera toxin complex (ciguatoxins [CTX], maitotoxin [MTX] (not shown), gambieric acids [GA]) and yessotoxin (YTX), are illustrated in Fig. 3. For ladder-frame polyethers, there are not even any remotely similar structures in nature, and the chemical architecture is apparently unique to biosynthetic products of marine dinoflagellates. Related metabolites found in other members of a marine food chains have likely been modified via biotransformation (e.g., CTX in the livers of moray eels). Thus some dinoflagellates must share a number of unique biosynthetic genes necessary for the production of these polyether analogues.

There are a few examples of microalgae other than dinoflagellates which are also capable of biosynthesizing polyether compounds. For example, prymnesin 2 (not illustrated), a toxin produced by the prymnesiophyte *Prymnesium parvum* (Igarishi *et al.* 1996) is particularly interesting in that it falls between being a linear polyether or a ladder-frame type. This metabolite possesses only short runs of ladder-frame structure (2-5 fused ether rings) linked together by a highly functionalized linear carbon chain.

2.4. Polyethers as chemotaxonomic or phylogenetic markers

Despite the consistent chemotaxonomic correlations among secondary metabolites that are observed for many species, such biochemical markers are seldom used to resolve problems in classification or phylogeny of marine microalgae. One concern is that the environmental or physiological conditions may not be conducive for the production of secondary metabolites. Furthermore, variations in growth conditions may have a profound effect upon the suite of secondary metabolites which is produced. In spite of these caveats, the unique biosynthetic pathways for dinoflagellate polyketides and the apparent clustering of polyethers into two major classes suggests that their distribution may provide insights into their eco-evolutionary significance.

Among the many attempts to establish potential phylogenetic relationships among dinoflagellates, the effort by Taylor (1987) to establish affinities based upon plate homology and ultrastructural features is among the most informative. In this scheme five basic morphotypes (one unarmoured) are recognized for the free-living vegetative stages; these are typified by representative genera: prorocentroid (*Prorocentrum*),

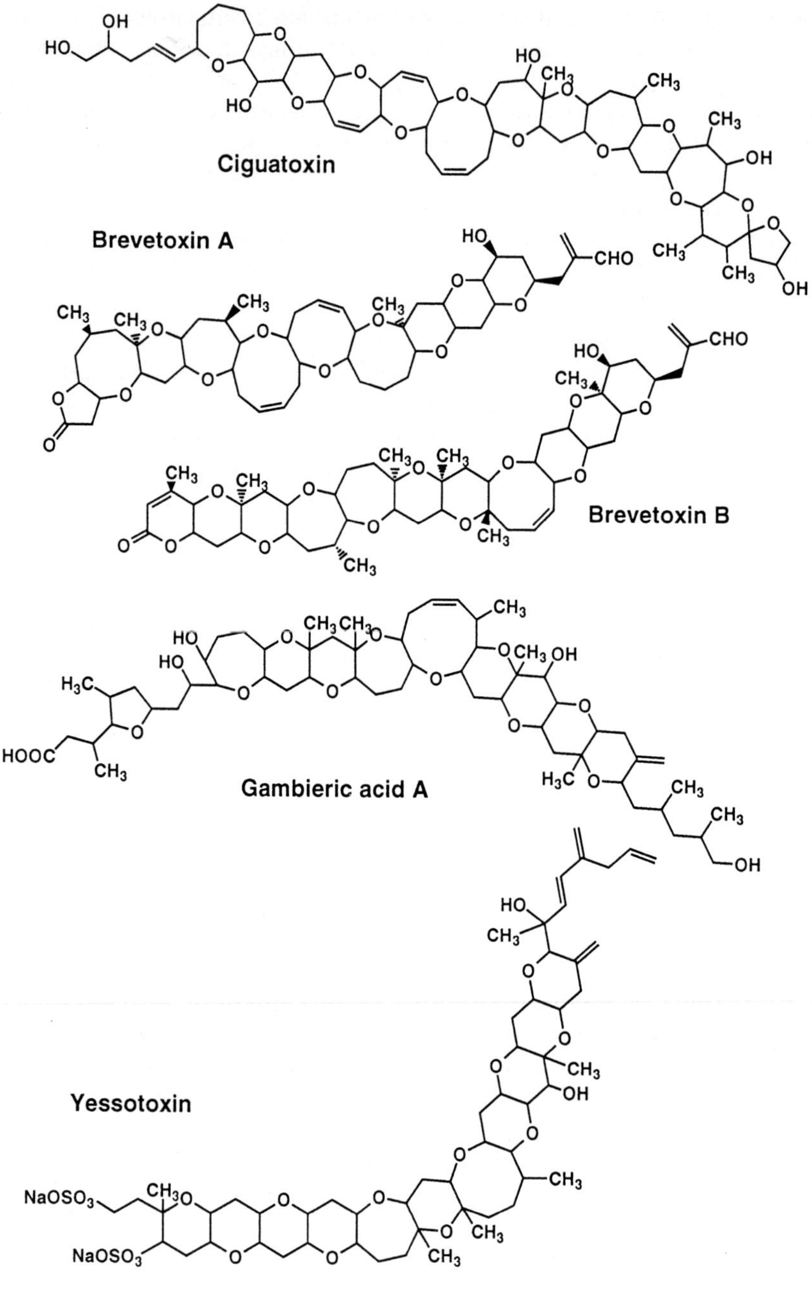

Ciguatoxin

Brevetoxin A

Brevetoxin B

Gambieric acid A

Yessotoxin

Figure 3. Selected examples of ladder-frame polyethers

dinophysoid (*Dinophysis*), gonyaulacoid (*Gonyaulax*, *Alexandrium, Pyrodinium*), peridinioid (*Peridinium*) and gymnodinoid (*Amphidinium*, *Gymnodinium*). Significantly, dinoflagellate species which produce polyether toxins are found within each of these groups; arrangement of the toxigenic dinoflagellate taxa which produce polyether compounds on a phylogenetic tree based upon deduced homologies of extant species (*sensu* Taylor 1987) is an attempt to examine potential affinities in biosynthetic pathways for these metabolites (Fig.4). Only a representative species is shown as the terminal taxon for each group of polyether compounds. It must be considered that this phylogenetic reconstruction is not universally accepted - recent molecular evidence (Zardoya *et al.* 1995) places the prorocentroids rather close to the gymnodinoids. A comprehensive list of the taxa known to produce particular biologically active polyether compounds (whether they are reported as toxic or not) is presented in Table 1.

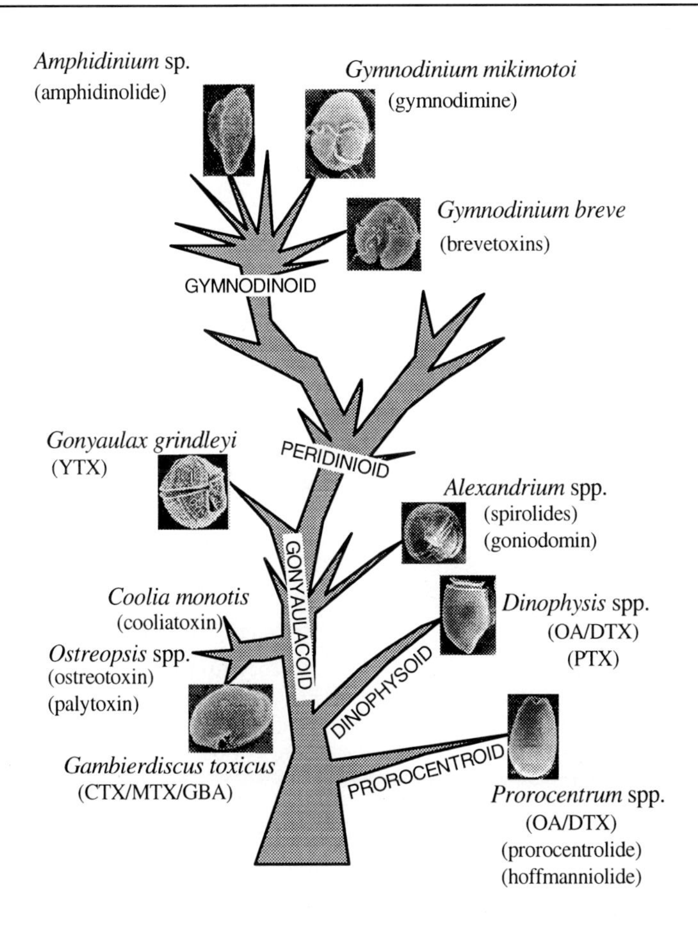

Fig.4. Arrangement of polyether-producing dinoflagellates on a hypothetical evolutionary tree (adapted from Taylor 1987).

Table 1. Biologically-active polyether compounds found in various eukaryotic microalgae. Only taxa where such metabolites were specifically identified are listed. Note that indication of a specific toxin does not indicate that it is present in all strains. OA=okadaic acid; DTX=dinophysistoxin; GBTX = brevetoxin; GA = gambieric acid; PTX = pectenotoxin; YTX = yessotoxin.

Species	Polyether Compound
Dinoflagellates	
Alexandrium ostenfeldii	spirolides
Alexandrium pseudogonyaulax (=*Goniodoma pseudogonyaulax*)	goniodomin
Amphidinium sp.	amphidinolides
Amphidinium klebsii	amphidinols
Coolia monotis	cooliatoxin
Dinophysis acuminata	OA/DTX
Dinophysis acuta	OA/DTX
Dinophysis caudata	OA
Dinophysis fortii	OA/DTX/PTX
Dinophysis mitra	DTX
Dinophysis norvegica	OA/DTX
Dinophysis rotundata	DTX
Dinophysis sacculus	OA
Dinophysis tripos	
Gambierdiscus toxicus	CTX/MTX/GA
Gonyaulax grindleyi (=*Protoceratium reticulatum*)	YTX
Gymnodinium breve (= *Ptychodiscus brevis*)	GBTX (=PbTx)
Gymnodinium mikimotoi (=*G. nagasakiense*)	gymnodimine
Ostreopsis lenticularis	ostreotoxin/ostreopsin
Ostreopsis siamensis	ostreocin (palytoxin)
Prorocentrum belizeanum	OA
Prorocentrum lima (=*Exuviella lima*)	OA/DTX/OA-diol ester; prorocentrolide
Prorocentrum hoffmannianum (= *Prorocentrum concavum*)	OA; hoffmanniolide
Prorocentrum maculosum (= *Prorocentrum concavum*)	OA/DTX/OA-diol ester
Symbiodinium sp.	zooxanthellotoxins
Other Taxa	
Chattonella marina [**Rhaphidophyte**]	GBTX
Prymnesium parvum [**Prymnesiophyte**]	prymnesins

Comparison of the hypothetical phylogenetic distribution of these polyether types among the genera containing polyether-producing species leads to some interesting observations (Fig.4). For example, there are links in structural homology of polyethers produced among the "primitive" dinoflagellates (*sensu* Taylor 1987) (prorocentroid, dinophysoid, and some gonyaulacoids, e.g., *Alexandrium* spp.) which produce linear or macrocyclic polyethers. Yet different gonyaulacoids, including the closely related group: *Gambierdiscus/Ostreopsis/Coolia* and *Gonyaulax* produce the fused ladder-frame type. Bacterial polyethers are generally smaller molecules than those produced by dinoflagellates, however there is little support for the idea that the level of molecular complexity (molecular weight or number of heterocycles) reflects the degree of primitiveness among dinoflagellates - one of the most elaborate structures (palytoxin) is produced within a relatively "primitive" genus (*Ostreopsis*). Both basic types of polyether structures are found among gymnodinoids, alternatively suggesting that one ancestral structural type was preferentially selected among different species during the long evolutionary history of this highly derived group or that these divergent structures have a polyphyletic origin.

2.5. The eco-evolutionary role of polyethers

Arguments have been made for a variety of roles for secondary metabolites in bacteria and fungi (e.g., Davies *et al.* 1992). For example, antibiotic production may be linked to critical events in the life cycle of the producing organism such as sporulation, when it is assumed the organism may be vulnerable. Although a defensive role for antibiotic production in bacteria is plausible, such a role does not apply to all secondary metabolites, and other functions such as signaling factors, hormones, cell regulators, etc., have been suggested. Such data for dinoflagellate metabolites is much more limited. Some dinoflagellate polyethers display potent antifungal activity, and it may be reasonable to argue a chemical defensive role for some of them (e.g., the phosphatase-inhibiting DSP toxins in *Prorocentrum* species), but less plausible for others (e.g., spirolides) where activity is directed towards the mammalian central nervous system (Hu *et al.* 1996).

3. Ecophysiology and polyether toxin production in dinoflagellates

3.1. Taxonomic and biogeographical distribution

One of the more intriguing features of the polyether compounds is their cosmopolitan distribution among a variety of taxa from widely diverse habitats. The occurrence of a particular sub-class of polyether compounds in widely divergent taxa, e.g., the presence of DSP toxins in the sponge *Halichondria*, as well as in several dinoflagellate species of epibenthic *Prorocentrum* and pelagic *Dinophysis*, are indications of possible polyphyletic and/or endosymbiotic origin. A brief survey of the chemical constituents of sponges strongly suggests that some other "sponge metabolites" (which are often polyethers) may be dinoflagellate products. Many dinoflagellates exist symbiotically with sponges, anemones, and corals (as zooxanthellae) and production of certain secondary metabolites may be partially or completely integrated.

The lipophilic polyether compounds originally associated with the DSP toxin complex include okadaic acid (OA) and dinophysistoxin derivatives (DTXs), pectenotoxins (PTXs), and yessotoxins (YTXs) (Yasumoto and Murata, 1990). Based upon our structural classification (Figs. 1-3), weak evidence of phosphatase inhibition and diarrheagenic activity, and its occurrence in *Gonyaulax grindleyi* (=*Protoceratium reticulatum*), it appears that YTX may not be a legitimate member of the DSP toxin complex (*sensu* Yasumoto and Murata 1990). The toxigenic dinoflagellates known to produce DSP toxins (certain *Dinophysis* spp. and *Prorocentrum* spp.) are considered to be primitive and close phylogenetic relatives (Fig.4), based primarily upon their common bilateral symmetry, apically-inserted flagellae and thick-walled thecae (Taylor 1987). Yet it is difficult to find two groups of free-living dinoflagellates with less apparent overlap in ecological niches - *Dinophysis* spp. are almost exclusively planktonic (see Maestrini, this volume) whereas toxigenic *Prorocentrum* spp. are epibenthic and are usually found attached to benthic substrates and debris, macrophytes or floating detritus (see Tindall and Morton, this volume). Toxic species of both genera have a cosmopolitan distribution, although the diversity of toxigenic epibenthic *Prorocentrum* species appears to be higher in tropical and sub-tropical waters. This begs the question of whether the expression of the common trait of toxin production merely represents a shared evolutionary artifact or if it is integral (perhaps with totally different functions) to ensuring survival in diverse habitats.

The epiphytic dinoflagellate *Gambierdiscus toxicus* is assumed to be the causative organism in most outbreaks of ciguatera fish poisoning (CFP). Although there is some inconclusive evidence that other dinoflagellates from *in situ* benthic assemblages in tropical waters (*Coolia*, *Ostreopsis*, *Amphidinium*, *Prorocentrum*) may play a subsidiary role in CFP (see Tindall and Morton, this volume), no true ciguatoxin analogues have been confirmed in any other species. Two sub-classes of cyclic polyethers which differ in mammalian potency and receptor binding affinity are specifically associated with ciguatera - the non-polar ciguatoxins (CTX) and the polar maitotoxins (MTX). Recent structural evidence (Holmes *et al.* 1991) indicates that certain CTX congeners found in finfish may represent oxidation products of related compounds synthesized by *G. toxicus* (now referred to as gambiertoxins, and related compounds: gambieric acid-A, -B,-C,-D and gambierol]).

The primary source of brevetoxins (NSP toxins) in marine ecosystems is the photoautotrophic red-tide dinoflagellate *Gymnodinium breve* (= *Ptychodiscus brevis*) (see Steidinger *et al.*, this volume), which occurs primarily off the coast of Florida and the Gulf of Mexico, but recent reports of NSP toxicity are linked to related morphotypes from New Zealand. The remarkable report of GBTX production by *Chattonella marina* (Ahmed *et al.* 1995) warrants further investigation, as the raphidophytes are not considered to be phylogenetically close to the dinoflagellates.

There remains the unresolved basic question of why these metabolites are produced by only some species of dinoflagellate genera, and why they can also be entirely absent from particular geographical populations belonging to the same toxigenic morphospecies. For example, the non-toxic *Dinophysis* populations or ecotypes present along the Canadian Atlantic coast (predominantly *D. norvegica*, *D.* cf. *acuminata* and *D. acuta*) are morphologically indistinguishable (within acceptable taxonomic variation) from their toxigenic counterparts from northern Europe and Japan. One is drawn to the conclusion that in these non-toxic species and populations, the biosynthetic genes are not being expressed, or that some or all of the biosynthetic genes are missing from the genomic DNA.

3.2. Factors affecting polyether toxin production

3.2.1. The brevetoxins

There have been only a few studies of genetic stability and physiological effects of environmental variables on brevetoxin production, because the early work focused only on structural elucidation of these notoriously labile derivatives. There is no information on the importance of life history events in toxigenesis; the sexual life cycle has been described for *Gymnodinium breve* (Steidinger *et al.*, this volume) but the induction of sexuality is rarely observed *in situ* and no corresponding toxin determinations are available.

Inconsistencies in the application of purification methods, low chemical stability of brevetoxin derivatives, and nomenclatural differences among various research groups has led to considerable confusion as to the number of naturally occurring analogues produced by *G. breve* (reviewed by Baden 1983). Brevetoxins isolated from *in situ* blooms off the Florida coast initially included BTX-B (=GB-2, T34) and the dihydro- derivative GB-3 (= T17, T46?), which is consistent with the production of these two neurotoxic derivatives by *G. breve* cultures, plus various hemolytic components. The brevetoxin analogues were subsequently re-named (as PbTx-1 to 8, or GBTX-A, GBTX-B, GBTX-*n*) with the discovery of several additional derivatives, and there are undoubtedly more derivatives produced by cultured isolates and natural populations which await further characterization.

Estimates of cellular toxicity and toxin content are subject to considerable uncertainty due to the lack of standardized culture conditions and methodologies for determining brevetoxin content, although almost all biochemical and physiological research has been carried out on the same isolate. Early work on brevetoxin purification (Spiegelstein *et al.* 1973) indicated that cellular toxin content in axenic batch cultures was roughly constant throughout exponential growth (mean: 1.4 ng crude toxin $cell^{-1}$, n=7). Subsequent work on multiple clones yielded a range of toxin content from 6.6-16.6 pg $cell^{-1}$ (Baden and Tomas 1989). Kim and Martin (1974) determined the effect of salinity on the synthesis of DNA, acidic polysaccharide and the ichthyotoxic components (presumably brevetoxins). The fact that the rate of toxin synthesis decreased linearly with the rate of DNA and polysaccharide synthesis suggests a direct negative correlation of toxin biosynthesis with active cell division, i.e., brevetoxins production may be stimulated under "stress" conditions.

The nutritional physiology of this species remains poorly understood. There is some indication of heterotrophic ability - after a long acclimation period, high cell yields and enhanced total brevetoxin yield may occur at high organic-N loading (glycine or urea at >0.5 mM) in the culture, whereas feeding of leucine or aspartate also results in a toxin profile shift and an increased toxin cell quota (Shimizu *et al.* 1995).

In a quantitative study on brevetoxin composition, the total cellular toxin content was found to be roughly constant in unialgal batch cultures throughout all growth phases but a clear shift in toxin profile was evident (Roszell *et al.* 1990). In logarithmic growth, two α-β unsaturated aldehyde derivatives of GBTX-A and GBTX-B (PbTx-1 and PbTx-2, respectively) were dominant, but four additional brevetoxins, including epoxide-, acetate- and primary alcohol-derivatives were produced in stationary phase. This shift in the ratios of -CHO:-CH_2OH to yield less potent primary alcohols is likely mediated by conventional enzymes (NADH/ NADPH-dependent alcohol dehydrogenase) and may function to reduce the auto-toxicity effect of brevetoxins in senescent cells.

Research on clonal variation of cultured isolates (n=6) from Florida and Texas has indicated considerable heterogeneity in brevetoxin profiles (Baden and Tomas 1989). The GBTX-B analogue (PbTx-2) was consistently the most abundant, but with large variations in the relative concentration of the corresponding alcohol (PbTx-3) and GBTX-A derivative (PbTx-1). Although these compounds are rather labile and interconversions can occur, evidence suggests that this heterogeneity has a genetic basis and may dramatically affect the toxicity of *in situ* blooms.

The brevetoxins are apparently well conserved within the cell and little leakage or excretion of toxin is found with healthy rapidly dividing populations. Erratic yields of brevetoxin from cultures in stationary growth phase (Sasner *et al.* 1972) indicate the probable loss of toxin through loss of membrane integrity or cell lysis. Some toxin increase in the cell-free medium may also result from the increased amount of more hydrophilic components in senescent cells (e.g., primary alcohol derivatives), which could be more readily excreted (Roszell *et al.* 1990). Only negligible amounts of brevetoxin are typically detected in the cell-free medium of cells in exponential growth phase, whereas in stationary phase the proportion of toxin in the culture medium increased to as much as 30% of the total (Spiegelstein *et al.* 1973). This effect can likely be attributed to the higher presence of disintegrating cells in mature cultures. This evidence supports the argument that brevetoxins are endotoxins which are not specifically designed as bacteriostatic, allelopathic, or ichthyotoxic agents.

3.2.2. The ciguatoxin complex

Ecophysiological studies on the production of CTX congeners by *G. toxicus* have been hampered by the low strain-specific toxin yields (often <1 MU [mouse unit] per 10^6 cells) in culture (Holmes *et al.* 1991; Satake *et al.* 1993), limited availability of chemical analytical methods, and difficulties in obtaining purified reference compounds. As is the case for many phycotoxins, cellular levels of “ciguatoxin” were often found to be substantially higher in material harvested from natural assemblages than from cultured cells (Holmes *et al.* 1991). Apparent wide variation in cellular toxin yield (an order of magnitude) and toxicity among wild populations harvested from the Caribbean (McMillan *et al.* 1986) may be indicative of substrate-related or genetic heterogeneity, but some differences could also be due to inconsistencies in sample preparation and toxin assays. Axenic cultures of *G. toxicus* from tropical Pacific sites with high detrital ciguatoxicity were found to produce extremely low yields of ciguatoxin (Yasumoto *et al.* 1979). This was also true for unialgal strains from Hawaii and the Caribbean which yielded little ciguatoxicity, but produced abundant mucilage and deformed cells (M. Leblanc, A. Cembella, F.J.R. Taylor, unpubl. obs.). Obviously, the growth conditions were sub-optimal; the exact nutritional requirements for optimal toxin production (presence of bacteria or microalgal cohort species, macroalgal exudates or other organic compounds as vitamins or chelators, etc.) are not yet completely defined.

Nevertheless, several detailed reports on clonal variation in toxin production by *G. toxicus* are available from culture experiments where apparently healthy cells and adequate toxin production were achieved. Unfortunately, many of these experimental studies have yielded apparently contradictory results regarding toxin stability and the specific effects of environmental factors. For example, Durand-Clement (1986) found little evidence of a large increase in toxicity with the age of the culture and cellular toxin yields were similar to those found for wild populations from the site of isolation in the southeast Pacific (in contrast to earlier findings of Yasumoto *et al.* 1979). These apparently discrepant results could be due to errors in cell enumeration, and/or to optimization of culture techniques and improvements in the preparation and assay of toxic extracts in the later studies. There is, nevertheless, clear evidence that manipulation of environmental variables can result in the almost complete suppression of toxin production even in a highly toxic clone (Morton *et al.* 1993).

Growth at high temperature and saturating irradiance, such as might be encountered during dispersal on drift macroalgae, enhanced cellular toxin content of *G. toxicus* from the Florida Keys (Bomber *et al.* 1988). However, Durand-Clement (1986) noted that toxicity per cell was largely independent of temperature and composition of the growth medium, although a two-fold increase in toxicity was found in a comparison of stationary phase versus freshly inoculated cells. There was a significant increase in the number of larger cells (max. diameter 100μm) with aberrant morphology in stationary phase - if the cellular toxicity was normalized per unit cell volume and these teratogenous morphotypes were considered as cells arrested in an advanced stage of the cell cycle (after toxin synthesis but prior to cytokinesis), it may be possible to explain this difference without invoking an increase in total toxin production.

In a detailed investigation of genetic differentiation, Bomber *et al.* (1988;1989) found stable clonal variation in toxicity among strains from the Florida Keys, Bahamas and the Caribbean harvested in early stationary phase. After a rigorous acclimation regime, cellular toxicity was found to be positively correlated with both cell division rates and the age of the isolate in culture (Bomber *et al.* 1989). This suggests that maximal toxin production is achieved only after adequate acclimation to the ambient regime. There is evidence that although cellular toxicity of contemporaneous toxic strains isolated from the same location may vary by up to two order of magnitude (Babinchak *et al.* 1986), and non-toxic *G. toxicus* isolates and natural populations may be more common than exceptional (Holmes *et al.* 1991), the toxin profile is relatively stable within a given isolate, especially when grown and harvested under standardized conditions (Babinchak *et al.* 1994; Holmes *et al.* 1991). The relative stability of the cellular toxicity of sub-clones (Durand-Clement (1986; Bomber *et al.* 1989) indicates both toxin profile and nominal cell potency are likely to be genetically determined. The ability for gambiertoxin production appears to be a trait which is neither readily acquired nor lost in long-term culture. There appears to a consistent difference in toxigenicity among clones from geographically disjunct populations (Holmes *et al.* 1991; Satake *et al.* 1993; Sperr and Doucette 1996) which applies to both the CTX precursors (gambiertoxins) and maitotoxin derivatives. Biogeographical analysis of strain potency of isolates (primarily from the sub-tropical Atlantic) revealed an inverse relationship with latitude (Bomber *et al.* 1989), which parallels the decrease in ciguatera with increasing latitude.

Few studies have examined the direct affect of macronutrient concentration and ratios on toxigenicity of *G. toxicus*. Examination of geographically distinct isolates from the Atlantic and Pacific showed that maximum growth rates and a shift up in cell toxicity from mid- to late exponential growth phase occurred within a wide range of N:P ratios (5:1 to 50:1) (Sperr and Doucette 1996). The toxicity peak in mid-logarithmic growth for three isolates occurred at an intermediate N:P ratio of 30:1 - approximately twice the Redfield ratio assumed to represent "balanced growth". Since Ca^{++} channel activity ("MTX-like") was more consistent among clones at all N:P ratios than Na^{+} channel responses ("CTX-like"), this suggests that the CTX-components may be more labile to catabolism or biotransformation than MTX. This also substantiates earlier reports indicating weak (or absent) CTX production by *G. toxicus* in culture and from natural populations, whereas MTX is usually dominant, particularly in cultured cells (Yasumoto *et al.* 1979). It has been suggested (Bomber 1991) that the increase in cellular ciguatoxicity with increasing NH_4^+ might indicate a functional role of these compounds in ion transport across the cell membrane. However, it is more likely that this pattern merely follows the increasing cell density in exponentially growing cultures; there is no specific N-requirement for biosynthesis of MTX or CTX or their immediate precursors.

Excretion of gambieric acid (GA) derivatives may serve an important role in sustaining the growth and viability of *G. toxicus* cells. These compounds have potent fungicidal activity (Nagai *et al.* 1993) and it has also been postulated that they may function as endogenous growth promoters (Sakamoto *et al.* 1996). Addition of CTX or MTX to *G. toxicus* cultures showed no stimulation of growth (Durand *et al. 1985*), whereas addition of GA enhanced cell growth following a dose-response up to a threshold, beyond which it was inhibitory (Sakamoto *et al.* 1996). Significantly, GA excretion increases suddenly as cells enter stationary phase, thus it is not clear that GA loss to the medium represents excretion or leakage due to loss of membrane integrity. This pattern of GA production and release to the medium may explain earlier observations (Bomber and Aikman 1989) that net toxicity decreased during senescence whereas reproductive rates were positively correlated with toxin production.

An early speculation (Steidinger and Baden 1984) that these polyether compounds may serve a function as photosynthetic modulators in low-light adapted epibenthic species such as *G. toxicus* was given additional credence with the discovery of a significant negative correlation between cellular chlorophyll levels and toxicity in multiple acclimated clones (Bomber *et al.* 1990). Nevertheless, higher growth rates (and less total photosynthetic pigments) were also exhibited at the higher light intensities, which could indicate that the response is related to optimized growth rather than a direct link to photosynthesis. In contrast, a detailed study of the response of a single clone to a gradient of light intensity (Morton *et al.* 1993) indicated no relationship between growth rate and cell potency, and toxicity was *positively* correlated with the content of major photosynthetic pigments (chlorophyll a and c and peridinin). Thus the relationship (if any) among growth rate, pigment biosynthesis and toxigenesis remains cryptic and elusive.

The influence of extracellular bacteria on toxin production is unclear, however no differences in toxicity of *G. toxicus* have been related to abundance of bacteria in the medium (Durand-Clement 1986; Bomber *et al.* 1989). Furthermore, TEM sections through toxigenic cells did not reveal any endosymbiotic bacteria (Tosteson *et al.* 1989). Ciguatera toxins produced by *G. toxicus* have been shown to inhibit the growth of certain marine microalgae (Bomber 1987) and there is some evidence that MTX may be a specific inhibitor of *in situ* cohort species of benthic diatoms co-occurring with *G. toxicus* (Bomber 1991). In nature, production of ectocrine allelochemicals by epi-benthic species would be a selective advantage for space competition, and may be particularly advantageous for mucus-producing species which may be subject to attack of the polysaccharide sheath by pathogenic bacteria or fungi.

3.2.3. *The diarrhetic shellfish toxin complex*

Until recently, research on the physiology of DSP toxin production in *Prorocentrum* spp. has been hampered by the limitations of analytical methods, incomplete characterizations of toxin analogues, and poor availability of reference standards. Dependence on liquid chromatography with fluorescence detection (LC-FD) restricts detection to DSP toxins with a free carboxylic acid group (e.g., OA, DTX1), and the discovery that the final biosynthetic products of *Prorocentrum* biosynthesis are actually the water-soluble sulfated compounds (e.g., DTX4 and DTX5a and 5b) (Hu *et al.* 1994, 1995), has led to misleading or erroneous conclusions regarding toxin production and cellular toxin content (discussed by Quilliam *et al.* 1996). Conventional extraction procedures result in a rapid enzymatic hydrolysis of DTX4 to the diol ester (see Fig. 1 for structures), which is more slowly hydrolyzed to OA, perhaps via a distinct esterase pathway (Hu *et al.* 1995). Since DTX4 and the diol ester are weak or inactive phosphatase inhibitors, it has been proposed that the DTX4-type sulfates are a means of safely harboring the lethal okadaic acid form in the cell (Hu *et al.* 1995), and this has important physiological consequences for auto-toxicity and sequestration of these metabolites.

3.2.3.1. Planktonic **Dinophysis** *spp.*

The failure to achieve "true cultures" of *Dinophysis* spp. (through serial transfers with significant biomass generation) has greatly impeded ecophysiological studies on DSP toxin production. Nevertheless a few attempts have tried to quantify cell toxin quotas from field populations either by bulk harvesting and fractionation of natural assemblages or by micropippete isolation and pooling of individual cells. The latter method was used to unequivocally demonstrate inter- and intra-specific differences in cellular toxicity (undetectable levels to 191.5 pg cell^{-1}) and toxin profile (OA/DTX1 ratios) of *Dinophysis* spp. (Lee *et al.* 1989a). Toxigenicity may be an ephemeral trait, since pronounced seasonal variation in toxicity is often expressed within a species from a given geographical region (Lee *et al.* 1989a; Reguera *et al.* 1993; Andersen *et al.* 1996; Sato *et al.* 1996). There is clear evidence of biogeographical divergence in DSP toxin profile among natural populations of the same species. In northern Japan, *D. fortii* was found to produce DTX1 but none was detected in southern populations (Yasumoto and Murata 1990). On the Atlantic coast of North America the toxicity of *Dinophysis* populations is rather enigmatic. Northern populations of *D. norvegica* and *D. acuminata* from the Gaspe coast, Quebec, Canada were moderately toxic (25-33 pg OA cell^{-1}) (Cembella 1989), but DSP toxins have not been confirmed in more southern populations from the Canadian Atlantic region (A. Cembella, M. Quilliam and J. Smith, unpublished data) nor from the Gulf of Maine along the New England coast. Initial hopes that toxic profiles could be used to establish clear inter-specific identities or to reflect geographical boundaries are not supported by current information. In northern Europe, *D. acuminata* and *D. acuta* were initially found to produce OA (but not DTX1), whereas DTX1 was dominant in *D. norvegica* from Norway (Lee *et al.* 1989a). The OA/DTX1 ratio in *D. acuminata* and *D. acuta* from Sweden were later found to be distinctive, but they are markedly influenced by nutrient-dependent growth (Johansson *et al.* 1996). The OA/DTX1 ratios in *D. fortii* and *D. acuminata* are also inconsistent within Japan (Sato *et al.* 1996).

The potential link between DSP toxin production and photosynthetic activity requires further investigation. With the exception of a suspect report of DTX1 in *D. rotundata* (Lee *et al.* 1989a) - a heterotrophic species which was found to produce little or no DSP toxins in other studies (Cembella 1989; Masselin *et al.* 1992), DSP toxins in *Dinophysis* have only been found in photoautrophic species (Table 1).

Inorganic macronutrient effects on toxin production in *Dinophysis* spp. are complex and there are insufficient data to define the relationships. Evidence from *Dinophysis*-enriched quasi-cultures of mixed natural assemblages (Johansson *et al.* 1996) indicated that nutrient-limited growth yields an increase in cell toxin quota, accompanied by a shift in the OA/DTX1 ratio - but the response is species-specific. For *D. acuminata,* the greatest increase in toxin content was associated with N-limitation of growth, but this cannot be directly compared with the P-dependent growth response, since growth rates were under low-P conditions were comparable to those in nutrient replete controls. Many toxigenic *Dinophysis* spp. are facultative heterotrophs (see Maestrini, this volume), therefore the potential for toxin acquisition via mixotrophy should not be excluded. There may be a relationship between the increased presence of inclusion bodies and higher toxin content under N-limitation (Granéli *et al.* 1997) - this could be linked to mixotrophic activity or to a shift from protein towards lipid and carbohydrate biosynthesis as cell-N quota is depleted.

3.2.3.2. Epibenthic **Prorocentrum** *spp.*

There is no information on the relationship between the induction of sexuality and the production of DSP toxins among *Prorocentrum* spp., although several benthic species are known to include a sexual stage in the life history. Most ecophysiological studies of DSP toxin production in prorocentroids have focused on *P. lima* which is found primarily in the benthos and attached to macroalgal debris, and less commonly in the plankton. The genetic diversity in DSP toxin production (OA/DTX1) among acclimated multiple clones from adjacent geographical regions has been clearly demonstrated (Morton and Tindall 1995). However, the direct effect of environmental cues versus genetic factors on DSP toxin production in *Prorocentrum* spp. remains to be fully elucidated. Multiple batch sub-cultures of *P. lima* grown under identical conditions and harvested at approximately the same cell density in late exponential growth phase were found to vary by as much as an order of magnitude in total DSP toxin content with no apparent systematic variation in the ratio of OA/DTX1/OA-diol ester (Bauder *et al.* 1996). A linear increase in the cell concentration of total DSP toxins throughout exponential growth and into early stationary phase (Quilliam *et al.* 1996) indicates the continuation of toxin biosynthesis and/or the ability to maintain the cell toxin quota as the growth rate declines. A recent detailed study of DSP toxin production by *P. lima* in a continuous culture (turbidostat) system revealed an approximately linear positive correlation of the cellular concentration of the predominant toxin [DTX4] and the growth rate, with a non-zero intercept (A. Windust, pers. comm.).

The direct effects of physical factors (light, temperature, salinity, turbulence, etc.) on toxin production in *Prorocentrum* spp. have yet to be dissected from their indirect effects on growth rate. For example, as would be expected in *P. lima*, the growth rate was a positive function of temperature (Jackson *et al.* 1993), but total cellular toxin content was inversely related to temperature, with no apparent effect on the OA/DTX1 ratio. It is likely that the higher toxicity at low temperature merely represents a higher cell toxin quota due to a reduced division rate rather an increase in the biosynthetic rate. In *P. hoffmannianum*, temperature- and light-dependent growth rates and cell OA content showed no consistent trends, whereas OA content was inversely correlated with salinity-dependent growth rate (Morton *et al.* 1994). The most consistent explanation is that increased cell toxin quota is a response to non-specific environmental stress modulated via the down-shift in growth.

A similar case may be made with respect to macronutrient limitation. The apparent higher cellular yield of OA when *P. lima* was grown on organic- versus inorganic-P enrichment may be related to the reduced growth rate and hence prolonged exponential growth phase with the organic source (Tomas and Baden 1993). Assuming the maintenance of a minimal cell-N quota from exogenous N, total toxin production in *P. lima* cells under N-dependent growth conditions was found to be highest in stationary phase (McLachlan *et al.* 1994). The negative correlation of macronutrient uptake (NH_4 and PO_4) and cell toxicity in *P. hoffmannianum* indicates a possible competitive interaction of the mechanisms of toxin biosynthesis and nutrient assimilation (Aikman *et al.* 1993).

Excretion (or leakage) of significant quantities of OA into the growth medium (*ca.* 20% of total) even in early exponential growth phase (Rausch de Traubenberg and Morlaix 1995) indicates the potential of *P. lima* for conditioning its immediate environment. The differential partitioning of DSP toxin components - with the phosphatase inhibitors OA (and some DTX1) present in the culture medium, whereas the phosphatase-inactive form DTX4 dominated in the cells of *P. lima* (Quilliam *et al.* 1996), may indicate enzymatic cleavage to the active forms at the cell periphery or in the periplasmic space. These potent phosphatase inhibitors may have an allelopathic role against potential competitors, as evidenced by the growth inhibitory response against certain microalgae (Windust *et al.* 1996).

The relationship of bacteria to DSP toxin production is extremely tenuous. Little toxin (*ca.* 1%) was associated with the free-living bacterial fraction in *P. lima* cultures (Rausch de Traubenberg and Morlaix 1995). An analysis of serial sections by TEM (Zhou and Fritz 1993) found no evidence of endocellular bacteria in either *P. lima* or *P. maculosum* - toxin production in these species is not likely due to prokaryotic endosymbionts.

The potentially significant relationship between the photosynthetic apparatus and DSP toxin production was highlighted by the discovery that a monoclonal OA-antibody probe was localized mainly to chloroplasts and pyrenoids, with some labeling of autodigestive lysosomes (Zhou and Fritz 1994). In addition to the storage of DSP analogues as water-soluble sulfates (Hu *et al.* 1995), this suggests a self-protection mechanism against this powerful phosphatase inhibitor through compartmentalization in the thylacoid membranes away from cytoplasmic protein phosphatases (PP1A and

PP2A). The high correlation between chlorophyll c_2 and DSP toxin content (Morton and Tindall 1995) supports this plausible linkage. The persistence of toxicity through long term culture and the relative stability of the toxin profile provides circumstantial evidence that toxin production in *Prorocentrum* species is genetically determined and tightly regulated, although none of the biosynthetic genes have been localized.

4. Conclusions

Chemical evidence that dinoflagellates biosynthesize unique polyether metabolites, which can be classed into two structural groups (linear- and macrocyclic- or fused ladder-frame-type), continues to emerge. However, the biosynthetic data indicates that polyethers from either group share many biosynthetic steps - they are first assembled from acetate units by a polyketide pathway and post-PKS modification results in the different structural types. Subsequent oxidative modification, methylation, and cyclization of the polyketide chain results in the highly functionalized polyether products. Although some of biosynthetic processes are reported to operate in other organisms (particularly bacteria), all of these unusual steps appear to operate simultaneously in dinoflagellates. As a result of their characteristic chemistry, and more particularly the unique biosynthetic pathways for their production, the polyether metabolites may serve as important ecophysiological and taxonomic markers.

As for most secondary metabolites, a definitive biological role for these compounds is difficult to establish, but obviously they were not selected for their latent activity as shellfish toxins. It is probably more than coincidental that - with the dubious exception of *D. rotundata* - all marine protist species confirmed to produce polyether compounds are facultative or obligate photoautotrophs. The correlation with cellular pigment levels and localization within the chloroplasts (DSP toxins) may indicate a role in modulation of photosynthesis. Many of these marine polyethers are ion channel effectors raising speculation as to whether they may play (or once played) a role as ion channel regulators. The hydrophobicity of these compounds, together with their considerable length and structural rigidity, which allows them to span large regions of an ion channel, are plausible supporting factors for this idea. The DSP toxins are notable exceptions in that they are not ion channel effectors, but rather they act as serine/threonine phosphatase inhibitors. They too may serve as enzyme regulators, though their allelopathic and antifungal properties also suggest a possible chemical defense role for these polyethers.

The reasons for variation in polyether metabolites among isolates and natural populations remain poorly defined. Presumably, exogenous environmental factors (light, temperature, salinity, nutrient availability) are modulated through endogenous mechanisms (circadian rhythms, cell cycle events) which are derived from a template of genetic diversity (presence and expression of genes and biosynthetic enzymes coding for toxin biosynthesis). In culure, biosynthesis of polyether derivatives appears to be constitutive, rather than facultative, although environmental factors can affect the rate of production. These compounds are not typical secondary metabolites synthesized *only* in response to short-term environmental stress. In many cases, rather non-specific exogenous factors which depress growth result in a higher cell quota (but not necessarily an increased rate of production) of these secondary metabolites. Use of molecular genetics and DNA analysis in biogeographical comparisons of genetic differentiation will greatly assist in determining the respective importance of clonal variation, habitat, and life cycle stages in the biosynthesis of these compounds. There are virtually no data on the heritability of genes for polyether biosynthesis or the role of sexual life histories. In the case of the polyether toxins, there is an absolute requirement for the determination of toxin components (not just net toxicity) because of potency differences. The coupling of biosynthetic studies with knowledge derived from experiments on phenotypic and genotypic expression of toxin production will ultimately help to establish the nature of the ecophysiological niche occupied by toxigenic species.

5. Acknowledgments

The authors thank Andrew Bauder and Nancy Lewis, Institute for Marine Biosciences, NRC for assistance in preparation of this manuscript. This publication is NRC (Canada) No. 39765.

6. References

Ahmed, M.S., Arakawa, O., Onoue, Y. (1995). Toxicity of cultured *Chattonella marina.* In: Lassus, P., Arzul, G., Erard-Le Denn, E., Gentien, P., Marcaillou-Le Baut, C. (eds.) *Harmful Marine Algal Blooms*, Lavoisier, Paris, pp. 499-504.

Aikman, K.E., Tindall, D.R., Morton, S.L. (1993). Physiology and potency of the dinoflagellate *Prorocentrum hoffmannianum* (Faust) during one complete growth cycle. In: Smayda, T.J., Shimizu, Y.(eds.) *Toxic Phytoplankton Blooms in the Sea*, Elsevier, N.Y., pp. 463-468.

Andersen, P., Hald, B., Emsholm, H. (1996). Toxicity of *Dinophysis acuminata* in Danish coastal waters. In: Yasumoto, T., Oshima, Y., Fukuyo, Y. (eds.) *Harmful and Toxic Algal Blooms* , IOC (UNESCO), Paris, pp. 281-284.

Babinchak, J.A., Jollow, D.G., Higerd, T.B. (1986). Toxin production by *Gambierdiscus toxicus* isolated from the Florida Keys. *Mar. Fish. Rev*. 48: 53-56.

Babinchak, J.A., Moeller, P.D.R., Van Dolah, F.M., Eyo, P.B., Ramsdell, J.S. (1994). Production of ciguatoxins in cultured *Gambierdiscus toxicus. Mem. Queensland Museum* 34: 447-453.

Baden, D.G. (1983). Marine food-borne dinoflagellate toxins. *Int. Rev. Cytol.* 82: 99-150.

Baden, D.G., Tomas, C.R. (1989). Variations in major toxin composition for six clones of *Ptychodiscus brevis.* In: Okaichi, T., Anderson, D.M., Nemoto, T. (eds.) *Red Tides Biology, Environmental Science and Toxicology*, Elsevier, N.Y., pp. 415-418.

Bauder , A.G., Cembella A.D., Quilliam M.A. (1996). Dynamics of diarrhetic shellfish toxins from the dinoflagellate, *Prorocentrum lima*, in the bay s*callop Argopecten irradians.* In: Yasumoto, T., Oshima, Y., Fukuyo, Y. (eds.) *Harmful and Toxic Algal Blooms*, IOC (UNESCO), Paris, pp. 433-436.

Bomber, J.W. (1987). Ecology, genetic variability and physiology of the ciguatera-causing dinoflagellate *Gambierdiscus toxicus* Adachi and Fukuyo, Ph.D. thesis, Florida Institute of Technology, Melbourne, Fl.

Bomber, J.W., (1991). Toxigenesis in dinoflagellates: genetic and physiological factors. In: Miller, D. M. (ed.) *Ciguatera Seafood Toxin*s, CRC Press, Boca Raton, Fl., pp. 135-170.

Bomber, J.W., Aikman, K.E. (l989). The ciguatera dinoflagellates. *Biol. Ocean.* 6: 291-311.

Bomber, J.W., Tindall, D.R., Miller, D.M. (l989). Genetic variability in toxin potencies among seventeen clones of *Gambierdiscus toxicus* (Dinophyceae). *J. Phycol.* 25: 617-625.

Bomber, J.W., Tindall, D.R., Venable, C.W., Miller, D.M. (1990). Pigment composition and low-light response of fourteen clones of *Gambierdiscus toxicus*. In: Granéli, E., Sundström, B., Edler L., Anderson, D.M. (eds.) *Toxic Marine Phytoplankton*, Elsevier, N.Y., pp. 263-268.

Bomber, J.W., Morton, S.L., Babinchak, J.A., Norris, D.R., Morton, J.G. (1988). Epiphytic dinoflagellates of drift algae-another toxigenic community in the ciguatera food chain. *Bull. Mar. Sci.* 43: 204-214.

Cembella, A.D. (1989). The occurrence of okadaic acid, a major diarrheic shellfish toxin, in natural populations of *Dinophysis* sp. From the eastern coast of North America. *J. Appl. Phycol.* 1: 307-310.

Chou, H-N, Shimizu, Y. (1987). Biosynthesis of brevetoxins: evidence for a mixed origin of the backbone carbon chain and the possible involvement of dicarboxylic acids. *J. Am. Chem. Soc.* 109: 2184-2185.

Davies, J., von Ahsen, U., Wank, H., Schroeder, R. (1992). Evolution of secondary metabolite production: potential roles for antibiotics as prebiotic effectors of catalytic RNA reactions. In: *Secondary Metabolites: Their Function and Evolution.* Ciba Foundation Symposium. J. Wiley, N.Y., pp. 24-32.

Durand, M., Squiban, A., Viso, A.-C., Pesando, D. (1985). Production and toxicity of *Gambierdiscus toxicus*. Effects of its toxins (maitotoxin and ciguatoxin) on some marine organisms. *Proc. Fifth Intl. Coral Reef Cong.*, Tahiti, 4: 483.

Durand-Clement, M. (1986). A study of toxin production by *Gambierdiscus toxicus* in culture. *Toxicon* 24: 1153-1157.

Garson, M. J. (1993). The biosynthesis of marine natural products. *Chem. Rev.* 93: 1699-1733.

Granéli, E., Johansson, N., Panosso, R. (1997). *Proceedings of the Eighth Intl. Conf. on Harmful Algae*, submitted.

Holmes, M.J., Lewis, R.J., Poli, M.A., Gillespie, N.C. (1991). Strain dependent production of ciguatoxin precursors (gambiertoxins) by *Gambierdiscus toxicus* (Dinophyceae) in culture. *Toxicon* 29: 761-775.

Hu, T., Curtis, J. M., Walter, J.A., Wright, J.L.C. (1994). Identification of DTX-4, a new water-soluble phosphatase inhibitor from the toxic dinoflagellate *Prorocentrum lima. J. Chem. Soc., Chem. Commun.* 597-599.

Hu, T., Curtis, J.M., Walter, J.A., Wright, J.L.C. (1995). Two new water-soluble DSP toxins from the dinoflagellate *Prorocentrum maculosum*: possible storage and excretion products. . *Tet. Letts.* 36: 9273-9276.

Hu, T., Curtis, J.M., Walter, J.A., Wright, J.L.C. (1996). Characterization of the biologically inactive spirolides E and F: Identification of the spirolide pharmacophore. *Tet. Letts.* 37: 7671-7674.

Igarishi, T., Satake, M., Yasumoto, T. (1996). Prymnesin-2: A potent ichthyotoxin and hemolytic glycoside isolated from the red tide alga *Prymnesium parvum. J. Am. Chem. Soc.* 118: 479-480.

Jackson, A.E., Marr, J.C., McLachlan, J.L. (1993). The production of diarrhetic shellfish toxins by an isolate of *Prorocentrum lima* from Nova Scotia, Canada. In: Smayda, T.J., Shimizu, Y. (eds.) *Toxic Phytoplankton Blooms in the Sea*, Elsevier, N.Y., pp. 513-518.

Johanson, N., Granéli, E., Yasumoto, T., Carlsson, P., Legrand, C. (1996). Toxin production by *Dinophysis acuminata* and *D. acuta* cells grown under nutrient sufficient and deficient conditions. In: Yasumoto, T., Oshima, Y., Fukuyo,Y. (eds.) *Harmful and Toxic Algal Blooms*, IOC (UNESCO), Paris, pp. 277-280.

Kim, Y.S., Martin, D.F. (1974). Effects of salinity on synthesis of DNA, acidic polysaccharide and ichthyotoxin in *Gymnodinium breve*. *Phytochem.* 13: 533-538.

Kobayashi, J., Ishibashi, M. (1993). Bioactive metabolites of symbiotic marine metabolites. *Chem. Rev.* 93: 1753-1770.

Kobayashi, J., Takahashi, M., Ishibashi, M. (1995). Biosynthetic studies of amphidinolide J: Explanation of the generation of the unusual odd-numbered macrocyclic lactone. *J. Chem. Soc. Chem. Commun.* 1639-1640.

Lee, J-S., Igarashi, T., Fraga, S., Dahl., E., Hovgaard P., Yasumoto, T. (1989a). Determination of diarrhetic shellfish toxins in various dinoflagellate species. *J. Appl. Phycol.* 1: 147-152.

Lee, M.S., Qin, G. Nakanishi, K., Zagorski, M. G. (1989b). . Biosynthetic studies of brevetoxins, potent neurotoxins produced by the dinoflagellate *Gymnodinium breve*. *J. Amemr. Chem. Soc.* 111: 6234-6241.

Masselin, P., Lassus, P., Bardouil, M. (1992). High performance liquid chromatography analysis of diarrhetic toxins in *Dinophysis* spp. from the French Coast. *J. Appl. Phycol.* 4: 385-389.

McLachlan, J.L., Marr, J.C., Conlon-Kelly, A., Adamson, A. (1994). Effects of nitrogen concentration and cold temperature on DSP-toxin concentrations in the dinoflagellate *Prorocentrum lima* (Prorocentrales, Dinophyceae). *Nat. Tox.* 2: 263-270.

McMillan, J.P., Hoffman, P.A., Granade, H.R. (1986). *Gambierdiscus toxicus* from the Caribbean: source of toxins involved in ciguatera. *Mar. Fish. Rev.* 48: 48-52.

Morton, S.L., Tindall, D.R. (1995). Morphological and biochemical variability of the toxic dinoflagellate *Prorocentrum lima* isolated from three locations at Heron Island, Australia. *J. Phycol.* 31: 914-921.

Morton, S.L., Bomber, J.W., Tindall, D.R. (1994). Environmental effects on the production of okadaic acid from *Prorocentrum hoffmannianum* Faust I. Temperature, light, and salinity. *J. exp. mar. Biol. Ecol.* 178: 67-77.

Morton, S.L., Bomber, J.W., Tindall, D.R., Aikman, K.E. (1993). Response of *Gambierdiscus toxicus* to light: cell physiology and toxicity. In: Smayda, T.J., Shimizu, Y. (eds.) *Toxic Phytoplankton Blooms in the Sea*, Elsevier, N.Y., pp. 541-546.

Nagai, H., Mikami, Y., Yazawa, K., Gonoit, T., Yasumoto, T. (1993). Biological activities of novel polyether antifungals, gambieric acids A and B from a marine dinoflagellate *Gambierdiscus toxicus*. *J. Antibiot.* 46: 520-522.

Needham, J., McLachlan, J.L., Walter, J.A., Wright, J.L.C. (1994). Biosynthetic origin of C-37 and C-38 in the polyether toxins okadaic acid and DTX-1. *J. Chem. Soc. Chem. Commun.* 2599-2600.

Needham, J., Hu, T., McLachlan, J.L., Walter, J.A., Wright, J.L.C. (1995). Biosynthetic studies of the DSP toxin DTX-4 and an okadaic acid diol ester. *J. Chem. Soc. Chem. Commun.* 1623-1624.

Nisbet, L.J. (1992). Useful functions of secondary metabolites. In: *Secondary Metabolites: Their Function and Evolution.* Ciba Foundation Symposium. J. Wiley, N.Y., pp. 215-225.

Quilliam, M.A., Hardstaff, W.R., Ishida, N., McLachlan, J.L., Reeves, A.R., Ross, N.W., Windust, A.J. (1996). Production of diarrhetic shellfish poisoning (DSP) toxins by *Prorocentrum lima* in culture and development of analytical methods. In: Yasumoto, T., Oshima, Y., Fukuyo, Y. (eds.) *Harmful and Toxic Algal Blooms*, IOC (UNESCO), Paris, pp. 289-292.

Rausch de Traubenberg, C., Morlaix, M. (1995). Evidence of okadaic acid release into extracellular medium in cultures of *Prorocentrum lima* (Ehrenberg) Dodge. In: Lassus, P., Arzul, G., Erard-Le Denn, E., Gentien, P., Marcaillou-Le Baut, C. (eds.) *Harmful Marine Algal Blooms*, Lavoisier, Paris, pp. 493-499.

Reguera, B., Bravo, I., Marcaillou-Le Baut, C., Masselin, P., Fernandez, M.L., Miguez, A., Martinez, A. (1993). Monitoring of *Dinophysis* spp. and vertical distribution of okadaic acid on mussel rafts in Ria de Ponteverda (NW Spain). In: Smayda, T.J., Shimuzu, Y. (eds.) *Toxic Phytoplankton Blooms in the Sea*, Elsevier, N.Y., pp. 553-558.

Robinson, J.A. (1991). Chemical and biological aspects of polyether-ionophore antibiotic biosynthesis. In: Herz, W., Kirby, G.W., Steglich, W., Tamm, C. (eds.) *Progress in the Chemistry of Organic Natural Products 58*, Springer-Verlag/Wien, N.Y., pp. 1-81.

Roszell, L.E., Sculman, L.S., Baden, D.G. (1990). Toxin profiles are dependent on growth stages in cultured *Ptychodiscus brevis*. In: Granéli, E., Sundström, B., Edler L., Anderson, D.M. (eds.) *Toxic Marine Phytoplankton*, Elsevier, N.Y., pp. 403-406.

Sakamoto, B., Nagai, H., Hokama, Y. (1996). Stimulators of *Gambierdiscus toxicus* (Dinophyceae) growth: the possible role of gambieric acid-A as an endogenous growth enhancer. *Phycologia* 3: 350-353.

Sasner, J.J., Ikawa, M., Thurberg, F., Alam, M. (1972). Physiological and chemical studies on *Gymnodinium breve* Davis toxin. *Toxicon* 10: 163-172.

Satake, M., Ishimaru, T., Legrand, A.M., Yasumoto, T. (1993). Isolation of a ciguatoxin analog from cultures of *Gambierdiscus toxicus*. In: Smayda, T.J., Shimuzu, Y. (eds.) *Toxic Phytoplankton Blooms in the Sea*, Elsevier, N.Y., pp. 575-579.

Sato, S., Koike, K., Kodama, M. (1996). Seasonal variation of okadaic acid and dinophysistoxin-1 in *Dinophysis* spp. in association with the toxicity of scallop. In: Yasumoto, T., Oshima, Y., Fukuyo, Y. (eds.) *Harmful and Toxic Algal Blooms*, IOC (UNESCO), Paris, pp. 285-288.

Shimizu, Y. (1993). Microalgal metabolites. *Chem. Rev.* 93: 1685-1698.

Shimizu, Y., Watanabe, N., Wrensford, G. (1995). Biosynthesis of brevetoxins and heterotrophic metabolism in *Gymnodinium breve*. In: Lassus, P., Arzul, G., Erard-Le Denn, E., Gentien, P., Marcaillou-Le Baut, C. (eds.) *Harmful Marine Algal Blooms*, Lavoisier, Paris, pp. 351-357.

Simpson, T. J. (1995). Polyketide biosynthesis. *Chem. Ind.* 407-411.

Sperr, A.E., Doucette, G.J. (1996). Variation in growth rate and ciguatera toxin production among geographically distinct isolates of *Gambierdiscus toxicus*. In: Yasumoto, T., Oshima, Y., Fukuyo, Y. (eds.) *Harmful and Toxic Algal Blooms*, IOC (UNESCO), Paris, pp. 309-312.

Spiegelstein, M.Y., Paster, Z., Abbott, B.C. (1973). Purification and biological activity of *Gymnodinium breve* toxins. *Toxicon* 11: 85-93.

Steidinger, K.A., Baden, D.G. (1984). In: Spector, D. (ed.) *The Dinoflagellates*, Academic Press, N.Y., pp. 201-261.

Taylor, F.J.R. (1987). *The Biology of Dinoflagellates*. Blackwell Scientific, Oxford, England. 785 pp.

Tomas, C.R., Baden, D.G. (1993). In: Smayda, T.J., Shimuzu, Y. (eds.) *Toxic Phytoplankton Blooms in the Sea*, Elsevier, N.Y., pp. 565-570.

Tosteson, T.R., Ballantine, D.L., Tosteson, H.V., Bardales, A.T (1989). Associated bacterial flora, growth and toxicity of cultured benthic dinoflagellates *Ostreopsis lenticularis* and *Gambierdiscus toxicus*. *Appl. Environ. Microbiol.* 55: 137.

Windust, A. J., Wright, J. L. C., McLachlan, J. L. (1996). The effects of the DSP toxins, okadaic acid and dinophysistoxin-1, on the growth of microalgae. *Mar. Biol.* 126: 19-23.

Wright, J.L.C., Hu, T., McLachlan, J.L., Needham, J., Walter, J.A. (1996). Biosynthesis of DTX4: Confirmation of a polyketide pathway, proof of a Baeyer-Villiger oxidation step, and evidence for an unusual carbon deletion step. *J. Amer. Chem. Soc.* 118: 8757-8758.

Yasumoto, T., Murata, M. (1990). Polyether toxins involved in seafood poisoning. In: Hall, S., Strichartz, G. (eds.) *Marine Toxins: Origin, Structure, and Molecular Pharmacology*, ACS Symposium Series, Washington D.C., pp. 20-132.

Yasumoto, T., Murata, M. (1993). Marine toxins. *Chem. Rev.* 93: 1897-1910.

Yasumoto, T., Torigoe, K. (1991). Seminar on marine chemistry: Biosynthesis of the bioactive compounds of *Prorocentrum lima*. *J. Nat Prods.* 54: 1486-1487.

Yasumoto, T., Nakajima, I., Oshima, Y., Bagnis, R. (1979). A new toxic dinoflagellate found in association with ciguatera. In: D.L. Taylor, D. L., H.H. Seliger, H.H. (eds.), *Toxic Dinoflagellate Blooms*, Elsevier-North Holland, N.Y., pp. 65-70.

Zardoya, R., Costas, E., López-Rodas, V., Garrido-Pertierra, A., Bautista, J.M. (1995). Revised dinoflagellate phylogeny inferred from molecular analysis of large-subunit ribosomal RNA gene sequences. *J. Mol. Evol.* 41: 637-645.

Zhou, J., Fritz, L. (1993). Ultrastructure of two toxic marine dinoflagellates, *Prorocentrum lima* and *Prorocentrum maculosum*. *Phycologia* 32: 444-450.

Zhou, J., Fritz, L. (1994). Okadaic acid antibody localizes to chloroplasts in the DSP-toxin-producing dinoflagellates *Prorocentrum lima* and *Prorocentrum maculosum*. *Phycologia* 33: 455-461.

Interactions Between Toxic Marine Phytoplankton and Metazoan and Protistan Grazers

Jefferson T. Turner[1], Patricia A. Tester[2] and Per Juel Hansen[3]

[1] Biology Department, University of Massachusetts Dartmouth, 285 Old Westport Road, North Dartmouth, MA 02747-2300, USA

[2] National Marine Fisheries Service, NOAA, Southeast Fisheries Science Center, Beaufort Laboratory, 101 Pivers Island Road, Beaufort, NC 28516-9722, USA

[3] Marine Biological Laboratory, University of Copenhagen, Strandpromenaden 5, DK-3000 Helsingør, DENMARK

1. Introduction

Interactions between toxic phytoplankton and their grazers are important but poorly-understood aspects of the ecology of harmful algal blooms. The importance of grazing becomes most apparent by its failure; if community grazing controls initial stages of toxic bloom development, there are no blooms, and the importance of grazing goes unnoticed. However, reduced grazing pressure may occur if toxicity makes bloom species unpalatable to grazers. Conversely, if grazers ingest toxic phytoplankton with impunity, and sequester phycotoxins, then vectorial intoxication of consumers beyond the grazer level in food webs can occur (White 1981).

There are considerable differences in chemistry and mode of action among phycotoxins (Baden and Trainer 1993; Hallegraeff 1993) and effects of toxic phytoplankters on their grazers. Some grazers are deleteriously affected by phycotoxins, whereas there are no apparent effects on others. Indeed, inconsistency seems to characterize many interactions between toxic phytoplankton and their grazers (reviewed by Turner and Tester 1997).

In this chapter the specific focus will be upon interactions between toxic marine phytoplankton and their direct metazoan and protistan grazers. Marine phytoplankton that are generally not "toxic" but are "harmful," in that they contribute to organic loading and/or anoxia (e.g., *Ceratium tripos* and gelatinous colonies of *Phaeocystis* spp.), will not be considered. We address effects of toxic phytoplankton on grazers, as well as impact of community grazing pressure upon toxic blooms. Transport of toxins through pelagic food webs, and effects of vectorial intoxication on upper-level consumers such as fish, birds or marine mammals have been reviewed by Anderson and White (1992) and Turner and Tester (1997).

We pose several generic questions concerning interactions between toxic phytoplankton and their grazers. These include:

- Are phycotoxins grazing deterrents?
- What types of deleterious effects do toxic phytoplankton have upon metazoan and protistan grazers?
- Why do grazers exhibit such varied responses to toxic phytoplankton?
- Is there "selective" grazing for or against toxic phytoplankton?

NATO ASI Series, Vol. G 41
Physiological Ecology of Harmful Algal Blooms
Edited by D. M. Anderson, A. D. Cembella and G. M. Hallegraeff

- What is the effect of metazoan and protistan zooplankton grazing on development and persistence of toxic phytoplankton blooms?
- What are potential influences of indirect trophic linkages on development of toxic phytoplankton blooms?
- What are potential interactions between toxic phytoplankters and the reproductive success of their grazers?

Answers to these questions are often uncertain, varied, and situation-specific. Thus, extrapolation of results from one scenario to another may be inappropriate. We limit citations to selected examples, and refer to Turner and Tester (1997) for more extensive documentation.

2. Questions on toxic phytoplankton-grazer interactions

2.1. Are phytoplankton toxins grazing deterrents?

The fact that some phytoplankton are toxic should not necessarily imply that toxicity evolved to deter grazers. Toxicity is often defined by effects on upper-trophic-level vertebrate carnivores that are not direct grazers of phytoplankton, including humans, birds or marine mammals (Anderson and White 1992). Effects of toxic phytoplankton upon many invertebrate zooplankton grazers often appear to be minimal (Table 1). Thus, while toxin production is often assumed to have evolved to deter grazers, if toxins do not inhibit feeding by direct grazers such as bivalves, zooplankters, or some fish, but rather affect mainly vertebrate carnivores such as humans, birds and whales (Dale and Yentsch 1978; Work *et al.* 1993; Gerachi *et al.* 1989), then the assumption that toxins evolved as grazing deterrents becomes questionable. However, if toxins did evolve as grazing deterrents, then they might work at cellular levels, such as on ion pumps in nervous systems or upon cell membranes. A consequence of such a generality of effect might be that phycotoxins could also intoxify non-grazing organisms such as fish, marine birds or marine mammals.

Table 1. Zooplankters that exhibit no apparent adverse effects from ingesting toxic phytoplankton.

Zooplankters	Toxic phytoplankter	Reference
Copepods -		
Acartia hudsonica	*Alexandrium tamarense*[a]	Turner & Anderson (1983)
A. erythraea	*Chattonella antiqua*	Uye (1986)
A. pacifica	"	"
Calanus sinicus	"	"
Centropages yamadi	"	"
Paracalanus parvus	"	"
Pseudiaptomus marinus	"	"
Calanus pacificus	*Heterosigma carterae*[b]	Sykes & Huntley (1987)
A. tonsa	*Gymnodinium breve*[c]	Turner & Tester (1989)
Oncaea venusta	"	"
Labidocera aestiva	"	"

Table 1. (continued)

Zooplankters	Toxic phytoplankter	Reference
Acartia omorii	*H. carterae*	Uye & Takamatsu (1990)
P. marinus	"	"
"	*Chattonella marina*	"
"	*Fibrocapsa japonica*	"
Calanus glacialis	*Pseudo-nitzschia multiseries*	Windust (1992)
Temora longicornis	"	"
A. tonsa	"	Lincoln, Turner, Bates & Léger (unpublished data)
A. tonsa	*Trichodesmium* sp. (healthy intact cells)	Guo & Tester (1994)
A. tonsa	*A. tamarense*	Teegarden & Cembella(1996)
Eurytemora herdmani	"	"
A. hudsonica	"	White (1981)
A. tonsa	*Pfiesteria pisicida*	Mallin *et al.* (1995)
Euphausiids-		
Thysanoessa raschii	*A. tamarense*	McClatchie (1988)
Planktonic polychaete larvae -	"	Turner & Anderson (1983)
Barnacle nauplii -	"	White (1981)
Rotifers -		
Brachionus plicatilis	*P. pisicida*	Mallin *et al.* (1995)
Tintinnids -		
Favella ehrenbergii	*A. tamarense*	Stoecker et al. (1981)
Heterotrophic Dinoflagellates -		
Gymnodinium dominans	*C. antiqua*	Nakamura *et al.* (1992)
Ciliates-		
Stylonichia putrina	*P. pisicida*	Burkholder & Glasgow (1995)

[a] previously known as *Gonyaulax tamarensis*

[b] previously known as *Olisthodiscus luteus* and *Heterosigma akashiwo*

[c] previously known as *Ptychodiscus brevis*

Most marine phycotoxins affect neurotransmission or inhibit enzymes, but different toxins have different physiological effects (reviewed by Baden and Trainer 1993). Paralytic shellfish poisoning (PSP) toxins produced by species of the dinoflagellate genera *Alexandrium, Pyrodinium,* and *Gymnodinium catenatum* are water-soluble sodium-ion-channel blockers that progressively inhibit nerve transmission, relaxing smooth muscle and causing paralysis and respiratory failure in humans. Conversely, brevetoxins produced by the dinoflagellate *Gymnodinium breve* are lipid-soluble sodium-channel activators, causing repetitive firing of nervous impulses, ultimately depleting acetylcholine at synapses. Amnesic shellfish poisoning (ASP) is caused by the water-soluble amino acid domoic acid (DA), produced by some species of the diatom genus *Pseudo-nitzschia.* Domoic acid acts as a surrogate for the excitatory neurotransmitter L-glutamic acid, and inability to regulate this substitute transmitter causes extensive neuronal depolarization, particularly in the vertebrate hippocampus which contains abundant glutamate receptors (Todd 1993). Thus, DA toxicity in humans and seabirds is caused by brain damage due to degeneration of the hippocampus. Diarrhetic shellfish poisoning (DSP) is caused by okadaic acid and several other lipid-soluble toxins produced by some species of the dinoflagellate genera *Dinophysis* and *Prorocentrum* (Yasumoto 1990). These toxins inhibit protein phosphatase and act upon specific enzyme subunits, affecting regulatory processes such as metabolism, membrane transport, secretion and cell division.

Various phytoplankton neurotoxins may affect vertebrate versus invertebrate nervous systems differently (Robineau *et al.* 1991a; 1991b; Turner and Tester 1997). Numerous studies reveal that various dinoflagellate toxins are lethal to a variety of larval, juvenile and adult fish of several species, either by direct exposure to dissolved toxin, ingestion of toxic dinoflagellates, or vectorial intoxication due to predation upon zooplankton that had ingested toxic dinoflagellates. In most cases, poisoned fish swam erratically, sank to bottoms of containers and remained paralyzed with still beating hearts until death. Conversely, feeding upon PSP toxins by lobster larvae, crab larvae and adults, bivalves, euphausiids, and copepods (references in Robineau *et al.* 1991a) produced no apparent adverse effects. The lack of effect of PSP toxins on some invertebrates may hinge on decentralization of invertebrate nervous systems, compared to those of vertebrates, and other differences such as binding site affinity or orientation.

If differences in the action of various phycotoxins on vertebrate versus invertebrate nervous systems explain many of the differential effects of various phytoplankton toxins on invertebrate grazers compared to their vertebrate carnivores, then the selective advantage of evolving toxins to poison organisms that do not directly graze on marine phytoplankton becomes obscured.

Obviously, toxicity is evidenced if zooplankton grazers die by direct intoxication from ingesting toxic phytoplankton, exposure to exudates, or if predators of these grazers die from vectorial intoxication. Although mortality due to direct ingestion of toxic algae is known to occur in copepods and fish larvae (Guo and Tester 1994; White *et al.* 1989), evidence for such mortality is limited. More frequently, effects of toxins on grazers may be sublethal, such as changes in behavior, or reductions in food intake or fecundity (Sykes and Huntley 1987; Hansen 1989; Uye and Takamatsu 1990; Gill and Harris 1987), or lethal if grazers simply starve rather than eat a toxic phytoplankter.

The mosaic of different effects of various phytoplankton toxins upon a variety of zooplankton grazers and phytoplanktivorous fish complicates the straightforward assumption that such toxins evolved solely as grazing deterrents.

2.2. What types of deleterious effects do toxic phytoplankton have upon metazoan and protistan grazers?

Ingestion of various toxic phytoplankters can produce a variety of adverse effects for some zooplankters (Table 2). With metazoans such as copepods there may be grazing-inhibitory effects from ingested or extracellular toxins (Huntley *et al.* 1986; Uye and Takamatsu 1990; Guo and Tester 1994). Copepods may exhibit reduced feeding upon toxic phytoplankton due to either behavioral rejection prior to ingestion, or physiological incapacitation after ingestion, whereby subsequent intoxication causes reduced feeding (Ives 1985; 1987). There is laboratory evidence for both types of rejection by copepods feeding on unialgal diets of toxic phytoplankton. Included are refusal of starved copepods to ingest toxic phytoplankters, with consequent low reproduction and high mortality due to starvation (Huntley *et al.* 1986; Uye 1986; Uye and Takamatsu 1990; Buskey and Stockwell 1993), regurgitation of unpalatable toxic dinoflagellates (Sykes and Huntley 1987), reduced development and survival of copepod nauplii fed upon toxic phytoplankters (Huntley *et al.* 1987; Buskey and Hyatt 1995), and paralysis or lethargic swimming after ingestion of toxic phytoplankton (Ives 1985; see below). Rejection or regurgitation of toxic phytoplankters is likely due to chemoreceptors such as those found on copepod mouthparts (Friedman 1980; Friedman and Strickler 1975). In many other experimental studies with copepods feeding upon toxic phytoplankton, there were no obvious deleterious effects on swimming behavior, or rates of ingestion, egg or fecal pellet production, and/or survival (Uye 1986; Uye and Takamatsu 1990; Sykes and Huntley 1987). In some cases, a given phytoplankton species such as *Heterosigma carterae* (= *Olisthodiscus luteus*) seemed to be suitable food for some copepod species (*Acartia omorii* - Uye and Takamatsu 1990; *Calanus pacificus* - Sykes and Huntley 1987), but not for others (*Acartia hudsonica* and *A. tonsa* - Tomas and Deason 1981; *Centropages hamatus* - Van Alstyne 1986).

As with metazoans, toxic effects on protistan grazers are variable. Some tintinnids appear to selectively prey upon dinoflagellates, including toxic ones, whereas other dinoflagellates are inadequate as food (Prakash *et al.* 1971; Stoecker *et al.* 1981). In some cases, protistan grazers are able to grow well when fed toxic phytoplankton at low concentrations, but are unable to sustain growth when exposed to high concentrations. This was the case with the tintinnid *Favella ehrenberghii* feeding upon *Alexandrium tamarense* (Hansen 1989) and the heliozoan *Heterophrys marina* feeding on *Chrysochromulina polylepis* (Tobisen 1991). In other cases, toxic phytoplankton have proven inadequate as food when fed to protists in monoculture, irrespective of algal concentrations (e.g., tintinnids feeding on *Heterosigma carterae*, *C. polylepis,* or *Gyrodinium aureolum,* Verity and Stoecker 1982; Carlsson *et al.* 1990; Hansen 1995; respectively). Such inadequacy was indicated by reduced growth and survival, relative to tintinnids fed non-toxic algae. Note, however, that nutritional inadequacy is not necessarily evidence for intoxication.

Table 2. Adverse effects in zooplankters from ingesting toxic phytoplankton.

No or low feeding, in some cases with low fecundity or survival

Zooplankter	Toxic phytoplankter	Reference
Copepods -		
Calanopia thompsoni	*Chattonella antiqua*	Uye (1986)
Labidocera rotundata	"	"
Tortanus forcipatus	"	"
Acartia omorii	*Gymnodinium nagasakiense*[a]	Uye & Takamatsu (1990)
Pseudodiaptomus marinus	"	"
"	*Heterosigma carterae*	"
Calanus helgolandicus	*G. nagasakiense*	Gill & Harris (1987)
Temora longicornis	"	"
A. tonsa	*Trichodesmium* sp. (aged cultures)	Guo & Tester (1994)
Calanus finmarchicus	*Alexandrium excavatum*	Turriff *et al.* (1995)
A. tonsa	*H. carterae*	Tomas & Deason (1981)
A. hudsonica	"	"
A. tonsa	Texas brown tide[b]	Buskey & Stockwell (1993)
Calanus pacificus	*Gymnodinium breve*	Huntley *et al.* (1986)
"	*Scrippsiella trochoidea*	"
"	*Protoceratium reticulatum*[c]	"
Paracalanus parvus	"	"
A. tonsa	*Aureococcus anophagefferens*	Durbin & Durbin (1989)
A. hudsonica	*Alexandrium tamarense*	Ives (1985), (1987)
Pseudocalanus sp.	"	"
Rotifers -		
Synchaeta cecilia	*H. carterae*	Egloff (1986)
Brachionus plicatilis	Texas brown tide	Buskey & Hyatt (1995)
Ciliates -		
Tintinnopsis tubulosoides	*H. carterae*	Verity & Stoecker (1982)
Favella sp.	"	"
Favella ehrenbergii	*Chrysochromulina polylepis*	Carlsson *et al.* (1990)
Strombidinopsis sp.	Texas brown tide	Buskey & Hyatt (1995)

Table 2. (continued)

Zooplankter	Toxic phytoplankter	Reference
Heterotrophic Dinoflagellates -		
Noctiluca scintillans	Texas brown tide	"
Regurgitation of toxic phytoplankters by copepods		
C. pacificus	*P. reticulatum*	Sykes & Huntley (1987)
"	*G. breve*	"
Reduced development & survival of copepod nauplii		
C. pacificus	*P. reticulatus*	Huntley *et al.* (1987)
"	*G. breve*	"
A. tonsa	Texas brown tide	Buskey & Hyatt (1995)
Lethargy or paralysis in copepods		
A. hudsonica	*A. tamarense*	Ives (1985)
Pseudocalanus sp.	"	"
A. tonsa	*G. breve*	Turner *et al.* (1996)
C. pacificus	"	Sykes & Huntley (1987)
A. tonsa	*Trichodesmium* sp. (aged cultures)	Guo & Tester (1994)

[a] formerly known as *Gyrodinium aureolum*
[b] this organism has been recently described as *Aureoumbra lagunensis*
[c] previously known as *Gonyaulax grindleyi*

In addition to effects upon growth, certain toxic phytoplankters have been shown to have acute toxic effects upon protistan grazers. When the tintinnid *Favella ehrenberghii* was exposed to high concentrations of *Alexandrium tamarense*, toxic effects included reversal of ciliary movement leading to backwards swimming, and swelling eventually leading to cell death (Hansen 1989). In this case it was demonstrated that the toxic effects were due to extracellular rather than ingested toxins.

Effects of ingested versus extracellular toxins cannot always be separated. In some cases where protistan grazers were fed ichthyotoxic algae such as *Heterosigma carterae*, *Chrysochromulina polylepis*, and *Gyrodinium aureolum*, protist survival decreased with increasing algal concentration, possibly indicating effects of toxins excreted to the medium (Verity and Stoecker 1982; Tobiesen 1991, Hansen 1989). However, no effects were found using filtrates from these cultures. This does not necessarily exclude the possibility that toxic exudates were not involved, because they may be quite labile. Toxic effects of filtrates from *A. tamarense* on tintinnids have been shown to disappear within hours (Hansen 1989). Thus, in cases where the toxin is labile and/or the potency of the toxin is low (Hansen *et al.* 1992), it can be hard to determine whether toxic effects are from ingestion of toxic prey or excretion of toxic or other exudates.

2.3. Why do grazers exhibit such varied responses to toxic phytoplankton?

There are often different responses to the same species of toxic phytoplankter by different, or the same species of grazers. For instance, feeding upon cultures of the toxic dinoflagellate *Gymnodinium breve* produced strong peristalsis in the copepod *Calanus pacificus* characterized by Sykes and Huntley (1987) as "retching," and reduced feeding and egg production rates with lethargy/paralysis in the copepod *Acartia tonsa* (Turner *et al.* 1996). Conversely, Turner and Tester (1989) noticed no adverse effects from ingesting *G. breve* for *A. tonsa* and several other species of copepods in during the 1987 North Carolina red tide.

Much of this variation in effects of phycotoxins on grazers probably relates to the considerable variation in toxin potency, composition and concentration between different blooms or clones of the same species (Anderson et al. 1990; 1994; Cembella *et al.* 1987; 1988; 1990; and numerous papers in Okaichi *et al.* 1989; Granéli *et al.* 1990; Smayda and Shimizu 1993; Lassus *et al.* 1995). Cellular toxin levels can also vary within a single clone with culture growth phase or nutrient conditions. For instance, *Pseudo-nitzschia multiseries* becomes more toxic when silicate or nitrogen limited, or with higher levels of temperature and light (Bates *et al.* 1991; 1993). *Alexandrium tamarense* has decreased toxicity when nitrogen limited, but increased toxicity when phosphate limited (Boyer *et al.* 1987). MacIntyre *et al.* (1997) found that *A. tamarense* had higher toxicity under nitrogen-replete, than nitrogen-limited conditions.

2.4. Is there "selective" grazing for or against toxic phytoplankters?

To evaluate patterns of "selective vs. nonselective" feeding, it is important to appreciate that selection compares the proportion of ingested diet comprised by a given food item to the proportion of that same item in the food assemblage available to be eaten. Thus, selection is not just dependent upon the choices by the grazer, but also on the relative proportions of various items in the available assemblage of food. Selection may be constrained by relative distributions of abundance, cell size and shape, chemical composition and toxicity of phytoplankton taxa within natural food assemblages.

Most studies of grazing on toxic phytoplankton have employed diets of unialgal cultures, but since toxic phytoplankters rarely if ever bloom in nature in the absence of other nontoxic species, results of such studies may not be representative. There appears to be only limited information on copepod grazing on various components of mixed natural assemblages of toxic and other non-toxic phytoplankton during blooms (e.g., Turner and Anderson 1983; Turner and Tester 1989; Turriff *et al.*, 1995), and results are mixed. For instance, during the 1987 North Carolina bloom of *Gymnodinium breve*, copepods ingested the toxic dinoflagellate when natural assemblages contained virtually nothing else, but when presented with alternative diatom diets, copepods switched to eating diatoms. Alternatively, Turner and Anderson (1983) found that copepods nonselectively ingested toxic *Alexandrium tamarense*, non-toxic *Heterocapsa triquetra*, and tintinnds, with these different food items being ingested in proportion to their relative abundances. Nonetheless, the relative paucity of definitive evidence for grazer selection for or against toxic phytoplankton should not imply that such selection does not occur during natural blooms.

There is relatively little information on what might drive selective feeding during natural blooms of toxic phytoplankton because there have been few investigations, even in the laboratory, quantifying concurrent ingestion of toxic as well as co-occurring non-toxic phytoplankters. In several recent laboratory studies of copepod grazing on mixtures of different toxic *Alexandrium* species and non-toxic dinoflagellates or diatoms, different copepods exhibited varied responses which were complicated and even somewhat contradictory (Turriff *et al* 1995; Teegarden and Cembella 1996).

Turriff *et al.* (1995) examined feeding by the copepod *Calanus finmarchicus* on unialgal cultures or mixtures of the toxic dinoflagellate *Alexandrium excavatum*, a non-toxic clone of the dinoflagellate *A. tamarense*, and the diatom *Thalassiosira weissflogii*. Copepods had reduced rates of feeding on unialgal diets of toxic compared to non-toxic dinoflagellates, irrespective of cellular toxin levels. In mixed food assemblages of *T. weissflogii* and toxic *A. excavatum*, copepods exclusively selected the diatom and avoided ingesting the toxic dinoflagellate, except at low diatom concentrations. Conversely, in similar mixtures of *T. weissflogii* and non-toxic *A. tamarense*, copepods selected the dinoflagellate. Despite low feeding rates on toxic *A. excavatum*, copepods exposed to this species accumulated toxins. Turriff *et al.* (1995) concluded that copepods were able to feed selectively upon non-toxic cells, and generally select against ingesting toxic cells.

Teegarden and Cembella (1996) examined feeding of the copepods *Eurytemora herdmani* and *Acartia tonsa* from the Gulf of Maine upon unialgal and mixed cultures of toxic and non-toxic dinoflagellates. There was no significant difference in grazing rates of either copepod on unialgal cultures of *Alexandrium tamarense* and *A. fundyense* of differing toxicity, although *E. herdmani* was a more effective grazer than *A. tonsa* on both cultures. *Eurytemora herdmani* accumulated continuously increasing levels of toxins over time, whereas *A. tonsa* exhibited more erratic patterns of toxin accumulation. Toxin retention efficiency based upon toxin accumulation relative to that ingested was an order of magnitude higher for *E. herdmani* than for *A. tonsa*. In mixtures of toxic *A. tamarense* and non-toxic *Lingulodinium polyedrum*, *E. herdmani* showed marked selectivity for the toxic species. Conversely, in mixtures of toxic *A. fundyense* and *L. polyedrum*, *E. herdmani* showed a complete reversal of behavior, selecting the non-toxic dinoflagellate. *Acartia tonsa* did not appreciably graze any of the toxic or non-toxic dinoflagellates in similar mixtures. Teegarden and Cembella (1996) concluded that grazing interactions with toxic phytoplankters are specific for various species of grazers and phytoplankters, and that the importance of zooplankton community grazing pressure likely depends more upon feeding preferences and selective behavior of grazers than upon putative chemical defenses of toxic phytoplankters.

2.5. What is the effect of metazoan and protistan zooplankton grazing on development and persistence of toxic phytoplankton blooms?

The role of zooplankton community grazing in development and persistence of natural blooms is variable, and outcomes are situation-specific. Turner and Anderson (1983) found that community grazing by copepods and polychaete larvae appeared incapable of preventing initiation of blooms of *Alexandrium tamarense* during the early spring in Cape Cod embayments. This was due to a combination of low grazing rates of individual grazers (due to low temperatures and low concentrations of *A. tamarense*) and low abundances of grazers. However, grazing may have contributed to the late spring decline of established blooms, due to higher grazing rates (associated with higher temperatures and higher *A. tamarense* concentrations) and dramatic increases in abundances of grazers, mainly from a seasonal pulse of planktonic polychaete larvae (Turner and Anderson 1983; Watras *et al.* 1985). Conversely, Uye (1986) concluded that copepod community grazing pressure could retard initial development of blooms of the toxic flagellate *Chattonella antiqua* in Japanese coastal waters, but that once established, such blooms were immune to grazing pressure. Grazing pressure by copepod adults, copepod nauplii, and/or rotifers can have either minimal (Sellner and Brownlee 1990; Sellner and Olson 1985), or substantial (Sellner *et al.* 1991) impact on non-toxic dinoflagellate blooms in the Chesapeake Bay. Such variation reflects differences in individual animal grazing rates due to temperature and food concentration, abundances of zooplankton populations at various stages of blooms, and rates at which abundances of bloom-formers increase due to growth and physical concentration.

The effect of protistan grazing on toxic phytoplankton blooms appears potentially substantial. There is circumstantial evidence that the 1985 blooms of the "brown tide" picoplankter *Aureococcus anophagefferens* in Narragansett Bay, Rhode Island and Long Island, New York embayments may have been triggered by breakdowns in grazing by heterotrophic flagellates during initial stages of bloom development (Smayda and Villareal 1989; Caron *et al.* 1989). After establishment of this initial population, bloom persistence may have been exacerbated by subsequent failures of grazing by copepods, marine cladocerans and bivalves. The 1985 Narragansett Bay brown tide declined in late summer, possibly in response to grazing by increasing populations of heterotrophic dinoflagellates. Similarly, heterotrophic dinoflagellate predation may be instrumental in causing the decline of toxic dinoflagellate and microflagellate blooms in Japanese waters (Nakamura *et al.* 1992; 1995).

Reduced zooplankton community grazing pressure by both metazoans and protists appears to have contributed to the persistence since 1990 of the brown tide along portions of the south Texas coast (Buskey and Stockwell 1993; Buskey and Hyatt 1995). Monitoring initiated prior to the bloom noted substantial declines in mesozooplankton populations, dominated by the copepod *Acartia tonsa* with the onset of the bloom. Reductions in *A. tonsa* body size, gut pigment content and fecundity suggested that the brown tide alga was poorly grazed by the copepod, due to unpalatability and/or nutritional inadequacy. Buskey and Hyatt (1995) confirmed that the Texas brown tide picoplankter (*Aureoumbra lagunensis*) was nutritionally inadequate to support growth of microzooplanktonic grazers such as ciliates, heterotrophic dinoflagellates, and rotifers, and was toxic to some ciliates and copepod nauplii.

2.6. What are the potential influences of indirect trophic linkages on bloom development?

There are several indirect trophic interactions that might affect toxic bloom development, but most have been suggested rather than demonstrated. For instance, Caron *et al.* (1989) speculated that heavy predation by larger zooplankters (such as copepods) upon protistan predators of brown tide picoplankters could have facilitated brown tide bloom development. French and Smayda (1995) proposed that *H. carterae* can bloom in Narragansett Bay even though the diatom *Skeletonema costatum* can outcompete it in growth rate. This can occur if copepod grazing on *H. carterae* is impaired by toxicity (Tomas and Deason 1981), but absence of ctenophore predation allows copepods to eat substantial amounts of *S. costatum* (Deason and Smayda 1982).

2.7. What are potential interactions between toxic phytoplankters and the reproductive success of their grazers? - An example from recent studies

Toxic phytoplankton blooms and copepod generations can both occur over similar time periods of days to weeks. A bloom of toxic phytoplankton implies inability of the grazer community to control the bloom. This may be due to toxicity depressing feeding rates of individual animals, but it could also be due to low abundances of grazers even if their individual feeding rates are high. Thus, dynamics of toxic phytoplankton blooms should be viewed within the context of the population dynamics of grazer communities.

Reproduction is an important component of population dynamics, and numerous studies have related copepod fecundity to quantity and composition of maternal diets (Kleppel 1993; Jónasdóttir *et al.* 1995). Further, although diets of certain types of phytoplankton may be sufficient foods for survival and/or egg production of adult copepods, these same foods may result in poor egg hatching success, due to inhibitors or nutritional inadequacy. Such has been suggested for several species of diatoms when fed to several species of copepods (Chaudron *et al.* 1996, and references therein).

Perhaps toxic phytoplankton can also "sabotage" future generations of their grazers by inhibiting reproductive success? If so, then bloom persistence could be enhanced by reducing abundances of future grazers, even though present grazers are unimpaired by toxins. This might allow a bloom to survive past the time when community grazing pressure might otherwise have caused bloom decline.

We examined this possibility with the copepod *Acartia tonsa* feeding upon diets of natural seawater, filtered seawater, and unialgal cultures of the toxic dinoflagellate*Gymnodinium breve* and its non-toxic congener *G. sanguineum.* Preliminary results were mixed and inconsistent, but serve as an example of the type of variability due to unassessed factors that may be encountered in such studies of interactions between toxic phytoplankton and their grazers.

Zooplankton were collected near Beaufort, North Carolina in March, 1995 and adult females of the copepod *Acartia tonsa* were isolated within 1-2 h of collection and held in natural sea water from the collection site, that had been sieved through 60-µm mesh to remove extraneous zooplankters. Salinity was 28 $^{o}/_{oo}$ and temperature was 17°C.

Because *G. sanguineum* cells were large (58.25 μm length x 38.5 μm width x 25 μm thickness, or 55,921 μm^3) compared to those of *G. breve* (38 μm length x 26 μm width x 16 μm thickness, or 15,808 μm^3), we prepared cultures at equivalent cell volume concentrations (total μm^3 ml^{-1}) rather than equal cell abundances. Volumetric concentrations were, for *G. sanguineum*, 55,921 μm^3 x the initial culture concentration of 78 cells ml^{-1}, = 4,361,838 μm^3 ml^{-1}, and for *G. breve*, 15,808 μm^3 x the initial culture concentration of 820 cells ml^{-1} x an added aliquot of 0.336 ml, = 4,355,420 μm^3 ml^{-1}. These virtually identical volumetric concentrations (4.36 x 10^6 μm^3 ml^{-1}) converted to projected cell abundances of 78 cells ml^{-1} for *G. sanguineum*, and 276 cells ml^{-1} for *G. breve*. Microscopically-determined initial experimental cell concentrations in feeding experiments with *G. breve* (actual concentration 271 cells ml^{-1}) were close to those projected by dilution (276 cells ml^{-1}) but for *G. sanguineum*, actual concentrations (171 cells ml^{-1}) were higher than projected ones (78 cells ml^{-1}).

We also compared diets of the toxic and non-toxic dinoflagellates with diets of natural phytoplankton and filtered sea water. The < 60 μm fraction from nearby waters was dominated by a bloom of the non-toxic dinoflagellate *Prorocentrum minimum*, present at natural abundances of 658 cells ml^{-1}. Dimensions of *P. minimum* cells were 22 μm length x 16.5 μm width x 9 μm thickness, producing cell volumes of 3,267 μm^3, which at 658 cells ml^{-1} resulted in volumetric concentrations of 2.15 x 10^6 μm^3 ml^{-1}.

Copepods were preconditioned at 18°C in 350-500 ml aliquots of experimental food suspensions for 13 h overnight, a period sufficient for egg production to reflect food used in preconditioning since *A. tonsa* converts food to eggs in < 10 h (Tester & Turner 1990). Because survival of *A. tonsa* declines rapidly in filtered sea water, we feared that many would die during experimental incubations if we preconditioned them for 13 h in filtered sea water. Thus, we preconditioned copepods in *P. minimum* /natural sea water before placing them in filtered sea water to serve as controls. Also, since *A. tonsa* fecundity quickly responds to variations in food (Dagg 1977), any eggs produced in filtered sea water would reflect residual egg production based on natural phytoplankton ingested near the end of the preconditioning period, allowing assessment of egg production in the absence of food over most of the same period as other copepods in different experimental food suspensions.

After preconditioning, copepods were placed in filtered seawater or experimental food suspensions at the same concentrations as in the preconditioning for experimental periods of 17 - 21 h. Ten adult female *A. tonsa* were placed in 100-ml food suspensions, replicated in triplicate for both egg production and feeding rate determinations. Experimental containers were 150-ml clear, wide-mouth plastic jars. These were used so that final egg counts could be made by placing these jars under a dissecting microscope with minimal manipulation of eggs that had settled to the bottoms of jars.

Feeding rate studies included, for each food type, triplicate experimental jars with copepods, initial samples preserved with Utermöhl's solution at the outset, and controls with no copepods which were preserved at the end. After 17-21 h of experimental incubation, samples for feeding rate determinations were preserved in Utermöhl's solution for subsequent microscopic counting of remaining phytoplankton, and determination of ingestion and clearance rates using the Frost (1972) equations.

Egg production experiments were terminated by pouring copepods and eggs from jars into petri dishes, with subequent pipetting of copepods out of the dishes while counting live versus dead copepods. Copepod mortality was minimal (3 of 120), and for the purpose of feeding rate and egg production rate calculations, dead copepods were assumed to have lived for half of the incubation period. Eggs were counted and gently pipetted back into experimental suspensions and left for another day for hatching. Approximately 24 h later, remaining eggs were counted, and hatching success was determined by subtraction.

Following initial preconditioning there were differences in behavior of copepods in different treatments. *A. tonsa* females in the *G. sanguineum* and *P. minimum*/natural sea water treatments were all active, but about a third of the copepods in the *G. breve* treatment were moribund. After pipetting these copepods into filtered sea water, most recovered quickly, suggesting intoxication by *G. breve.*

Rates of ingestion of both *G. breve* (121.3 $\pm$ 10.2 cells copepod^{-1} h^{-1}) and *G. sanguineum* (25.6 $\pm$ 31.9 cells copepod^{-1} h^{-1}) by *A. tonsa* were low, compared to ingestion of *P. minimum* (380.5 $\pm$ 27.9 cells copepod^{-1} h^{-1}) from natural sea water. Rates for ingestion of *G. breve* by both copepods were comparable to the low rates at similar *G. breve* concentrations found for *A. tonsa* during the 1987 North Carolina red tide (Turner & Tester 1989).

Egg production and hatching data for *A. tonsa* in March were surprising (Fig. 1). Rates of egg production on diets of toxic *G. breve* were low, but most of these eggs hatched. These were essentially the same results as for the treatments with filtered sea water. Conversely, egg production on diets of *G. sanguineum* was high, but none of these eggs hatched. On diets of natural sea water dominated by *P. minimum*, both egg production and hatching success were high.

Because we were suspicious of high *A. tonsa* egg production but zero hatching on a diet of non-toxic *G. sanguineum* in March, an abbreviated experiment to re-examine egg production and hatching rates was performed in October, 1995. In this case, there was 20 h of preconditioning at 20°C, followed by experimental incubations in which 200 *A. tonsa* females in each of 2 one-liter containers had *G. sanguineum* and *G.breve* concentrations of 78 cells ml^{-1} and 280 cells ml^{-1}, respectively. These were virtually identical to the concentrations used in the March, 1995 experiment. Hatching success was determined after 24 h. Over experimental incubations of 17 h, 200 *A. tonsa* reduced *G. sanguineum* concentrations from 78 cells ml^{-1} to zero, but *G. breve* concentrations from only 280 to 254 cells ml^{-1}, indicating that *G. breve* cultures were essentially ungrazed.

Results from October were not the same as for March (Fig. 2). As in March, October egg production on a diet of *G. breve* was low but hatching success was high. However, on a diet of *G. sanguineum*, both egg production and hatching success were high.

Rates of egg production and hatching success appear to have been influenced by concentration and type of food, but results were variable. With unialgal diets of *G. breve* in both March and October, egg production rates were low, but hatching success was high. With unialgal diets of *G. sanguineum* rates of *A. tonsa* egg production were high in both March and October, but hatching was zero in March and high in October. Rates of egg production and hatching success were most consistently high when copepods fed upon natural sea water.

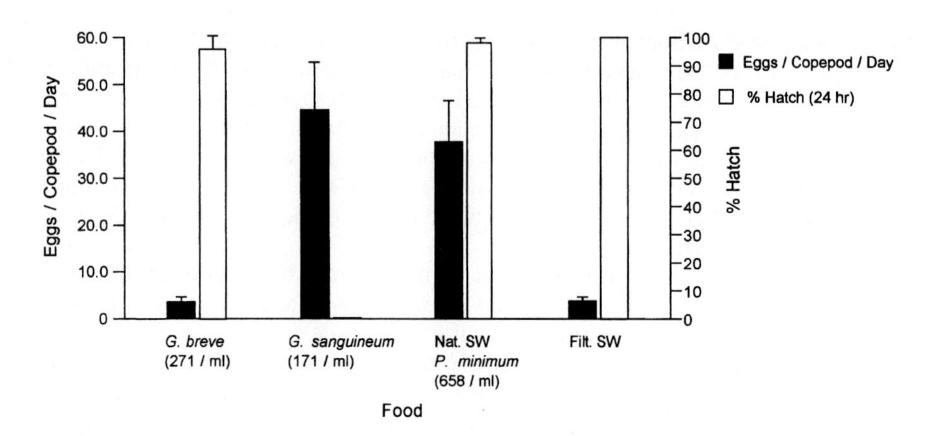

Fig. 1. Egg production and hatching rates for the copepod *Acartia tonsa* feeding upon diets of toxic *Gymnodinium breve*, non-toxic *G. sanguineum*, natural seawater dominated by *Prorocentrum minimum*, and filtered seawater, Beaufort, North Carolina, March, 1995. Microscopically-determined cell concentrations for food suspensions are given in parentheses.

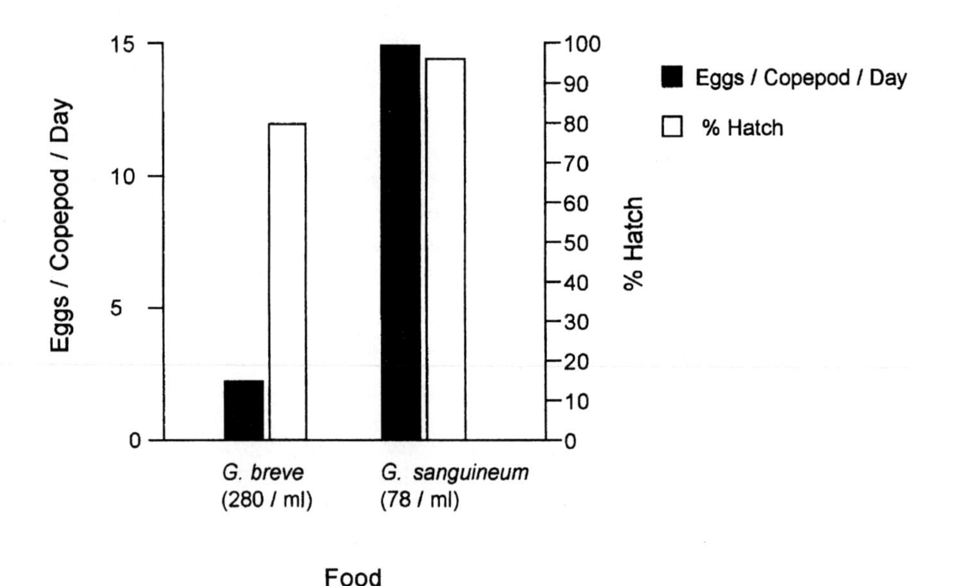

Fig. 2. Egg production and hatching rates for the copepod *Acartia tonsa* feeding upon diets of toxic *Gymnodinium breve*, non-toxic *G. sanguineum*, Beaufort, North Carolina, October, 1995. Microscopically-determined cell concentrations for food suspensions are given in parentheses.

Why did we obtain no egg hatching with *A. tonsa* on a diet of *G. sanguineum* in March, but near complete egg hatching when the experiment was repeated in October? Most likely the difference in egg hatching could relate to differences in the unialgal cultures used in the two experiments. Perhaps *G. sanguineum* produces an unknown metabolite that inhibits copepod egg development? Although this possibility is intriguing, there is no evidence for it.

We also thought that since undiluted *G. sanguineum* culture was used as food in March, that culture medium chemicals might have caused egg hatching inhibition. Accordingly, the cultures used in October were grown to higher density (3,702 cells ml^{-1}), and greatly diluted with filtered natural sea water so that one liter of food suspension contained only 21 ml of *G. sanguineum* culture (= experimental concentration of 78 cells ml^{-1}). To further examine this possibility, we performed egg production hatching replicates with half filtered sea water:half dense *G. sanguineum* culture, for a final culture concentration of 2,100 cells ml^{-1}. A day later 765 of 804 eggs had hatched (95% success), and there were numerous apparently healthy nauplii present.

Although we do not know why the two *G. sanguineum* experiments produced opposite results in terms of egg hatching, our results highlight potential problems and suggest caution in using unialgal cultures as diets in such studies. Since, for toxic phytoplankters, toxicity levels can vary considerably with culture growth phase, nutrient levels, and such factors as temperature and irradiance levels, then perhaps egg hatching inhibitors and/or essential nutrients for copepod reproduction can also vary between different cultures of the same species. There is also recent field evidence for such variation in the seasonal succession of phytoplankton assemblages, and copepod diets, fecundity and egg hatching success from the English Channel (Laabir et al. 1995).

Can *G. breve* toxins sabotage the reproduction of future generations of their potential grazers? Our results suggest that they can. In both experiments with *A. tonsa*, while rates of egg hatching were high, rates of egg production were low. Copepods produced similar numbers of eggs in *G. breve* treatments and in filtered sea water. Ingestion rate data for March, and cell reduction data for October reveal that *A. tonsa* was grazing at comparatively low rates on *G. breve*. Thus, the low egg production could be due to starvation.

Such starvation of *A. tonsa* in the presence of *G. breve* appears due to intoxication. Many of the copepods exposed to *G. breve* in March were lethargic or moribund until they were transferred to filtered sea water where they quickly recovered. This suggests neurotoxicity. If so, then continued ingestion of *G. breve* was precluded by physiological incapacitation, similar to that noted by Ives (1985, 1987) for other copepods after ingesting *Alexandrium tamarense*. Ingestion of *G. breve* has also been shown to produce paralysis and elevated heartbeat rate, as well as strong peristalsis and regurgitation in other copepods (Huntley et al. 1987, Sykes & Huntely 1987). Thus, if *G. breve* toxins inhibit copepod feeding by behavioral rejection and/or physiological incapacitation, this could contribute to bloom persistence through immediate reduction of grazing pressure, as well as by diminishing future generations of grazers through starvation-induced reduction of egg production.

Ingestion of *G. breve* by copepods is not always harmful, however. During the 1987 North Carolina bloom of *G. breve*, three species of copepods (*A. tonsa, Oncaea venusta*, and *Labidocera aestiva*) fed heavily upon high natural concentrations (up to 19,567 cells ml^{-1}) of *G. breve* (Turner & Tester 1989). Further, these copepods exhibited no apparent adverse effects in terms of reduced activity or mortality after 17.5-21.75 h of exposure.

How can cultures versus a natural bloom of *G. breve* be alternatively both strong and weak copepod feeding deterrents? The answer probably lies in the aforementioned variation in toxicity of different clones or blooms of the same dinoflagellate species. These can be influenced by numerous chemical and physical factors (reviewed by Turner & Tester 1996).

If *A. tonsa* does not appreciably eat *G. breve*, then how can it become accumulated in copepods, the ingestion of which causes vectorial intoxication in fish (Tester et al. 1996)? Since the copepods in the Tester et al. study were starved for lengthy periods in order to facilitate toxin accumulation, perhaps this is why the copepods ingested *G. breve* for a time, accumulating toxin, when less hungry copepods would normally avoid ingesting this species. If this is the case, it would explain the apparent contradiction between the present and the Tester et al. (1996) study, as well as why copepods ingested *G. breve* during the 1987 North Carolina red tide, when surface water samples used as food were virtually monospecific cultures of *G. breve* (Turner & Tester 1989). This would also explain extremely similar recent results of Turriff et al. (1995) which showed that the copepod *Calanus finmarchicus* from the Gulf of St. Lawrence generally avoided ingesting toxic *Alexandrium excavatum*, but nonetheless still accumulated toxins from it. If cultures used in such studies contain high cellular levels of toxins, copepods may become intoxicated, even by ingesting relatively few cells.

Our preliminary results suggest that *G. breve* may inhibit reproduction of its grazer *A. tonsa* by reducing egg production, even if not by reducing egg hatching. The mechanism seems to be one of causing the copepods to reduce ingesting the dinoflagellate, with subsequent dimunition of fecundity by intoxication and/or starvation. Such effects for other copepods fedding on diets of toxic phytoplankters are not so apparent (Turner *et al.* 1996; 1997). Thus, interactions between copepod egg production, egg hatching, and types of phytoplankton food appear to lie within a wide envelope of variability. Accordingly, extrapolation of results on zooplankton interactions with toxic phytoplankton from one bloom or culture study to another may be inappropriate without sufficient attention to details of variations between species of grazers and culture or bloom toxicity in different scenarios.

3. Conclusions

Interactions between toxic phytoplankton and their grazers are complex, variable, and situation-specific. An overall synthesis of these interactions is elusive and premature because present results are still too disparate. Much of this mosaic is due to a plethora of toxins which have different physiological effects on vertebrate versus invertebrate nervous systems, and to toxin potency or intracellular concentrations which vary with culture or bloom growth phase, nutrient levels, irradiance, or other factors. Different grazers also exhibit different responses to different toxic species, or to different clones or blooms of the same species. Additional complications can arise from effects of organisms at other trophic levels that, while not direct grazers of toxic phytoplankton, may affect their blooms by preying upon their grazers. Accordingly, information from one experimental study or natural bloom should be extrapolated to another with caution.

Improved understanding of dynamics of toxic phytoplankton blooms will emerge with more effort to define the role of zooplankton grazing in these blooms. Much can be learned from mesocosm experiments which examine phytoplankton growth under different regimes of nutrient addition and zooplankton grazing (e.g., Granéli *et al.* 1993). Further, zooplankton monitoring should be incorporated into phytoplankton monitoring programs focused on harmful blooms. In the very few cases where ongoing phytoplankton and zooplankton monitoring preceeded development of a harmful bloom (e.g., Falkowski et al. 1980; Buskey and Stockwell 1993), zooplankton dynamics appear to be important components of toxic phytoplankton dynamics.

Perhaps most importantly, we recommend that studies of interactions between toxic phytoplankton and their grazers be conducted within the context of known, or even manipulated intracellular toxin concentrations. Also, attention should be given to ingestion of naturally co-occurring non-toxic as well as toxic phytoplankters. Only through collaboration between specialists on phytoplankton, zooplankton and phycotoxin chemistry will better understanding of interactions between toxic algae and their grazers emerge.

4. References

Anderson, D. M., White, A. W. (1992). Marine biotoxins at the top of the food chain. *Oceanus* 35(3): 55-61.

Anderson, D. M., Kulis, D. M., Sullivan, J. J., Hall, S., Lee, C. (1990). Dynamics and physiology of saxitoxin production by the dinoflagellates *Alexandrium* spp. *Mar. Biol.* 104: 511-524.

Anderson, D. M., Kulis, D. M., Doucette, G. J., Gallagher, J. C., Balech, E. (1994). Biogeography of toxic dinoflagellates in the genus *Alexandrium* from the northeastern United States and Canada. *Mar. Biol.* 120: 467-478.

Baden, D. G., Trainer, B. L. (1993). Mode of action of toxins of seafood poisoning. In: Falconer, I. R. (ed.), *Algal Toxins in Seafood and Drinking Water.* Academic, London.

Bates, S. S., Worms, J., Smith, J. C. (1993). Effects of ammonium and nitrate on growth and domoic acid production by *Nitzschia pungens* in batch culture. *Can. J. Fish. Aquat. Sci.* 50: 1248-1254.

Bates, S. S., deFreitas, A. S. W., Milley, J. E., Pocklington, R., Quilliam, M. A., Smith, J. C., Worms, J. (1991). Controls on domoic acid production by the diatom *Nitzschia pungens* f. *multiseries* in culture: nutrients and irradiance. *Can. J. Fish. Aquat. Sci.* 48: 1136-1144.

Boyer, G. L., Sullivan, J. J., Andersen, R. J., Harrison, P. J., Taylor, F. J. R. (1987). Effects of nutrient limitation on toxin production and composition in the marine dinoflagellate *Protogonyaulax tamarensis*. *Mar. Biol.* 96: 123-128.

Burkholder, J. M., Glasgow, H. B. Jr. (1995). Interactions of a toxic estuarine dinoflagellate with microbial predators and prey. *Arch. Protistenkd.* 145: 177-188.

Buskey, E. J., Hyatt, C. J. (1995). Effects of the Texas (USA) 'brown tide' alga on planktonic grazers. *Mar. Ecol. Prog. Ser.* 126: 285-292.

Buskey, E. J., Stockwell, D. A. (1993). Effects of a persistent "brown tide" on zooplankton populations in the Laguna Madre of south Texas. In: Smayda, T. J., Shimizu, Y. (eds.), *Toxic phytoplankton blooms in the sea.* Elsevier, Amsterdam, pp. 659-666.

Carlsson, P., Granéli, E., Olsson, P. (1990). Grazer elimination through poisoning: One of the mechanisms behind *Chrysochromulina polylepis* blooms? In: Granéli, E., Sundström, B., Edler, L., Anderson, D. M. (eds.) *Toxic marine phytoplankton.* Elsevier, Amserdam, pp. 116-122.

Caron, D. A., Lim, E. L., Kunze, H., Cosper, E. M., Anderson, D. M. (1989). Trophic interactions between nano- microzooplankton and the "brown tide." In: Cosper, E. M., Bricelj, V. M., Carpenter, E. J., (eds.), *Novel phytoplankton blooms. Causes and impacts of recurrent brown tides and other unusual blooms.* Springer-Verlag, Berlin, pp. 265-294.

Cembella, A. D., Therriault, J. C., Béland, P. (1988). Toxicity of cultured isolates and natural populations of *Protogonyaulax tamarensis* from the St. Lawrence Estuary. *J. Shellfish Res.* 7: 611-621.

Cembella, A. D., Sullivan, J. J., Boyer, G. L., Taylor, F. J. R., Andersen, R. J. (1987). Variation in paralytic shellfish toxin composition within the *Protogonyaulax tamarensis/catenella* species complex; red tide dinoflagellates. *Biochem. Syst. Ecol.* 15: 171-186.

Cembella, A., Destombe, C., Turgeon, J. (1990). Toxin composition of alternative life history stages of *Alexandrium*, as determined by high-performance liquid chromatography. In: Granéli, E., Sundström, B., Edler, L., Anderson, D. M. (eds.) *Toxic marine phytoplankton.* Elsevier, Amsterdam, pp. 333-338.

Chaudron, Y., Poulet, S. A., Laabir, M., Ianora, A., Miralto, A. (1996). Is hatching success of copepod eggs diatom density-dependent? *Mar. Ecol. Prog. Ser.* 144: 185-193.

Dagg, M. J. (1977). Some effects of patchy food environments on copepods. *Limnol. Oceanogr.* 22: 99-107.

Dale, B., Yentsch, C. M. (1978) Red tide and paralytic shellfish poisoning. *Oceanus* 21(3): 41-49.

Deason, E. E., Smayda, T. J. (1982). Ctenophore-zooplankton-phytoplankton interactions in Narragansett Bay, Rhode Island, during 1972-1977. *J. Plankton Res.* 4: 203-217.

Durbin, A. G., Durbin, E. G. (1989). Effect of the "brown tide" on feeding, size and egg laying rate of adult female *Acartia tonsa.* In: Cosper, E. M., Bricelj, V. M., Carpenter, E. J., (eds.), *Novel phytoplankton blooms. Causes and impacts of recurrent brown tides and other unusual blooms.* Springer-Verlag, Berlin, pp. 625-646.

Egloff, D. A. (1986). Effect of *Olisthodiscus luteus* on the feeding and reproduction of the marine rotifer *Synchaeta cecilia. J. Plankt. Res.* 8: 263-274.

Falkowski, P. G., Hopkins, T. S., Walsh, J. J. (1980). An analysis of factors affecting oxygen depletion in the New York Bight. *J. Mar. Res.* 38: 479-506.

French, D. P., Smayda, T. J. (1995). Temperature regulated responses of nitrogen limited *Heterosigma akashiwo*, with relevance to its blooms. In: Lassus, P., Arzul, G., Erard, E., Gentien, P., Marcaillou, C. (eds.), *Harmful marine algal blooms.* Lavoisier, Paris, pp. 585-590.

Friedman, M. M. (1980). Comparative morphology and functional significance of copepod receptors and oral structures. In: Kerfoot, W. C. (ed.), *Evolution and ecology of zooplankton communities.* University Press of New England, Hanover, New Hampshire, pp. 185-197.

Friedman, M. M., Strickler, J. R. (1975). Chemoreceptors and feeding in calanoid copepods (Arthropoda: Crustacea). *Proc. Nat. Acad. Sci. USA* 72: 4185-4188.

Frost, B. W. (1972). Effects of size and concentration of food particles on the feeding behavior of the marine planktonic copepod *Calanus pacificus*. *Limnol. Oceanogr.* 17: 805-815.

Gerachi, J. R., Anderson, D. M., Timperi, R. J., St. Aubin, D. J., Early, G. A., Prescott, J. H., Mayo, C. A. (1989). Humpback whales (*Megaptera novaeangliae*) fatally poisoned by dinoflagellate toxin. *Can. J. Fish. Aquat. Sci.* 46: 1895-1898.

Gill, C. W., Harris, R. P. (1987). Behavioural responses of the copepods *Calanus helgolandicus* and *Temora longicornis* to dinoflagellate diets. *J. Mar. Biol. Assoc. U. K.* 67: 785-801.

Granéli, E., Sundström, B., Edler, L., Anderson, D. M. (eds.). (1990). *Toxic marine phytoplankton*. Elsevier, Amsterdam, 554 pp.

Granéli, E., Olsson, P., Carlsson, P., Granéli, W., Nylander, C. (1993). Weak 'top-down' control of dinoflagellate growth in the coastal Skagerrak. *J. Plankton Res.* 15: 213-237.

Guo, C., Tester, P. A. (1994). Toxic effect of the bloom-forming *Trichodesmium* sp. (Cyanophyta) to the copepod *Acartia tonsa*. *Nat. Tox.* 2: 222-227.

Hallegraeff, G. M. (1993). A review of harmful algal blooms and their apparent global increase. *Phycologia* 32: 79-99.

Hansen, P. J. (1989). The red tide dinoflagellate *Alexandrium tamarense*: effects on behaviour and growth of a tintinnid ciliate. *Mar. Ecol. Prog. Ser.* 53: 105-116.

Hansen, P. J. (1995). Growth and grazing response of a ciliate feeding on the red tide dinoflagellate *Gyrodinium aureolum* in monoculture and in mixture with a non-toxic alga. *Mar. Ecol. Prog. Ser.* 121: 65-72.

Hansen, P. J., Cembella, A. D., Moestrup, Ø. (1992). The marine dinoflagellate *Alexandrium ostenfeldii:* Paralytic shellfish toxin concentration, composition, and toxicity to a tintinnid ciliate. *J. Phycol.* 28: 597-603.

Huntley, M. E., Ciminello, P., Lopez, M. D. G. (1987). Importance of food quality in determining development and survival of *Calanus pacificus* (Copepoda: Calanoida). *Mar. Biol.* 95: 103-113.

Huntley, M. E., Sykes, P., Rohan, S., Marin, V. (1986). Chemically-mediated rejection of dinoflagellate prey by the copepods *Calanus pacificus* and *Paracalanus parvus:* mechanism, occurrence and significance. *Mar. Ecol. Prog. Ser.* 28: 105-120.

Ives, J. D. (1985). The relationship between *Gonyaulax tamarensis* cell toxin levels and copepod ingestion rates. In: Anderson, D. M., White, A. W., Baden, D. G. (eds.), *Toxic dinoflagellates*. Elsevier, Amsterdam, pp. 413-418.

Ives, J. D. (1987). Possible mechanisms underlying copepod grazing responses to levels of toxicity in red tide dinoflagellates. *J. Exp. Mar. Biol. Ecol.* 112: 131-145.

Jónasdóttir, S. H., Fields, D., Pantoja, S. (1995). Copepod egg production in Long Island Sound, USA, as a function of the chemical composition of seston. *Mar. Ecol. Prog. Ser.* 119: 87-98.

Jeong, H. J., Latz, M. I. (1994). Growth and grazing rates of the heterotrophic dinoflagellates *Protoperidinium* spp. on red tide dinoflagellates. *Mar. Ecol. Prog. Ser.* 106: 173-185.

Kleppel, G. S. (1993). On the diets of calanoid copepods. *Mar. Ecol. Prog. Ser.* 99: 183-195.

Laabir, M. Poulet, S. A., Ianora, A., Miralto, A., Cueff, A. (1995). Reproductive response of *Calanus helgolandicus*. II. *In situ* inhibition of embryonic development. *Mar. Ecol. Prog. Ser.* 129: 97-105.

Lassus, P., Arzul, G., Erard, E., Gentien, P., Marcaillou, C. (eds.), (1995) *Harmful marine algal blooms.* Lavoisier, Paris, 877 pp.
MacIntyre, J. G., Cullen, J. J., Cembella, A. D. (1997). Vertical migration, nutrition and toxicity in the dinoflagellate *Alexandrium tamarense. Mar. Ecol. Prog. Ser.* 148: 201-216.
Mallin, M. M., Burkholder, J. M., Larsen, L. M., Glasgow, H. B., Jr. (1995). Response of two zooplankton grazers to an ichthyotoxic estuarine dinoflagellate. *J. Plankton Res.* 17: 351-363.
McClatchie, S. (1988). Functional response of the euphausiid *Thysanoessa raschii* grazing on small diatoms and toxic dinoflagellates. *J. Mar. Res.* 46: 631-646.
Nakamura, Y., Suzuki, S.-y., Hiromi, J. (1995). Growth and grazing of a naked heterotrophic dinoflagellate, *Gyrodinium dominans. Aquat. Microb. Ecol.* 9: 157-164.
Nakamura, Y., Yamazaki, Y., Hiromi, J. (1992). Growth and grazing of a heterotrophic dinoflagellate, *Gyrodinium dominans*, feeding on a red tide flagellate, *Chattonella antiqua. Mar. Ecol. Prog. Ser.* 82: 275-279.
Okaichi, T., Anderson, D. M., Nemoto, T. (eds.), (1989). *Red tides. Biology, environmental science, and toxicology.* Elsevier, Amsterdam, 489 pp.
Prakash, A., Medcof, J. C., Tennant, A. D. (1971). Paralytic shellfish poisoning in eastern Canada. *Fish. Res. Bd. Can. Bull.* 177: 1-87.
Robineau, B., Gagne, J. A., Fortier, L., Cembella, A. D. (1991a). Potential impact of a toxic dinoflagellate (*Alexandrium excavatum*) bloom on survival of fish and crustacean larvae. *Mar. Biol.* 108: 293-301.
Robineau, B., Fortier, L., Gagne, J. A., Cembella, A. D. (1991b). Comparison of the response of five larval fish species to the toxic dinoflagellate *Alexandrium excavatum* (Braarud) Balech. *J. Exp. Mar. Biol. Ecol.* 152: 225-242.
Sellner, K. G., Brownlee, D. C. (1990). Dinoflagellate-microzooplankton interactions in Chesapeake Bay. In: Granéli, E., Sundström, B., Edler, L., Anderson, D. M. (eds.) *Toxic marine phytoplankton.* Elsevier, Amsterdam, pp. 221-226.
Sellner, K. G., Olson, M. M. (1985). Copepod grazing in red tides of Chesapeake Bay. In: Anderson, D. M., White, A. W., Baden, D. G. (eds.),*Toxic dinoflagellates.* Elsevier, Amsterdam, pp. 245-250.
Sellner, K. G., Lacoture, R. V., Cibik, S. J., Brindley, A., Brownlee, S. G. (1991). Importance of a winter dinoflagellate-microflagellate bloom in the Patuxent River Estuary. *Est. Coast. Shelf Sci.* 32: 27-42.
Smayda, T. J., Villareal, T. A. (1989). The 1985 'brown-tide' and the open phytoplankton niche in Narragansett Bay during summer. In: Cosper, E. M., Bricelj, V. M., Carpenter, E. J., (eds.), *Novel phytoplankton blooms. Causes and impacts of recurrent brown tides and other unusual blooms.* Springer-Verlag, Berlin, pp. 159-187.
Smayda, T. J., Shimizu, Y. (eds.), (1993). *Toxic phytoplankton blooms in the sea.* Elsevier, Amsterdam, 952 pp.
Stoecker, D. K., Guillard, R. R. L., Kavee, R. M. (1981). Selective predation by *Favella ehrenbergii* (Tintinnia) on and among dinoflagellates. *Biol. Bull.* 160: 136-145.
Sykes, P. F., Huntley, M. E. (1987). Acute physiological reactions of *Calanus pacificus* to selected dinoflagellates: direct observations. *Mar. Biol.* 94: 19-24.

Teegarden, G. J., Cembella, A. D. (1996). Grazing of toxic dinoflagellates, *Alexandrium* spp., by adult copepods of coastal Maine: Implications for the fate of paralytic shellfish toxins in marine food webs. *J. Exp. Mar. Biol. Ecol.* 196: 145-176.

Tester, P. A., Turner, J. T. (1990). How long does it take copepods to make eggs? *J. Exp. Mar. Biol. Ecol.* 141: 169-182.

Tester, P. A., Turner, J. T., Shea, D. (1996). Brevetoxin transfer in the marine food web. *Eos* 76(3): OS12G-1.

Tobiesen, A. (1991). Growth rates of *Heterophrys marina* (Heliozoa) on *Chrysochromulina polylepis* (Prymnesiophyceae). *Ophelia* 33: 205-212.

Todd, E. C. D. (1993). Domoic acid and amnesic shellfish poisoning - A review. *J. Food Protect.* 56: 69-83.

Tomas, C. R., Deason, E. E. (1981). The influence of grazing by two *Acartia* species on *Olisthodiscus luteus* Carter. *P.S.Z.N.I: Marine Ecology* 2: 215-223.

Turner, J. T., Anderson, D. M. (1983). Zooplankton grazing during dinoflagellate blooms in a Cape Cod embayment, with observations of predation upon tintinnids by copepods. *P.S.Z.N.I: Marine Ecology* 4: 359-374.

Turner, J. T., Tester, P. A. (1989). Zooplankton feeding ecology: Copepod grazing during an expatriate red tide. In: Cosper, E. M., Bricelj, V. M., Carpenter, E. J., (eds.), *Novel phytoplankton blooms. Causes and impacts of recurrent brown tides and other unusual blooms.* Springer-Verlag, Berlin, pp. 359-374.

Turner, J. T., Tester, P. A. (1997). Toxic marine phytoplankton, zooplankton grazers, and pelagic food webs. *Limnol. Oceanogr.* (in press).

Turner, J. T., Lincoln, J. A., Cembella, A. D. (1997). Effects of toxic and non-toxic dinoflagellates on copepod grazing, egg production and egg hatching success. *Abstracts of papers presented at VII International Conference on Harmful Algae, Vigo, Spain, 25-29 June, 1997.*

Turner, J. T., Lincoln, J. A., Tester, P. A., Bates, S. S., Léger, C. (1996). Do toxic phytoplankton reduce egg production and hatching success of the copepod *Acartia tonsa?* Eos 76(3): OS12G-2

Turriff, N., Runge, J. A., Cembella, A. D. (1995). Toxin accumulation and feeding behaviour of the planktonic copepod *Calanus finmarchicus* exposed to the red-tide dinoflagellate *Alexandrium excavatum. Mar. Biol.* 123: 55-64.

Uye, S. (1986). Impact of copepod grazing on the red-tide flagellate *Chattonella antiqua. Mar. Biol.* 92: 35-43.

Uye, S., Takamatsu, K. (1990). Feeding interactions between planktonic copepods and red-tide flagellates from Japanese coastal waters. *Mar. Ecol. Prog. Ser.* 59: 97-107.

Van Alstyne, K. L. (1986). Effects of phytoplankton taste and smell on feeding behavior of the copepod *Centropages hamatus*. *Mar. Ecol. Prog. Ser.* 34: 187-190.

Verity, P. G., Stoecker, D. K. (1982). Effects of *Olisthodiscus luteus* on the growth and abundance of tintinnids. *Mar. Biol.* 72: 79-87.

Watras, C. J., Garcon, V. C., Olson, R. J., Chisholm, S. W., Anderson, D. M. (1985). The effect of zooplankton grazing on estuarine blooms of the toxic dinoflagellate *Gonyaulax tamarensis*. *J. Plankton Res.* 7: 891-908.

White, A. W. (1981). Marine zooplankton can accumulate and retain dinoflagellate toxins and cause fish kills. *Limnol. Oceanogr.* 26: 103-109.

White, A. W., Fukuhara, O., Anraku, M. (1989). Mortality of fish larvae from eating toxic dinoflagellates or zooplankton containing dinoflagellate toxins. In: Okaichi, T., Anderson, D. M., Nemoto, T. (eds.), *Red tides. Biology, environmental science, and toxicology*. Elsevier, Amsterdam, pp. 395-398.
Windust, A. (1992). The responses of bacteria, microalgae and zooplankton to the diatom *Nitzschia pungens* f. *multiseries* and its toxic metabolite domoic acid. M. S. thesis, Dalhousie Univ., 107 p.
Work, T. M., Beale, A. B., Fritz, L. Quilliam, M. A., Silver, M., Buck, K., Wright, J. L. C. (1993). Domoic acid intoxication of brown pelicans and comorants in Santa Cruz, California. In: Smayda, T. J., Shimizu, Y. (eds.), *Toxic Phytoplankton Blooms in the Sea*. Elsevier, Amsterdam.
Yasumoto, T. (1990). Marine microorganisms toxins - an overview. In: Granéli, E., Sundström, B., Edler, L., Anderson, D. M. (eds.) *Toxic marine phytoplankton*. Elsevier, Amsterdam, pp. 3-8.

Species Composition of Harmful Algal Blooms in Relation to Macronutrient Dynamics

R. Riegman

Netherlands Institute for Sea Research, P.O. Box 59, 1790 AB Den Burg, The Netherlands.

1. Introduction

Eutrophication, the enrichment of natural waters with plant nutrients, was initially only known as a phenomenon in freshwater environments. At present, many coastal areas have experienced an order of magnitude enrichment with nitrogen (N) and phosphorus (P) compared to several decades ago. Notably, the global nitrogen cycle is extraordinary to the extent in which it has been modified by human activity. Industrial N-fixation for fertilizer production, and by internal combustion engines, and the increased farming of legume crops introduce more newly fixed N into the biosphere than the estimated annual natural background fixation of 100 Tg N (Vitousek 1994). A part of this newly fixed and mobilised N is transported to the atmosphere, but most may be expected to be allocated in solution and consequently discharged in continental coastal areas. Coastal marine environments, being the most productive in biological and economical terms, are likely to be dramatically affected by anthropogenic nutrification. In addition to algal biomass related effects, changes in marine vegetation types may lead to shifts in biogeochemical cycles, reduced productivity of coastal areas (due to mortality at higher trophic levels related to toxification or anoxia) and associated economical damage. A global increase of harmful algal blooms has been attributed in part to eutrophication (Smayda 1990).

Adequate water quality measures will be necessary to achieve sustainability of coastal ecosystems. Insight into the mechanisms that determine the algal species composition is needed to predict the possible effect of water quality measures.

Anthropogenic nutrification usually involves an increase in N and/or P but not silicate. Consequently, under eutrophic conditions non-siliceous phytoplankton species can reach enhanced biomass levels whereas diatom growth remains limited by the availability of silicate (van Bennekom *et al.* 1975). The speciation of available N in coastal areas may also be altered as a result of anthropogenic activities. Deforestation may induce discharge of humic acids. Waste water treatment to restore the oxic conditions in rivers can cause a chemical conversion from ammonium to nitrate (Schaub and Gieskes 1991). An increase in sedimentation of organic matter, induced by eutrophication, may cause sediments to act as an ammonium rather than a nitrate source, due to an increase in anoxic conditions (Sloth *et al.* 1995). At high nitrate concentrations (±10-30 μM) in the overlaying waters, anoxic sediments may act as a sink for N as a result of denitrification. At low nitrate concentrations in the overlying waters, N_2 production and consequent N-removal becomes restricted due to nitrate-limited denitrification under anoxic conditions (Rysgaard *et al.* 1994).

With respect to P, eutrophication-induced sedimentation of organic matter may create anoxic conditions which will reduce the P-storage and P-removal capacity of the sediments (Van Cappellen and Ingall 1996), introducing more P to the pelagic zone. Alternatively, there may be occasions where the release of reduced iron from

NATO ASI Series, Vol. G 41
Physiological Ecology of Harmful Algal Blooms
Edited by D. M. Anderson, A. D. Cembella and G. M. Hallegraeff

the sediments becomes enhanced (Canfield et al. 1993; Slomp et al. in press) facilitating a transition of dissolved phosphate into the particulate fraction within the water phase. Hypothetically, this could create a temporal habitat for mixotrophs (such as *Chrysochromulina polylepis*) prior to mass sedimentation of iron-phosphate precipitates.

These biogeochemical processes act at different magnitudes at various geographical locations, which may explain the lack of uniformity in algal species composition in eutrophic areas. Finally, the issue is complicated by the natural differences existing among coastal areas with regard to physical factors such as temperature, irradiance, and vertical stability of the water column.

In many coastal areas eutrophication causes a shift in the ratio of macronutrient supply rates in the photic zone. Due to competition for the limiting nutrient, this might lead to a shift in species composition. It has been suggested that the apparant global increase of harmful blooms is related to enhanced N:Si and P:Si supply ratios favouring non-siliceous phytoplankton (Smayda 1990), some of which are toxic or poorly edible by zooplankton. However, there are also various siliceous species which produce toxins or are poorly edible for herbivorous predators. Whether nutrification in the presence of silicate would also lead to a selection towards harmful species is a valid question. A shift in N:P supply ratios can facilitate a shift in species composition as was reported for the Dutch continental coast (Cadée, 1990). Summer blooms of *Phaeocystis* appeared after a shift from P-limitation to N-limitation in the area, accompanied with a shift from ammonium towards nitrate in the river Rhine discharge (Riegman *et al.* 1992).

The indirect effect of enrichment with macronutrients on algal species composition is related to rigorous changes in the modulation of resource limitation and predation in the pelagic zone, and the enhanced biogeochemical and biological impact of the sediments on the pelagic system.

In this review the initial focus will be on the direct impact of nutrient supply rates on algal species composition. Second, the alteration of system intrinsic factors and its effect on harmful algal blooms will be discussed.

2. Competition for macronutrients

2.1. Nutrient supply ratios and speciation

Even during the establishment of a bloom, algae usually do not grow at their temperature-dependent maximum growth rate (Riegman *et al.* 1993). This indicates that either light or nutrients are non-saturating for growth. Under these conditions, algal species compete for the limiting resource where the affinity of each species for the controlling resource determines the differences in their growth rate response. The ratio between resource supply rates determines the character of the growth rate limiting factor (Tilman *et al.* 1982). The nutrient supply ratio theory of Tilman has not always been applied with great care. In theory, the nutrient supply ratio only has an effect on algal species composition if one of the nutrients is actually limiting the growth rate, i.e. if irradiance levels and temperature are sufficient to support a higher growth rate. Recently the theory on competition has been extended for light-limiting growth conditions (Huisman and Weissing 1995). Basically, there is no essential difference between competition for light (spring bloom conditions) or nutrients (summer blooms). In both cases, an algal species will outcompete others when it is able to reduce the availability of the limiting factor to such an extent that the other competing species are not able to obtain enough of it to compensate for their natural population loss rates. This ability is related to the affinity for a limiting resource and the efficiency at which it is converted into new cell material.

Mixed algal populations can be limited by more than one nutrient, as was theoretically predicted (Tilman *et al.* 1982) and also observed in a natural system such as the North Sea (Riegman *et al.* 1990). Species-dependent variations in nutrient requirement lead to multiple nutrient limitations. Silicate limiting the diatom growth rate and phosphate limiting non-diatoms is a classic example. Co-existence may also be expected when different forms of the same nutrient are used by different species. Nitrogen and phosphorus are available in both the inorganic dissolved fraction and in the particulate (primarily bacterial) fraction. This can lead to co-existence of mixotrophs (feeding mainly on bacteria) and obligatory photoautotrophs (specialised in dissolved organic and/or inorganic nutrient uptake).

Dissolved N is usually available in the form of ammonium, nitrate, nitrite, urea, humic acids and dissolved amino acids (at low concentrations). Depending on the nitrogen speciation some species will be good competitors for N or not. Fig. 1 (Riegman, unpubl. results) shows an example of growth in mixed algal cultures maintained on different N sources which were supplied at similar rates. The diatom *Ditylum brightwelli* was a good competitor for nitrate but not for other N-sources, whereas the diatom *Cymatosira belgica* outcompeted other species on ammonium, urea or humic acids but not on nitrate. The prymnesiophyte *Chrysochromulina polylepis* grew fast, relative to other species, on ammonium and humic acids. The cryptomonad *Rhodomonas* did not grow on urea in mixed cultures. The cyanobacterium *Synechococcus* grew fast under ammonium and urea limitation, and to a lesser extent under nitrate and humic acid limitation.

This example illustrates that not only nutrient ratios but also nutrient speciation can have an effect on the performance of species in mixed populations. It has been suggested that humic acids favour dinoflagellates over diatoms (Granéli and Moreira 1990). This conclusion was based on algal growth response to different mixtures of riverwater from forests and agricultural areas. Unfortunately, silicate levels were not recorded. Therefore, the possibility that diatoms were limited by silicate during growth on the humic enriched forest water, enabling the dinoflagellates to grow faster due to reduced N-depletion during the experiments cannot be excluded. Using commercially available humic acids (Fluka Chemie, Buchs, Germany), we were not able to confirm these results when silicate was in excess (unpubl. results).

Another reason for co-existence in natural populations is the time factor. Simple calculations show that if the net specific growth rate of two species (after correction for grazing and sedimentation losses) differs by 0.1 d^{-1} and both species start to grow at equal biomass, it will take 24 days before 90% dominance is achieved. When competition experiments in mixed algal cultures are performed, the growth rate of several species frequently differ by only 10-20 percent. For example, in one of our experiments, the diatom *Lauderia borealis* was found to grow 1.4 times faster than the diatom *Chaetoceros socialis* under light-limiting conditions in a mixed culture of six algal species. If this would also happen in the natural environment, and both species would experience a specific loss rate of 0.2 d^{-1}, the net increase of *L. borealis* would be three times faster than that of *C. socialis*. Therefore, relatively small differences in growth rate become much more important at significantly high natural mortality rates, resulting in a larger proportion of the limiting nutrient accumulating in the population with the highest net specific growth rate.

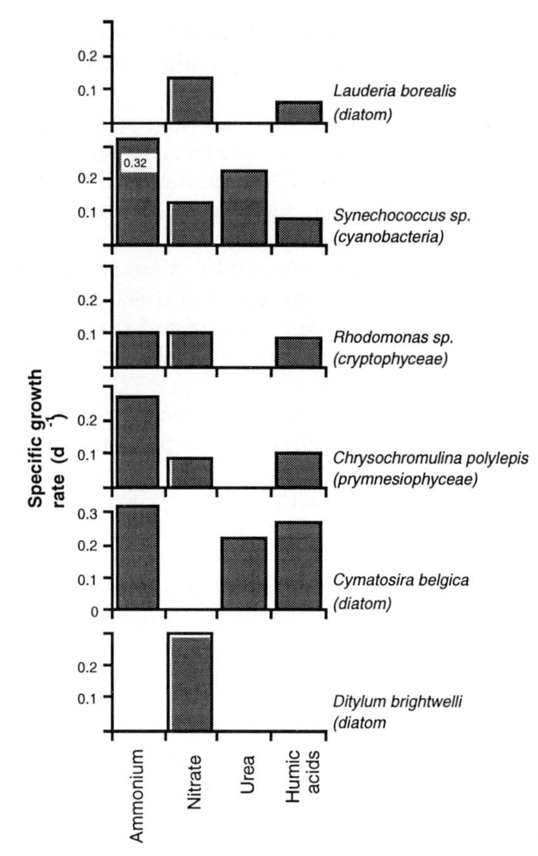

Fig. 1. Response of different algae to different nitrogen sources under N-limiting growth conditions, in discontinuously diluted mixed algal cultures at a dilution rate of 0.1 d^{-1}.

2.2. Algal cell size

One constraining factor with respect to algal competition is interspecific differences in cell size. Unicellular algal species cover a cell size range from 4 to 4 x10^6 μm^3. The competitive advantage of small cell size of microalgae under nutrient sufficient conditions at both light-saturating and light-limiting growth rates is due to enhanced catalytic efficiencies of growth and light absorption (Geider *et al.* 1986). Observations along a transect from the Dogger Bank to the Shetland Islands during spring 1994 are illustrative. The specific growth rate (calculated from specific ^{15}N-ammonium, -nitrate and -urea uptake experiments and corrected for potential bacterial uptake; Riegman and Kuipers, unpubl. results) of the algal size fraction <5µm was significantly higher than that of the larger size fraction (Fig. 2). Yet the bloom which was observed at the Dogger Bank consisted mainly of algae >5µm as indicated by chlorophyll-*a*. Short generation times of microzooplankton often results in effective biomass control of the small size fraction (Riegman *et al.* 1993). Although larger algae are growing slower than their smaller counterparts, their size reduces grazing losses and enables them to accumulate biomass.

Under nutrient-limiting growth conditions small cells are better competitors than large ones (Munk and Riley 1952. This was evident from a theoretical analysis of allometric relationships between minimum nutrient quota, maximum growth rate, and cell volume (Stolte and Riegman 1995). An exception to this general statement is the growth of diatoms on nitrate under pulsed N-limiting conditions. Nitrate is a nutrient which, in contrast to ammonium, can be stored in the vacuole of plant cells. Since the vacuole volume/ cytoplasm ratio increases with cell volume in diatoms, larger cells will have a relatively higher cell N-specific storage capacity. Consequently, larger diatoms are better competitors in nitrate-controlled environments than smaller species (Stolte *et al.* 1994). Competition experiments with dinoflagellates growing on nitrate did not show a similar trend (Riegman, unpubl. res.). This is likely due to a much lower vacuole/cytoplasm ratio which reduces the importance role of the vacuole as a storage reservoir for macronutrients.

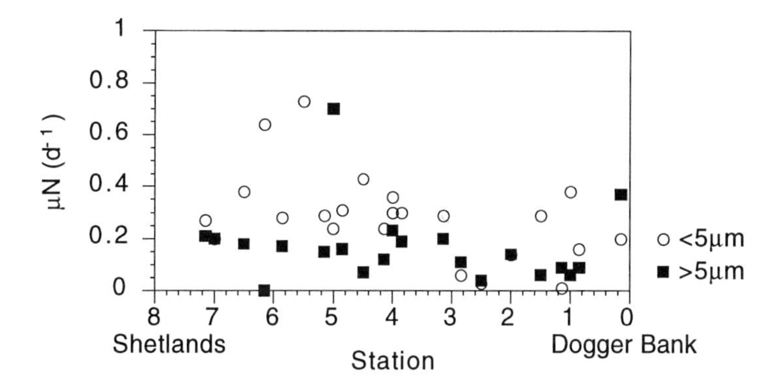

Fig. 2. Growth rate of phytoplankton along a transect from the Dogger Bank to The Shetlands (North Sea) in April 1994, measured as specific ammonium + urea + nitrate uptake per unit of particulate N, indicating higher growth rates for the <5μm fraction. Data were corrected for bacterial uptake which was estimated from leucine incorporation rates (Riegman, unpubl. results).

Experimental evidence for size differential competitive ability was provided from a set of 26 different competition experiments under N-, P-, and light-limited growth conditions with mixtures of dinoflagellates, haptophytes and diatoms. Experimental design and results are published in detail elsewhere (Riegman *et al.* 1996). The observed average specific growth rate of the different species under nutrient limiting growth conditions correlated negatively to their cell volume, confirming that smaller algae are better competitors for nutrients than larger ones (Fig. 3). Since smaller species also grow faster under saturating irradiance and nutrient levels (Banse 1976), many large-celled algal species must exist for reasons other than their growth characteristics.

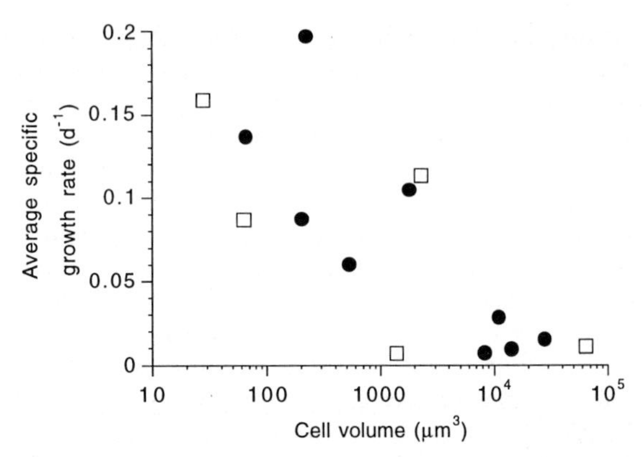

Fig. 3. Specific growth rates of dinoflagellates and haptophytes (circles), and diatoms (squares), grown under different nutrient-limiting growth conditions (for a detailed description: Riegman *et al.* 1996), in relation to their cell size. Calculated for each species as average specific growth rate during growth at different nutrient limitations in mixed algal cultures.

2.3. Taxonomic differences

Competition experiments with mixtures of dinoflagellates, haptophytes and diatoms indicated that dinoflagellates were poor competitors compared to haptophytes. If sufficient silicate was available, diatoms always outcompeted non-diatoms under nutrient- and/ or light-limiting growth conditions. (Riegman *et al.* 1996). A summary of these results, illustrated by the ecological niche of the different groups of organisms, is presented in Fig. 4. Especially under light-limiting growth conditions (in center of diagram) diatoms were the fastest growing organisms. Under saturating light conditions (around centre of circle in diagram), the character of the growth rate-limiting nutrient determined the species composition. Under N-limitation with nitrate as N source, large diatom species such as *Ditylum brightwelli,* became dominant. Haptophytes such as *Emiliania huxleyi* and *Phaeocystis* (Riegman *et al.* 1992) were good competitors for nitrate as well. When nutrients were in excess, smaller diatoms were the fastest growing organisms. At non-limiting nutrient concentrations and photo-inhibiting irradiance levels, *Emiliania huxleyi,* the raphidophyte *Heterosigma carterae*, and the dinoflagellates *Heterocapsa triquetra*, and *Gymnodinium simplex* became dominant. The cyanobacterium *Synechococcus* performed very well under P-limiting growth conditions. This was also the case for *Chrysochromulina polylepis*. Presumably, *C. polylepis* was growing at least partly mixotrophically, using bacteria as an additional P-source (Jones *et al.* 1993). Under N-limiting growth conditions with ammonium as N-source, diatoms grew as fast as non-diatoms. Dinoflagellates were more capable of competing for ammonium than for nitrate.

As discussed above, in natural field populations other factors have a selective effect as well. Each algal species has its own temperature range for growth which restricts its performance to water masses with a temperature within this range. An illustrative example was presented by Karentz and Smayda (1984). Species succession in Narragansett Bay could partly be explained by seasonal changes in temperature and the optimum temperature for growth of the individual species. Additionally, it must be mentioned that within light- or nutrient-controlled populations the vertical stability of the water column may also act as a selective factor (Margalef, 1978). This topic is discussed in detail by Estrada and Berdalet (this volume).

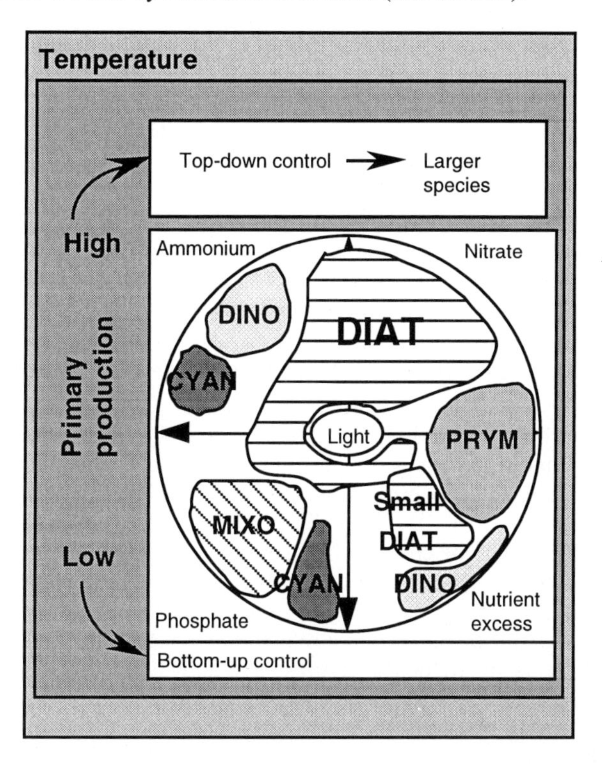

Fig. 4. Ecological niche of dinoflagellates (DINO), diatoms (DIAT), haptophyceae (PRYM), cyanobacteria (CYAN) and mixotrophic algae (MIXO), according to their performance in competition experiments as described in Riegman *et al.* (1996). Centre of circle indicates light limiting conditions; border of circle indicates saturating irradiance levels and nutrient limitation or nutrient excess. Within the indicated taxonomical groups, primary production in the ecosystem may have, depending on its magnitude, its effect on species composition *via* top-down control and corresponding size selective grazing.

The question remains why one should expect to observe any taxonomically related differences in competitive ability of algal species? It is believed that each taxonomic group has a common ancestor. Later, new species developed from this common ancestor. It would be likely that these newly evolved species would differ from each other in their ability to compete under ammonium-, nitrate-, phosphate- or light limiting conditions. In other words, one would expect that within each taxonomic group specialists in nitrate-, ammonium-, phosphate-, light-limited growth have evolved. This is what we actually observed most clearly for diatoms (Riegman *et al.* 1996). However, each species has its own restrictions and specific ecophysiological properties, some of which are shared with other members of the same taxonomic group. These taxonomically distributed properties have their effect on the ecophysiogy of the particular species. It seems that in addition to evolutionary morphological and genetical clustering in the phytoplankton world, we also should be able to observe associated ecological clusters, like diatoms, having a relatively high growth rate under low irradiance conditions but are dependant on silicate, dinoflagellates having a lower intrinsic growth rate (which is only partly explained by their lower chlorophyll-*a*/carbon ratio; Tang 1996), and haptophytes having high maximum growth rates under optimum conditions.

Some species (such as many dinoflagellates; Riegman *et al.* 1996) are poor competitors for nutrients and light. These species probably exist due to their specific life cycle strategies. Resting spore formation followed by sedimentation when the photic zone becomes a hostile, non-productive environment, and germination of the resting spores when conditions again favour vegetative growth, are genetically programmed. The timing of variations in environmental factors at a geographical location determines which species with corresponding life cycle strategy will be selected.

3. Eutrophication and the occurrence of harmful algal blooms

Since nutrient discharges usually have an N:P ratio which differs from the receiving coastal waters, shifts in the controlling resource may occur which consequently will lead to a shift in species composition. At a local site, this may result in the appearance (Riegman *et al.* 1992), or disappearance of a particular harmful algal species as a consequence of changes in resource competition. Since harmful algal blooms (HABs) are known from both P- and N-controlled environments (Nielsen and Ærtebjerg 1984; Rosenberg *et al.* 1988; Smayda 1990) it is very likely that the apparent global increase in HABs is not related to a general shift in supply ratios. In other words, there is no evidence at present that P-limitation leads to a higher frequency of HABs than N-limitation, or *vice versa*.. More likely, the global increase of HABs is indirectly related to a macronutrient enrichment-induced increase in phytoplankton biomass.

Algal species selection is partly related to the intrinsic properties of the biological components of the pelagic foodweb. Hunter and Price (1992) addressed this topic by raising the question: "What factors modulate resource limitation and predation in a system, determining when and where predators or resources will dominate in regulating populations?". According to a theoretical multi-algal species foodweb model (Riegman and Kuipers 1994) increasing eutrophication would lead to a replacement of competitive specialists by rapidly growing generalists (mesotrophic conditions) and finally by poorly edible primary producers (hypertrophic conditions). The basic assumption in this mathematical model was that algae which invest part of their energy in grazing resistance (for example by making toxic substances, poorly digestable cell walls, or spines) are not able to invest the same energy in producing new cells. Consequently, poorly edible algal species are expected to be poor competitors compared to related species which are edible. In fact, at higher total nutrient levels (eutrophicated foodweb) the carrying capacity enhances zooplankton biomass and selective grazing pressure channels the excess of nutrients into poorly edible algal species. The general conclusion from this exercise was (assuming that the basic assumption of poorly edible species as poor competitors is correct) that modulation between bottom-up and top-down control is determined by the trophic state of the foodweb. This hypothesis is supported by field observations: eutrophic and hypertrophic waters harbour a stable community with low turn-over rates, mainly dominated by poorly edible phytoplankton. This type of community is observed in both marine and freshwater systems. Marine red tides (e.g. Smayda 1990) and eutrophic shallow lakes are dominated by populations of dinoflagellates and cyanobacteria respectively. Ecological studies show that the dominant species from these taxonomic groups have a low maximum growth rate (Van Liere and Mur 1979; Falkowski *et al.* 1985; Langdon 1987) and are, relative to other species, poorly edible for zooplankton (Nizan *et al.* 1986; Huntley *et al.* 1987; Lampert 1987). A remarkable example is the estuarine Comacchio Lagoons (Ferrara, Italy), where anthropogenic nutrient discharges in stagnant waters allowed the phytoplankton to develop completely towards the typical hypertrophic state. The population was dominated by chain-forming, poorly edible, cyanobacteria. Carbon fixation rates remained high, but the majority of the fixed carbon was respired during the night by the same cyanobacteria. A drastic reduction of zooplankton, zoobenthos and fish occurred (Sorokin *et al.* 1996). Unfortunately no measurements were made during the establishment of this bloom but it is very likely that grazing on edible species caused a channelling of nearly all the P into the biologically "inert" biomass of cyanobacteria, ending up with an elimination of higher trophic levels. A highly stable bloom developed. No P was recycled, allowing edible species to take over again since the grazers were diminished due to the poor food quality.

More than one fundamental biological principle may be responsible for eutrophication-induced shifts towards harmful species within the different functional groups in the ecosystem. However, the same principle of modulation of regulation factors can be considered in a broader context. Simplifying the evolutionary design of species, two fundamentally different types of organisms can be distinguished: species specialised in conversion of harvested irradiance energy either into new cells (reproduction) or into investment of the fixed carbon in protection mechanisms (survival) such as toxin production, spines, resting stages, etc.

MacArthur and Wilson (1967), reviewing stages of colonisation of islands by terrestrial species, recognised selection for species with high intrinsic rates of increase (r- selection) versus selection for competitive survival (K-selection) in more mature systems. Eutrophication prolongs the bloom period of algae. More and more populations of different species develop and feedback mechanisms such as nutrient-controlled reproduction and top-down control become more important. Occasionally, eutrophication introduces additional stress factors such as high pH (Hinga 1992), anoxic sediments or water layers inducing elevated concentrations of potentially growth rate-limiting chemicals such as S^{2-} and nitrite which reduces the reproductive potential of a functional group. An increase in feedback mechanisms and/or growth rate reducing environmental factors diminishes the competitive advantage of r-specialists and will replace them by K-specialists. Stronger selection towards specialists in survival will occur. The presence of a harmful algal species in a certain eutrophic area could be explained primarily by its ability to survive rather than by its reproductive capacity.

The ecological role of algal toxins is still unclear. The major role of toxins might be protection of populations against predation losses. On the species level typical K-selection species such as dinoflagellates (poor competitors; Riegman *et al.* 1996) are also poorly eaten by microzooplankton compared to non- dinoflagellates (Sellner and Brownlee 1990). This might be related to their toxin production. On the cellular level, it is often observed that reduction of the growth rate by nutrient depletion or non- saturating irradiances elevates the toxin content of cells. The most extreme example of a survival strategy amongst algae is the production of resting spores. The formation of resting spores is induced by nutritional deficiencies (Doucette *et al.* 1989; Blanco 1995). Resting spores of diatoms are protected against grazing losses in the sediments by a heavy silicon cell wall (Drebes 1974), some dinoflagellate cysts have spines, and some cysts (e.g. *Alexandrium* spp.) are toxic. Resting spores of dinoflagellates may contain more toxin than reproducing vegetative cells (Lirdwitayaprasit *et al.* 1990; Oshima *et al.* 1992).

A eutrophication-associated increase in the organic load of the sediments stimulates the microbial foodweb in the sediments. This might enhance the biological selection pressure on resting stages of algal species, depending on their toxicity. In areas with (eutrophication induced) high biological activity in the sediments, strains reinoculating the water column could be more toxic compared to strains in areas where no selection on toxicity of cysts in the sediments occurs. This also implies that toxicity events due to cyst forming dinoflagellates in the photic zone during summer periods are dependent on the biological activity in the sediments during the preceding resting stage period. The hypothesis that, especially in eutrophicated areas, sediments act as a selective enrichment factor of toxic algal species in coastal areas needs further exploration.

The economic value of marine ecosystems is closely related to the productivity of their predators, rather than to the phytoplankton directly. Moderate anthropogenic eutrophication, which brings an oligotrophic ecosystem into the mesotrophic state will favour rapidly growing and edible phytoplankton (Riegman and Kuipers 1993), which might explain the enhancement of, for example, shellfish productivity (Beukema and Cadée 1986, 1991). However, excessive nutrient discharges drive the system into a eutrophic or even hypertrophic state. Persistence leads to K-selection, favouring toxic or poorly edible algae, as is observed most clearly in stagnant waters in marine (Sorokin *et al.* 1996), as well as in freshwater ecosystems (Sommer *et al.* 1986; Post and McQueen 1987; Vanni 1987). Once the dominance of K-selected algal species is established, mass mortalities amongst heterotrophic organisms by toxification may occur, or the productivity of certain higher organisms becomes reduced since the phytoplankton have become less suitable for consumption. This shift in algal species composition may be caused by herbivores which are not necessarely the same as the species which are economically beneficial.

Adequate measures to reduce economic damage caused by eutrophication are i) a reduction of nutrient discharges and/or ii) facilitation of nutrient discharges into water masses with a low residence time such that horizontal dispersion and dilution reduce the effect.

The global increase of HABs is not only attributed to eutrophication (Smayda 1990), it is also understandable from a mechanistic point of view, this being the consequence of the anthropogenic effect on the global N- and P-budget. The coastal zone of the oceans occupies only 18 percent of the surface of the globe. However, it is the area where 60% of the human population lives (65% of the world's cities with populations of over 1.6 million people are located in this area) and it supplies approximately 90 percent of world fish catch (Turner and Adger 1996). Considering the economical importance of coastal areas and the need for sustainable ecosystems, the expansion of HABs in these waters should be a serious concern for the entire international community.

4. Acknowledgements

This work was supported by BEON, contract nr. NIOZ 95 E 02-RKZ-173. A.M.H. van der Heyden, M. de Boer, and L. de Senerpont Domis are gratefully acknowledged for their practical assistence. I am indebted to my colleague Wim van Raaphorst for his valuable information on the role of sediments in macronutrient dynamics.

5. References

Banse, K. (1976). Rates of growth, respiration and photosynthesis of unicellular algae as related to cell size- a review. *J. Phycol.* 12: 135-140.

Bennekom, A. J. van, Gieskes, W. W. C., Tijssen, S. B. (1975). Eutrophication of Dutch coastal waters. *Proc. R. Soc.* (Ser. B) 189: 359-374.

Beukema, J.J., Cadée, G.C. (1991). Growth rates of the bivalve *Macoma baltica* in the Wadden Sea during a period of eutrophication: relationships with concentrations of pelagic diatoms and flagellates. *Mar. Ecol. Prog. Ser.* 68: 249-256.

Beukema, J.J., Cadée, G.C. (1986). Zoobenthos responses to eutrophication of the Dutch Wadden Sea. *Ophelia* 26: 55-64.

Blanco, J. (1995). Cyst production in four species of neritic dinoflagellates. *J. Plankton Res.* 17: 165-182.

Cadée, G.C. (1990). Increase of *Phaeocystis* blooms in the westernmost inlet of the Wadden Sea, the Marsdiep, since 1973. In: Lancelot, C., Billen, G., Barth, H. (eds.) *Water Pollution Research Report* 12: 105-112.

Cadée, G. C. (1992). Phytoplankton variability in the Marsdiep, The Netherlands. *ICES mar. Sci. Symp.* 195: 213-222.

Canfield, D.E., Thamdrup, B. , Hansen, J.W. (1993). The anaerobic degradation of organic matter in Danish coastal sediments: iron reduction, manganese reduction, and sulfate reduction. *Geochimica Cosmochimica Acta* 57: 3867-3883.

Doucette, G.J., Cembella, A.D., Boyer, G.L. (1989). Cyst formation in the red tide dinoflagellate *Alexandrium tamarense* (Dinophyceae): Effects of iron stress. *J. Phycol.* 25: 721-731.

Drebes, G. (1974). Marines Phytoplankton. Georg Thieme Verlag, Stuttgart, p. 12.

Falkowski, P.G., Dubinski, Z., Wyman, K. (1985). Growth irradiance relationships in phytoplankton. *Limnol. Oceanogr.* 30(2): 311-321.

Geider, R.J., Platt, T., Raven, J.A. (1986). Size dependence of growth and photosynthesis in diatoms: a synthesis. *Mar. Ecol. Prog. Ser.* 30: 93-104.

Granéli, E., Moreira, M.O. (1990). Effects of river water of different origin on the growth of marine dinoflagellates and diatoms in laboratory cultures. *J. Exp. Mar. Biol. Ecol.* 136: 89-106.

Hinga, K.R. (1992) Co-occurance of dinoflagellate blooms and high pH in marine enclosures. *Mar. Ecol. Prog. Ser.* 86: 181-187.

Huisman, J., Weissing, F.J. (1995). Competition for nurients and light in a mixed water column: a theoretical analysis. *Am. Nat.* 146: 536-564.

Hunter, M. D., Price, P. W.. (1992). Playing chutes and ladders: heterogeneity and the relative roles of bottom-up and top-down forces in natural communities. *Ecology* 3: 724-732.

Huntley, M.E., Ciminiello, P. , Lopez, M.D.G. (1987). Importance of food quality in determining development and survival of *Calanus pacificus* (Copepoda: Calanoida). *Mar. Biol.* 95: 103-113.

Jones, H.L.J.. Leadbetter, B.S.C., Green, J.C. (1993). Mixotrophy in main species of *Chrysochromulina* (Prymnesiophyceae): digestion of small green flagellates. *J. Mar. Biol. Ass.* U.K. 73: 283-294.

Karentz, D., Smayda, T.J. (1984). Temperature and seasonal occurance patterns of 30 dominant phytoplankton species in Narragansett Bay over a 22-year period (1959-1980). *Mar. Ecol. Prog. Ser.* 18: 277-293.

Kodama, M. (1990). Possible links between bacteria and toxin production in algal blooms. In: Granéli, E., Sundström, B., Edler, L., Anderson, D.M. (eds.) *Toxic Marine Phytoplankton*:. Elsevier, Amsterdam, pp. 52-61.

Lampert, W. (1987). Laboratory studies on zooplankton-cyanobacteria interactions. *N.Z.J. Mar. Fresh/Water Res.* 21: 483-490.

Langdon, C. (1987). On the causes of interspecific differences in the growth-irradiance relationship for phytoplankton. Part I. A comparative study of the growth-irradiance relationship of three marine phytoplankton species: *Skeletonema costatum, Olisthodiscus luteus* and *Gonyaulax tamarensis. J. Plankton Res.* 9(3): 459-483.

Lirdwitayaprasit, T., Nishio, S., Montani, S., Okaichi, T. (1990). The biochemical processes during cyst formation in *Alexandrium catenella.* In: Granéli, E., Sundström, B., Edler, L., Anderson, D.M. (eds.). *Toxic Marine Phytoplankton.* Elsevier, Amsterdam, pp. 294-299.

MacArthur, R., Wilson, E.O. (1967). *The Theory of Island Biogeography.* Princeton University Press, Princeton, N.J., pp. 1-203.

Margalef, R. (1978). Life-forms of phytoplankton as survival alternatives in an unstable environment. *Oceanol. Acta* 1: 493-509.

Middelburg, J.J., Soetaert, K., Herman, P.M.J. (submitted). Denitrification in marine sediments: A model study. *Global Biochem. Cycles.*

Munk, W.H., Riley, G.A. (1952). Absorption of nutrients by aquatic plants. *J. Mar. Res.* 11: 215-240.

Nielsen, A., Ærtebjerg, G. (1984). Plankton blooms in Danish waters. *Ophelia* suppl. 3: 181-188.

Nizan, S., Dimentman, C., Shilo, M. (1986). Acute effects of the cyanobacterium *Microcystus aeruginosa* on *Daphnia magma. Limnol. Oceanogr.* 31: 497-502.

Oshima, Y., Bolch, C.J., Hallegraeff, G.M. (1992). Toxin composition of resting cysts of *Alexandrium tamarense* (Dinophyceae). *Toxicon* 30: 1539-1544.

Post, J.R. and McQueen, D.J. (1987). The impact of planktivoruous fish on the structure of a planktonic community. *Freshwater Biol.* 17: 79-89.

Riegman, R., Colijn, F., Malschaert, J.F.P., Kloosterhuis, H.T., Cadée, G.C. (1990). Assesment of growth rate limiting nutrients in the North Sea by the use of nutrient-uptake kinetics. *Neth. J. Sea Res.* 26: 53-60.

Riegman, R., Noordeloos A. A. M., Cadée, G. C. (1992). *Phaeocystis* blooms and eutrophication of the continental coastal zones of the North Sea. *Mar. Biol.*112: 479-484.

Riegman, R., Kuipers, B.R., Noordeloos, A.A.M., Witte, H.J. (1993). Size-differential control of phytoplankton and the structure of planktonic communities. *Neth. J. Sea Res.* 31: 255-265.

Riegman, R., Kuipers, B.R. (1994). Resource competition and selective grazing of plankton in a multispecies pelagic food web model. *Mar. Ecol.* 15: 153-165.

Riegman, R. (1995). Nutrient related selection mechanisms in marine phytoplankton communities and the impact of eutrophicaion on the planktonic foodweb. *Wat. Sci. Technol.* 32: 63-75.

Riegman, R., de Boer, M., de Senerpont Domis, L. (1996). Growth of harmful marine algae in multispecies cultures. *J. Plankton Res.* 18(9): 1851-1866.

Rosenberg, R., Lindahl, O., Blanck, H. (1988). Silent spring in the sea. *Ambio* 17: 289-290.

Rysgaard, S., Risgaard-Petersen, N., Sloth, N.P., Jensen, K., Nielsen, L.P. (1994). Oxygen regulation of nitrification and denitrification in sediments. *Limnol. Oceanogr.* 39: 1643-1652.

Schaub, B. E. M., Gieskes, W. W. C. (1991). Eutrophication of the North Sea: the relation between Rhine river discharge and chlorophyll a concentration in Dutch coastal waters. In: Elliot, M., Ducrotoy, J.P. (eds.) *Estuaries and coasts: spatial and temporal intercomparisons.* Elsevier, Amsterdam. pp. 85-90.

Sellner, K.G. , Brownlee, D.C. (1990). Dinoflagellate-microooplankton interactions in Chesapeake Bay, in: Granéli, E., Sundström, B., Edler, L., Anderson, D.M. (eds.) *Toxic Marine Phytoplankton*:. Elsevier, Amsterdam. pp. 221-226.

Slomp, C. P., van Raaphorst, W. (1993). Forms of phosphorus in North Sea sediments and fluxes across the sediment-water interface. In: van Raaphorst, W. , Boon, J.P. (eds.) The Integrated North Sea Programme 1991-1992, preliminary results. *NIOZ-Report* 9. NIOZ, The Netherlands. pp. 20-23.

Slomp, C. P., Malschaert, J.F.P., Lohse, L., van Raaphorst, W. (in press). Iron and manganese cycling in different sedimentary environments on the North Sea continental margin. *Cont. Shelf Res.*.

Sloth, N.P., Blackburn, H., Hansen, L.S., Risgaard-Petersen, N., Lomstein, B. Aa. (1995). Nitrogen cycling in sediments with different organic loading. *Mar. Ecol. Prog. Ser.* 116: 163-170.

Smayda, T. J. (1990). Novel and nuisance phytoplankton blooms in the sea: evidence for a global epidemic. In: Granéli, E., Sundström, B., Edler, L., Anderson, D.M. (eds.) *Toxic Marine phytoplankton*. Elsevier, Amsterdam. pp. 29-41.

Sommer,U., Gliwicz, Z.M., Lampert, W., Duncan, A. (1986). The PEG-Model on seasonal successional events in fresh waters. *Archiv für Hydrobiol.* 106: 433-471.

Sorokin, Y.I., Dallocchio, F., Gelli, F., Pregnolato, L. (1996). Phosphorus metabolism in anthropogenically transformed lagoon ecosystems: the Comacchio Lagoons (Ferrara, Italy). *J. Sea Res.* 35(4): 243-250.

Stolte, W., McCollin, T, Noordeloos, A. A. M., Riegman, R. (1994). Effect of nitrogen source on the size distribution within marine phytoplankton populations. *J. Exp. Mar. Biol. Ecol.* 184: 83-97.

Stolte, W., Riegman, R. (1995). The effect of phytoplankton cell size on transient state nitrate and ammonium uptake kinetics. *Microbiology* 141: 1221-1229.

Tang, E.P.Y. (1996). Why do dinoflagellates have lower growth rates? *J. Phycol.* 32: 80-84.

Tilman, D., Kilham, S. S., Kilham, P. (1982). Phytoplankton community ecology: the role of limiting nutrients. *Annu. Rev. Ecol. Syst.* 13: 349-372.

Turner, R.K., Adger, W.N. (1996). Coastal Zones Resources Assessment Guidelines. *LOICZ Reports & Studies* 4, LOICZ, NIOZ, Texel, The Netherlands: 101 pp.

Van Cappellen, Ingall, E.D. (1996). Redox stabilization of the Atmosphere and Oceans by Phosphorus-limited Marine Productivity. *Science* 271: 493-496.

Van Liere, L., Mur, L.R. (1979). Growth kinetics of *Oscillatoria agardhii* Gomont in continuous culture, limited in its growth by the light-energy supply. *J. Gen. Microbiol.* 115: 153-160.

Vanni, M.J. (1987). Effects of nutrients and zooplankton size on the structure of a phytoplankton community. *Ecology* 68: 624-635.

Vitousek, P.M. (1994). Beyond global warming: ecology and global change. *Ecology* 75: 1861-1876.

Trace Elements and Harmful Algal Blooms

Gregory L. Boyer[1] and Larry E. Brand[2]

[1] Faculty of Chemistry, State University of New York, College of Environmental Science and Forestry, Syracuse, NY, 13210-2786, USA.
[2] Division of Marine Biology and Fisheries, Rosenstiel School of Marine and Atmospheric Sciences, University of Miami, 4600 Rickenbacker Causeway, Miami, FL, 33149-1099, USA.

1. Introduction

Recent progress on the concentration and speciation of trace elements in seawater has increased the evidence that trace elements may influence phytoplankton population dynamics. Our understanding of how trace elements affect phytoplankton is best developed in the oceanic environment. John Martin and his colleagues, using "trace metal-clean techniques", studied areas of the ocean where high concentrations of nitrogen and phosphorus remain in the water column (Martin *et al.* 1989; 1990). Both small scale bottle enrichments and large scale open water enrichments (Coale *et al.* 1996) provided evidence for iron (Fe) limitation in these areas. Addition of iron in these studies affected a variety of physiological parameters including chlorophyll fluorescence, Fe uptake rates, specific growth rate and species composition (reviewed in Hutchins 1995). However most harmful algal blooms (HABs) occur in coastal waters where trace element concentrations are higher, sometimes by orders of magnitude, than the concentrations in oceanic systems. Could these higher concentrations of trace elements still be limiting? In this chapter, we examine what is known about the concentrations of trace elements in marine waters, their distribution and changes over time, their chemical speciation and availability to phytoplankton. The effects of complexation, multiple metal limitations, trace metal antagonisms, and the variation in individual requirements among different species are all considered as factors that could lead to trace metal limitation in coastal waters despite the higher concentrations found there. Several specific examples of HABs where trace element limitation or toxicity has been implicated are discussed in light of these controlling factors.

1.1. Trace Metal Concentrations in Marine Waters

In the central gyres of the open ocean, trace element concentrations are low and fairly constant. By contrast, estuarine trace element concentrations are often high and extremely variable due to differences in watershed biogeochemistry and run-off. Various biogeochemical scavenging processes can, under some circumstances, reduce estuarine trace element concentrations to levels similar to those found in the open ocean. The range of trace element concentrations found in seawater is summarized in Table 1. Careful measurements of trace element concentrations have

NATO ASI Series, Vol. G 41
Physiological Ecology of Harmful Algal Blooms
Edited by D. M. Anderson, A. D. Cembella and G. M. Hallegraeff

been made in relatively few estuarine and coastal waters, so these data may underestimate the total concentration range of trace metals in these environments.

TABLE 1. The range of trace element concentrations (nM) in surface waters.

Trace Element	*Oceanic*	*Coastal*	*Estuarine*
Iron (Fe)	0.2 - 1	5 - 50	0.5 - 5,000
Manganese (Mn)	0.5 - 3	2 - 30	5 - 10,000
Zinc (Zn)	0.1 - 0.2	0.5 - 4.5	4 - 500
Cobalt (Co)	0.004 - 0.03		0.2 - 10
Copper (Cu)	0.5 - 1.3	2 - 40	4 - 60
Cadmium (Cd)	0.001 - 0.01	0.05 - 0.07	0.02 - 0.85
Selenium (Se)	0.5		0.6 - 4.5

Oceanic concentrations are from Donat and Bruland (1995). Coastal and estuarine concentrations are from Cutter (1991) and Sunda *et al.* (1990).

Most trace elements increase by approximately one order of magnitude in going from the surface of the open ocean onto the continental shelf but individual differences in the elements are observed. Selenium concentrations do not substantially change from oceanic to neritic waters; its concentration is at most ten times higher in some estuaries than in the open ocean. In contrast, Fe and Mn can increase several orders of magnitude over the same geographic distance. The concentrations of Fe, Mn and Zn show an additional increase as one moves from coastal to estuarine waters, whereas there is less change in either the Cu or Cd concentrations. Within estuaries, the concentration of different elements can vary from one to three orders of magnitude, with Fe, Mn and Zn being the most variable, and Cu, Cd and Se the most conservative. Variations due to heavy metal pollution can further magnify these differences. Thus, it is clear that coastal and estuarine phytoplankton are exposed to a wide range of trace element concentrations.

1.2. Complexation

Current dogma suggests that the trace metal uptake rate is dependent on the free metal concentration and not the total metal concentration (Sunda 1994). While some exceptions to this rule may exist, (e.g., Fe uptake may depend on the sum of the labile Fe hydroxy-complexes), total dissolved trace metal concentrations are generally not particularly useful in determining trace element availability to phytoplankton. Complexation of trace metals by inorganic and organic ligands can make most of the total metal concentration unavailable to phytoplankton. If such complexation occurs in coastal waters, the trace metal ion concentrations actually experienced by phytoplankton can be orders of magnitude less than the total measured concentrations. Evidence for strong, primarily organic, complexation in seawater has been observed for Fe, Cu, Cd, Co and Zn (for a review, see Sunda 1991; 1994).

Copper offers an excellent example on the effects of complexation on algal uptake since free cupric ions are toxic to phytoplankton. Early researchers advanced the hypothesis that the high concentration of ubiquitous dissolved organic matter in seawater complexes most of the free Cu (Sunda *et al.* 1981). Recently it has been reported that Cu speciation in the photic zone is controlled by low concentrations of organic molecules with extremely high affinities for Cu (Coale and Bruland 1990; Moffett *et al.* 1990). As a result, free ionic Cu decreases from around 10^{-10} M in deep water where the strong chelators are largely absent to 10^{-13} M in the photic zone. This has led to the hypothesis that strong upwelling, while providing nutrients, also delivers toxic Cu to the photic zone (Barber and Ryther 1969). An observed decline in cyanobacterial abundance after a storm in the North Pacific may be the result of such upwelled Cu (Ditullio and Laws 1991).

The concentrations of organic compounds that chelate Cu are usually only slightly higher than the total concentration of Cu in the seawater. Cyanobacteria, but not eukaryotic phytoplankton, produce strong Cu complexing compounds in culture with almost identical stability constants to those found in the photic zone (Moffett and Brand 1996). These cyanobacteria use a feedback mechanism to produce the minimal amount of ligand needed to complex essentially all the Cu. This results in a metal buffered system in which free ionic Cu is maintained at a nontoxic concentration (around 10^{-13} M), but one with a low buffering capacity. Addition of a relatively small amount of Cu can over-titrate the organic buffer. If this happens, free ionic Cu concentrations increase by orders of magnitude and can become toxic. Field studies in selected embayments on Cape Cod (Moffett *et al.* 1996) showed that in two embayments with relatively low total Cu concentrations (around 5 nM), cyanobacteria were as abundant inside the embayments as in the adjacent coastal waters. Essentially all the Cu was complexed by organic ligands resulting in a free ionic concentration around 10^{-13} M, a level nontoxic to cyanobacteria. A large numbers of boats were present in the other two embayments examined and total Cu was around ten times higher. The flux of Cu into these embayments (probably from antifouling paint on boat hulls) was faster than the rate the cyanobacteria could excrete the strong Cu complexing compounds, resulting in a thousand fold increase in free ionic Cu concentration. The Cu buffer system was over-titrated, resulting in toxic concentrations of free ionic Cu and a resultant large decrease in the cyanobacterial populations.

It is significant that the free ionic Cu concentrations in this system appear to be controlled primarily by cyanobacteria. As discussed in Section 2, Cu toxicity has been cited as a controlling factor for toxic dinoflagellate blooms. However, the degree of Cu toxicity that dinoflagellates experience appears to be controlled by another group of organisms within the community. Where cyanobacteria are unable to grow or become abundant, the free ionic Cu concentrations might be expected to be higher. Therefore, factors that affect cyanobacteria ultimately could affect dinoflagellates as well. Dinoflagellates and other HAB species may be able to excrete their own Cu complexing compounds (Robinson and Brown 1991), although these are weaker than those produced by cyanobacteria. Thus, the free ionic Cu concentrations will vary during the course of species succession and species composition is both the cause and an effect of this variation. Not only can trace metal chemistry affect phytoplankton, but phytoplankton in turn affect the trace metal chemistry. Subtle changes in these dynamics may have dramatic effects on the phytoplankton community.

Less is known about biotic interactions with the other trace metals. Zinc in oceanic seawater is complexed by organic ligands that have very high affinity constants (Bruland 1989), resulting in a 1000-fold decrease in the free Zn ion concentrations in the photic zone. Others (Gledhill and van den Berg 1994; Rue and Bruland 1995) have shown that over 99% of the Fe is complexed by organic ligands. Most of the Cd (Bruland 1992) and 45-100% of the Co (Zhang *et al.* 1990) in seawater is also organically complexed. The availability of these organic complexes to phytoplankton is unknown. Algal studies using Zn suggest that these complexes are not available for uptake (Sunda and Huntsman 1992; 1995a). However biological feedback mechanisms would prevent the production of a large surplus of ligands, limiting the buffering capacity of the system. In weakly metal-buffered systems such as these, the free ionic concentrations of the trace metals can vary over several orders of magnitude as a result of relatively small changes in metal or ligand concentrations. Such dramatic changes in trace metal chemistry may explain some of the large and rapid changes observed in phytoplankton species composition in estuarine and coastal waters.

1.3. Multiple Limitation

A significant artifact in most laboratory experiments on nutrient limitation is that the effects of low concentrations of one nutrient are examined while keeping all other nutrients at high concentrations. While this is an appropriate experimental design to ensure attributing the observed effects to one particular nutrient, the experimental environment is clearly not an accurate reflection of the environment phytoplankton normally experience. Biogeochemical cycles in the ocean that result in low concentrations of a particular nutrient typically result in low concentrations of other nutrients as well. Trace metals can substitute for others in the active sites of enzymes, e.g., both Cd and Co can substitute nutritionally for Zn (Price and Morel 1990; Sunda and Huntsman 1995a). As a result, an experiment in which only one nutrient is reduced could erroneously indicate little nutrient limitation because unrealistically high concentrations of other nutrients substitute for the experimental one. Experiments in which Fe, Zn, and Mn concentrations were all reduced at the same time (Brand, unpublished data), using an experimental design parallel to earlier experiments in which Fe, Zn, and Mn were reduced individually in separate experiments (Brand *et al.* 1983), indicate that simultaneous limitation by Fe, Zn, and Mn is much more severe than limitation by any one metal alone. Concentrations of trace metals that can induce nutritional limitation in combination may be orders of magnitude less than what has been estimated from experiments where only individual metals were varied.

1.4. Trace Metal Antagonism

Most essential metals (e.g. Fe, Zn) are toxic at some concentration and toxic metals (e.g. Cu) also have a nutritional role. Relative concentrations are important: high levels of Cu and Cd compete with Fe, Zn or Mn for the active sites of uptake or incorporation into intracellular enzymes (Sunda 1991). Copper (Sunda *et al.* 1981; Sunda and Huntsman 1983), Zn and Cd (Sunda and Huntsman 1996) inhibit Mn uptake and growth on low concentrations of Mn. High levels of Cu are more inhibitory to uptake and growth at low concentrations of Zn (Rueter and Morel 1981) or Fe (Stauber 1995), and Cd is inhibitory at low levels of Fe (Harrison and Morel 1983). This competition between toxic and nutritional roles of different metals means that trace metal ratios are often more important than the concentrations of individual metals. Coastal waters, with their higher concentrations of nutritional trace metals, are also likely to have higher concentrations of competitive "toxic" trace metals. The competitive inhibition of uptake and incorporation of the nutritional metals by toxic concentrations of Cu and Cd could therefore induce trace metal nutritional limitation even in coastal waters with normally sufficient concentrations of Fe, Zn, Mn and Co.

1.5. Species Differences and Species Composition

If large differences exist among species in their adaptations to different trace metal regimes, changes in environmental conditions can have a large effect on species composition of the community. The differences among species are the result of their individual evolutionary histories in response to different ecological selective forces, combined with fundamental biochemical and genetic constraints. The influence of natural selection can be seen in comparing the ability of coastal and oceanic phytoplankton species to grow at low metal concentrations. Brand *et al.* (1983) found that oceanic species of eukaryotic phytoplankton can grow at substantially lower concentrations of Fe, Zn and Mn than species from coastal or estuarine waters. The cellular requirements of oceanic eukaryotic species for Fe and Zn are considerably lower than for coastal and estuarine species (Brand 1991; Sunda and Huntsman 1992; 1995a; 1995b). Thus, coastal species have difficulty competing for nutritional trace metals in the open ocean and are restricted to coastal waters.

In contrast, no significant differences are observed between most coastal and oceanic species of eukaryotic phytoplankton in their ability to resist Cu toxicity (Brand *et al.* 1986). Total Cu concentrations are generally around four times higher in coastal waters than oceanic waters. Brand *et al.* (1986) speculated that free ionic Cu may not be any higher in coastal waters because the higher concentration of dissolved organic matter may complex and detoxify it. Recent Cu-complexation studies have confirmed this speculation (Sunda and Huntsman 1991; Moffett *et al.* 1996). Some species, such as *Skeletonema costatum,* appear to be particularly resistant to Cu. These species are often found in newly upwelled waters where free ionic Cu concentrations are expected to be highest. However it can be argued both ways: that these species have evolved a high tolerance to trace metals because they live in high trace metal environments, or that these species live in high trace metal environments because they have a high tolerance to trace metals. The wide range in adaptations to trace metals found among coastal phytoplankton (Brand *et al.* 1983; 1986) may result from the wide temporal and spatial variations in trace metal concentrations and chemical speciation found in these coastal ecosystems.

These adaptations reflect not only the results of natural selection, but also the biochemical and genetic constraints that prevent other adaptations from evolving. The redox biogeochemical history of the earth has had a major influence on the adaptations of phytoplankton with respect to trace metals. Dramatic changes in trace metal chemistry in the ocean resulted from the biosphere shifting from anaerobic to aerobic conditions. Prokaryotic cyanobacteria, for the first half of their evolutionary history, were exposed to high concentrations of Fe, Mn, and Co, and low concentrations of Cu, Zn and Cd relative to the ocean today. By contrast, eukaryotic phytoplankton have always lived in an aerobic ocean, and thus have been exposed to relatively low concentrations of Fe, Mn, and Co, and high concentrations of Cu, Zn and Cd throughout their evolutionary history. These differences in evolutionary history are reflected in the different abilities of prokaryotic cyanobacteria and eukaryotic phytoplankton to grow at low concentrations of essential nutrients and resist the toxic levels of others (Brand 1991; Sunda and Huntsman 1995b).

2. Individual Case Studies on Micronutrient Limitation or Metal Toxicity in Harmful Algal Blooms

For several decades, scientists have postulated the importance of trace metals in the formation of HABs. Ingle and Martin (1971) reported on the use of an Fe index to predict the occurrence of *Gymnodinium breve* blooms in southern Florida. Similarly, Fe has been reported to limit blooms of *Alexandrium tamarense* in the Gulf of Maine, *Chattonella antiqua* in the Seto Inland Sea and *Heterosigma akishiwo* in Osaka Bay. Iron is not the only trace metal implicated in the occurrence of HABs. Table 2 presents a brief summary of those cases where the growth of harmful algal species were reported to be dependent on trace metal input.

The importance of environmental factors (temperature, salinity, water stratification) and macronutrients (nitrogen, phosphate) in the control of harmful algal blooms cannot be denied. However some HABs are sporadic in nature, failing to occur even under apparently optimal conditions of temperature, salinity and macronutrient availability. Trace metals are often implicated in either stimulating or inhibiting harmful algal blooms, but there are few examples where this has been examined in detail. In reviewing these examples, it is useful to keep in mind Koch's postulates from microbial pathology. Using an analogous set of rules, a claim for nutrient limitation by trace metals in harmful algal blooms should meet the following criteria:

(1) Deficiency of the metal should exist prior to the bloom period;

(2) Addition of the metal should stimulate growth of the organism;

(3) The magnitude of the response observed *in situ* should agree with the amount of metal added to the system and its availability to the organism.

TABLE 2. Trace Metals Implicated in the Growth of Harmful Algal Species.

Metal	Species	Reference
IRON		
	Alexandrium tamarense	Wells *et al.* 1991a
	Aureococcus anophagefferens	Gobler 1995
	Chattonella antiqua	Okaichi *et al.* 1989
		Nakamura 1990
	Gymnodinium breve	Ingle and Martin 1971
		Kim and Martin 1974
	Gymnodinium sanguineum	Doucette and Harrison 1990; 1991
	Heterosigma sp.	Yamochi 1983; 1989
COPPER		
	Alexandrium tamarense	Anderson and Morel 1978
		Schenck 1984
	Chattonella antiqua	Nakamura *et al.* 1987
		Nakamura 1990
	Gymnodinium breve	Martin and Olander 1971
	Gymnodinium sanguineum	Robinson and Brown 1991
	Prorocentrum minimum	Granéli *et al.* 1986
SELENIUM		
	Aureococcus anophagefferens	Cosper *et al.* 1993
	Chattonella verruculosa	Imai (this volume)
	Chrysochromulina polylepis	Dahl *et al.* 1989
	Gymnodinium nagasakiense	Koike *et al.* 1993
		Ishimaru *et al.* 1989
	Pyrodinium bahamense	Usup and Corrales (this volume)
COBALT		
	Chrysochromulina polylepis	Granéli *et al.* 1993
		Granéli and Risinger 1994

The effects of trace metals on the growth of HABs will be discussed in a series of case studies. The first two studies deal with the importance of Fe. Iron is a key nutrient for the growth of phytoplankton and it is required at concentrations higher than other "trace" elements. Iron is essential for a number of biochemical processes and is found in the cytochromes and iron-sulfur proteins needed for the photosynthetic and respiratory electron transport chains, nitrate and nitrite reductase important for nitrogen utilization, sulfate reductase and the enzymes of chlorophyll biosynthesis. Phytoplankton limited by Fe often show signs of N-limitation and may be unable to effectively use certain nitrogen sources (Price *et al.* 1991). They utilize light energy poorly and show increased chlorophyll fluorescence (Green *et al.* 1994). To meet their cellular Fe needs, selected phytoplankton utilize Fe-sparing mechanisms such as the replacement of the Fe-containing protein ferredoxin with the riboflavin-dependent flavodoxin in the electron transport chain (La Roche *et al.* 1995). Induction of a high-affinity Fe uptake system may also allow uptake from sources of Fe normally unavailable to the cell (Soria-Dengg and Horstmann 1995). The mechanisms for Fe uptake in harmful algal species are poorly known as most studies on Fe uptake in phytoplankton have utilized diatoms (Anderson and Morel, 1982; Hudson and Morel 1990; 1993). However total Fe is generally a bad predictor of Fe availability. The use of dissolved Fe (that which passes through an 0.45 μm filter) as an estimate of biologically available Fe is compromised by the fact that dissolved Fe colloids (Wells *et al.* 1983; Rich and Morel 1990) and soluble Fe complexes with synthetic chelators (Anderson and Morel 1982) have been found to be unavailable to some phytoplankton. Free Fe concentrations are extremely low in seawater and Fe uptake has been proposed to be dependent on the concentration of soluble inorganic hydrolysis species (Hudson and Morel 1990; 1993). The $Fe(OH)^{2+}$ ion is the most abundant Fe hydroxy species in seawater. It can readily shed its outer layers of water of hydration and is a likely species for uptake. Free Fe^{3+}, $Fe(OH)^{2+}$ and other forms of the ferric hydroxides can rapidly interconvert. Theoretical calculations comparing the total amount of Fe required for growth to the solubility of the ferric hydroxide species suggest that at least some Fe must also be supplied by dissolution of colloidal particles. Iron availability would depend in part on the speed of dissolution of these particles (Wells 1989). Iron-sparing mechanisms may dramatically change the cellular Fe quota and induction of a high-affinity Fe uptake system could solubilize forms (e.g. ferric hydroxides) that are otherwise unavailable for use by the cell. Both responses would change the apparent need for Fe by the cell and may be important in that adaptation by phytoplankton to low-Fe conditions.

Case 1: Iron Limitation of Dinoflagellates along the Eastern Coast of the United States.

Iron limitation has often been suggested as a controlling factor for HABs along the eastern seaboard of North America. In the Fe index of Ingle and Martin (1971), the level of Fe discharged in the Peace River over a three month interval was strongly correlated with the occurrence of *Gymnodinium breve* blooms in southern Florida. While having limited predictive value, this index was unable to differentiate between the importance of Fe versus humic acids and rain-fall, two other variables strongly correlated with the riverine Fe content (Kim and Martin 1974). Riverine humic input into coastal systems can affect phytoplankton productivity by attenuating light, providing essential nutrients such as N or P, or act as a chelator that may decrease the toxicity of metals such as Cu. More information on the nutrient status of the organisms is needed before assigning a causative role to the riverine input.

More detailed studies have been conducted on *Alexandrium tamarense* blooms in the Gulf of Maine. Earlier studies by Glover (1978) observed increased Fe in the freshwater plume associated with river run-off entering the Gulf of Maine near Monhegan Island. This freshwater plume was followed by a dinoflagellate bloom (*Gymnodinium* sp.), suggesting riverine input may stimulate primary productivity in the area. This area also serves as a seed population for the recurrent blooms of *A. tamarense* that occur in the region. To test if these populations of *A. tamarense* were limited by Fe, Wells *et al.* (1991a) studied the distribution and bioavailability of Fe in this region.. They developed an extraction technique using 8-hydroxyquinoline or "oxine" to estimate the concentration of labile Fe species in sea water and the freshwater plumes of estuaries (Wells *et al.* 1991b). The affinity of the oxine chelator for Fe is similar to that calculated for the proposed Fe uptake receptors; thus, Fe available to oxine should be representative of Fe available for cellular uptake. Careful calibration of Fe-limited cultures of *A. tamarense* using this oxine technique showed that at least 3 nM "oxine" Fe was necessary to support dense (10^7 cells l^{-1}) *Alexandrium* blooms. Because of the problems of luxury consumption and changes in the stability of ferric hydrite with time, this number represents an upper limit to the amount of Fe required. Blooms below the 10^7 cells l^{-1} threshold could easily occur at concentrations less that 3 nM oxine Fe. Detailed field studies along the Damariscotta River region of Maine and out to Georges Bank showed that for the inland coastal regions, the levels of oxine Fe were consistently above that 3 nM threshold, ranging from 5-50 nM (*i.e.* 25-1,445 nM total Fe). For Fe to initiate the formation of red tides, the cells must be Fe-limited prior to bloom initiation. The levels of Fe in these coastal waters averaged 8 nM throughout much of the time period when blooms are expected to be initiated and should never have been limiting to *Alexandrium*. Oxine Fe was not correlated to the occurrence of PSP toxins in shellfish, a crude indicator of the presence of *Alexandrium* in the water column.

These results suggest that Fe *per se* was not controlling bloom initiation or cell densities in the nearshore waters of the Gulf of Maine. However, Fe may be important in limiting that bloom to the coastal margin. Transects run from the coast offshore indicated the levels of oxine Fe dropped below the 3 nM threshold as one moved into deeper waters. This condition would be unfavorable for the growth of *A. tamarense* and would limit its growth to the near-shore environment. Levels of oxine Fe rebounded above the 3 nM threshold once over Georges Bank, a region that has experienced blooms of *A. tamarense* based on PSP toxicity records. While these studies are strongly suggestive of the importance of Fe, the use of the oxine technique does have its limitations. Photochemical processes dramatically affect the lability of Fe (Finden *et al.* 1984) and its availability to phytoplankton. Bacterial Fe transport compounds, similar to those produced by marine bacteria in oceanic waters (Haygood *et al.* 1993), may solubilize particulate Fe and increase its bioavailibility to *Alexandrium* in these off-shore waters. Thus more Fe may be available to phytoplankton in the off-shore environment than estimated using the oxine technique. Knowledge of the cellular uptake mechanisms available to *Alexandrium* and a metabolic marker for the intracellular Fe status of the cell are needed to understand the importance of Fe in this system.

Case 2. Importance of Iron in Raphidophyte Blooms in the Seto Inland Sea.

A second example where Fe has been implicated in HABs can be found in the study of red tides due to *Chattonella antiqua* (Raphidophyceae). For years, these blooms have sporadically occurred during summer in the Seto Inland Sea of Japan. Heavy rains often occur in mid-June to mid-July, resulting in large amounts of freshwater run-off with its associated nutrients flowing into the Inland Sea. Nitrate and phosphate concentrations are generally stratified in the water column by mid-July. The levels of macronutrients in the surface waters are often insufficient to support maximal growth of *Chattonella.* However, the raphidophyte is capable of diel vertical migration and, if a shallow nutricline is imposed, they migrate to the bottom waters to utilize the excess nutrients found there (Watanabe *et al.* 1995). Temperature, salinity, light intensity and vitamin B_{12} were generally optimal for a *Chattonella* bloom, even in years that the bloom did not occur (Nakamura *et al.* 1988; 1989). Since earlier blooms of *Chattonella* (e.g. 1983) were strongly correlated with river run-off, total organic carbon, and Fe and Cu concentrations were monitored around the Ie-shima Islands to see if they contributed to bloom formation (Montani *et al.* 1989). Soluble Fe concentrations in the surrounding waters were generally in the range of 4-10 nM with total Fe as high as 160 nM (Nakamura 1990). The Fe concentrations reported for maximum growth of *C. antiqua* in culture were ~1 μM Fe (as FeEDTA), suggesting the organisms would either be limited by the available Fe, or have uptake mechanisms that would allow it to use particulate Fe (Okaichi *et al.* 1989). Initial attempts reported the isolation of desferrioxamine B, a common terrestrial siderophore produced by numerous *Streptomyces* spp., in iron-limited cultures of *C. antiqua.* Siderophore production is as part of a high-affinity Fe-uptake system and is common in marine bacteria (Trick 1989). It has yet to be established for marine eukaryotes. These results, therefore, need to be viewed cautiously until they have been repeated using axenic cultures. More likely, the Fe is absorbed to the glycocalyx layer where it can be solubilized and absorbed by the cell. A positive correlation between chlorophyll-*a* and total Fe suggested that much of the Fe was associated with cells. High concentrations (~6-10 nM) of filterable Fe (< 0.8 μm) observed in both bloom and non-bloom years were not correlated with chlorophyll or the occurrence of *Chattonella.* These observations suggest that Fe was not a factor in bloom formation. In contrast, evidence supporting a role for Fe was obtained from the bioassay experiments. Addition of 100 nM Fe to bottle cultures stimulated growth of *C. antiqua* in the non-bloom (1986, 1988) years, and in bloom years (1987) prior to development of the bloom, but not during the course of the bloom. Thus, the levels of Fe may be insufficient to support maximal growth except in bloom years. Cognizant of the fact that the observed levels of Fe were not significantly different between the two time periods, Nakamura (1990) suggested that Fe bioavailability may be different in non-bloom years. One possibility is that filtered Fe did not represent the labile Fe actually available for uptake. Experimental reasons dictated the use of an 0.8 μm filter, and a major fraction of this Fe could have been as Fe colloids, unavailable Fe chelates, or in picoplankton and bacteria. The oxine or a substitute technique was not used, so it is uncertain what fraction of the filterable Fe was actually available for uptake. The cellular quota for Fe was not reported, making it difficult to estimate if the levels of Fe found in the system matched the needed biochemical demand for the observed bloom density. With only one of three criteria fulfilled, the current evidence for Fe limitation in this system is inconclusive.

Case 3. Copper Limitation of Harmful Algal Blooms.

The role of Cu on HABs is different from that of Fe. Copper is an essential metal needed for growth, being found in plastocyanin in the photosynthetic electron transport system, cytochrome C oxidase in the respiratory electron transport system and in amine oxidases important in amino acid catabolism. However, there is little evidence for Cu limitation other than in laboratory cultures. More commonly, Cu exerts its effects by being toxic to phytoplankton. As discussed earlier, this toxicity is a function of the free metal concentration whose concentration is controlled by Cu-binding ligands present in the water column. Whatever the source of these ligands, either by riverine input or through *in situ* production by cyanobacteria, this Cu - ligand buffer can easily be overwhelmed by inputs of additional Cu. As a result, small changes in Cu concentration can result in large differences in the free metal concentration.

Laboratory studies of the marine flagellates *Alexandrium tamarense*, *Gymnodinium sanguineum* and *Chattonella antiqua* (Table 2) have all suggested that the levels of inorganic Cu found in natural environments during bloom events should have been toxic to these organisms if only inorganic complexation was considered. In the case of *C. antiqua* in the Seto Inland Sea, the total and soluble Cu concentrations were similar during bloom and non-bloom events and did not change significantly throughout the course of the year. Organic complexation of Cu must have been present to allow the bloom to occur, but there was no evidence to suggest that the level of complexation changed between bloom and non-bloom years. Hence the authors concluded that Cu was probably not an important regulator of these blooms (Nakamura 1990). In contrast, the levels of complexed Cu as measured by anodic stripping voltammetry increased during a bloom of *G. sanguineum* (Robinson and Brown 1991). These results suggested that chelation may be important in controlling the bloom initiation for this organism. New techniques developed to measure low concentrations of high affinity Cu chelators in seawater should be applied to better evaluate the role of cupric ions in harmful bloom formation.

Case 4. Role of Cobalt in Chrysochromulina Blooms in the Kattegat - Skagerrak.

Cobalt is an important component in vitamin B_{12}. This vitamin is produced by many marine bacteria and these bacteria probably serve as net producers of the vitamin in the marine environment. Most algae readily absorb vitamin B_{12} and function as consumers of this essential compound. Recent studies indicate that Co also has other functions in the cell and the cellular requirements for Co may vary depending on environmental conditions. Cobalt and Zn are at least partially interreplaceable in marine algae, and thus, Zn limitation may increase the cell's requirement for Co (Price and Morel 1990; Sunda and Huntsman 1995a). The cell quota for Co in the red tide species *Heterosigma akashiwo (=H. carterae)* depended on the phosphate status of the organism (Watanabe *et al.* 1989). Thus low phosphate levels, coupled with the ratio of other metals, could combine to increase the number of situations where Co limitation is important.

The Kattegat-Skagerrak area of the Norwegian Sea was the site of an unusual bloom of the prymnesiophyte *Chrysochromulina polylepis* in 1988. This bloom turned out to have unexpected toxic activity towards trout and salmon farms in the area. The causes of the bloom are uncertain but the nitrogen and phosphate levels prior to the event did not appear extraordinary, suggesting a potential role for micronutrients. Bioassay results hinted that the *C. polylepis* bloom in the Kattegat in spring 1988 may have been due to the leaching of Co into coastal waters due to soil acidification and river run-off (Granéli *et al.* 1993; Granéli and Haraldsson 1993). To test this hypothesis, Granéli and Risinger (1994) enriched seawater samples with Co acetate, CoEDTA, or cyanocobalamin (Vitamin B_{12}). All three forms of Co gave similar results. The growth rate did not change, but cell yields increased with increasing concentrations of Co up to 1 nM. Cobalt concentrations measured in the Kattegat averaged between 0.2 - 0.4 nM during the bloom. Using an estimate of the cell quota (0.55-0.69 fg Co $cell^{-1}$) and the maximum bloom population of 10^9 cells per liter from the 1988 bloom, these Co concentrations would be just below the amount sufficient to support the bloom. Thus, this study fulfills three major criteria: that the nutrient (Co) be limiting to the organisms, that addition of the nutrient stimulated the bloom, and that the magnitude of the response was proportional to the concentration of metal added. Only missing are studies on the levels of Co during non-bloom years to establish that these conditions are bloom specific.

Case 5. The Importance of Selenium in Harmful Algal Blooms.

Selenium is an essential nutrient for the growth of a wide range of phytoplankton (Harrison *et al.* 1988) and is routinely added to culture media for sensitive or axenic species. Selenium is present in natural waters as selenate (Se^{VI}), selenite (Se^{IV}) and organoselenium compounds. Most of the Se present in coastal waters is probably in the form of organoselenium, a common component of humic materials present in river run-off. Laboratory experiments suggest that the addition of soil extract to cultures can be replaced by the addition of Se in some cases. While all three Se forms (selenite, selenate and organoselenium) are available for phytoplankton uptake (Cutter and Bruland 1984), species-specific differences in Se uptake do exist. Selenocysteine, the primary form of cellular Se, functions as an antioxidant in enzymes such as glutathione peroxidase. Selenocysteine can also degrade peroxides nonenzymatically (Gennity *et al.* 1985).

Several harmful algal species have been examined for their Se requirements. It is an essential nutrient for *Aureococcus anophagefferens* (Cosper *et al.* 1993), *Chattonella verruculosa* (Imai, this volume), *Gymnodinium nagasakiense* (Ishimaru *et al.* 1989; Koike *et al.* 1993), *Pyrodinium bahamense* (Usup and Corrales, this volume) and species of *Chrysochromulina*. Oceanic strains of *Chrysochromulina polylepis* require very low levels of Se (Harrison *et al.* 1988) whereas *C. polylepis* isolated from Scandinavian waters showed a marked growth increase upon addition of 10^{-8} M Se as selenite (Dahl *et al.* 1989). Perhaps the most detailed studies on the effects of Se in harmful algal bloom formation have been with blooms of *Gymnodinium nagasakiense* from the Wakayama Prefecture in Japan. Nutrient addition experiments using unialgal cultures and seawater from Temma Bay indicated that only Se in the form of selenite significantly enhanced the growth of test cultures of *G. nagasakiense* (Ishimaru *et al.* 1989). Field experiments during the 1988 *G. nagasakiense* bloom indicated that during the month immediately preceding the bloom, the Se levels remained low (<0.1 nM). During the month with the maximum cell density, the Se concentrations were at maximum. The peak Se concentration during the bloom was 2.8 nM Se^{IV} (~3.2 nM total Se) at the surface. Using a cellular quota of 0.044 fmol Se $cell^{-1}$, these levels should have supported a bloom density near 72,000 cells ml^{-1}, in good agreement with the 100,000 cells ml^{-1} actually observed. Selenium concentrations above 0.4 nM were linearly related to cell concentration. These results suggest that an influx of Se may be essential to achieve high cell densities. They do not necessarily support the hypothesis that Se input triggered the bloom. The initial rise in cell numbers was observed prior to an increase in Se and the pre-bloom Se levels (0.1 nM) should have supported a cell concentration of nearly 2,000 cells ml^{-1}, well above the 100-400 cells actually observed. This suggests that other factors, such as temperature, may play a more important role in bloom initiation. The source of Se to this system is unknown. The concentrations of Se in the Aizu river flowing into the Tanabe Bay have not been determined. Selenium is common in bottom sediments and the pulse in Se observed during the bloom period may simply reflect a mixing event. More studies on the changes in Se during the early bloom events, as well as the possible sources to the system are needed the clarify its role in red tide formation.

Other Trace Elements.

In addition to Fe, Cu, Co and Se, numerous other trace elements have been suggested to limit phytoplankton growth in natural environments. Ishimaru *et al.* (1989), in their study on the effects of trace metals on the growth of *G. nagasakiense*, looked at the addition of 19 different trace elements (Cu, Fe, Se, Cr, Co, Mn, Mo, Rb, V, W, Zn, Zr, B, Ba, Li, Si, Br, I, F) to seawater collected from Temma Bay. While only selenite showed any noticeable effect in that particular study, inorganic elements other than Fe, Cu, Co and Se can limit the growth of phytoplankton. Three likely candidates include Mn, Ni, and Zn. These metals are briefly discussed below.

Manganese:

Manganese is an essential cofactor in both photosystem II and superoxide dismutase. It is required for the growth of all algae. Manganese is present in the marine environment as soluble Mn^{II} and as the insoluble Mn^{III} and Mn^{IV} oxides which are lost from the system by precipitation and settling. This oxidation is kinetically slow, allowing the more soluble Mn^{II} to persist for some time in seawater despite its thermodynamic instability. While the abiotic oxidation of Mn^{II} is slow, Mn oxidizing bacteria can considerably increase the rates of oxidation in natural environments (Sunda and Huntsman 1987). Once formed, the oxides readily absorb humic and non-living organic mater in the water column. A pronounced photochemical reduction of Mn and photoinhibition of Mn oxidizing bacteria leads to a surface maximum of Mn^{II} observed in ocean waters (Sunda and Huntsman 1988). It is estimated that the turnover time for Mn in coastal waters is on the order of days to weeks, providing an excellent opportunity for its uptake by phytoplankton (Sunda and Huntsman 1987). Laboratory studies have shown coastal coccolithophores and diatoms are limited by free Mn concentration below $\sim 10^{-9}$ M (Brand *et al.* 1983). While coastal waters often have total Mn concentrations two to three orders of magnitude above this value (Table 1), antagonisms between inhibitory metals (Cu, Zn and Cd) and Mn can greatly increase concentrations of Mn^{II} needed to support maximum growth rates (Sunda and Huntsman 1983; 1996). Addition of Mn increased the growth of *Heterosigma* sp. and *Gymnodinium* sp. in natural assemblages taken from Osaka Bay in Japan (Takahashi and Fukazawa 1982). Further studies on the importance of this metal in HABs are warranted.

Nickel:

Nickel is also an essential nutrient for many phytoplankton. It is a cofactor in urease, the enzyme responsible for the cleavage of urea. Cells grown on other nitrogen sources such as nitrate or ammonia show a much lower uptake rate for Ni and an apparent loss of their cellular requirement for Ni (Oliveira and Antia 1986; Price and Morel 1991). Recent experiments, however, have suggested that Ni may have a cellular role in enzymes other than urease in phytoplankton (Soeder and Engelmann 1984; Brand unpublished). The importance of Ni in other enzymes other than urease, and its specific role in HABs, is unknown.

Zinc:

Zinc is an essential cofactor for carbonic anhydrase and acid and alkaline phosphatase, enzymes that play important roles in the carbon and phosphorus nutrition of many phytoplankton. In addition, Zn can be found in DNA-RNA ligases and in DNA regulatory proteins in the form of Zn finger proteins. Cadmium and Co, two nutrient analogs, can substitute for Zn in many circumstances (Price and Morel 1990; Sunda and Huntsman 1995a). Zinc speciation in oceanic waters is determined by the presence of a relatively long lived organic chelates (Bruland 1989), and Zn concentrations appear to be controlled by biological uptake and regeneration processes (Sunda and Huntsman 1992). Zinc levels are often much higher in coastal environments (Table 1) and coastal phytoplankton show a correspondingly higher ($>10^{-11}$ M) Zn requirement (Brand *et al* 1983; Sunda and Huntsman 1995a). Zinc, along with Fe, may play an important role in the evolution and spatial distribution of phytoplankton species (Brand *et al.* 1983). In spite of this, the evidence for Zn limitation in coastal phytoplankton in general and harmful algal blooms in particular, is lacking. Interestingly, *Cochlodinium* type 78, a red tide forming dinoflagellate from the Yatsushiro Sea, Japan, forms a Zn complex with its PSP toxins (Onoue and Nozawa 1989). It is uncertain at this time if this represents a native toxin or is simply an artifact formed during isolation.

3. Summary

While there is some evidence for the influence of Fe, Se, Co, and Cu on the growth of harmful algal species, the exact role played by trace elements and chelators in the initiation and maintenance of most HABs remains unknown. Despite decades of work, key physiological parameters such as the cellular metal quotas, uptake kinetics, and transport mechanisms have yet to be determined for the majority of harmful algal species. The over-riding question is not if the harmful organism has a general requirement for a particular metal. The nutritional physiology of harmful algal species is often very similar to their non-toxic counterparts. Trace elements are essential for the growth of all phytoplankton, regardless of their origin or ecological significance and harmful algal species are not likely to be any different. Rather the question is whether harmful algal species have unique trace element requirements, physiology, metal uptake or detoxification mechanisms that would allow them to dominate non-toxic species. Coastal environments have higher concentrations of essential micronutrients than oceanic systems. Yet compensating factors such as metal complexation and speciation, multiple nutrient limitation, and trace metal antagonism all combine to create a system where, despite these higher levels, a given micronutrient may be limiting for a particular species. The question of whether a harmful algal bloom will develop then becomes: what controls species composition in these systems and how are trace elements involved? The differences among species of phytoplankton in their ability to grow at low nutritional trace element concentrations and resist high concentrations of toxic metals needs to be examined in much greater detail for all species, toxic or not, before the importance of individual metals on specific harmful algal blooms is suitably evaluated.

4. Acknowledgments

Preparation of this review was supported by the Suffolk County Department of Health Services and by a National Oceanic and Atmospheric Administration award #NA46RG0090 from the New York Sea Grant Association to G.L.B., and by an Office of Naval Research award #N000149310893 to L.E.B. We would like to thank William G. Sunda for his helpful review of the manuscript.

5. References

Anderson, D.M., Morel, F.M.M. (1978). Copper sensitivity of *Gonyaulax tamarensis*. *Limnol. Oceanogr.* 23:283-295.

Anderson, M.A., Morel, F.M.M. (1982). The influence of aqueous iron chemistry on the uptake of iron by the coastal diatom *Thalassiosira weissflogii*. *Limnol. Oceanogr.* 27: 789-813.

Barber, R.T., Ryther, J.H. (1969). Organic chelators: factors affecting primary production in the Cromwell current upwelling. *J. Exp. Mar. Biol. Ecol.* 3:191-199.

Brand, L.E. (1991). Minimum iron requirements of marine phytoplankton and the implications for the biogeochemical control of new production. *Limnol. Oceanogr.* 36:1756-1771.

Brand, L.E., Sunda, W.G., Guillard, R.R.L. (1983). Limitation of phytoplankton reproductive rates by zinc, manganese and iron. *Limnol. Oceanogr.* 28:1182-1198.

Brand, L.E., Sunda, W.G., Guillard, R.R.L. (1986). Reduction of marine phytoplankton reproduction rates by copper and cadmium. *J. Exp. Mar. Biol. Ecol.* 96:225-250.

Bruland, K.W. (1989). Complexation of zinc by natural organic ligands in the central North Pacific. *Limnol. Oceanogr.* 34:269-285.

Bruland, K.W. (1992). Complexation of cadmium by natural organic ligands in the central North Pacific. *Limnol. Oceanogr.* 37:1008-1017.

Coale, K.H., Bruland, K.W. (1990). Spatial and temporal variability in copper complexation in the North Pacific. *Deep Sea Res.* 34:317-336.

Coale, K.H., Johnson, K.S., Fitzwater, S.E., Gordon, R.M., Tanner, S., *et. al.* (1996). A massive phytoplankton bloom induced by an ecosystem-scale iron fertilization in the equatorial Pacific Ocean. *Nature* 383:495-501.

Cosper, E.M., Garry, R.T., Milligan, A.J., Doall, M.H. (1993). Iron, selenium, and citric acid are critical to the growth of the brown tide microalga, *Aureococcus anophagefferens.* In: Smayda, T.J., Shimizu, Y. (eds.), *Toxic Phytoplankton Blooms in the Sea.* Elsevier, Amsterdam, pp. 667-673.

Cutter, G.A. (1991). Trace elements in estuarine and coastal water - U.S. studies from 1986-1990. *Reviews of Geophysics, Supplement, Am. Geophys. Union.* pp. 639-644.

Cutter, G.A., Bruland, K.W. (1984). The marine biogeochemistry of selenium: A re-evaluation. *Limnol. Oceanogr.* 29:1179-1192.

Dahl, E., Lindahl, O., Paasche, E., Throndsen, J. (1989). The *Chrysochromulina polylepis* bloom in Scandinavian waters during spring 1988. In: Cosper, E.M., Bricelj, V.M., Carpenter, E.J. (eds.), *Novel Phytoplankton Blooms.* Springer-Verlag, Berlin, pp. 383-405.

Ditullio, G.R., Laws, E.A. (1991). Impact of an atmospheric-oceanic disturbance on phytoplankton community dynamics in the North Pacific central gyre. *Deep Sea Res.* 38:1305-1329.

Donat, J.R., Bruland, K.W. (1995). Trace elements in the oceans. In: Salbu, B., Steinnes, E. (eds.), *Trace Elements in Natural Waters.* CRC Press, Boca Raton, FL., pp. 247-281.

Doucette, G.J., Harrison, P.J. (1990). Some effects of iron and nitrogen stress on the red tide dinoflagellate *Gymnodinium sanguineum.* *Mar. Ecol. Prog. Ser.* 62:293-306.

Doucette, G.J., Harrison, P.J. (1991). Aspects of iron and nitrogen nutrition in the red tide dinoflagellate *Gymnodinium sanguineum.* II. Effects of iron depletion and nitrogen source on iron and nitrogen uptake. *Mar. Biol.* 110:175-182.

Finden, D.A.S., Tippings, E., Jaworski, G.H.M., Reynolds, C.S. (1984). Light-induced reduction of natural iron (III) oxide and its relevance to phytoplankton. *Nature* 309:783-784.

Gennity, J. M., Bottino, N.R., Zingaro, R.A. Wheeler, A.E., Irgolic, K.J. (1985). A selenium-induced peroxidation of glutathione in algae. *Phytochemistry* 24:2817-2821.

Gledhill, M., van den Berg, C.M.G. (1994). Determination of complexation of iron(III) with natural organic complexing ligands in seawater using cathodic stripping voltammetry. *Mar. Chem.* 47:41-54.

Glover, H.E. (1978). Iron in Maine coastal waters; seasonal variation and its apparent correlation with a dinoflagellate bloom. *Limnol. Oceanogr.* 23:534-537.

Gobler, C.J. (1995). The role of iron in the occurrence of *Aureococcus anophagefferens* blooms. M.Sc. thesis, State Univ. New York-Stony Brook, 127 p.

Granéli, E., Haraldsson, C. (1993). Can increased leaching of trace metals from acidified areas influence phytoplankton growth in coastal waters? *Ambios* 22:308-311.

Granéli, E., Persson, H., Edler, L. (1986). Connection between trace metals, chelators and red tide blooms in the Laholm Bay, SE Kattegat - An experimental approach. *Mar. Environ. Res.* 18:61-78.

Granéli, E., Paasche, E., Maestrini, S.Y. (1993). Three years after the *Chrysochromulina polylepis* bloom in Scandinavian waters in 1988: Some conclusions of recent research and monitoring. In: Smayda, T.J., Shimizu, Y. (eds.), *Toxic Phytoplankton Blooms in the Sea.* Elsevier, Amsterdam, pp. 23-32.

Granéli, E., Risinger, L. (1994). Effects of cobalt and vitamin B_{12} on the growth of *Chrysochromulina polylepis* (Prymnesiophyceae). *Mar. Ecol. Prog. Ser.* 113:177-183.

Green, R.M., Kolber, Z.S., Swift, D. G., Tindale, N.W., Falkowski, P.G. (1994). Physiological limitation of phytoplankton photosynthesis in the eastern equatorial Pacific determined from variability in the quantum yield of fluorescence. *Limnol. Oceanogr.* 39:1061-1074.

Harrison, G.I., Morel, F.M.M. (1983). Antagonism between cadmium and iron in the marine diatom *Thalassiosira weissflogii*. *J. Phycol.* 19:495-507.

Harrison, P.J., Yu, P.W., Thompson, P.A., Price, N.M., Phillips, D. J. (1988). Survey of selenium requirements in marine phytoplankton. *Mar. Ecol. Prog. Ser.* 47:89-96.

Haygood, M.G., Holt, P.D., Butler, A. (1993). Aerobactin production by a marine planktonic marine *Vibrio* sp.. *Limnol. Oceanogr.* 38:1091-1097.

Hudson, R.J.M., Morel, F.M.M. (1990). Iron transport in marine phytoplankton: kinetics of cellular and medium coordination reactions. *Limnol. Oceanogr.* 35: 1002-1020.

Hudson, R.J.M., Morel, F.M.M. (1993). Trace metal transport by marine microorganisms: implications of metal coordination kinetics. *Deep Sea Res.* 40:129-150.

Hutchins, D.A. (1995). Iron and the marine phytoplankton community. *Prog. Phycol. Res.* 11:1-49.

Ingle, R.M., Martin, D.F. (1971). Prediction of the Florida red tide by means of the iron index. *Environ. Lett.* 1:69-74.

Ishimaru, T., Takeuchi, T., Fukuyo, Y., Kodama, M. (1989). The selenium requirement of *Gymnodinium nagasakiense*. In: Okaichi, T., Anderson, D.M., Nemoto, T. (eds.), *Red Tides; Biology, Environmental Science and Toxicology.* Elsevier, New York, NY, pp. 357-360.

Kim, Y.S., Martin, D.F. (1974). Interrelationship of Peace River parameters as a basis of the iron index: a predictive guide to the Florida red tide. *Water Res.* 8:607-616.

Koike, Y., Nakaguchi, Y., Hiraki, K., Takeuchi, T., Kokubo, T., Ishimaru, T. (1993). Species and concentrations of selenium and nutrients in Tanabe Bay during red tide due to *Gymnodinium nagasakiense*. *J. Oceanogr.* 49:641-656.

La Roche, J., Murray, H., Orellana, M., Newton, J. (1995). Flavodoxin expression as an indicator of iron limitation in marine diatoms. *J. Phycol.* 31:520-530.

Martin, D. F., Olander, K.W. (1971). Effects of copper, titanium and zirconium on the growth rates of the red tide organism, *Gymnodinium breve*. *Environ. Lett.* 2:135-142.

Martin, J.H., Gordon, R.M., Fitzwater, S., Broenkow, W.W. (1989). VERTEX: Phytoplankton/iron studies in the Gulf of Alaska. *Deep Sea Res.* 36:649-680.

Martin, J.H., Gordon, R.M., Fitzwater, S.E. (1990). Iron in Antarctic waters. *Nature* 345:156-158.

Moffett, J.W., Brand, L.E. (1996). Production of strong, extracellular Cu chelators by marine cyanobacteria in response to Cu stress. *Limnol. Oceanogr.* 41:388-395.

Moffett, J.W., Brand, L.E., Croot, P.L., Barbeau, K.A. (1997). Copper speciation and cyanobacterial distribution in harbors subject to anthropogenic Cu inputs. *Limnol. Oceanogr. (in press).*

Moffett, J.W., Zika, R.G., Brand, L.E. (1990). Distribution and potential sources and sinks of copper chelators in the Sargasso Sea. *Deep Sea Res.* 37:27-36.

Montani, S., Tokuyasu, M., Okaichi, T. (1989). Occurrence and biomass estimation of *Chattonella marina* red tides in Harima Nada, the Seto Inland Sea, Japan. In: Okaichi, T., Anderson, D.M., Nemoto, T., (eds.), *Red Tides: Biology, Environmental Science and Toxicology.* Elsevier, New York, NY, pp. 197-200.

Nakamura, Y. (1990). Chemical environment for red tides due to *Chattonella antiqua*. Part 3. Roles of iron and copper. *J. Oceanogr. Soc. Jpn.* 46: 84-95.

Nakamura, Y., Sawai, K., Watanaba, M. (1987). Growth inhibition of a red tide flagellate, *Chattonella antiqua* by copper. *J. Oceanogr. Soc. Jpn.* 42:481-486.

Nakamura, Y., Takashima, J., Watanabe, M. (1988). Chemical environment for red tides due to *Chattonella antiqua* in the Seto Inland Sea, Japan. Part 1. Growth bioassay of the seawater and dependence of growth rate on nutrient concentrations. *J. Oceanogr. Soc. Jpn.* 44:113-124.

Nakamura, Y., Umemori, T., Watanabe, M. (1989). Chemical environment for red tides due to *Chattonella antiqua*. Part 2. Daily monitoring of the marine environment throughout the outbreak period. *J. Oceanogr. Soc. Jpn.* 45:116-128.

Okaichi, T., Montani, S., Hiragi, J., Hasui, A. (1989). The role of iron in the outbreaks of *Chattonella* red tides. In: Okaichi, T., Anderson, D.M., Nemoto, T. (eds.), *Red Tides: Biology, Environmental Science and Toxicology.* Elsevier, New York, NY, pp. 353-356.

Oliveira, L., Antia, N.J. (1986). Nickel ion requirements for autotrophic growth of several marine microalgae with urea serving as nitrogen source. *Can. J. Fish. Aquati. Sci.* 43:2427-2433.

Onoue, Y., Nozawa, J. (1989). Zinc-bound PSP toxins separated from *Cochlodinium*. *Bioact. Mol. (Mycotoxins Phycotoxins* 1988) 10:359-366.

Price, N.M., Andersen, L.F., Morel, F.M.M. (1991). Iron and nitrogen nutrition of equatorial Pacific plankton. *Deep Sea Res.* 38:1361-1378.

Price, N.M., Morel, F.M.M. (1990). Cadmium and cobalt substitution for zinc in marine diatom. *Nature* 344:658-660.

Price, N.M., Morel, F.M.M. (1991). Colimitation of phytoplankton growth by nickel and nitrogen. *Limnol. Oceanogr.* 36:1071-1077.

Rich, H.W., Morel, F.M.M. (1990). Availability of well-defined iron colloids to the marine diatom *Thalassiosira weissflogii*. *Limnol. Oceanogr.* 35:652-662.

Robinson, M.G., Brown, L.N. (1991). Copper complexation during a bloom of *Gymnodinium sanguineum* Hirasaka (Dinophyceae) measured by ASV. *Mar. Chem.* 33:105-118.

Rue, E.L., Bruland, K.W. (1995). Complexation of iron(III) by natural organic ligands in the Central North Pacific as determined by a new competitive ligand equilibration/adsorptive cathodic stripping voltammetric method. *Mar. Chem.* 50:117-138.

Rueter, J.G.,Jr., Morel, F.M.M. (1981). The interaction between zinc deficiency and copper toxicity as it affects the silicic acid uptake mechanism in *Thalassiosira pseudonana*. *Limnol. Oceanogr.* 26:67-73.

Schenck, R.C. (1984). Copper deficiency and toxicity in *Gonyaulax tamarensis* (Lebour). *Mar. Biol. Lett.* 5:13-19.

Soeder, C.J., Engelmann, G. (1984). Nickel requirement in *Chlorella emersonii*. *Arch. Microbiol.* 137:85-87.

Soria-Dengg, S., Horstmann, U. (1995). Ferrioxamines B and E as iron sources for the marine diatom *Phaeodactylum tricornutum*. *Mar. Ecol. Prog. Ser.* 127: 269-277.

Stauber, J.L. (1995). Toxicity testing using marine and freshwater unicellular algae. *Austral. J. Ecotoxicol.* 1:15-24.

Sunda, W.G. (1991). Trace metal interactions with marine phytoplankton. *Biol. Oceanogr.* 6:411-442.

Sunda, W.G. (1994). Trace metal / phytoplankton interactions in the sea. In: Bidoglio, G., Stumm, W., (eds.), *Chemistry of Aquatic Systems: Local and Global Perspectives*. ECES, EEC, EAEC, Brussels, Netherlands, pp. 213-247.

Sunda, W.G., Barber, R.T., Huntsman, S.A. (1981). Phytoplankton growth in nutrient rich seawater: Importance of copper-manganese cellular interactions. *J. Mar. Res.* 39:567-586.

Sunda, W.G., Huntsman, S.A. (1983). Effect of competitive interactions between manganese and copper on cellular manganese and growth in estuarine and oceanic species of the diatom *Thalassiosira*. *Limnol. Oceanogr.* 28:924-934.

Sunda, W.G., Huntsman, S.A. (1987). Microbial oxidation of manganese in a North Carolina estuary. *Limnol. Oceanogr.* 32: 552-564.

Sunda, W.G., Huntsman, S.A. (1988). Effect of sunlight on redox cycles of manganese in the southwestern Sargasso Sea. *Deep Sea Res.* 35:1297-1317.

Sunda, W.G., Huntsman, S.A. (1991). The use of chemiluminescence and ligand competition with EDTA to measure copper concentration and speciation in seawater. *Mar. Chem.* 36: 137-163.

Sunda, W.G., Huntsman, S.A. (1992). Feedback interactions between zinc and phytoplankton in seawater. *Limnol. Oceanogr.* 37:25-40.

Sunda, W.G., Huntsman, S.A. (1995a). Cobalt and zinc interreplacement in marine phytoplankton: biological and geochemical implications. *Limnol. Oceanogr.* 40:1404-1417.

Sunda, W.G., Huntsman, S.A. (1995b). Iron uptake and growth limitation in oceanic and coastal phytoplankton. *Mar. Chem.* 50: 189-206.

Sunda, W.G., Huntsman, S.A (1996) Antagonisms between cadmium and zinc toxicity and manganese limitation in a coastal diatom. *Limnol. Oceanogr.* 41:373-387.

Sunda, W.G., Tester, P.A., Huntsman, S.A. (1990). Toxicity of trace metals to *Acartia tonsa* in the Elizabeth River and southern Chesapeake Bay. *Estuar. Coastal Shelf Sci.* 30: 207-221.

Takahashi, M., Fukazawa, N. (1982). A mechanism of "red tide" formation. II. Effect of selective nutrient stimulation on the growth of different phytoplankton species in natural water. *Mar. Biol.* 70:267-273.

Trick, C.G. (1989). Hydroxamate-siderophore production and utilization by marine eubacteria. *Curr. Microbiol.* 18:375-378.

Watanabe, M., Takamatsu, T., Kohata, K., Kunugi, M., Kawashima, M., Koyama M. (1989). Luxury phosphate uptake and variation of intracellular metal concentrations in *Heterosigma akashiwo* (Raphidophyceae). *J. Phycol.* 25:428-436.

Watanabe, M., Kohata, K., Kimura T., Takamatsu, T., Yamaguchi, S., Ioriya, T. (1995). Generation of a *Chattonella antiqua* bloom by imposing a shallow nutricline in a mesocosm. *Limnol. Oceanogr.* 40:1447-1460.

Wells, M.L. (1989). The availability of iron in seawater: A perspective. *Biol. Oceanogr.* 6:463-476.

Wells, M.L., Mayer, L.M., Guillard, R.R.L. (1991a). Evaluation of iron as a triggering factor for red tide blooms. *Mar. Ecol. Prog. Ser.* 69:93-102.

Wells, M.L., Mayer, L.M., Guillard, R.R.L. (1991b). A chemical method estimating the availability of iron to phytoplankton in seawater. *Mar. Chem.* 33:23-40.

Wells, M.L., Zorkin, N.G., Lewis, A.G. (1983). The role of colloid chemistry in providing a source of iron to phytoplankton. *J. Mar. Res.* 41:731-746.

Yamochi, S. (1983). Mechanisms for an outbreak of *Heterosigma-akashiwo* red tide in Osaka Bay, Japan. 1. Nutrient factors involved in controlling the growth of *Heterosigma-akashiwo*. *J. Oceanogr. Soc. Jpn.* 39:310-316.

Yamochi, S. (1989). Mechanisms for outbreak of *Heterosigma akashiwo* red tide in Osaka Bay, Japan. In: Okaichi, T., Anderson, D.M., Nemoto, T., (eds.), *Red Tides: Biology, Environmental Science and Toxicology.* Elsevier, New York, NY, pp. 253-256.

Zhang, H., van den Berg, C.M.G., Wollast, R. (1990). The determination of interactions of cobalt (II) with organic compounds in seawater using cathodic stripping voltammetry. *Mar. Chem.* 28:285-300.

Utilization of Dissolved Organic Matter (DOM) by Phytoplankton, Including Harmful Species

P. Carlsson and E. Granéli

University of Kalmar, Dept. of Natural Sciences, PO Box 905, S-391 29 Kalmar, Sweden

1. Introduction

Many phytoplankton species, including some harmful species, are known to be partly autotrophic and partly heterotrophic and are therefore called mixotrophic. The mixotrophic behavior can be manifested either as ingestion of food particles (termed phagotrophy, see chapters by Hansen, and Granéli and Carlsson, this volume) or by active uptake of dissolved/colloidal organic substances by osmotrophy/pinocytosis, processes which are believed to play a minor role in phytoplankton nutrition. For example, *Protoperidinium* species without chloroplasts were considered to be osmotrophic for a long time. However, these species have now been shown to capture phytoplankton cells and perform extracellular digestion (Gaines and Taylor 1984; Jacobson and Anderson 1986). This can be considered as "extracellular osmotrophy". Mixotrophy is a means for some phytoplankton to obtain energy, macronutrients or certain vitamin(s) necessary for growth (Gaines and Elbrächter 1987; Caron *et al.* 1990). Most dinoflagellates, both marine and freshwater, are known to require small amounts of some specific externally produced organic compound(s), usually the vitamins B_{12} (cyanocobalamine), biotin and/or thiamine (Provasoli and Carlucci 1974; Gaines and Elbrächter 1987). Thus, most photosynthetic dinoflagellates are "auxotrophs", i.e. they cannot solely rely on photosynthesis for growth.

The ecological importance of phytoplankton utilizing dissolved organic matter (DOM) as a significant part of their nutrition is still an open question. That some phytoplankton can grow heterotrophically on dissolved organic carbon compounds (DOC) in the dark has been known for some time (Droop 1974). However, growth of phytoplankton in the dark based on DOC as the carbon (C) source requires very high concentrations of substrate that do not occur in natural waters (Droop 1974; Richardson and Fogg 1982). Furthermore, utilization of organic C compounds in the light can increase the growth rate of phytoplankton species (Cheng and Antia 1970; Combres *et al.* 1994), but utilization of DOC in the light also requires high concentrations of substrate. Experiments with natural levels of substrates have shown no significant stimulation of either growth or survival of axenic algal species (Richardson and Fogg 1982). These results lead to the conclusion that microalgal utilization of C in DOM can be considered as insignificant (Richardson and Fogg 1982).

Soil extract has often been used to increase success in culturing phytoplankton (e.g. Provasoli *et al.* 1957). Prakash and Rashid (1968) and Prakash *et al.* (1973)

NATO ASI Series, Vol. G 41
Physiological Ecology of Harmful Algal Blooms
Edited by D. M. Anderson, A. D. Cembella and G. M. Hallegraeff

found that DOM in the form of humic substances increased both yield and growth rates of marine dinoflagellates and diatoms. Although the reason for the growth-stimulating effect could not be adequately explained, the authors suggested that the humic substances acted as chelators and made essential trace metals available for the algae. The content of nitrogen (N) and phosphorus (P) in the humic substances were considered to be of minor importance (Prakash and Rashid 1968). The amount of N and P in the humic substances used in the experiments of Prakash and Rashid (1968) were also negligible compared to the N and P present in the medium used. Later experiments have shown that the enhancement of phytoplankton growth when soil extract/humic substances are added may be due to the ability of DOM to chelate essential metals and make them available for phytoplankton (e.g. Anderson and Morel 1982).

However, the N present in organic form in soil extract can also be beneficial for phytoplankton. Morrill and Loeblich (1979) found that N-limited axenic cultures of the dinoflagellate *Kryptoperidinium foliaceum* increased in growth rate and biomass yield when supplied with sterile soil extract, probably because of the N content. Granéli *et al.* (1985) suggested that the dinoflagellate *Prorocentrum minimum* was able to use N in humic substances since the biomass yield increased considerably when humic substances and phosphate were added to the medium. The cellular N content of *P. minimum* grown with addition of humic substances was also comparable to cells grown with inorganic N (Granéli *et al.* 1985).

Direct utilization of small organic molecules, such as amino acids, can also be an important mechanism for phytoplankton to obtain macronutrients (see review by Flynn and Butler 1986). Note that the traditional view is that phytoplankton use inorganic N forms, while bacteria use dissolved organic nitrogen (DON) (Billen 1984).

Concentrations of inorganic N usually become low (0.1-2 μmol l^{-1}, Flynn and Butler 1986; Hansell 1993) in many coastal surface waters during summer months, while concentrations of DON are higher (around 5-10 μmol l^{-1}, Hansell 1993), and even higher if freshwater is influencing the seawater (Antia *et al.* 1991, and references therein). Thus, a large fraction of the N in coastal seawater occurs in the form of organic molecules. It would be beneficial for phytoplankton if they could use the pool of DON to some extent when inorganic N limits production. Organically-bound P can be utilized by phytoplankton through the action of phosphatases (see review by Cembella *et al.* 1984), but the utilization of organically bound N other than urea or amino acids is less well known. See recent review of phytoplankton utilization of DON by Antia *et al.* (1991).

This chapter will focus on phytoplankton utilization of DOM as a source of N since phytoplankton production in most coastal waters are considered to be N-limited and DON makes up a potentially large N pool in these areas. Moreover, riverine input of DON is a large potential N source in coastal waters. Direct utilization of low- and high-molecular-weight (HMW) organic N compounds by phytoplankton will be discussed, as will the possibility for phytoplankton to utilize organically-bound N indirectly via the action of the "microbial loop".

2. Properties of dissolved organic matter (DOM)

Dissolved organic matter can be operationally defined as the organic fraction that passes a 0.2 μm pore-size filter (Benner *et al.* 1992). In addition to truly dissolved compounds, DOM therefore also consists of ultraparticles/colloids, small bacteria and viruses that pass such filters.

Most DOM of terrestrial origin is composed of C (about 50% by weight), but it also contains N (about 1-3%) and P (about 0.2%) (Thurman 1985; Hedges 1987). The largest portion of the DOM (80-95%) in freshwater aquatic systems, such as rivers reaching coastal areas occurs as dissolved HMW (>1 kDa) polymeric compounds, such as polypeptides, polysaccharides and polyphenolic humic substances (HS) (Allen 1976; Thurman 1985). The rest is composed of dissolved amino acids and smaller carbohydrates (Münster and Chróst 1990). About 50-75% of the DOM in freshwater waters consists of polymeric organic acids, also considered as HS (Thurman 1985). HS can be operationally defined as the fraction of DOM that can be isolated by hydrophobic adsorption chromatography on nonionic resins (Thurman and Malcolm 1983). The riverine HS have generally been described as terrestrially derived organic matter that has leached from soils and considered refractory to degradation (Beck *et al.* 1974).

In the sea, however, DOM is mainly composed (65-80%) of small (<1 kDa) molecules (Carlson *et al.* 1985a; Benner *et al.* 1992) and consequently a smaller part is humic or HMW than in freshwater systems. The concentration of HMW DOM is higher in coastal areas influenced by river runoff than in offshore waters. The concentration of DOC is in the range of 50-200 $\mu mol\ l^{-1}$ in marine waters (Moran *et al.* 1991; Amon and Benner 1994) with the highest concentrations in nearshore waters.

2.1. DOM reaching coastal waters via rivers

River runoff carries about 2.0×10^{14} g of DOC into the ocean annually (Deuser 1988). Thus, coastal waters adjacent to estuaries receive large quantities of DOM, including organically bound nutrients, especially nutrients bound to humic substances (Fleischer and Stibe 1989).

Using the occurrence of lignin phenols (that are only found in terrestrial plants) in coastal seawater, Moran *et al.* (1991) showed that a substantial part of coastal DOC is derived from terrestrial ecosystems. Between 5-36% of nearshore DOC (0-20 m depth) in marine waters outside the southeastern US was estimated to be derived from terrestrial plants (Moran *et al.* 1991).

Dissolved humic acids precipitate to some extent when they reach saline coastal waters (Mulholland 1981), a process that causes part of the humic substances to sediment in estuaries before reaching marine waters. However, it is only a small fraction of the DOM that precipitates and is removed from the pelagic in estuaries (Mantoura and Woodward 1983). Precipitated DOM formed after addition of humic substances to seawater can also disappear after 2-4 days, simultaneously with increased bacterial production (P. Carlsson, unpubl.).

2.2. Importance of DOM as a nutrient source

The pool of DOM is composed of a large number of known and unknown components; Only about 50% of the DON has been characterized (Sharp 1983), being dominated by urea and amino acids (free and combined).

Dissolved organic nitrogen is the dominant fraction of the flux of N from the land to the sea, generally making up 60-90% of the total dissolved N-export (Meybeck 1982). Nitrogen content of riverine humic substances is usually 1-3 % (by weight) and the P content is about 0.2% (Thurman 1985; Hedges 1987). The DON concentrations in oceanic waters are considered to be 3-5 $\mu mol\ l^{-1}$ (Sharp 1983). Suzuki *et al.* (1985), however, reported values of 20-40 $\mu mol\ l^{-1}$ in surface oceanic waters using a new high-temperature catalytic oxidation method. Since then, Hansell (1993) have measured DON concentrations in nearshore and open-ocean waters consistently lower than 10 $\mu mol\ l^{-1}$ also using high-temperature catalytic oxidation. Irrespective of the exact amount of DON present in marine waters, the DON pool is usually substantially larger than the pool of inorganic nitrogen during the summer in many coastal areas (Flynn and Butler 1986; Mantoura *et al.* 1988) (Fig. 1).

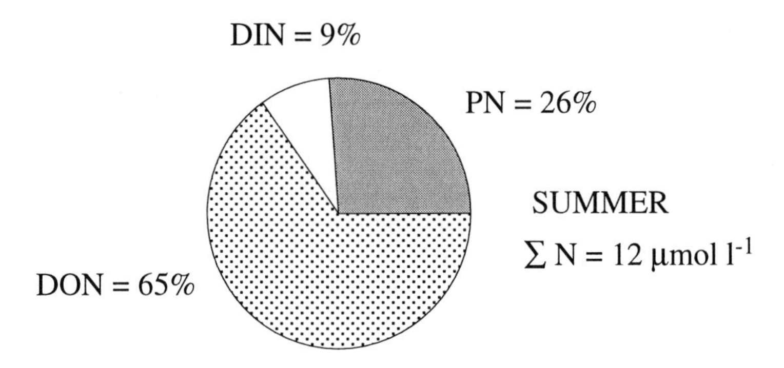

Figure 1. Nitrogen speciation during summer in the relatively shallow (average depth 17 m) Carmarthen Bay, Bristol Channel. The bay is influenced by two river systems but the circulation and dispersion of water is dominated by tidal exchange with the Bristol Channel rather than river discharge (Mantoura *et al.* 1988). PN = particulate nitrogen (phytoplankton, zooplankton and detritus), DIN = dissolved inorganic nitrogen (nitrate+ammonium+nitrite), DON = dissolved organic nitrogen (After Mantoura *et al.* 1988).

Dissolved organic nitrogen is usually not considered as an available nutrient for phytoplankton when management of eutrophic coastal areas is discussed. However, several experiments have shown that DON stimulates phytoplankton production. Phytoplankton in a stratified system were found to increase more in particulate N than could be explained by the uptake of inorganic N and urea (Price *et al.* 1985), presumably by utilization of organic N compounds.

Carlsson *et al.* (1993) found a stimulation of the microbial food web (bacterial and ciliate numbers increased) and increased regeneration of inorganic N which in turn increased phytoplankton production when riverine humic substances were added to a natural plankton community.

2.3. How refractory is DOM?

The average age of DOM in the deep ocean is about 6000 years, which implies that DOM cycles very slowly on average (Williams and Druffel 1987). Nevertheless, there is also a significant fraction that is used by bacteria as substrate (Kirchman *et al.* 1991). Some portion of the DON (amino acids, proteins and urea) is rapidly metabolized by bacteria and has short turnover times (Hollibaugh and Azam 1983). Gardner *et al.* (1996) shows that HMW DOM with a high C:N ratio can be a carbon source for bacteria, whereas the bacteria form a sink for ammonium.

In the sea, low-molecular-weight (LMW) compounds are thought to be rapidly processed by bacteria, whereas a smaller fraction consisting of HMW compounds are considered to be more refractory (Thurman 1985). However, Amon and Benner (1994) have shown that bacterial growth and respiration of HMW DOM (>1 kDa) was three and six times greater respectively than for LMW DOM (<1 kDa).

Even if most DOM in marine waters is LMW, less than 25-33% of the DON consists of LMW compounds (Benner *et al.* 1992). Thus, the dominant fraction is HMW DON, and belongs to the part that recently has been shown to be more available for bacteria.

In coastal waters receiving river inputs, a large amount of HMW DOM is supplied in a form which can be processed in the pelagic system.

Irradiance, especially energy-rich UV, breaks down DOM into smaller molecules (Kieber *et al.* 1989). When river water containing high concentrations of organic compounds reaches coastal waters, the lower salinity of this water causes it to float above the marine water where it will be exposed to high irradiance levels (Kieber *et al.* 1990). Half-life of riverine DOC reaching coastal waters has been estimated to be 1-5 years due to photodegradation (Kieber *et al.* 1990). This is much shorter than the half-life of oceanic DOC, estimated to be several thousands of years.

3. Direct uptake of DOM by phytoplankton

High-molecular-weight polymeric compounds are too large to pass the cytoplasmic membrane (Payne 1980). Therefore, only that part of the DOM that occurs in low molecular form (e.g. urea and amino acids) can be taken up through the cell membranes. Uptake of urea and amino acids can take place as passive diffusion through the cell membrane, but this process requires high concentrations in the water, e.g. >70 $\mu mol\ l^{-1}$ for urea and >1 $\mu mol\ l^{-1}$ for amino acids (concentrations that do not normally occur in seawater). At normal environmental concentrations, active, energy consuming transport systems involving specific transport enzymes (permeases), will instead be used for the uptake of both urea and amino acids (see review by Antia *et al.* 1991).

Pinocytosis is a mechanism to take up HMW compounds whereby the plasma membrane extends and form a vesicle enclosing liquid containing the HMW compounds. This process is studied only to a minor extent in phytoplankton (Kivik and Vesk 1974; Klut *et al.* 1987).

3.1. Direct utilization of low-molecular-weight organic compounds

Heterotrophic growth of microalgae in the dark on LMW carbon compounds has been demonstrated for many species (see review by Hellebust and Lewin, 1977 for diatoms), but very high substrate concentrations are needed to support growth. These conditions can be fulfilled only in sediments and during hypertrophic conditions in the water column, and consequently, pure heterotrophic growth in the dark is probably only of importance for benthic diatoms and plankton inhabiting hypertrophic waters (Bonin and Maestrini 1981). However, amino acids can be used directly by phytoplankton to obtain N (e.g. Flynn and Butler 1986). Concentrations of free amino acids in marine waters are generally <0.01 to >5 $\mu mol\ l^{-1}$ (Mopper and Lindroth 1982; Flynn and Butler 1986). Amino acids can also be bound abiotically to dissolved HMW compounds (Carlson *et al.* 1985b).

Some harmful phytoplankton have been shown to use amino acids as N sources. The toxin-producing prymnesiophyte *Prymnesium parvum* has been shown to grow solely on the amino acids ethionine or methionine as the N-source (Rahat and Hochberg 1971). Baden and Mende (1979) showed that the toxic dinoflagellate *Gymnodinium breve* had half-saturation (K_s) values of 110 and 150 $\mu mol\ l^{-1}$ for uptake of the amino acids glycine and valine, respectively. These K_s values for amino acid uptake make it probable that *G. breve* may not use amino acids in its natural environment. However, other experiments have shown much lower K_s values (0.6-2 $\mu mol\ l^{-1}$) for phytoplankton growth on amino acids as exclusive N source (Flynn and Syrett 1986).

Phytoplankton should potentially compete with bacteria for the utilization of amino acids, and bacteria are supposed to be better competitors for nutrients at low concentrations because of their larger surface to volume ratio (Bratbak and Thingstad 1985). However, any argument against the use of amino acids by phytoplankton based on competition with bacteria should apply equally to the use of regenerated ammonium which is also used as a N source by bacteria (Laws *et al.* 1985).

The incorporation rate of amino acids into protein is generally much higher in the light than in the dark (see review by Flynn and Butler 1986). Low concentrations of dissolved free amino acids in seawater do not necessarily mean that they are not important for phytoplankton growth. A tight coupling between production of amino acids by enzymatic degradation and uptake by bacteria and phytoplankton could well be the explanation. If the enzymatic hydrolysis of peptides is high, then a substantial amount of N could be used by phytoplankton as amino acids, even if concentrations are low.

3.2. Direct uptake of high-molecular-weight compounds

Larger organic molecules can be taken up via the pusule system by some flagellates, although the major function of this organelle is still unknown. *Dunaliella tertiolecta*, *Amphidinium carterae* and *Prorocentrum micans* have been shown to take up macromolecular markers such as labeled lectins and horseradish peroxidase presumably by pinocytosis using the pusule system, accumulated in vesicles inside the cell membrane (Klut *et al.* 1987). This process has not been studied to any significant extent among phytoplankton however, and it still remains an open question whether or not pinocytosis is of any significance in phytoplankton nutrition.

For heterotrophic nanoflagellates, significant uptake of "colloidal DOM" has been shown (Sherr 1988; Tranvik *et al.* 1993). In these experiments, fluorescent (FITC-labeled) macromolecules (carbohydrates and proteins) spanning a large range of molecular weights (50-2000 kD) at concentrations similar to the ones occurring in marine waters, were ingested by heterotrophic flagellates and supported growth. No uptake of FITC-labeled macromolecules have so far been observed in flagellates exhibiting chlorophyll autofluorescence.

Results by Carlsson *et al.* (in prep.) show that the toxic dinoflagellate *Alexandrium catenella* is able to grow well on N bound to humic substances isolated from river water. *Alexandrium catenella* showed equivalent growth without bacteria or with a natural bacterial community present (Fig. 2). Aminopeptidase activity was negligible in *A. catenella* cultures grown with humic substances without bacteria, but high in the treatment with bacteria. Thus, in the treatments with bacteria, *A. catenella* might have used amino acids as an N source, helped by the bacterial aminopeptidases, but the high growth of *A. catenella* in the axenic treatment cannot be explained in this way. Perhaps *A. catenella* was able to use the HMW DOM directly, as indicated by the presence of FITC-labeled dextran molecules of 2000 kDa size in vacuoles of *A. catenella* (Legrand and Carlsson, in prep.).

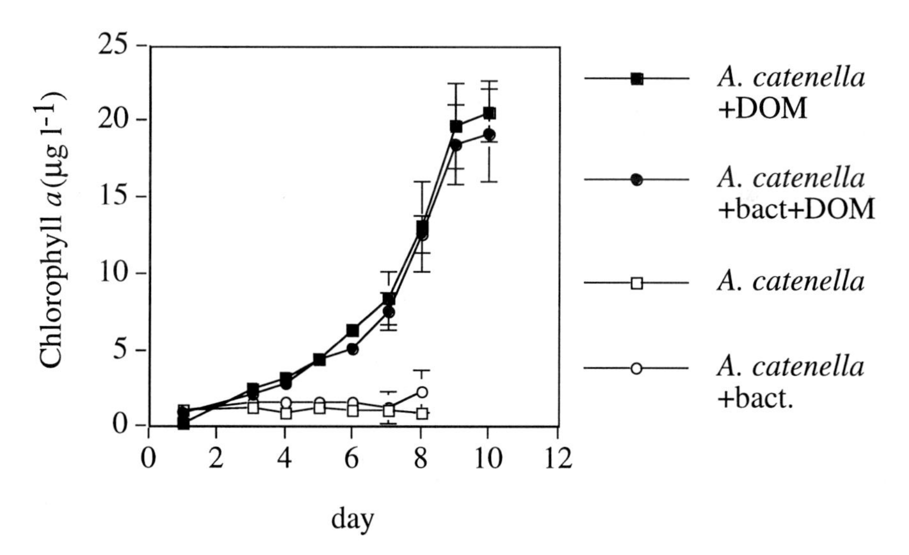

Figure 2: Growth of *Alexandrium catenella* in axenic cultures or together with bacteria with organic nitrogen in the form of riverine humic substances (DOM) as the major N source present (Carlsson *et al.*, in prep.).

4. Enzymatic breakdown of DOM

The bacterial production observed in aquatic systems can usually not be explained by utilization of the rather small pool of directly assimilable LMW DOM (Riemann and Søndergaard 1986). Therefore it is apparent that bacteria can use a part of the HMW DOM (Tranvik 1988). These macromolecules are enzymatically hydrolyzed by bacterial exoenzymes (e.g. Chróst 1990).

Both bacteria and phytoplankton produce extracellular phosphatases that hydrolyze organic phosphorus compounds and releases orthophosphate which is taken up by the cells (Stewart and Wetzel 1982; Cembella *et al.* 1984; 1985).

Degradation of macromolecules with peptide bonds can occur by the action of leucine aminopeptidase, an enzyme that is widely distributed in aquatic environments and hydrolyses a large number of peptides and amide linkages of amino acids (Hoppe *et al.* 1988). Cell surface or extracellular peptidases are always associated with bacteria (Rosso and Azam 1987) and there are no reports of phytoplankton utilization of these enzymes. Bacterial exoenzymes are associated with the cell membrane in the periplasmic space (Chróst 1991) and the products from their activity should be mainly available for the bacteria and not phytoplankton. Accordingly, aminopeptidase activity is often low in 0.2 μm filtrates where bacteria are not present (e.g. Cróst *et al.* 1989). However, extracellular enzymes can be washed away from the periplasmic space, liberated by lysis or damage of cells by grazers, and intracellular enzymes may also become dissolved by cell lysis or grazing (Chróst 1991). High aminopeptidase activities (10-90 % of total activity) has also been measured in 0.2 μm filtrates (Jacobsen and Rai 1991) which would also produce free amino acids for phytoplankton uptake. It is possible that phytoplankton use amino acids produced by the action of bacterial peptidases to some extent.

Three phytoplankton species, among them the toxic prymnesiophyte *Prymnesium parvum*, have been shown to have cell-surface L-amino acid oxidases which oxidize amino acids and primary amines to ammonium for cellular uptake, yielding peroxide and either a-keto acids (from amino acids) or aldehydes (from primary amines) (Palenik and Morel 1990) (Fig. 3). None of the phytoplankton species tested by Palenik and Morel (1990) possesed cell-surface or extracellular peptidases and consequently they could not cleave peptides into amino acids.

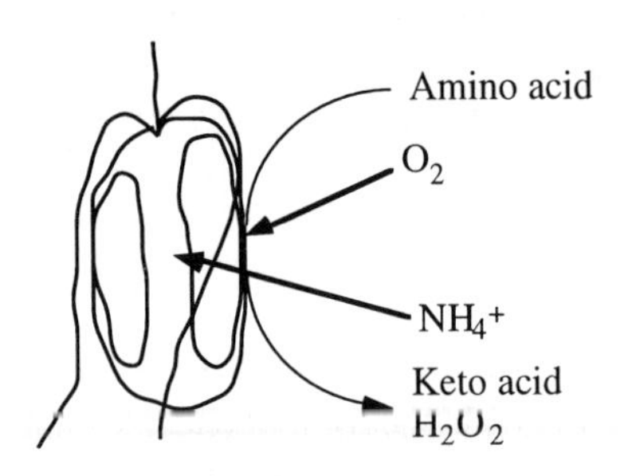

Figure 3. The enzymatic mechanism whereby some phytoplankton use cell-surface amino acid oxidases to convert amino acids to ammonium that is taken up and used for growth (After Palenik and Morel 1990).

5. Indirect utilization of DOM by phytoplankton - the role of bacteria and regeneration of inorganic nutrients

Inorganic nutrients are regenerated by the grazing activity of heterotrophic flagellates on bacteria (Caron and Goldman 1990). The regenerated inorganic N and P are then available both for bacterial and phytoplankton uptake. A transfer of nutrients from organic to inorganic form occurs in this way.

Bacteria are considered to be the main organisms using DOM as a substrate for growth. Since the bacterial C:N ratio is lower than for phytoplankton, they need more N per unit biomass than phytoplankton. In marine waters DOM has a high C:N ratio, about 15 (by weight) (Benner *et al.* 1992), while riverine DOM that enters coastal waters has an even higher C:N ratio (20- 50) (Malcolm 1985; Gardner *et al.* 1996). Bacteria have a much lower C:N ratio (about 3-7, Bratbak 1985; Nagata 1986), therefore they will tend to keep the N and not regenerate it in inorganic form available for phytoplankton (Goldman and Caron 1985). However, when bacteria are grazed by heterotrophs, such as nanoflagellates and ciliates, inorganic N is released (Caron and Goldman 1990). The amount of N that is regenerated has been estimated to be between 10-50% of the bacterial N ingested (Andersson *et al.* 1985, Goldman *et al.* 1985) and will be available for phytoplankton (Fig. 4). The rest of the N is retained in the heterotrophic flagellate biomass and is available for higher trophic levels (Azam *et al.* 1983).

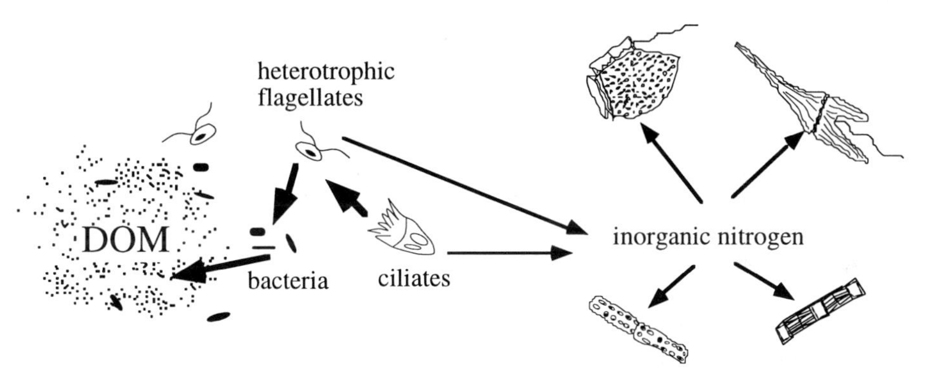

Figure 4. Tentative pathway for indirect utilization of organic bound N by phytoplankton via heterotrophic grazing activity on bacteria which have used DOM as substrate. Thick arrows = utilization of substrate/ingestion of food; Thin arrows = excretion/uptake of inorganic N.

Ietswaart *et al.* (1994) showed that axenic cultures of the prymnesiophyte *Emiliania huxleyi* were able to grow on amino acids as the only available N source, whereas the diatom *Thalassiosira pseudonana* was not. When a marine bacterial strain was present, both phytoplankton species grew well with amino acids as the only

N source. Thus, bacteria both assimilated amino acids and regenerated ammonium that could be used by the phytoplankton.

Dark bottle experiments performed by Gardner *et al.* (1996) showed large remineralization rates of organic matter by bacteria in the Mississippi River plume water, and high accumulation rates of ammonium at intermediate salinity stations, suggesting that the C:N ratio of the available substrates were low. However, it is also possible that bacterial regeneration of inorganic nitrogen will take place even when bacteria uses organic matter with a high C:N ratio, if the relative availability of the N is higher than for the C.

6. Possible connections between terrestrial supply of DOM and occurrence of red tides

There is no direct evidence that a large supply of DOM has induced a harmful algal bloom. However, toxic blooms composed of *A. tamarense* form regularly in the outflow of two rivers in the Gulf of Maine (USA) and can persist in the plume for up to a month (Franks and Anderson 1992), indicating that some property of the river water may be important for the development and persistence of these blooms.

Baden and Mende (1979) found that *G. breve* utilized amino acids even when inorganic N sources were present and speculated that even though amino acid concentrations in marine environments are usually low, the concentrations of free amino acids can become substantial in areas of low water turnover and high productivity, areas which coincidentally are sites for red tide outbreaks. Granéli and Moreira (1990) have shown that rivers draining agricultural soil, thereby rich in inorganic N and P, stimulates growth of diatoms, whereas river water from forested areas, rich in humic substances, increase growth of dinoflagellates (Fig. 5).

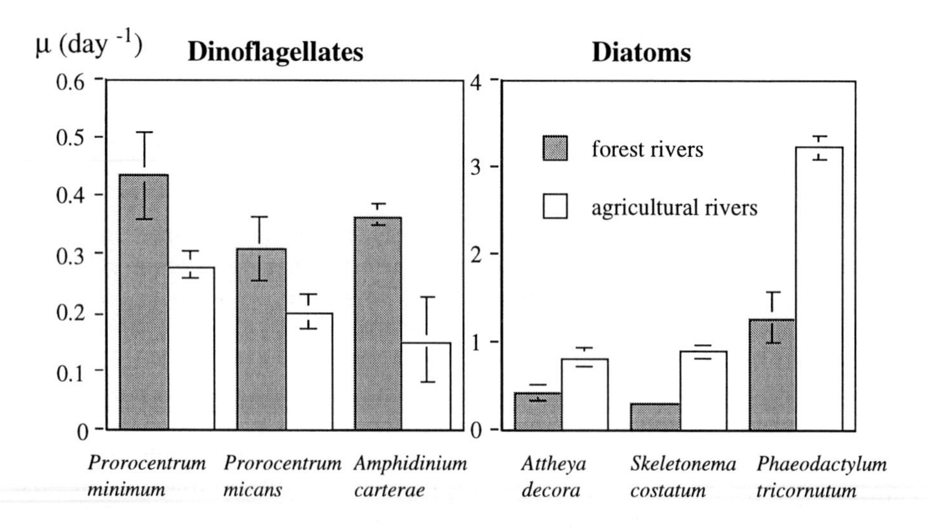

Figure 5. Growth rates for some dinoflagellates and diatoms grown in seawater with additions of river water from either forested areas or rivers draining agricultural areas (After Granéli and Moreira 1990).

A large increase of the discharge of humic substances by rivers into Swedish coastal waters during the last 20 years (Andersson *et al.* 1991) coincides with an increased number and intensity of dinoflagellate blooms in the same area (Nielsen and Ærtebjerg 1984; Granéli *et al.* 1989). There may be a connection between these two events.

7. Conclusions

Substantial experimental evidence shows that phytoplankton do not use DOM as a major C source in natural environments, except perhaps during extreme conditions such as in hypertrophic environments. It is, however, evident that many phytoplankton species benefit from organic substances present in seawater or reaching coastal zones with rivers. The growth stimulating effect can be caused via trace metal complexation by the organic molecules (Anderson and Morel 1982), but utilization of organically bound N is also important. The organic N can be used directly if the molecules are small, such as urea and amino acids (Flynn and Butler 1986). Some phytoplankton also possess cell-surface amino acid oxidases (Palenik and Morel 1990) which can degrade amino acids to ammonium. The occurrence of free, bacterially produced, amino peptidases in the water produces free amino acids which could be available for phytoplankton uptake. Indirect utilization of organically bound N via remineralization by heterotrophic grazers cropping bacteria which previously used DOM as a growth substrate is also a mechanism for phytoplankton utilization of N in DOM (Carlsson *et al.* 1993; 1995).

Direct ingestion of HMW organic molecules by some species of phytoplankton might also be a mechanism, so far overlooked.

DOM in river water entering coastal waters are subjected to two major breakdown processes: 1) bacterial degradation (Chin-Leo and Benner 1992, Carlsson *et al.* 1993) and 2) photochemical transformations that increase their availability to bacteria (Kieber *et al.* 1989), and perhaps also to phytoplankton.

Therefore, coastal waters influenced by river runoff may show increased growth of phytoplankton, including blooms of harmful algal species as an effect of direct or indirect utilization of N bound to DOM.

8. Acknowledgments

Financial support from NATO, SCOR, the Swedish Natural Science Research Council (NFR), and the Swedish Environmental Protection Agency (SNV) is gratefully acknowledged. We also thank C. Legrand for the permission to quote her unpublished data.

9. References

Allen, H.L. (1976). Dissolved organic matter in lake water: characteristics of molecular weight size fractions and ecological implications. *Oikos*. 27: 64-70.

Amon, R.M.W., Benner, R. (1994). Rapid cycling of high-molecular-weight dissolved organic matter in the marine carbon cycle. *Nature*. 369: 549-552.

Anderson, M.A., Morel, F.M.M. (1982). The influence of aqueous iron chemistry on the uptake of iron by the coastal diatom *Thalassiosira weissflogii*. *Limnol. Oceanogr*. 27: 789-813.

Andersson, A., Lee, C., Azam, F., Hagstrom, Å. (1985). Release of amino acids and inorganic nutrients by heterotrophic marine microflagellates. *Mar. Ecol. Prog. Ser.* 23: 99-106.

Andersson, T., Nilsson, Å., Jansson, M. (1991). Coloured substances in Swedish rivers-temporal variation and regulating factors. In: Allard, B. Borén, H., Grimvall, A. (eds.). *Humic substances in the aquatic and terrestrial environment.* Lecture notes in earth sciences. vol. 33. Springer-Verlag, Berlin-Heidelberg, pp. 243-253.

Antia, N.J., Harrison, P.J., Oliviera, L. (1991). The role of dissolved organic nitrogen in phytoplankton nutrition, cell biology and ecology. *Phycologia*. 30: 1-89.

Azam, F., Fenchel, T., Field, J.G., Gray, J.S., Meyer-Reil, L.A., Thingstad, F. (1983). The ecological role of water-column microbes in the sea. *Mar. Ecol. Prog. Ser.* 10: 257-263.

Baden, D.G., Mende, T.J. (1979). Amino acid utilization by *Gymnodinium breve. Phytochem.* 18: 247-251.

Beck, K.C., Reuter, J.H., Perdue, E.M. (1974). Organic and inorganic geochemistry of some coastal plain rivers of the southeastern United States. *Geochim. Cosmochim. Acta*. 38: 341-364.

Benner, R., Pakulski, J.D., McCarthy, M., Hedges, J.I., Hatcher, P.G. (1992). Bulk chemical characteristics of dissolved organic matter in the ocean. *Science*. 255: 1561-1564.

Billen, G. (1984). Heterotrophic utilization and regeneration of nitrogen. In: Hobbie, J. E., Williams, P.J.leB. (eds.). *Heterotrophic activity in the sea.* Plenum Press, New York, pp. 313-355.

Bonin, D.J., Maestrini, S.Y. (1981). Importance of organic nutrients for phytoplankton growth in natural environments: implications for algal species succession. In: Platt, T. (ed.). *Physiolological bases of phytoplankton ecology.* Can. Bull. Fish. Aquat. Sci. Ottawa. Bull. No 210, pp. 279-291.

Bratbak, G. (1985). Bacterial biovolume and biomass estimation. *Appl. Environ. Microbiol.* 49: 1488-1493.

Bratbak, G., Thingstad, T.F. (1985). Phytoplankton-bacteria interactions: an apparent paradox? Analysis of a model system with both competition and commensalism. *Mar. Ecol. Prog. Ser.* 25: 23-30.

Carlson, D.J., Brann, M.L., Mague, T.H., Mayer, L.M. (1985a). Molecular weight distribution of dissolved organic materials in seawater determined by ultrafiltration: a re-examination. *Mar. Chem*. 16: 155-171.

Carlson, D.J., Mayer, L.M., Brann, M.L., Mague, T.H. (1985b). Binding of monomeric compounds to macromolecular dissolved organic matter in seawater. *Mar. Chem*. 16: 141-153.

Carlsson, P., Segatto, A.Z., Granéli, E. (1993). Nitrogen bound to humic matter of terrestrial origin - a nitrogen pool for coastal phytoplankton? *Mar. Ecol. Prog. Ser.* 97: 105-116.

Carlsson, P., Granéli, E, Tester, P., Boni, L. (1995). Influences of riverine humic substances on bacteria, protozoa, phytoplankton, and copepods in a coastal plankton community. *Mar. Ecol. Prog. Ser.* 127: 213-221.

Caron, D.A., Goldman, J.C. (1990). Protozoan nutrient generation. In: Capriulo, G.M. (ed.). *Ecology of marine protozoa.* Oxford University Press, New York, pp. 283-306.

Caron, D.A., Porter, K.G., Sanders, R.W. (1990). Carbon, nitrogen and phosphorus budgets for the mixotrophic phytoflagellate *Poterioochromonas malhamensis* (Chrysophyceae) during bacterial ingestion. *Limnol. Oceanogr.* 35: 433-443.

Cembella, A.D., Antia, N.J., Harrison, P.J. (1984). The utilization of inorganic and organic phosphorus compounds as nutrients by eucaryotic microalgae: A multidisciplinary perspective. Part 1. *CRC Critical Reviews in Microbiology*. 10: 317-391.

Cembella, A.D., Antia, N.J., Harrison, P.J. (1985). The utilization of inorganic and organic phosphorus compounds as nutrients by eucaryotic microalgae: A multidisciplinary perspective. Part 2. *CRC Critical Reviews in Microbiology*. 11: 13-81.

Cheng, U.Y., Antia, N.J. (1970). Enhancement by glycerol of phototrophic growth of marine planktonic algae and its significance to the ecology of glycerol pollution. *J. Fish. Res. Board. Can.* 27: 335-346.

Chin-Leo, G., Benner, R. (1992). Enhanced bacterioplankton production and respiration at intermediate salinities in the Mississippi River plume. *Mar. Ecol. Prog. Ser.* 87: 87-103.

Chróst, R.J. (1990). Microbial enzymes in aquatic environments. In: Overbeck, J., Chróst, R. J. (eds.). *Aquatic microbial ecology: Biochemical and Molecular Approaches*. Springer-Verlag, New York, pp. 47-78.

Chróst, R.J. (1991). Environmental control of the synthesis and activity of aquatic microbial enzymes. In: Chróst, R.J. (ed.). *Microbial enzymes in aquatic environments*. Springer-Verlag, New York, pp. 29-59.

Chróst, R.J., Münster, U., Rai, H., Albrecht, D., Witzel, K.P., Overbeck, J. (1989). Photosynthetic production and exoenzymatic degradation of organic matter in the euphotic zone of an eutrophic lake. *J. Plankton. Res.* 11: 223-242.

Combres, C., Laliberté, G., Sevrin Reyssac, J., de la Noüe, J. (1994). Effect on acetate on growth and ammonium uptake in the microalga *Scenedesmus obliquus. Physiol. Plant*. 91: 729-734.

Deuser, W.G. (1988). Whither organic carbon? *Nature*. 332: 396-397.

Droop, M.R. (1974). Heterotrophy of carbon. In: Stewart, W. D. P. (ed.). *Algal physiology and biochemistry*. Univ. California Press, Los Angeles, pp. 530-559.

Fleischer, S., Stibe, L. (1989). Agriculture kills marine fish in the 1980s. Who is responsible for fish kills in the year 2000? *Ambio*. 6: 347-350.

Flynn, K.J., Butler, I. (1986). Nitrogen sources for the growth of marine microalgae: role of dissolved free amino acids. *Mar. Ecol. Prog. Ser.* 34: 281-304.

Flynn, K.J., Syrett, P.J. (1986). Characteristics of the uptake system for L-lysine and L-arginine in *Phaeodactylum tricornutum*. *Mar. Biol.* 90: 151-158.

Franks, P.J.S., Anderson, D.M. (1992). Alongshore transport of a toxic phytoplankton bloom in a buoyancy current: *Alexandrium tamarense* in the Gulf of Maine. *Mar. Biol.* 112: 153-164.

Gaines, G., Elbrächter, M. (1987). Heterotrophic nutrition. In: Taylor, F. J. R. (ed.). *The Biology of Dinoflagellates*. Blackwell Scientific Publications, Oxford, pp. 224-268.

Gaines, G., Taylor, F.J.R. (1984). Extracellular digestion in marine dinoflagellates. *J. Plankton. Res*. 6: 1057-1061.

Gardner, W.S., Benner, R., Amon, R.M.W., Cotner, J.B.Jr, Cavaletto, J.F., Johnson, J.R. (1996). Effects of high-molecular-weight dissolved organic matter on nitrogen dynamics in the Mississippi River plume. *Mar. Ecol. Prog. Ser.* 133: 287-297.

Goldman, J.C., Caron, D.A. (1985). Experimental studies on an omnivorous microflagellate: Implications for grazing and nutrient regeneration in the marine microbial food chain. *Deep Sea Res.* 8: 899-915.

Goldman, J.C., Caron, D.A., Andersen, O.K., Dennett, M.R. (1985). Nutrient cycling in a microflagellate food chain: I. nitrogen dynamics. *Mar. Ecol. Prog. Ser.* 24: 231-242.

Granéli, E., Edler, L., Gedziorowska, D., Nyman, U. (1985). Influence of humic and fulvic acids on Peorocentrum minimum Pav.) J. Schiller. In: (Anderson, D.M., White, A.W., Baden, D.G. (eds). *Toxic dinoflagellates*. Elsevier Science Publishing Co. Inc. New York, pp. 201-206.

Granéli, E., Carlsson, P., Olsson, P., Sundström, B., Granéli, W., Lindahl, O. (1989). From anoxia to fish poisoning: the last ten years of phytoplankton blooms in Swedish marine waters. In: Cosper, E.M., Bricelj, V.M., Carpenter, E.J. (eds.). *Novel phytoplankton blooms - causes and impacts of recurrent brown tides and other unusual blooms*. Springer-Verlag. New York, pp. 407-427.

Granéli, E., Moreira, M.O. (1990). Effects of river water of different origin on the growth of marine dinoflagellates and diatoms in laboratory cultures. *J. Exp. Mar. Biol. Ecol.* 136: 89-106.

Hansell, D. (1993). Results and observations from the measurement of DOC and DON in seawater using a high-temperature catalytic oxidation technique. *Mar. Chem*. 41: 195-202.

Hedges, J.I. (1987). Organic matter in sea water. *Nature*. 330: 205-206.

Hellebust, J.A., Lewin, J. (1977). Heterotrophic nutrition. In: Werner, D. (ed.). *The biology of diatoms*. Blackwell Scientific Publ., pp. 169-197.

Hollibaugh, J.T., Azam, F. (1983). Microbial degradation of dissolved proteins in seawater. *Limnol. Oceanogr.* 28: 1104-1116.

Hoppe, H.G., Kim, S.J., Gocke, K. (1988). Microbial decomposition in aquatic environments: combined processes of extracellular enzyme activity and substrate uptake. *Appl. Environ. Microbiol.* 54: 784-790.

Ietswaart, T., Schneider, P.J., Prins, R.A. (1994). Utilization of organic nitrogen sources by two phytoplankton species and a bacterial isolate in pure and mixed cultures. *Appl. Environ. Microbiol*. 60: 1554-1560.

Jacobsen, T.R., Rai, H. (1991). Aminopeptidase activity in lakes of differing eutrophication. In: Chróst, R. J. (ed.). *Microbial enzymes in aquatic environments.* Springer-Verlag, New York, pp. 155-164.

Jacobson, D.M., Anderson, D.M. (1986). Thecate heterotrophic dinoflagellates: Feeding behaviour and mechanisms. *J. Phycol*. 22: 249-258.

Kieber, D.J., McDaniel, J., Mopper, K. (1989). Photochemical source of biological substrates in sea water: implications for carbon cycling. *Nature*. 341: 637-639.

Kieber, R.J., Zhou, X., Mopper, K. (1990). Formation of carbonyl compounds from UV-induced photodegradation of humic substances in natural waters: Fate of riverine carbon in the sea. *Limnol. Oceanogr*. 35: 1503-1515.

Kirchman, D.L., Suzuki, Y., Garside, C., Ducklow, H.W. (1991). High turnover rate of dissolved organic carbon during a spring phytoplankton bloom. *Nature*. 352: 612-614.

Kivic, P.A., Vesk, M. (1974). Pinocytotic uptake of protein from the reservoir in *Euglena*. *Arch. Microbiol* 96: 155-159.

Klut , M.E., Bisalputra, T., Antia, N.J. (1987). Some observations on the structure and function of the dinoflagellate pusule. *Can. J. Bot*. 65: 736-744.

Laws, E.A., Harrison, W.G., DiTullio, G.R. (1985). A comparison of nitrogen assimilation rates based on 15N uptake and protein synthesis. *Deep Sea Res.* 32: 85-95.

Malcolm, R.L. (1985). Geochemistry of stream fulvic and humic substances. In: Aiken, G.R., McKnight, D.M., Wershaw, R.L. (eds.). *Humic substances in soil, sediment, and water*. John Wiley & Sons, New York, pp. 181-210.

Mantoura, R.F.C., Woodward, M.S. (1983). Conservative behaviour of the riverine dissolved organic carbon in the Severn estuary: Chemical and geochemical implications. *Geochim. Cosmochim. Acta*. 47: 1293-1309.

Mantoura, R.F.C., Owens, N.J.P., Burkill, P.H. (1988). Nitrogen biochmistry and modelling of Carmarthen Bay. In: Blackburn, T. H., Sørensen, J. (eds.). *Nitrogen cycling in coastal marine environments*. Wiley. New York, pp. 415-441.

Meybeck, M. (1982). Carbon, nitrogen and phosphorus transport by world rivers. *Am. J. Sci*. 282: 401-450.

Moran, M.A, Pomeroy, L.R., Sheppard, E.S., Atkinson, L.P., Hodson, R.E. (1991). Distribution of terrestrially derived dissolved organic matter on the southeastern U. S. continental shelf. *Limnol. Oceanogr*. 36: 1134-1149.

Morrill, L.C., Loeblich, A.R. (1979). An investigation of heterotrophic and photoheterotrophic capabilities in marine Pyrrhophyta. *Phycologia*. 18: 394-404.

Mulholland, P.J. (1981). Formation of particulate organic carbon in water from a southeastern swamp-stream. *Limnol. Oceanogr*. 26: 790-795.

Münster, U., Chróst, R.J. (1990). Origin, composition and microbial utilization of dissolved organic matter. In: Overbeck, J., Chróst, R.J. (eds.). *Aquatic Microbial Ecology. Biochemical and Molecular Approaches*. Springer-Verlag, New York, pp. 8-46.

Nagata, T. (1986). Carbon and nitrogen content of natural planktonic bacteria. *Appl. Environ. Microbiol*. 52: 28-32.

Nielsen, A., Ærtebjerg, G. (1984). Plankton blooms in Danish waters. *Ophelia. Suppl.*. 3: 181-184.

Palenik, B., Morel, F.M.M. (1990). Comparison of cell-surface L-amino acid oxidases from several marine phytoplankton. *Mar. Ecol. Prog. Ser*. 59: 195-201.

Payne, J.W. (1980). *Microorganisms and nitrogen sources*. John Wiley and Sons, New York. 764 pp.

Prakash, A., Rashid, M.A. (1968). Influence of humic substances on the growth of marine phytoplankton: dinoflagellates. *Limnol. Oceanogr.* 13: 598-606.

Prakash, A., Rashid, M.A., Jensen, A., Subba Rao, D.V. (1973). Influence of humic substances on the growth of marine phytoplankton: diatoms. *Limnol. Oceanogr.* 18: 516-524.

Price, N.M., Cochlan, W.P., Harrison, P.J. (1985). Time course of uptake of inorganic and organic nitrogen by phytoplankton in the Strait of Georgia: comparison of frontal and stratified communites. *Mar. Ecol. Prog. Ser.* 27: 39-53.

Provasoli, L., McLaughlin, J.J.A., Droop, M.R. (1957). The development of artificial media for marine algae. *Arch. Microbiol.* 25: 392-428.

Provasoli, L., Carlucci, A.F. (1974). Vitamins and growth regulators. In: Stewart, W. D. P. (ed.). *Algal Physiology and Biochemistry*, Blackwell Scientific Publications, Oxford, pp. 741-787.

Rahat, M., Hochberg, A. (1971). Ethionine and methinine metabolism by the chrysomonad flagellate *Prymnesium parvum*. *J. Protozool.* 18: 378-382.

Richardson, K., Fogg, G.E. (1982). The role of dissolved organic material in the nutrition and survival of marine dinoflagellates. *Phycologia*. 21: 17-26.

Riemann, B., Søndergaard, M. (1986). Bacteria. In: Riemann, B., Søndergaard, M. (eds.). *Carbon dynamics in eutrophic, temperate lakes*. Elsevier, Amsterdam, pp. 127-197.

Rosso, A.L., Azam, F. (1987). Proteolytic activity in coastal oceanic waters: depth distribution and relationship to bacterial populations. *Mar. Ecol. Prog. Ser.* 41: 231-240.

Sharp, J.H. (1983). The distributions of inorganic nitrogen and dissolved and particulate organic nitrogen in the sea. In: Carpenter, E.J., Capone, D.G. (eds.) *Nitrogen in the marine environment.* Academic Press, New York, pp. 1-35.

Sherr, E.B. (1988). Direct use of high molecular weight polysaccharide by heterotrophic flagellates. *Nature*. 335: 348-351.

Suzuki, Y., Sugimura, Y., Itoh, T. (1985). A catalytic oxidation method for the determination of total nitrogen dissolved in seawater. *Mar. Chem.* 16: 83-97.

Stewart, A.J., Wetzel, R.G. (1982). Phytoplankton contribution to alkaline phosphatase activity. *Arch. Hydrobiol.* 93: 265-271.

Thurman, E.M. (1985). *Organic geochemistry of natural waters*. Nijhoff/Junk, Boston, 687 pp.

Thurman, E.M., Malcolm, R.L. (1983). Structural study of humic substances: New approaches and methods. In: Christman, R. F., Gjessing, E. T. (eds.). *Aquatic and terrestrial humic materials*. Ann Arbor Science, Ann Arbor, pp. 1-23.

Tranvik, L.J. (1988). Availability of dissolved organic carbon for planktonic bacteria in oligotrophic lakes of differing humic content. *Microb. Ecol.* 16: 311-322.

Tranvik, L.J., Sherr, E.B., Sherr, B.F. (1993). Uptake and utilization of "colloidal DOM" by heterotrophic flagellates in seawater. *Mar. Ecol. Prog. Ser.* 92: 301-309

Williams, P.M., Druffel, E.R.M. (1987). Radiocarbon in dissolved organic matter in the central North Pacific Ocean. *Nature*. 330: 246-24.

Phagotrophic Mechanisms and Prey Selection in Mixotrophic Phytoflagellates

Per Juel Hansen

Marine Biological Laboratory, University of Copenhagen, Strandpromenaden 5, DK-3000 Helsingør, Denmark

1. Introduction

Planktonic phytoflagellates are often considered autotrophs, but species that are exclusively heterotrophic or mixotrophic are common. The term mixotrophy is used for organisms that combine photosynthesis and phagotrophy (ingestion and digestion of particulate prey). Mixotrophic species are found within several groups of phytoflagellates, including cryptophytes, chrysophytes, dinoflagellates and prymnesiophytes. Within these phytoflagellates, toxic species or species causing other harmful effects in marine environments have so far been found exclusively among prymnesiophytes and dinoflagellates. Granéli and Carlsson (this volume) discuss the methods used in studies of mixotrophy and the ecological significance of mixotrophy. This chapter focuses on feeding mechanisms and prey selection among prymnesiophytes and dinoflagellates.

2. Prymnesiophytes

2.1. Phagotrophy and feeding mechanisms in prymnesiophytes

Phagotrophy has only been recorded in three photosynthetic genera of prymnesiophytes (Green 1991), namely, *Chrysochromulina, Prymnesium* and *Coccolithus*. In the phototrophic genus *Chrysochromulina*, mixotrophy is widely distributed and about 25 species of *Chrysochromulina* have been described as mixotrophs (see Green 1991). Mixotrophy is rarely found in either *Prymnesium* or *Coccolithus*, and is so far only reported in *P. saltans* and the motile stage of *C. pelagicus* (Conrad 1941, Parke and Adams 1960).

The feeding mechanisms of prymnesiophytes have been studied in great detail in two species of *Chrysochromulina, C. hirta* Manton and *C. spinifera* (Fournier) Pienaar et Norris (Kawachi *et al.* 1991, Kawachi and Inouye 1995). First, an introduction to the two organisms is warranted. *Chrysochromulina hirta* (size 5-9 μm) has 3 distinctive types of scales: long (12-16 μm in length), short (length 4-6 μm) and plate scales. The flagella are equal or subequal in length and measure 15-18 μm. The haptonema is 45 - 55 μm long and can coil and bend. *Chrysochromulina spinifera* (size 5-8 μm) is covered with two types of scales. Closest to the plasmalemma is a layer of scales ca. 1 μm in diameter. These are covered with spine scales, which are 10-18 μm long. The two flagella are unequal in length. The haptonema is fairly short (8-12 μm), and can coil and bend.

NATO ASI Series, Vol. G 41
Physiological Ecology of Harmful Algal Blooms
Edited by D. M. Anderson, A. D. Cembella and
G. M. Hallegraeff

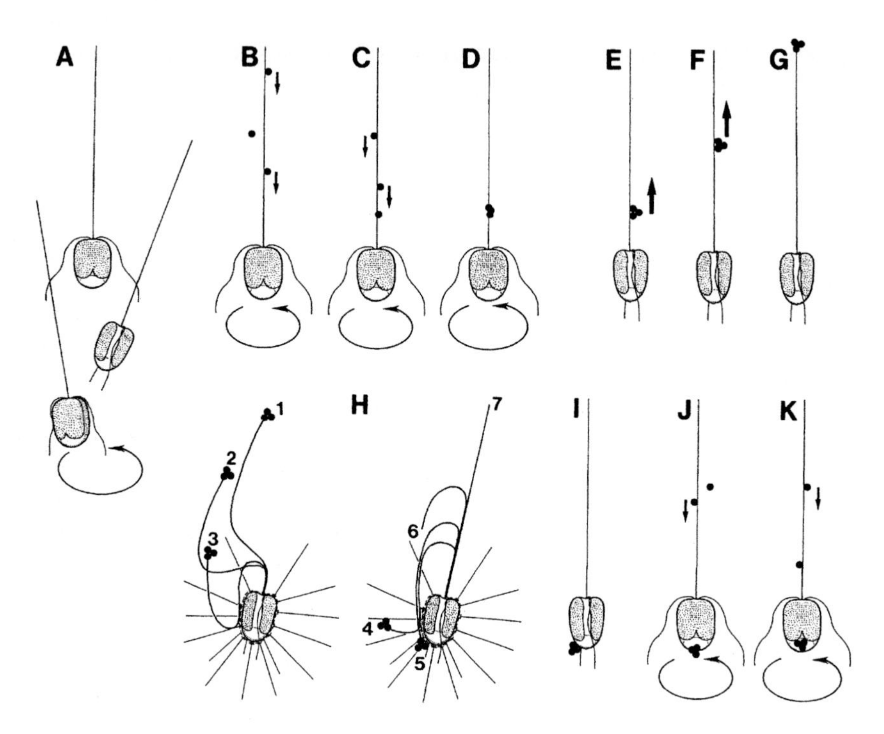

Fig. 2.1. Processes of food capture and phagotrophy in *Chrysochromulina hirta*. See text for a detailed description. Periplast scales are only shown in H. From Kawachi *et al.* 1991.

In *Chrysochromulina hirta*, the haptonema plays an important role in capture and transport of food in the process of phagotrophy (Fig. 2.1). The capture of food and its transportation by the haptonema has been observed only in cells which were swimming with the haptonema extended and pointing forward, while the two flagella undulate antapically alongside the cell body (Fig. 2.1A). Food particles become attached and adhere to the haptonema. They move down to a particular point on the haptonema, the particle aggregation center (PAC), where they accumulate. The cell stops swimming, and the aggregated food moves toward the tip of the haptonema (Fig. 2.1B-F). After the food aggregate reaches the tip, the haptonema bends and delivers the aggregate to the lateral posterior surface of the cell. The particle aggregate is then translocated at the posterior end of the cell, where it is taken into the cell by phagocytosis (Fig. 2.1 G-K). Food vacuoles are found in the posterior end of the cell. The haptonema bends even when it does not carry any particles, suggesting that the same motion is repeated at certain programmed intervals.

The feeding mechanism of *Chrysochromulina spinifera*, while resembling that of *C. hirta*, exhibits significant differences. In *C. spinifera*, a water current is generated by flagellar movements, which brings food particles to the cell. These particles adhere to the long spine scales that cover the cell, then move to the cell surface, and even-

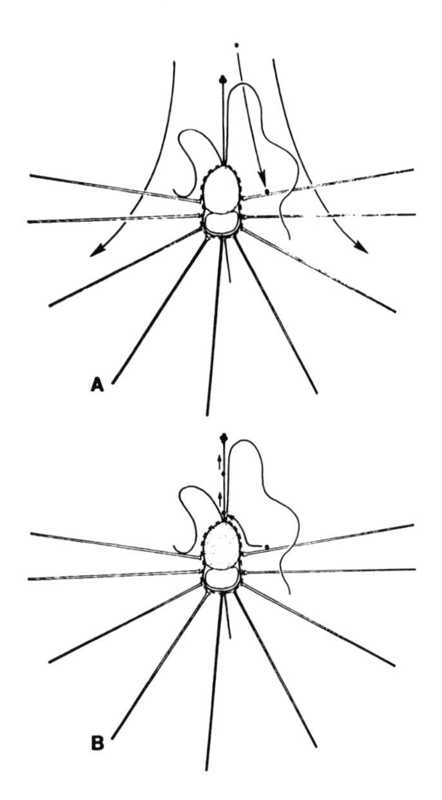

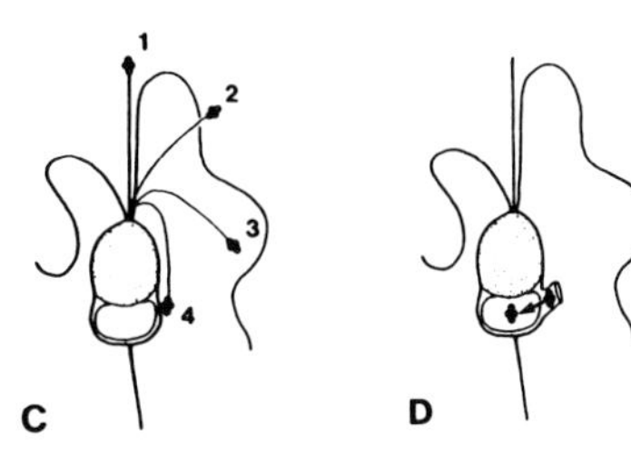

Fig. 2.2. Food capture and feeding process in *Chrysochromulina spinifera*. A. Delivery of food particles to spine scales by a water current (indicated by arrows). B. Translocation of captured particles to the tip of the haptonema and aggregate formation of particles (indicated by arrows). C1-4. Delivery of the aggregate at the cell surface by bending of the haptonema. D. Uptake of the aggregate into a food vacuole. The periplast scales are omitted in C and D. From Kawachi and Inouye (1995)

tually move to the tip of the haptonema, where they form an aggregate. The haptonema bends to transfer the aggregate to the posterior end of the cell body, where the food is ingested by phagocytosis. Thus in *C. spinifera*, both the spine scales and the haptonema participate in the feeding process. The spine scales play a role as a food-capturing device and the haptonema participates in aggregate formation and transport to the cell body. Only the haptonema is involved in those processes in *C. hirta*. Adhesion and movement of particles on the surface of spine scales has not been observed in other species of *Chrysochromulina* examined thus far (Kawachi and Inouye, 1995).

The surface motility observed on the haptonema of both species is most likely menbrane-mediated. However, the transport of captured particles on the surface of scales and from the scale base to the base of the haptonema is an unusual phenomenon. The plasma membrane, except for that of the flagella and haptonema, is covered with spine and plate scales so that movement of particles cannot be a membrane-mediated process.

Some interesting differences exist in the haptonematal transport mechanisms in the two species. In *C. hirta*, the particle aggregating center is located close to the base of the haptonema, whereas in *C. spinifera*, it is located at the tip. Also, the motility of particles on the surface is strictly unidirectional in *C. spinifera*, whereas the direction of surface motility in *C. hirta* differs depending on the stage of particle capture. In *C. hirta*, particles that are captured initially by the haptonema move toward the particle aggregating center, but once an aggregate is formed, the direction of the movement changes and the aggregate moves towards the tip.

2.2. Prey selection in prymnesiophytes

Among prymnesiophytes prey selection has only been investigated in a few cases within the genus *Chrysochromulina*. Species of *Chrysochromulina* range in size from 2-20 μm. They typically ingest a broad range of live prey and inert particles in the size range of 0.33-6 μm. In rare cases, pennate diatoms of 10 μm in length can be ingested (see Green 1991). The ability to feed on a variety of different particles has been demonstrated in a few species of *Chrysochromulina*.

In *C. spinifera*, Kawachi and Inouye (1995) demonstrated uptake of fluorescent microspheres in the size range of 0.30 to 1.66 μm. Ingestion was also observed when *C. spinifera* was fed a rod-shaped bacterium (0.8 μm) and the prasinophyte, *Micromonas pusilla* (2.3 μm). Large microspheres (3.06 μm) were not ingested, although they adhered to the spine scales and were transported to the haptonemal tip and the cell body. This suggests that the rejection of particles too large to be ingested is associated with the cell cytostome. However, a rejection mechanism was also observed that involved passing an aggregate from the haptonemal tip to the long flagellum, translocating the aggregate towards the tip of the haptonema, and vigorously beating the flagellum, thus discarding the aggregate. In healthy cells, this activity was only observed with an aggregate larger than about 3 μm in diameter.

Parke *et al.* 1956 presented *C. ericina* with a wide variety of prey. Some small cryptophytes, green algae, diatoms, a red alga, and graphite particles, all within the size range 1-6 μm, were ingested. However, not all prey presented to *C. ericina* were ingested. This prymnesiophyte failed to ingest the green algae *Dunaliella* spp. (6-12 μm) and *Nannochloris atomus* (2-3 μm), and the diatom *Phaeodactylum tricornutum* (8-35 μm). Jones *et al.* (1993) presented a variety of particles to *C. brevifilum*. While it fed on inert particles (carmine and fluorescent particles), a single green alga, and two yeasts, it failed to feed on other green algae and yeasts. In both cases, size selection alone could not account for the observed difference in particle uptake. The basis for selection is unknown, but it may relate to differences in the ability of different particles to stick to the haptonema.

3. Dinoflagellates

3.1. Phagotrophy and feeding mechanisms in dinoflagellates

Phagotrophy appears to be widespread among the photosynthetic dinoflagellates and has been described in detail for several naked and thecate genera (Table 3.1). The large majority of reports are based on recognition of what has been interpreted as prey inside food vacuoles. However, food vacuoles may be confused with degradation bodies (PAS-bodies) or in fact they may be protistan parasites that have infected the dinoflagellate. Nevertheless, recent papers have documented beyond doubt that particles found inside food vacuoles are indeed prey cells (Bockstahler and Coats 1993,

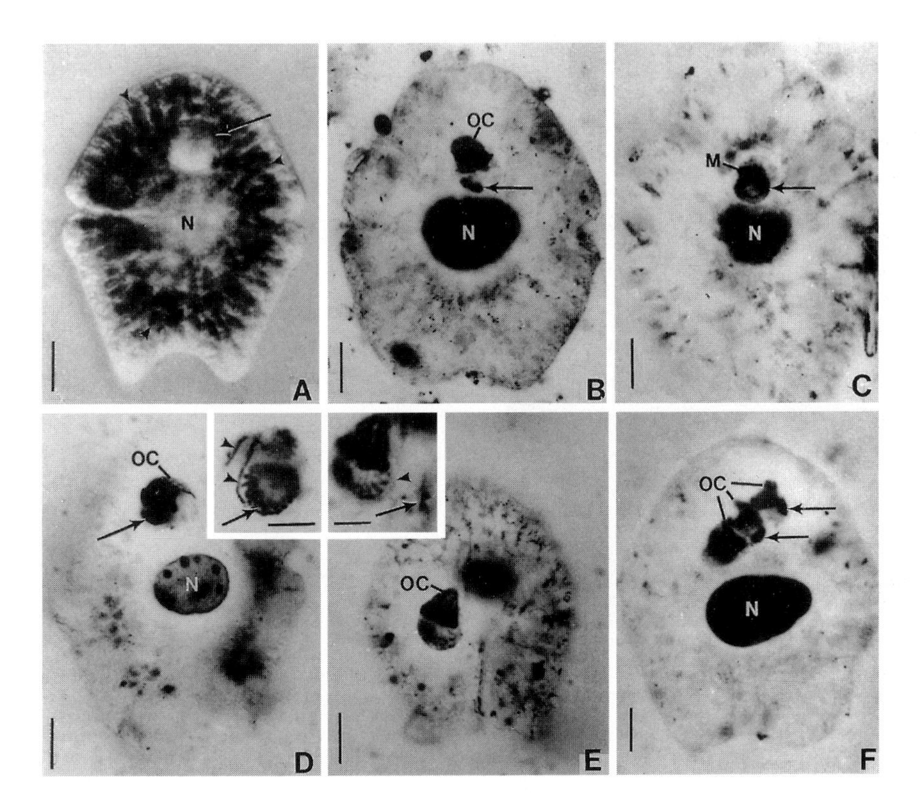

Fig. 3.1. *Gymnodinium sanguineum* and representative prey. A. A large inclusion body (arrow) in the anterior end of a live cell; the nucleus (N) is surrounded by chloroplats (arrowheads). B-F. Protargol stained *G. sanguineum* showing inclusion bodies which can be identified as oligotrichous ciliates; food vacuoles (arrows), somatic cilature (arrow heads); OC, oral cilia; M, ciliate macronucleus; N, dinoflagellate nucleus. Scale bar 10μm. From Bockstahler and Coats (1993).

Jacobson and Anderson 1996, Li *et al.* 1996). A few species have been found which are able to sequester chloroplasts from other phytoplankters. These species lack their own chloroplasts, but are able to photosynthetically utilize the "stolen" chloroplasts. This kind of mixotrophy has only been reported among species belonging to the naked genera *Amphidinium* and *Gymnodinium* (Table 3.1). This behaviour is also reported for the ambush predator dinoflagellate *Pfiesteria* (Burkholder, this volume).

Actual feeding mechanisms have only been observed in a few species (Table 3.1). Among the naked genera feeding mechanisms have been described in *Amphidinium*, *Gymnodinium* and *Gyrodinium*. Some species have been shown to feed by directly engulfing the prey (Fig. 3.1), while other species belonging to the same genera use a feeding tube (peduncle or phagopod). Among thecate species, food uptake using a peduncle has been directly observed in two heterotrophic species of the genus *Dinophysis* (Fig. 3.2; Elbrächter 1991, Hansen 1991), and this is suggested to be the

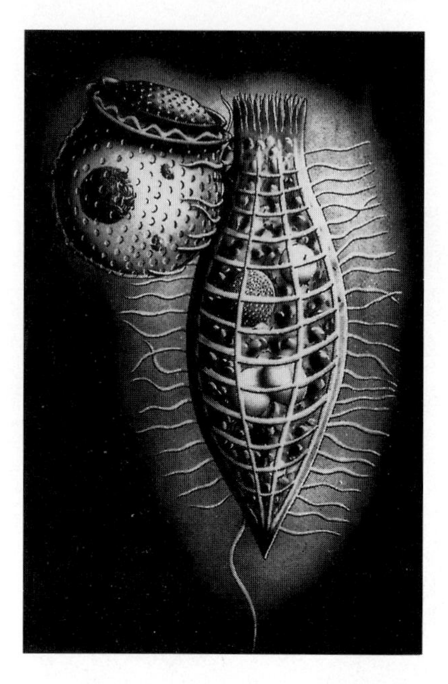

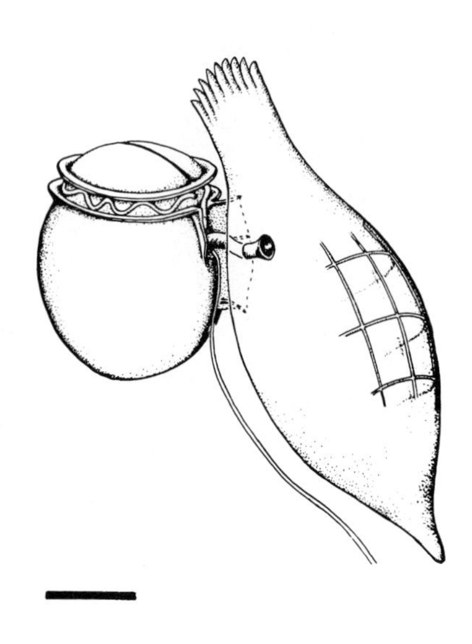

Fig. 3.2. The heterotrophic dinoflagellate *Dinophysis rotundata* feeding on the prostomatid ciliate *Tiarina fusus*. The ciliate has just been caught and the feeding tube (peduncle) connects the dinoflagellate with the ciliate. Schematic drawing to the right illustrates the physical connection between the dinoflagellate and the ciliate. Scale bar 20 µm. From Hansen (1991).

feeding mechanism in phototrophic species of *Dinophysis* (Jacobson and Andersen 1994). In a number of cases, large food vacuoles containing recognizable prey have been found in thecate phototrophic dinoflagellate species, suggesting direct engulfment (Table 3.1; Fig. 3.3). Direct engulfment has been reported in the thecate mixotrophic dinoflagellate *Fragilidium* (= *Helgolandinium*) *subglobosum* (Skovgaard, 1996). *Fragilidium subglobosum* feeds on *Ceratium* species which are ingested through the sulcus (Fig. 3.4). The prey theca is gradually dissolved during engulfment. During the process, the theca of *F. subglobosum* remains intact until engulfment is completed. The feeding cell is able to increase its volume approximately three-fold. This is possible because the individual thecal plates detach from one another, making the theca more flexible. Direct engulfment of prey has also been reported in thecate heterotrophic *Peridinium gargantua* (Biecheler 1936; 1956). Among thecate heterotrophic dinoflagellates, a third feeding mechanism, pallium-feeding is common. The prey is surrounded by a pseudopodium, the pallium, originating from the flagellar pore and digestion takes place "outside" the main cell body (e.g. *Protoperidium*, the *Diplopsalis* group; Jacobson and Anderson 1986). However, reports of mixotrophic dinoflagellates using this feeding mechanism are lacking.

Table 3.1. Prey items and feeding mechanisms of naked and thecate mixotrophic dinoflagellates.

Naked species		Prey items	Feeding mechanism	Reference
Amphidinium	*poecilochroum* [b]	cryptophytes	peduncle	1
	cryophilum [a]	dinoflagellates	phagopod	2
	steinii [c]	?	nd	3
Gymnodinium	*aeruginosum* [b]	*Chroomonas* sp	peduncle	4
	fulgens [c]	?	nd	3
	galatheanum [a]	bacteria?	nd	5
	ravenescens [c]	?	nd	3
	sanguineum [a]	ciliates, *Pyrenomonas salina*	direct engulfment	6
Gyrodinium	*estuariale* [a]	*Cryptomonas* sp., *Isochrysis* sp.	nd	21
	instriatum [a]	tintinnid ciliates	direct engulfment	24
	melo [c]	?	nd	3
	pavillardii [a]	ciliates	direct engulfment	7, 8
	uncatenum [a]	ciliates	nd	6
	vorax [a]	dinoflagellates	peduncle	7,8
Nematodinium	sp.[a]	diatoms	direct engulfment	9
Noctiluca	*scintillans* [b]	*Pedinomonas* sp.	direct engulfment	10, 11, 20
Pfiesteria	*piscicida* [b] (dinospore)	cryptophytes, prymnesiophytes,	peduncle	22
Torodinium	*teredo*.[a]	?	?	9

[a], Plastidic species. [b], Species which retain chloroplasts or endosymbionts. [c], existence of own chloroplasts unknown. [d], doubts about species identification. ?, unknown prey items. ?text: refers suggested prey item or feeding mechanism. nd: not described.

Table 3.1. continued.

Thecate species		Prey items	Feeding mechanism	Reference
Prorocentrum	*micans*[a]	cyst/parasite?	nd	12
	minimum[a]	*Cryptomonas* sp. *Strobilidium* sp	nd	21
Dinophysis	*acuminata*[a]	ciliates?	peduncle?	13
	norvegica[a]	ciliates?	peduncle?	13
Alexandrium	*tamarense*[a]	bacteria?	nd	5
	ostenfeldii[a]	*Dinophysis* sp. (ciliates)	direct engulfment?	12
Fragilidium	*heterolobum*[a]	dinoflagellates	nd	17
	mexicanum[a]	dinoflagellates	nd	18
	subglobosum[a]	*Ceratium* spp.	direct engulfment	19
Gonyaulax	*grindleyi*[a]	?	nd	12
	spinifera[a]	?	nd	12
Scrippsiella	sp.[a]	?	nd	12
Ceratium	*declinatum*[a]	?	nd	16
	furca[a]	*Strobilidium*, other ciliates	direct engulfment?	6,21
	fusus[a]	?	nd	16
	hirundinella[a]	?	nd	14, 15
	longipes[a]	?	nd	16
	teres[a]	?	nd	16
Peridinium[d]	*brevipes*[a]	*Cryptomonas* sp.	nd	21
Peridinium	*inconspicuum*[a]	bacteria?	nd	23

Larsen 1988; 2. Wilcox and Wedemayer 1991; 3. Fields and Rhodes 1991; 4. Kofoid and Swecy 1936; 5. Nygaard and Tobisen 1993; 6. Bockstahler and Coats 1993; 7. Biecheler 1935; 8. Biecheler 1952; 9. Elbrächter 1991; 10. Sweeney 1971; 11. Sweeney 1978; 12. Jacobson and Ander-son 1996; 13. Jacobson and Andersen 1994; 14. Dodge and Crawford 1970; 15. Chapman *et al.* 1981; 16. Chang and Carpenter 1994; 17. Balech 1988; 18. Balech 1990; 19. Skovgaard (1996); 20. Sweeney 1976; 21. Li *et al.* 1996; 22. Burkholder and Glasgow 1995; 23. Porter 1988; 24. Uchida et al. 1997.

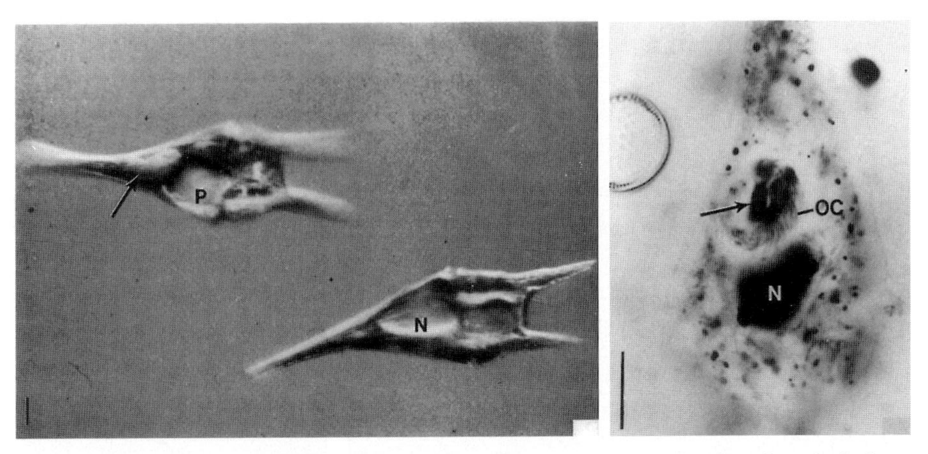

Fig. 3.3. *Ceratium furca*. Differential interference contrast image of two freshly collected *C. furca* cells. One of the specimens contains a triangular inclusion body (arrow). P. pusule; N, nucleus. Protargol stain showing the remains of a partially digested ciliate is shown to the right. OC, oral cilia; N, dinoflagellate nucleus; Arrow, macronucleus of the prey. Scale bars = 10 μm. From Bockstahler & Coats (1993).

3.2. Prey selection in dinoflagellates

3.2.1. Prey size selection

Mixotrophic dinoflagellates typically range in size from about 8 to 500 μm and have been found to ingest live prey in the size range of about 4 to 300 μm. Nanoflagellates, other dinoflagellates, and ciliates are among the prey items most often reported. Thus, the feeding mechanisms found among dinoflagellates all allow ingestion of large prey (Gaines and Elbrächter 1987, Jacobson and Anderson 1986, Schnepf and Elbrächter 1992). The efficiency of prey capture by planktonic protists depends on size, and often this prey size spectrum is bell-shaped (Hansen *et al.*, 1994) Very small particles are not detected by the protist, while particles too big for the protist to engulf are rejected. A prey size spectrum determined for the heterotrophic dinoflagellate, *Gyrodinium spirale*, revealed that it is most efficient on prey of its own size (Hansen 1992).

A few studies have indicated that even large mixotrophic dinoflagellates are able to feed substantially on small (0.6 μm) suspended bacteria (Porter 1988, Nygaard and Tobisen 1993), but the results of these studies are questionable for the following reasons. Porter (1988) carried out experiments in which the photosynthetic *Peridinium inconspicuum* was fed fluorescent polystyrene beads for a short time. Despite the fact that the beads only were observed on the surface of *P. inconspicuum* and not in food vacuoles inside the cell, this was interpreted as ingestion. Nygaard and Tobisen (1993) studied the uptake of radioactively-labeled bacteria (RLB) by a number of red tide phytoflagellates including the dinoflagellates *Gymnodinium galatheanum* and *Alexandrium tamarense*. Using short term incubation, a linear uptake of radioactivity was observed with time. As a control for leakage of radioactivity

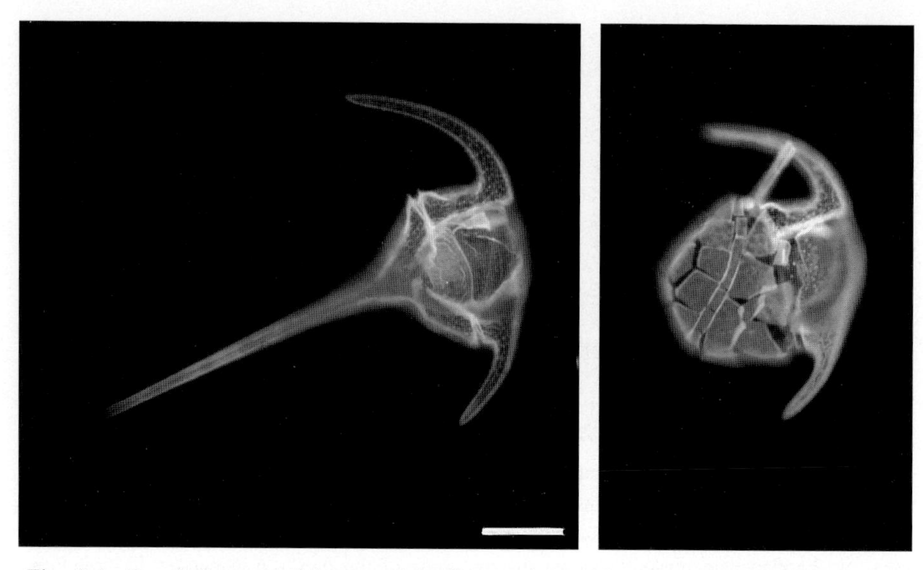

Fig. 3.4. *Fragilidium subglobosum*. Calcofluor staining of a *F. subglobosum* cell in the process of engulfing the dinoflagellate *Ceratium tripos*. Note that the plates of *F. subglobosum* detach from each other in order to ingest the large prey. Calcofluor staining of a *Ceratium tripos* cell for comparison is shown to the left. Scale bar 20 µm. Skovgaard, original.

from the bacteria, some cultures were fixed prior to the addition of labeled prey. First, the RLB method always gave positive uptake rates in the phytoflagellates tested. Thus, a "control" alga that did not take up the radioactivity is lacking from their data set. If the RLB in some way attaches to the cell surface, this would also be a time-dependent process and lead to a false conclusion. Such a time-dependent attachment of bacteria to *A. tamarense* has been observed in laboratory cultures exposed to natural bacterial communities (R. Jovine, unpubl. data). Second, no evidence of bacteria inside food vacuoles was provided by either light microscopy or transmission electron microscopy. Finally, the use of fluorescently labeled bacteria (FLB) failed to document uptake of bacteria in food vacuoles in both dinoflagellates studied. Thus, in these cases, the use of FLB and RLB techniques may have led to artifacts or misinterpretations.

3.2.2. Prey specificity

The literature on prey selection due to prey specificity in mixotrophic dinoflagellates is sparse. *Ceratium furca* has been shown to primarily ingest small ciliates of the genera *Strobilidium* and *Mesodinium*, but will not ingest species of the genus *Strombidium* although they are in similar size to cells of the former ciliate genera (Bockstahler and Coats 1993). *Prorocentrum minimum* apparently is capable of ingesting the cryptophyte *Cryptomonas* sp. and the ciliate *Strobilidium* sp., but not another cryptophyte *Pyrenomonas salina* (Li *et al.* 1996). *Fragilidium subglobosum* only ingests species of *Ceratium,* when presented with a variety of organisms of different sizes and taxonomical groups (Skovgaard, 1996). Also, growth and grazing responses of *F. subglobosum* are significantly higher when fed *C. tripos* and *C. lineatum* compared to *C. furca* and *C. fusus.* This observation cannot be explained solely by size selection, because *C. tripos* and *C. lineatum* are the largest and the smallest of the *Ceratium* species tested, respectively (Skovgaard, 1996, Hansen and Nielsen, in press). How *F. subglobosum* is able to recognize a *Ceratium* species is unknown, but it may relate to surface properties.

Thus one of the difficulties in studying mixotrophy in phytoflagellates is related to the specificity/selectivity of prey items. The reason for the high degree of specificity in some species is unknown. While it seems a poor "strategy" when the species diversity is high, it may be quite advantageous when the prey forms monospecific blooms.

4. Conclusions

Mixotrophy appears to be widespread among prymnesiophytes and dinoflagellates. A large variety of feeding mechanisms within these groups have been found, all allowing the ingestion of relatively large prey. Some degree of selectivity which cannot be explained by size selection apparently exists. We need to know more about the feeding mechanisms of mixotrophic phytoflagellates and to understand the mechanisms of prey selection in order study the role of mixotrophy in the bloom dynamics of harmful algae.

5. Acknowledgments

This work was funded by the Danish Natural Research Council grant no 9502163-28808. I thank Alf Skovgaard for the use of photographs illustrating the feeding mechanism of *Fragilidium subglobosum.*

6. References

Balech, E. (1988). Una especie nueva del género *Fragilidium* (Dinoflagellata) de la Bahía de Chamela, Jalisco, Mexico. *Anales Inst. Biol.* 58: 479-486.

Balech, E. (1990). Four new dinoflagellates. *Helgoländer Meeresuntersuch.* 44: 387 -396.

Biecheler, B. (1936a). Des conditions et du méchanism de la predation chez un dinoflagellé envelope tabulée, *Peridinium gargantua* n.sp. *C.R. Séanc. Soc. Biol.* 121: 1054-1057.

Biecheler, B. (1936b). Observation de la capture et de la digestion des proies chez un peridinien vert. *C.R. Séanc. Soc. Biol.* 122: 1173-1175.

Biecheler, B. (1952). Recherces sur les péridiniens. *Bull. Biol. Fr. Belg.* (suppl.) 36: 1-149.

Bockstahler, K.R., Coats, D.W. (1993). Spatial and temporal aspects of mixotrophy in Chesapeake Bay dinoflagelates. *J. Eur. Microbiol.* 40: 49-60.

Chang. J., Carpenter, E.J. (1994). Inclusion bodies in several species of *Ceratium*

Bockstahler, K.R., Coats, D.W. (1993). Spatial and temporal aspects of mixotrophy in Chesapeake Bay dinoflagelates. *J. Eur. Microbiol.* 40: 49-60.

Chang. J., Carpenter, E.J. (1994). Inclusion bodies in several species of *Ceratium* Schrank (Dinophyceae) from Caribbean Sea examined with DNA-specific staining. *J. Plankton Res.* 16: 197-202.

Chapman, D.V., Livingstone, D., Dodge J.D. (1981). An electron microscope study of the encystment and early development of the dinoflagellate *Ceratium hirundinella*. *Br. Phycol. J.* 16: 183-194.

Conrad, W. (1941). Sur les Chrysomonadines à trois fouets. Apercu synoptique. *Bull. Musée Royal d'Histoire Naturelle de Belg.* 17: 1-16.

Dodge, J., Crawford, R.M. (1970). The morphology and fine structure of *Ceratium hirundinella* (Dinophyceae). *J. Phycol.* 6: 137-149.

Elbrächter, M. (1991). Faeces production by dinoflagellates and other small flagellates. *Mar. Microbial Food Webs* 5: 189-204.

Fields, S.D., Rhodes R.G. (1991). Ingestion and retention of *Chroomonas* sp. (Cryptophyceae) by *Gymnodinium acidotum* (Dinophyceae). *J. Phycol.* 27: 525-529

Gaines, G. and Elbrächter, M. (1987). Heterotrophic nutrition. *In*: Taylor, F.J.R (ed) *The biology of dinoflagellates*. Blackwell, Oxford, pp. 224-268.

Green, J.C. (1991). Phagotrophy in prymnesiophyte flagellates. In: Patterson, D.J. and Larsen, J. (eds) *The biology of free-living heterotrophic flagellates*. Systematics Association Special 45. Clarendon Press, Oxford, pp. 401-414.

Hansen, P.J. (1991). *Dinophysis* - a planktonic dinoflagellate genus which can act both as a prey and a predator of a ciliate. *Mar. Ecol. Prog. Ser.* 69: 201-204.

Hansen, B., Bjørnsen, P.K., Hansen, P.J. (1994). The size ratio between planktonic predators and their prey. *Limnol. Oceanogr.* 39: 395-403.

Hansen, P.J., Nielsen, T.G. (1997). Mixotrophic feeding of *Fragilidium subglobosum* (Dinophyceae) on *Ceratium* spp: effects of prey concentration prey species and light intensity. *Mar. Ecol. Prog. Ser.* 147: 187-196.

Jacobson D.M., Anderson, D.M. (1986). Thecate heterotrophic dinoflagellates: feeding behavior and mechanism. *J. Phycol.* 22: 249-258.

Jacobson, D.M., Andersen, R.A. (1994). The discovery of mixotrophy in photosynthetic species of *Dinophysis* (Dinophyceae): light and electron microscopical observations of food vacuoles in *Dinophysis acuminata, D. norvegica* and two heterotrophic dinophysoid dinoflagellates. *Phycologia* 33: 97-110.

Jacobson, D.M., Anderson D.M. (1996). Widespread phagocytosis of ciliates and other protists by marine mixotrophic and heterotrophic thecate dinoflagellates. *J. Phycol.* 32: 279-285.

Jones H.L.J., Leadbeater, B.S.C., Green, J.C. (1993). Mixotrophy in marine species of *Chrysochromulina* (Prymnesiophyceae): ingestion, and digestion of a small green flagellate. *J. mar. biol. Ass. UK.* 73: 283-296.

Kawachi, M., Inouye, I., Maeda, O., Chihara, M. (1991). The haptonema as a food-capturing device: observations on *Chrysochromulina hirta* (Prymnesiophyceae). *Phycologia* 30: 563-573.

Kawachi, M., Inouye, I. (1995). Functional roles of the haptonema and the spine scales in the feeding process of *Chrysochromulina spinifera* (Fournier) Pienaar et Norris (Haptophyta = Prymnesiophyta). *Phycologia* 34: 193-200.

Kofoid, C.A., Swezy, O. (1921). The free-living unarmoured dinoflagellata. *Mem. Univ. Calif.* 5: 1-564.

Larsen, J. (1988). An ultrastructural study of *Amphidinium poecilochroum* (Dinophyceae), a phagotrophic dinoflagellate feeding on small species of crypto-

Norris (Haptophyta = Prymnesiophyta). *Phycologia* 34: 193-200.
Kofoid, C.A., Swecy, O. (1921). The free-living unarmoured dinoflagellata. *Mem. Univ. Calif.* 5: 1-564.
Larsen, J. (1988). An ultrastructural study of *Amphidinium poecilochroum* (Dinophyceae), a phagotrophic dinoflagellate feeding on small species of cryptophytes. *Phycologia* 27: 366-377.
Li, A., Stoecker, D.K., Coats, D.W., Adam, E.J. (1996). Ingestion of fluorescently labeled and phycoerythrin-containing prey by mixotrophic dinoflagellates. *Aquat. Microbial Ecol.* 10: 139-147.
Nygaard, K., Tobisen, A. (1993). Bacterivory in algae: a survival strategy during nutrient limitation. *Limnol. Oceanogr.* 38: 273-279.
Parke, M., Adams, I. (1960). The motile (*Crystallolithus hyalinus* Gaarder & Markali) and non-motile phases in the life history of *Coccolithus pelagicus* (Wallich) Schiller. *J. mar. biol. Ass. UK.* 39: 21-35.
Parke, M., Manton, I., Clarke, B. (1956). Studies of marine flagellates. III. Three further species of *Chrysochromulina. J. mar. biol. Ass. UK.* 35: 387-414.
Porter, K. G. (1988). Phagotrophic phytoflagelates in microbal food webs. *Hydrobiologia* 159: 89-97.
Schnepf, E., Elbrächter, M. (1992). Nutritional strategies in dinofagellates. *Europ. J. Protistol.* 28: 3-24.
Skovgaard, A. (1996). Engulfment of *Ceratium* spp. (Dinophyceae) by the thecate photosynthetic dinoflagellate *Fragilidium subglobosum. Phycologia 35: 400-409.*
Sweeney, B.M. (1971). Laboratory studies of a green *Noctiluca* from New Guinea. *J. Phycol.* 7: 53-58.
Sweeney, B.M. (1976). *Pedinomonas noctilucae* (Prasinophyceae), the flagellate symbiotic in *Noctiluca* (Dinophyceae) in South-East Asia. *J. Phycol.* 12: 460-464.
Sweeney, B.M. (1978). Ultrastructure of *Noctiluca miliaris* (Pyrrhophyta) with green flagellate symbionts. *J. Phycol.* 14: 116-120.
Uchida, T., Kamiyama, T., Matsuyama, Y. (1997). Predation by a photosynthetic dinoflagellate *Gyrodinium instriatum* on loricated ciliates. J. Plankton Res. 19: 603-608
Wilcox, L.W., Wedemayer, G.J. (1991). Phagotrophy in the freshwater photosynthetic dinoflagellate *Amphidinium cryophilum. J. Phycol.* 27: 600-609.

The Ecological Significance of Phagotrophy in Photosynthetic Flagellates

E. Granéli and P. Carlsson

University of Kalmar, Dept.. of Natural Sciences, PO Box 905, S-391 29, Kalmar, Sweden

1. Introduction

Mixotrophy is the ability of an organism to be both phototrophic and heterotrophic, in the latter case by utilizing either organic particles (phagotrophy) or dissolved organic substances (osmotrophy). This review deals with phagotrophic algae, i.e. potentially phototrophic organisms with the capacity to "graze" on particles, e.g. bacteria. For more information on the uptake of dissolved organic substances by phytoplankton see Carlsson and Granéli (this volume). The idea that photosynthetic organisms might be able to utilize organic material is not new. Older observations of phagotrophy typically relied on conventional light microscopy. For example, Hofeneder (1930) reported that the freshwater dinoflagellate *Ceratium hirundinella* engulfed prey through the ventral area using a pseudopodium (see Gaines and Elbrächter 1987). Although such descriptive microscopic observations of algal phagotrophy have been known for decades, attempts to quantify the ecological significance of phagotrophy in the marine environment are relatively new (not more than 5 years). Recent studies of mixotrophy have been stimulated by new methods for the quantification of grazing by small plankton and by the recent reconceptualization of the pelagic food chain to include the "bacterial loop". The new view of the role of pelagic bacteria also stressed the importance of bacterivory through heterotrophic and mixotrophic microflagellates. Methodological innovations that have been especially important are the development of techniques based on fluorescent- or radioactively-labelled particles (Lessard and Swift 1985, Rublee and Gallegos 1989, Sherr *et al.* 1991). Using such methods (latex beads), Bird and Kalff (1986) detected remarkable phagotrophy in the freshwater chrysophyte *Dinobryon*. This investigation has spurred many subsequent studies of phagotrophy by algae in both marine and freshwater pelagic communities. Hansen (this volume), provides a comprehensive review of the mechanisms by which algae ingest particles (bacteria, other algae or even zooplankton).

The significance of phagotrophy in the ecology of "phytoplankton" is still largely unknown. Boraas *et al.* (1988), in considering the direction of future research, stated: "Many questions (numerical and biomass abundance, carbon flux, feeding rates, nutrient regeneration rates, etc.) are directly analogous to those asked by zooplankton ecologists". Phagotrophy can be seen as an important factor regulating population dynamics. Thus we may ask, for example: What part of grazing on bacteria is due to phagotrophy by phytoplankton? However, this review focus on the grazer, i.e. the phagotrophic phytoplankton. Some important questions are: How can we detect phagotrophy? Among which members of the different phytoplankton taxonomic groups is phagotrophy most common? What environmental conditions trigger phagotrophic behaviour? Why are photosynthetic organisms phagotrophic, i.e. is there a nutritional advantage? Can phagotrophs outcompete autotrophs and/or heterotrophs?

NATO ASI Series, Vol. G 41
Physiological Ecology of Harmful Algal Blooms
Edited by D. M. Anderson, A. D. Cembella and G. M. Hallegraeff

Is phagotrophy an alternative way, to addition to photosynthesis and uptake of dissolved inorganic or organic substances, to acquire supplementary carbon, macronutrients (N, P, Si), trace elements or special compounds, e.g. vitamins, that the algae cannot synthesise ? Or is phagotrophy only a remnant of an ancient, formerly important behaviour among planktonic organisms, that has little ecological significance? These are some of the questions that remain largely unanswered.

Phagotrophy among algae is presumably regulated through external abiotic or biotic factors that influence cell physiological state. If light is too low to allow for sufficient CO_2 fixation to meet metabolic demands of the cell, phagotrophy can supplement or even substitute for photosynthesis as a source of organic carbon (Andersson *et al.* 1989, Sanders *et al.* 1990). A similar argument may be used to explain phagotrophy with respect to nutrients, i.e., phagotrophy may supply the organism with N or P or some micronutrient if dissolved sources have been exhausted (Kimura and Ishida 1985, Caron *et al.* 1993, Nygaard and Tobiesen 1993). For phagotrophy to be effective, there must be a sufficient supply of suitable organic particles (prey organisms). Phagotrophy might thus be induced in environments where the organism encounters a high concentration of prey, e.g. bacteria. The three most important triggering/regulating factors for phagotrophy might then be light, nutrient availability, and prey abundance.

It is natural to assume that algae use phagotrophy to obtain macro- (N, P) and micro-nutrients (e.g. vitamins), when dissolved inorganic or organic nutrients are growth-limiting (Aaronson 1974, Kimura and Ishida 1985, Sanders and Porter 1988). Mixotrophy may thus be a primitive trait, a notion that is supported by the fact that groups with many phagotrophs are evolutionarily old (Porter 1988). Jones (1994) has proposed that some phagotrophic algae have evolved from primitive heterotrophs, while others are "secondary" phagotrophs, where the character has evolved from strict phototrophy. He further suggested that mixotrophy should not be viewed as a single strategy developed by planktonic organisms placed between the two extremes of nutrition: autotrophy and heterotrophy. There is probably a continuous gradient between truly autotrophic and heterotrophic organisms (Fig. 1). Some organisms will initiate phagotrophy only in the presence of high quantities of prey (Andersson *et al.* 1989, Sanders *et al.* 1990), whereas for others phagotrophy appears to be more dependent on abiotic factors, such as light (Caron *et al.* 1993, Jones *et al.* 1993, 1995, Keller *et al.* 1994). Some algae may be efficient phagotrophs but poor phototrophs (Caron *et al.* 1990), whereas others may be obligate photoautotrophs, still capable of phagotrophy (Caron *et al.* 1993). From an ecological point of view, there is an apparent trade-off between phagotrophy and photosynthesis. Both modes of nutrition have metabolic costs, and the simultaneous ability to perform both has high costs, but enables the organism to outcompete heterotrophic or strictly photosynthetic competitors under certain environmental conditions (Rothhaupt 1996 a, Thingstad *et al.* in press).

Phagotrophy among photosynthetic plankton has been studied mainly for phytoflagellates, including dinoflagellates, in the marine environment (except for a few studies on chrysophytes and prymnesiophytes in marine/estuarine waters and for chrysophytes in freshwater environments). Among dinoflagellates there are species which are strictly heterotrophic (lacking photosynthetic pigments), whereas closely related species are mixotrophic or photosynthetic, e.g. in the genus *Gyrodinium* (Gaines and Elbrächter 1987). Also there are species which after ingesting their prey keep their actively working chloroplasts= "cleptochloroplasts" (see Burkholder *et al.*, this volume). Due to the scarcity of literature on phagotrophy in toxic or harmful phytoplankton, we will consider examples of phagotrophy for both non-toxic and toxic/harmful phytoplankton species, including marine, brackish and freshwater organisms.

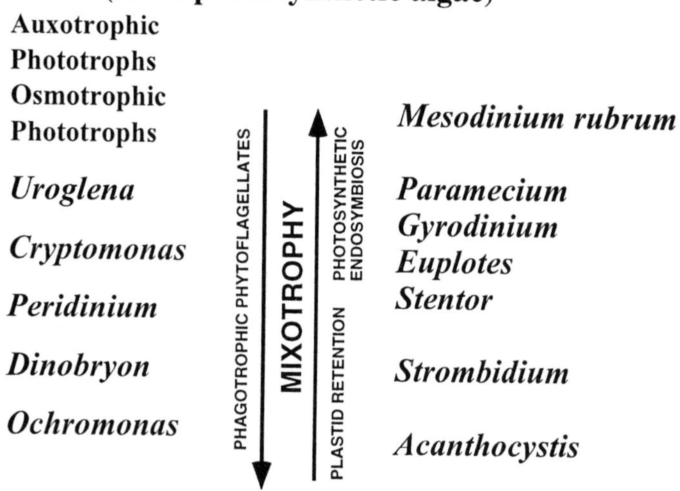

Fig. 1. Schematic representation of the continuum of nutritional strategies amongst planktonic protists. Examples are shown of genera which include mixotrophic forms amongst their species. (redrawn from Jones 1994).

2. Methods and Approaches - Progress and Limitations

Several methods can be used to detect and quantify phagotrophy in phototrophic algae, but all have both limitations and advantages. A direct approach for investigating phagotrophy is to quantify the proportion of a population of cells that contain food vacuoles and to identify the content of the vacuoles. Use of protargol staining and high magnification (500-1250X) optical microscopy enables identification of ciliates, e.g. inside food vacuoles of the phagotrophic organism (Bockstahler and Coats 1993 a, b). Chang and Carpenter (1994) used DAPI (DNA stain) and epifluorescence microscopy to show the presence of prey inside food vacuoles. A similar approach with better resolution is the use of transmission electron microscopy (TEM) for identification of food vacuole content (Jacobson and Andersen 1994, Jacobson and Anderson 1996). These methods can be directly applied to natural plankton communities, without experimental manipulations. Such studies, however, only detect the occurrence of phagotrophy, and do not quantify grazing rates. An example of a study based on this approach is by Bockstahler and Coats (1993 b), who reported that in water samples from the Chesapeake Bay taken between June and September, as much as 20 to 30% of the red-tide dinoflagellates *Gymnodinium sanguineum, Ceratium furca* and *Gyrodinium uncatenum* possessed food vacuoles containing nanociliates (diameter < 20 μm).

In some instances, it may be difficult to interpret the significance of inclusions contained within vacuoles. Studies of vacuole content or occurrence of inclusion bodies give an indication of the existence of phagotrophy, but do not strictly prove that active grazing has taken place. This is due in part to digestion in the vacuoles.

Another difficulty lies in determining if the particle inside the organism is a prey, a parasite or a gamete. For example *Heterocapsa triquetra* cells exposed to darkness were found to contain cells resembling smaller individuals of the same species. This observation suggests that darkness induces cannibalism by *H. triquetra* (Legrand *et al.* submitted). While cannibalism has been reported in phyto- and zooflagellates (Aaronson 1974, Jeong and Latz 1994) the presence of smaller cells found inside the "predator" are often gametes (Walker 1984, Berland *et al.* 1995 , Maestrini *et al.* 1996). In one case, Berland *et al.* (1995) reported that small flagellate cells were liberated into the medium and swam freely after disruption of the cell wall of the toxic dinoflagellate *Dinophysis acuminata.* These small motile flagellates were interpreted as *D. acuminata* gametes but the possibility that they were parasites cannot be excluded.

A wide variety of particles have been used to test phagotrophic capabilities and estimate grazing rates of planktonic organisms. A population can be exposed to inert particles, e.g. (i) fluorescent plastic latex beads (Bird and Kalff 1987, Porter 1988, Jones *et al.* 1993), or (ii) live or killed prey, e.g. unlabelled green flagellates (Jones *et al.*, 1993, 1995), or ^{14}C-radioactively- or fluorescently-labelled bacteria (RLB, FLB) or algae (RLA, FLA) (Lessard and Swift 1985, Rublee and Gallegos 1989, Nygaard and Hessen 1990, Nygaard and Tobiesen 1993), and naturally fluorescing live prey (Li *et al.* 1996). The prerequisite is that the particle or some constituent of the particle can be detected inside the grazer. The decrease in prey particles in the medium during incubation with the grazer can also be taken as a measure of phagotrophy, just as in studies of zooplankton grazing, if other prey loss factors are insignificant or can be quantified.

Use of artificial particles or manipulated (dead) natural prey may cause bias or underestimation compared to phagotrophy on live, natural prey. However, Nygaard and Hensen (1990) showed that there was only a small difference (10%) between ingestion rates of small flagellates fed FLB and RLB. Similarly Jones and Rees (1994) found almost no discrimination between FLB and FM (fluorescent microspheres) by *Dinobryon divergens.* Heating bacteria for FLB labelling seems to cause minor problems (Sherr *et al.* 1990). However, other studies have indicated that use of artificial particles and labelled and/or dead prey is not without problems, underestimating the process. Jones *et al.* (1993) studied phagotrophy in the prymnesiophyte *Chrysochromulina brevifilum* using a variety of particles and living cells of different sizes (Table 1). Not all types of cells were ingested, but the inorganic particles were; however, ingestion of carmine particles was much lower than ingestion of a small green flagellate. After one hour of incubation 51.6% of *C. brevifilum* cells were found to contain the green flagellate but only 8% contained carmine particles. According to Nygaard and Tobiesen (1993) grazing rates for marine flagellates (including dinoflagellates) based on FLB were much lower than those based on RLB. Ingestion of FLB for a toxic *C. polylepis* was also very low (Legrand *et al.* 1996), similar to results of Nygaard and Tobiesen (1993) for this species, indicating that methodological problems in the use of RLB may produce artificially high estimates of grazing rates.

Table 1. *Chrysochromulina brevifilum* screened for the ingestion of a variety of particles and cells (From Jones *et al.* 1993).

	Species/Particles	Size, µm	Ingestion
Inert particles	Carmine particles	1-4	+
	Fluorescent microspheres	0.5	+
Green algae	*Micromonas pusilla*	1.5-3	-
	Chlorella sp.	-	-
	C. stigmatophora	2-5	-
	C. salina	4-8	-
	Small green flagellate	3-4	+
	Stichococcus bacillaris	5-8	-
Yeasts	*Sporobolomyces roseus*	2.5-5	+
	Saccharomyces cerevisae	4.5-7	+
	Rhodosporidium diboratum	2.5-5	-

3. Phagotrophy in relation to taxonomic affiliations

Phagotrophy has been reported for pigmented phototrophic flagellated species in the classes Chrysophyceae, Euglenophyceae, Prymnesiophyceae, Xanthophyceae, Dinophyceae, Raphidophyceae and Cryptophyceae (Bird and Kalff 1986, Porter *et al.* 1985, Sanders and Porter1988, Jones *et al.* 1993, 1995, Nygaard and Tobiesen 1993, Keller *et al.* 1994, Legrand *et al.* 1996), but has not been reported and may not exist among certain phototrophs such as diatoms (see Table 2). Pioneering work on phagotrophy was conducted in freshwater systems. The freshwater chrysophyte genera *Dinobryon* and *Poteriochromonas* contain chloroplast-bearing species among which phagotrophy is frequently found (Bird and Kallf 1986, 1987, Berninger *et al.* 1992, Sanders *et al.* 1990). For a comprehensive review of phagotrophy among photosynthetic freshwater species see Porter (1988) and Sanders and Porter (1988). For marine waters there are two reviews on phagotrophic feeding mechanisms among heterotrophic and mixotrophic dinoflagellates (Gaines and Elbrächter 1987; Schnepf and Elbrächter 1992; see also chapter by Hansen, this volume).

4. Phagotrophy in relation to environmental conditions

4.1 Prey abundance

If there are no or only few organic particles in a suitable size range, phagotrophy should not be induced. Conversely, if the potential for phagotrophy exists, ingestion of particles should be high when prey of appropriate size are abundant. There might be a simple numerical predator-prey response similar to that found for zooplankton grazing on microalgae. Jones and Rees (1994) found a strong relation between ingestion rate for *Dinobryon divergens* and density of fluorescent microspheres. The amount of FM ingested per cell $^{-1}$· min^{-1} increased proportionally to FM densities in the medium from 10^5 up to a maximum of 10^9 FM ml^{-1}. A threshold was found at 10^{10} FM ml^{-1}, above which no further increase in ingestion was observed. For the chrysophyte *Ochromonas* sp., bacterivory increased hyperbolically with bacterial (FLB) density (Rothhaupt 1996a). Maximal ingestion rate was only achieved at $3x10^7$ bacteria ml^{-1}. Similar results have also been found with live prey (Jones *et al.* 1993, Li *et al.* 1996). Hansen and Nielsen (submitted) found that phagotrophy in the marine dinoflagellate *Fragilidium subglobosum* on *Ceratium tripos* was related to prey concentration with a threshold around 10 *C. tripos* cells ml^{-1}. *Fragilidium subglobosum* growth rates were 0.16 and 0.5 d^{-1} for cells growing phototrophically and phagotrophically, respectively. Thus phagotrophy seems to be the preferred mode of nutrition for this dinoflagellate. *Fragilidium subglobosum* could feed on the three *Ceratium* species offered as prey (*C. tripos, C. furca* and *C. fusus*), but growth rate was higher when *C. tripos* (the largest of the species) was the prey.

Table 2. Phytoflagellate genera reported to utilize particulate food (redrawn from Porter 1988 and *Nygaard and Tobiesen 1993).

Division Chrysophyta:		Division Pyrrophyta:
Class Chrysophyceae:		Class Dinophyceae:
Catenochrysis	*Dinobryon*	*Amphidinium*
Chromulina	*Ochromonas*	*Ceratium*
Chrysococcus	*Palatinella*	*Gymnodinium*
Chrysosphaerella	*Pedinella*	*Gyrodinium*
Chrysamoeba	*Phaeaster*	*Massartia*
Chrysostephanosphaera	*Poterioochromonas*	*Peridinium*
Cyrtophora	*Pseudopodinella*	
	Uroglena	Division Cryptophyta:
Class Prymnesiophyceae:		
Chrysochromulina		Class Cryptophyceae:
Coccolithus		*Cryptomonas*
Prymnesium		
Class Xantophyceae:		Class Raphidophyceae*
Chlorochromonas		*Heterosigma*

Light, nutrient availability, and prey concentration may interact in enhancing or suppressing phagotrophy in more or less complicated ways. It is also possible that there is some physiological adaptation involved when switching from one mode of nutrition to another. If this is true, then the mode of nutrition may not immediately track changes in environmental conditions, as is the case for physiological changes taking place when a phytoplankter is exposed to different light regimes.

An example of the effect of previous growth conditions has been described by Sanders *et al.* (1990). The chrysophyte *Poterioochromonas malhamensis* was pre-cultivated phototrophically (low and high light intensities without prey). After pre-cultivation, algae previously grown without bacteria were inoculated into a bacteria-containing medium. The mode of nutrition then changed. Chlorophyll *a* content decreased drastically during the first 10 h of incubation as the organisms switched to phagotrophic behaviour independent of light conditions. The threshold concentration of bacteria, (around 10^6 bacteria ml^{-1}, a common concentration in nature), induced the algae to switch from autotrophy to heterotrophy. Above this concentration, *P. malhamensis* was phagotrophic, and at lower bacterial densities it was autotrophic in sufficient light. Another example of the decoupling of light and prey availability in phagotrophy of a photosynthetic organism was found by Rothhaupt (1996b) for *Ochromonas* sp. (Fig 2, 3). This species increased its ingestion rate with increasing bacterial density, independent of light conditions. Growth rate also increased in proportion to an increase in bacterial density, independent of light regime. However, growth rates were slightly higher when *Ochromonas* sp. was growing in light, especially at low bacterial densities. Prey concentrations had no effect on the ingestion rates or chlorophyll concentrations of *C. brevifilum*, while low light intensity increased chlorophyll per cell and the ingestion rates for this species (Jones *et al.* 1995; Table 3).

The toxic prymnesiophyte *C. polylepis* grew to higher cell numbers at high than at low light intensities in batch cultures (Legrand *et al.* in prep.). This was explained by elevated bacterial growth and increased ingestion rates of bacteria under high light (250 μmol m^{-2} s^{-1}) compared to low light (20 μmol m^{-2} s^{-1}). As expected, cellular chlorophyll levels were low at the high light intensity. *Chrysochromulina polylepis* cells growing at high light, but in a medium containing humic acids, contained less than half of the chlorophyll of cells grown in an inorganic nutrient medium (Table 4). In the high light plus humic treatment *C. polylepis* bacterial ingestion rates were 5- to 60- fold higher than in any other treatment, bacteria being found in greater quantities in this treatment. The authors concluded that humic substances were a substrate for bacteria, and as algae had more bacteria to ingest, they switched to phagotrophy. As a consequence, cellular chlorophyll decreased. Phagotrophic behaviour of *C. polylepis* was similar to that of *Poterioochromonas malhamensis* (Sanders *et al.* 1990), which apparently preferred the phagotrophic mode of nutrition over phototrophy in the presence of abundant bacterial prey, independent of light regime and chlorophyll concentrations.

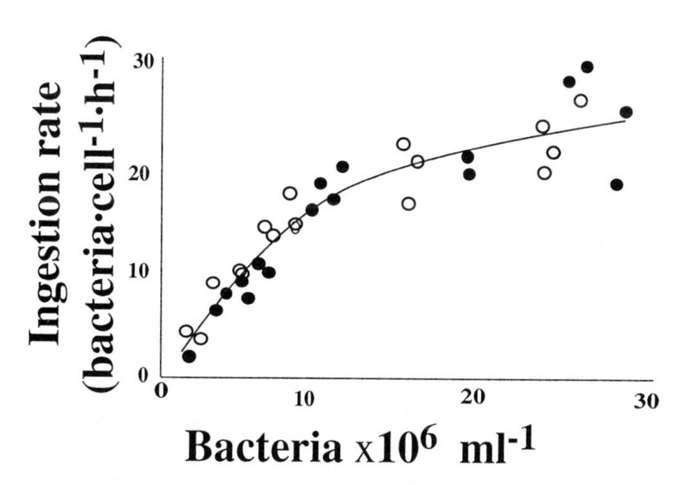

Fig. 2. Ingestion rate as a function of bacterial density in the dark (filled circles) and in the light (open circles). A Michaelis-Menten model was fit to the data (I_{max}=34.8± 2.4 bacteria $cell^{-1}$ h^{-1}; K_1=12.6±1.9 x 10^6 bacteria ml^{-1} ; R^2= 0.92). (Redrawn from: Rothhaupt 1996b)

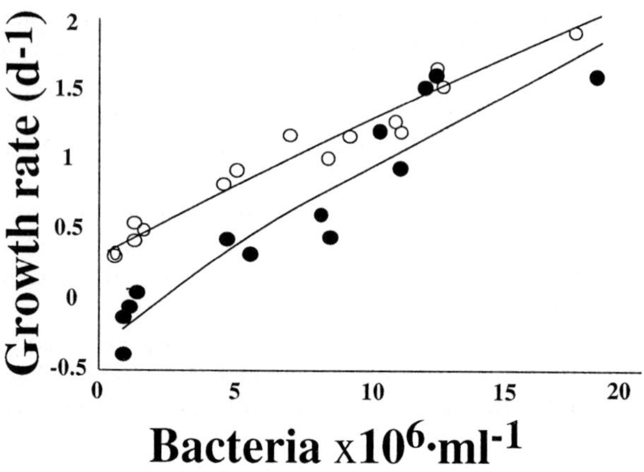

Fig. 3. Growth rates (μ) of *Ochromonas* as functions of bacterial density in the dark (filled circles) and in the light (open circles). Lines are the fits of modified Monod models. (Redrawn from Rothhaupt 1996b)

Table 3. Chlorophyll *a* per cell of *Chrysochromulina brevifilum* with and without *Marsupiomonas pelliculata*, at different prey/predator ratios, at photon flux densities of 100 μmol $m^{-2} s^{-1}$ & 25 μmol $m^{-2} s^{-1}$. (Adapted from Jones *et al.* 1995).

Prey/predator ratio	25 μmol m^{-2} s^{-1} (fg chl *a* cell^{-1})±SE.[1]	100 μmol m^{-2} s^{-1} (fg chl *a* cell-1)±SE[3]
(*C.brevifilum* alone)[2]	722±50	281±15
2:1	-	270±4.0
5:1	751±2.0	302±26

[1]Concentration of *C. brevifilum* =4x10^5 cells ml^{-1} (25 μmol $m^{-2} s^{-1}$) or 2x10^5 cells ml^{-1} (100 μmol $m^{-2} s^{-1}$).

[2]72 hours after the introduction of *M. pelliculata* to cultures of *C. brevifilum*.

[3]Six days after the introduction of *M. pelliculata* to cultures of *C. brevifilum*.

Table. 4. *Chrysochromulina polylepis* chlorophyll *a* (pg cell^{-1}) and ingestion of FLB (fluorescent labelled bacteria) per cell (FLB cell^{-1} h^{-1}) in low and high light (20 and 250 μmol m^{-2} s^{-1} respectively). (Redrawn from Legrand *et al.* in prep.)

	Low light		High light	
	FLB	Chl *a*	FLB	Chl *a*
- humic substances	0-0.01	1.0-1.6	0.05-0.1	0.6-0.7
+ humic substances	0-0.01	0.8-1.7	0.20-0.6	0.2-0.3

One might intuitively expect low light to be a major triggering factor for the induction of heterotrophy in phytoplankton (Jones *et al.* 1993, 1995, McKenzie *et al.* 1995, Legrand *et al.* submitted), although, as was shown above, for some phagotrophic species grazing is independent of light. There may be a stimulation of either osmotrophy (Pintner and Provasoli 1968, Rivkin and Putt 1987) or phagotrophy (Bird and Kalff 1986, 1987, Jones *et al.*, 1993, 1995, Legrand *et al.* submitted) in low light/darkness. The opposite has also been found, however, e.g. inhibition of bacterivory in a freshwater chrysophyte exposed to darkness or low light (Caron *et al.* 1993). Maximum ingestion rates and high rates of photosynthesis occurred simultaneously at 40 µmol m^{-2} s^{-1} for the chrysophyte *P. malhamensis* (Porter 1988). The dinoflagellate *Heterocapsa triquetra* significantly increased phagotrophy in darkness relative to cells grown in light (100 µmol m^{-1} s^{-1}) (Legrand *et al.* submitted). This dinoflagellate ingested radioactively- and fluorescently-labelled algal cells (the cyanobacterium *Synechococcus* (=1µm), a phototrophic flagellate (=3µm) and the diatom *Thalassiosira pseudonana* (=6 µm) (Figs 4a,b). Jones *et al.* (1993) tested 16 species of the prymnesiophyte genus *Chrysochromulina* for phagotrophy under low and high light, and with live and dead prey. Only 5 among the 16 species did not become phagotrophic, at least in response to some of the offered prey. Among the phagotrophic species was the toxic prymnesiophyte *C. polylepis.* The authors found that *C. brevifilum* showed an inverse relation between phagotrophy and light, which was re-confirmed in further experiments (Jones *et al.* 1993). Not only did phagotrophy increase at low light intensity, but the number of *C. brevifilum* cells grown at low light (45 µmol m^{-2} s^{-1}) was much higher when algal prey (a small green flagellate) was provided as food (Fig. 5). This is an example where phagotrophy seems to have substituted for photosynthesis as a source of organic carbon at low light intensities.

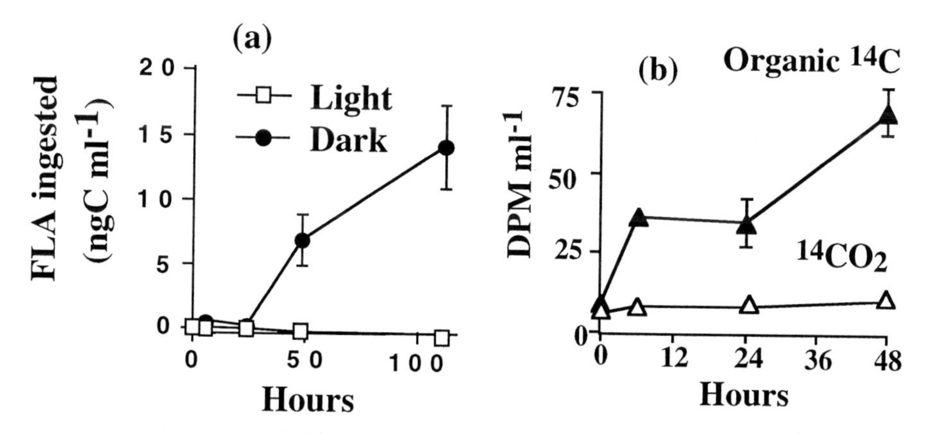

Fig. 4. Ingestion of fluorescently (FLA) (a) or radioactively- (RLA) (4 b) -labelled algae by *Heterocapsa triquetra* at a light intensity of 100 µmol m^{-2} s^{-1} and in darkness. (Error bars, standard error.) (Fig. 4a. Redrawn from Legrand *et al.* 1996) (Fig. 4b. Redrawn from Legrand *et al.*, submitted)

Chlorophyll concentrations usually increase in phytoplankton cells exposed to low light as chloroplasts become larger and more numerous compared to cells growing under high light (Coats and Harding 1988). As a consequence, the photosynthetic efficiency increases, as was found e.g. for the red-tide dinoflagellate *Prorocentrum marie-lebouriae* (=*P. minimum*) (Harding 1988) and for the mixotrophic prymnesiophyte *C. brevifilum* (Jones *et al.* 1995). For the latter species, prey ingestion also increased at low light intensity (Jones *et al.* 1995), although prey concentrations had no effect on ingestion rates. Increases in chlorophyll per cell and phagotrophy can thus be adaptations to low light, increasing the availability of organic carbon sources for grow and respiration.

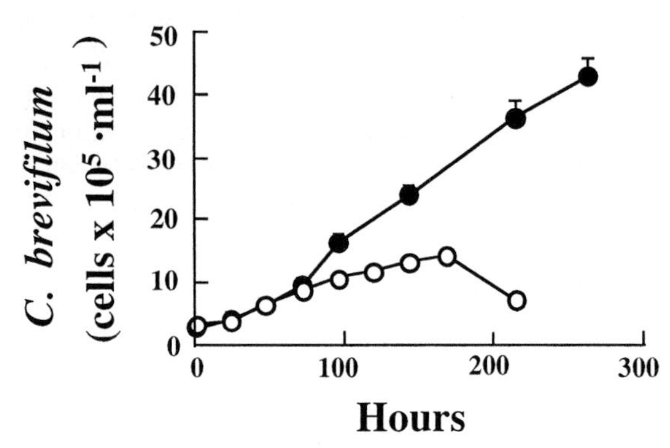

Fig. 5. Growth of *Chrysochromulina brevifilum* at a light intensity of 45 µmol m^{-2} s^{1} either with (●) or without (O) a food source. (Error bars, standard error) (Redrawn from Jones *et al.* 1993)

4.2 Macronutrients

Nitrogen (N) and phosphorus (P), the most important inorganic nutrients for algal growth, are often in low concentrations in marine waters. Some phytoplankton species that are not good competitors for the limiting nutrient at low concentrations (see Schöllhorn and Granéli 1995) may have retained or re-developed the capacity to prey on other planktonic organisms to obtain the required nutrients. The ability to use an alternative nutrient source may enable these species to coexist with, or outcompete, species that rely strictly on dissolved nutrients (Rothhaupt 1996b).

An increase in bacterivory in several species of toxic flagellates was observed when the cells were grown in P-deficient medium (Nygaard and Tobiesen 1993). Ingestion of smaller algae by the toxic prymnesiophyte *C. polylepis* increased during the last hours of incubation under P-deficiency while the larger algae were ingested only during the first hours of incubation (Legrand *et al.* 1996). Using a flow cytometer, we found that *C. polylepis* ingested more prey (FLA) under N-deficient conditions than under nutrient sufficient or P-deficient conditions. However, the degree of phagotrophy of *C. polylepis* was several fold lower than that found by Legrand *et al.* (1996). An explanation seems to be that flow cytometric detection of ingested FLAs underestimated microscope-based phagotrophy, as confirmed during an exercise during the NATO ASI workshop in Bermuda performed by T. Cucci, C. Legrand and M Sieracki.

Nygaard and Tobiesen (1993) concluded that ingestion of bacteria, which are rich in P, may be a mechanism for phagotrophic dinoflagellates/flagellates to obtain this critical element. Rothhaupt (1996a) showed that the chrysophyte *Ochromonas* sp. switched between uptake and excretion of SRP (soluble reactive phosphorus) when the phagotrophic species to coexist with phototrophic species, which must rely on SRP or dissolved organic P (Rothhaupt 1996b). Although the chrysophyte *Uroglena* sp. depends on the assimilation of inorganic P through phototrophy, it is also able to take up phospholipids via ingestion of bacteria (Kimura and Ishida 1985). According to Keller *et al.* (1994), phagotrophy in *Ochromonas sp.* is used to supplement its nutrition when light or inorganic nutrients are limiting (Fig. 6). The same authors found that phagotrophy on bacteria by the phytoflagellate *Ochromonas* sp. increased when light and nutrients became limiting (Fig. 6). However, there was an apparent threshold of response, as phagotrophy decreased or even ceased after periods of prolonged darkness.

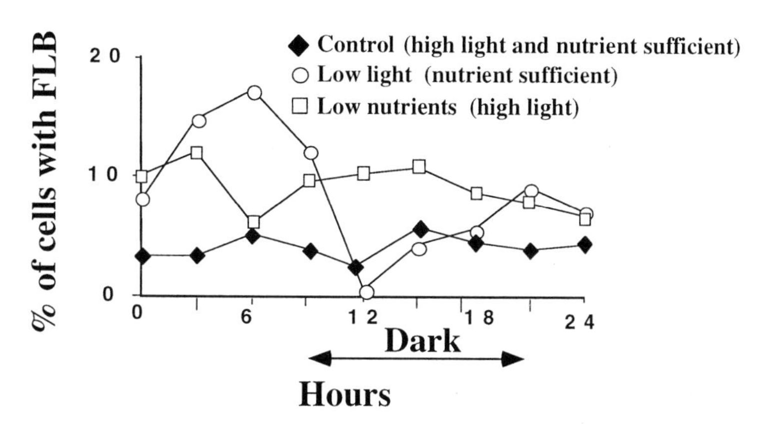

Fig. 6. 24-hour time course of FLB uptake by *Ochromonas* cells. Subsamples from unlabelled cultures were taken every 3 h and FLB were added. Samples were incubated for 30 min. and then FLB uptake was assessed. (Redrawn from Keller *et al.* 1994)

The freshwater chrysophyte *Dinobryon cylindricum* was shown to supply 25% of its C, N and P requirements through ingestion of bacteria (Caron *et al.* 1993).

Bird and Kalff (1989) estimated that 50% of the C- demand by *Dinobryon sertularia* was met by phagotrophy. Assuming an assimilation efficiency of 60% for C (Calow 1977) and close to 100% for N, Bird and Kalff (1989) estimated that nearly 100% of the N demand by *Dinobryon* could be met by phagotrophy. For the marine red-tide dinoflagellate *Gymnodinium sanguineum*, Bockstahler and Coats (1993 b) calculated that natural populations of this algae in the Chesapeake Bay were able to balance nitrogen requirement by ingesting small nanociliates (< 20 μm).

5. Phagotrophy among toxic/harmful species

5.1. Occurrence and regulation

Phagotrophy has been found in several harmful marine phytoplankton species. The toxic species *Chrysochromulina polylepis, Heterosigma akashiwo* and *Alexandrium tamarense* were shown by Nygaard and Tobiesen (1993) to ingest radioactively-labelled bacteria at high rates, indicating that phagotrophy may be an important mode of nutrition for these toxic algae. Most of the eight investigated species responded within minutes to the addition of radioactively labelled bacteria, with only the dinoflagellate *Gyrodinium galateanum* showing a delayed response.

Chrysochromulina polylepis was grown under N- or P- deficiency and under nutrient sufficient conditions (Legrand *et al.* 1996). The disappearance of FLA from the medium in P-deficient treatments was appreciably higher than for other treatments. Thus *C. polylepis* cells might have been using FLA as a P source. The same authors found that phagotrophy in *C. polylepis* accounted for a maximum of 5% of the carbon intake, but since dead cells were presented as prey, the ingestion rates may have been significantly underestimated. Legrand *et al.* (submitted) showed that the carbon provided by the ingestion of fluorescently-labelled flagellates by *Heterocapsa triquetra* in darkness could have supported 0.7 divisions per day, more than half of the maximum reported for autotrophic growth. However, *H. triquetra* did not divide but rather increased in size (C-content).

Until recently, only the non-pigmented *Dinophysis rotundata* among the toxic (DSP) *Dinophysis* species has been found to prey on other cells (a ciliate; Hansen 1991). However, the toxic, pigmented *D. norvegica* and *D. acuta* have been shown to be capable of mixotrophy in darkness (Granéli *et al* in press; Fig. 7). In this study it was not possible to differentiate between mixotrophy from ingestion of (radioactively-labelled) dissolved organic substances, versus phagotrophy on radioactively-labelled prey. Y. Fukuyo has shown photographs of a *D. acuminata* cell with a cell of *Cryptomonas* sp. inside (pers. communication). Food vacuoles were observed in *D. norvegica* and *D. acuminata* by Jacobson and Andersen (1994) using electron microscopy. Although these authors could not identify the prey, the resemblance of the particles to those in the obligate heterotrophic *Oxyphysis oxytoxoides* suggests that *D. acuminata* was preying on ciliates.

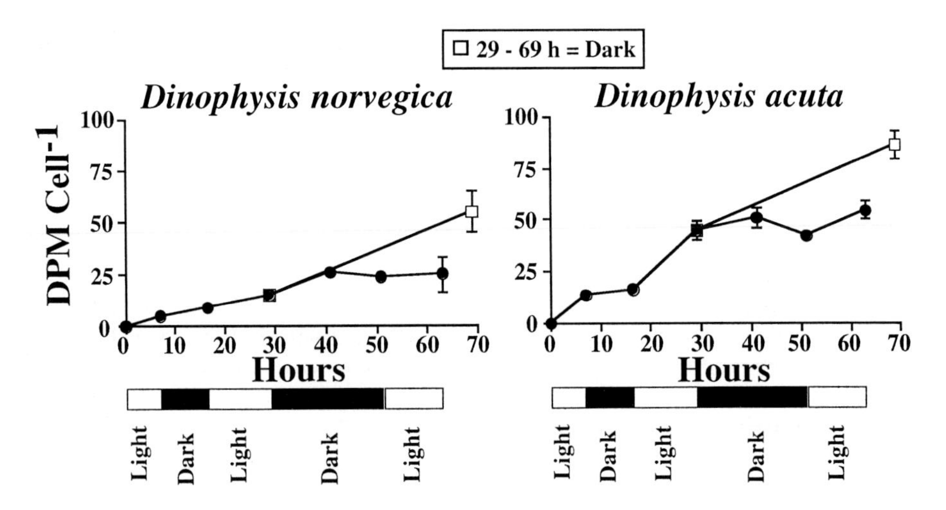

Fig. 7. ^{14}C uptake by *Dinophysis norvegica* and *D. acuta* cells exposed to light and dark periods (filled circles). Empty squares: cells that were incubated in complete dark from the 29 to 69 h. (Redrawn from Granéli *et al.*, 1997).

During summer months *D. norvegica* was in high concentrations at 12-15 m depths in the Baltic Sea (Carpenter *et al.* 1995). Measurements of photosynthesis showed that the cells would take 4 to 11 months to double their cellular carbon content (0.004 to 0.001 doublings day^{-1}) if cells depended on inorganic carbon fixation only. The *in situ* growth rate for this species has been estimated to be 0.2 to 0.7 divisions day^{-1} (Subba Rao *et al.* 1993. Granéli *et al.,* 1995, 1997). Mixotrophy seems to be the only explanation for *D. norvegica* growth (Carpenter *et al.* 1995), as photosynthesis was not providing enough carbon to support reasonable growth rates. In the Kattegat, *D. norvegica* increased from 700-1000 cells l^{-1} by the beginning of October to 3500-9000 cells l^{-1} in the beginning of December (1-5 m depth integrated water samples, P. Olsson, pers. comm) when light had decreased to 5 μmol m^{-2} s^{-1}, corroborating the assumption of mixotrophy for this species at low light/darkness.

Heterotrophy of *Dinophysis* species found by Granéli *et al.* (1997) might also help to explain previous lack of success in cultivating these algae. *Dinophysis* may need some compound found only in dissolved organic substances or in organic particles to support sustained growth, whereas only few divisions were possible in attempts to culture them in inorganic media (Sampayo 1993, Maestrini *et al.*1995).

Little is known about the connection between mixotrophy and toxin production. Okadaic acid (OA) concentrations in cells of *Prorocentrum lima* were higher (11.2-14.2 pg OA cell^{-1}) when organic P(glycerol-PO_4) was supplied instead of inorganic P (7.5-8.9 pg OA cell^{-1}) (Tomas and Baden 1993). Toxin concentration (OA) in *D. acuminata* was related to occurrence of inclusion bodies (Legrand *et al.* unpublished) (Fig. 8). These cells had been growing under nutrient sufficient and N- or P-deficient conditions. Under N-deficient conditions, 20-40% of *D. acuminata* and *D. acuta* cells contained 1 to 5 inclusion bodies while 10% in the other two treatments (P-deficient or nutrient sufficient conditions). Nitrogen deficient cells increased their toxin content from 1 to 26 pg OA cell^{-1}, whereas in the nutrient sufficient treatments there was a suppression of toxin production. It was not possible, however, to distinguish if the higher toxin levels found in the N-deficient treatment were due to the ingestion of prey or due to nutrient deficiency *per se*. Of course nutrient deficiency may be the triggering cause of phagotrophy.

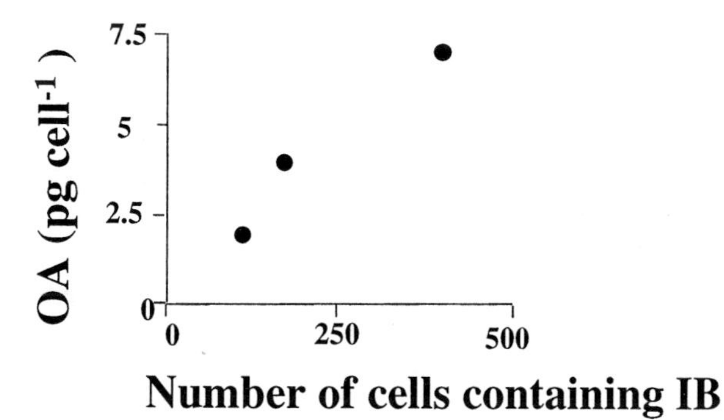

Fig. 8. Relation between okadaic acid (OA) and *Dinophysis acuminata* cells containing at least one inclusion body (IB) (Legrand *et al.* unpublished)

6. The ecological significance of phagotrophy

Until recently phytoplankton were regarded as strictly phototrophic, with a well defined position at the base of the food webs of lakes and oceans. Recently we have learned that the nutritional demands of a growing number of phytoplankton species appears to have the potential to be met at least partly through heterotrophic nutrition (Rothhaupt 1996b). This finding has direct implications for our view of algal survival strategies, and also for the ability of phagotrophic species to out-compete other algae under less favourable growth conditions, e.g. in waters poor in inorganic nutrients or under low light. It also affects the traditional view of the "microbial loop" (Azam *et al.* 1983), where DOC is thought to be channelled from algal photosynthesis to bacteria and then up the food chain through heterotrophic flagellates, ciliates and mesozooplankton. Are phagotrophic phytoplankton that feed on bacteria "taking back" some of the lost (excreted) photosynthetic DOC? How can we estimate the fluxes of carbon and nutrients between different trophic levels in the plankton food web when phagotrophic algae are involved?

Phagotrophy has been considered a significant grazing process with respect to ecosystem loss rates. Phagotrophic C uptake as a percentage of total (photosynthesis plus phagotrophy) C incorporation varies widely in the literature, from a few percent (Tranvik *et al.* 1989, Legrand *et al.* 1996) to more than 50% (Porter 1988, Bird and Kalff 1989, Caron *et al.* 1993). This variation can occur within the same algal group (genus). Bird and Kalff (1989) showed that C assimilation by algal phagotrophy (chrysomonads) in the deeper waters (below 6 m) of Lac Gilbert can be equal to or higher than photosynthetic fixation. Tranvik *et al.* (1989), however, reported that bacterivory by *Cryptomonas* sp. was not an important C source for this phytoflagellate, representing not more than 2% of the C intake.

Control of bacteria in lakes and marine waters has, until recently, been attributed mostly to heterotrophic flagellates. Some phytoplankton species have been shown to have similar or sometimes higher grazing rates compared to heterotrophic flagellates, controlling not only bacteria but even heterotrophic organisms of the same size as the algae themselves (Sanders and Porter 1988, Bockstaller and Coats 1993 a, b, Hansen and Nielsen submitted).

Phagotrophic photoflagellates are reported to be responsible for a major part of the grazing pressure on bacteria in some lakes (Bird and Kalff 1986, Berningher *et al.* 1992). Sanders and Porter (1988) reported clearance rates between 2.5-8.4 nl hr^{-1} for photoflagellates in Oglethorpe Lake. Similar values were found by Nygaard and Tobiesen (1993) for marine flagellate/dinoflagellate species such as *Chrysochromulina polylepis.* The occurrence of food vacuoles in *Gymnodinium sanguineum* was positively correlated to ciliate densities (Bockstahler and Coats 1993b). The daily removal of these nanociliates by *G. sanguineum* accounted for 6 to 67% of the <20 µm ciliate standing stock (Bockstahler and Coats 1993a).

These contradictory results show that the contribution by phagotrophy of mixotrophic algae to the control of biomass of bacteria or other plankton organisms is highly variable, depending on the algal species and the environmental conditions in which the algae is living. However, it is clear that losses due to pigmented flagellates must be considered on grazing estimates of bacteria, algae and microzooplankton in lakes or marine waters.

7. Summary

Phagotrophy is widespread in certain groups of photosynthetic organisms, especially among various phytoflagellates including several important toxic or harmful species, such as *Chrysochromulina polylepis, Pfisteria piscicida , Heterosigma akashiwo, Alexandrium tamarense, Gymnodinium galateanum* and *Heterocapsa triquetra* . Phagotrophy most probably occurs in the *Dinophysis* species responsible for diarrhetic shellfish poisoning (DSP). Phagotrophic algae have been shown to prey on bacteria, other algae, and even microzooplankton. Phagotrophy may substitute for photosynthesis, and thus may be an alternative way of acquiring reduced carbon. Phagotrophy may also enable the organism to obtain macronutrients (P, N), micronutrients, vitamins or other organic substances that the organism cannot synthesise itself. Both low light and nutrient deficiency have been shown to promote phagotrophy, with grazing rate generally dependent on prey concentration. However, some phagotrophic algae graze independent of light conditions. There is a wide range in degree of mixotrophy, from species that can only supplement their nutrition with phagotrophy to species that are able to grow phagotrophically in complete darkness. Mixotrophic organisms may have a competitive advantage over strict heterotrophs or strict photoautotrophs under specific environmental conditions. However, there is most likely a cost attached to being mixotrophic that makes phagotrophic algae photosynthetically less efficient than obligate autotrophs and less efficient grazers than heterotrophic unicellular organisms. Under certain conditions phagotrophic algae can be the primary bacterivores of the microbial food web. Phagotrophy among algae was "rediscovered" only about 10 years ago, and the ecological significance of this mode of nutrition for the organisms themselves, as well as for the plankton ecosystem, is still poorly known.

8. Acknowledgments

Financial support from NATO, SCOR, the Swedish Natural Science Research Council (NFR), the Swedish Board for Agricultural and Forestry Research and the Swedish Environmental Protection Agency (SNV) is gratefully acknowledged. We also thank C. Legrand for the permission to quote her unpublished data.

9. References

Aaronson, S. (1974). The biology and ultrastructure of phagotrophy in *Ochromonas danica* (Chrysophyceae: Chrysomonadida). *J. Microbiol.* 83: 21-9.

Andersson, A., Falk, S., Samuelsson, G., Hagström, Å. (1989). Nutritional characteristics of mixotrophic nanoflagellate, *Ochromonas* sp. *Microb. Ecol.* 17: 251-262.

Azam, F., Fenchel, T., Field, J. G., Gray, J. S., Meyer-Reil, L. A., Thingstad, F. (1983). The ecological role of water-column microbes in the sea. *Mar. Ecol. Prog. Ser.* 10: 257-263.

Berland, B.R., Maestrini, S.Y., Grzebyk, D. (1995). Observations on possible life cycle stages of the dinoflagellate *Dinophysis* cf. *acuminata*, *Dinophysis acuta* and *Dinophysis pavillardi*. *Aquat. Microb. Ecol.* 9: 183-189.

Berninger, U. G., Caron, D. A., Sanders, R. W. (1992). Mixotrophic algae in three ice-covered lakes of the Poconc Mountains, U.S.A. *Freshwater Biol.* 28: 263-272.

Bird, D. F., Kalff, J. (1986). Bacterial grazing by planktonic lake algae. *Science* 231: 493-495.

Bird, D. F., Kalff, J. (1987). Algal phagotrophy: regulating factors and importance relative to photosynthesis in *Dinobryon* (Chrysophyceae). *Limnol Oceanogr* 32: 277-284.

Bird, D. F., Kalff, J. (1989). Phagotrophic sustenance of a metalimnetic phytoplankton peak. *Limnol. Oceanogr.* 34: 155-162.

Bockstahler, K. R., Coats, D. W. (1993 a). Grazing of the mixotrophic dinoflagellate *Gymnodinium sanguineum* on ciliate populations of Chesapeake Bay. *Mar. Biol.* 116: 477-487.

Bockstahler, K. R., Coats, D. W. (1993 b). Spatial and temporal aspects of mixotrophy in Chesapeake Bay dinoflagellates. *J. Euk. Microbiol.* 40: 46-60.

Boraas, M. E., Estep, K. W., Johnson, P. W., Sieburth, J. M. (1988). Phagotrophic phototrophs: the ecological significance of Mixotrophy. *J. Protozool* 35: 249-252.

Calow, P. (1977). Conversion efficiencies in heterotrophic organisms. *Biol. Rev.* 52: 385-409.

Caron, D. A., Porter, K. G., Sanders, R. W. (1990). Carbon, nitrogen and phosphorus budgets for the mixotrophic phytoflagellate *Poterioochromonas malhamensis* (Chrysophyceae) during bacterial ingestion. *Limnol. Oceanogr.* 35: 433-443.

Caron, D. A., Sanders, R. W., Lim, E.L. (1993). Light-dependent phagotrophy in the freshwater mixotrophic Chrysophyte *Dinobryon cylindricum. Microb. Ecol.* 25: 93-111.

Carpenter, E. J., Jansson, S., Boje, R., Pollehne, F., Chang, J. (1995).The dinoflagellate *Dinophysis norvegica* : biological and ecological observations in the Baltic Sea. *Eur. J. Phycol.* 30: 1-9.

Chang, F., Carpenter E. J. (1994). Inclusion bodies in several species of Ceratium Schrank (Dinophyceae) from the Caribbean Sea examined with DNA-specific staining. *J. Plankton Res.* 16: 197-202.

Coats, D. W., Harding, L. W. (1988). Effect of light history on the ultrastructure ansd physiology of *Prorocentrum mariae-lebouriae* (Dinophyceae). *J. Phycol.* 24: 8-14.

Gaines, G., Elbrächter, M. (1987). Heterotrophic nutrition. In: Taylor, F. J. R. (ed.). *The Biology of Dinoflagellates*. Blackwell Scientific Publications, Oxford. pp. 224-268.

Granéli, E., Anderson, D. M., Carlsson, P., Finenko, G., Maestrini, S. Y. (1997). Light and dark carbon fixation by *Dinophysis*-species in comparison to other photosynthetic and heterotrophic dinoflagellates . *Aquat. Microb. Ecol.*

Granéli, E., Anderson, D. M., Carlsson, P., Finenko, G., Maestrini, S. Y., Sampayo, M. A. de M., Smayda, T. M. (1995) . Nutrition, growth rate and sensibility to grazing for the dinoflagellates *Dinophysis acuminata, D. acuta* and *D. norvegica. La Mer* 33: 81-88.

Hansen, P. J. (1991). Dinophysis - a planktonic dinoflagellate genus which can act both as prey and a predator of a ciliate. *Mar. Ecol. Prog. Ser.* 69: 201-204.

Hansen, P. J., Nielsen, T. G. (submitted). Mixotrophic feeding of *Fragilidium subglobosum* (Dinophyceae) on *Ceratium* spp: effects of prey concentration, prey species and light intensity. *J. Plankton Res.*

Harding, L. W. (1988). The time-course of photoadaptation to low-light in *Prorocentrum mariae-lebouriae* (Dinophyceae). *J. Phycol.* 24: 274-281.

Havskum, H., Riemann , B. (1996). Ecological importance of bacterivorous, pigmented flagellates (mixotrophs) in the Bay of Aarhus.*Mar. Ecol. Prog. Ser.* 137: 251-263.

Hofeneder, H. (1930). Über die animalische Ernährung von *Ceratium hirundinella* O.F. Muller und über die Rolle des kernes bei dieser Zellfunktion. *Arch. Protistenkd.* 71: 1-32.

Jacobson, D. M., Andersen, R. A. (1994). The discovery of mixotrophy in photosynthetic species of *Dinophysis* (Dinophyceae): light end electron microscopical observations of food vacuoles in *Dinophysis acuminata*, *D. norvegica* and two heterotrophic dinophysoid dinoflagellates. *Phycologia* 33:97-110.

Jacobson, D. M., Anderson, D. M. (1996). Widespread phagocytosis of ciliates and other protists by marine mixotrophic and heterotrophic thecate dinoflagellates. *J. Phycol.* 32, 279-285.

Jeong, H. J., Latz, M. I. (1994). Growth and grazing rates of the heterotrophic dinoflagellates *Protoperidinium* spp. on red tide dinoflagellates. *Mar. Ecol. Prog. Ser.* 106: 173-185.

Jones, R. I. (1994). Mixotrophy in planktonic protists as a spectrum of nutritional strategies. *Mar. Microb. Food. Webs* 8: 87-96.

Jones, R. I., Rees, S. (1994). Characteristics of particle uptake by the phagotrophic phytoflagellate, Dinobryon divergens . *Mar. Microb. Food Webs* 8: 97-110.

Jones: H. L. J., Leadbeater: B. S. C., Green: J. C. (1993). Mixotrophy in marine species of *Chrysochromulina* (Prymnesiophyceae): ingestion and digestion of a small green flagellate. *J. mar. biol. Ass. U.K.* 73: 283-296.

Jones, H. L. J., Durjun, P., Leadbeater, B. S. C., Green, J. C. (1995). The relationship between photoacclimation and phagotrophy with respect to chlorophyll a, carbon and nitrogen content, and cell size of *Chrysochromulina brevifilum* (Prymnesiophyceae). *Phycologia* 34: 128-134.

Keller, M. D., Shapiro, L. P., Haugen, E. M., Cucci, T. L., Sherr, E. B. (1994). Phagotrophy of fluorescently labelled bacteria by an oceanic phytoplankter. *Microb. Ecol* 28: 39-52.

Kimura, B., Ishida, Y. (1985). Photophagotrophy in *Uroglena americana*, Chrysophyceae. *Jpn. J. Limn.* 46: 315-318.

Legrand, C., Granéli E., Carlsson P. (submitted). Induced mixotrophy in the photosynthetic dinoflagellate *Heterocapsa triquetra. Aquat. Microb. Ecol.*

Legrand, C., Sæmundsdottir, S., Granéli E. (1996). Phagotrophy in *Chrysochromulina polylepis* (Prymnesiophyceae): ingestion of fluorescent labelled algae (FLA) under different nutrient conditions. In: T. Yasumoto, Oshima Y., Fukuyo Y. (eds.), *Harmful and Toxic Algal Blooms.* Intergovernmental Oceanographic Commission of UNESCO, pp. 339-342.

Lessard, E. J., Swift, E. (1985). Species-specific grazing rates of heterotrophic dinoflagellates in oceanic waters, measured with a dual-label radioisotope technique. *Mar. Biol.* 87: 289-296.

Li, A., Stoecker, D. K., Coats, W., Adam, E. J. (1996). Ingestion of fluorescently labelled and phycoerythrin-containing prey by mixotrophic dinoflagellates. *Aquat. Microb. Ecol.* 10: 139-147.

Maestrini, S. Y., Berland, B. R., Carlsson, P., Granéli, E., Pastoureaud, A. (1996). Recent advances in the biology and ecology of the toxic dinoflagellate genus *Dinophysis*: the enigma continues. In: T. Yasumoto, Oshima Y., Fukuyo Y. (eds.), *Harmful and Toxic Algal Blooms.* Intergovernmental Oceanographic Commission of UNESCO, pp. 339-342.

Maestrini, S. Y., Berland, BR, Grzebyk, D, Spano, A-M (1995). *Dinophysis* spp. cells concentrated from nature for experimental purposes, using size fractionation and reverse migration. *Aquat Microb Ecol.* 9:177-182

McKenzie, C. H., Deibel, D., Paranjape, M. A., Thompson, R. J. (1995). The marine mixotroph *Dinobryon balticum* (Chrysophyceae): phagotrophy and survival in a cold ocean. *J. Phycol.* 31: 19-24.

Nygaard, K., Hessen, D. O. (1990). Use of ^{14}C-protein-labelled bacteria for estimating clearance rates by heterotrophic and mixotrophic flagellates. *Mar. Ecol. Prog. Ser.* 68: 7-14.

Nygaard, K., Tobiesen, A. (1993). Bacterivory in algae: a survival strategy during nutrient limitation. *Limnol. Oceanogr.* 38: 273-279.

Pintner, I. J., Provasoli, L. (1968). Heterotrophy in subdued light of 3 *Chrysochromulina* species. *Bull. Misaki Mar. Biol. Inst.,* Kyoto Univ., Proceedings of the U.S.-Japan seminar on Marine Microbiology, 12: 25-31.

Porter, K. G.(1988). Phagotrophic phytoflagellates in microbial food webs. *Hydrobiologica* 159: 89-97.

Porter, K. G., Sherr, E. B., Sherr, B. F., Pace, M., Sanders, R. W. (1985). Protozoa in planktonic food webs. *J. Protozool.* 32: 409-415.

Rivkin, R.B., Putt M. (1987). Heterotrophy and photoheterotrophy by Antarctic microalgae: light-dependent incorporation of amino acids and glucose. *J. Phycol.* 23: 442-452.

Rothhaupt, K. O. (1996 a). Laboratory experiments with a mixotrophic chrysophyte and obligately phagotrophic and phototrophic competitors. *Ecology* 77: 716-724.

Rothhaupt, K. O. (1996 b). Utilization of substitutable carbon and phosphorus sources by the mixotrophic chrysophyte *Ochromonas* sp. *Ecology.* pp. 706-715.

Rublee, P. A., Gallegos, C. L. (1989). Use of fluorescently labelled algae (FLA) to estimate microzooplankton grazing. *Mar. Ecol. Prog. Ser.* 51: 221-227.

Sampayo, M. A. de M. (1993). Trying to cultivate *Dinophysis* spp. In: T.J.Smayda, Y.Shimizu (eds.), *Toxic Phytoplankton Blooms in the Sea.* Elsevier Sci. Pub.., Amsterdam, pp. 807-810.

Sanders, R.W., Porter, K. G. (1988). Phagotrophic phytoflagellates. *Aquat. Microb.* Ecol. 10: 167-192.

Sanders, R.W., Porter, K. G., Caron, D. A. (1990). Relationship between phototrophy and phagotrophy in the mixotrophic chrysophyte *Poterioochromonas malhamensis* . *Microb. Ecol.* 19: 97-109.

Schnepf, E., Ellbrächter, M. (1992). Nutritional strategies in dinoflagellates: a review with emphasis on cell biological aspects. *Europ. J. Protistol.* 28: 3-24.

Schöllhorn, S., Granéli, E. (1996). Influence of different nitrogen to silica ratios and artificial mixing on the structure of a summer phytoplankton community from the Swedish West Coast (Gullmar Fjord). *J. Sea Res.*, 35: 159-167

Sherr, B. F., Sherr, E. B., McDaniel, J. (1991). Clearance rates of <6 μm fluorescently labeled algae (FLA) by estuarine protozoa: Potential grazing impact of flagellates and ciliates. *Mar. Ecol. Prog. Ser.* 69: 81-92.

Sherr, B. F., Sherr, E. B., Pedrós-Alió, C. (1990). Simultaneous measurement of bacterioplankton production and protozoan bacterivory in estuarine water. *Mar. Biol. Prog. Ser.* 54: 209-219.

Subba Rao, D. V. and Pan, Y., (1993). Diarrhetic shellfish poisoning (DSP) associated with a subsurface bloom of *Dinophysis norvegica* in Bedford Basin, eastern Canada. *Mar. Ecol .Prog .Ser.* 97: 117-126.

Thingstad, TF., Havskum, H., Garde, K., Riemann, B. (1996). On the strategy of "eating your competitor". A mathematical analysis of algal mixotrophy. *Ecology* 77: 39-49

Tomas, C. R., Baden, D. G. (1993). The influence of phosphorus source on the growth and cellular toxin content of the benthic dinoflagellate *Prorocentrum lima*. In: T.J.Smayda, Y.Shimizu (eds.), *Toxic Phytoplankton Blooms in the Sea.* Elsevier Sci. Pub., Amsterdam, pp. 565-570.

Tranvik, L., Porter, K. G., Sieburth, J. M. (1989). Occurrence of bacterivory in *Cryptomonas* , a common freshwater phytoplankter. *Oecologia.* 78: 473-476.

Walker, L. M. (1984). Life histories, dispersal and survival in marine planktonic dinoflagellates. In: K.A Steidinger, L.M. Walker (eds.), *Marine Plankton Life Strategies,* CRC Press.19-34.

Behavior, Physiology and the Niche of Depth-Regulating Phytoplankton

John J. Cullen and J. Geoffrey MacIntyre

Center for Environmental Observation Technology and Research, Department of Oceanography, Dalhousie University, Halifax, Nova Scotia, B3H 4J1, Canada

1. Introduction

To understand the population dynamics of phytoplankton species contributing to harmful algal blooms (HABs), we focus on physiological, biochemical, and behavioral features of phytoplankton that might influence ecological selection in different hydrographic regimes. Each species of phytoplankton has a different combination of adaptive characteristics that in many ways defines its niche, *i.e.*, the suite of ecological factors that determines its distribution and activities. By understanding the adaptations of different phytoplankton species, it should be possible to describe patterns of species abundance as functions of hydrographic processes, nutrient distributions and the potential impact of herbivores. This understanding could be applied to determine the principal causes of HABs (*e.g.*, eutrophication, upwelling, stratification, advection), to achieve better skill at forecasting their occurrence, and to predict the consequences for phytoplankton of environmental changes in coastal waters, such as might be associated with aquaculture, coastal engineering, or climate change. Because the ecological and commercial effects of algal blooms depend on the particular species involved, it is essential to focus on characteristics of phytoplankton that are likely to distinguish one species from another with respect to ecological selection during development and maintenance of a bloom.

Recognizing that many harmful algae can move vertically, we discuss depth regulation in phytoplankton as an ecological strategy, emphasizing physiological and biochemical adaptations that can strongly influence the growth and survival of phytoplankton species in different hydrographic regimes. We identify adaptive features that can be quantified experimentally, so that species can be compared and predictive models can be developed. This discussion is focused on marine systems; a comprehensive presentation, including a thorough review of the extensive literature on freshwater cyanobacteria (Zevenboom 1986; Oliver 1994) is beyond the scope of this chapter.

2. Hydrographic regimes and corresponding strategies for survival

Without denying the relevance of physiological adaptations to light, temperature, or nutrients, Margalef (1978) suggested that, with respect to ecological selection of phytoplankton species, the most important factor is the mechanical energy of the water column: "Water movement controls plankton communities." Because nutrients are incorporated into particles, and particles tend to sink, the fate of a permanently stratified water column with no input of mechanical energy is complete segregation of

NATO ASI Series, Vol. G 41
Physiological Ecology of Harmful Algal Blooms
Edited by D. M. Anderson, A. D. Cembella and G. M. Hallegraeff

the factors necessary for the growth of phytoplankton: light, but no nutrients near the surface; nutrients, but no light at depth. The productivity of such systems is minimal. Mechanical energy from winds, tides, and some types of currents displaces water and delivers nutrients to the photic zone through mixing and upwelling, leading to enhanced productivity (Yentsch 1980; Kiørboe 1993). There are limits, however: deep vertical mixing reduces mean irradiance for surface-layer phytoplankton, thereby diminishing primary productivity (Sverdrup 1953); in other places, upwelling can be too intense for the assimilatory responses of phytoplankton to keep pace (Dugdale and Wilkerson 1989). The phytoplankton that survive under specific regimes of turbulence have developed "adaptation syndromes...to recurrent patterns of selective factors" (Fig. 1; Margalef 1978). The dominant selective factor is turbulence. A principal mode of action is the delivery of nutrients to phytoplankton cells (Kiørboe 1993).

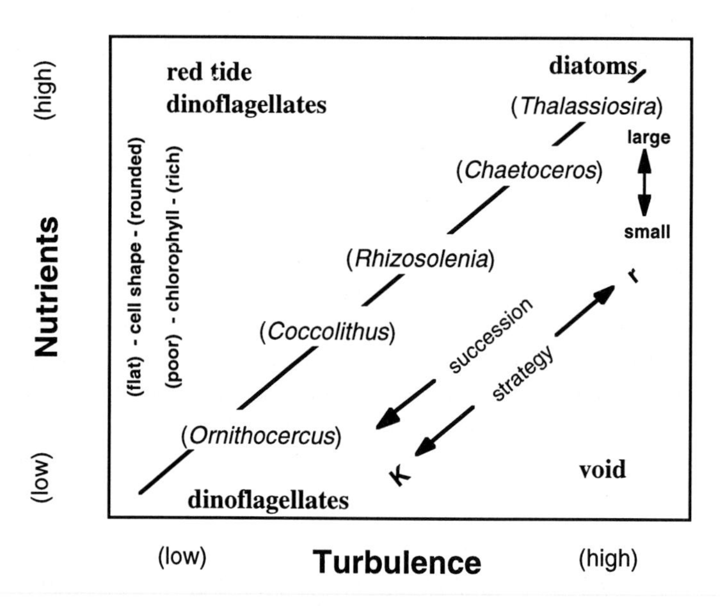

Fig. 1. Margalef's (1978) view of how life-forms of phytoplankton, represented by exemplary species, correspond to different regimes of turbulence and availability of nutrients. The trends in cell size, shape, pigment content, and ecological strategy are consistent with a strong influence of nutrient availability on selection. Depth regulation and vertical migration were not explicitly considered, although the dominance of red-tide dinoflagellates in low-turbulence, high-nutrient waters can be associated with swimming behavior (redrawn from Margalef, 1978).

Accepting that life forms of phytoplankton species reflect in large part alternative strategies for survival in different regimes of turbulence (Kiørboe 1993; Estrada and Berdalet, this volume), we describe in this chapter how our understanding of niche space defined by turbulence regimes is expanded when strategies of depth regulation by phytoplankton are considered. Some adaptations of depth-regulating phytoplankton are identified, and an attempt is made to describe adaptation syndromes that could enhance the success of particular phytoplankton species in particular regimes of turbulence or nutrients. One reason for doing this is to see if potentially harmful phytoplankton species have developed adaptations that would be selected for in response to changes in coastal waters, such as eutrophication (Smayda 1990; Hallegraeff 1993) or hydrographic shifts associated with climate change (Fraga and Bakun 1993; Tester *et al.* 1993).

2.1 Classification of adaptive strategies

Selective patterns of phytoplankton can be associated with water columns classified as high-energy, high-nutrients (deep vertical mixing or active upwelling), moderate-energy, moderate nutrients (transiently stratified water, nutricline well within the photic zone), and low-energy, low-nutrients (highly stratified, nutricline very deep in the photic zone or below it). The persistence of high nutrients in stratified marine waters can be viewed as an anomaly associated with external inputs of nutrients, conducive to dinoflagellate blooms (Fig. 1); because of coastal eutrophication, such conditions have become more common in recent decades (Smayda 1990).

With respect to depth regulation in different regimes of turbulence, there are three general strategies that require different combinations of physiological and behavioral adaptations (behavior is very loosely defined to include the regulation of buoyancy): mixing, migrating and layer-formation.

2.1.1 Mixers

In regimes of high turbulence, displacement of phytoplankton cells (m s^{-1}) is much greater than maximum sinking or swimming speeds (see Kamykowski *et al.*, this volume), thus, the phytoplankton ("mixers") go with the flow. Because elevated nutrients are generally associated with enhanced turbulence, mixers should have adaptations to exploit variable irradiance (Legendre and Demers 1984; Demers *et al.* 1991; Ibelings *et al.* 1994), rather than to acquire nutrients efficiently.

How important is depth regulation to mixers? In well-mixed surface layers (ignoring upwelling), losses due to sinking (l_s, d^{-1}) can be quantified: $l_s = w_s / z_m$, where w_s is sinking rate (m d^{-1}) and z_m (m) is mixed layer depth. Regulation of sinking rate is important to the extent that l_s makes a significant contribution to net growth rate (*i.e.*, the specific growth rate, μ (d^{-1}) minus specific loss rates to grazing, mixing and sinking). For example, a diatom with a sinking rate of 1 m d^{-1}, blooming in a mixed layer of 50 m, will not be affected much by sinking losses. See the review by Walsby and Reynolds (1980) for a comprehensive treatment.

When turbulence decays, for example when a mixed layer shoals and near-surface nutrients are depleted as phytoplankton grow and accumulate, sinking losses become more important (z_m decreases and μ decreases). One strategy for a mixer is capitulation by sinking from the system and entering a life-cycle that culminates in re-introduction to the surface layer when conditions are once again favorable (Smetacek 1985). This strategy could be facilitated by an increase in cell density in response to nutrient depletion (Richardson and Cullen 1995 and references therein) and also the production of mucilage, setae or microfibrils, which would promote aggregation, thereby accelerating the sinking rate by increasing the Stokes radius (Jackson 1990; Kiørboe 1993). For mixers, an alternative adaptation to decaying turbulence is to minimize sedimentation by reducing sinking rate when entering the nutrient-rich thermocline, thereby increasing the probability of resuspension into the mixed layer (Lande and Wood 1987). This strategic trend of nutrition-influenced vertical movement culminates in vertical migration, which is the next category.

2.1.2 Migrators

In lower-energy water columns, nutrients in the well-lighted surface layer can be depleted. By migrating vertically in such systems, phytoplankton can acquire nutrients at depths where light would limit growth rate, but also exploit saturating irradiance near the surface, where depletion of nutrients would otherwise restrict the accumulation of biomass (Holmes *et al.* 1967; Cullen 1985). Diel vertical migration (DVM) is the best studied and probably most common form of vertical migration (Blasco 1978; Heaney and Eppley 1981; Watanabe *et al.* 1991). However, cyclical vertical movements of phytoplankton likely occur over other temporal scales (Rivkin *et al.* 1984; Villareal *et al.* 1996). The environment of migrators is characterized by extreme, but in large part predictable variability in light and nutrients. Important adaptations for vertical migration include: the capacity to take up nutrients in low light or in the dark; photosynthetic physiology tuned to exploiting predictably varying irradiance; and behavioral responses (*i.e.*, changes in swimming behavior or cellular buoyancy) regulated effectively by environmental feedback (Richardson *et al.* 1996).

2.1.3 Layer-formers

Migration is but one adaptation for survival in low-energy water columns. Many species of phytoplankton segregate vertically in stratified waters, remaining in distinct strata (Venrick 1988). This layering can be described partly on the basis of population dynamics, *i.e.*, different local optima for net growth rates (Lande *et al.* 1989), but for some species, buoyancy regulation and active swimming certainly play a role. Some motile species aggregate and migrate vertically within a restricted range (Sommer 1985) but others aggregate and show little or no evidence of vertical migration during phases of their life cycles: examples include *Gyrodinium aureolum* (*Gymnodinium mikimotoi*) (Holligan 1978) and *Ceratium tripos* (Falkowski *et al.* 1980; Eppley *et al.* 1984). Such aggregating phytoplankton can be considered as layer-formers. They are found in environments characterized by maximum predictability of irradiance and nutrient supply. Efficiency of light-utilization would be an important selective factor for phytoplankton that form deep layers, whereas efficient nutrient utilization is very important for layer-formers that aggregate near the surface. Other adaptations to layer formation would include effective physiological controls on depth regulation, and the development of mechanisms to repel grazers (Turner *et al.*, this volume).

The three strategies — mixing, migrating, and layer formation are neither totally distinct nor relevant to all phytoplankton. These categories should be useful, however, for discussing adaptations of depth-regulating phytoplankton. In this chapter, we will focus on photosynthesis, nutrient utilization, behavior, and their interactions. Several other adaptive processes can be explored fruitfully without considering in detail the movements of phytoplankton in the water column. For example, special nutrient requirements (iron, vitamins, ammonium), allelopathy, mixotrophy, cyst formation and germination, as well as parasitism have all been studied in the context of HABs; they are discussed elsewhere in this volume.

3. Adaptations to environmental variability

Turbulence, solar radiation and depth-regulation behavior interact to determine the irradiance experienced by phytoplankton. Phytoplankton respond to changes in irradiance through a suite of adjustments called photoacclimation (Harding *et al.* 1987; Falkowski and LaRoche 1991). Adaptations of different taxa to irradiance have been compared in an analysis of strategies which revealed broad differences between dinoflagellates, diatoms, and green algae that were generally consistent with their patterns of dominance in aquatic environments (Richardson *et al.* 1983). In the hopes of finding key properties that might explain the dominance of phytoplankton species in particular hydrographic regimes, we build upon that analysis, looking at aspects of photoacclimation that might relate closely to strategies of depth regulation.

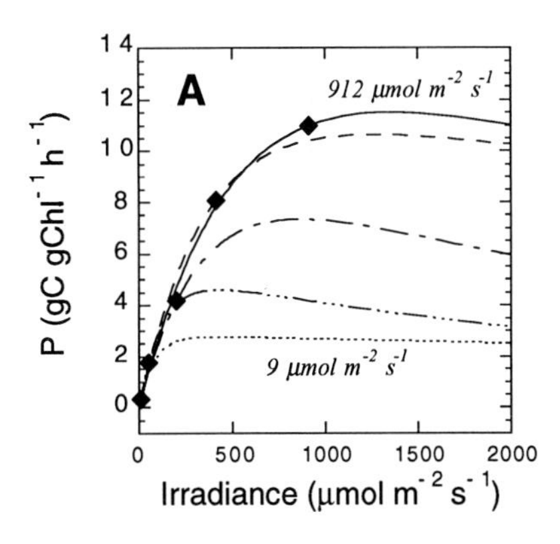

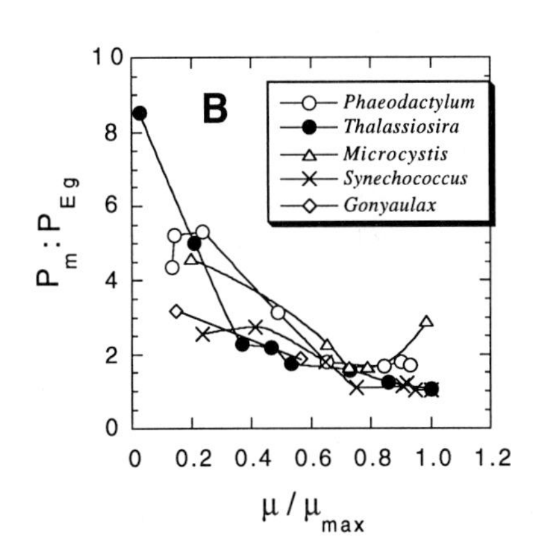

Fig. 2. Adaptations to exploit variable irradiance. A. Photosynthesis (gC $gChl^{-1}$ h^{-1}) vs E for *Thalassiosira pseudonana* (3H) in semi-continuous cultures adapted to different irradiance (9, 50, 200, 410, and 912 μmol m^{-2} s^{-1}; lowest to highest curves, respectively) during the photoperiod of a 12h:12h light:dark cycle at 20°C. Photosynthesis was measured as ^{14}C-bicarbonate uptake for 24 subsamples over 20 min (Cullen *et al.* 1989), fit to the model of Platt *et al.* (1980). Closed diamonds mark photosynthesis at growth irradiance for each culture, P_{Eg}. B. Excess photosynthetic capacity, P_m:P_{Eg}, as a function of light-limited growth rate, relative to the maximum at that temperature (μ / μ_{max}). Cultures of *T. pseudonana* from A (•), along with data presented by Kana and Glibert (1987b) for *Phaeodactylum tricornutum* (o), *Microcystis aeruginosa* (Δ), *Synechococcus* WH7803 (x), and *Alexandrium* (*Gonyaulax*) *tamarense* (◊).

3.1. Excess photochemical capacity

The relationship between photosynthesis and irradiance (*P-E*; here we refer to photosynthesis normalized to chlorophyll, P, gC $gChl^{-1}$ h^{-1}, but we omit the superscript B) is an adaptive feature of phytoplankton; it varies with growth irradiance (Fig. 2A, see also Falkowski 1980); nutrition (Cullen *et al.* 1992), and temperature (Li and Morris 1982; Maxwell *et al.* 1995). The *P-E* relationship can be characterized by an initial slope, α (gC $gChl^{-1}$ h^{-1} (μmol m^{-2} $s^{-1})^{-1}$), a maximum rate, P_m (gC $gChl^{-1}$ h^{-1}), and a term characterizing the susceptibility to photoinhibition (*e.g.*, β, gC $gChl^{-1}$ h^{-1} (μmol m^{-2} $s^{-1})^{-1}$; Platt *et al.* 1980). For cultures grown under constant irradiance (E_g)

during the photoperiod, the ratio of maximal to adapted rate of photosynthesis, P_m:P_{Eg} (where P_{Eg} is P at growth irradiance) is an index of excess photosynthetic capacity (Kana and Glibert 1987b), *i.e.*, the capacity to exploit irradiance greater than the mean. Excess photosynthetic capacity (EPC) represents genetically determined capability to exploit variable irradiance, especially exposures to bright light during vertical mixing.

The enzymes (principally ribulose bisphosphate carboxylase) and the reaction center proteins (Sukenik *et al.* 1987) required to maintain EPC represent a capital cost that carries with it a reduction in growth rate when irradiance does not vary much (Richardson *et al.* 1983). Hence, we would expect mixers to have maximal EPC, and layer-forming phytoplankton to have minimal excess capacity.

It is premature to generalize, but one can speculate that excess photochemical capacity indeed reflects depth-regulation strategy. A preliminary survey of available data (Fig. 2B; Kana and Glibert 1987b) suggests that the genetically determined photosynthetic responses of diatoms are consistent with what we would expect for mixers: high EPC for cultures grown in low irradiance. Two species have lower values of P_m/P_{Eg}: *Synechococcus* and *Alexandrium* (Fig. 2B). *Synechococcus* , too small to move very far in a day, seems adapted to stable water columns. *Alexandrium* can migrate vertically, but at least one strain forms a persistent layer near the surface when nutrients are replete (MacIntyre *et al.* 1997). A comprehensive survey of photosynthetic responses might reveal if EPC indicates the hydrographic regime to which different phytoplankton species are best adapted. The diagnostic is potentially useful for characterizing environmental preferences of harmful phytoplankton, and their potential responses to altered hydrography.

Differences in EPC, reflecting ecological selection linked to turbulence, could influence the interpretation of P_m in nature as a measure of photoacclimation (Falkowski 1981): for example, a relatively high P_m for phytoplankton in optically deep mixed layers with low mean irradiance has been interpreted as an adaptation to the highest irradiance encountered during vertical mixing (Vincent *et al.* 1994). In a sense, the mixing phytoplankton would indeed be adapted to exploit the highest irradiance encountered, but it may be the consequence of genotypic selection for mixers with high EPC rather than phenotypic photoacclimation to highest irradiance experienced during mixing.

3.2. Capacity for photoacclimation

Many species of phytoplankton cannot survive exposures to full sunlight. Hence, species of phytoplankton that form blooms at the surface should have adaptations for tolerating potentially harmful solar radiation. In this chapter, only photosynthetically available radiation (PAR) will be considered. Ultraviolet radiation (UV) can also be a factor; effects of UV and adaptive responses are reviewed elsewhere (Vincent and Roy 1993; Cullen and Neale 1994).

A first step in characterizing the growth responses of a phytoplankter (say, a species that forms harmful blooms) is to examine its growth rate as a function of irradiance. One can determine growth-irradiance relationships of phytoplankton (Langdon 1988) and generalize about tolerance of bright light or adaptation to low light, but caution is warranted: for example, the marine *Synechococcus* was originally thought to require low light regimes for growth (Glover 1986) but an oceanic clone was later shown to grow very well at irradiance levels approaching full sunlight (Kana and Glibert 1987a). To obtain high growth rates in bright light, it was crucial for Kana and Glibert (1987a) to introduce high irradiance gradually, because even though low-light-adapted *Synechococcus* had some excess photosynthetic capacity (Fig. 2B), it did not survive abrupt increases to supersaturating (*i.e.*, inhibiting) irradiance (Barlow and Alberte 1985). Oceanic *Synechococcus* are now known to be abundant over a broad range of depths in the photic zone (Campbell and Vaulot 1993). We can infer that adaptation of phytoplankton to variable environmental irradiance includes not only the capacity to exploit irradiance higher than the mean, but also mechanisms to tolerate excess irradiance long enough for effective adaptation to take place.

3.2.1. Short-term responses to excess irradiance

Phytoplankton respond to excess irradiance through a variety of mechanisms that operate over a range of time-scales (Harris 1986). An operating principle is that when irradiance is high enough to saturate photosynthetic systems, additional photons absorbed by photosynthetic pigments, especially those directed to photosystem II (PSII) reaction centers, are potentially damaging (Ibelings *et al.* 1994 and references therein). In this section, we discuss short-term diversion of excess absorbed irradiance away from sensitive sites, with concomitant changes in characteristics of *in vivo* fluorescence. Longer-term adjustments are examined below.

Rapid (seconds to minutes) responses to excessive irradiance involve processes that reduce the efficiency with which absorbed photons (excitons) are transferred to PSII reaction centers. The absorbed but diverted energy is dissipated as heat; the concomitant reduction in fluorescence yield is called nonphotochemical quenching. Nonphotochemical quenching can be effected by several processes (Krause and Weis 1991; Arsalane *et al.* 1994). In studies of phytoplankton, quenching through xanthophyll cycling (Demers *et al.* 1991; Olaizola *et al.* 1994) is recognized as being important in photoprotection; it is responsive to growth conditions (Arsalane *et al.* 1994), and variable among species (Demers *et al.* 1991) . The xanthophyll cycle involves the physiologically harmless conversion of pigments with enzymatic reversion to complete the cycle. Activity of xanthophyll cycling is manifest in rapid (minutes), reversible reductions in chlorophyll fluorescence during exposure to bright light (Demers *et al.* 1991). Although quenching associated with xanthophyll cycling may provide only partial protection from excess irradiance (Olaizola *et al.* 1994), it seems likely that physiological regulation of xanthophyll pools is an adaptive feature of phytoplankton that would reflect the different strategies of mixers, migrators, and layer-formers (Demers *et al.* 1991; Ibelings *et al.* 1994).

If excess irradiance is not quenched or otherwise diverted, PSII reaction centers can be inactivated, leading to photoinhibition. Photoinhibition is defined here as a decrease in the rate of photosynthesis *during* exposure to supersaturating irradiance — *i.e.*, a reduction in the capacity for photosynthesis, concomitant with decreases in efficiency; (see Fig. 3 and Neale 1987). Strong, slowly reverting nonphotochemical quenching processes can also reduce the efficiency of photosynthesis (Ting and Owens 1994), but not necessarily the rate of photosynthesis in bright light (see Fig. 6C in Cullen *et al.*

1988). While such strong quenching is active, rates of photosynthesis would be maintained if cells remained at the surface, but the exposure to bright light would reduce subsequent photosynthesis if cells were mixed vertically (Ibelings *et al.* 1994). Mechanisms such as strong quenching and inactivation of PSII reaction centers are reversible and thus can protect the long-term integrity of PSII (Öquist *et al.* 1992; Ibelings *et al.* 1994), even though they reduce short-term photosynthesis in nature. If such "protective" mechanisms are not effective, more severe and possibly irreversible damage could occur. Hence, when trying to describe the photophysiology of a phytoplankter, it is important to explore long-term adaptability to bright light, as well as short-term responses. Note that a low-light adapted diatom (*Thalassiosira pseudonana*, presumably a mixer) showed strong inhibition of short-term photosynthesis during exposure to bright light, but it survived and adapted (Fig. 3A), whereas a layer-former, *Oscillatoria agardhii*, showed no inhibition of photosynthesis during short-term exposures to irradiance much higher than saturating irradiance, but would not survive (Fig. 3B).

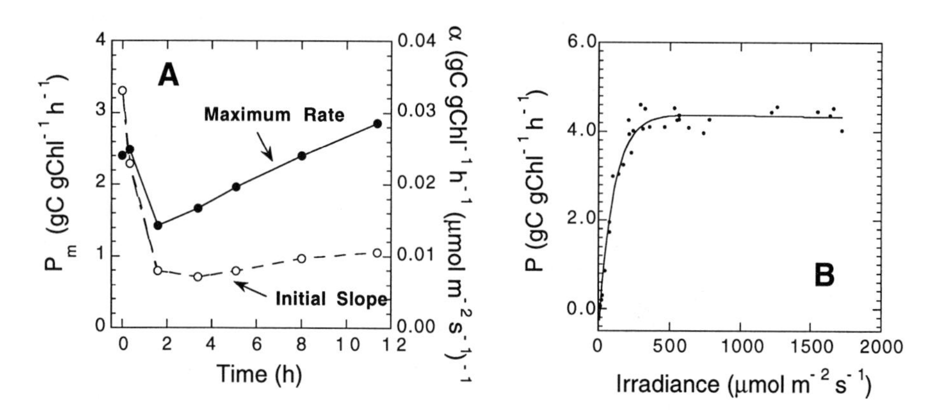

Fig. 3. Photosynthetic responses to increased irradiance of phytoplankton suited to different regimes of turbulence. A. *Thalassiosira pseudonana*, a coastal diatom, presumably adapted to turbulent environments: a culture adapted to low light, 20 μmol m^{-2} s^{-1} was shifted to 2000 μmol m^{-2} s^{-1} (Cullen and Lewis 1988); both the initial slope of the *P-E* curve (α) and the maximum rate P_m declined rapidly, consistent with nonphotochemical quenching and inactivation of PSII reaction centers, causing photoinhibition. After 2 hours, photosynthesis continued, chemical composition changed consistent with photoacclimation, photosynthetic capacity recovered, and the culture eventually acclimated to high light. B. *Oscillatoria agardhii*, a depth-regulating cyanobacterium, sampled from a low-light metalimnetic peak in an experimental mesocosm: *P* as a function of *E* (15 min incubations with ^{14}C bicarbonate) showed no short-term photoinhibition, even at $E >$ 10 times the irradiance at which cells had aggregated. Cells did not survive after transfer to bright light, however (J.J. Cullen and A.R. Klemer, unpublished).

It thus seems possible that dynamic physiological responses to bright light, including those that might reduce short-term photosynthesis, are adaptations that promote survival of phytoplankton that move through the water column. Depth-regulating phytoplankton need not have these adaptations, if they can descend rapidly enough to avoid damaging conditions (Heaney and Talling 1980; Häder and Worrest 1991; Neale *et al.* 1991). Perhaps the existence of strong circadian rhythms in photosynthetic performance (Prézelin 1992) with maximal resistance to photoinhibition near midday (J.J. Cullen, unpublished data on *Heterocapsa niei*), is an adaptation to predictable exposures to bright light.

Experiments on the dynamic responses of phytoplankton to bright light should thus be helpful for inferring how species might compete in different turbulence environments. For example, research on freshwater bloom formers, especially cyanobacteria, has been very successful (Ibelings *et al.* 1994 and references therein). These studies have been noteworthy not only for the development and testing of hypotheses, but also for the application of powerful fluorescence techniques (Schreiber *et al.* 1986) to describe dynamic changes in photosynthetic physiology in an environmental context. Research on short-term photosynthetic responses of marine and estuarine phytoplankton has focused more on the oceanographic problem of describing photosynthetic responses to fluctuating light (Neale and Marra 1985; Kirkpatrick *et al.* 1990; Franks and Marra 1994) than on comparing species. However, Sakshaug *et al.* (1987) were able to identify differences in bright-light responses between two congeneric diatoms that could be related to the oceanic vs coastal environments from which they were isolated.

3.2.2. Longer-term photoacclimation through unbalanced growth

Aspects of the longer-term (hours-days) acclimation of phytoplankton to changes in irradiance have been studied very well (reviewed by Falkowski and LaRoche 1991). For example, the ratio of cellular C to chlorophyll (C:Chl) increases with growth irradiance (Geider 1987), along with the capacity for photosynthesis, P_m (Falkowski 1981). These changes are consistent with adaptive shifts in the allocation of photosynthate between three major cellular pools: light harvesting, maintenance, and storage (Shuter 1979; Lancelot *et al.* 1991; Geider *et al.* 1996). Rates of acclimation depend on growth conditions, magnitude and direction of the change, species, which property is measured, and what constituent is used in the denominator of a ratio (*e.g.*, C:Chl; Geider and Platt 1986; Cullen and Lewis 1988). In general, the time scales of change for C:Chl and P_m range from hours to a generation time or so (Prézelin and Matlick 1980; Post *et al.* 1984; Cullen and Lewis 1988). These changes are accomplished through unbalanced growth, the synthesis (or more precisely, the net accumulation) of some cellular components at different rates than others (Shuter 1979). An understanding of the regulation of unbalanced growth leads directly to an understanding of long-term photoacclimation (Geider *et al.* 1996).

Cellular chemical composition is sensitive to irradiance, but other factors also have a strong influence. In general, light harvesting capability is reduced (in essence, diluted; C increases relative to Chl) and storage products are accumulated in relatively bright light, under relatively low temperature, and in response to nutrient limitation. These are the conditions under which the absorption of light by photosystems (the *source* of photosynthetic electron flow: a function of pigmentation and irradiance) exceeds the cell's ability to utilize the light (the *sink* for photosynthetic electron flow: a function of assimilatory enzymes, temperature, and nutrients available for protein synthesis). In other words, the capacity to harvest light is reduced and photosynthate is stored when the source-sink ratio is high (Cullen 1985). When light is limiting (sink exceeds source), storage is reduced and photosynthate flows to light harvesting components. Using fluorescence induction methods, the balance between light absorption and utilization (source:sink) can be quantified as "excitation pressure", the reduction state of the electron acceptor pool in PSII (Maxwell *et al.* 1995). This measurement, as well as a related analytical representation (Geider *et al.* 1996), can be extremely useful in studies of photoacclimation through unbalanced growth.

Photoacclimation involves trade-offs, and thus ecological strategies probably exist. For example, adaptation to low light involves processes that maximize light harvesting (*e.g.*, accumulation of pigment) and minimize losses to respiration (Geider *et al.* 1986). Adaptation of mixers to low mean irradiance is thus an interesting problem. Excess photosynthetic capacity facilitates use of variable irradiance in deep mixed layers, but the adaptations to support exploitation and tolerance of bright light (high levels of photosynthetic enzymes, active photoprotective and repair mechanisms) come with a cost (elevated respiration, inefficient photosynthesis after exposure to bright light). The consequences of these kinds of physiological trade-offs have been considered by Ibelings et al. (1994); the topic merits careful examination in future studies of adaptation to different mixing regimes.

Deep layer-formers are expected to be small (internal self-shading of pigments is minimized, hence light absorption efficiency is maximized, Morel and Bricaud 1986) and to have inherently low respiration rates (Richardson *et al.* 1983). The existence of big cells in deep layers suggests that efficient photosynthesis is not always necessary: processes such as phagotrophy, osmotrophy or resistance to grazing are likely important to their ecology (discussed elsewhere in this volume). Vertical migrators that can reach the surface need not adapt to low light, unless the nutricline is so deep that saturating irradiance cannot be reached during a diurnal ascent. Size and swimming speed (or rates of sinking or floating) would then be factors (see Kamykowski *et al.*, this volume). Note that the energetic costs of swimming are small (Richardson *et al.* 1983), and not considered to be a major ecological factor for phytoplankton.

For ecological modeling, as well as for understanding adaptations to environmental variability and the physiological trade-offs that are involved, it is critical to understand the dynamic regulation of biosynthesis during photoacclimation. Some progress has been made (Shuter 1979; Lancelot *et al.* 1991; Geider *et al.* 1996). Source-sink regulation is central to the process of photoacclimation, and as we show in the following sections, it may be important to the nutrition and buoyancy regulation of vertically migrating phytoplankton.

4. Unbalanced growth, behavior, and nutrient assimilation

It has long been recognized that motility of marine phytoplankton allows them to exploit nutrients over a relatively large part of the water column (Gran 1929), thereby achieving concentrations greater than could be attained through growth at any one depth in stratified waters (Holmes *et al.* 1967). In coastal waters, this exploitation of nutrients is achieved principally through diel vertical migration. For a time, the ability of dinoflagellates to support growth by taking up nitrate during nocturnal descent of DVM was controversial (reviewed by Cullen 1985). Now, it is well established that vertically migrating flagellates can and do take up deep nutrients (N and/or P) to support growth in stratified water columns (Watanabe *et al.* 1995; MacIntyre *et al.* 1997 and references therein). Interest has been focused primarily on nitrate and phosphate uptake by flagellates during DVM, but there have been demonstrations (either direct or indirect) of deep nutrient acquisition by vertically cycling marine diatoms (cycle time of several days, Villareal *et al.* 1996) and marine cyanobacteria (Karl *et al.* 1992). Also, it has been suggested that through buoyancy regulation, mixed-layer diatoms can exploit deep nutrients obtained during periodic encounters with the thermocline (Lande and Wood 1987; Richardson and Cullen 1995). Buoyancy reversals in freshwater cyanobacteria, and regulation through light-nutrient interactions have been long recognized and well studied over the years (reviewed by Oliver 1994 and discussed by Klemer *et al.* 1996). For all examples, the strategy of deep-nutrient acquisition through vertical cycling involves close coupling between physiological condition and changes in buoyancy or swimming direction (see Kamykowski *et al.*, this volume).

4.1 Physiological adaptations for assimilation of deep nutrients

What are the physiological adaptations for the exploitation of deep nutrients? One is the ability to take up nutrient (*e.g.*, nitrate) in the dark, or in very low light. There is much to be learned by comparing the abilities of different species to take up nitrate in the dark (Paasche *et al.* 1984); one important observation is that for some dinoflagellates, the capability is induced only after a short period of N-starvation (Paasche *et al.* 1984; Cullen 1985).

Assimilation of nitrate into protein requires products of photosynthesis for energy, reductant, and carbon skeletons (Syrett 1981); when N-starved dinoflagellates were fed nitrate in the dark (thereby greatly increasing the sink for photosynthate), these requirements were satisfied by the mobilization of carbohydrate that had been stored during the light period. The stoichiometry (about 6 mol carbohydrate-C mobilized for each mol nitrate-N taken up and assimilated) is consistent with calculated demands (Cullen 1985). This mechanism of N-acquisition is apparently sustainable: a dinoflagellate could be maintained on nitrate supplied only in the dark (Harrison 1976).

Assimilation of nitrate and synthesis of protein at the expense of accumulated carbohydrate is one aspect of a general dependence of nocturnal protein synthesis on carbohydrate dynamics (Cuhel *et al.* 1985; Zenvenboom 1986). It can thus be concluded that uptake and assimilation of nitrate during nocturnal descent of DVM requires few, if any, special physiological adaptations. Accumulation of storage products is the consequence of being in nutrient-depleted surface layers during the day, consistent with photoacclimation to high excitation pressure (see 3.2.2). Mobilization of storage products for assimilation of nitrate in the dark is likewise consistent with source-sink regulation: exposure to inorganic nitrogen creates a large demand for photosynthate to provide carbon skeletons for protein synthesis. Under similar conditions, diatoms are capable of nocturnal nitrate uptake (Fig. 4), suggesting that the ability to take up nutrients in the dark is not a special adaptation of migrators.

Not all vertical migrators do so every day. For example, large diatoms (Villareal *et al.* 1996), the dinoflagellate *Pyrocystis* (Rivkin *et al.* 1984) and cyanobacteria of the genus *Trichodesmium* (Karl *et al.* 1992) in the open ocean are thought to migrate between the nutricline and surface waters with a period longer than a day. The large diatoms may cycle on the order of a generation time (Richardson *et al.* 1996). Mixed-layer diatoms might also cycle vertically through nutrient-dependent sinking and resuspension from the thermocline (Lande and Wood 1987); this would happen on an irregular basis. Buoyancy regulation in freshwater cyanobacteria is likewise tied more to physiological state than to the diel cycle (Oliver 1994). Adaptations for all these non-motile migrators include the ability to take up large amounts of nutrient during periodic exposures, along with the capacity to accumulate large amount of storage products to support nutrient assimilation at depth (Fig. 4B; Richardson *et al.* 1996). As shown in Fig. 4B, the accumulation of storage products acts as a buffer for environmental variability (Gibson and Jewson 1984). Species might vary in the amount they can store and in the time they can remain nutrient starved before survival is threatened (Fig. 5). Can some species of depth-regulating phytoplankton buffer

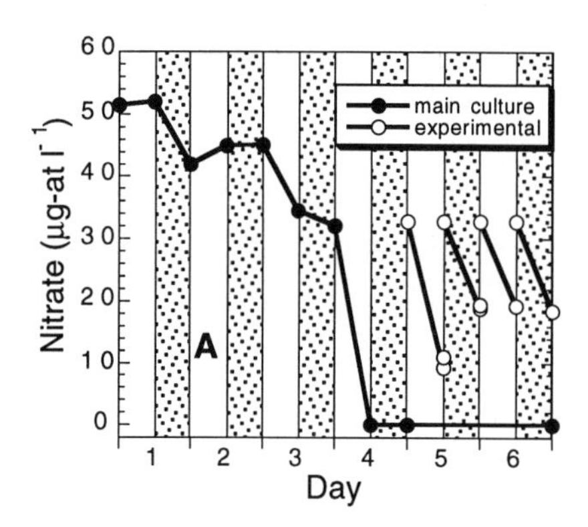

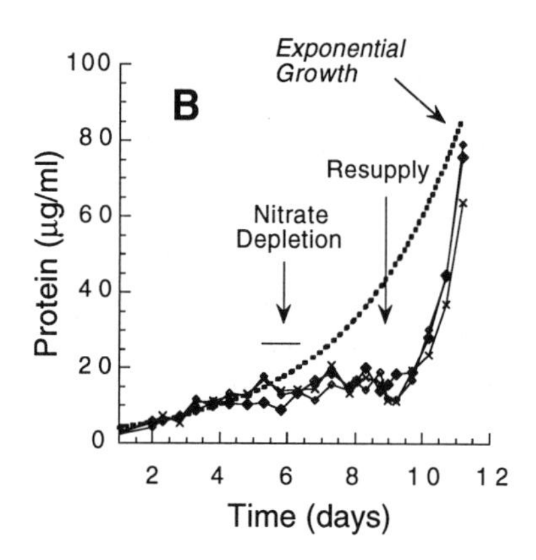

Fig. 4. Uptake and assimilation of nitrate after N-starvation. A. The diatom *Skeletonema costatum*, grown at 20°C on a 12h:12h light-dark cycle with nutrients in excess except for nitrate. Filled circles, nitrate in the culture; open circles, connected by lines, show responses of subsamples to the addition of 32 µg-at l^{-1} nitrate at the beginning or end of the light period. Conditions and manipulations are essentially the same as reported by Cullen (1985) for the dinoflagellate *Heterocapsa niei*, and results are very similar, showing capacity for the uptake of nitrate in the dark after brief N-starvation. Like in *Heterocapsa*, carbohydrate was mobilized during dark nitrate uptake in the stoichiometry appropriate for nitrate reduction and protein synthesis. However, *Skeletonema* took up only about half as much nitrate during the dark period. (M. Zhu and J.J. Cullen, unpublished). B. Changes of protein in triplicate cultures of the diatom *Thalassiosira weissflogii* during N-starvation and resupply of nitrate. From day 6 through 9, protein did not increase because N was unavailable. Photosynthate was accumulated as carbohydrate; changes in chemical composition were sufficient to explain increased sinking rate of the diatom. After nitrate was supplied on day 9, the uptake of nitrate proceeded in the light and in the dark; carbohydrate was mobilized to support protein synthesis, decreasing cell density in the process. By day 11, protein in the cultures was almost fully restored to what would have accumulated during unperturbed exponential growth. Carbohydrate acted as an environmental buffer (results from Richardson and Cullen 1995).

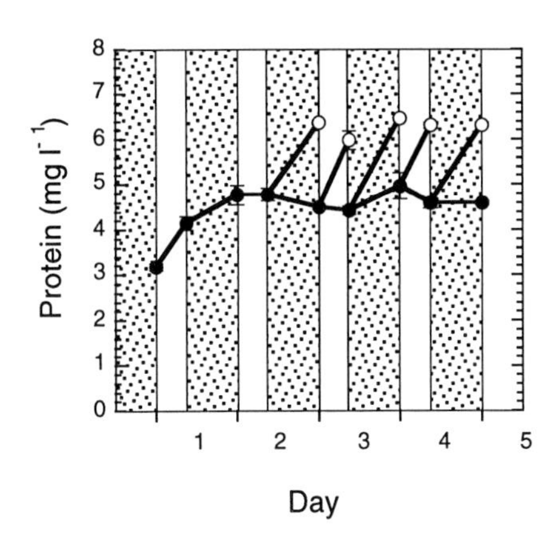

Fig. 5. Maintenance of the ability to assimilate nitrate during N-starvation. Protein concentration (closed circles ± s.e. of triplicates; method of Dorsey *et al.* 1978) for *Heterocapsa niei* grown to nitrate depletion on day 1 (conditions as in Fig. 4A, initial nitrate = 50 μM, but on a 9h:15h light-dark cycle). Open circles, increase of protein (±s.e.) after addition of 30 μM nitrate to subsamples. The ability to take up nitrate in the dark was undiminished after three days: essentially all added nitrate was consumed during the last dark period, and 79% of the nitrate appeared as protein (assuming protein is 16% N by weight; Dorsey *et al.* 1978). It is not known how long the capability for nitrate assimilation would persist or how other species might differ in this respect. Unpublished data of M. Zhu and J.J. Cullen.

more than one cell division? Can some buffer much less? Does it matter for ecological selection? These questions deserve further study.

4.2. Behavioral adaptations for deep nutrient assimilation

The preceding discussion has shown that physiological adaptations for deep nutrient acquisition are widespread. Thus, it may be the ability to cycle between deep nutrient pools and the surface layer, rather than the capacity to take up nutrients in the dark, that is key to exploitation of deep nutrient pools. Consequently, a variety of behavioral adaptations for deep nutrient acquisition are likely to exist.

4.2.1 Motile phytoplankton

For motile phytoplankton, vertical migration, though phased to solar irradiance, is governed by factors other than phototactic swimming, including temperature, cell size, cell density, and environmental feedback on swimming behavior (Kamykowski 1995). Not surprisingly, species differ in the depth or light level of aggregation during the day, extent of nocturnal descent, and timing of both descent and ascent. Further, migration patterns can be strongly influenced by nutrition (Heaney and Eppley 1981; MacIntyre *et al.* 1997) so that each species probably has a unique suite of nutrient- and light-dependent vertical migration patterns, each suited to a particular hydrographic regime (Cullen 1985). Indeed, each dinoflagellate studied to date has displayed a distinct suite of responses to altered regimes of nutrients or stratification. The challenge is to infer from field observation and experimental results where and why a particular species will bloom, or when its strategy might fail. For some species, bloom dynamics have been described as a function of the interactions between behavior, physiology, and regional hydrography (*e.g.*, Tyler and Seliger 1981; Watanabe *et al.* 1995; Figueiras *et al.* 1996). More studies using different species in experimental water columns may be effective in revealing physiologically-influenced behavioral and physiological adaptations that are suited for growth and survival in other hydrographic regimes.

4.2.2 Non-motile phytoplankton

Depth regulation in vacuolate freshwater cyanobacteria has been well described (Oliver and Walsby 1984). The synthesis and collapse of gas vesicles strongly influences buoyancy, but the accumulation of ballast (*e.g.*, polymeric carbohydrate) also affects sinking and floating. Vesicle synthesis, turgor collapse, and carbohydrate accumulation are influenced by interactions of light and nutrients, and the general tendency is for nutrient-rich, low-light cells to float and nutrient-starved, high-light cells to sink (Oliver 1994). The concept of nutritional effects on the accumulation and mobilization of carbohydrate ballast has been applied to marine cyanobacteria (Romans *et al.* 1994) and diatoms (Richardson and Cullen 1995) as well. Clearly, physiological responses to nutrients and light have a direct influence on vertical movements of non-motile phytoplankton. Research on marine forms will benefit from careful consideration of the extensive literature on freshwater cyanobacteria.

4.3. The mechanistic links between nutrition and vertical migration

Because nutrition influences vertical migration of motile phytoplankton, swimming behavior must be under physiological control. But how does light absorption and nutrition influence swimming? A direct link has been suggested (see Lieberman *et al.* 1994; Kamykowski 1995): accumulation of dense carbohydrate in the posterior region of a dinoflagellate, perhaps with lipid accumulation anteriorly, would direct swimming upward. The importance in nature of this mechanism of orientation is unknown and deserving of further study (Häder *et al.* 1991 and references therein).

Even if the exact mechanism of physiologically-induced behavior modification is unknown, there is a need to describe the environmental influences on swimming trajectories of motile phytoplankton so that the vertical movements of phytoplankton can be incorporated into hydrodynamic or ecological models. Phototaxis alone cannot explain migratory behavior of phytoplankton; modeling exercises show that even simple physiological control on migration behavior greatly improves predicted growth (Kamykowski *et al.*, this volume). However, the interaction of light and nutrients in influencing cellular orientation and consequently vertical distribution is poorly described. This is an important topic for future research. One approach would be to relate orientation patterns to internal stores of storage compounds. Another would be to calculate something akin to excitation pressure (Maxwell *et al.* 1995; Geider *et al.* 1996; see 3.2.2) and relate it to swimming orientation. Considering the rapid migratory response of a nitrogen-starved dinoflagellate to N addition (MacIntyre *et al.* 1997), we might focus on the source-sink balance as reflected in the enzymes of carbohydrate synthesis and mobilization (Cullen 1985) rather than on PSII, site for excitation pressure. A principal objective would be to provide appropriate algorithms for models of vertical migration. Another product, of course, would be a deeper appreciation of how migrating phytoplankton respond to their environment, and what response patterns are best suited to particular hydrographic regimes.

5. Summary and conclusions

We began this chapter by relating behavioral strategies of depth-regulating phytoplankton to different regimes of turbulence (see also Estrada and Berdalet, this volume). Recognizing that interactions of behavior and hydrography determine the temporal variability of irradiance and nutrient supply, we extended Margalef's (1978) search for "adaptation syndromes...to recurrent patterns of selective factors," looking for adaptations to dominant modes of environmental variability that are specific to different behavioral strategies. Distinct syndromes were identified: mixers should be adapted to highly variable irradiance, with some influence of nutrition on buoyancy; migrators should have a well-developed capacity for unbalanced growth and effective physiological control of vertical movements; and layer-formers should develop efficient systems for utilizing light or nutrients at the expense of tolerating environmental variability. Ecological trade-offs are associated with each adaptation, so there is wide latitude for differentiation between species. We suggest that differences in adaptation patterns can be related to ecological success in particular hydrographic regimes.

Quantifiable properties are associated with several key adaptations. Excess photochemical capacity, an index of the ability to exploit variable irradiance, should be high in mixers and low in layer-formers. Short-term photoprotective responses, detectable with fluorescence techniques, can have a strong influence on the survival of phytoplankton exposed to bright light during mixing or vertical migration. Photoacclimation and nutrient acquisition through unbalanced growth can be related to the ratio of light absorbed to light (*i.e.*, photosynthate) utilized: regulation of the allocation of photosynthate is subject to selection, so we expect adaptive differences between species. The process of unbalanced growth can be examined through a number of simple experiments (Figs. 4, 5; also Cullen and Lewis 1988) and described quantitatively in the context of recent models (Lancelot *et al.* 1991; Geider *et al.* 1996). Such models should also serve to describe nutrient uptake during vertical migration and the physiological control of buoyancy (Kromkamp and Walsby 1990) as well as vertical migration behavior (Kamykowski *et al.*, this volume).

Because adaptation syndromes to different hydrographic regimes include physiological controls on depth regulation, many niches are available, associated with a range of behavioral responses to light and nutrients. By characterizing environmental influences on depth regulation for individual species, and the consequences for population dynamics, we may be able to define the niches of different phytoplankton species and identify the selective factors that promote their growth in particular environments. Progress has been made for freshwater cyanobacteria (Oliver 1994) and a few marine species (*e.g.*, Tyler and Seliger 1981; Richardson *et al.* 1996; Villareal *et al.* 1996), but appropriate experimental data for marine HAB phytoplankton is scarce (*e.g.*, Watanabe *et al.* 1995; MacIntyre *et al.*, 1997). We conclude that to describe the growth of depth-regulating phytoplankton in nature, the interaction of physiology and behavior must be considered. Because many HAB species can move vertically, new approaches, some of which have been outlined here, should be useful in trying to understand the physiological ecology of harmful algal blooms.

6. Acknowledgments

Supported by the Natural Sciences and Engineering Research Council of Canada. Thanks to Patrick Neale and Michele DuRand for helpful information, and to the editors for many good suggestions. CEOTR Publication 12.

7. References

Arsalane, W., Rousseau, B., Duval, J.C. (1994). Influence of the pool size of the xanthophyll cycle on the effects of light stress in a diatom - competition between photoprotection and photoinhibition. *Photochem. Photobiol.* 60:237-243.

Barlow, R.G., Alberte, R.S. (1985). Photosynthetic characteristics of phycoerythrin-containing marine *Synechococcus* sp. 1. Responses to growth flux density. *Mar. Biol.* 86:63-74.

Blasco, D. (1978). Observations on the diel migration of marine dinoflagellates off the Baja California coast. *Mar. Biol.* 46:41-47.

Campbell, L., Vaulot, D. (1993). Photosynthetic picoplankton community structure in the subtropical North Pacific Ocean near Hawaii (station ALOHA). *Deep-Sea Res.* 40:2043-2060.

Cuhel, R.L., Ortner, P.B., Lean, D.R.S. (1985). Night synthesis of protein by algae. *Limnol. Oceanogr.* 29:731-744.

Cullen, J.J. (1985). Diel vertical migration by dinoflagellates: roles of carbohydrate metabolism and behavioral flexibility. *Contr. Mar. Sci.* 27 (Suppl.):135-152.

Cullen, J.J., Lewis, M.R. (1988). The kinetics of algal photoadaptation in the context of vertical mixing. *J. Plankton Res.* 10:1039-1063.

Cullen, J.J., MacIntyre, H.L., Carlson, D.J. (1989). Distributions and photosynthesis of phototrophs in sea-surface films. *Mar. Ecol. Prog. Ser.* 55:271-278.

Cullen, J.J., Neale, P.J. (1994). Ultraviolet radiation, ozone depletion, and marine photosynthesis. *Photosyn. Res.* 39:303-320.

Cullen, J.J., Yang, X., MacIntyre, H.L. (1992). Nutrient limitation of marine photosynthesis. In: Falkowski P.G., Woodhead A., (eds.) *Primary Productivity and Biogeochemical Cycles in the Sea.* Plenum, New York, pp. 69-88.

Cullen, J.J., Yentsch, C.S., Cucci, T.L., MacIntyre, H.L. (1988). Autofluorescence and other optical properties as tools in biological oceanography. *Proc. SPIE Int. Soc. Opt. Eng.* 925:149-156.

Demers, S., Roy, S., Gagnon, R., Vignault, C. (1991). Rapid light-induced changes in cell fluorescence and in xanthophyll-cycle pigments of *Alexandrium excavatum* (Dinophyceae) and *Thalassiosira pseudonana* (Bacillariophyceae): a photo-protection mechanism. *Mar. Ecol. Prog. Ser.* 76:185-193.

Dorsey, T.E., McDonald, P., Roels, O.A. (1978). Measurements of phytoplankton-protein content with the heated biuret-folin assay. *J. Phycol.* 14:167-171.

Dugdale, R.C., Wilkerson, F.P. (1989). New production in the upwelling center at Point Conception, California: temporal and spatial patterns. *Deep-Sea Res.* 36:985-1007.

Eppley, R.W., Reid, F.M.H., Cullen, J.J., Winant, C.D., Stewart, E. (1984). Subsurface patch of dinoflagellate (*Ceratium tripos*) off Southern California: patch length, growth rate, associated vertically migrating species. *Mar. Biol.* 80:207-214.

Falkowski, P.G. (1980). Light-shade adaptation in marine phytoplankton. In: Falkowski P.G., (ed.) *Primary Productivity in the Sea.* Plenum Press, New York, p. 99–119.

Falkowski, P.G. (1981). Light–shade adaptation and assimilation numbers. *J. Plankton Res.* 3:203–216.

Falkowski, P.G., Hopkins, T.S., Walsh, J.J. (1980). An analysis of factors affecting oxygen depletion in the New York Bight. *J. Mar. Res.* 38:479-506.

Falkowski, P.G., LaRoche, J. (1991). Acclimation to spectral irradiance in algae. *J. Phycol.* 27:8-14.

Figueiras, F.G., Gómez, E., Nogueira, E., Villarino, M.L. (1996). Selection of *Gymnodinium catenatum* under downwelling conditions in the Ria del Vigo. In: Yasumoto T., Oshima Y., Fukuyo Y., (eds.) *Harmful and Toxic Algal Blooms.* Intergovernmental Oceanographic Commission of UNESCO, Paris, pp. 215-218.

Fraga, S., Bakun, A. (1993). Global climate change and harmful algal blooms: The example of *Gymnodinium catenatum* on the Galician coast. In: Smayda T.J., Shimizu Y., (eds.) *Toxic Phytoplankton Blooms in the Sea.* Elsevier, Amsterdam, pp. 59-65.

Franks, P.J.S., Marra, J. (1994). A simple new formulation for phytoplankton photoresponse and an application in a wind-driven mixed-layer model. *Mar. Ecol. Prog. Ser.* 111:145-153.

Geider, R.J. (1987). Light and temperature dependence of the carbon to chlorophyll a ratio in microalgae and cyanobacteria: implications for physiology and growth of phytoplankton. *New Phytol.* 106:1-34.

Geider, R.J., MacIntyre, H.L., Kana, T.M. (1996). A dynamic model of photoadaptation in phytoplankton. *Limnol. Oceanogr.* 41:1-15.

Geider, R.J., Osborne, B.A., Raven, J.A. (1986). Growth, photosynthesis and maintenance metabolic cost in the diatom *Phaeodactylum tricornutum* at very low light levels. *J. Phycol.* 22:39-48.

Geider, R.J., Platt, T. (1986). A mechanistic model of photoadaptation in microalgae. *Mar. Ecol. Prog. Ser.* 30:85-92.

Gibson, C.E., Jewson, D.H. (1984). The utilisation of light by microorganisms. In: Codd G.A., (ed.) *Aspects of microbial metabolism and ecology.* Academic Press, London, pp. 97-127.

Glover, H.E. (1986). The physiology and ecology of marine cyanobacteria, *Synechococcus* spp. In: Jannasch H.W., Williams P.J.L., (eds.) *Advances in aquatic microbiology. 3* Academic Press, New York, pp. 49-107.

Gran, H.H. (1929). Investigation of the production of plankton outside the Romsdalfjord 1926-1927. *J. Cons.* 56:1-112.

Häder, D.-P., Liu, S.-M., Kreuzberg, K. (1991). Orientation of the photosynthetic flagellate, *Peridinium gatunense*, in hypergravity. *Curr. Microbiol.* 22:165-172.

Häder, D.-P., Worrest, R.C. (1991). Effects of enhanced solar ultraviolet radiation on aquatic ecosystems. *Photochem. Photobiol.* 53:717-725.

Hallegraeff, G.M. (1993). A review of harmful algal blooms and their apparent global increase. *Phycologia.* 32:79-99.

Harding, L.W., Jr., Fisher, T.R., Jr. , Tyler, M.A. (1987). Adaptive responses of photosynthesis in phytoplankton: specificity to time-scale of change in light. *Biol. Oceanogr.* 4:403-437.

Harris, G.P. (1986). *Phytoplankton ecology: structure, function and fluctuation*, Chapman and Hall, New York.

Harrison, W.G. (1976). Nitrate metabolism of the red tide dinoflagellate *Gonyaulax polyedra* Stein. *J. Exp. Mar. Biol. Ecol.* 21:199-209.

Heaney, S.I., Eppley, R.W. (1981). Light, temperature and nitrogen as interacting factors affecting diel vertical migrations of dinoflagellates in culture. *J. Plankton Res.* 3:331-344.

Heaney, S.I., Talling, J.F. (1980). Dynamic aspects of dinoflagellate distribution patterns in a small productive lake. *J. Ecol.* 68:75-94.

Holligan, P.M. (1978). Patchiness in subsurface phytoplankton populations on the northwest European continental shelf. In: Steele J.H., (ed.) *Spatial patterns in phytoplankton communities.* Plenum Press, New York, pp. 222-238.

Holmes, R.W., Williams, P.M., Eppley, R.W. (1967). Red water in La Jolla Bay, 1964-1966. *Limnol. Oceanogr.* 12:503-512.

Ibelings, B.W., Kroon, B.M., Mur, L.R. (1994). Acclimation of photosystem II in a cyanobacterium and a eukaryotic green alga to high and fluctuating photosynthetic photon flux densities, simulating light regimes induced by mixing in lakes. *New Phytol.* 128:407-424.

Jackson, G.A. (1990). A model of the formation of marine algal flocs by physical coagulation processes. *Deep-Sea Res.* 37:1197-1211.

Kamykowski, D. (1995). Trajectories of autotrophic marine dinoflagellates. *J. Phycol.* 31:200-208.

Kana, T.M., Glibert, P.M. (1987a). Effect of irradiances up to 2000 $\mu E\ m^{-2}\ s^{-1}$ on marine *Synechococcus* WH7803–I. Growth, pigmentation, and cell composition. *Deep-Sea Res.* 4:479-495.

Kana, T.M., Glibert, P.M. (1987b). Effect of irradiances up to 2000 $\mu E\ m^{-2}\ s^{-1}$ on marine *Synechococcus* WH7803–II. Photosynthetic responses and mechanisms. *Deep-Sea Res.* 4:497-516.

Karl, D.M., Letelier, R., Hebel, D.V., Bird, D.F., Winn, C.D. (1992). *Trichodesmium* blooms and new nitrogen in the North Pacific Gyre. In: Carpenter E.J., Capone D.G., Reuter J.G., (eds.) *Marine Pelagic Cyanobacteria: Trichodesmium and other diazotrophs.* Kluwer Academic, Boston, pp. 219-237.

Kiørboe, T. (1993). Turbulence, phytoplankton cell size, and the structure of pelagic food webs. *Adv. Mar. Biol.* 29:1-72.

Kirkpatrick, G.J., Curtin, T.B., Kamykowski, D., Freezor, M.D., Sartin, M.D., Reed, R.E. (1990). Measurement of photosynthetic response to euphotic zone physical forcing. *Oceanography Mag.* 3:18-22.

Klemer, A.R., Cullen, J.J., Mageau, M.T., Hanson, K.M., Sundell, R.A. (1996). Cyanobacterial buoyancy regulation: The paradoxical roles of carbon. *J. Phycol.* 32:47-53.

Krause, G.H., Weis, E. (1991). Chlorophyll fluorescence and photosynthesis: the basics. *Annu. Rev. Plant Physiol. Plant Mol. Biol.* 42:313-349.

Kromkamp, J., Walsby, A.E. (1990). A computer-model of buoyancy and vertical migration in cyanobacteria. *J. Plankton Res.* 12:161-183.

Lancelot, C., Veth, C., Mathot, S. (1991). Modelling ice-edge phytoplankton bloom in the Scotia-Weddell sea sector of the Southern Ocean during spring 1988. *J. Mar. Syst.* 2:333-346.

Lande, R., Li, W.K.W., Horne, E.P., Wood, A.M. (1989). Phytoplankton growth rates estimated from depth profiles of cell concentration and turbulent diffusion. *Deep-Sea Res.* 36:1141-1159.

Lande, R., Wood, A.M. (1987). Suspension times of particles in the upper ocean. *Deep-Sea Res.* 34:61-72.

Langdon, C. (1988). On the causes of interspecific differences in the growth-irradiance relationship for phytoplankton. II. A general review. *J. Plankton Res.* 10:1291-1312.

Legendre, L., Demers, S. (1984). Towards dynamic biological oceanography and limnology. *Can. J. Fish. Aquat. Sci.* 41:2-19.

Li, W.K.W., Morris, I. (1982). Temperature adaptation in *Phaeodactylum tricornutum* Bohlin: photosynthetic rate compensation and capacity. *J. Exp. Mar. Biol. Ecol.* 58:135-150.

Lieberman, O.S., Shilo, M., van Rijn, J. (1994). The physiological ecology of a freshwater dinoflagellate bloom population: vertical migration, nitrogen limitation, and nutrient uptake kinetics. *J. Phycol.* 30:964-971.

MacIntyre, J.G., Cullen, J.J., Cembella, A.D. (1997). Vertical migration, nutrition and toxicity of the dinoflagellate, *Alexandrium tamarense*. *Mar. Ecol. Prog. Ser.* in press.

Margalef, R. (1978). Life forms of phytoplankton as survival alternatives in an unstable environment. *Oceanol. Acta.* 1:493-509.

Maxwell, D.P., Falk, S., Huner, N.P.A. (1995). Photosystem II excitation pressure and development of resistance to photoinhibition. *Pl. Physiol.* 107:687-694.

Morel, A., Bricaud, A. (1986). Inherent properties of algal cells including picoplankton: Theoretical and experimental results. In: Platt T., Li W.K.W., (eds.) *Photosynthetic Picoplankton.* pp. 521-559.

Neale, P.J. (1987). Algal photoinhibition and photosynthesis in the aquatic environment. In: Kyle D.J., Osmond C.B., Arntzen C.J., (eds.) *Photoinhibition.* Elsevier, Amsterdam, pp. 35- 65.

Neale, P.J., Heaney, S.I., Jaworski, G.H.M. (1991). Responses to high irradiance contribute to the decline of the spring diatom maximum. *Limnol. Oceanogr.* 36:761-768.

Neale, P.J., Marra, J. (1985). Short-term variation of Pmax under natural irradiance conditions: a model and its implications. *Mar. Ecol. Prog. Ser.* 26:113-124.

Olaizola, M., La Roche, J., Kolber, Z., Falkowski, P.G. (1994). Non-photochemical fluorescence quenching and the diadinoxanthin cycle in a marine diatom. *Photosyn. Res.* 41:357-370.

Oliver, R.L. (1994). Floating and sinking in gas-vacuolate cyanobacteria. *J. Phycol.* 30:161-173.

Oliver, R.L., Walsby, A.E. (1984). Direct evidence for the role of light-mediated gas vesicle collapse in the buoyancy regulation of *Anabaena flos-aquae* (cyanobacteria). *Limnol. Oceanogr.* 29:879-886.

Öquist, G., Chow, W.S., Anderson, J.M. (1992). Photoinhibition of photosynthesis represents a mechanism for long-term regulation of photosystem II. *Planta.* 186:450-460.

Paasche, E., Bryceson, I., Tangen, K. (1984). Interspecific variation in dark nitrogen uptake by dinoflagellates. *J. Phycol.* 20:394-401.

Platt, T., Gallegos, C.L., Harrison, W.G. (1980). Photoinhibition of photosynthesis in natural assemblages of marine phytoplankton. *J. Mar. Res.* 38:687–701.

Post, A.F., Dubinsky, Z., Wyman, K., Falkowski, P.G. (1984). Kinetics of light-intensity adaptation in a marine planktonic diatom. *Mar. Biol.* 83:231-238.

Prézelin, B.B. (1992). Diel periodicity in phytoplankton productivity. *Hydrobiologia.* 238:1-35.

Prézelin, B.B., Matlick, H.A. (1980). Time-course of photoadaptation in photosynthesis-irradiance relationship of dinoflagellate exhibiting photosynthetic periodicity. *Mar. Biol.* 58:85-96.

Richardson, K., Beardall, J., Raven, J.A. (1983). Adaptation of unicellular algae to irradiance: an analysis of strategies. *New Phytol.* 93:157-191.

Richardson, T.L., Ciotti, A.M., Cullen, J.J., Villareal, T.A. (1996). Physiological and optical properties of *Rhizosolenia formosa* (Bacillariophyceae) in the context of open-ocean vertical migration. *J. Phycol.* 32:741-757.

Richardson, T.L., Cullen, J.J. (1995). Changes in buoyancy and chemical composition during growth of a coastal marine diatom: Ecological and biogeochemical consequences. *Mar. Ecol. Prog. Ser.* 128:77-90.

Rivkin, R.B., Swift, E., Biggley, W.H., Voytek, M.A. (1984). Growth and carbon uptake by natural populations of oceanic dinoflagellates *Pyrocystis noctiluca* and *Pyrocystis fusiformis*. *Deep-Sea Res.* 31:353-367.

Romans, K.M., Carpenter, E.J., Bergman, B. (1994). Buoyancy regulation in the colonial diazotrophic cyanobacterium *Trichodesmium tenue*: Ultrastructure and storage of carbohydrate, polyphosphate, and nitrogen. *J. Phycol.* 30:935-942.

Sakshaug, E., Demers, S., Yentsch, C.M. (1987). *Thalassiosira oceanica* and *T. pseudonana* : two different photoadaptational responses. *Mar. Ecol. Prog. Ser.* 41:275-282.

Schreiber, U., Schliwa, U., Bilger, B. (1986). Continuous recording of photochemical and nonphotochemical chlorophyll fluorescence quenching with a new type of modulation fluorometer. *Photosyn. Res.* 10:51-62.

Shuter, B. (1979). A model of physiological adaptation in unicellular algae. *J. theor. Biol.* 78:519–552.

Smayda, T.J. (1990). Novel and nuisance phytoplankton blooms in the sea: evidence for a global epidemic. In: Granéli E., Sundström B., Edler L., Anderson D.M., (eds.) *Toxic Marine Phytoplankton.* Elsevier, New York, pp. 29-40.

Smetacek, V.S. (1985). Role of sinking in diatom life-history cycles: ecological, evolutionary and geological significance. *Mar. Biol.* 84:239-251.

Sommer, U. (1985). Differential migration of cryptophyceae in Lake Constance. *Cont. Mar. Sci.* 27 Suppl.: 166-175.

Sukenik, A., Bennett, J., Falkowski, P.G. (1987). Light–saturated photosynthesis - limited by electron transport or carbon fixation? *Biochim. Biophys. Acta.* 891:205-215.

Sverdrup, H.U. (1953). On conditions for the vernal blooming of phytoplankton. *J. Cons. Cons. Int. Explor. Mer.* 18:287-295.

Syrett, P.J. (1981). Nitrogen metabolism of microalgae. In: Platt T., (ed.) *Physiological Bases of Phytoplankton Ecology. 210* pp. 182-210.

Tester, P.A., Geesey, M.A., Vukovich, F.M. (1993). *Gymnodinium breve* and global warming: What are the possibilities? In: Smayda T.J., Shimizu Y., (eds.) *Toxic Phytoplankton Blooms in the Sea.* Elsevier, Amsterdam, pp. 67-72.

Ting, C.S., Owens, T.G. (1994). The effects of excess irradiance on photosynthesis in the marine diatom *Phaeodactylum tricornutum. Pl. Physiol.* 106:763-770.

Tyler, M.A., Seliger, H.H. (1981). Selection for a red tide organism: physiological responses to the physical environment. *Limnol. Oceanogr.* 26:310-324.

Venrick, E.L. (1988). The vertical distributions of chlorophyll and phytoplankton species in the North pacific central environment. *J. Plankton Res.* 10:987-998.

Villareal, T.A., Woods, S., Moore, J.K., Culver-Rymsza, K. (1996). Vertical migration of *Rhizosolenia* mats and their significance to NO_3^- fluxes in the central North Pacific gyre. *J. Plankton Res.* 18:1103-1121.

Vincent, W.F., Bertrand, N., Frenette, J.-J. (1994). Photoadaptation to intermittent light across the St. Lawrence Estuary freshwater-saltwater transition zone. *Mar. Ecol. Prog. Ser.* 110:283-292.

Vincent, W.F., Roy, S. (1993). Solar ultraviolet-B radiation and aquatic primary production: damage, protection and recovery. *Environ. Rev.* 1:1-12.

Walsby, A.E., Reynolds, C.S. (1980). Sinking and floating. In: Morris I., (ed.) *Physiological Ecology of Phytoplankton.* Blackwell Scientific, Oxford, pp. 371-412.

Watanabe, M., Kohata, K., Kimura, T. (1991). Diel vertical migration and nocturnal uptake of nutrients by *Chattonella antiqua* under stable stratification. *Limnol. Oceanogr.* 36:593-602.

Watanabe, M., Kohata, K., Kimura, T., Takamatsu, T., Yamaguchi, S., Ioriya, T. (1995). Generation of a *Chattonella antiqua* bloom by imposing a shallow nutricline in a mesocosm. *Limnol. Oceanogr.* 40:1447-1460.

Yentsch, C.S. (1980). Phytoplankton growth in the sea, a coalescence of disciplines. In: Falkowski P.G., (ed.) *Primary Productivity in the Sea.* Plenum, New York, pp. 17-31.

Zevenboom, W. (1986). Ecophysiology of nutrient uptake, photosynthesis and growth. In: Platt T., Li W.K.W., (eds.) *Photosynthetic Picoplankton.* pp. 391-422.

A Comparison of How Different Orientation Behaviors Influence Dinoflagellate Trajectories and Photoresponses in Turbulent Water Columns

D. Kamykowski[1], H. Yamazaki[2], A. K. Yamazaki[3], and G. J. Kirkpatrick[4]

[1] Department of Marine, Earth & Atmospheric Sciences, Box 8208, North Carolina State University, Raleigh NC 27695, USA
[2] Department of Ocean Sciences, Tokyo University of Fisheries, 4-5-7 Konan, Minato-ku, Tokyo 108 Japan
[3] Department of Information Sciences, Nippon Institute of Technology, 4-1 Gakuen-dai, Miyashiro-cho, Saitama 345 Japan
[4] Mote Marine Laboratory, 1600 Thompson Parkway, Sarasota FL 34236, USA

1. Introduction

Single species dinoflagellate blooms, including those termed 'Harmful Algal Blooms' or 'HABs' (Anderson 1995) can attain concentrations of 10^9 cells L^{-1} (Taylor and Pollingher 1987). These blooms, however, are composed of single cells or chains of cells that act as individuals. One mechanism of bloom formation is when physical dispersal mechanisms are sufficiently weak that motility-aided photosynthesis and biosynthesis yield a large accumulation of cells in a given location due to many cell divisions. Alternately, populations of cells that undergo fewer cell divisions but that initially occupy a large area may accumulate in a smaller bloom area due to physical convergence mechanisms that result from interaction between cell behavior and water motion. With both mechanisms, the growth environment of individual cells can be influenced by the interdependent influences of swimming behavior, small scale physical processes in the vertical dimension, and time-dependent photosynthetic responses. A literature review will set a context for each of these topics. A biophysical model then will be used to explore relationships between the biological vector due to swimming and the vertical physical vector represented by a turbulence approximation and between different orientation mechanisms as they influence and are influenced by photosynthate use and accumulation. One purpose of this modeling is to provide a base for detailed conceptual and eventual mathematical models of HAB initiation, development, aggregation and dissipation that more fully consider the role of individual cells within their environment.

1.1. Swimming behavior

Swimming behavior can be separated into two components: speed and orientation. Kamykowski (1995) reviewed both topics for autotrophic marine dinoflagellates, but see Jones (1993) for additional discussion of phytoplankton orientation in freshwater. Typical average swimming speeds for different dinoflagellate species

NATO ASI Series, Vol. G 41
Physiological Ecology of Harmful Algal Blooms
Edited by D. M. Anderson, A. D. Cembella and G. M. Hallegraeff

that range between 50-600 μm s^{-1} (Kamykowski and McCollum 1986, Levandowsky and Kaneta 1987) place this group among the fastest phytoflagellates (Raven and Richardson 1984). The instantaneous swimming speed exhibited by a given species depends on environmental conditions such as gravity, temperature, salinity and light intensity (Hand *et al.* 1965, Kamykowski *et al.* 1988). Other environmental influences are likely but remain unquantified. Swimming orientation often is described as positively phototactic in daylight and positively geotactic in the dark. However, deviations from these simple generalized movements have been observed (Sournia 1974). Accumulating evidence based on different taxa suggests that dinoflagellate orientation responds to gravity, light intensity, temperature, salinity, oxygen and inorganic nutrients (Lebert and Hader 1996; Eppley *et al.* 1968, Kamykowski 1981, Harris *et al.* 1979, Cullen and Horrigan 1981). Again, other environmental influences are likely but remain unquantified. The synthesis of information in Kamykowski (1995) concerning circadian rhythms (Chisholm *et al.* 1984), gyrotaxis (Pedley and Kessler 1992), dinoflagellate organelles (Rizzo 1987) and storage product placement (Dodge and Gruet 1987), dinoflagellate sensory capabilities (Levandowsky and Kaneta 1987), and biochemical fluxes (Cullen 1985) provide a compelling case for a more dynamic interpretation of cell orientation in terms of internal cellular characteristics. The biophysical model discussed below includes a swimming speed submodel containing gravity, temperature and light dependence based on Kamykowski *et al.* (1988). The model also involves a swimming orientation submodel, one version of which uses photosynthesis and respiration thresholds as proxies for biochemical changes within the cell.

1.2. Small scale physical processes

Denman and Gargett (1983) considered turbulent mixing, internal waves and Langmuir circulation as physical processes that are especially pertinent to phytoplankton vertical displacement in the upper ocean. The continued accumulation of information on these (Caldwell and Moum 1995, Thorpe 1995) and other (Thorpe 1995) physical processes in the upper ocean have encouraged continued exploration of their ecological implications (Mann and Lazier 1991, Franks 1995). Similarly, increased knowledge concerning planktonic motility (Denman and Gargett 1995) has stimulated new interest in how the biologically determined vector influences organism trajectories in natural water columns (Denman 1994). Representative modeling efforts that include organism behavior for small scale turbulence (Yamazaki and Kamykowski 1991, Franks and Marra 1995, Kamykowski *et al.* 1995), Langmuir circulation (Evans and Taylor 1980, Watanabe and Harashima 1986), internal waves (Kamykowski 1979) and fronts (Franks 1992) demonstrate complex, often non-intuitive, patterns in space and time. However, the need for more realistic physical models in terms of water column stratification (MacIntyre 1993) and realistic vertical water motion (Li and Garrett 1995, Kamykowski 1995) is recognized. Denman (1994) and Denman and Gargett (1995) suggested that measurements by several recently developed physical oceanographic instruments including freefall probes, upgraded current meters, acoustics and neutral floats need to be incorporated into community models of the upper ocean. For example, the biophysical model discussed below is based on a simple wind-related random walk that decays exponentially with depth down through the Ekman layer (the water column below the sea surface that is significantly influenced by the wind). Brainard and Gregg (1995) summarized information collected with the Advanced Microstructure Profiler that clearly demonstrated the limits of this model condition. The purpose here is to examine simple biophysical relationships based on increased behavioral complexity in anticipation of the more advanced upper ocean models to come.

1.3. Time-dependent photosynthetic responses

Lagrangian models of phytoplankter trajectories through representative water columns often include a diurnal light/dark cycle with incident radiation that varies sinusoidally during the light period and that attenuates exponentially with depth (*e.g.* Kamykowski 1974). The potential for rapid excursions through the strong light gradient that can coincide with an energetic upper ocean requires the consideration of time-dependent photosynthetic responses (Falkowski and Wirick 1981, Cullen and Lewis 1988). Lande and Lewis (1989) incorporated such time-dependence in terms of first-order kinetics in Eulerian (particle motion inferred from water motion measured at a point through time) and Lagrangian (particle motion measured by following the particle) models in response to the Lagrangian approach of Woods and Onken (1982) that used integral exponential reaction kinetics. McGillicuddy (1995) suggested that the discrepancy between the results obtained in these two papers was due to the different photosynthesis formulations and favored the Lande and Lewis (1989) approach. Barkmann and Woods (1996) recently applied first order kinetics in their study of *in situ* incubations for primary production. Janowitz and Kamykowski (1990) expanded the Eulerian approach by incorporating photosynthetic time-dependence in a modified version of the robust upper ocean model of Price *et al.* (1986). Kamykowski *et al.* (1994) and Franks and Marra (1995) used different time-dependent photosynthesis formulations in the context of the simple Larangian model suggested by Yamazaki and Kamykowski (1991) to examine the effects of turbulence on the statistics of cell-specific photosynthesis with depth. Geider *et al.* (1996) pointed out that most biophysical studies of the upper mixed layer emphasize the more rapid photoresponses and provided a longer term photoacclimation point-of-view in the context of cell biochemistry. Kamykowski and Yamazaki (1996) considered the three time scales of photoresponse that also are included here representing photoinhibition, diel variability in maximum photosynthetic rate, and sun-shade photoacclimation.

2. Methods

As in Kamykowski and Yamazaki (1996), the biophysical model used in this paper began with the upper ocean, wind-driven turbulence model described in Yamazaki and Kamykowski (1991), the photoacclimation model described in Janowitz and Kamykowski (1991) and Kamykowski *et al.* (1994), and the swimming behavior model described in Yamazaki and Kamykowski (1991). The turbulence model, criticized for producing artificial patchiness (Holloway 1994) but applied as a dispersive mechanism in the original context (Yamazaki and Kamykowski 1994), is still used awaiting better alternative formulations (Denman and Gargett 1995). The model was run in two different modes. In the single-day mode, 10 cells were followed over a night\day period; in the multi-day mode, one cell was followed over 10 days. Both modes calculated the cell's position at 0.2 hour intervals, but only the hourly values are reported here. Some units in Table 1 are based on this 0.2 hour increment.

Though the biophysical model used here was based on Kamykowski and Yamazaki (1996), some important changes, briefly described next, were made concerning turbulence intensity, the photoinhibition model, the behavioral response to sun-shade photoacclimation, and the depth threshold for descent.

Table 1: A list of symbols used in the text.

Symbol	Definition	Units
C	initial average percent cloud cover	%
H_c	cumulative photoinhibition at present time step	
H_{cp}	cumulative photoinhibition at previous time step	
H_i	instantaneous photoinhibition	
I_3	average three day PAR exposure	$\mu mol\ m^{-2}\ s^{-1}$
I_c	PAR after cloud effect	$\mu mol\ m^{-2}\ s^{-1}$
I_h	PAR threshold for photoinhibition	$\mu mol\ m^{-2}\ s^{-1}$
I_k	saturation PAR equal to P_m/α	$\mu mol\ m^{-2}\ s^{-1}$
I_m	maximum noon PAR	$\mu mol\ m^{-2}\ s^{-1}$
I_s	PAR at time t_i	$\mu mol\ m^{-2}\ s^{-1}$
I_z	PAR at the depth of the cell	$\mu mol\ m^{-2}\ s^{-1}$
P_m	maximum photosynthetic rate at sunrise	pg-at O_2 cell^{-1} 0.2h^{-1}
P_{md}	P_m plus the diel increment	pg-at O_2 cell^{-1} 0.2h^{-1}
$\mathcal{R}_1,\mathcal{R}_2,\mathcal{R}_3$	independent random numbers	
R_a	respiration enhancement (sun-shade acclimation)	pg-at O_2 cell^{-1} 0.2h^{-1}
R_i	respiration enhancement (photoinhibition)	pg-at O_2 cell^{-1} 0.2h^{-1}
R_m	minimal respiration	pg-at O_2 cell^{-1} 0.2h^{-1}
t_d	hours of daylight	h
t_i	time of day	h
t_s	length of time step	h
w_a	initial wind speed for a computer run	m s^{-1}
w_d	daily wind speed after random effect	m s^{-1}
α	inital slope of the PI curve	(pg-at O_2 cell^{-1} 0.2h^{-1}) ($\mu mol\ m^{-2}\ s^{-1}$)$^{-1}$
Γ	decay time constant for photoinhibition	h
π	pi	
ϕ	phase adjustment for daylight relative to t_i	h

The instantaneous (t_i; see Table 1 for a summary of symbol definitions) photosynthetically active radiation or PAR (I_s) resulted from the multiplication of the maximum noon PAR (I_m) by a sinusoidal function (adjusted by ϕ so that sunrise occurred 12 hours after the model run began, where t_d is half daylength) such that

$$I_s=I_m(\mathrm{SIN}(t_i+\phi)\pi/t_d).$$

The cloud-influenced PAR (I_c) resulted from

$$I_c=I_s(1.0\text{-}C),$$

where C represented an initial average cloud cover of 30% that randomly varied at each time step by $\mathcal{R}_1/10$, where $\mathcal{R}_1$ is a random number between 0 and 1. The cloud cover remained within a range between 10 to 50%. The incident radiation attenuated exponentially with water column depth based on a predetermined coefficient (0.1 m^{-1}).

The wind speed during the single-day model runs equaled the average wind speed. However, small day-to-day random variation ($\mathcal{R}_2$) altered the daily wind speed (w_d) for multi-day model runs to yield deviations from the average wind speed (w_a) by

$$w_d=w_a+((\mathcal{R}_2\,(w_a))/2).$$

The water column instantaneously responded to wind speed changes between days. The wind-driven turbulence yielded a random walk for each cell that was 2.8x less intense for a given wind speed than that in Kamykowski and Yamazaki (1996) and that decreased in magnitude with increasing depth in the Ekman layer. No turbulence existed below the Ekman layer.

Three different time scales of photoresponse were included. Photoinhibition remained as in Kamykowski and Yamazaki (1996) except that the decay constant for the influence of previous PAR exposure in this model decayed at faster rates for low-to-high (1 h^{-1}) PAR trends than for high-to-low (0.33 h^{-1}) PAR trends. This change made the respiration penalty (R_i), based on the influence of residual inhibition (H_c) that extended into the night

$$R_i=R_m(H_c)^{1/2}$$

and added on to a randomly modified respiration rate (R_m; defined below), more significant. A diel variation was added by making the instantaneous maximum photosynthetic rate (P_{md}) a function of the initial maximum photosynthetic rate (P_m) at sunrise incremented by a sine-based, time-of-day variable (notation as above) that gave the highest value at local noon,

$$P_{md}=P_m+8.0((\mathrm{SIN}((t_i+\phi)\pi)/t_d)^3).$$

Sun-shade photoacclimation was added by calculating the value of P_m and the saturation intensity (I_k) on a given day based on the three day running average (Savidge 1988) of the total incident radiation (I_3) experienced by a cell. See Kamykowski and Yamazaki (1996) for details of P_m and I_k ranges under different PAR conditions. The respiration term based on maximum photosynthesis was altered from Kamykowski and Yamazaki (1996) by adding a penalty for shade acclimation (R_a) using the term

$$R_a=(500/I_k)R_m.$$

Only multi-day model runs examined sun-shade acclimation. This term increased respiration rate in shade-adapted cells so that ascent began sooner.

Swimming speed remained a function of temperature, light intensity and gravity effects. Orientation was either taxis-directed (TD) with descent at night and ascent during the day or metabolism-influenced (MI). The latter case used proxies for each cell's internal state at a given stage of its diel vertical migration. Beginning at sunset, a cell descended for 12 hours and then ascended for the following 12 hours as in the taxis-directed case unless some factor intervened. The proxy for the nutricline changed from that used in Kamykowski and Yamazaki (1996). If the thickness of the Ekman layer was less than 10 m, then the cell stopped its descent at 10 m. However, if the Ekman layer extended below 10 m, the cell continued to descend toward the bottom of the Ekman layer. The cell potentially could hold the attained depth (10 m or the bottom of the Ekman layer) for the remainder of the 12 hour period before it ascended. However, each cell also responded to a randomly modified ($\mathcal{R}_3$) respiration rate (R_m) based on a percentage of the maximum potential photosynthetic rate (P_m) for a day given by

$$R_m=-(P_m/25)+((P_m\ (\mathcal{R}_3))/50)$$

that served as a hypothetical proxy for metabolic repairs associated with the previous day's PAR exposure. Photoinhibition (R_i) and shade-acclimation (R_a) increased the respiration rate and thus the approach to the predetermined respiration threshold. When the cell's cumulative respiration exceeded the assigned threshold, a proxy for the full utilization of the photosynthate pools generated during the previous light period, ascent began even if less than 12 hours had elapsed since sunset. Once ascent began, the cell continued to ascend until the next sunset unless additional factors intervened. If the cell moved through the water column until it reached the surface, it then stopped its ascent and potentially could remain near the surface until sunset when it again descended. However, other factors could cause the descent to begin early. If the cell's cumulative photoinhibition, a proxy for exposure to excessive light, exceeded 0.5 on a scale of 0 to 1, it descended until the cumulative photoinhibition equaled or fell below 0.5. The present cumulative photoinhibition (H_c) was calculated by summing the instantaneous inhibition (H_i) at each time step from

$$H_i=1.0\text{-}EXP(-(((I_z-I_h)/I_h)^2)),$$

where I_d is the PAR at the depth of the organism and I_h is the PAR threshold above which photoinhibition began, in the equation

$$H_c=H_{cp}+((H_i-H_{cp})t_s/\Gamma,$$

where H_{cp} is the cumulative photoinhibition from the previous time step, t_s is the length of the time step, and Γ is the time constant by which photoinhibition decays from the time of initial exposure. The cell then ascended until it reached the surface or until photoinhibition again exceeded 0.5. Also, each cell responded to a daily total photosynthesis threshold as a proxy for newly filled photosynthate pools. Each cell's light exposure influenced the time required to reach the predetermined photosynthesis threshold value. When the cell's cumulative photosynthesis exceeded the assigned threshold, descent began even if sunlight was still available. If the cell's cumulative photosynthesis never exceeded the assigned threshold, then descent began at sunset.

Table 2: An input list entered in response to inquiries from the computer program.

Factor		Value	Units
Maximum Incident PAR		2000	µmol m^{-2} s^{-1}
Clouds:	Initial	30	%
	Maximum	50	%
	Minimum	10	%
Water Column PAR Attenuation		0.1	m^{-1}
Wind Speed		3	m s^{-1}
Respiration Threshold			
	Taxis-directed (not attained)	-10	pg-atO_2 cell^{-1} night^{-1}
	Metabolism-influenced (attained)	-4	pg-atO_2 cell^{-1} night^{-1}
Photosynthesis Threshold			
	Taxis-directed (not attained)	120	pg-atO_2 cell^{-1} day^{-1}
	Metabolism-influenced (attained)	50	pg-atO_2 cell^{-1} day^{-1}
Mode			
	Organism Number	25	
	Days	10	days
Random Number Seed		7894	

The single-day and multi-day modes, written as parallel programs initiated by screen calls for input (see Table 2 for the values used) and distinguished by the cycling patterns through the subroutines, both included paired runs that examined TD and MI orientation under the same physical forcing conditions. Since orientation mechanism influenced the depth occupied by the individual cell, exposure to all depth-dependent factors like turbulence intensity, PAR and nutrients was affected. The single-day mode will be discussed first in a comparison of the relative influence of biological and physical motion on cell trajectories. The multi-day mode then will be discussed briefly in reference to biological *versus* physical motion and then more extensively in a specific comparison of the impacts of the different orientation mechanisms.

3. Results

3.1. The relationship between biologically and physically influenced cell trajectories

The biophysical model provides an opportunity to examine how well the modeled cells are able to determine their motion under increasing intensity of physical forcing. This relationship presently ignores cellular damage that may result from increased turbulence as discussed by Thomas and Gibson (1990a,b). However, the biophysical model readily can incorporate a threshold of turbulence impact and a time course of effect (flagellar repair?) on swimming speed and orientation in

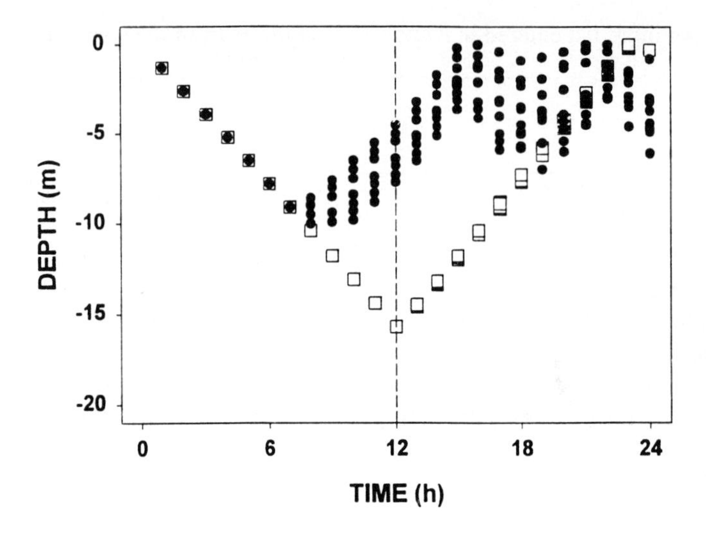

Fig. 1: A comparison between taxis-directed (white squares) and metabolism-influenced (black circles) orientation behavior in the absence of any wind-derived turbulence.

analogy to the photoresponse submodel. The differences in cell trajectories that result under the two orientation mechanisms when no physical forcing is applied (Fig. 1) provide the probe used to monitor the success of biological effort under different physical forcing conditions. As wind speed increases, the differences between the two orientation mechanisms should decrease as swimming becomes less effective. Increasing wind stress due to wind speeds from 3 to 18 m s^{-1} incremented at 3 m s^{-1} intervals causes increases in the ranges of water particle displacement under the modeled conditions. In order to place this water motion in perspective, if these displacements are divided by cell swimming speeds, this ratio increases to a maximum of about 15 at the highest wind speed used. A scatterplot with the depths attained by TD cells on the x-axis and by MI cells on the y-axis examines the state of the chosen probe as wind speed increases (Fig. 2). The regression line labeled ‘0’ (right axes on the two plots) represents the trend of black circles and sets the baseline for the plot as presented in Fig. 1. As wind speed increases, the regression lines representing successive wind speeds rise toward the 1:1 line given by the dashed line. This rise is interpreted as decreasing effectiveness of cell motility. If cell behavior were totally ineffective, the regression line would be indistinguishable from the dashed line. The family of regression lines in the left plot for the multi-cell, single day data suggests that over the short term the influence of the two different behaviors weakens with increasing wind speed, but the rate of change between the regression lines decreases as wind speed increases. The family of regression lines in the right plot for the single cell, multi-day data suggests that intermediate wind speeds have the greatest effect over 10 days. However, a comparison of the white circles (representing the 18 m s^{-1} wind

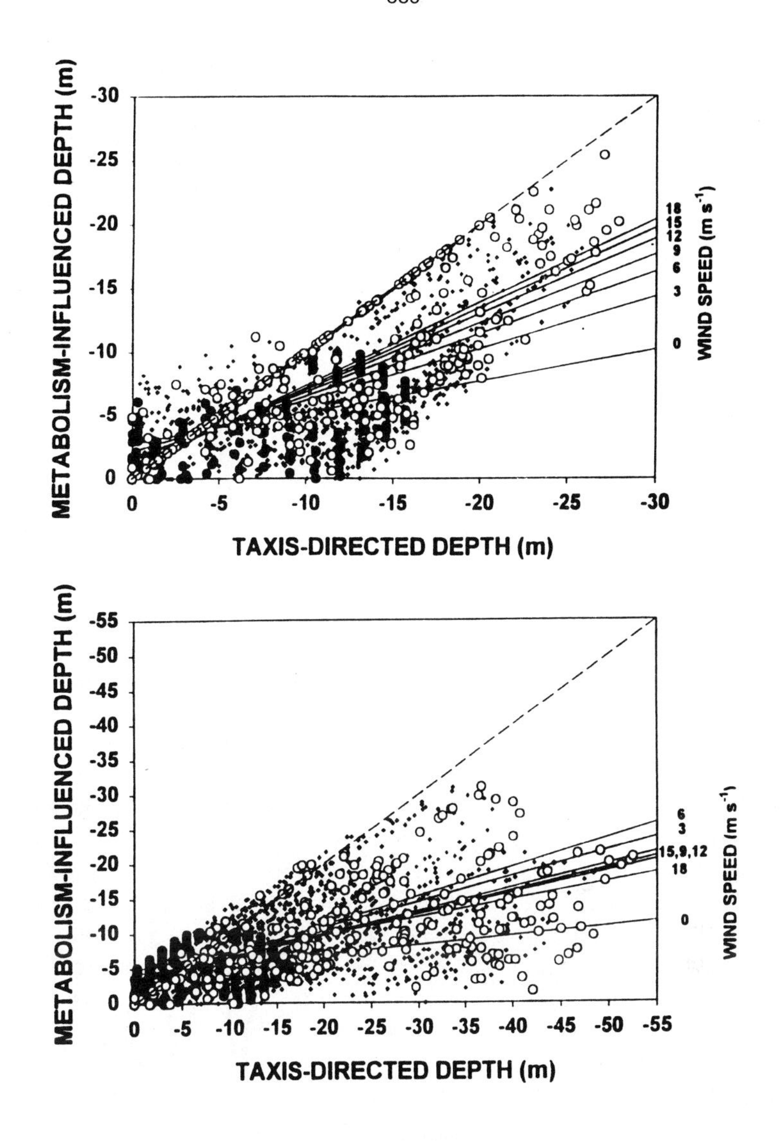

Fig. 2: Upper: The depths attained by 10 cells over a single-day with taxis-directed behavior compared to cells with metabolism-influenced behavior under increasing wind forcing. When wind speed is zero (black circles), the depths are very different as represented by a regression line that deviates from a slope of 1; as the wind speed increases (dots for all other wind speeds except 18 m s^{-1} (white circles)), the depths become more similar as the regression lines approach a slope of 1. Lower: Same except that single cell, multi-day data were used in this figure.

speed) between the two plots demonstrates the enhanced deepening of the TD cell (28 m to 54 m) over the MI cell (25 m to 30 m) in the 10 day plot. In this 10 day case, higher wind speed actually is making the cell trajetories more different with time due to the effectiveness of the MI behavior. The deviations of the regression lines from the dashed lines and the depth stability of the MI cells for both modeled conditions, suggest that persistent cell behavior influences cell trajectories under a broad range of natural wind speeds that contribute to relatively large intensity, disorganized water motion. Recall, however, that the present model does not include organized flows (*e.g.* waves, Langmuir cells).

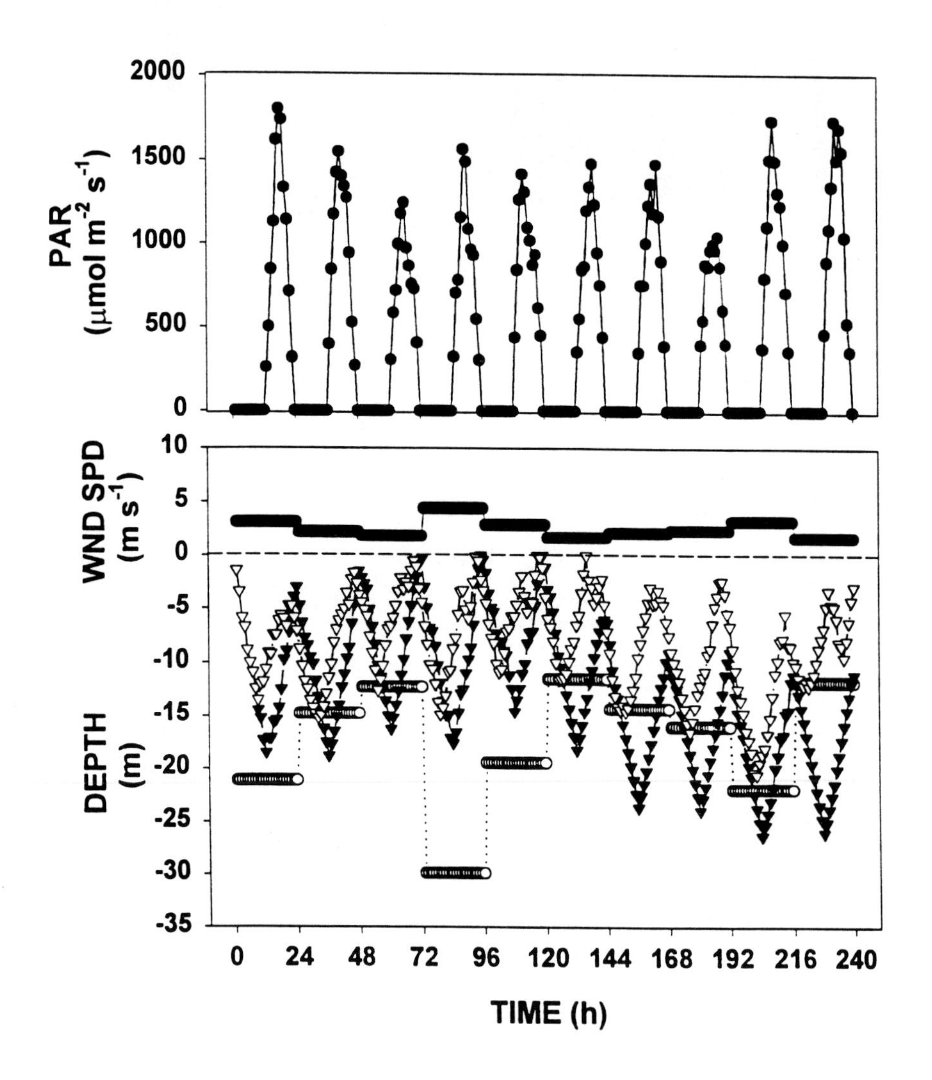

Fig. 3: A comparison of the incident surface PAR (top) that changes day-to-day due to cloud cover, the wind speed (middle) and the associated Ekman depth (bottom; white circles) that change randomly from day-to-day, and the trajectories of the two cells (bottom) that follow either taxis-directed (black triangles) or metabolism-influenced (white triangles) behavior.

3.2. A comparison of taxis-directed and metabolism-influenced orientation mechanisms

Several changes were made to the model used in Kamykowski and Yamazaki (1996) to correct for unrealistic components that became evident when the multi-day model was examined in more detail. As mentioned in the 'Methods', these changes included a modified wind-turbulence relationship, reversed decay time constants for photoinhibition induction and recovery, a respiration penalty for

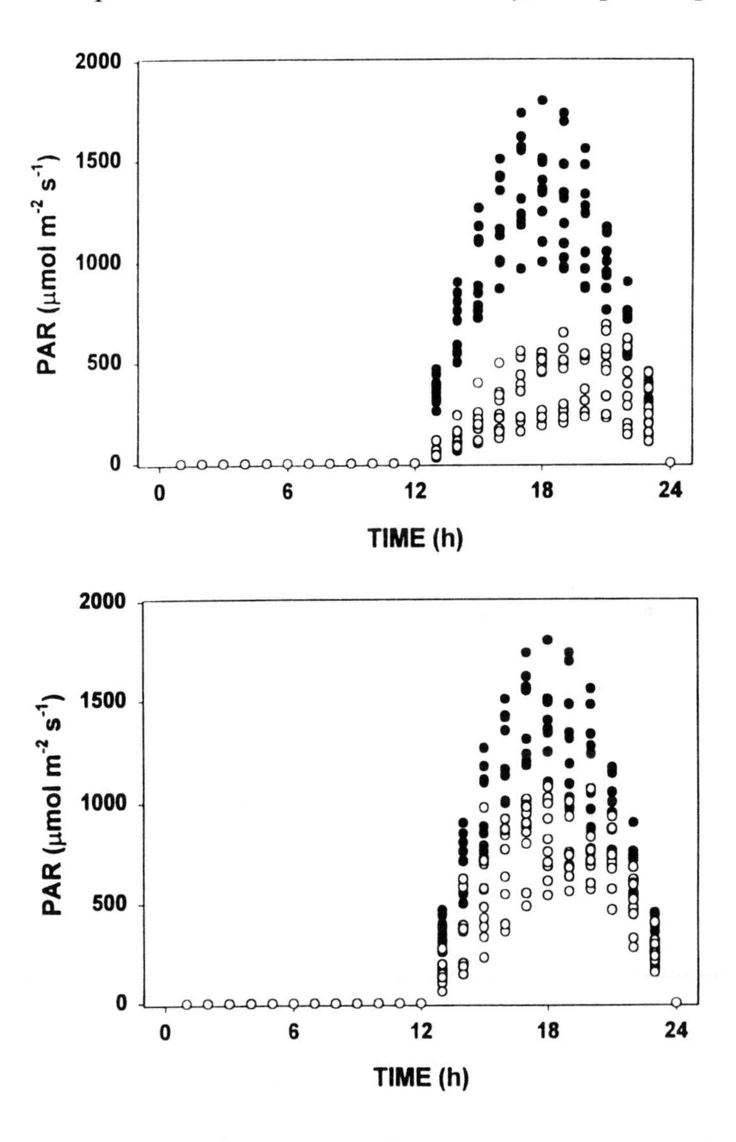

Fig. 4: A comparison of the surface PAR (black circles) and the PAR at the depth of the organism (white circles) for the taxis-directed (upper) and the metabolism-influenced (lower) cells.

shade acclimation, and a nutricline defined relative to the deeper of either 10 m depth or the Ekman depth.

For the purposes of this comparison, the 3 m s^{-1} wind speed from the multi-day mode is examined in detail. A composite summary (Fig. 3) of the 10-day variation of surface PAR, wind speed and the associated Ekman depth, and the cell trajectories for the two types of orientation mechanisms clearly exhibits the random variation that occurs in PAR and wind speed. More significantly in terms of nutrient exposure, the TD cell often swims through the Ekman depth inefficiently, going well beyond the depth necessary to experience increased nutrient concentrations. In contrast, the MI cell efficiently stops at the Ekman depth once higher nutrient concentrations are reached and thus has less distance to cover in order to return to higher PAR during the next daylight period.

Both types of cells may live under the same surface PAR (Fig. 3 and 4), but the TD cell typically experiences lower PAR than the MI cell (Fig. 4), primarily because the latter attains a respiratory threshold that initiates ascent prior to sunrise. Furthermore, the timing of maximum PAR is skewed toward mid-afternoon for the TD cell, while the MI cell experiences a more symmetrical PAR exposure around noon. The Productivity-Intensity (PI) response curves associated with the PAR exposure at depth (Fig. 5) demonstrates that the MI cell spends more time at saturating PAR conditions. This cell also is less likely to experience debilitating photoinhibition due to the behavior that triggers descent as long as the cell's cumulative photoinhibition exceeds a pre-assigned threshold. The cumulative effect of the 10-day exposure to different PAR levels results in a gradual shift to the shade-acclimated state, as measured by $I_k = P_m / \alpha$, by the TD cell (I_k begins at 500 but ends near 100 μmol m^{-2} s^{-1}), while the MI cell generally remains sun-acclimated (I_k begins at 500 and ends near 400 μmol m^{-2} s^{-1}). In the present model, the TD cell's decreased I_k is due a decreasing P_m and an increasing α compared to the MI cell. The overall effect of these changes, results in about 30% higher cumulative productivity by the MI cell at the end of 10 days.

4. Discussion

4.1. SUPA application

The present biophysical model provides a synthesis of cell trajectories based on representative physical water motion acting on a representative cell's swimming capability. In the MI cell, the latter includes feedback loops that tie cell orientation to the photosynthetic experience of a cell over different time scales. Although these components yield an interesting range of complexity, is the end result realistic? The easy answer is 'no' for many of the reasons stated in the 'Introduction'. However a qualitative comparison to the output of the Self-contained Underwater Photosynthesis Apparatus (SUPA; Kirkpatrick *et al.* 1990) provides a more graphic illustration that the approach used in the biophysical model at least is moving in the right direction. SUPA records temperature, PAR, oxygen flux and pH changes converted to carbon dioxide flux every minute during a deployment. SUPA, filled with a culture (≈30 μg Chl *a* l^{-1}) of the diatom *Thalassiosira weisflogii*, was

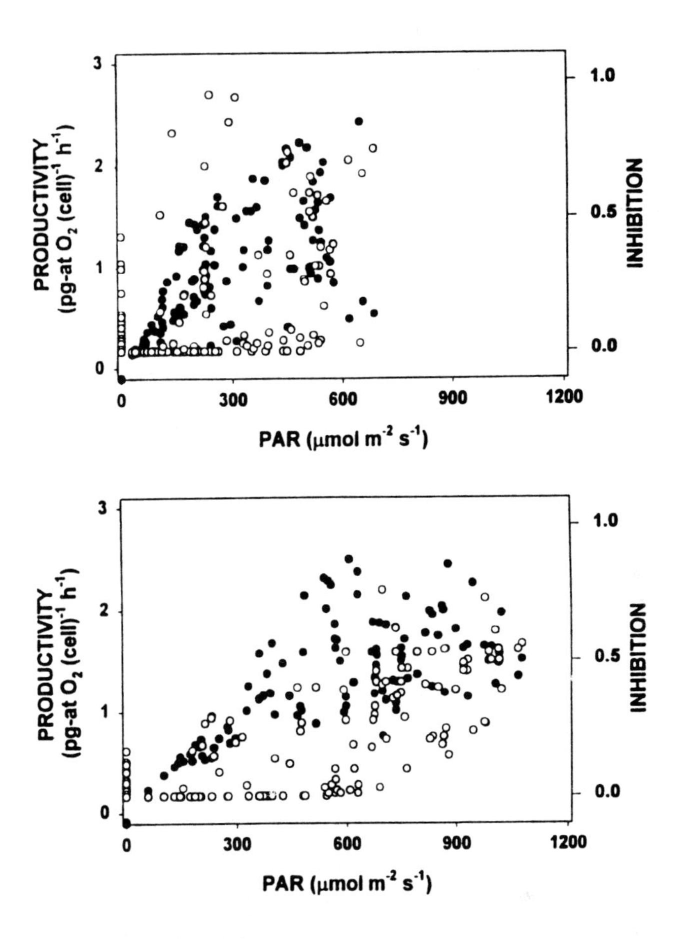

Fig. 5: PI (black circles) and inhibition-intensity (white circles) plots for the taxis-directed (upper) and the metabolism-influenced (lower) cells.

deployed on a *R/V Pelican* cruise in the Gulf of Mexico on 25 June, 1993. The instrument initially was placed with the culture at 10 m depth at about 0645 (EDT) and then was raised 2 m every 0.5 hr starting at 0700 until 0830 when the culture was at 2 m depth. For the next 6.5 hours, the instrument was lowered 2 m every 0.5 hr until reaching 28 m depth at 1500. For the next 5 hours, SUPA was raised at 2 m every 0.5 hr until it reached 8 m depth at 2000. Although the subsequent comparison is between a SUPA diatom measured at minute intervals over one day and a biophysical model dinoflagellate monitored hourly over 10 days and the vertical movements in the two cases have different phase relationships with daylight, both cases share a Lagrangian experience in a vertical PAR gradient. The SUPA PAR signal has more detail, but the larger scale features resemble the structure provided by the cloud submodel (Fig. 6, top). The productivity comparison (Fig. 6 bottom) is more direct probably because phytoplankton cell physiology and the measurement of bulk oxygen and pH in the SUPA container combine to smooth

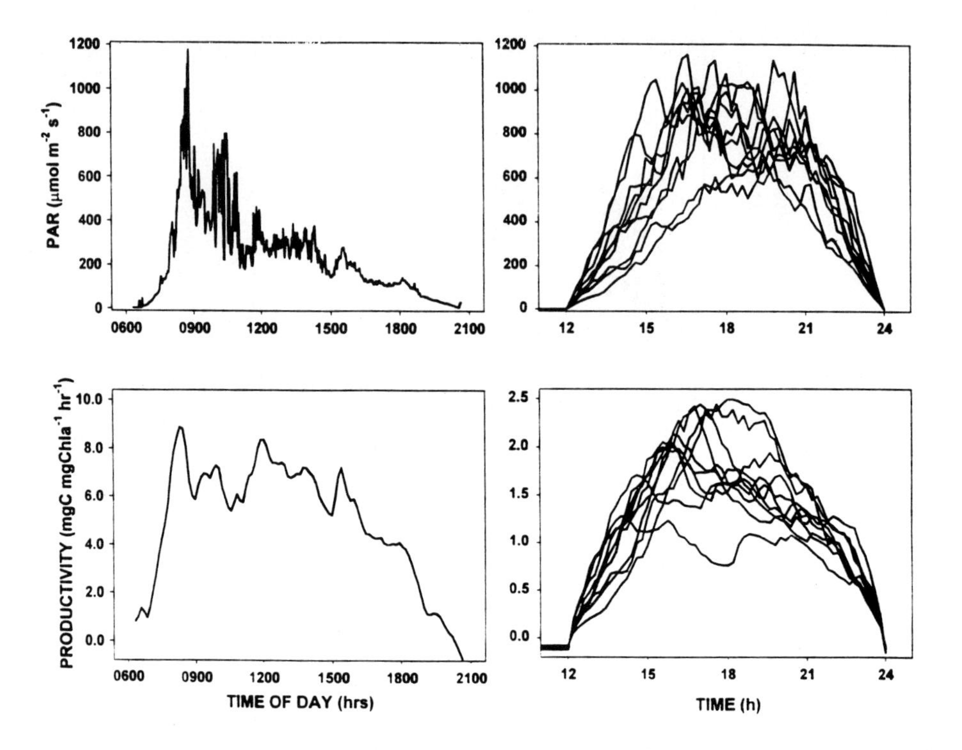

Fig. 6: A qualitative comparison of the PAR exposure (top) and productivity response (bottom) between a SUPA population of *Thalassiosira weisflogii* (left) and the metabolism-influenced cell from the biophysical model (right).

the measured response to high frequency PAR variability. This comparison is not merely a qualitative vindication of the modeling approach, but also a preview of expected results from a planned application of SUPA to a *Gymnodinium breve* bloom off the Florida coast. The present biophysical model will be tuned to *G. breve* based on measurements of population swimming behavior and physiology to support a more quantitative comparison with an actual HAB species.

4.2. Flow charts

The present biophysical model is composed of metabolic proxies for cellular state included as modulated photosynthetic and respiratory responses to environmental exposure. The first flow diagram labelled 'proxy' (Fig. 7) is discussed in more detail in Kamykowski and Yamazaki (1996) and depicts the control exerted by light (through photoinhibition [INH], diel variation of photosynthetic parameters [DIEL] and sun-shade acclimation [SUN-SHD]), gravity and nutrients on these rate processes and how these rate processes in turn influence the decisions related to ascent and descent. Kamykowski (1995) provided a case for the control of orientation decisions in terms of the biochemical pools that are influenced by photosynthesis and respiration. In this scenario, as represented in the second flow diagram labelled 'organic synthesis' (Fig. 7), some orientation control can be determined either passively by how a cell's center of mass changes relative to the cell's geometric center based on changing organelle size and/or location and

METABOLISM-INFLUENCED DINOFLAGELLATE: PROXY

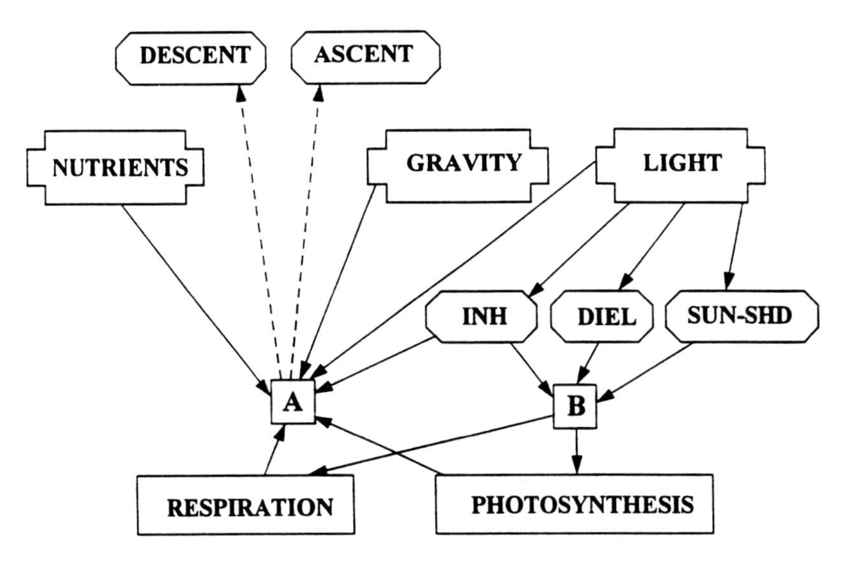

METABOLISM-INFLUENCED DINOFLAGELLATE: ORGANIC SYNTHESIS

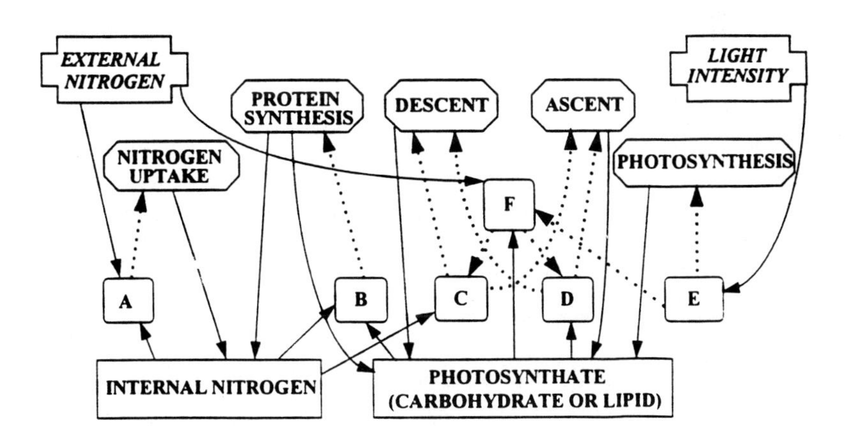

Fig. 7: The top flow diagram represents orientation control based on photosynthesis and respiration as presently included in the biophysical model. The bottom flow diagram respresents a future orientation control that includes a consideration of biochemical pools.

storage pool state with time, or actively by how the cell interprets sensory signals based on some chosen optimization of instantaneous biochemical state. The changes in cell biochemical state are related to the cell's age in the division cycle and to alternating biosynthetic pathways (like the accumulation of carbohydrate and lipid during the day and the utilization of these constitutents together with available nitrogen at night to form protein) based on the day/night cycle and environmental gradients. This second flow diagram looks to the future and provides an experimental outline of the information required in the next stage of biophysical models. Ideally, the planned Florida SUPA experiment focusing on *Gymnodinium breve* should be accompanied by measurements that monitor the changing state of at least some of the identified nutrient and biochemical pools to support model development.

5. Conclusions

Although the present model provides a step toward increased biological realism in biophysical models that include dinoflagellate behavior and physiology, improvements in both the physics and the biology are required. The physical improvements are underway based on improved measurements and the growing call for integration of these results into new physical models of the upper ocean. The biological improvements also are underway but are complicated by the many different species that need to be considered. Present day biological models inevitably depend on information from many sources (Wood and Latham 1992). At best, the information often is compiled on different clones of the same species from different regions; more likely, information also must be included from different taxa, ranging from closely related species to different phytoplankton classes, because the complete set of required measurements are not available for the selected species. The future application of biophysical modeling to HAB species requires a dedicated effort to obtain detailed behavioral and physiological information on selected species in order to make biophysical modeling a useful tool that relates to natural occurrences of those species.

6. Acknowledgements

Dr. G. S. Janowitz formulated the initial version of the time-dependent photoinhibition model. Mr. Robert Reed contributed to the development of the swimming speed model and to the conceptualization of the orientation model. This work was supported by NSF OCE-9503253 and NASA grant NAGW 3575-2 (DK, J. M. Morrison, G. S. Janowitz and GJK).

7. References

Anderson, D. 1995. Toxic red tides and harmful algal blooms: A practical challenge in coastal oceanography. *Rev. Geophys. (Suppl.)* p. 1189-1200, July 1995.

Barkmann, W. and Woods, J. D. 1996. On using a Lagarangian model to calibrate primary production determined from *in vitro* incubation measurements. *J. Plankton Res.* 18:767-788.

Brainard, K. E. and Gregg, M. C. 1995. Surface mixed and mixing layer depths. *Deep Sea Res.* 42:1521-1543.

Caldwell, D. R. and Moum, J. N. 1995. Turbulence and mixing in the ocean. *Rev. Geophys. (Suppl.)* p.1385-1394, July 1995.

Chisholm, S. W., Vaulot, D. and Olsen, R. J. 1984. Cell cycle controls in phytoplankton. In: Edmunds, L. N., Jr. (ed.) *Cell Cycle Clocks*. Marcel Dekker, New York, pp. 365-394.

Cullen, J. J. 1985. Diel vertical migration by dinoflagellates: roles of carbohydrate metabolism and behavioral flexibility. In: Rankin, M. A. (ed.) *Migration: Mechanisms and Adaptive Significance. Cont. Mar. Sci.* 27:135-152.

Cullen, J. J. and Horrigan, S. G. 1981. Effects of nitrate on the diurnal vertical migration, carbon to nitrogen ratio, and the photosynthetic capacity of the dinoflagellate, *Gymnodinium splendens. Mar. Biol.* 62:81-89.

Cullen, J. J. and M. R. Lewis. 1988. The kinetics of algal photoadaptation in the context of vertical mixing. *J. Plankton Res.* 10:1039-1063.

Denman, K. L. 1994. Scale-determining biological-physical interactions in oceanic food webs. In: Giller, P. S., Hildrew, A. G. and Raffaelli, D. G. (eds.) *Aquatic Ecology: Scale, Patterns and Processes*. Blackwell Scientific, New York, pp. 377-402.

Denman, K. L. and Gargett, A. E. 1983. Vertical mixing and advection of phytoplankton in the upper ocean. *Limnol. Oceanogr.* 28:801-815.

Denman, K. L. and Gargett, A. E. 1995. Biological-physical interactions in the upper ocean: the role of vertical and small scale transport processes. *Annu. Rev. Fluid Mech.* 27:225-255.

Dodge, J. D. and Gruet, C. 1987. Dinoflagellate ultrastructure and complex organelles. In: Taylor, F. J. R. (ed.) *The Biology of Dinoflagellates*. Blackwell Scientific, New York, pp. 92-119.

Evans, G. T. and Taylor, F. J. R. 1980. Phytoplankton accumulation in Langmuir cells. *Limnol. Oceanogr.* 25:840-845.

Eppley, R. W., Holm-Hansen, O. and Strickland, J. D. H. 1968. Some observations on the vertical migration of dinoflagellates. *J. Phycol.* 4:333-340.

Falkowski, P. G. and Wirick, C. D. 1981. A simulation model of the effects of vertical mixing on primary productivity. *Mar. Biol.* 65:69-75.

Franks, P. J. S. 1992. Sink or swim: accumulation of biomass at fronts. *Mar. Ecol. Prog. Ser.* 82;1-12.

Franks, P. J. S. 1995. Coupled physical-biological models in oceanography. *Rev. Geophys. (Suppl)* p. 1177-1187.

Franks, P. J. S. and Marra, J. 1995. A simple formulation for phytoplankton photoresponse and an application in a wind-driven mixed-layer model. *Mar. Ecol. Prog. Ser. 111*:143-153.

Geider, R. J., MacIntyre, H. L. and Kana, T. M. 1996. A dynamic model of photoadaptation in phytoplankton. *Limnol. Oceanogr.* 41:1-15.

Hand, W. G., P. A. Collard and Davenport, D. 1965. The effects of temperature and salinity changes of the swimming rate in the dinoflagellates, *Gonyaulax* and *Gymnodinium. Biol. Bull* 128:90-101.

Harris, G. P., Heaney, S. I. and J. F. Talling. 1979. Physiological and environmental constraints in the ecology of the planktonic dinoflagellate *Ceratium hirudinella. Freshwater Biol.* 9:413-428.

Holloway, G. 1994. On modeling vertical trajectories of phytoplankton in a mixed layer. *Deep-Sea Res.* 41: 957-959.

Janowitz, G. S. and Kamykowski, D. 1991. An Eulerian model of phytoplankton photosynthetic response in the upper mixed layer. *J. Plankton Res.* 13: 983-1002.

Jones, R. I. 1993. Phytoplankton migrations: Patterns, processes and profits. *Ergeb. Limnol.* 39:67-77.

Kamykowski, D. 1974. Possible interactions between phytoplankton and semidiurnal internal tides. *J. Mar. Res.* 32:67-89.

Kamykowski, D. 1979. The growth response of a model *Gymnodinium splendens* in stationary and wavy water columns. *Mar. Biol.* 50:289-303.

Kamykowski, D. 1981. Laboratory experiments on the diurnal vertical migration of marine dinoflagellates through temperature gradients. *Mar. Biol. 62*:81-89.

Kamykowski, D. 1995. Trajectories of autotrophic marine dinoflagellates. *J. Phycol.* 31:200-208.

Kamykowski, D. and McCollum, S. A. 1986. The temperature acclimatized swimming speed of selected marine dinoflagellates. *J. Plankton Res.* 8:275-287.

Kamykowski, D., McCullom, S. A. and G. J. Kirkpatrick. 1988. Observations and a model concerning the translational velocity of a photosynthetic marine dinoflagellate under variable environmental conditions. *Limnol. Oceanogr.* 33:66-78.

Kamykowski, D., Yamazaki, H. and Janowitz, G. S. 1994. A Lagrangian model of phytoplankton photosynthetic response in the upper mixed layer. *J. Plankton Res.*16:1059-1069.

Kamykowski, D. and Yamazaki, H. 1996. A study of metabolism-influenced orientation in marine dinoflagellate diel vertical migration. *Limnol. Oceanogr. (In Press).*

Kirkpatrick, G. J., Curtin, T. B., Kamykowski, D., Feezor, M. D., Sarton, M. D. and Reed, R. E. 1990. Measurement of photosynthetic response to euphotic zone physical forcing. *Oceanography* 3:18-22.

Lande, R. and Lewis, M. R. 1989. Models of photoadaptation and photosynthesis by algal cells in a turbulent mixed layer. *Deep-Sea Res.* 36:1161-1175.

Lebert, M. and Hader, D.-P. 1996. How Euglena tells up from down. *Nature* 379:590.

Levandowsky, M. and Kaneta, P. J. 1987. Behavior in dinoflagellates. In: Taylor F. J. R. (ed.) *The Biology of Dinoflagellates.* Blackwell Scientific, New York, pp. 360-397.

Li, M. and Garrett, C. 1995. Large eddies in the surface mixed layer and their effects on mixing, dispersion and biological cycling. *IUTAM Symposium on Physical Limnology*, Broome, Australia, p. 57-77.

MacIntyre, S. 1993. Vertical mixing in a shallow, eutrophic lake: Possible consequences for the light climate of phytoplankton. *Limnol. Oceanogr.* 38:798-817.

Mann, K. H. and Lazier, J. R. N. 1991. *Dynamics of Marine Ecosystems*. Blackwell Scientific, New York, 466 p.
McGillicuddy Jr., D. J. 1995. One-dimensional numerical simulation of primary production: Lagrangian and Eulerian formulations. *J. Plankton Res.* 17:405-412.
Pedley, T. J. and Kessler, J. O. 1992. Hydrodynamic phenomena in suspensions of swimming microorganisms. *Annu. Rev. Fluid Mech.* 24:313-358.
Price, J. F., Weller, R. A. and Pinkel, R. 1986. Diurnal cycling: observations and models of the upper ocean response to diurnal heating, cooling and wind mixing *J. Geophys. Res.* 91:8411-8427.
Raven, J. A. and Richardson, K. 1984. Dinophyte flagella: a cost benefit analysis. *New Phytopl.* 98:259-276.
Rizzo, P. J. 1987. Biochemistry of the dinoflagellate nucleus. In: Taylor, F. J. R. (ed.) *The Biology of Dinoflagellates*. Blackwell Scientific, New York, pp. 143-173.
Savidge, G. 1988. Influence of inter- and intra-daily light-field variablity on photosynthesis of coastal phytoplankton. *Mar. Biol.* 100: 127-133.
Sournia, A. 1974. Circadian periodicities in natural populations of marine phytoplankton. *Mar. Biol.* 12:325-389.
Taylor, F. J. R. and Pollingher, U. 1987. Ecology of dinoflagellates. In: Taylor, F. J. R. (ed.) *The Biology of Dinoflagellates*. Blackwell Scientific, New York, pp. 398-502.
Thomas, W. H. and Gibson, C. H. 1990a. Effects of small-scale turbulence on microalgae. *J. Appl. Phycol.* 2:71-77.
Thomas, W. H. and Gibson, C. H. 1990b. Quantified small-scale turbulence inhibits a red tide dinoflagellate, *Gonuaulax polyedra* Stein. *Deep-Sea Res.* 37:1583-1593.
Thorpe, S. A. 1995. Dynamical processes of transfer at the sea surface. *Prog. Oceanog. 35*:315-352.
Watanabe, M. and Harashima, A. 1986. Interaction between motile phytoplankton and Langmuir circulation. *Ecol. Modelling* 31:175-183.
Wood, A. M. and Latham, T. 1992. The species concept in phytoplankton ecology. *J. Phycol.* 28:723-729.
Woods, J. and Onken, R. 1982. Diurnal variation and primary production in the ocean-preliminary results of a Lagrangian ensemble model. *J. Plankton Res.* 4:735-756.
Yamazaki, H. and Kamykowski, D. 1991. The vertical trajectories of motile phytoplankton in a wind-mixed water column. *Deep-Sea Res.* 38: 219-241.
Yamazaki, H. and Kamykowski, D. 1994. Reply to Holloway. *Deep-Sea Res.* 41: 961-963.

Effects of Turbulence on Phytoplankton

Marta Estrada and Elisa Berdalet

Institut de Ciències del Mar. CSIC, P. Joan de Borbó, s/n, 08039 Barcelona, Spain

1 Introduction

The physical properties of the aquatic environment play a fundamental role in driving the dynamics of planktonic communities. Physical forcing not only shapes the structure of the pelagic environment but also affects biological processes in many direct and indirect ways (Mann and Lazier 1991). Advective and turbulent flows with typical dimensions of a few to tens of meters transport organisms around the water column, changing their light regime and photosynthetic environment and influencing physiological processes. In the case of motile cells, swimming behaviour combines with water motion to modulate the trajectories of the organisms in the water column. In addition, small-scale turbulence may exert direct effects on plankton, for example through interaction with nutrient uptake or by mechanical disturbance or disruption of cells.

Relationships between turbulence, photosynthetic activity and cell motility will be dealt with in other chapters of this volume (Cullen and MacIntyre, Kamykowski *et al.*). The aim of this contribution is to review information on some direct effects of turbulence with potential ecophysiological importance for phytoplankton. The topic is relevant in connection with harmful algal events because dinoflagellates, which are often responsible for noxious effects, exhibit particularly strong responses to turbulence. The subsequent presentation will begin with a brief consideration of basic fluid dynamic concepts relevant to the subject, followed by a discussion of major functional types of phytoplankton organization. Next, a survey of studies on effects at the organismal and population level will be presented. Finally, a consideration of how these findings may apply in the aquatic environment will be made.

2 Characteristics of turbulent motion

There is no generally accepted definition of turbulence. Turbulent flow patterns have been characterized by expressions like "a subtle mixture of order and chaos" (Nelkin 1992) or "a puzzling blend of order and disorder" (Vassilikos 1995). Turbulence is a property of the motion, not of the fluid, and two of its characteristics are randomness and diffusivity (Tennekes and Lumley 1972). Several theoretical approaches have proved useful for our understanding of turbulence and its effects, but this subject continues to be an unsolved problem, in the sense that there is not a clear understanding of the observed phenomena (Nelkin 1992). However, recent improvements in computer performance have allowed realistic direct simulation of turbulent flows at low Reynolds numbers (defined below). Appropriate references for

NATO ASI Series, Vol. G 41
Physiological Ecology of Harmful Algal Blooms
Edited by D. M. Anderson, A. D. Cembella and G. M. Hallegraeff

more information on turbulence theory or its ecological implications are Tennekes and Lumley 1972 or Mann and Lazier 1991.

3. Basic fluid dynamics concepts

The space and time scales of motion in aquatic environments cover a wide range (Denman and Gargett 1995). Kinetic energy is introduced in the ocean through heat transfer, wind and tides and is transmitted from large to progressively smaller scales, down to the domain where molecular viscosity smoothes out the flow and converts turbulent kinetic energy into heat. A fluid element within a flow is subjected to both inertial and viscous forces. Inertial forces are those that produced the acceleration of the fluid element until its present velocity or that would be necessary to stop it from travelling under its own inertia (Mann and Lazier 1991). Viscous forces are those related to the internal resistance of the fluid molecules. The relative importance of the inertial versus viscous forces acting on a fluid element can be estimated by the Reynolds number (Re), a dimensionless parameter which can be calculated using the expression:

$$Re=UL/\nu$$

where U is the velocity of the flow, L is a characteristic dimension and ν the kinematic viscosity of the fluid. In a pipe flow, for example, L would be the pipe diameter.

Reynolds numbers can be applied not only to fluid elements, but also to other bodies of interest, such as organisms. In this case, the length scale may be a characteristic dimension of the organism and the velocity a typical swimming or sinking speed. In this manner, Okubo (1987) calculated Reynolds numbers for living organisms ranging from large mammals to bacteria (Fig. 1).

One of the important breakthroughs in turbulence theory was the introduction by Kolmogorov of the idea that, in fully developed turbulence, kinetic energy is transferred through a cascade of eddies

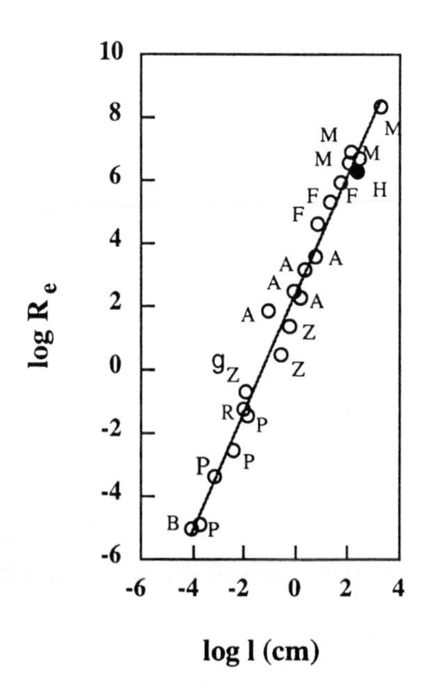

Figure 1. Reynolds number (Re) versus organism size (l). M: mammals, F: fish, A: amphipod, Z: zooplankton, R: protozoa, P: phytoplankton, B: bacteria, H: Man. Redrawn from Okubo (1987), with permission from the editor.

of decreasing size and velocity, down to dissipation scales at which molecular viscosity erases velocity gradients more rapidly that they can be produced. Within a range of scales, the input of turbulent kinetic energy into large eddies is balanced by its rate of dissipation at the smallest scales. The key dynamic quantity in this process is ε, the average rate of energy dissipation per unit mass. Kolmogorov assumed that the smallest length scale of the energy cascade (known as the Kolmogorov length scale) depended only on ε and ν, the viscosity of the fluid. Based on dimensional analysis, the Kolmogorov length scale (λ) can be expressed as:

$$\lambda_\nu \approx C(\nu^3/\varepsilon)^{1/4}.$$

In this expression, the factor C (which cannot be determined by dimensional analysis) is often taken as 1. Nevertheless, a C factor of 2π (introduced for mathematical convenience) appears to result in more realistic values for the length scale (Mann and Lazier 1991; Gargett, 1997). In natural aquatic media, the smallest values of $(\nu^3/\varepsilon)^{1/4}$ are of the order of mm (Table 1). The large scales are very important from the point of view of transport, but direct effects on phytoplankton occur at small scales. It has been shown by measurements that the maximum shear (or velocity gradient) resides at relatively small scales of the order of (3-4)λ_ν.

In general, mathematical treatment of turbulence considers Newtonian fluids, with constant molecular viscosity. However, viscosity and elasticity of sea water may vary in response to biological factors such as mucus excretion by phytoplankton. These changes could have theoretical and ecological implications (Jenkinson amd Biddanda 1995). It has also been pointed out that movement of animals could increase turbulent dissipation rates (Farmer *et al.* 1987).

4 The turbulent environment of phytoplankton cells

Phytoplankton organisms are typically smaller than the Kolmogorov length in aquatic environments, so that the velocity field in their vicinity should present a nearly linear velocity gradient or shear varying randomly in time, with a typical shear magnitude determined by (Lazier and Mann 1989, Jiménez 1997, Gargett 1997):

$$du/dz = C_{sh}(\varepsilon/\nu)^{1/2}$$

where du/dz is the velocity gradient and C_{sh} is a proportionality factor.

The shear environment for a particular cell depends not only upon its size, which determines the length scales for which velocity gradients will be relevant for the organism, but also on ε.

Turbulent shear is not evenly distributed in space. Strong gradients are more common than they would be in the case of a Gaussian distribution. This constitutes the basis of the intermittency of turbulence.

Oceanic turbulence appears to be strongly intermittent. This property is important regarding the effects of water motion on planktonic organisms. At small scales, intermittency is a general feature of turbulent flows and is related to the presence of strong coherent vortices, with diameters of the order of ten times the Kolmogorov scale, but much longer lengths. These events should be very intense from the point

Table 1. A: Rates of turbulent energy dissipation (ε), Kolmogorov length scales (λ_v) and strain rates (τ) for natural systems and for experimental systems. Average turbulent energy dissipation generated by wind in the upper 10 m (ε_{10}) and corresponding λ_v and τ (according to the model referred to in Kiørboe and Saiz, 1995). References: 1, Reynolds (1994); 2, Kiørboe and Saiz (1995); 3, Lakhotia and Papoutsakis (1992); 4, Thomas and Gibson (1990b); 5, Dempsey (1982).

Natural systems	ε ($cm^2\ s^{-3}$)	λ_v (mm)	τ (s^{-1})	Ref
Lakes	0.014 10^{-2}- 4 10^{-2}	0.7-2.9	0.1-2.1	1
Open ocean	10^{-6} -10^{-2}	1-0.10	0.01-1.	2
Shelf	10^{-3} -10^{-2}	0.18-0.10	0.32-1.	2
Coastal zone	10^{-3}-10^{0}	0.18-0.03	1.0-10.	2
Tidal front	10^{-1}	0.06	3.16	2
Tidal estuary (Severn)	2.2-8 10^{-2}	1.3 10^{-1}- 4.3 10^{2}	0.2-0.67	1
Experimental systems				
Animal cell cultures	0.6-1.	0.36-0.31	7.7-11	3
Couette cylinders				
Range	0.045-164	0.08-0.66	2.2-132	
Effect threshold	0.18	0.48	4.4	4
Paddle stirrer	0.096-0.14	0.62-0.52	3.1-3.7	5
Model Wind speed ($m\ s^{-1}$)				
5	1.7 10^{-3}	0.16	0.4	--
10	1.5 10^{-2}	0.09	1.2	--
15	4.9 10^{-2}	0.07	2.2	--
20	8.4 10^{-2}	0.06	2.9	--

of view of plankton, but calculations show that their probability is small. At relatively large scales, intermittency may be induced by factors like variable wind forcing and breaking of surface waves (George *et al.* 1994).

In terms of classic fluid dynamic parameters, the relevant characteristics of the velocity field, concerning direct effects on algae, can be described by the viscous dissipation rate, ε ($L^2\ T^{-3}$), the rate of strain parameter, γ (T^{-1}) and the shear stress,

$\tau \approx \mu du/dz$ (MLT^{-2}), where $\mu = \rho\nu$ is the dynamic viscosity (Thomas and Gibson 1990). The rate of strain parameter is proportional to a velocity gradient or shear and represents the magnitude of the deformation rate due to mean velocity gradients in the flow fields; it is dimensionally equivalent to $(\varepsilon/\nu)^{1/2}$. A formal treatment of this subject can be found in Tennekes and Lumley (1972).

There is an important body of bioengineering literature addressing the mechanical effects of turbulence on cells cultured in bioreactors. Intense stirring used to supply oxygen to bacterial or yeast cultures does not harm them. However, animal cells have much lower tolerance of fluid forces (Cherry 1993) and reduced growth rates can occur under agitation conditions within the range used in long term cultures. Some experiments excluding headspace air from the vessels have shown that the bulk liquid may not be the site where most deletereous effects occur and that in normal cultures many cells can be killed by bubble bursting. However, there are also indications that fluid motion affects cell physiology with sublethal or lethal effects.

The forces acting on a spherical particle suspended in a turbulent flow have been modelled using expressions which include terms representing buoyancy, pressure gradients (implying fluid acceleration), Stokes drag and other forces. Such equations have been used, with some simplifications, in simulation models of particle motion and particle-fluid interactions. A similar approach could be used to describe the mechanical forces acting on living cells suspended in a fluid (Cherry 1993). Other possibilities include experimental modelling using nylon capsules filled with an indicator liquid to simulate cells and particle tracking velocimetry.

5 Functional morphology of phytoplankton as related to external energy

Phytoplankton organisms can be classified into a number of ideal categories or "life-forms", based on their apparent adaptations to recurrent combinations of environmental factors. According to Margalef (1978), the supply of nutrients and intensity of turbulence are key factors determining the morphological and physiological properties of phytoplankton. These life-forms express apparent adaptations to recurrent patterns of selective factors and tend to coincide with taxonomical groups based on genetic affinities. This view of phytoplankton strategies was summarized (Margalef 1978; Margalef *et al.* 1979) in a conceptual graph using nutrient concentration and intensity of turbulence as coordinates. In principle, high nutrient concentrations, a situation generally associated with relatively high turbulence and mixing, would allow the development of forms with high growth rates and large sizes. In a strongly turbulent medium, lack of motility would not be a disadvantage. In marine environments, diatoms tend to be the dominant group under these conditions. However, species adapted to persist under low nutrient concentrations would have to be adapted also to low turbulence. In this case, motility may help to exploit the resources of different levels of the water column. Dinoflagellates, with their often peculiar shapes, motility, and migratory behaviour, offer a typical example of this kind of strategy. Coccolithophorids tend to occupy an intermediate position between diatoms and dinoflagellates. An intermediate or close analogy to the dinoflagellate situation could be expected for flagellates, although their taxonomical diversity and the scarcity of reliable information on their field distributions do not allow for generalizations. Within this general scheme, properties of the organisms, such as size, may represent extreme niches within major groups. Thus, small diatoms like *Nitzschia closterium* or some *Thalassiosira*, with low sedimentation rates may thrive in intermediate stages of the succession in microcosms in conditions of stagnant water and high nitrate concentrations.

Quantitative attempts to interpret these relationships have been put forward by Bowman *et al.* (1981), who plotted the abundances of diatoms and microflagellates in Long Island and Block Island Sounds in a diagram using the stratification index (s) and depth-scaled-by-light (kh, where h is depth and k the diffuse extinction coefficient) as coordinates. These parameters, which can be directly measured or derived from convenient models, were used to express water motion (s) and growth conditions of the organisms (kh). A similar approach was proposed by Jones and Gowen (1990).

Attempts to examine the spectrum of morphological features of phytoplankton, to understand their ecophysiological significance, have been reviewed by several authors. Sournia (1982) considered the diversity of morphological and cytological features of marine phytoplankton, the environmental requirements regarding planktonic life and the implications of cell size. Elbrächter (1984) discussed the most commonly accepted functional types of primary producers with particular regard to their use by modellers. Physical properties of the phytoplankton organisms such as size, shape and density can be expected to interact with water motion and turbulence (Fogg 1991; Kiørboe 1993; Kamykowski *et al.*, this volume). Regarding cell size, at least some of its implications for physical-biological interactions in the planktonic ecosystem seem to be well understood (Kiørboe 1993). For example, larger cells sink faster than smaller cells of similar shape. However, in spite of repeated efforts, many other proposed relationships between "form" and "function" in phytoplankton remain speculative (Sournia 1982).

6 Effects of turbulence on nutrient uptake.

A motionless phytoplankton cell suspended in the water and taking up nutrients will tend to create a layer of decreased nutrient concentration around it which increases the possibility of diffusion limitation of nutrient uptake. Relative motion of cell and fluid, either originated by active swimming or sinking of the organism or by motion of the fluid will have an effect on renewing the depleted zone. Munk and Riley (1952) derived formulae for the rates of nutrient absorption by phytoplankton cells approximating the shapes of spheres, discs, cylinders and plates and considered the effect of the movement of the water relative to sinking cells. Their results were not essentially modified by turbulence. The influence of diffusion transport and its interaction with biological properties of the organisms in determining nutrient uptake rates were examined by Pasciak and Gavis (1974) and Gavis (1976) using Michaelis-Menten dynamics. Gavis (1976) described the effect of relative motion between an ideal spherical organism and the medium. He concluded that motion could reduce diffusion transport limitation, but not eliminate it completely. Lazier and Mann (1989) and Mann and Lazier (1991) reviewed studies addressing the effect of turbulence on diffusive layers around small organisms and concluded that relative motion by swimming or sinking would have a significant effect for cells larger than about 5 μm, and that strong turbulence would produce only a 2% increase in flux for stationary cells 100 μm in diameter. The effects of fluid motion on nutrient fluxes to planktonic cells were revisited by Karp-Boss *et al.* (1996) who proposed improvements and extensions of previous solutions. According to these authors, a minimum cell radius of about 20 μm was necessary in order for sinking or swimming to produce a relevant increase of nutrient flux with respect to the stagnant-water case. Karp-Boss *et al.* (1996) stated that turbulence effects could be an order of magnitude greater than previously postulated, with a minimum cell size of 60 μm needed to experience substantial gain. Chains of diatoms or filamentous cyanobacteria, with lengths approaching the Kolmogorov scale (the smallest eddy size in the fluid) could be expected to experience enhanced relative motion. Another conclusion, of potential importance with respect to motile cells such as dinoflagellates, was that rotation, whether caused by active swimming or by shear from fluid motion, would reduce the rate of nutrient transfer relative to a non-rotating cell, except when the axis of rotation was parallel to the direction of flow (i.e. the direction of swimming or of shear). Shape of the cells was another important factor, and its dependence on cell size varied for different motion environments. These observations, although in need of experimental testing, provide further basis for suggestions concerning the possible ecophysiological role of spines, horns and other complex shapes.

There have been few experimental studies of the effect of fluid motion on nutrient uptake by phytoplankton. Pasciak and Gavis (1975), using a Couette device, found that nitrate uptake by *Ditylum brightwellii*, a relatively large diatom, increased about 5% at 5 rad s^{-1}. Canelli and Fuhs (1976) examined the effect of the sinking rate of two diatoms (*Thalassiosira* spp.) spp. on phosphorus uptake using, as an analogy, a constant flow of medium through a filter in which the cells were held. However, according to Karp-Boss *et al.* (1996), this approach is not valid to derive conclusions concerning natural fluid motions in the water column. Savidge (1981) examined the effect of turbulence, obtained by means of an oscillating grid, on nutrient uptake by the diatom *Phaeodactylum tricornutum*. In this case, nitrate uptake was enhanced by stirring, but phosphate uptake and carbon fixation (as measured by the ^{14}C method) decreased. These results show that effects of turbulence on nutrient uptake may be more complex than predicted by relatively simple theoretical considerations.

7 Microzones

The same reasoning applied to the problem of nutrient uptake can be applied to metabolite excretion and the formation of microzones or microenvironments around plankton organisms (Currie 1984, Mann and Lazier 1991). Lehman and Scavia (1982a) used autoradiography to show that *Chlamydomonas* was able to take up labelled phosphorus excreted by *Daphnia*. They presented model results showing that excretion by *Daphnia* could be a significant source of nutrient for the alga (Lehman and Scavia 1982b). These results were challenged by Currie (1984), who concluded that typical nutrient concentrations in oligotrophic areas were sufficient to meet the growth needs of phytoplankton in these areas. Jackson (1990) showed that nutrient plumes from microzooplankton would diffuse too fast to be used by primary producers. According to Mann and Lazier (1991), most calculations suggest that micronutrient patches may be too few and disperse too fast to be of great significance for phytoplankton. The occurrence of relatively persistent microenvironments may be more likely in connection with macroscopic particles (Alldredge and Cohen 1987).

An interesting experimental demonstration of the influence of cell size has been reported by Richardson and Stolzenbach (1995). These authors investigated the developments of extracellular microenvironments of pH produced by individual photosynthesizing phytoplankton, using a blue dye to detect extracellular oxidation of Mn (II) to Mn O_x.

Small-scale turbulence has been advocated as an explanation of the low rate of N_2 fixation in estuaries, coastal marine waters and many open-ocean waters. Carpenter and Price (1976) suggested that turbulence could disrupt the colonies of *Trichodesmium.* Paerl and Bebout (1988) proposed that higher turbulence and lower dissolved organic matter in marine environements relative to fresh-water ones reduced the likelihood of oxygen-depleted microzones around cyanobacterial cells, increasing the possibility of nitrogenase inactivation. This hypothesis is currently the object of a lively discussion (Howarth *et al.* 1995).

8 Encounter rate and predator-prey interactionss

The effects of turbulence on particle distribution have been examined using experimental methods and numerical models. Hill *et al.* (1992) used neutrally buoyant spheres in grid turbulence to examine interparticle velocities. Squires and Yamazaki (1996) carried out simulations showing that particles were preferentially concentrated by turbulence. Jiménez (1997) proved that the conclusions of Squires and Yamazaki (1996) were valid for heavy particles (relative to the fluid), but that no appreciable concentrations could develop for particles with density close to that of the fluid (the case of most biological material in aquatic media). Turbulent shear is hypothesized to increase particle encounter rates and enhance aggregation of small living or non-living particles into larger and presumably faster sinking ones. Jackson (1990) applied coagulation theory to study the formation of aggregates by collision between suspended phytoplankton cells. He showed that coagulation rates increased rapidly above a cell concentration threshold which was inversely related to fluid shear, whether laminar or turbulent, algal size and a so-called coefficient of stickiness, which could be particularly important for diatoms possessing spines or setae. As pointed out by Jackson (1990) and Kiørboe *et al.* (1990), physical coagulation processes could place an upper boundary on the accumulation of algal blooms. Predator-prey interactions among planktonic organisms may be influenced by the existence of relative motion between predator and prey or by other mechanisms related to water turbulence (Alcaraz *et al.* 1988, Osborn 1996, Marrasé *et al.* 1990). A simple model of predator-prey encounter rates was formulated by Kiørboe and Saiz (1995), who concluded that the effect of turbulence was minor for suspension feeding copepods but benefited ambush feeding ones. Osborn (1996) proposed an alternative model of copepod feeding based on the diffusion of food particles towards the predators and concluded that feeding currents interacted with turbulence to increase the flux of food.

Turbulence was found to increase grazing rates of protozoa on bacteria by Peters *et al.* (1996).

9 Effects of turbulence on growth and cell division

Agitation by means of aeration, paddle wheels, magnetic stirrers or orbital shakers is often used as a way of enhancing growth of algal cultures. Up to a certain intensity, mixing should improve light and nutrient transport regimes. Very high levels of agitation can be expected to produce mechanical cell damage. In some cases, even presumably moderate levels of agitation have been found to inhibit growth rates or to cause cell damage; however, few studies have tested the effect of different agitation rates on the growth of particular phytoplankton strains. In general, agitation conditions are described as rotation speeds or oscillations per unit time; only in very few cases have the characteristics of agitation been expressed in fluid mechanics terms such as the rate of turbulent energy dissipation (ε), the rate of strain or the shear stress (Thomas and Gibson 1990). Even in these rare occasions it is difficult to compare results obtained using different devices (Peters and Redondo, 1997).

An overview of experimental studies concerning the effects of turbulence on plankton growth rates has been prepared by F. Peters and C. Marrasé (unpublished.). A summary of their phytoplankton data is presented in Fig. 2. The ε values were taken from the papers or estimated on the basis of other data provided. For the data shown in the figure, the threshold for deletereous effects appears to be about 0.1 cm^2s^{-3}. Most negative effects were evidenced in dinoflagellate cultures.

Dinoflagellates appear to be particularly sensitive to water turbulence, but relatively moderate agitation may also have effects on other algal groups. Fogg and Than-Tun (1960) found that growth of the cyanobacterium *Anabaena cylindrica* was stimulated up to a shaking rate of 90 oscillations per minute and decreased at higher rates. Volk and Phinney (1968) described inhibitory effects of agitation for *Anabaena spiroides.*

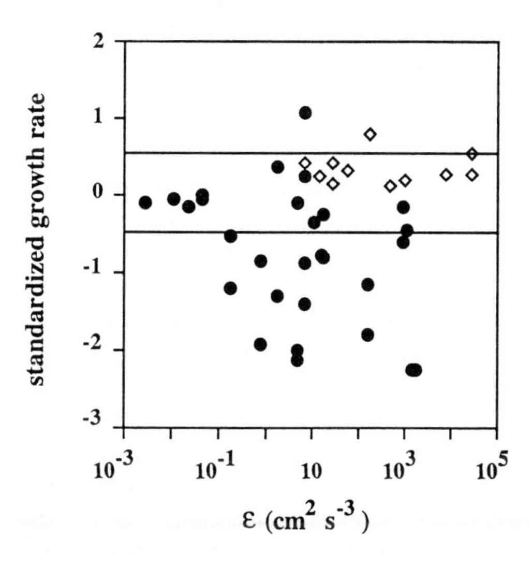

Figure 2. Growth rates (μ) under different turbulence levels normalized to growth rates under still water conditions. Data from F. Peters and C. Marrasé (unpublished). ● Dinophyceae. ◊ Other phytoplankton.

Savidge (1981) reported that cell division times of *Phaeodactylum tricornutum* in the exponential phase decreased with increasing agitation in phosphate-limited cultures, but increased in nitrate-limited cultures. Schöne (1970) showed that intensity of the motion of the sea surface (in a categorical scale) was inversely related to chain length of several colony-forming diatoms like *Chaetoceros curvisetus* and *Skeletonema costatum*; the diminution of chain length with increasing motion was apparently due to mechanical breakage of chains. He could produce similar effects using air bubbling of *S. costatum* cultures in the laboratory. Other effects of agitation were more subtle, such as the synchronization of cell division in *S. costatum* populations, after only 5 min of agitation per day. Bakus (1973) carried out competition experiments between *Scenedesmus* (chlorophyte) and *Stichococcus* (generally classified as a chlorophyte) under varying conditions of irradiance and aeration and found that the advantage in the final yield of *Scenedesmus* over *Stichococcus* could be reversed in the presence of a combination of turbulent water movements and decreased illumination.

Mention of inhibitory effects of aeration on dinoflagellates was made by Tuttle and Loeblich (1975) and Galleron (1976). However, Siegelman and Kycia (1979) could grow large-scale cultures of dinoflagellates under strong aeration. White (1976) reported that continuous rotary shaking of *Alexandrium fundyense* (= *Gonyaulax excavata*) cultures at speeds of 125 rpm and greater caused death and disintegration of the cells. Intermittent shaking, even for only 30 min d^{-1} also caused growth inhibition. Pollingher and Zemel (1981) found that even intense winds blowing during the day did not affect the division rate of *Peridinium gatunense* (= *Peridinium cinctum* forma *westii*) in Lake Kinneret. However, an inhibitory effect was noted when wind episodes (speed exceeding 3.5 m s^{-1}) occurred between 18:00 and 02:00 h, a period which corrresponded to the premitotic and mitotic phases of cell division. The relationship between water turbulence and division rate of *P. gatunense* was tested experimentally using rotary shaking at 100 rpm in the laboratory. Continuous shaking increased cell mortality and decreased division rates with respect to the controls. Intermittent shaking of 2 h d^{-1} during the dark inhibited cell division, while shaking during the light period did not affect the division process.

Berdalet (1992) placed *Gymnodinium nelsonii* cultures on an orbital shaker at 100 rpm and showed that cell division was prevented, cellular volume increased up to 1.5 times that of unperturbed controls, nuclear morphology was modified and RNA and DNA concentrations per cell increased up to 10 times those of controls. The effects of shaking were reversible after 10 days of treatment, but after 20 days there was total cell death and disintegration. Berdalet and Estrada (1993) reported results of shaking experiments using dinoflagellates and representatives of other phytoplankton groups. No significant differences were found between cultures of *Dunaliella tertiolecta* (chlorophyte), *Isochrysis galbana* (chrysophyte), *Heterosigma akashiwo* (raphidophyte; strain from the Plymouth culture collection) and *Thalassiosira weissflogii* (diatom) unshaken or shaken at speeds between 100 and 130 rpm. In contrast, shaking at similar intensities caused growth inhibition of several dinoflagellates (Table 2). Further experiments showed similar results with *A. minutum* and *Prorocentrum triestinum* (Berdalet and Koumandu, in prep.), but a small *Gymnodinium* sp. was not affected by shaking. Several points should be noted from Table 2, besides the

Table 2. Effects of orbital and vertical Netlon grid shaking on growth of several dinoflagellate species. Volume: volume of culture vessel. Diameter: mean diameter of the organisms for each experimental condition. Growth rates (divisions d^{-1}) were calculated from cell counts except for *S. trochoidea* where growth was followed by measuring *in vivo* fluorescence.

Species	Treatment	Volume	Diameter	Division rate	Reference
Gymnodinium nelsonii	Still	4 L	35 µm	0.42	Berdalet 1992
	100 rpm	4 L	41 µm	no growth	id.
	Netlon grid	1 L		death	id.
Gymnodinium sp.	Still	4 L	10 µm	0.35	Berdalet & Koumandou in prep.
	100 rpm	4 L	10 µm	0.35	id.
Alexandrium minutum	Still	4 L	17 µm	0.65	Berdalet & Koumandou in prep.
	110 rpm	4 L	19 µm	0.1	id.
	Netlon grid	1 L	17 µm	No effect	Berdalet & Estrada 1993
Prorocentrum triestinum	Still	4 L	11 µm	0.66	Berdalet & Koumandou in prep.
	110 rpm	4 L	14.5 µm	0.16	id.
Procentrum micans	80-90 rpm	4 L	18 µm	reduced growth	Berdalet & Estrada 1993
	100 rpm	4 L	18 µm	reduced growth	id.
Scrippsiella trochoidea	100 rpm	4 L	20 µm	reduced growth	Berdalet & Estrada 1993
	Still	10 mL	20 µm	0.12*	id.
	100 rpm	10 mL	20 µm	no growth	id.

differences between dinoflagellates and other groups. First, for the same speed of the orbital shaker, effects within a species are stronger in smaller vessels. Second, the only non-affected dinoflagellate is the smallest of the group. These relationships are based on a very limited data set and could be fortuitous. However, they are what could be expected from fluid dynamic considerations. The energy imparted by the orbital shaker is dissipated into a relatively greater fluid volume in the larger vessels, thus leading to smaller energy dissipation rates and rates of strain. For the same characteristic value of the velocity gradient, velocity differences across a small cell will be lower than across a large cell.

An early attempt to reproduce natural levels of turbulence in the laboratory and measure their effects on phytoplankton was carried out by Dempsey (1982). She used a six paddle stirrer, but she had several difficulties with the apparatus, ranging from toxic effects of the stainless steel paddles to problems in maintaining the low rotation rates needed (6-7 rpm). She found that *Heterocapsa triquetra* and *Skeletonema costatum* appeared to show some inhibition of growth rate at turbulence levels of the order of 10^{-2} watts m^{-3}. In order to mix the cells prior to sampling, Dempsey had to apply, for a short time, rotation speeds that were more than 10 times greater than the experimental ones. She applied a presumably similar agitation to the control cultures, using a magnetic stirrer, but was unable to make precise comparisons. This kind of problem may plague many experimental studies.

Thomas and Gibson (1990a,b) used Couette devices to expose cultures of *Lingulodinium polyedrum* (= *Gonyaulax polyedra*) to known shear rates simulating the small-scale shear encountered by the organisms in turbulent waters. They found that growth inhibitory levels of turbulent kinetic energy dissipation (ε) and rate of strain (γ) were about 0.18 cm^2 s^{-3} and 4.4 rad s^{-1}, respectively. Motile cells in the shear-inhibited cultures lost their longitudinal flagella. Thomas *et al.* (1995) used the inhibitory response of *L. polyedrum* to compare the turbulence level in culture bottles with that in Couette devices; the conclusion was that photosynthesis, although inhibited at high turbulence levels, was less sensitive than cell division. To account for possible effects of turbulence intermittency, Gibson and Thomas (1995) subjected cultures of *L. polyedrum* in Couette devices, to various rotation rates maintained during different fractions of the day. The authors reported that the daily averaged rate of strain threshold for zero growth, was reduced by two orders of magnitude by intermittency of the simulated turbulence.

In natural ecosystems, the effects of water turbulence cannot be separated from those of other factors such as light and nutrients; this makes it difficult to investigate the direct effects of turbulence on phytoplankton in natural conditions. An experimental approach to this problem has been based on the use of meso- or microcosms (Eppley *et al.* 1978) enclosing natural plankton communities. Estrada *et al.* (1987) studied the response of marine phytoplankton populations enclosed in 30 dm^3 microcosms and exposed to different conditions of water stirring obtained by means of oscillating or rotating grids. Several dinoflagellate taxa were abundant throughout the experiment in two microcosms with grids oscillating vertically at 20 rpm, but dinoflagellates disappeared soon from the water column in non stirred or in more strongly stirred containers. This selective effect was probably a result of direct effects of turbulence on the physiology of the dinoflagellate cells, combined with alterations of their swimming patterns and migratory behaviour.

10 Mechanisms of cellular damage in dinoflagellates

Small-scale turbulence may produce deletereous effects on dinoflagellates by mechanical damage to cell integrity, by influencing physiological processes like cell division, or by disorienting the organisms and interfering with their movements or migrations (White 1976). It can be speculated that the peculiar sensitivity of dinoflagellates to water motion and their adaptations for living in stratified water columns are related to their evolutionary history of diversification in mesozoic seas, which are supposed to have been less turbulent than seas in later periods (Kennett 1982).

The observation that nucleic acid content increases and cells become larger under inhibitory turbulence levels (Berdalet 1992) suggests that the processes linked to nutrient uptake, biosynthetic metabolism and cell division are affected to different degrees by water turbulence. The DNA concentration increase in shaken cells suggests an impairment of chromosome separation, after the duplication of DNA (Berdalet 1992). Dinoflagellate mitosis is extranuclear and characterized by an entirely cytoplasmatic spindle. In free-living forms, the spindle is subdivided into bundles enclosed in cytoplasmatic tunnels piercing the nucleus. These tunnels contain both kinetochore and non-kinetochore microtubules (Raikov 1995). The blockage of dinoflagellate division by shaking is hypothesized to occur by physical disturbance of the microtubule assemblage and/or the mechanisms responsible for chromosome separation (Karentz 1987, Berdalet 1992). A comparable impairment of chromosome separation has been observed in mutants of the yeast *Schizosaccharomyces plombe* subjected to starvation or sublethal temperature (Broeck *et al.* 1991). Ultrastuctural studies of dinoflagellate cells affected by shaking should help to clarify this point.

As proposed by White (1976), inhibition of dinoflagellate growth by turbulence can also be related to more subtle mechanisms than those causing cell damage or impairing cell division. These effects could include alteration of swimming patterns, phototaxis and migration (White 1976, Estrada *et al.* 1987). Evidence of active response of dinoflagellates to water turbulence is provided by the common observation that dinoflagellates concentrate in preferred zones of containers with non-uniform agitation intensity (White 1976, Berdalet and Estrada 1993). Other possible effects are suggested by Karp-Boss *et al.* (1996), who concluded that water motion may decrease nutrient uptake if it hinders the ability of swimming cells to maintain a rotational axis parallel to the direction of swimming or of the shear flow. A promising approach to explore physical-biological interactions concerning dinoflagellate motility consists in the combining experimental observation with the use of numerical simulation models (Kamykowski *et al.* this volume).

11 Comparison of experimental and natural conditions

It is difficult to ascertain whether laboratory results showing inhibition of dinoflagellate growth by turbulence are significant for natural conditions. In most cases, the turbulence used in the experiments has not been defined in fluid dynamic terms; even in the occasions in which turbulence parameters were given, comparison with field conditions must be made with caution. Couette flows, for example, may not be very representative of field conditions. Furthermore, it is likely that not only turbulence intensity, but also differences in the flow field at sufficiently large scales.

As a first approximation, rates of turbulent kinetic energy dissipation (ε) applied in laboratory experiments can be compared with conditions at sea. The range of ε used in experiments with dinoflagellates can be seen in Fig. 2. Additional data for experimental systems are given in Table 1. In most cases, viscous dissipation rates used in experiments with phytoplankton are larger than those typically reported for natural conditions (Table 1). However, it is possible that processes like intermittent wave breaking could enhance dissipation rates in surface layers above the threshold needed to cause cellular effects on dinoflagellates (Gibson and Thomas 1995). The role of bubbles could be important, as suggested by experiments on animal cells. In particular, environments like the surf zone have been shown to present ε values exceeding 10^2 cm^2 s^{-3} (George *et al.* 1994). In view of these observations, it could be speculated that direct effects of turbulence on phytoplankton cells could be significant in areas such as shallow estuaries subjected to intense winds.

In spite of the possibility of inhibitory effects, small-scale turbulence is not likely to be *per se* a major factor controlling bloom formation. Algal blooms can be better explained as a result of the combination of mesoscale circulation features and physiological responses of the organisms, basically their motility (Wyatt and Horwood 1973). Bloom termination is generally associated with storms or strong winds inducing circulation patterns that tend to disperse the cells. Under these weather conditions, bloom disappearance could be favoured by direct effects of turbulence on the organisms. However, in natural environments, it is difficult to separate direct effects of turbulence from those related to organism transport. The relationship between the dominance of major life-forms like diatoms or dinoflagellates and turbulence can be explained through the interactions between organism entrainment within the water column, motility and physiological responses to nutrient and light availability (Margalef 1978). A new examination of these concepts in the context of harmful algal events is given in Riegman (this volume).

12 Conclusion

Turbulence is a key environmental property for plankton and plays a central role in controlling or modulating the effects of other factors such as light and nutrients. However, interaction between hydrodynamics and phytoplankton life is poorly understood, especially over small scales. Progress in this field of work will require improvements in characterization of the physical conditions of the environment and the organism response at appropriate scales. This is a challenging and exciting field in which recent instrumentation and computing developments have opened a host of research possibilities. These advancements and the increasing emphasis on interdisciplinary collaboration lead to expect that the near future will bring significant advances in our knowledge of the relevance of turbulence for the planktonic ecosystem and, in particular, for harmful algal events.

13. Acknowledgements

This work was supported by EU grant 950033, CICYT (Spain) grant AMB94-0853 and the CSIC. We thank all colleagues whose comments improved the final version of the manuscript. Dr. Mikel Latasa provided valuable help with editing.

14. References

Alcaraz, M., Saiz, E., Marrasé, C., Vaqué, D. (1988). Effects of turbulence on the development of phytoplankton biomass and copepod populations in marine microcosms. *Mar. Ecol. Prog. Ser.* 49: 117-125.

Alldredge, A L., Cohen, Y. (1987). Can microscale chemical patches persist in the sea? Microelectrode study of marine snow, fecal pellets. *Science* 235: 689-691.

Bakus, G.J. (1973). Some effects of turbulence and light on competition between two species of phytoplankton. *Inv. Pesq.* 37: 87-99.

Berdalet, E. (1992). Effects of turbulence on the marine dinoflagellate *Gymnodinium nelsonii*. *J. Phycol.* 28: 267-272.

Berdalet, E., Estrada, M. (1993). Effects of turbulence on several dinoflagellate species. In: Smayda, T. J., Shimizu, Y. (eds.) *Toxic phytoplankton blooms in the sea.* Elsevier, New York, pp. 737-740.

Bowman, M.J., Esaias, W.E., Schnitzer, M.B. (1981). Tidal stirring and the distribution of phytoplankon in Long Island and Block Island Sounds. *J. Mar. Res.* 39: 587-603.

Broeck, D., Bartlett, R., Crawford, K., Nurse, P. (1991). Involvement of $p34^{cdc2}$ in establishing the dependency of S phase on mitosis. *Nature* 349: 388-393.

Canelli, E., Fuhs, G.W. (1976). Effect of the sinking rate of two diatoms (*Thalassiosira* spp.) on uptake from low concentrations of phosphate. *J. Phycol.* 12: 93-99.

Carpenter, E., Price, C. (1976). Marine *Oscillatoria* (*Trichodesmium*) explanation for aerobic nitrogen fixation without heterocysts. *Science* 191: 1278-1280.

Cherry, R S. (1993). Animal cells in turbulent fluids: details of the physical stimulus and the biological response. *Biotech. Adv.* 11: 279-299.

Cullen, J.J., MacIntyre, J.G. (This volume). behavior, physiology and the niche of depth-regulating phytoplankton.

Currie, D.J. (1984). Phytoplankton growth and the microscale nutrient patch hypothesis. *J. Plank. Res.* 6: 591-599.

Dempsey, H.P. (1982). The effects of turbulence on three algae *Skeletonema costatum, Gonyaulax tamarensis, Heterocapsa triquetra.* S. B. Thesis. Massachusetts Institute of Technology, Cambridge, Massachusetts.

Denman, K.L., Gargett, A E. (1995). Biological-physical interactions in the upper ocean: the role of vertical and small scale transport processes. *Ann. Rev. Fluid Mech.* 27: 225-255.

Elbrächter, M. (1984). Functional types of marine phytoplankton primary producers and their relative significance in the food web. In: Fasham, J. R. (ed.) *Flows of energy and materials in marine ecosystems.* Plenum Publ. Corp. pp. 191-221.

Eppley, R.W., Koeller, P., Wallace Jr., G.T. (1978). Stirring influences the phytoplankton species composition within enclosed columns of natural sea water.

J. exp. mar. Biol. Ecol. 32: 219-239.

Estrada, M., Alcaraz, M., Marrasé, C. (1987). Effects of turbulence on the composition of phytoplankton assemblages in marine microcosms. *Mar. Ecol. Prog. Ser.* 38: 267-281

Farmer, D.D., Crawford, G B., Osborn, T R. (1987). Temperature and velocity microstructure caused by swimming fish. *Limnol. Oceanogr.* 290: 390-392.

Fogg, G.E. (1991). Tansley Review No. 30. The phytoplanktonic ways of life. *New Phytol.* 118: 191-232.

Fogg, G.E., Than-Thun. (1960). Interrelations of photosynthesis and assimilation of elementary nitrogen in a blue green alga. *Proc. R. Soc.* London B 153: 111-127.

Galleron, C. (1976). Synchronization of the marine diinoflagellate *Amphidinium carteri* in dense cultures. *J. Phycol.* 12: 69-73.

Gargett, A.E. (1997). "Theories" and techniques for observing turbulence in the ocean euphotic zone. In: Marrasé, C., Saiz, E., Redondo, J. M. (eds.) Lectures on plankton and turbulence. *Scientia marina* 61 (Suppl. 1): 25-45.

Gavis, J. (1976). Munk and Riley revisited: nutrient diffusion transport and rates of phytoplankton growth. *J. Mar. Res.* 34: 161-179.

George, R., Flick, R.E., Guza, R.T. (1994). Observations of turbulence in the surf zone. *J. Geophy. Res.* 99: 801-810.

Gibson, C.H., Thomas, W.H. (1995). Effects of turbulence intermittency on growth inhibition of a red tide dinoflagellate, *Gonyaulax polyedra* Stein. J. *Geophys. Res.* 100: 24.841-24.846.

Hill, P S., Nowell, A R M., Jumars, P.A. (1992). Encounter rate by turbulent shear of particles similar in diameter to the Kolmogorov scale. *J. Mar. Res.* 50: 643-668.

Howarth, R.W., Swaney, D., Marino, R., Butler, T. (1995). Turbulence does not prevent nitrogen fixation by plankton in estuaries and coastal seas (reply to the comment by Paerl *et al.*). *Limnol. Oceanogr.* 40: 639-643.

Jackson, G.A. (1990). A model of the formation of marine algal flocs by physical coagulation processes, *Deep-Sea Res.* 37: 1197-1211.

Jenkinson, I.R., Biddanda, B.A. (1995). Bulk-phase viscoelastic properties of seawater: relationship with plankton components. *J. Plank. Res.* 17: 2251-2274.

Jiménez, J. (1997). Oceanic turbulence at millimetre scales. In: Marrasé, C., Saiz, E., Redondo, J. M. (eds.) Lectures on plankton and turbulence. *Scientia Marina* 61 (Suppl. 1): 47-56.

Jones, K.J., Gowen, R.J. (1990). Influence of stratification and irradiance regime on summer phytoplankton composition in coastal and shelf areas of the British Isles. *Estuar. Coast. Shelf Sci.* 30: 557-567.

Kamykowski, D., Yamazaki, H., Yamazaki, A. Kirkpatrick, G.J. (This volume). A comparison of how different orientation behaviors influence dinoflagellate trajectories and photoresponses in turbulent water columns.

Karentz, D. (1987). Dinoflagellate cell cycles. In: Kumar, D.H. (ed.). *Phycotalk.* Vol. I, Print House, India, pp. 377-397.

Karp-Boss, L., Boss, E., Jumars, P.A. (1996). Nutrient fluxes to planktonic osmotrophs in the presence of fluid motion. *Oceanogr. Mar. Biol. Ann. Rev.* 34:71-107.

Kennett, J. P. (1982). Marine Geology. Prentice Hall, Englewood Cliffs, pp. 813.
Kiørboe, T., Andersen, K.P., Dam, H. (1990). Coagulation efficiency and aggregate formation in marine phytoplankton. *Mar. Biol.* 107: 235-245.
Kiørboe, T. (1993). Turbulence, phytoplankton cell size, and the structure of pelagic food webs. *Adv. Mar. Biol.* 29: 1-72.
Kiørboe, T., Saiz, E. (1995). Planktivorous feeding in calm and turbulent environments, with emphasis on copepods. *Mar. Ecol. Prog. Ser.* 122: 135-145.
Lakhotia, S., Papoutsakis, E.T. (1992). Agitation induced cell injury in microcarrier cultures. Protective effect of viscosity is agitation intensity dependent: Experiments and modeling. *Biotech. Bioeng.* 39: 95-107.
Lazier, J.R.N., Mann, K H. (1989). Turbulence and the diffusive layers around small organisms. *Deep-Sea Res.* 36: 1721-1733.
Lehman, J.T., Scavia, D. (1982a). Microscale patchiness of nutrients in plankton communities. *Science* 216: 729-730.
Lehman, J.T., Scavia, D. (1982b). Microscale nutrient paches produced by zooplankton. *Proc. Natl. Acad. Sci.* 79: 5001-5005.
Mann, K H., Lazier, J. R. N. (1991). *Dynamics of marine ecosystems. Biological-Physical interactions in the ocean.* Blackwell Scientific Publications, Oxford, pp. 466.
Margalef, R. (1978). Life-forms of phytoplankton as survival alternatives in an unstable environment. *Oceanol. Acta* 1: 493-509.
Margalef, R., Estrada, M., Blasco, D. (1979). Functional morphology of organisms involved in red tides, as adapted to decaying turbulence. In: R. Taylor, Seliger, (eds.) *Toxic dinoflagellate blooms.* Elsevier North Holland, Inc., pp. 89-94.
Marrasé, C., Costello, J.H., Granata, T., Strickler, J.R. (1990). Grazing in a turbulent environment: energy disipation, encounter rates and efficacy of feeding currents. *Proc. Natl. Acad. Sci.* USA 87: 1653-1657.
Munk, W.H., Riley, G. A. (1952). Absorption of nutrients by aquatic plants. *J. Mar. Res.* 11: 215-240.
Nelkin, M. (1992). In what sense is turbulence an unsolved problem? *Science* 255: 566-570.
Okubo, A. (1987). Fantastic voyage into the deep: Marine biofluid nechanics. In: Teramoto, E., Yamaguti, M. (eds.) *Mathematical Topics in Population Biology*, Lecture Notes in Biomathematics, vol. 17. Springer Verlag, pp. 32-47.
Osborn, T. (1996). The role of turbulent diffusion for copepods with feeding currrents. *J. Plankton Res.* 18: 185-195.
Paerl, H.W., Bebout, B.M. (1988). Direct measurement of O_2-depleted microzones in marine *Oscillatoria*: relation to N_2 fixation. *Science* 241: 442-445.
Pasciak, W J., Gavis, J. (1974) Transport limitation of nutrient uptake on phytoplankton. *Limnol. Oceanogr.* 19: 881-888.
Pasciak, W.J., Gavis, J. (1975). Transport limited nutrient uptake rates in *Ditylum brightwellii. Limnol. Oceanogr.* 20: 604-617.
Peters, F., Choi, J.W., Gross, T. (1996). *Paraphysomonas imperforata* (Protista, Chrysomonadina) under different turbulence levels: feeding, physiology and energetics. *Mar. Ecol. Prog. Ser.* 134: 235-245

Peters, F., Redondo, J. M. (1997). Turbulence generation and measurement: application to studies on plankton. In: Marrasé, C., Saiz, E., Redondo, J. M. (eds.) Lectures on plankton and turbulence. Scientia Marina 61 (Suppl.1): 205-228.

Pollingher, U., Zemel, E. (1981). *In situ* and experimental evidence of the influence of turbulence on cell division processes of *Peridinium cinctum* forma *westii* (Lemm.) Lefèvre. *Br. Phycol. J.* 16: 281-287.

Raikov, I.B. (1995). The dinoflagellate Nucleus and Chromosomes: Mesokaryote Concept Reconsidered. *Acta Protozoologica* .34: 239-247.

Reynolds, C. S. (1994). The role of fluid motion in the dynamics of phytoplankton in lakes and rivers. In: Giller, P.S., Hildrew, A.G., Raffaelli, D.G. (eds.) *Aquatic ecology. Scale, pattern and process.* British Ecological Society. Blackwell Science, Oxford, pp. 141-187.

Richardson, L.L., Stolzenbach, K.D. (1955). Phytoplankton cell size and the development of microenvironments. *FEMS Microbiol. Ecol.* 16: 185-192.

Riegman, R. (This volume). Macronutrient dynamics related to harmful algal blooms.

Savidge, G. (1981). Studies of the effects of small-scale turbulence on phytoplankton. *J. Mar. Biol. Ass. U. K.* 61: 477-488.

Schöne, H. (1970). Untersuchungen zur ökologischen Bedeutung des Seegangs für das Plankton mit besonderer Berücksichtigung mariner Kieselalgen. *Int. Revue ges. Hydrobiol.* 55: 595-677.

Siegelman, H W., Kycia, J.H. (1979). Large scale cultures of dinoflagellate algae. In: Taylor, D. L., Seliger, H. H. (eds) *Toxic dinoflagellate blooms.* Elsevier, New York, pp. 115-120.

Sournia, A. (1982). Form and function in marine phytoplankton. *Biol. Rev.* 57: 347-394.

Squires, K.D., Yamazaki, H. (1996). Preferential concentration of marine particles in isotropic turbulence. *Deep-Sea Res.* 42: 1989-2004.

Tennekes, H., Lumley, J.L. (1972). A first course in turbulence. The Massachusetts Institute of Technology Press, pp. 300.

Thomas, W.H., Gibson, C.H. (1990a). Effects of small-scale turbulence on microalgae. *J. Appl. Phycol.* 2: 71-77.

Thomas, W.H., Gibson, C.H. (1990b). Quantified small-scale turbulence inhibits a red tide dinoflagellate, *Gonyaulax polyedra* Stein. *Deep-Sea Res.* 37: 1583-1593.

Thomas, W H., Vernet, M., Gibson, C H. (1995). Effects of small-scale turbulence on photosynthesis, pigmentation, cell division, and cell size in the marine dinoflagellate *Gonyaulax polyedra* (Dinophyceae). *J. Phycol.* 31: 50-59.

Tuttle, R.C., Loeblich, A R. (1975). An optimal growth medium for the dinoflagellate *Crypthecodinium cohnii. Phycologia* 14: 1-8.

Vassilikos, J C. (1995). Turbulence and intermittency. *Nature* 374: 408-409.

Volk, S.L., Phinney, H.K. (1968). Mineral requirements for the growth of *Anabaena spiroides in vitro. Can. J. Bot.* 46: 619-630.

White, A.W. (1976). Growth inhibition caused by turbulence in the toxic marine dinoflagellate *Gonyaulax excavata.. J. Fish. Res. Board Can.* 33: 2598-2602.

Wyatt, T., Horwood, J. (1973). Model which generates red tides. *Nature* 244: 238-240.

Bacterial Interactions with Harmful Algal Bloom Species: Bloom Ecology, Toxigenesis, and Cytology

G.J. Doucette[1], M. Kodama[2], S. Franca[3], S. Gallacher[4]

[1] Marine Biotoxins Prog., National Marine Fisheries Svc., Charleston, SC, 29412, USA
[2] School of Fisheries Sciences, Kitasato University, Sanriku, Iwate 022-01, Japan
[3] Instituto Nacional Saude, Av-Padre Cruz, 1699 Lisbon, Portugal
[4] SOAEFD, The Marine Laboratory, Victoria Rd., Aberdeen, AB11 9DB, UK

1. Introduction

The apparent global increase in the occurrence of harmful algal blooms (HABs) over the past decade (e.g., Hallegraeff 1993) has been accompanied by enhanced efforts to identify factors ultimately controlling the population and toxin dynamics of HABs. As the processes of bloom initiation, maintenance, and decline, as well as toxin production, are dissected into their most basic elements, bacteria and their interactions with HAB species are among the components and processes increasingly cited as potentially important regulators of algal growth and toxicity. Indeed, although still an emerging area of study, it appears that now the question is not *whether* bacterial-algal interactions play a role in HAB ecology, but instead *how* these interactions are manifested in terms of population and toxin dynamics, and *what* mechanisms are involved (Doucette 1995).

The relationship between bacteria and algae is considered to play a critical role in such important oceanic processes as carbon flux (e.g., the 'microbial loop') and nutrient regeneration. The literature is thus replete with studies detailing the nature of bacteria-phytoplankton trophic interactions. By comparison, much of the work on bacterial-algal associations in the context of HAB ecology is driven by concerns unique to HABs and their frequently negative environmental and human health effects. Rather than treating bacteria and algae simply as adjacent trophic compartments, issues such as selective growth inhibition of HAB species by bacteria, and bacterial production of phycotoxins have received considerable attention (see reviews by Rausch de Traubenberg and Lassus 1991, Doucette 1995). Many studies are also beginning to evaluate the potential use of bacteria as a HAB mitigation strategy, and the possibility that phycotoxin-producing bacteria are capable of contaminating fisheries resources.

Another distinctive feature of HAB-related work on bacterial-algal relationships is its focus on interactions between bacteria and individual HAB species, rather than the aggregate phytoplankton community (see Doucette 1995). The reasons for this approach appear to be two-fold: first, many HAB-related bacterial strains studied in the laboratory originate from cultures of HAB species; and, second, most HAB field investigations target the dynamics of a single algal taxon. Interactions with associated bacteria are viewed in the latter context rather than in relation to the phytoplankton assemblage. Yet, bacterial effects on co-occurring, potentially competing, algal species are likely to be important in determining the success of a HAB taxon in achieving bloom concentrations.

There are special considerations imposed upon studies of bacteria and HAB species, yet many of the fundamental processes and mechanisms involved are also common to bacterial interactions with more benign, non-bloom forming algae. The techniques and

NATO ASI Series, Vol. G 41
Physiological Ecology of Harmful Algal Blooms
Edited by D. M. Anderson, A. D. Cembella and
G. M. Hallegraeff

experimental approaches to address specific questions are broadly applicable, whether or not the alga is a HAB species. Therefore, the aims of this chapter are to emphasize the unique aspects of bacterial-HAB species associations, while identifying the commonalities linking these relationships with more 'generic' interactions between bacteria and phytoplankton. Although many of the points considered below are closely related in terms of their net effects on HAB population and toxin dynamics, the material is organized into three areas: 1) the impact of bacteria on HAB dynamics and algal growth, 2) the involvement of bacteria in phycotoxin production, and 3) the cytology of bacterial associations with HAB species. Finally, we attempted to identify critical issues that warrant immediate attention by researchers.

2. Effect of bacteria on HAB dynamics and algal growth

The population dynamics of HABs, including the initiation, growth, maintenance, and decline of a bloom, are governed by a series of complex interactions among physico-chemical and biological variables. The *in situ* population growth of a phytoplankton species follows the equation:

$$N_t = N_o e^{(K + K_i - K_a - K_m - K_s - K_g)\,t}$$

where the rate of population increase (N_t) is a function of the difference between cellular growth rate (K) and immigration of cells into the population (K_i), and the net losses from the population due to washout or flushing (K_a), mortality (K_m), sinking (K_s), and grazing (K_g) (Smayda 1996). In the case of bacterial effects on phytoplankton population growth, K and K_m may be affected by extracellular growth-inhibiting and algicidal compounds (e.g., Ishio *et al.* 1989), as well as direct attack of algal cells by bacteria (Imai *et al.* 1993). Bacterial metabolites can act in a species-specific fashion to interfere with or terminate algal sexual reproduction (Sawayama *et al.* 1993), although in other instances algicidal activities can be effective against a broad range of taxa (Ishida *et al.* in press). It is these cellular level interactions between bacteria and algae which are among the factors contributing to the regulation of HAB population growth, as well as influencing the successional dominance of the algal community by a given species.

2.1. Role of bacterial-algal interactions in species succession and HAB dynamics

The close spatial and temporal coupling of bacterial and algal populations in the ocean is well documented, as is the tendency for both groups of organisms to synthesize metabolites that may be beneficial or harmful to one another (reviewed by Doucette 1995). Such capabilities imply a potential for feedback mechanisms which may ultimately play an important role in controlling both bacterial and algal species succession and population dynamics. A case in point is the work of Furuki and Kobayashi (1991) which suggested the presence of a bacterial assemblage promoting the growth of *Chattonella antiqua* when this bloom-forming raphidophyte was attaining maximum concentrations. Conversely, as the *C. antiqua* bloom declined, the ambient bacterial flora inhibited growth of this alga. In fact, a gliding bacterium, *Cytophaga* sp., affecting specifically *C. antiqua* (see Ishida *et al.* in press), was isolated during this phase of the bloom. The data of Fukami *et al.* (1991a, 1996) also documented negative correlations between concentrations of a given HAB species and bacteria inhibitory to its growth, as well as shifts in the bacterial population with the dominant phytoplankton taxon. For example, bacterial strains inhibiting the growth of *Skeletonema costatum* were present during a shift in species dominance from this diatom to the raphidophyte *Heterosigma carterae (=akashiwo)*. While many of these bacteria exhibited relatively broad spectrum algicidal effects, there was no growth inhibition of *H. carterae*.

In contrast, other investigators (Ishida *et al.* in press) have isolated bacteria lethal to *H. carterae* from other Japanese waters, and documented a sharp increase in the frequency of these bacteria immediately upon and after the peak of a bloom. Similarly, Fukami's group isolated a bacterium (*Flavobacterium* sp., strain 5N-3) that has specific algicidal effects on a red tide dinoflagellate, *Gymnodinium mikimotoi* (Fukami *et al.* 1991b). This bacterium, obtained at the end of a *G. mikimotoi* bloom, has exhibited no effect on other potential bloom formers occurring in local waters. This work strongly suggests that under certain conditions, species-specific inhibition of algal growth by bacteria, or avoidance of bacterial effects by algae, may play a role in driving the phytoplankton assemblage toward dominance by a different algal taxon.

Changes in microbial assemblages associated with blooms occur not only through the initial phases of population decline, but continue during decomposition of the algal biomass. Examining community composition during the break down of a *S. costatum* bloom, Fukami *et al.* (1985) observed that the bacterial flora attached to particulate material in the water column shifted from a *Pseudomonas-Alcaligenes* (*Ps*) group, to an *Acinetobacter-Moraxella* group, to a chromogenic bacterial group, and finally to a slower growing *Ps* group. To demonstrate that the initial *Ps* group comprised an assemblage of bacteria with heterotrophic activity well-suited to growing on the complex DOM characteristic of early stages of decomposition, this *Ps* group was re-established by adding fresh plankton cell material during the latter phase of decomposition.

2.2. Taxon-specific relationships between bacteria and HAB species

It has been well-documented that the amount and nature of dissolved organic matter (DOM) released by algae exhibit marked changes as a phytoplankton bloom proceeds through its phases of initiation, maintenance, and decline (e.g., Fuhrman *et al.* 1980; see also Carlsson and Graneli, this volume). This variation in DOM, which can include substances of nutritive value as well as bioactive compounds (e.g., antibiotics), appears to be a principal driving force behind temporal shifts in bacterial number, diversity, and functional groups during blooms (Fukami *et al.* 1985, Romalde *et al.* 1990). The spatial association between bacteria and algae is also dynamic, and can differ greatly among HAB species in the abundance and diversity of bacteria (see Doucette 1995). Although the majority of bacteria in the ocean are free-living, attached forms tend to increase proportionally as a function of particle surface availability (see Fukami *et al.* 1985), as would be the case during an algal bloom. Furthermore, attachment of bacteria to surfaces can elicit profound changes in phenotype (Costerton *et al.* 1995), which would influence the bacterial-algal interaction. Algal cells can thus be viewed as a source of a given suite of dissolved organic compounds and as a surface with a high potential for bacterial attachment. Such attributes suggest that an algal species may be capable of promoting the growth and possibly the attachment of a unique bacterial flora, components of which may affect algal reproduction and thus population dynamics.

The important role of DOM in modulating the activity of algicidal bacteria against species of harmful algae is demonstrated by the *Gymnodinium mikimotoi*-killing bacterium E401 (Ishida in press), isolated at the end of a *G. mikimotoi* bloom in Japan and described as a new genus within the γ-proteobacteria subdivision (Yoshinaga and Ishida 1995). Cultures of this dinoflagellate are completely lysed within 24-48 h of introducing strain E401, mediated via a high molecular weight (10 kD) heat labile compound excreted by the bacterium. Further study showed that this 'killing substance' was produced in response to excreted organic matter (EOM) from *G. mikimotoi* in a specific manner, in that EOM from a variety of other HAB species representing different algal groups did not elicit production of this substance. The algicidal activity of the compound was restricted to two closely related dinoflagellates, *G. mikimotoi* and *Gymnodinium catenatum*, but had no effect on other dinoflagellates, diatoms, or raphidophytes tested. Another example of a specific interaction mediated by algal organics is Furuki and Kobayashi's (1990) report that a *Cytophaga* sp. present during the decline of a *C. antiqua* bloom could be cultivated from seawater samples only when provided autolyzed cells of the raphidophyte as a growth substrate and not on more conventional isolation media. Later work demonstrated that the lethal effects of *Cytophaga* sp. on *C. antiqua* are species-specific, with no observable impact on the growth of other raphidophytes and dinoflagellates tested (M. Furuki unpubl. results).

Microbial assemblages associated with, and thus possibly 'selected for' by, HAB species are now being studied at the molecular level. The most extensive molecular-based analysis of the bacterial flora associated with a HAB event is the work of Ishida and co-workers (in press) on a bloom of the raphidophyte *Heterosigma carterae* in Hiroshima Bay, Japan. These authors documented maximum abundances of bacteria capable of killing *H. carterae* co-incident with peak and subsequently declining concentrations of the alga, and isolated 96 bacterial strains during this period. These strains were then characterized based on RFLP analysis and sequencing of the 16S rRNA gene. Three dominant RFLP types were identified and determined to be members of the *Cytophaga* class and the γ-proteobacteria subdivision (Ishida in press). The temporal and spatial correspondence of these bacteria with the sudden dissipation of the *H. carterae* red tide led to the suggestion that these microbes may have been closely involved with the termination of this bloom. It is interesting to note that the *Cytophaga* and γ-proteobacteria groups are among the dominant bacterial flora attached to macroagreggates or marine snow (DeLong *et al.* 1993). Moreover, in the case of the predominant *Cytophaga* component, representatives of this group exhibit a tendency to associate with surfaces, show surface-dependent gliding motility, and produce exoenzymes allowing them to degrade a number of complex, high molecular weight compounds (e.g., proteins, polysaccharides, nucleic acids, etc.). The findings of Ishida and his colleagues are thus consistent with the idea that specific bacterial assemblages characterized by members whose phenotypic traits are conducive to particle interaction and decomposition may play a role in the termination phase of HABs.

It has been suggested that algicidal bacteria may frequently cause the collapse of algal blooms (Ishida in press). While this scenario is possible, more definitive evidence is needed to establish algicidal bacteria as a unique mechanism of bloom termination. Such bacteria could certainly contribute to bloom decline by negatively affecting HAB species, thereby increasing the algae's susceptibility to other environmental 'pressures' such as interspecific nutrient competition, grazing, and allelopathy. Although data are few, it also appears that certain bacterial groups are more likely to be associated with bloom decline. Yoshinaga and Ishida (1995) classified the bacterium E401 within the γ-proteobacteria. Members of this group, as well as *Cytophaga* spp., were among the dominant bacteria present upon termination of a *H. carterae* outbreak (Ishida *et al.* in press). Further evidence for the algicidal tendencies of *Cytophaga* is provided by Imai *et al.*'s (1993) report of a *Cytophaga* strain lethal to 10 of 11 algal species tested. Since the traits of these microbes are consistent with an ability to acquire and utilize organics associated with algal cells, representatives of these bacterial groups may be involved with the decline and decomposition of algal blooms in a general context and show specific relationships with blooms of individual HAB species.

While characterization of marine bacterioplankton community structure, dynamics, and processes has long eluded microbial ecologists, molecular techniques based largely on bacterial rRNA genes (see Stahl 1995) now provide a means to address these issues. Recent investigations employing rRNA probe- and sequencing-based approaches, as well as analyses of low-molecular-weight RNA profiles, indicate not only that oceanic bacterioplankton communities consist of diverse assemblages of previously unknown bacteria (Mullins *et al.* 1995), but that genetically classified bacterial groups can be common to geographically distinct regions (e.g., Atlantic vs. Pacific Oceans). It has also been reported (e.g., Rehnstam *et al.* 1993) that actual 'blooms' of individual bacterial taxa or rRNA sequence type are not uncommon and that, otherwise, relatively few taxa numerically dominate the bacterioplankton community. The possibility that a certain bacterial taxon or phylogenetic group may dominate HAB-associated bacterioplankton communities, and may exhibit strong, interactive associations with particular HAB species, should be examined. Indeed, a comparison of free-living and particle (> 3 μm)-associated bacteria in the Chesapeake Bay estuary revealed a remarkable consistency in genetic level composition of the latter assemblage (and not the former) across a wide range of habitats, implying a more uniform set of conditions 'selecting' for members of the attached community (Bidle and Fletcher 1995). Therefore, during a HAB event in which cells of one algal species may represent a considerable portion of the surfaces available for bacterial attachment as well as a primary source of DOM, the local 'particle environment' could be quite constant and possibly promote the growth of bacteria uniquely capable of responding to those conditions.

2.3. Mechanisms and models for bacterial influence of HAB dynamics

A useful way of grouping the various effects of bacteria on HAB species, whether beneficial or detrimental, is according to whether direct contact with the algae is required. Such information provides important insights into the overall mechanism of interaction, and can be used to guide attempts at identifying the substance(s) mediating a particular effect.

It is apparent from Table 1.1 (from Ishida in press) that the direct or indirect nature of algicidal interactions between bacteria and algae is closely tied to the bacterial group involved, and thus likely reflects the phenotypic traits characteristic of the respective

Table 1.1. A summary of algicidal bacteria isolated from seawater (adapted from Ishida in press). ● = direct bacterial attack; ○ = indirect bacterial attack; ▬ = no effect.

	Cytophaga* strain			***Alteromonas* strain***		***Flavo-bacterium* strain***		***Vibrio* strain***
HOST ALGAE	**1**	**2**	**3**	**4**	**5**	**6**	**7**	**8**
DIATOMS								
Skeletonema costatum	●	●			▬	▬	▬	▬
Chaetoceros didymum	●			○				
Ditylum brightwellii	▬	●		▬			▬	▬
Eucampia zodiacus	●							
Thalassiosira sp.	●	●					▬	▬
RAPHIDOPHYTES								
Chattonella antiqua	▬	●	●	○	▬		▬	
Chattonella marina		●	▬	○	▬	▬		
Heterosigma carterae		●	▬	▬	▬	▬	▬	○
Fibrocapsa japonica		●	▬					
DINOFLAGELLATES								
Gymnodinium mikimotoi	▬	●	▬	○	○	○	○	○
Gymnodinium catenatum							○	
Scrippsiella trochoidea		▬						
Alexandrium catenella							▬	○

* Strain designations and source: 1. Strain Y-6 (Mitsutani *et al.* 1992); 2. Strain J18/MO1 (Imai *et al.* 1993); 3. M. Furuki (unpubl. data); 4. Strains K, D (Imai *et al.* 1995); 5. Strain 6/6-46 (Yoshinaga *et al.* 1995b); 6. Strain 5N-3 (Fukami *et al.* 1991b); 7. E401 (Yoshinaga *et al.* 1995a); 8. Strains B46, C1, T27 (Yoshinaga *et al.* 1995b).

group. As noted above, certain groups of bacteria show a tendency toward particle attachment and synthesis of exoenzymes, while others generally occur within the free-living component of the bacterioplankton community. Thus, HAB-associated bacteria with phylogenetic affinities to the former group appear most likely to depend on physical interaction with algal cells in order to exert their effect, and may be less apt to affect algae through excretion of highly specific bioactive metabolites (e.g., dinoflagellate mating inhibitors, Sawayama *et al.* 1993) into the surrounding water. Nonetheless, the biosynthesis of certain bacterial metabolites can be closely regulated by surface attachment, even at the level of gene transcription (see Cooksey and Wigglesworth-Cooksey 1995). It is therefore essential to include cells growing on a surface, in addition to the liquid bacterial cultures commonly used for laboratory study, when screening for biologically active substances produced by these microbes.

In most cases where bacteria influence the growth of HAB species, the effect is of an inhibitory rather than stimulatory nature (see Doucette 1995). However, bacterial metabolites can also promote algal growth as reported by Keshtacher-Liebson *et al.* (1995), who observed the enhancement of algal reproductive rates under iron-limiting conditions by bacteria. In their experiments, a bacterium (*Halomonas* sp.) mediated an increase in Fe solubility, apparently through excretion of siderophores, thereby allowing the chlorophyte *Dunaliella* to continue growing at maximum rates at Fe levels that otherwise limit its growth. While Fe-limited growth of marine phytoplankton is generally associated with oligotrophic environments (see Chisholm and Morel 1991), the potential importance of this element in regulating the population dynamics of HAB species in coastal regimes has been suggested by several authors (e.g., Gobler and Cosper 1996; see Boyer and Brand this volume). Thus, although it has yet to be empirically verified, bacterial siderophores may provide a competitive advantage to those HAB taxa capable of acquiring the iron chelated by these metabolites.

It is now possible and desirable to begin formulating simple, conceptual models describing how bacteria may influence HAB dynamics. Clearly, there exists a broad spectrum in the degree of specificity exhibited by bacteria and their effects on HAB species, as well as potentially co-occurring phytoplankton taxa. The data compiled by Ishida (in press) and shown in Table 1.1, while not comprehensive, can be used to illustrate this point. Whether or not the algicidal activity is mediated by direct or indirect means, bacterial strains can be identified that affect taxa belonging to either multiple or individual groups of algae, or even to single algal species. There are also cases (see Fig. 4, Fukami *et al.* 1996) in which all but one of the algal species tested were susceptible to the algicidal effects of a bacterial strain. Situations such as the latter show a high potential for influencing successional changes in the phytoplankton community which may lead to blooms of harmful species. By way of illustration, Fukami

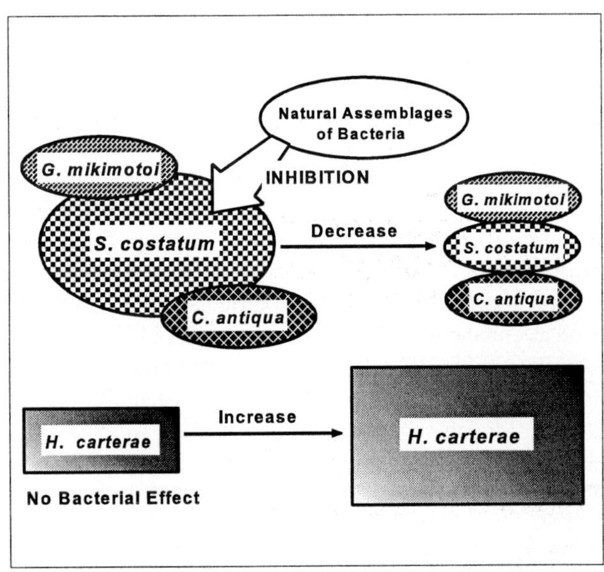

Figure 1.1. Conceptual model for bacterial effects on phytoplankton species succession (adapted from Fukami *et al.* 1996).

et al. (1996) isolated five bacterial strains which inhibited the growth of the *S. costatum* (and several other algal taxa), but showed no effect on *H. carterae* during a shift in species dominance from the diatom to the raphidophyte. Based on these observations a model was proposed to explain the role of bacteria in mediating this successional change (Fig. 1.1). Their model suggests that through selective growth inhibition of the dominant species, other unaffected taxa have an increased probability of achieving enhanced population growth. In cases where the unaffected species exhibits allelopathic effects against other phytoplankton taxa and is known to be capable of bloom formation (e.g., *H. carterae*), the potential for initiation of a bloom event would likely be enhanced.

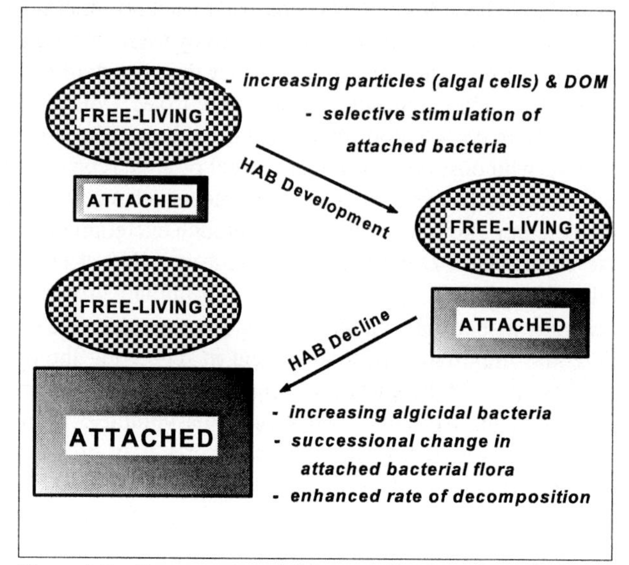

Figure 1.2. Conceptual model for contribution of attached bacteria to the decline and decomposition of HABs.

An analogous framework within which to examine the contribution of bacteria to bloom decline and decomposition can also be useful (Fig. 1.2). This model is based on data indicating that the proportion of attached bacteria increases with particle availability (as would occur during HAB development) and that an algal species, via excretion of a unique suite of DOM, may selectively stimulate bacteria with a tendency toward surface attachment, algicidal activity, and/or decomposition of complex algal organics. As a result, bacteria adapted to the environment provided by cells of the algal species comprising a bloom may both contribute to the start of bloom termination processes, as well as enhance the rate of decomposition during the declining phase. Note that while box sizes in Fig. 1.2 reflect changes in the proportion of attached vs. free-living bacteria, total bacterial abundance can also be expected to increase over the course of an algal bloom. Models such as those outlined here, while neither quantitative nor predictive in nature, can aid in designing studies to identify and ultimately measure processes by which bacteria may influence HAB dynamics.

3. Role of bacteria in phycotoxin production

The longstanding issue of bacterial involvement in phycotoxin production, first proposed over three decades ago (Silva 1962) and, has attracted increasing interest as well as controversy in recent years. Whether referring to the autonomous bacterial synthesis of these toxins normally attributed to certain HAB species, or to the possibility that bacteria may directly/indirectly enhance toxicity levels associated with algal cells, this area of research has important implications for how phycotoxin dynamics in natural systems are viewed. Data supporting such ideas have been slow to accumulate; nonetheless, the evidence is now compelling enough to justify a rigorous examination of the mechanisms involved and how they are controlled. Since most work in this area has dealt with two groups of sodium channel blockers, saxitoxin (STX) and its derivatives, causative agents of paralytic shellfish poisoning (PSP), and the tetrodotoxin analogues (TTXs), also referred to as 'puffer fish toxins', these toxins will be emphasized. Studies concerned with the relationship of bacteria to production of other phycotoxins (e.g., domoic acid, brevetoxin) have been reviewed by Doucette (1995), and will be mentioned only briefly.

3.1. Bacterial production of phycotoxins

In some toxic algal cultures, bacteria living outside or inside the algal cells are either directly or indirectly associated with phycotoxin production (e.g., Kodama 1990, Bates *et al.* 1995, Doucette 1995, Franca *et al.* 1995, Gallacher *et al.* 1997), although no clear picture of the relationship between bacteria and algal toxin production has emerged. Neither have questions of how or why either organism produces these toxins been fully elucidated. Complete details of the biosynthetic pathways remain unknown, as do the likely metabolic function(s) of these toxins. A potentially key issue here is the autonomous bacterial production of algal toxins, including the PSP toxins (Kodama *et al.* 1988, Doucette and Trick 1995, Gallacher *et al.* 1996, 1997) and TTXs (Noguchi *et al.* 1986, Yasumoto *et al.* 1986, Simidu *et al.* 1987).

Bacterial production of sodium channel blocking (SCB) toxins was first demonstrated for the tetrodotoxins (Noguchi *et al.* 1986, Yasumoto *et al.* 1986), which bind to the same biological receptor as the PSP toxins. Tetrodotoxin is an exogenous substance to puffer (Matsui *et al.* 1982), but its origin remained uncertain for many years until two laboratories (Noguchi *et al.* 1986, Yasumoto *et al.* 1986) showed that bacteria isolated from TTX-bearing crab and algae could synthesize this toxin, suggesting a bacterial origin of TTX in such organisms. TTX-producing bacteria have since been reported from various sources (reviewed by Tamplin 1990, Gallacher *et al.* 1996). Interestingly, the characteristics of and conditions for TTX production by bacteria are very similar to those now being reported for PSP toxin synthesis by bacteria.

A bacterial origin of the PSP toxins was first suggested by Silva (1979, 1982a), who observed bacteria-like particles inside dinoflagellate cells exhibiting mouse-based toxicity. Silva reported further that bacterial strains isolated from toxic dinoflagellates, while unable to elicit a toxic response in mice, could confer toxicity to non-toxic algal strains; nevertheless, these findings remain to be confirmed by other investigators. While studying the toxin production of a dinoflagellate, *Alexandrium tamarense*, Ogata *et al.* (1987) observed that toxin yields of subclones isolated from a clonal culture and grown under identical conditions exhibited a 20-fold difference, suggesting that toxin production in this species is not a hereditary characteristic of the alga. As this finding appeared consistent with Silva's idea of a bacterial source of the PSP toxins, attempts were made to isolate bacteria from this dinoflagellate. A single toxigenic bacterium, referred to as "*Moraxella* sp." (later classified as a *Pseudomonas/Alteromonas* species (Doucette and Trick 1995) and most recently as a new genus within the α-proteobacteria (Kopp *et al.* submitted)), was isolated when cells of an axenic *A. tamarense* culture (produced by antibiotic treatments) were homogenized lightly and inoculated to agar plates (Kodama *et al.* 1988, 1989). The number of bacterial colonies was small, and since the isolated bacterium was sensitive to those antibiotics used to treat the culture, *Moraxella* sp. was considered to be of intracellular origin. A later study revealed the presence of bacteria in the nucleus of this *A. tamarense* strain (Kodama 1990); however, subsequent attempts to demonstrate endocellular bacteria in the same dinoflagellate species by this and other research groups have been unsuccessful, leaving in question the precise *in vivo* location of bacteria associated with *A. tamarense.*

Evidence for the autonomous bacterial synthesis of PSP toxins has been based primarily on chromatographic behavior and biological activity (i.e., receptor binding assay) of both crude and purified preparations of *Moraxella* sp. (Kodama *et al.* 1988, Doucette unpubl. data). Doucette and Trick (1995) and Franca *et al.* (1995) employed high performance liquid chromatography (HPLC) methods to examine toxin production characteristics of *Moraxella* sp. and another bacterium obtained from the toxic dinoflagellate *A. lusitanicum* (= *minutum*), respectively. Both groups reported PSP toxins in bacterial cell extracts. Recently, Gallacher and co-workers have made an important step in confirming the synthesis of PSP toxins by bacteria isolated from five *Alexandrium* cultures, including the *A. lusitanicum* strain noted above (Gallacher *et al.* 1997), using a capillary electrophoresis-mass spectrometry (CE-MS) method.

In general, toxicity of *Moraxella* sp. tended to increase under starved conditions (Kodama *et al.* 1990a) and when exposed to P-limiting conditions in chemically defined medium (Doucette and Trick 1995). Studies of other bacterial isolates producing SCB toxins (either PSPs or TTXs) have also found that maximum toxin levels coincide with entry of batch cultures into stationary phase, frequently with growth limited by phosphorus (Yasumoto *et al.* 1986, Gallacher and Birkbeck 1993, Gallacher *et al.* 1997). Moreover, the greatest proportion of toxin produced by the bacteria at this point seems to occur in the culture medium, rather than inside the cells. One possible interpretation of the above findings is that since toxin accumulates when the anabolic processes of bacteria are depressed, the PSPs (and also TTXs) may represent catabolites of some unknown bacterial substance(s). Alternatively, the apparently close relationship of bacterial toxin yields with P availability may imply either the regulation of toxin gene expression by phosphorus, as has been demonstrated for several secondary metabolites (Vining 1990), or simply that excess nitrogen is available for toxin synthesis (STX and TTX contain six and three N atoms per molecule, respectively).

Given the close similarities between the PSPs and TTXs, ranging from chemical structure to what is known of their production characteristics, it is perhaps not surprising that both of these SCB toxin groups have been detected in the same organism. A variety of marine invertebrates, including xanthid crabs, horseshoe crabs, and snails (e.g., Yasumura *et al.* 1986, see Tamplin 1990), as well as puffer fish (Kodama *et al.* 1983, Sato *et al.* 1997) have all been shown to accumulate PSP toxins and tetrodotoxins. Although yet to be confirmed independently, both toxins have also been detected in a strain of the dinoflagellate, *A. tamarense* (Kodama *et al.* 1996b), and scallops can accumulate considerable amounts of TTX during a bloom of this algal species (Kodama *et al.* 1993). While tempting to ascribe such observations to a single source of both toxins, evidence for the existence of such an organism has only recently been obtained. The bacterium, *Shewanella alga*, originally described as a TTX-producer and isolated from the red alga *Jania* sp., has been reported to synthesize PSPs in addition to TTXs

(Gallacher *et al.* submitted). Moreover, a second strain of the same bacterial species, isolated from a culture of *Alexandrium affine*, has also been demonstrated via CE-MS to produce both classes of these sodium channel blockers (H. Hines pers. comm.).

Virtually all the examples of autonomous bacterial synthesis of SCB toxins outlined above involve purified cultures of strains isolated from close associations with other organisms, such as dinoflagellates or marine invertebrates. Nonetheless, evidence suggesting that such bacteria may actively produce these toxins in natural systems is now beginning to appear in the literature. The initial studies of Kodama and co-workers in Ofunato Bay, Japan (Kodama *et al.* 1990b) indicated that PSP toxins could be detected in bacterial size fractions of seawater collected after the disappearance of *A. tamarense* cells following a bloom event, although definitive proof of a bacterial toxin source was lacking. Recently, the potential production of saxitoxin congeners by bacteria in the St. Lawrence estuary, Canada (Levasseur *et al.* 1996) was examined. Bacterial strains isolated from the St. Lawrence (48 total) were screened for PSPs by HPLC, yielding four positive strains and indicating the presence of putatively (pending confirmation by mass spectrometry) toxigenic bacteria in the region. Attempts were then made to measure ambient toxin levels in various particle size fractions during and after an *Alexandrium* bloom, as well as production of toxin by the heterotrophic component of the plankton community (e.g., bacteria) during dark incubation of size-fractioned material. While ambient PSP toxins were detected by HPLC only in the two largest size fractions (> 21 μm and 5-21 μm) obtained during the bloom, *de novo* toxin synthesis was detected (following dark incubation) in most size fractions, including the 0.22-5 μm range, regardless of sampling time relative to the bloom. Although free-living and particle-associated bacteria capable of producing the PSPs appear to exist in the St. Lawrence estuary, only toxin associated with larger size fractions accounted for shellfish toxicity during this *Alexandrium* bloom. However, a contribution to toxicity of bacteria attached to these larger particles (including algal cells) could not be ruled out. Whether a close spatial relationship with particulates in general, or specifically with algal cells and their related DOM, facilitates toxin production by HAB-associated bacteria remains in question. A similar follow-up study (Levasseur *et al.* unpubl. data) reproduced results from previous *de novo* toxin synthesis experiments, but also revealed ambient PSP toxins in the 0.22-5 μm size fraction as well as larger fractions. The latter findings support further the idea that bacteria may directly contribute to PSP toxin levels in this region.

The distribution of SCB producing bacteria in the marine environment appears quite widespread (e.g., Gallacher and Birkbeck 1995) and not necessarily restricted to associations with HABs. An increasing number of investigators are now reporting the isolation of such bacteria from locations that are spatially and temporally distinct from the cells of harmful algal species. For example, the four putatively toxic bacterial strains isolated by Levasseur *et al.* (1996) from the St. Lawrence estuary originated from sites exhibiting no concurrent PSP activity, and three of these locations had no prior history of bloom formation. A recent report by Tiecco *et al.* (1996), documenting production of SCB toxins by 7 of 85 bacterial strains isolated from marine sediments, has provided additional evidence of the ubiquitous presence of such bacteria in marine ecosystems with no apparent relationship to recognized toxic organisms. Furthermore, direct analysis of sediment samples provided evidence for the co-occurrence of these toxins in sediments from which 5 of the 7 toxigenic bacteria were obtained. Based on these and other similar data (e.g., Do *et al.* 1990), it may be argued that the feature of bacterial microscale distribution most important in controlling toxin production by these microbes is simply attachment to particle surfaces, rather than a direct association with eukaryotic organisms (i.e., algae and marine invertebrates), although the influence of benthic fauna cannot be discounted. This viewpoint is supported further by the work of Sakamoto *et al.* (1992) who noted that detritus and silt particles, favorable sites for bacterial attachment, often showed high PSP toxin levels in the absence of toxic dinoflagellates.

3.2. Bacterial effects on algal toxicity

The role bacteria play in determining the levels of phycotoxins associated with algal cells is complex. On one hand, there is now good evidence that bacteria isolated from cultures of toxic dinoflagellates are, themselves, capable of synthesizing algal toxins such as the PSPs and could thus directly contribute to algal toxicity. Indirect enhancement of algal toxin yield by bacteria not considered to be toxigenic has also been demonstrated in the case of a domoic acid-producing diatom, *Pseudo-nitzschia multiseries* (Bates *et al.* 1995). On the other hand, several studies indicate that axenic, or bacteria-free, dinoflagellate cultures retain the ability to produce these toxins at levels similar to those of non-axenic cultures (e.g., Kim *et al.* 1993). Still other investigators argue against an extra-chromosomal location of PSP toxin genes, based on observations that inheritance of STX derivative profiles in dinoflagellates follows a Mendelian pattern (Ishida *et al.* 1993). The picture is still far from clear since it is most difficult to establish that an algal culture is truly axenic at a molecular level (i.e., containing no intact, actively transcribed bacterial genome whether or not it is contained within a 'recognizable' bacterial cell), and possession of the genetic machinery needed for toxin synthesis by both algae and bacteria is not necessarily a mutually exclusive proposition. While the evidence is still largely circumstantial, we suggest that bacteria can modulate the toxicity of algal cells, at least in laboratory cultures of PSP toxin producing dinoflagellates.

Several recent studies have examined both the quality and quantity of PSP toxins produced by bacteria originating from cultures of toxic *Alexandrium* spp. As is the case for dinoflagellates, bacterial toxin composition appears to vary among strains (Franca *et al.* 1996, Gallacher *et al.* 1997) and change within a strain as a function of nutritional status (Doucette and Trick 1995). Doucette and Trick (1995) reported that while P-limitation enhanced the toxin per unit protein of *Moraxella* sp. by about 3-fold, a response comparable to that reported for *Alexandrium* (Anderson *et al.* 1990), actual toxin levels per unit protein were still only ca. 5% of those in the dinoflagellate cells. By comparison, when the toxin content was normalized on a per unit cell volume basis, Gallacher *et al.* (1997) concluded that bacteria and algae showed roughly equivalent levels of the PSP toxins. They also noted that the potential bacterial contribution to algal toxicity will obviously depend on the number of bacteria associated with the algae. Based on their enumeration of 10^6 to 10^7 bacteria per ml of algal culture, the maximum possible contribution to toxicity was about 3%. These authors emphasized, however, that their data should be interpreted with care, since bacterial concentrations represent only those culturable in the growth medium used, and the proportion of dinoflagellate-associated bacteria capable of toxin synthesis is unknown. In addition, the population dynamics and concentrations of bacteria associated with *Alexandrium* blooms have yet to be critically examined. Finally, several authors (Doucette 1995, Gallacher *et al.* 1997) have correctly pointed out that the conditions for bacterial growth in laboratory experiments differ considerably from that experienced by bacteria in natural systems, especially with regard to the availability of algal-derived organic material and the attachment of bacteria to algal cell walls which can facilitate marked changes in bacterial phenotype (Costerton *et al.* 1995), possibly including toxin production.

Attempts are now being made to assess contributions of toxigenic bacteria to algal toxicity under conditions which take into account the presence of algal cells and their excreted DOM, as well as the possibility of bacterial adhesion to the algae. Experiments have been performed (G. Doucette unpubl. data) in which a toxigenic bacterium (*Pseudomonas stutzeri*, see Franca *et al.* 1995) was added to an axenic culture of the dinoflagellate from which it was isolated (*Alexandrium lusitanicum*). While toxicity associated with algal cells in axenic cultures was ca. half that in the original xenic cultures, bacterial additions yielded a roughly two-fold increase in algal toxicity relative to controls receiving no bacteria, thereby restoring toxicity levels to those of the xenic cultures. However, when the bacterial supplement was contained inside dialysis tubing (300,000 mol. wt. cutoff), thus preventing physical contact with the algal cells, no enhancement occurred. These results suggest that the toxicity of an algal species may be modified via attachment of bacteria, although the mechanism remains uncertain.

The taxonomic specificity in bacterial-algal associations has been examined recently (Gallacher *et al.* in prep) for five *Alexandrium* isolates using 16S rRNA signatures of the associated (culturable) bacteria. Their data revealed that a limited number of bacterial taxa were associated with each algal culture, and that some strains were unique to a dinoflagellate species while others were common to all algal isolates. The bacterial strains included species of *Roseobacter*, *Caulobacter*, *Alteromonas* and *Shewanella*, as well as several unknown taxa. Three of the strains isolated were definitively classified as PSP toxin producers. One of these strains was associated with cultures of *Alexandrium tamarense* (two cultures) and *A. lusitanicum* (one culture), while the other two strains occurred in respective cultures of *A. tamarense* and *A. affine*. The toxigenic bacteria were all members of the γ-proteobacteria. Continued molecular characterization of toxigenic bacteria associated with a variety of *Alexandrium* spp. isolates originating from different locations is needed to better define the taxonomic specificity of bacterial-algal relationships as it affects toxicity.

The final body of evidence consistent with the idea that toxigenic bacteria may influence algal toxicity focuses on the spatial relationship and cellular level dynamics of the bacterial-algal association. Several authors have reported the direct observation of endocellular bacteria in toxic species of dinoflagellate using electron microscopy (Silva 1979, Kodama 1990, Franca *et al.* 1995), although bacterial numbers are generally few and such bacteria are not always discernable even within cells of the same algal culture. While the same researchers have successfully isolated toxigenic bacteria from these dinoflagellates, unequivocal proof at the molecular level (e.g., immuno- or nucleic acid probe labeling) that the isolated bacterial strains and the corresponding endocellular bacteria are the same is lacking. Kodama *et al.* (1996a) have used an antibody against *Moraxella* sp. strain to probe western blots of the 'host' *A. tamarense* isolate (axenic cultures), and observed increased cross-reactivity with the apparent disruption of residual (= accumulation) bodies contained in the algal cells. These results, and the fact that bacteria could be isolated from an axenic culture of this dinoflagellate only upon disruption of residual bodies, led the authors to contend that endocellular bacteria existed predominantly in residual bodies and retained the ability to reproduce within the algal cell. They also hypothesized that bacterial growth inside residual bodies may be controlled by digestive processes, thereby 'harmonizing' bacterial growth with that of the dinoflagellate and allowing extended maintenance of this endocellular bacterial flora. Consistent with this idea are the electron microscopic observations of partially digested bacteria in the residual body of several toxic dinoflagellates (Franca *et al.* 1995), as well as the apparently stable association of certain bacteria with an algal isolate over many years (see below). Kodama and co-workers, citing recently published evidence of mixotrophy among thecate, photosynthetic dinoflagellates, including toxic *Alexandrium* species (Jacobson and Anderson 1996; Ogata *et al.* 1996), also suggested that bacteria co-existing inside dinoflagellate cells may represent a 'self-sustaining' source of organic nutrition to the algae in a mutualistic fashion. The issue of taxonomic specificity is important as to whether a single taxon or multiple bacterial taxa are able to reside and reproduce within the cells of a given dinoflagellate species. Application of molecular probes to verify the identity of endocellular bacteria is required to unambiguously resolve whether these organisms are the toxigenic bacteria isolated from the algal cells. Such evidence would strengthen the argument for a direct bacterial contribution to algal toxicity. Nevertheless, a more indirect biochemical mechanism, such as the supply of toxin precursor molecules by bacteria, cannot be discounted at present.

3.3. Phycotoxins in marine food webs: a bacterial source?

It is important to determine whether toxigenic bacteria represent an alternative source of phycotoxins in marine food webs, accumulating in and/or affecting other organisms either directly through ingestion or indirectly through trophic transfer. Again, the PSP toxins will be emphasized due to their predominance in the literature. The accumulation of PSP toxins in plankton feeders following ingestion of toxic dinoflagellates is well-documented and generally accepted as the primary mechanism of toxin entry into local food webs. However, in the case of bivalve toxicity, several lines of evidence suggest a non-algal source of these toxins. For example, shellfish toxicity, which generally declines to some extent in the absence of toxic dinoflagellates, has been reported to reach a peak as late as one week after the corresponding peak in abundance of the causative dinoflagellate (Ogata *et al.* 1982). After toxin levels in shellfish began to decrease, a further rise in toxicity was noted even though the causative dinoflagellates were no longer detected. Possible explanations include the spatial patchiness of algal populations or the biotransformation of sulfocarbamate PSP toxin derivatives to their corresponding, and far more potent, carbamate forms. Alternatively, these observations may indicate *de novo* toxin synthesis within the bivalves (e.g., undigested, viable algal cells, or bacteria) and/or the presence of a non-algal PSP toxin source (e.g., free-living or particle associated bacteria). One laboratory study does indicate that ingestion of free-living, toxigenic bacteria can result in low level mussel toxicity (Gallacher and Birkbeck 1995). Although evidence from natural populations remains largely circumstantial, studies aimed at explaining cryptic patterns in the relationship between shellfish toxicity and toxigenic organisms, including bacteria, should be pursued.

As outlined above (see section 3.1), both PSP toxins and TTXs are associated with a variety of marine invertebrates other than shellfish, and also the dinoflagellate, *A.*

tamarense (Kodama *et al.* 1996b). The association of a PSP- and TTX-producing bacterium, *Shewanella alga*, with both a red alga and a dinoflagellate (see Gallacher *et al.* submitted) originating from widely separated locations indicates a broad phylogenetic association and geographic distribution, suggesting strongly that additional studies will identify other bacteria with similar toxigenic properties. It is plausible that the direct acquisition of such bacteria by certain eukaryotic organisms may represent one mechanism by which the SCB PSPs and TTXs can simultaneously enter and be transferred through marine food webs leading to accumulation in certain trophic compartments. A recent survey of numerous common marine animals revealed low levels of PSPs and TTXs in viscera samples (Sato *et al.* 1993). These findings, together with the fact that TTX-bearing animals are highly resistant to TTX (Koyama *et al.* 1983, Saito *et al.* 1985), apparently support this food web theory. Nevertheless, the levels of TTX found in common animals seem far too low to explain the elevated concentrations of this toxin group in highly toxic animals, such as the puffer fish.

In the case of the puffer fish, discovery of TTX-producing bacteria in the intestinal flora of toxic specimens (Noguchi *et al.* 1986) suggested that over an extended period puffer accumulated the toxin (predominantly in the liver) produced by these intestinal bacteria. Yet subsequent studies revealed no significant difference in the intestinal flora of toxic versus nontoxic puffer (Sugita *et al.* 1988). Low levels of TTX in the intestine of both juvenile and adults of non-toxic cultured puffer (with no accumulation of TTX in the liver regardless of age) (Sato *et al.* 1990), indicate that TTX in puffer is not associated with environmental nor intestinal bacteria. Recent efforts (Kodama *et al.* 1995) have yielded isolates of SCB toxin producing bacteria from liver tissues of highly toxic, but not of non-toxic, puffer fish maintained in aquaria for over 6 months. This suggests that endocellular bacteria are maintained in liver cells for extended periods and indicates that they may be the TTX source in puffer fish. Since puffer fish contain both PSPs and TTXs, it is also possible that these bacteria can synthesize both classes of SCB toxins, although this remains to be proven. While the source of these bacteria, as well as the mechanisms by which they invade and are maintained in the puffer's liver tissue are uncertain, a high degree of 'host specificity' would seem to be required. Such relationships between toxigenic bacteria and certain marine fauna, if determined to be widespread in occurrence (e.g., digestive gland of bivalves), may implicate these bacteria in the introduction and trophic transfer of SCB toxins in marine food webs.

A final point is the bacterial contribution of SCB toxins to marine benthic communities. Several investigators have isolated such toxigenic bacteria from various marine sediments (e.g., Do *et al.* 1990, Tiecco *et al.* 1996), but of particular note is the accumulation of a SCB toxin 'pool' in this environment (Tiecco *et al.* 1996) and in components of the benthic faunal assemblage (Kogure *et al.* 1989). Certainly, examples of gastropods, including predators, scavengers, and grazers, containing PSP toxins and/or TTX are common (Sato *et al.* 1993, see review by Shumway 1995). A bacterial origin of these SCB toxins appears possible, yet other sources such as bivalve prey species and dinoflagellate resting cysts may also be important in certain cases. In the former case, sediment-associated toxigenic bacteria may, themselves, be incorporated into benthic consumers and then transferred (in a viable state) to higher trophic compartments where they could continue to elaborate SCB toxins in 'host' organisms, as discussed above. Certainly, a source of SCB toxins and the bacterial flora capable of producing them do exist in marine benthic communities, suggesting a route of entry into marine food webs apart from well-established planktonic sources such as toxic dinoflagellates.

4. Cytology of bacterial associations with HAB species

Some of the earliest accounts of bacteria-like particles in cells of cultured dinoflagellates were provided by Silva (1962). Subsequent light microscopic (LM) studies (Silva 1990) revealed bacteria in external cytoplasmic regions or immediately beneath the cell wall of dinoflagellates from both cultures and natural samples. However, she noted that under the transmission electron microscope (TEM) intracellular bacteria were observed neither as frequently nor as clearly as expected based on LM evidence. In any case, detailed examination of the relationship between bacteria and dinoflagellates (Silva 1978, later references in Silva 1990) yielded numerous LM and/or TEM micrographs with bacteria in extra- or intracellular association with these algal cells (Table 4.1). The principal aim of this section is to summarize the sub-cellular aspects of bacterial associations with three acknowledged bloom forming dinoflagellates, *Alexandrium lusitanicum* Balech, *Gyrodinium instriatum* Freudenthal and Lee and *Gymnodinium catenatum* Graham.

Table 4.1. List of the dinoflagellate species with bacteria in extra- or intracellular association with the algal cells and references where LM and/or TEM micrographs are shown.

Dinoflagellate species	LM & TEM extra-cellular	LM intra-cellular	TEM cyto-plasm	TEM nucleus	ref*
Gonyaulax spinifera		+			1
G. tamarensis (= *Alexandrium lusitanicum*)	+	+	+		1,4
Scrippsiella trochoidea		+			1
Cochlodinium heterolobatum		+			2,4
Gyrodinium instriatum		+	+	+	2,4,5
Gymnodinium splendens			+	+	3,4
Glenodinium foliaceum		+	+	+	3,4,5
Prorocentrum minimum		+			4
P. scutellum		+			4
P. micans	+				4

* 1. Silva 1962; 2. Silva 1967; 3. Silva 1978; 4. Silva 1982a; 5. Silva and Franca 1985

4.1. *Alexandrium lusitanicum* Balech

The Portuguese strain of *Alexandrium lusitanicum* Balech was identified as *Gonyaulax tamarensis* prior to its re-classification by Balech (1985). Blooms of this species were associated with the occurrence of PSP toxins in shellfish in 1958, 1959, and 1962 in Obidos Lagoon, Portugal (Silva 1962). The clone of *A. lusitanicum* selected for detailed study was isolated in 1962 from "red water" of high toxicity in Obidos Lagoon, causing human intoxications. This clone has been maintained in Provasoli's ASP7 medium, with frequent addition of Provasoli's AM9 antibiotics. The toxin profile of this isolate was first investigated using HPLC methods by Cembella *et al.* (1987) and again by Mascarenhas *et al.* (1995), who confirmed that the same toxin composition was conserved in the cultured cells over this *ca.* 10 year period.

Spatial aspects. The LM study of *A. lusitanicum* showed maintenance of bacteria inside the cell, apparently without damaging the normal growth of the host. TEM observations indicated that bacteria were confined to the cell periphery (Silva 1962, 1982a). Cells examined under LM and stained with methylene blue, or DAPI (DNA stain) for epi-fluorescence, showed few distinct bacteria beneath the theca. Most DAPI fluorescence was concentrated in the nucleus, and it was difficult to ascribe fluorescent points to bacteria without a complete extraction of chlorophyll. No label was seen in the nucleolus nor general cytoplasmic space, while a few isolated points beneath the theca could be discerned and were interpreted as representing bacteria. DAPI fluorescence was not observed exterior to the cell, so the medium was considered to be bacteria free.

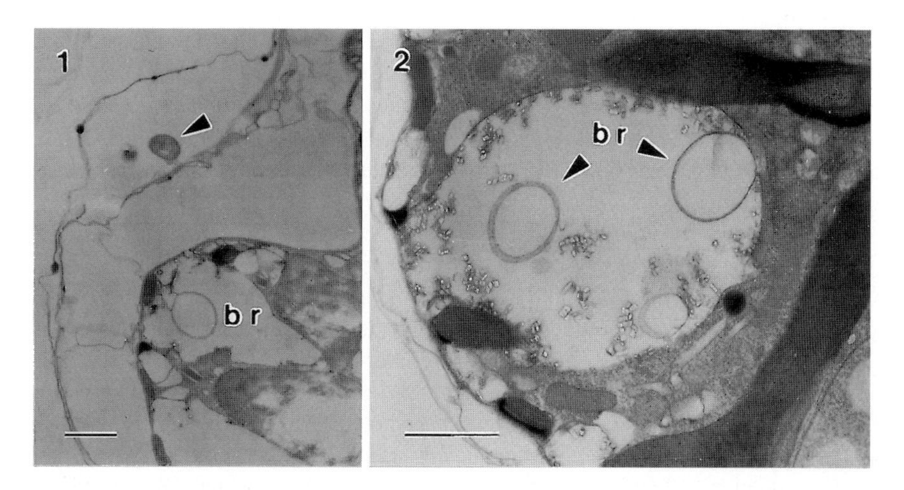

Figure 4.1. *A. lusitanicum.* 1. Intracellular bacterium (arrowhead) in space between outer cellular membrane and thecal complex. 2. Structures with appearance of bacterial remnants (br) inside cytoplasmic vesicles. Note similar structure in Fig. 4.1.1. (br). Scale bars = 1μm.

Ultrastructural features. TEM observation of cell sections showed bacteria and possibly cyanobacteria in low numbers immediately external to some cells, in particular near constrictions of paired cells. High bacterial abundances were limited to sections of degrading dinoflagellates. Well-preserved intracellular bacteria were rarely seen, but when apparent, they occurred only between the outer cellular membrane and the thecal complex (Fig. 4.1.1), especially near the flagellar-cingular region but not deeper in the cytoplasm nor in the nucleus. Some cytoplasmic vesicles appeared to contain degraded structures of concentric membrane, possibly remnants of bacteria (Fig. 4.1.1, 4.1.2).

Temporal stability. In spite of frequent AM9 antibiotic treatments needed to maintain axenic cultures, bacteria have remained associated with cells of *A. lusitanicum* more than 30 years after the original isolation, albeit in low numbers. Bacterial isolations from these dinoflagellate cells using two different procedures yielded similar results (Franca *et al.* 1995). The predominant isolate, *Pseudomonas stutzeri*, showed an ultrastructure consistent with that observed for bacteria occurring inside and outside the *A. lusitanicum* cell periphery. Other isolates included *P. fluorescens* and a gram positive chromogenic bacterium yet to be identified. A permanent association of *A. lusitanicum* with these bacteria, described both as intracellular and as isolated from culture medium, has been confirmed by other reports (Silva 1962, 1982a, 1990; Dimanlig and Taylor 1985).

4.2. *Gyrodinium instriatum* Freudenthal & Lee

Gyrodinium instriatum Freudenthal and Lee is a species often abundant in seawater lagoons or inlets along the Portuguese coast, where it has never been described as toxic. Clones of this dinoflagellate isolated from spatially and temporally separate populations, have been described as either with or without endonuclear bacteria (Silva and Franca 1985). Silva (1982b) reported that *G. instriatum* clones with endonuclear bacteria were toxic. However, preliminary chromatograms from HPLC analysis of PSP toxins in the same clones conducted ten years later failed to show any toxicity (Franca 1991). The strain of *G. instriatum* chosen for study as a positive control, based on previous evidence of intracellular bacteria, was found to possess bacteria both in the nucleus and in the cytoplasm (Silva and Franca 1985). This strain was isolated from Santo-André Lagoon, Portugal in 1982, and has been maintained in culture using a mixture of ASP1+ASP2 (1:1) Provasoli media, with occasional addition of AM9 antibiotics.

Spatial aspects. By staining with methylene blue and DAPI the nucleus was clearly observed, as were numerous points, thought to represent bacteria, scattered within the cells and also some external to the cells. This culture, which received no regular antibiotic treatments, was obviously not axenic, and the fluorescent points outside the cells were considered to be free-living extracellular bacteria.

Ultrastructural features. New ultrastructural details of the relationship between *G. instriatum* and its endocellular bacteria have been noted. Bacteria were previously (Silva and Franca 1985) observed within the nucleoplasm, not only encircled by transparent halos, but sometimes showing connections throughout these halos with DNA fibrils from chromosomes. Detailed observations now indicate a close morphological relationship between the outer layer of the bacterial wall and the nucleoplasm. In addition, endonuclear bacteria appear to be transferred into the cytoplasm via evaginations of the nuclear envelope, which protrude into the cytoplasm while enclosing bacteria and are ultimately pinched off and released from the nuclear envelope (Fig. 4.2.1, 4.2.2). Once in the cytoplasm, the structure of these bacteria can appear as follows (Fig. 4.2.3-5): 1) groups of undamaged bacteria enveloped by a host membrane; 2) dense clusters of bacteria, also enveloped by sack-like showing laminated arrangements of outer layers of the bacterial cell wall; 3) apparently free bacteria in the cytoplasm, isolated by a halo from the surrounding cytoplasmic organelles, which appear morphologically normal.

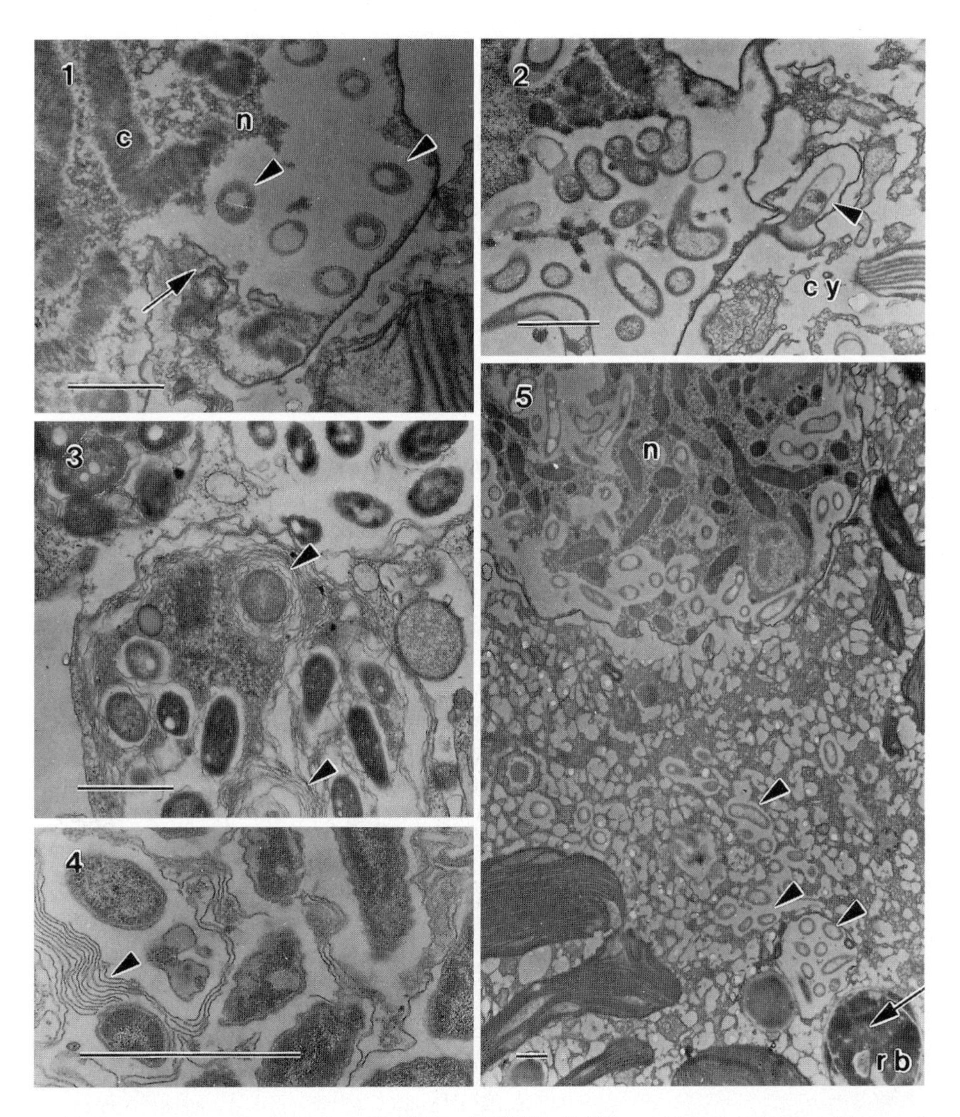

Figure 4.2. *G. instriatum*. 1. Endonuclear bacteria within dividing nucleus (n) showing V-shaped chromosomes (c) and microtubule cross-section (arrow) along invagination of nuclear envelope. Bacteria (arrowheads) are concentrated in evagination of envelope 2. Endonuclear bacterium (arrowhead) located in evagination of nuclear envelope, possibly undergoing transfer into cytoplasm (cy). 3,4. Clusters of degrading bacteria enveloped by sack-like formations of host membrane. Note laminated arrangement of outer bacterial cell wall layers (arrowheads). 5. Overview of cell showing endonuclear bacteria (n = nucleus) and clusters of undamaged bacteria throughout cytoplasm (arrowheads). Note presence of apparent bacterial cross-section (arrow) contained within residual body (rb). Scale bars = 1 µm.

Temporal stability. The long term relationship between bacteria and *G. instriatum* cells of cultures established over 10 years ago was confirmed recently by TEM observations. The vitality of these dinoflagellate cultures was clearly indicated by average division rates of 0.4-0.5 div.day^{-1} based on data from 18 cultures established from three separate inocula. Morphological development of cells in the presence of bacteria appeared to be normal. One bacterial strain, *Alcaligenes paradoxus*, was isolated from these dinoflagellate cells and maintained on ZoBell nutrient marine agar. The fine structure of its wall was similar to that of other gram-negative bacteria.

4.3. *Gymnodinium catenatum* Graham

The first recognized PSP contamination of bivalves in Portugal occurred in 1986, and was attributed to *Gymnodinium catenatum* Graham (Franca and Almeida 1989). The two Portuguese clones of *G. catenatum* considered here (a gift of M.A. de M. Sampayo) were obtained during PSP outbreaks in 1986 from the northeast coast (Aguda), and in 1992 from the south coast (Sagres). Culture conditions after isolation from water samples were described by Sampayo (1985, 1993); maintenance of cultures utilized GPM medium (Franca *et al.* 1993). Cell division rates were 0.26 ± 0.05 div.day-1.

Spatial aspects. Using direct LM and methylene blue staining, bacteria were clearly observed at the periphery of some cells. The bacterial presence within the dinoflagellate cells seemed very irregular. For example, two or three bacilli were often free in the peripheral cellular space beneath the cell wall at connection between cells in a chain. However, some cells showed no evidence of associated bacteria. With DAPI staining, some scattered fluorescence points, considered to be bacteria, were present beneath the peripheral membrane of some cells. No extracellular fluorescence was observed.

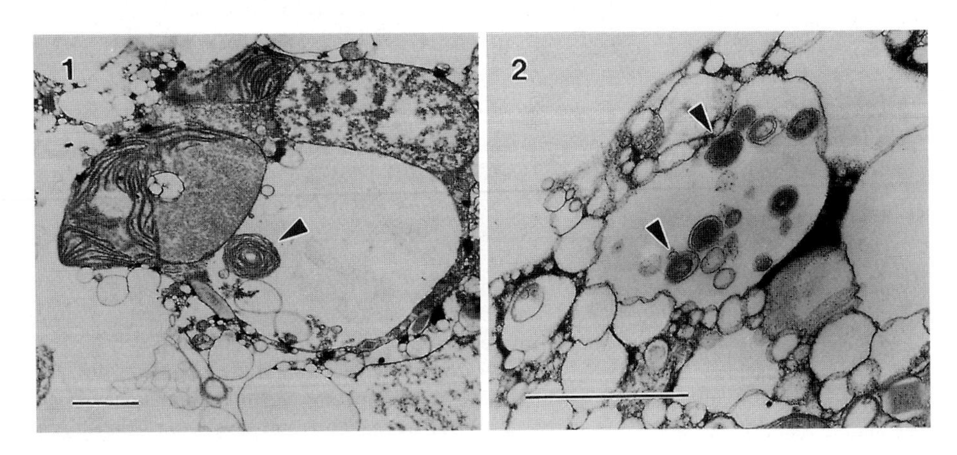

Figure 4.3. *G. catenatum.* 1, 2. Concentric laminated structures having appearance of degraded bacteria located within cytoplasmic vesicles (arrowheads). Scale bars = 1 µm.

Ultrastructural features. Earlier attempts using TEM to detect endocellular bacteria in sections of *G. catenatum* cells were unsuccessful (Franca 1987, Rees and Hallegraeff 1991, Franca *et al.* 1993). However, more recent TEM observations of *G. catenatum*

(Franca *et al.* 1996) revealed well-preserved bacterial cross sections near the periphery in a few specimens, although intracellular bacteria with fine micromorphology were not generally apparent. In certain cases, bacteria-like structures were observed as inclusions at the cell periphery, and the cytoplasm occasionally exhibited extreme vesiculation containing what appeared to be degraded bacteria (Fig. 4.3).

Temporal stability. The first attempts to isolate bacteria yielded no discernable colonies (Franca *et al.* 1993). Recently, however, two bacterial strains forming white and yellow colonies, respectively, were isolated consistently from *G. catenatum* (Franca *et al.* 1996). The dominant white colonies were classified as *Pseudomonas diminuta*, while the less abundant yellow colonies have not yet been identified. It is not clear why the initial isolation attempts were unsuccessful.

5. Critical issues

It is evident from the information presented above that progress is being made in characterizing the association between bacteria and HAB species. Yet it is equally clear that we are still in the early stages of this developing field, with many important issues remaining unresolved. There is compelling evidence to suggest that the most 'influential players' in microbial assemblages associated with HABs likely belong to particular classes or subdivisions of bacteria; however, the genetic identity of these bacteria and how the taxonomic structure of the bacterioplankton community is modified during blooms events must be elucidated. It is also uncertain to what extent an algal species can mediate changes in the co-occurring microbial assemblage, and whether these alterations occur on temporal and spatial scales required to influence bloom dynamics. Attempts to move beyond qualitative/descriptive studies in order to acquire quantitative data on the taxonomic structure and metabolic activity of bacterial communities, while essential, have been constrained by an inability to culture a large fraction of these microbes. The approaches required for quantitative evaluation of changes in bacterial community structure at the level of individual taxa, or at least bacterial groups, are clearly those based on molecular techniques. As is now the rule for microbial ecologists describing the dynamics of unique bacterial communities (e.g., microbial mats, marine snow, sediments), sequencing, profiling, and probing the genetic signatures of these microbes are the methods of choice, yielding a wealth of new information.

Cellular level interactions between bacteria and HAB species, affecting algal cell growth and ultimately population growth, can take many forms and be effected by a variety of bacterial, as well as algal, metabolites. Thus, it is important that attempts be made to isolate and characterize bacterial and algal bioactive metabolites, describe their effects on bacterial and algal growth, and elucidate patterns of synthesis in natural communities. Classical chemical and biochemical approaches are satisfactory for initial laboratory and field investigations; however, molecular techniques allowing individual cells of selected taxa to be screened for the genetic machinery needed to synthesize a given substance and for the active expression of these genes will ultimately be required to resolve bacterial-algal interactions at the cellular level. Once the genetic basis for production of a metabolite by a HAB species or its associated bacteria is established, techniques such as *in-situ* PCR and reporter gene assays, currently used in studies of marine prokaryotes (e.g., Hodson *et al.* 1995, Cooksey and Wigglesworth-Cooksey 1995), will be effective tools in the study of bacterial interactions with harmful algae.

The available evidence demonstrates that certain bacteria are capable of synthesizing one or both of the sodium channel blocking PSP toxins and tetrodotoxins. These bacteria not only show wide-ranging geographic distributions, but also appear to be common to both planktonic and benthic communities. However, simply demonstrating an ability to produce toxin in laboratory cultures does not constitute proof of bacterial toxigenesis in natural systems. Herein lies perhaps the most challenging issue surrounding this topic. Much of what we presently know suggests that physical and/or biochemical interactions between toxic bacteria and living organisms (e.g., dinoflagellates, puffer fish) or non-living particulate matter (e.g., detritus, marine snow, silt), may be crucial to inducing and/or modulating toxin production by bacteria. More direct approaches aimed at identifying and characterizing such interactions would appear to be a logical next step. Laboratory-based study of the effects of surface attachment on bacterial production of certain metabolites, including SCB toxins, employing both classical and molecular techniques, may yield important insights that could be applied to field-oriented research. The utility of this approach can be seen in the rapid advances being made by microbial ecologists in their study of bacterial biofilms and their complex interactions with surfaces in the marine environment. Researchers must also pursue the description of toxin genetics and biosynthesis in bacterial model systems. Elucidation of toxin genes and the enzymes involved in toxin synthesis will ultimately lead to an understanding of the environmental control of toxin production at the molecular level.

Another aspect of bacterial involvement in phycotoxin production which warrants attention is the issue of endocellular bacteria and their interactions with 'host' organisms. Whether referring to associations with algae, invertebrate animals, or fish, the possibility exists that a bacterial metabolite(s) acts as either a direct or indirect precursor for toxin synthesis by the host, and that both organisms contribute to the process yielding 'typical' levels of PSP toxins and TTXs measured in nature. The apparent maintenance of bacteria inside these eukaryotic organisms, especially in critical organs such as the puffer fish liver, for extended periods shows the potential for such a 'mutualistic' relationship, although such occurrences may simply represent bacterial 'infections'. Further work is also needed to characterize at a genetic level the taxonomic specificity of bacterial associations with toxic eukaryotes, and molecular, as well as biochemical, approaches will be required to identify bacterial metabolites that may serve as toxin precursors.

It is apparent that the spatial component of bacterial-algal associations is highly variable both within and between dinoflagellate species. Whether this variability reflects the limitations of the techniques employed (e.g., chemical fixations, viewing individual cross-sections of cells, etc.), or the actual patterns of bacterial distribution within and among algal cells *in vivo*, must be discerned. The most obvious approaches to this task involve the use of nucleic acid or antibody probes for bacteria in conjunction with whole cell imaging techniques such as confocal microscopy, which allows visualization of a probe via high resolution 'optical' sectioning of the entire cell. The images may then be used to locate sites of probe binding (i.e., bacteria) within a three-dimensional reconstruction of an algal cell, thereby allowing precise mapping of intracellular bacterial distributions. From a temporal viewpoint, although bacterial cells show a dynamic relationship with algal cells (i.e., movement within and among cellular compartments), long term stability of a bacterial taxon's association with an algal culture is not unusual and should be examined. Data obtained will aid not only in establishing the degree of taxonomic specificity between bacteria and host algae, but also in describing the 'handling' of bacteria by algae. Indeed, observations of bacteria within membrane-bound vacuoles having a lysosome-like structure, suggest that bacteria may contribute metabolically to the algal cell. In addition, release of bacteria into the growth medium and the apparent maintenance of an attached bacterial assemblage may be other mechanisms by which algal cells 'control' the number and type of associated bacteria.

To conclude, it is apparent that there co-exist with algal communities containing HAB species microbial assemblages which undergo changes in species composition, exhibit competitive interactions, and produce bioactive compounds, including toxins. These and other microbial processes are influenced by the algal community and can potentially influence the population dynamics, cellular growth, and toxin characteristics of the algae. Bacterial-algal associations operate on a number of levels (i.e., population, cellular, sub-cellular) and in a variety of ways potentially affecting both growth (i.e., stimulatory, inhibitory) and toxicity (i.e., modifying toxin content and/or composition) of HAB species. Now our challenge is to begin piecing together this puzzle, and defining where it fits within the ecology of harmful algal blooms.

6. References

Anderson, D.M., Kulis, D.M., Sullivan, J.J., Hall, S., Lee, C. (1990). Dynamics and physiology of saxitoxin production by the dinoflagellates *Alexandrium* spp. *Mar. Biol.* 104: 511-524.

Balech, E. (1985). The genus *Alexandrium* or *Gonyaulax* of the Tamarensis Group. In Anderson D.,White, A., Baden, D. (eds.) *Toxic Dinoflagellates*, Elsevier Publ., NY, pp. 33-38.

Bates, S.S., Douglas, D.J., Doucette, G.J. Leger, C. (1995). Enhancement of domoic acid production by reintroducing bacteria to axenic cultures of the diatom *Pseudo-nitzschia multiseries*. *Nat. Toxins* 3: 428-435.

Bidle, K.D., Fletcher, M. (1995). Comparison of free-living and particle-associated bacterial communities in the Chesapeake Bay by stable low-molecular-weight RNA analysis. *Appl. Environ. Microbiol.* 61: 944-952.

Cembella, A. D., Sullivan J.J., Boyer, G.L., Taylor F.J.R., Anderson, R.J. (1987). Variation in paralytic shellfish toxin composition within the *Protogonyaulax tamarensis/catenella* species complex: red tide dinoflagellates. *Biochem. System. Ecol.* 15: 171-186.

Chisholm, S.W., Morel, F.M.M. (eds.) (1991). What Controls Phytoplankton Production in Nutrient-rich Areas of the Open Sea? *Limnol. Oceanogr.* 36: 1507-1970.

Cooksey, K.E., Wigglesworth-Cooksey, B. (1995). Adhesion of bacteria and diatoms to surfaces in the sea: a review. *Aquat. microb. Ecol.* 9: 87-96.

Costerton, J.W., Lewandowski, Z., Caldwell, D.E., Korber, D.R., Lappin-Scott, H.M. (1995). Microbial Biofilms. *Ann. Rev. Microbiol.* 49: 711-745.

DeLong, E.F., Franks, D.G., Alldredge, A.L. (1993). Phylogenetic diversity of aggregate-attached vs. free-living marine bacterial assemblages. *Limnol. Oceanogr.* 38: 924-934.

Dimanlig, M.N.V., Taylor, F.J.R. (1985). Extracellular bacteria and toxin production in *Protogonyaulax species*. In: Anderson D.,White, A., Baden, D. (eds.) *Toxic Dinoflagellates*, Elsevier Publ., NY, pp. 103-108.

Do, H.K., Kogure, K., Simidu, U. (1990). Identification of deep-sea-sediment bacteria which produce tetrodotoxin. *Appl. Environ. Microbiol.* 56: 1162 -1163.

Doucette, G.J. (1995). Interactions between bacteria and harmful algae: a review. *Nat. Toxins* 3: 65-74.

Doucette, G.J., Trick, C.G. (1995). Characterization of bacteria associated with different isolates of *Alexandrium tamarense*. In: Lassus, P., Arzul, G., Erard, E., Gentien, P., Marcaillou-LeBaut, C. (eds.) *Harmful Marine Algal Blooms*, Lavoisier Science Publ., Paris, pp. 33-38.

Franca, S. (1987). Ultrastructural study of *Gymnodinium catenatum*, toxic dinoflagellate-preliminary results. Abst.14, XXII R, A, SPME, Évora.

Franca, S., Almeida, J.F. (1989). Paralytic shellfish poisons in bivalve molluscs on the Portuguese coast caused by a bloom of the dinoflagellate *Gymnodinium catenatum*. In: Okaichi, T., Anderson, D.M., Nemoto, T. (eds.) *Red Tides: Biology, Environmental Science, and Toxicology*, Elsevier Sci. Pub. Co., Inc., NY, pp. 93-96.

Franca, S. (1991). Dinoflagellate toxicity: harmful effects, warning measures and research trends in Portugal. In: Frémy, J.M. (ed.) *Proceedings of Symposium on Marine Biotoxins*, (CNEVA), Paris, pp. 229-233.

Franca, S., Alvito, P., Sousa, I., Mascarenhas, V. (1993). The dinoflagellate *Gymnodinium catenatum* isolated from the coast of Portugal: observations on development, toxicity and ultrastructure. In: Smayda, T.J., Shimizu, Y. (eds.) *Toxic Phytoplankton Blooms in the Sea*, Elsevier, NY, pp. 869-874.

Franca, S., Viegas, S., Mascarenhas, V., Pinto, L., Doucette, G.J. (1995). Prokaryotes in association with a toxic A*lexandrium lusitanicum* in culture. In: Lassus, P., Arzul, G., Erard, E., Gentien, P., Marcaillou-LeBaut, C. (eds.) *Harmful Marine Algal Blooms*, Lavoisier Science Publ., Paris, pp. 44-51.

Franca, S., Pinto, L., Alvito, P., Sousa, I., Vasconcelos, V., Doucette, G.J. (1996). Studies on prokaryotes associated with PSP producing dinoflagellates. In: Yasumoto, T., Oshima, T., Fukuyo, Y. (eds.) *Harmful and Toxic Algal Blooms*, Intergov. Oceanogr. Comm. of UNESCO, pp. 347-350.

Fuhrman, J.A., Ammerman, J.W., Azam, F. (1980). Bacterioplankton in the coastal euphotic zone: distribution, activity and possible relationships with phytoplankton. *Mar. Biol.* 60: 201-207.

Fukami, K., Simidu, U., Taga, N. (1985). Microbial decomposition of phyto- and zooplankton in seawater. II. Changes in the bacterial community. *Mar. Ecol. Prog. Ser.* 21: 7-13.

Fukami, K., Nishijima, T., Murata, H., Doi, S., Hata, Y. (1991a). Distribution of bacteria influential on the development and the decay of *Gymnodinium nagasakiense* red tide and their effects on algal growth. *Nippon Suisan Gakkaishi* 57: 2321-2326.

Fukami, K., Yuzawa, A., Nishijima, T., Hata, Y. (1991b). Isolation and properties of a bacterium inhibiting the growth of *Gymnodinium nagasakiense*. *Nippon Suisan Gakkaishi* 58: 1073-1077.

Fukami, K., Sakaguchi, K., Kanou, M., Nishijima, T. (1996). Effect of bacterial assemblages on the succession of blooming phytoplankton from *Skeletonema costatum* to *Heterosigma akashiwo*. In: Yasumoto, T., Oshima, T., Fukuyo, Y. (eds.)

Harmful and Toxic Algal Blooms, Intergov. Oceanogr. Comm. of UNESCO, pp. 335-338.

Furuki, M., Kobayashi, M. (1991). Interaction between *Chattonella* and bacteria and prevention of this red tide. *EMECS '90* 23: 189-193.

Gallacher, S., Birkbeck, T.H. (1993). Effect of phosphate concentration on production of tetrodotoxin by *Alteromonas tetraodonis*. *Appl. Environ. Microbiol.* 59: 3981-3983.

Gallacher, S., Birkbeck, T.H. (1995). Isolation of marine bacteria producing sodium channel blocking toxins and the seasonal variation in their frequency in sea water. In: Lassus, P., Arzul, G., Erard, E., Gentien, P., Marçaillou-LeBaut, C. (eds.) *Harmful Marine Algal Blooms*, Lavoisier Science Publ., Paris, pp. 445-450.

Gallacher, S., Flynn, K.J., Leftley, J., Lewis, J., Munro, P.D., Birkbeck, T.H. (1996). Bacterial production of sodium channel blocking toxins. In: Yasumoto, T., Oshima, T., Fukuyo, Y. (eds.) *Harmful and Toxic Algal Blooms*, Intergov. Oceanogr. Comm. of UNESCO, pp. 355-358.

Gallacher, S., Flynn, K.J., Franco, J.M., Brueggemann, E.E., Hines, H.B. (1997). Evidence for the production of paralytic shellfish toxins by bacteria associated with *Alexandrium* spp. (Dinophyta) in culture. *Appl. Environ. Microbiol.* 63: 239-245.

Gallacher, S., Flynn, K.J., Franco, J.M. (submitted). Production of paralytic shellfish toxins by the tetrodotoxin producing bacterium *Shewanella alga*. *FEMS Microbial Ecology*.

Gallacher, S., Mass, E., Moore, E., Hold, G. (in prep). Diversity and identity of bacteria isolated from PST associated dinoflagellates: *Alexandrium* spp.

Gobler, C.J., Cosper, E.M. (1996). Stimulation of "brown tide" blooms by iron. In: Yasumoto, T., Oshima, T., Fukuyo, Y. (eds.) *Harmful and Toxic Algal Blooms*, Intergov. Oceanogr. Comm. of UNESCO, pp. 321-324.

Hallegraeff, G.M. (1993). A review of harmful algal blooms and their apparent global increase. *Phycologia* 32: 79-99.

Hodson, R.E., Dustman, W.A., Garg, R.P., Moran, M.A. (1995). In situ PCR for visualization of microscale distribution of specific genes and gene products in prokaryotic communities. *Appl. Environ. Microbiol.* 61: 4074-4082.

Imai, I., Ishida, Y., Hata, Y. (1993). Killing of marine phytoplankton by a gliding bacterium *Cytophaga* sp., isolated from the coastal sea of Japan. *Mar. Biol.* 116: 527-532.

Imai, I., Ishida, T., Sakaguchi, K., Hata, Y. (1995). Algicidal marine bacteria isolated from Northern Hiroshima Bay, Japan. *Fish. Sci.* 61: 628-636.

Ishida, Y. (in press). Microbial impact on occurrence of harmful algal red tides. In: *Proceedings of 1st Korea-Japan Marine Biotechnology Symposium*.

Ishida, Y., Kim, C.-H., Sako, Y., Hirooka, N., Uchida, A. (1993). PSP toxin production is chromosome dependent in *Alexandrium* spp. In: Smayda, T.J., Shimizu, Y. (eds.) *Toxic Phytoplankton Blooms in the Sea*, Elsevier, NY, pp. 881-887.

Ishida, Y., Yoshinaga, I., Mu-Chan, K., Uchida, A. (in press). Possibility of bacterial control of harmful algal blooms. In: *Proceedings of VII International Symposium of Microbial Ecology*. Santos, Brazil, 1995.

Ishio, S., Mangindaan, R.E., Kuwahara, M., Nakagawa, H. (1989). A bacterium hostile to flagellates: identification of species and characters. In: Okaichi, T., Anderson,

D.M., Nemoto, T. (eds.) *Red Tides: Biology, Environmental Science, and Toxicology*, Elsevier Sci. Pub. Co., Inc., NY, pp. 205-208.

Jacobson, D.M., Anderson, D.M. (1996). Widespread phagocytosis of ciliates and other protists by marine mixotrophic and heterotrophic thecate dinofagellates. *J. Phycol.* 32: 279-285.

Keshtacher-Liebson, E., Hadar, Y., Chen, Y. (1995). Oligotrophic bacteria enhance algal growth under iron-deficient conditions. *Appl. Environ. Microbiol.* 61: 2439-2441.

Kim, C.-H., Sako, Y., Ishida, Y. (1993). Variation of toxin production and composition in axenic cultures of *Alexandrium catenella* and *A. tamarense. Nippon Suisan Gakkaishi* 59: 633-639.

Kodama, M. (1990). Possible links between bacteria and toxin production in algal blooms. In: Granéli, E., Sundström, B., Edler, L., Anderson, D.M. (eds.) *Toxic Marine Phytoplankton*, Elsevier, NY, pp. 52-61.

Kodama, M., Ogata, T., Noguchi, T., Maruyama, J., Hashimoto, K. (1983). Occurrence of saxitoxin and other toxins in the liver of the pufferfish *Takifugu pardalis. Toxicon* 21: 897-900.

Kodama, M., Ogata, T., Sato, S. (1988). Bacterial production of saxitoxin. *Agric. Biol. Chem.* 52: 1075-1077.

Kodama, M., Ogata, T., Sato, S. (1989). Saxitoxin-producing bacterium isolated from *Protogonyaulax tamarensis*. In: Okaichi, T., Anderson, D.M., Nemoto, T. (eds.) *Red Tides: Biology, Environmental Science, and Toxicology*, Elsevier, NY, pp. 363-366.

Kodama, M., Ogata, T., Sakamoto, S., Sato, S., Honda, T., Miwatani, T. (1990a). Production of paralytic shellfish toxins by a bacterium *Moraxella* sp. isolated from *Protogonyaulax tamarensis. Toxicon* 28: 707-714.

Kodama, M., Ogata, T., Sato, S., Sakamoto, S. (1990b). Possible association of marine bacteria with paralytic shellfish toxicity of bivalves. *Mar. Ecol. Prog. Ser.* 61: 203-206.

Kodama, M., Sato, S., Ogata, T. (1993). *Alexandrium tamarense* as a source of tetrodotoxin in the scallop *Patinopecten yessoensis*. In: Smayda, T.J., Shimizu, Y. (eds.) *Toxic Phytoplankton Blooms in the Sea*, Elsevier, NY, pp. 401-406.

Kodama, M., Shimizu, H., Sato, S., Ogata, T., Terao, K. (1995). The infection of bacteria in the liver cells of toxic puffer --- A possible cause for organisms to be made toxic by tetrodotoxin in association with bacteria. In: Lassus, P., Arzul, G., Erard, E., Gentien, P., Marcaillou-LeBaut, C. (eds.) *Harmful Marine Algal Blooms*, Lavoisier Science Publ., Paris, pp. 457-462.

Kodama, M., Sakamoto, S., Koike, K. (1996a). Symbiosis of bacteria in *Alexandrium tamarense*. In: Yasumoto, T., Oshima, T., Fukuyo, Y. (eds.) *Harmful and Toxic Algal Blooms*, Intergov. Oceanogr. Comm. of UNESCO, pp. 351-354.

Kodama, M., Sato, S., Ogata, T. (1996b). Occurrence of tetrodotoxin in *Alexandrium tamarense*, a causative dinoflagellate of paralytic shellfish poisoning. *Toxicon* 34: 1101-1106.

Kogure, K., Do, H.K., Simidu, U. (1989). Production of tetrodotoxin by marine bacteria and its accumulation through food-web in marine ecosystems. Program 1st Int. Marine Biotechnol. Conf. (IMBC `89), 1989. pp. 11.

Kopp, M., Doucette, G.J. Kodama, M., Gerdts, G., Schutt, C., Medlin, L. (Submitted). Phylogenetic analysis of selected toxic and non-toxic bacterial strains isolated from the toxic dinoflagellate *Alexandrium tamarense*. *FEMS Lett.*

Koyama, K., Noguchi, T., Uzu, A., Hashimoto, K. (1983). Resistibility of toxic and nontoxic crabs against paralytic shellfish poison and tetrodotoxin. *Bull. Japan. Soc. Sci. Fish.* 49: 481-484.

Levasseur, M., Monfort, P., Doucette, G.J., Michaud, S. (1996). Preliminary study of the impact of bacteria as PSP producers in the Gulf of St. Lawrence. In: Yasumoto, T., Oshima, T., Fukuyo, Y. (eds.) *Harmful and Toxic Algal Blooms*, Intergov. Oceanogr. Comm. of UNESCO, pp. 363-366.

Mascarenhas,V., Alvito, P, Franca, S., Sousa, I., Martinez, A.G., Rodriguez-Vazquez, J.A. (1995). The dinoflagellate *Alexandrium lusitanicum* isolated from the coast of Portugal: observations on toxicity and ultrastructure during growth phases. In: Lassus,P., Arzul, G., Erard, E., Gentien, P., Marcaillou-LeBaut, C. (eds.) *Harmful Marine Algal Blooms*, Lavoisier Science Publ., Paris, pp. 71-76.

Matsui, T., Sato, H., Hamada, S., Shimizu, C. (1982). Comparison of toxicity of the cultured and wild puffer fish *Fugu niphobles*. *Bull. Jpn. Soc. Sci. Fish.* 48: 253.

Mitsutani, A., Takesue, K., Kirita, M., Ishida, T. (1992). Lysis of *Skeletonema costatum* by *Cytophaga* sp. isolated from the coastal water of Ariake Sea. *Nippon Suisan Gakkaishi* 58: 2159-2167.

Mullins, T.D., Britschgi, T.B., Krest, R.L., Biovannoni, S.J. (1995). Genetic comparisons reveal the same unknown bacterial lineages in Atlantic and Pacific bacterioplankton communities. *Limnol. Oceanogr.* 40: 148-158.

Noguchi, T. Jeon, J., Arakawa, O., Sugita, H., Deguchi, T., Shida, T., Hashimoto, K. (1986). Occurrence of tetrodotoxin and anhydrotetrodotoxin in *Vibrio* sp. isolated from the intestine of a xanthid crab, *Atergatis floridus*. *J. Biochem.* 99: 311-314.

Ogata, T., Kodama, M., Fukuyo, Y., Inoue, T., Kamiya, H., Matsuura, F., Sekiguchi, K., Watanabe, S. (1982). The occurrence of *Protogonyaulax* spp. in Ofunato Bay, in association with the toxification of the scallop *Patinopecten yessoensis*. *Bull. Jpn. Sci. Soc. Fish.* 48: 563-566.

Ogata, T., Kodama, M., Ishimaru, T. (1987). Toxin production in the dinoflagellate *Protogonyaulax tamarensis*. *Toxicon* 25: 923-928.

Ogata, T., Kodama, M., Komaru, K., Sakamoto, S., Sato, S., Simidu, U. (1990). Production of paralytic shellfish toxins by bacteria isolated from toxic dinoflagellates. In: Granéli, E., Sundström, B., Edler, L., Anderson, D.M. (eds.) *Toxic Marine Phytoplankton*, Elsevier, NY, pp. 311-315.

Ogata, T., Koike, K., Nomura, S., Kodama, M. (1996). Utilization of organic substances for growth and toxin production by *Alexandrium tamarense*. In: Yasumoto, T., Oshima, T., Fukuyo, Y. (eds.) *Harmful and Toxic Algal Blooms*, Intergov. Oceanogr. Comm. of UNESCO, pp. 343-345.

Provasoli, L. (1963). Growing marine seaweeds. Proc. 4th Int. Seaweed Symp., Biarritz. Pergamon Press, pp. 9-17.

Rausch de Traubenberg, C., Lassus, P. (1991). Dinoflagellate toxicity: are marine bacteria involved? Evidence from the literature. *Mar. Microbial Food Webs* 5: 205-226.

Rees, A.J.J., Hallegraeff, G.M. (1991). Ultrastructure of the toxic, chain forming dinoflagellate *Gymnodinium catenatum* (Dinophyceae). *Phycologia* 30: 90-105.
Rehnstam, A.-S., Backman, S., Smith, D.C., Azam, F., Hagstrom, A. (1993). Blooms of sequence-specific culturable bacteria in the sea. *FEMS Microbiol. Ecol.* 102: 161-166.
Romalde, J.L., Toranzo, A.E., Barja, J.L. (1990). Changes in bacterial populations during red tides caused by *Mesodinium rubrum* and *Gymnodinium catenatum* in north west coast of Spain. *J. Appl. Bacteriol.* 68: 123-132.
Saito, T., Noguchi, T., Harada, T., Murata, O., Abe, T., Hashimoto, K. (1985). Resistibility of toxic and non-toxic pufferfish against tetrodotoxin. *Bull. Jpn. Soc. Sci. Fish.* 51: 1371.
Sakamoto, S., Ogata, T., Sato, S., Kodama, M., Takeuchi, T. (1992). Causative organism of paralytic shellfish toxins other than toxic dinoflagellates. *Mar. Ecol. Prog. Ser.* 89: 229-235.
Sampayo, M.A. de M. (1985). Encystment and excystment of a Portuguese isolate of *Amphidinuim carterae* in culture. In: Anderson D.,White, A., Baden, D. (eds.) *Toxic Dinoflagellates*, Elsevier Publ., NY, pp. 125-130.
Sampayo, M.A. de M. (1993). Trying to cultivate *Dinophysis* spp. In: Smayda, T.J., Shimizu, Y. (eds.) *Toxic Phytoplankton Blooms in the Sea*, Elsevier, NY, pp. 807-810.
Sato, S., Kodama, M., Ogata, T., Saitanu, K., Furuya, M., Hirayama, K., Kakinuma, K. (1997). Saxitoxin as a toxic principle of a freshwater puffer *Tetraodon fangi* in Thailand. *Toxicon* 35: 137-140.
Sato, S., Komaru, K., Ogata, T., Kodama, M. (1990). Occurrence of tetrodotoxin in cultured puffer. *Nippon Suisan Gakkaishi* 56: 1129-1131.
Sato, S., Ogata, T., Kodama, M. (1993). Wide distribution of toxins with sodium channel blocking activity similar to tetrodotoxin and paralytic shellfish toxns in marine animals. In: Smayda, T.J., Shimizu, Y. (eds.) *Toxic Phytoplankton Blooms in the Sea*, Elsevier, NY, pp. 429-434.
Sawayama, S., Sako, Y., Ishida, Y. (1993). Bacterial inhibitors for the mating reaction of *Alexandrium catenella* (Dinophyceae). In: Smayda, T.J., Shimizu, Y. (eds.) *Toxic Phytoplankton Blooms in the Sea*, Elsevier Sci. Publ. BV, Amsterdam, pp. 177-181.
Shumway, S.E. (1995). Phycotoxin-related shellfish poisoning: bivalve molluscs are not the only vectors. *Rev. Fisheries Sci.* 3: 1-31.
Silva, E.S. (1959). Some observations on marine dinoflagellate cultures. I. *Prorocentrum micans* Ehr. and *Gyrodinium* sp. *Notas e Estudos do Inst. Biol. Mar.* 12: 1-15.
Silva, E.S. (1962). Some observations on marine dinoflagellate cultures. III. *Gonyaulax spinifera* (Clap. and Lach.) Dies., *Gonyaulax tamarensis* Leb., and *Peridinium trochoideum* (Stein) Lemm. *Notas e Estudos do Inst. Biol. Mar.* 26: 1-21.
Silva, E.S. (1978). Endonuclear bacteria in two species of dinoflagellates. *Protistologica* 14: 113-119.
Silva, E.S. (1979). Intracellular bacteria, the origin of the dinoflagellates toxicity. In: *Proc. IVth IUPAC Symposium on Mycotoxins and Phycotoxins, Lausane*, Pathotox Publ., Lausane, pp. 8.
Silva, E.S. (1980). As grandes populações de dinoflagelados tóxicos na Lagoa de Obidos. *Arquivos do Instituto Nacional de Saúde (Lisboa).* 4: 253-262.

Silva, E.S. (1982a). Relationship between dinoflagellates and intracellular bacteria. In: Hoppe, H.A., Levring, T. (eds.) *Marine Algae in Pharmaceutical Science, vol. 2*, Walter de Gruyter & Co, Berlin, pp. 269-288.

Silva, E.S. (1982b). Toxic clones of *Gyrodinium instriatum* with endonuclear bacteria. *Proc. Vth Int. IUPAC Symp.on Mycotoxins and Phycotoxins*. pp. 216-219.

Silva, E.S. (1990). Intracellular bacteria: the origin of dinoflagellate toxicity. *JEPTO* 10: 124-128.

Silva, E.S., Franca, S. (1985). Ultrastructural relationships between two dinoflagellates and their intracellular bacteria. *Protistologica* 21: 429-446.

Simidu, U., Noguchi, T., Hwang, D.F., Shida, Y., Hashimoto, K. (1987). Marine bacteria which produce tetrodotoxin. *Appl. Environ. Microbiol.* 53: 1714-1715.

Smayda, T.J. (1996). Dinoflagellate bloom cycles: what is the role of cellular growth rate and bacteria? In: Yasumoto, T., Oshima, T., Fukuyo, Y. (eds.) *Harmful and Toxic Algal Blooms*, Intergov. Oceanogr. Comm. of UNESCO, pp. 331-334.

Stahl, D.A. (1995). Application of phylogenetically based hybridization probes to microbial ecology. *Mol. Ecol.* 4: 535-542.

Sugita, H., Noguchi, T., Furuta, M., Harada, T., Murata, O., Hashimoto, K., Deguchi, Y. (1988). The intestinal microflora of cultured specimens of a puffer *Fugu rubripes*. *Nippon Suisan Gakkaishi* 54: 733.

Tamplin, M.L. (1990). A bacterial source of tetrodotoxins and saxitoxins. In: Hall, S., Strichartz, G. (eds.) *Marine Toxins: Origin, Structure, and Molecular Pharmacology*, ACS Symp. Ser. 418. Amer. Chem. Soc., Washington, D.C., pp. 78-85.

Tiecco, G., Ianieri, A., Francioso, E., Todaro, M.P., Buonavoglia, D. (1996). Isolamento di batteri marini produttori di sostanze ad azione neurotossica. *Industrie Alimentari* 35: 134-135.

Vining, L.C. (1990). Functions of secondary metabolites. *Ann. Rev. Microbiol.* 44: 395-427.

Yasumoto, T., Yasumura, D., Yotsu, M., Mishishita, T., Endo, A., Kotaki, Y. (1986). Bacterial production of tetrodotoxin and anhydrotetrodotoxin. *Agric. Biol. Chem.* 50: 793-795.

Yasumura, D., Oshima, Y., Yasumoto, T., Alcala, A.C., Alcala, L.C. (1986). Tetrodotoxin and paralytic shellfish toxins in Philippine crabs. *Agric. Biol. Chem.* 50: 593-598.

Yoshinaga, I., Kawai, T., Ishida, T. (1995a). Lysis of *Gymnodinium nagasakiense* by marine bacteria. In: Lassus, P., Arzul, G., Erard, E., Gentien, P., Marcaillou-LeBaut, C. (eds.) *Harmful Marine Algal Blooms*, Lavoisier Science Publ., Paris, pp. 687-692.

Yoshinaga, I., Kawai, T., Takeuchi, T., Ishida, Y. (1995b). Distribution and fluctuation of bacteria inhibiting the growth of a marine red tide phytoplankton *Gymnodinium mikimotoi* in Tanabe Bay (Wakayama Pref., Japan). *Fish. Sci.* 61: 780-786.

Yoshinaga, I., Ishida, Y. (1995). Taxonomical studies on a marine bacterium, E401, isolated from Tanabe Bay, Wakayama Pref., killing a harmful microalga *Gymnodinium mikimotoi*, and analysis of its killing mechanisms. *Mem. Interdiscipl. Res. Inst. Environ. Sci.* 14: 47-58.

Ecophysiological Processes and Mechanisms: Towards Common Paradigms for Harmful Algal Blooms

A.D. Cembella
Institute for Marine Biosciences, National Research Council, Halifax, Nova Scotia, Canada B3H 3Z1

The key to every biological problem must finally be sought in the cell
-E.B. Wilson

1. Introduction

In response to the mandate of SCOR Working Group #97: "To review the data and existing knowledge on the ecology and physiology of harmful algal species", members chose to convene a NATO Advanced Study Institute, with international experts in their respective disciplines as lecturers. To fill the gaps in our ecophysiological knowledge of HABs, the NATO-ASI was structured around two major themes: AUTECOLOGY (what do we know about individual HAB species in relation to their environment?) and PROCESSES AND MECHANISMS (what niche-defining strategies can be identified among taxonomically disjunct HAB species?). The task of the lecturers on the latter theme was to define common strategies which might explain the success of types of HAB species and their community interactions. Fortunately, as authors of the review chapters, the lecturers have taken their task seriously, such that instead of a mere bibliographic review and data compilation, we now have a conceptual synthesis for each key ecophysiological topic we had identified.

2. Processes and Mechanisms

For several decades, the scientific community working on HABs has attempted the transition from a purely descriptive approach to HAB dynamics (magnitude, species composition, range extension, dispersion, etc.) to an interpretation of the causative processes and mechanisms underlying bloom initiation and development. The aim is to shift from this descriptive approach to conceptual models and ultimately to mechanistic/dynamic models which would give us predictive capabilities - in much the same way as weather forecasting and climatological research have evolved in the past century. In spite of remarkable advances in diagnostic technologies, from sub-cellular (e.g., nuclear and immunological probes; Scholin, this volume) to macro-scale surveys (e.g., satellite and airborne remote sensing), it is fair to state that we remain far from this goal. Indeed, it could be argued that progress in defining the critical parameters regulating bloom dynamics, many of which were suggested more than 20 years ago during the First International Conference on Toxic Dinoflagellate Blooms in Boston, Massachusetts, has been disappointing. Although we generally accept the importance of alternative life history strategies (e.g., the cyst "seed bed" hypothesis), and the role of turbulence, stratification, vertical migration, and macronutrient loading in the ecophysiology of HAB species, we remain generally perplexed at how the interplay of these key factors can lead to a sudden shift in community structure, with concomitant emergence of our "species of interest". How

NATO ASI Series, Vol. G 41
Physiological Ecology of Harmful Algal Blooms
Edited by D. M. Anderson, A. D. Cembella and G. M. Hallegraeff

do these species emerge from relative obscurity within the background matrix to dominate our attention, if not a plankton community?

The identification of key ecophysiological processes and mechanisms which permit a given "species" to occupy a defined ecological niche is a crucial step in modeling bloom dynamics and succession. There are a number of ways in which types of harmful species have been classified: a) ichthyotoxic species vs. toxic species affecting human health via vectorial toxin transfer; b) high biomass-formers (r-strategists) vs. low density background species (K-strategists); c) pelagic vs. epiphytic/epibenthic species; d) photoautotrophic vs. heterotrophic/mixotrophic species; e) species stimulated by high organic enrichment vs. those which are not; f) "layer formers" vs. "mixers"; and g) life history strategists vs. temporally persistent species. Some of these dichotomies are useful constructs, whereas others now appear to be fortuitous or artificial assemblages.

Harmful algal species are capable of exploiting a bewildering array of ecological niches, survival strategies, and nutritional modes. This is why attempts at dynamic modeling using conventional input parameters (nutrient uptake kinetics, grazing rates, specific growth rates, etc.), and based upon a simple trophic structure paradigm (primary production ≈ phytoplankton ≈ chlorophyll biomass; secondary production ≈ zooplankton; tertiary production ≈ ichthyoplankton/ctenophores) have been of little utility for HABs. For example, where does *Noctiluca* (see Elbrächter and Qi, this volume) fit into this heirachy?

Among the toxigenic microalgal species, the photosynthetic flagellates and particularly the dinoflagellates are overwhelmingly represented; this flagellate dominance of toxic species is poorly counterbalanced by only a few closely related species of pennate diatoms (Bates, this volume). The phylogenetic distribution of PSP toxins among diverse taxa from both tropical and temperate latitudes (Cembella, this volume), strongly suggests a polyphyletic origin, likely derived from a prokaryotic source (Doucette *et al.*, this volume). The hypothesized interaction between dinoflagellates and eubacteria in PSP toxin production, either as free-living cohorts or as endosymbionts, and food web toxin transfer via bacteria must now be considered plausible mechanisms. In contrast to the PSP toxins, the biosynthesis of polyether toxins derived via polyketide biosynthesis (Wright and Cembella, this volume) are restricted to phylogenetically clustered dinoflagellate species - with the exception of the recent report of brevetoxins in the rhaphidophyte, *Chattonella*.

The ecophysiological significance of phycotoxins has engendered much speculation, e.g., allelopathy or chemical defense by toxin-producing species against grazers. The apparent deleterious effects of toxigenic microalgal species on protistan and metazoan grazers do not in themselves constitute proof that toxin production is an eco-evolutionary adaptation to discourage predation. Nevertheless, the expression of sub-lethal effects towards some predators, including grazing avoidance, physiological impairment, and reduced egg production and viability, do lend credence to the chemical defense hypothesis. As pointed out by Turner *et al.* (this volume) the evidence for toxin-modulated grazing by zooplankton in natural ecosystems is highly species-specific and resistant to generalization.

The relative importance of "top down" (e.g., grazing) vs. "bottom up" (e.g. nutrient dynamics) mechanisms in the regulation of HABs remains largely unresolved. The fact that most harmful algal species are photosynthetic and can be grown on defined mixtures of macro- and micro-nutrients, originally led to a degree of hubris, whereby HAB events would be explained and predicted by a classic nutrient dynamic paradigm. Perhaps if we only knew ambient nutrient concentrations and supply ratios in natural ecosystems, species-specific nutrient uptake and assimilation parameters (K_s, V_{max}), and nutrient-dependent growth rates, we could predict the outcome of species competition and hence predict HABs. Alas, if it were only that simple. Reigman (this volume) has noted, nutrient supply ratio theory only affects competitive outcomes if one nutrient actually limits the growth rate of a species. In mixed natural populations, multiple nutrient limitation of differentspecies, or even co-existence of species capable of utilizing alternative forms of a given nutrient (dissolved organic, particulate, or inorganic) may occur. Reigmann has provided a useful ecological niche model for broad taxonomic groupings (dinoflagellates, diatoms, prymnesiophytes, etc.) based upon competitive outcomes which is driven primarily by light and macronutrients. Such a structure does not, however, allow one to predict which species will emerge as dominant from among each group - a critical question where HAB species are concerned.

Emerging knowledge of the utilization of dissolved organic matter (DOM) via heterotrophic uptake (osmotrophy) (Carlsson and Granéli, this volume) by various harmful microalgal species has complicated the interpretation of the nutrient dynamic paradigm. For most microalgae, DOM may be regarded as a nutritional supplement or growth stimulant, rather than a replacement for inorganic macronutrients. In the form of humic acids, DOM may be an important factor regulating the availability of trace metals in phytoplankton. The complex biogeochemistry of coastal ecosystems provides a fluctuating and highly variable supply of essential trace metals, some of which could be growth-rate limiting for certain phytoplankton including HAB species (Boyer and Brand, this volume). Definitive evidence that particular HAB species have been growth-rate or biomass limited, or subject to metal toxicity effects in coastal ecosystems, remains inconclusive.

Mixotrophy is increasingly recognized as crucial to the nutritional physiology of certain HAB species (Granéli and Carlsson, and Hansen; this volume), by providing fixed carbon and nitrogen in forms which are bioenergetically favorable. Phagotrophic ingestion and retention of exogenous chloroplasts (cleptochloroplastidy) even yields a bonus by functioning as an auxiliary source of photosynthate (Burkholder, Glasgow and Lewitus, this volume). Since phagotrophy is not restricted to HAB organisms - although it is widely reported among diverse phytoflagellates - it is difficult to determine to what extent this alternative nutritional strategy contributes to the development and maintenance of bloom events. Could protistan phagotrophy be a viable means of toxin transfer in marine food webs via the microbial loop?

Microparasites are well known from marine habitats, but their specific interactions with HAB species remain poorly characterized (Elbrachter and Schnepf, this volume). According to evidence presented at this NATO ASI, microparasites have been linked to such diverse HAB phenomena as: 1) the collapse of chrysophycean (brown tide) blooms due to viral infection; 2) catastrophic declines in raphidophyte populations caused by algicidal bacteria; 3) infection of toxigenic dinoflagellates (*Alexandrium* spp.) by the endocytic/endonuclear parasite *Amoebophyra*; and 4) invasion of *Pseudo-nitzschia* cells by parasitic chytrid fungi. Identification of infective organisms and determination of their respective effects on HAB population dynamics are challenging tasks due to the episodic and often cryptic nature of the infestation. Nevertheless, the exploration of host-specific parasites of HAB organisms is a worthwhile pursuit in view of their long range potential in bloom mitigation.

Micro- and meso-scale physical phenomena (turbulence, thin layer formation, etc.), and behavioral interactions of HAB species with physical structure of the water column, are key ecophysiological factors. In this context, the early statement by

Margalef that "water movement controls phytoplankton communities" is an over-simplification, but one that should not be ignored. Kamykowski *et al.* (this volume) contributed a bio-physical model which couples behavior (swimming, orientation), physiological response (photosynthesis, photoinhibition, nutrient assimilation) with physical parameters (turbulence). The specific effectsof turbulence on phytoplankton are multivariate and interactive, but it is clear that many bloom-forming dinoflagellates are particularly susceptible to the deleterious effects of turbulence (Estrada and Berdalet, this volume). Cullen and McIntyre (this volume) successfully characterized a suite of niche-defining strategies among phytoplankton, related to depth regulation via mixing, layer formation, or vertical migration, which can be used to model the behavior of HAB species.

In conclusion, the search for a "unified field theory" which will yield a mechanistic model to interpret all HAB phenomena is a quixotic and probably fruitless quest. Nevertheless, through the contributions to this NATO-ASI, we have significantly expanded our understanding of the fundamental ecophysiological mechanisms which govern bloom dynamics. This information must now be integrated with our increasingly sophisticated knowledge of hydrodynamic interactions (advection, stratification, upwelling) if we are to achieve a predictive capability for HABs.

PARTICIPANTS LIST

AUSTRALIA
Blackburn, Susan
Hallegraeff, Gustaaf

BRAZIL
Garcia, Virginia

CANADA
Black, Edward
Cembella, Allan D.
Cullen, John
Cochlan, William, P.
Lawrence, Janice
Levasseur, Maurice
Pan, Youlian
Parrish, Christopher
Taylor, F. J. R. (Max)
Trick, Charles

CHILE
Clement, Alejandro

DENMARK
Hansen, Per Juel
Jenkinson, Ian
Kaas, Hanne

FRANCE
Gentien, Patrick
Legrand, Catherine
Maestrini, Serge Y.

GERMANY
Elbrächter, Malte
Lenz, Jürgen
Luckas, Bernd, B.
Medlin, Linda
Taroncher-Oldenburg, Gaspar

ICELAND
Gudmundsson, Kristinn

ITALY
Honsell, Giorgio
Jovine, Rafffael
Pistocchi, Rossella

JAPAN
Fukuyo, Yasuwo
Imai, Ichiro
Ishida, Yuzaburo
Kodama, Masaaki

LATVIA
Balode, Maija

MALAYSIA
Anton, Ann
Usup, Gires

NEW ZEALAND
Chang, F. Hoe

NORWAY
Edvardsen, Bente

PERU
Cordova, José

PHILIPPINES
Corrales, Rhodora

PORTUGAL
Franca, Susana, Maria
Sampayo, Maria Antonia de Mello

RUSSIA
Vershinin, Alexander

SOUTH AFRICA
Pitcher, Grant

SPAIN
Berdalet, Elisa
Bravo, Isabel
Estrada, Marta
Fraga, Santiago
Garcés, Esther
Reyero, Isabel
Zapata, Manuel

SWEDEN
Carlsson, Per
Granéli, Edna

THE NETHERLANDS
Maas, Els
Peletier, Harry
Riegman, Roel

TURKEY
Kideys, Ahmet E.

UNITED KINGDOM
Gallacher, Susan
John, Eurgain Haf
Lewis, Jane M.
Wood, Gareth J.

UNITED STATES
Anderson, Donald, M.
Doucette, Gregory, J.
Boyer, Gregory
Brand, Larry
Burkholder, JoAnn M.
Carmichael, Wayne W.
Coats, D. Wayne
Dortch, Quay
Dyhrman, Sonya
Gallagher, Jane
Kamykowski, Daniel
Keller, Maureen
McDonnell, Tracey A.
Plumley, F. Gerald
Scholin, Christopher, A.
Sellner, Kevin
Sieracki, Michael
Smayda, Theodore, J.
Tester, Patricia
Tindall, Donald R.
Turner, Jefferson, T.
Van Dolah, Fran

URUGUAY
Méndez, Silvia M.

INDEX

A

Acartia tonsa 456, 460
aerosol 133
agglutinins 31
airborne toxin 133
Alexandrium iii, 3-8, 13-23, 29-44, 116, 340, 342, 353, 382, 383, 456, 546
  acatenella 3, 29
  affine 16, 20, 631
  andersonii 16, 20, 384
  catenella 3-8, 13-23, 29, 36, 39, 39, 52, 59, 344, 346, 389, 515
  cohorticula 29
  excavatum 6, 386, 357, 468
  fundyense 6, 14, 16, 17, 20, 29-44, 226-228, 392, 610
  insuetum 15
  lusitanicum 49, 628, 634, 636
  minutum 5, 16, 29, 33, 37-39, 395, 610
  monilatum 4, 20, 29
  ostenfeldii 5, 29, 39, 229, 384, 387
  pseudogonyaulax 7, 16
  tamarense 3-8, 13-23, 29-44, 49, 226-228, 344, 346, 371, 384, 386, 387, 391, 393,397, 459, 461, 467, 494, 496, 518, 533, 549, 627
  tamiyavanichi 29
algicidal activity 625
alkaline phosphatase 90, 117, 411
allelopathy 124, 164, 220, 252-259, 282, 375, 399, 421, 454
allozyme profiles 65, 227
Alsidium corallinum 268
Amansia glomerata 268
ambush-predator dinoflagellate 175
amino acids 514, 518
aminopeptidase 516
ammonia 315, 411, 475
Amnesic Shellfish Poisoning (ASP) 267, 405-422
amoebae 179, 356
Amoebophrya ceratii 3, 353
Amphidinium 293, 353, 529
  carterae 294, 297, 514
  klebsii 294, 297
  toxins 293
Amphora coffeaeformis 269, 278
Anabaena
  circinalis 387
  cylindrica 609
  flos-aquae 387
  spiroides 609
anchovies 270
anoxia 33, 180, 328
antibiotic production 437
antibody probes 55, 156, 640
antifeedant (see allelopathy)
antifungal activity 433
Aphanizomenon flos-aquae 257, 383, 387, 397
Aureococcus anophagefferens 351, 500

B

bacteria 49, 167, 236, 315, 323, 352, 389, 412, 413, 517, 539, 544-546, 548, 552, 619-641
ballast water 23, 62
Bay of Fundy 274, 275
biogeochemical cycles 492
biogeography, 4, 120, 198, 211, 232, 278, 317
biological control 353
bioluminescence 83, 316, 318
biosynthetic pathways 414, 427-447
bloom development 220
bloom initiation 29-44, 135, 167
bloom termination 145, 166, 614
brevetoxin 133, 425, 438-440
buoyancy 213
buoyant plume 41, 156

C

C:N 517
CFP (see ciguatera fish poisioning)

Calanus
finmarchicus 461, 468
glacialis 417
helgolandicus 164
cadmium 490, 492
Cardigan Bay 271-274, 281
cell division 613
cell size 160, 474, 606
cell cycle 160-162, 391, 392, 409
Centropages hamatus 453
Ceratium
furca 534, 542
fusus 535
lineatum 535
tripos 544, 562
Chaetoceros 108, 164, 422
socialis 213, 214, 219
Chaetomorpha 294
chain-formation *62*
Chattonella 95-108, 498
antiqua 95, 96, 98, 101, 108, 114, 134, 352, 494, 498, 499, 620
globosa 95
marina 95, 96, 98, 101, 134, 438
minima 95
ovata 95
subsalsa 95
verruculosa 95
chelator 304
chemical defense 457
chemosensis attraction 187
chemostat culture (see continuous culture)
chlorophyll fluorescence 40
chloroplast DNA 15, 233
chloroplast 114
Chondria 168, 421
Chrysochromulina 193-203
brevifilum 197, 542, 545, 547
breviturrita 198
birgeri 198, 200
ericina 528
hirta 193, 525
kappa 193, 197
leadbeateri 198, 200, 202
polylepis 193, 197, 198, 200, 201, 203, 230, 352, 457, 459, 477, 499, 545, 548, 549, 550
spinifera 526, 527
strobilus 197
Chrysochromulina toxin 198
ciguatera fish poisoning (CFP) 293-309, 441
ciguatoxin 293, 423-445
circannual rhythm 31, 34, 321
cladistic analysis 232
cobalt 492, 494, 499
coastal current 41, 157
Cochlodinium
polykrikoides 301
catenatum 62
colony formation 211, 212, 219
complexation (see chelation)
continuous culture 412
Coolia
monotis 294, 294, 308
tropicalis 296
copper 490, 491, 493, 494, 499
cormorants 270
Coscinodiscus wailesii 355
Cryothecomonas 355
Crypthecodinium 175, 177, 188
cohnii 49, 229
cyanobacteria 491
cyst 24, 29, 31-44, 59, 66, 69, 74, 82-85, 98, 99, 102, 104, 108, 123, 136, 176, 194, 251, 284, 368, 397, 482
Cytophaga 620

D

DA (see domoic acid)
DAP (see amnesic shellfish poisoning)
DNA 49, 114, 392, 438, 613
DOC 509-519
DOM 509-519, 551, 621, 628
DON 509-519, 539-553
DSP (see diarrhetic shellfish poisoning)
dark uptake 569
deforestation 475
depth regulation (see vertical migration)

diagnostic kits 347
diarrhetic shellfish poisoning (DSP) 437, 443-446
Dictyota 294
dimethyl sulfide 210
Dinobryon
divergens 542, 543
cylindricum 549
Dinophysis 243-265, 344, 353, 434, 452, 529
acuminata 243-265, 434, 440, 542, 550, 551
acuta 243, 248, 249, 252, 257, 434, 550
caudata 243, 251
dens 252
fortii 243, 251, 255, 439
norvegica 243, 244, 252, 256, 438, 443, 550
rotundata 243, 257, 444, 550
sacculus 243, 244, 257
skagii 243, 244
tripos 243
toxins (see diarrhetic shellfish poisoning)
dissolved organic carbon (see DOC)
dissolved organic nitrogen (see DON)
dissolved organic matter (see DOM)
division rates (see growth rate)
DNA synthesis 408
domoic acid (DA) 267-285, 405-422
domoic acid poisoning (DAP) (see amnesic shellfish poisoning)
dormancy 31, 101
downwelling 41, 135, 145, 248
Dubosquella 354
Dunaliella tertiolecta 610
Dungeness crabs 268, 277

E

ecological role of algal toxins 399, 416-418, 428, 437, 483
Ectrogella 357
egg production 463, 465, 468
El Nino/Southern Oscillation (ENSO) 43, 86, 121
Emiliania huxleyi 232, 517
encystment 38
endogenous clock 31, 34
environmental control 35, 180, 195, 250, 279, 280, 305-307, 319, 373
enzymes 49, 516
Eurytemora herdmani 457
eutrophication 121, 221, 327, 475-484
evolution 234, 398
excystment 38
exopolysaccharides 212
Exuviaella iv

F

Favella sp. 85
ehrenberghii 165, 457, 459
feeding mechanisms 525-535
Fibrocapsa japonica 123, 134
Firth of Forth 39
fish kills 133, 195, 200, 372
fish mortality 103
flow cytometry 341
fluorescence 35
food webs 632
food vacuoles 528, 530, 541
Fragilidium subglobosum 8, 535, 544
fronts 41, 135, 147, 156, 158, 221, 315
fungi 274, 356

G

Galician rias 71, 75
Gambierdiscus toxicus 293-309, 438, 440
gambier toxins 441
gambieric acid 438, 442
gambierol 438
gamete 30, 38, 66, 136, 179, 351
gas vesicles 572
genes 55, 389
genetic diversity (see genetic variability)
genetic variability 13-23, 43, 225-238, 270, 405, 440
genetic analysis 49, 228
genospecies 13-23, 225-238
German Bight 325

Gessnerium 5
Gloeodinium montanum 49
glutamate 419
glutamate receptors 267
glycocalyx 114, 117
glycoproteins 115
Goniodoma 7, 8
Gonyaulax 4, 5, 353
  grindleyi 295, 308
  polyedra 49, 295
grazer avoidance 124, 421
grazing 124, 165, 187, 203, 211, 214, 257-259, 322, 324, 375, 453-469, 552, 568, 608
grazing rates 165, 453, 458, 609
grazing deterrents (see allelopathy)
green *Noctiluca* 316, 319
growth rate 43, 85, 119, 143, 160, 203, 244, 259, 283, 390, 393, 407, 408, 413, 476, 478, 483, 609, 613
growth inhibition 282
Gulf of Alaska 211
Gulf Loop Current 138, 145, 148
Gulf of Maine 39-43, 211, 385, 397
Gulf of Mexico 133-148
Gulf of St. Lawrence 156, 385, 386
Gymnodinium 353
  breve 28, 133-148, 368, 425, 434, 435, 452, 456, 461, 467, 468, 494, 518, 594
  catenatum iv, 32, 52, 59-76, 228, 368, 382, 384, 387, 389- 391, 397, 622, 634, 638, 639
  mikimotoi iv, 133, 155-169, 562, 622
  nagasakiense iv, 38, 155-169, 500
  nelsonii 610
  sanguineum 310, 464-467, 542, 549
Gyrodinium
  aureolum iv, 133, 155-169, 459, 562
  fissum 301
  galatheanum 523
  impudicum 62
  instriatum 636
  spirale 523
  uncatenum 542

H

haptonema 526, 527
Hemistasia phaeocysticola
Helgolandinium 8
Heterocapsa triquetra 547, 550,612
Heterosigma iii, 340
  akashiwo (see *H. carterae*)
  carterae 113-124, 457, 459, 463, 494, 549, 610, 621-623
heterotrophy 304, 539-553
horizontal gene flow 235
humic substances 510-519, 545
hypnozygote (see cyst)

I

immunofluorescence 271
initiation zone 35, 39-42, 123, 278
intracellular bacteria 634-641
interannual variability 217
iron 215, 490, 494, 496-498, 625
irradiance 33, 87, 96, 98, 108, 114, 143, 144, 157, 159, 180, 215, 280, 306, 320, 393, 415, 444, 476, 513, 547, 563-565
Isochrysis galbana 610
isozyme electrophoresis 15

J

Jania sp. 294, 382

K

K-selection 140, 483, 484
Katodinium fungiforme 175-177, 188
kleptochloroplastidy 181, 183
Ks 117, 124, 142, 162

L

ladder-frame polyethers 433
Lagenisma 356
latent toxicity 203
layer-formers 155-170, 247, 252, 253, 562,

life cycle 29-44, 66, 68, 81, 98, 100, 119, 136, 137, 175, 178, 179, 209, 229, 250, 284, 315, 316, 561, 561
light (see irradiance)
linear polyethers 431, 432
linear sorting 234, 235
Lingulodinium polyedrum 612
lipid content 410
Lyngbya wollei 612

M

maitotoxin 293, 438, 441
mandatory dormancy 32, 100
manganese 490, 492, 494, 501
mating compatibility 15, 30, 50, 56, 227-229
maturation 100
memory loss 267
Mesodinium rubrum 254
microbial food web 187, 209. 513
microbial loop 510
Micromonas pusilla 528
microsatellite 65
microzones 607
microzooplankton 220
"*minutum*" group 382, 387
mixotrophy 42, 193, 202, 256, 257, 304, 375, 509, 539-553
modeling 103, 123, 167, 568, 581-596, 625
molecular probes 67, 337-347
monoclonal antibodies 15
monophyletic radiation 19
monsoon 86
Monterey Bay 275, 277
Moraxella 627-632
morphospecies 6, 13-23
motility 115
mucoid web 323
mucopolysaccharide 210
mucus 114

N

Neurotoxic Shellfish Poisoning (NSP) 133
Neuse River Estuary 180
new production 211, 212
nickel 502
nitrogen (N) 90, 96, 117, 140, 142, 162, 163, 181-186, 211, 244, 256, 304, 305, 394, 477, 510, 148, 569-571
nitrogen limitation 116, 161, 395, 495, 551, 569
nitrogen storage 411
Noctiluca iv, 315-329, 353
scintillans 315-329
miliaris 315
Nodularia spumigena 257
nucleic acid probes 15, 337-347, 622
nucleotide sequences 56, 271, 337-347
nutrient limitation 31, 85, 98, 161, 384, 395, 406, 412
nutrients 38, 42, 67, 85, 88, 96, 116, 124, 139, 141, 143, 146, 161, 162, 181-186, 200, 211, 216, 256, 272, 278, 281, 282, 300, 304, 305, 394, 445, 475-485, 548, 568-571, 605
nutrient pools 31
nutrient ratios 217, 476-477, 482

O

Ochromonas 548, 549
okadaic acid (OA) 429, 437, 551
Olisthodiscus luteus iii
Olpidium 357
organic N 435
organic complexation 489-503
organic nutrients 142, 146, 181-186, 282
Oscillatoria
agardhii 566
morigeotii 387
osmotrophy 509, 539, 568
Ostreopsis 293
heptagona 293, 296
labens 296
lenticularis 293, 296, 298, 302
mascarenensis 296
ovata 293, 296
siamensis 293, 296, 308, 433
oxygen depletion 33
Oxyphysis oxytoxoides 550

P

PCR 344
pallium 530
palmelloid stages 136
palytoxin 426
paralytic shellfish poisoning (PSP) 13, 49-56, 59, 71, 81-83, 381-400, 426, 672, 630-633
parasites 274, 351- 358
pectenotoxins 438
Pedinomonas noctilucae 316
peduncle 177
pelicans 270
Peridinium
  quinquecorne 295
  cinctum 610
Pfiesteria piscicida 175-188, 340
Pfiesteria-like species 184
Pfiesteria toxins 187
Phaeocystis iv, 209-221, 230, 231, 453
  antarctica 210, 220, 231
  globosa 210, 220, 231
  pouchetii 210, 220, 231
  scrobiculata 211
Phaomyxa 357
phagotrophy 176, 539-553, 568
pheromones 399
phosphate 96, 146, 202, 281
phosphatase inhibition 438, 445
phosphorus 117, 140, 141, 143, 146, 162, 181-186, 194, 304, 316, 407, 510
phosphorus limitation 146, 161, 194, 281, 395, 411, 477, 548
photoacclimation 564, 567, 568, 581-596
photoinhibition 565, 581-596
photosynthetic efficiency 116, 144, 159, 213, 563, 564
phylogenetic relationships 19, 233, 381, 435
physical/biological coupling 37
pinocytosis 513
planomeiocyte 35, 36, 40, 67
planozygote 30, 38, 39, 66, 136, 176
pollution 42, 118, 121, 147, 211, 221, 300, 327, 475-485
polyamines 161, 162
polyether toxins 427-447
polyketide biosynthesis 429
polyphyletic radiation 17
Poterioochronomonas malhamensis 544, 547
precipitation 138
predator-prey interactions 608
prey selection 525, 528
Prince Edward Island (PEI) 271-274, 405
probe (see antibody or nucleic acid probe)
Prorocentrum iv, 293, 353, 480, 438
  belizeanum 295
  concavum 294, 296, 302
  emarginatum 294, 296
  hoffmannianum 295
  lima 294, 296, 429, 444, 551
  mexicanum 294, 296, 302, 308
  micans 229, 253, 514
  minimum 460, 534, 548
  triestinum 610
Protogonyaulax (see *Alexandrium*)
Protoperidinium 354, 530
Prymnesium 193-203
  calathiferum 195
  parvum 194, 195, 514, 516
  patelliferum 194, 195
Pseudo-nitzschia 214, 267-285, 340, 342, 405-422
  australis 251, 275-278, 282, 345, 405, 407
  delicatissima 214, 269, 405
  fraudulenta 412
  multiseries 230, 268, 269, 272, 276, 277, 280, 345, 460
  pseudodelicatissima 269, 274, 276, 345, 405
  pungens 230, 268, 269, 272, 276, 277, 280, 345, 412
  seriata 269, 405, 407
  turgidula 268
Pseudocalanus acuspes 421

Pseudomonas 627
  stutzeri 636
  diminuta 639

pseudopod 177
PSP (see paralytic shellfish poisoning)
PSP toxin/s (see saxitoxin)
pycnocline 157, 158, 167
Pyrocystis 141
Pyrodinium bahamense 4, 7, 8, 59, 81-91, 368, 382, 387, 391, 397, 500

Q

quiescence 31, 67, 101

R

RAPD 65
RFLP 344
r-selection 483
razor clams 268, 277
rDNA 15, 17, 56, 65
recombination 52
red tide 107, 315
red *Noctiluca* 316
remote sensing 139
reporter gene assays 639
respiration rates 328
resting cysts (see cyst)
RFLP 622
Rhizophydium 357
Rhynchopus coscinodiscivorus 354
Ria de Vigo 71-75
ribosomal DNA (see rDNA)
ribosomal RNA (see rRNA)
ribotype 17, 22
river runoff 121, 124, 216, 511
Ross Sea 215
rRNA 15, 340, 622

S

salinity 65, 69, 87, 119, 135, 139, 156, 180, 195, 197, 202, 251, 278, 282, 283, 301, 305, 320, 294, 411, 439
sandwich assay 339, 344
saxitoxin (STX) 49-56, 65, 381-400, 456, 627, 268, 630, 640
Scrippsiella trochoidea 301
seabirds 270, 276
secondary metabolites 416, 427
seedbeds (see initiation zone)
selenium 89, 202, 500
Seto Inland Sea 99, 497
sexual compatibility (see mating compatibility)
sexual reproduction 30, 234, 274, 284
silicate limitation 285, 408, 411, 472
silicate 218, 281
sinking 255, 561
Skeletonema costatum 116, 120, 232, 610, 612, 621, 626
soil extract 88, 509
species composition 480, 492, 493
species concept 15, 65, 225, 236, 237, 426
spirolides 433
St. Lawrence Estuary 39, 40
stratification 115, 124, 145, 158, 163, 164, 167, 200, 215, 243, 244, 247, 249, 251, 276, 284, 559-573
STX (see saxitoxin)
subsurface blooms 157
sulfotransferase 52-54
SUPA 592
swimming speed 581-596
swimming behavior 453, 559-573, 581-596, 601
Symbiodinium microadriaticum 49
Synechococcus 547, 564

T

"*tamarensis*" group 15, 382, 384, 395
Tasmania 69
Temora sp. 417
temperature 32, 65, 67, 69-71, 74, 87, 96, 99, 100, 103, 119, 139, 157, 195, 197, 202, 210, 218, 220, 247, 249, 251, 276, 281, 306, 320, 393, 414, 480
temperature window 32, 35
temporary cyst 30
Thalassiosira pseudonana 566
tetrodotoxin (TTX) 627
toxin biosynthesis 414, 630

toxin cell quota 385-389, 435, 439, 440
toxin composition 15, 49, 50, 65, 226, 384-389, 439, 440
toxin content (see toxin cell quota)
toxin genes 55, 389, 392, 410, 424, 426, 438, 442, 630
toxin profile (see toxin composition)
toxin production 13, 203, 229, 381-400, 384-389, 405-418, 439, 483, 626-634
trace metals 88, 89, 117, 489-503
trace metal complexation 88, 89, 489, 503
turbulence 85, 165, 166, 219, 220, 255, 294, 297, 303, 375, 560, 561, 567, 587, 601-614

U

unbalanced growth 567
upwelling 41, 67, 73, 74, 86, 118, 121, 135, 138, 143, 145-147, 248, 277, 560

V

vertical migration 37, 42, 67, 105, 106, 115, 124, 144, 158, 162, 252, 308, 375, 559-574, 581-596
Vidalia obtusiloba 268
viruses 236, 351
vitamins 509, 540

W

wind stress 69

X

Y

yessotoxin 434

Z

zinc 492, 502
zoospores 175, 176, 179, 186, 316
zooxanthella 430

NATO ASI Series G

Vol. 1: **Numerical Taxonomy.** Edited by J. Felsenstein. 644 pages. 1983. (out of print)

Vol. 2: **Immunotoxicology.** Edited by P. W. Mullen. 161 pages. 1984.

Vol. 3: **In Vitro Effects of Mineral Dusts.** Edited by E. G. Beck and J. Bignon. 548 pages. 1985.

Vol. 4: **Environmental Impact Assessment, Technology Assessment, and Risk Analysis.** Edited by V. T. Covello, J. L. Mumpower, P. J. M. Stallen, and V. R. R. Uppuluri. 1068 pages.1985.

Vol. 5: **Genetic Differentiation and Dispersal in Plants.** Edited by P. Jacquard, G. Heim, and J. Antonovics. 452 pages. 1985.

Vol. 6: **Chemistry of Multiphase Atmospheric Systems.** Edited by W. Jaeschke. 773 pages. 1986.

Vol. 7: **The Role of Freshwater Outflow in Coastal Marine Ecosystems.** Edited by S. Skreslet. 453 pages. 1986.

Vol. 8: **Stratospheric Ozone Reduction, Solar Ultraviolet Radiation and Plant Life.** Edited by R. C. Worrest and M. M. Caldwell. 374 pages. 1986.

Vol. 9: **Strategies and Advanced Techniques for Marine Pollution Studies: Mediterranean Sea.** Edited by C. S. Giam and H. J.-M. Dou. 475 pages. 1986.

Vol. 10: **Urban Runoff Pollution.** Edited by H. C. Torno, J. Marsalek, and M. Desbordes. 893 pages. 1986.

Vol. 11: **Pest Control: Operations and Systems Analysis in Fruit Fly Management.** Edited by M. Mangel, J. R. Carey, and R. E. Plant. 465 pages. 1986.

Vol. 12: **Mediterranean Marine Avifauna: Population Studies and Conservation.** Edited by MEDMARAVIS and X. Monbailliu. 535 pages. 1986.

Vol. 13: **Taxonomy of Porifera from the N. E. Atlantic and Mediterranean Sea.** Edited by J. Vacelet and N. Boury-Esnault. 332 pages. 1987.

Vol. 14: **Developments in Numerical Ecology.** Edited by P. Legendre and L. Legendre. 585 pages. 1987.

Vol. 15: **Plant Response to Stress. Functional Analysis in Mediterranean Ecosystems.** Edited by J. D. Tenhunen, F. M. Catarino, O. L. Lange, and W. C. Oechel. 668 pages. 1987.

Vol. 16: **Effects of Atmospheric Pollutants on Forests, Wetlands and Agricultural Ecosystems.** Edited by T. C. Hutchinson and K. M. Meema. 652 pages. 1987.

Vol. 17: **Intelligence and Evolutionary Biology.** Edited by H. J. Jerison and I. Jerison. 481 pages. 1988.

NATO ASI Series G

Vol. 18: Safety Assurance for Environmental Introductions of Genetically-Engineered Organisms. Edited by J. Fiksel and V.T. Covello. 282 pages. 1988.

Vol. 19: Environmental Stress in Plants. Biochemical and Physiological Mechanisms. Edited by J. H. Cherry. 369 pages. 1989.

Vol. 20: Behavioural Mechanisms of Food Selection. Edited by R. N. Hughes. 886 pages. 1990.

Vol. 21: Health Related Effects of Phyllosilicates. Edited by J. Bignon. 462 pages. 1990.

Vol. 22: Evolutionary Biogeography of the Marine Algae of the North Atlantic. Edited by D. J. Garbary and G. R. South. 439 pages. 1990.

Vol. 23: Metal Speciation in the Environment. Edited by J. A. C . Broekaert, Ş. Güçer, and F. Adams. 655 pages. 1990.

Vol. 24: Population Biology of Passerine Birds. An Integrated Approach. Edited by J. Blondel, A. Gosler, J.-D. Lebreton, and R . McCleery. 513 pages. 1990.

Vol. 25: Protozoa and Their Role in Marine Processes. Edited by P. C. Reid, C. M . Turley, and P. H. Burkill. 516 pages. 1991.

Vol. 26: Decision Support Systems. Edited by D. P Loucks and J. R. da Costa. 592 pages. 1991.

Vol. 27: Particle Analysis in Oceanography. Edited by S. Demers. 428 pages. 1991.

Vol. 28: Seasonal Snowpacks. Processes of Compositional Change. Edited by T. D. Davies, M . Tranter, and H . G. Jones. 484 pages. 1991.

Vol. 29: Water Resources Engineering Risk Assessment. Edited by. J. Ganoulis. 551 pages. 1991.

Vol. 30: Nitrate Contamination. Exposure, Consequence, and Control. Edited by I. Bogárdi and R. D. Kuzelka. 532 pages. 1991.

Vol. 31: Industrial Air Pollution. Assessment and Control. Edited by A. Müezzinoğlu and M. L. Williams. 245 pages. 1992.

Vol. 32: Migration and Fate of Pollutants in Soils and Subsoils. Edited by D. Petruzzelli and F. G. Helfferich. 527 pages. 1993.

Vol. 33: Bivalve Filter Feeders in Estuarine and Coastal Ecosystem Processes. Edited by R. F. Dame. 584 pages. 1993.

Vol. 34: Non-Thermal Plasma Techniques for Pollution Control. Edited by B. M. Penetrante and S. E. Schultheis.
Part A: Overview, Fundamentals and Supporting Technologies. 429 pages. 1993.
Part B: Electron Beam and Electrical Discharge Processing. 433 pages. 1993.

Vol. 35: Microbial Mats. Structure, Development and Environmental Significance. Edited by L. J. Stal and P. Caumette. 481 pages. 1994.

NATO ASI Series G

Vol. 36: **Air Pollutants and the Leaf Cuticle.**
Edited by K. E. Percy, J. N. Cape, R. Jagels and C. J. Simpson. 405 pages. 1994

Vol. 37: **Azospirillum VI and Related Microorganisms.**
Edited by I. Fendrik, M. del Gallo, J. Vanderleyden, and M. Zamaroczy. 588 pages. 1995.

Vol. 38: **Molecular Ecology of Aquatic Microbes.**
Edited by I. Joint. 423 pages. 1995.

Vol. 39: **Biological Fixation of Nitrogen for Ecology and Sustainable Agriculture.**
Edited by A. Legocki, H. Bothe, and A. Pühler. 339 pages. 1997.

Vol. 40: **National Parks and Protected Areas. Keystones to Conservation and Sustainable Development.**
Edited by J. G. Nelson and R. Serafin. 1997.

Vol. 41: **Physiological Ecology of Harmful Algal Blooms.**
Edited by D. M. Anderson, A. D. Cembella and G. M. Hallegraeff. 1998.

Printing: Druckhaus Beltz, Hemsbach
Binding: Buchbinderei Schäffer, Grünstadt